RED WINE TECHNOLOGY

RED WINE TECHNOLOGY

Edited by

ANTONIO MORATA

Department of Chemistry and Food Technology, Technical University of Madrid, Madrid, Spain

Academic Press is an imprint of Elsevier
125 London Wall, London EC2Y 5AS, United Kingdom
525 B Street, Suite 1650, San Diego, CA 92101, United States
50 Hampshire Street, 5th Floor, Cambridge, MA 02139, United States
The Boulevard, Langford Lane, Kidlington, Oxford OX5 1GB, United Kingdom

Copyright © 2019 Elsevier Inc. All rights reserved.

No part of this publication may be reproduced or transmitted in any form or by any means, electronic or mechanical, including photocopying, recording, or any information storage and retrieval system, without permission in writing from the publisher. Details on how to seek permission, further information about the Publisher's permissions policies and our arrangements with organizations such as the Copyright Clearance Center and the Copyright Licensing Agency, can be found at our website: www.elsevier.com/permissions.

This book and the individual contributions contained in it are protected under copyright by the Publisher (other than as may be noted herein).

Notices
Knowledge and best practice in this field are constantly changing. As new research and experience broaden our understanding, changes in research methods, professional practices, or medical treatment may become necessary.

Practitioners and researchers must always rely on their own experience and knowledge in evaluating and using any information, methods, compounds, or experiments described herein. In using such information or methods they should be mindful of their own safety and the safety of others, including parties for whom they have a professional responsibility.

To the fullest extent of the law, neither the Publisher nor the authors, contributors, or editors, assume any liability for any injury and/or damage to persons or property as a matter of products liability, negligence or otherwise, or from any use or operation of any methods, products, instructions, or ideas contained in the material herein.

British Library Cataloguing-in-Publication Data
A catalogue record for this book is available from the British Library

Library of Congress Cataloging-in-Publication Data
A catalog record for this book is available from the Library of Congress

ISBN: 978-0-12-814399-5

For Information on all Academic Press publications
visit our website at https://www.elsevier.com/books-and-journals

Publisher: Charlotte Cockle
Acquisition Editor: Patricia Osborn
Editorial Project Manager: Tasha Frank
Production Project Manager: Denny Mansingh
Cover Designer: Mark Rogers
Cover Pictures: Courtesy of Rafael Cuerda García-Junceda, Winemaker

Typeset by MPS Limited, Chennai, India

Contents

List of Contributors — ix
Prologue — xi

1. Grape Maturity and Selection: Automatic Grape Selection — 1
SUSANA RÍO SEGADE, SIMONE GIACOSA, VINCENZO GERBI AND LUCA ROLLE

1.1 Physicochemical Characteristics of Enological Interest — 1
1.2 Vineyard Approaches to Grape Selection and Harvest Date Determination — 3
1.3 Grape Selection in Winery — 9
References — 13

2. Acidification and pH Control in Red Wines — 17
PIERGIORGIO COMUZZO AND FRANCO BATTISTUTTA

2.1 Importance of Acidic Fraction and pH Control in Red Wines — 17
2.2 Main Organic Acids in Must and Wine — 17
2.3 Total Acidity and Wine pH — 19
2.4 Acid–Base Equilibrium and Wine Buffer Capacity — 22
2.5 Traditional Strategies for Chemical Acidification — 23
2.6 Traditional Strategies for Chemical Deacidification — 25
2.7 New Technologies for pH Control — 27
2.8 Laboratory Techniques for Measuring pH and Acidic Fraction — 30
Acknowledgments — 32
References — 32

3. Maceration and Fermentation: New Technologies to Increase Extraction — 35
ANTONIO MORATA, CARMEN GONZÁLEZ, WENDU TESFAYE, IRIS LOIRA AND JOSE A. SUÁREZ-LEPE

3.1 Introduction — 35
3.2 Tank Design for Red Winemaking — 35
3.3 Vessel Materials in Red Winemaking — 36
3.4 Kinetics of Extraction: The Effect of Temperature — 38
3.5 Mechanical Processes During Maceration — 40
3.6 New Extraction Technologies — 44
3.7 Conclusions — 47
References — 48

4. Use of Non-*Saccharomyces* Yeasts in Red Winemaking — 51
MAURIZIO CIANI AND FRANCESCA COMITINI

4.1 Introduction — 51
4.2 Yeast Ecology of Grape Berry — 51
4.3 Controlled Fermentation: The Role of *Saccharomyces Cerevisiae* — 55
4.4 Non-*Saccharomyces* Yeasts Features in Red Wine — 56
References — 64

5. Yeast Biotechnology for Red Winemaking — 69
KARINA MEDINA, VALENTINA MARTIN, EDUARDO BOIDO AND FRANCISCO CARRAU

5.1 Introduction — 69
5.2 Yeast Diversity in Red Grapes and Musts — 70
5.3 Influence of Red Wine Technology on *Saccharomyces* Strains — 71
5.4 *Saccharomyces Cerevisiae* and Flavor Compounds — 74
5.5 Practical Red Winemaking and Yeast Performance — 76
Acknowledgments — 80
References — 80
Further Reading — 83

6. Malolactic Fermentation — 85
IRENE GIL-SÁNCHEZ, BEGOÑA BARTOLOMÉ SUÁLDEA AND M. VICTORIA MORENO-ARRIBAS

6.1 Introduction — 85
6.2 Lactic Acid Bacteria in Winemaking — 85
6.3 Factors Impacting LAB at Winery — 87
6.4 Technological Strategies for Managing the MLF Performance — 88
6.5 Impact of MLF on Wine Organoleptic Properties — 89
6.6 Production of Off-Flavors by Lactic Acid Bacteria — 91
6.7 Implications of LAB and MLF in Wine Safety — 93
6.8 Conclusion — 94
Acknowledgments — 95
References — 95

7. Yeast-Bacteria Coinoculation — 99
ISABEL PARDO AND SERGI FERRER

7.1 Introduction — 99
7.2 Objectives — 101
7.3 Interactions Between Wine Microorganisms — 107

Acknowledgments 108
References 108

8. Molecular Tools to Analyze Microbial Populations in Red Wines 115
KAROLA BÖHME, JORGE BARROS-VELÁZQUEZ AND PILAR CALO-MATA

8.1 Introduction 115
8.2 Classical and Phenotypic Methods 116
8.3 DNA-Based Methods 116
8.4 Matrix-Assisted Laser Desorption/Ionization–Time of Flight Mass Spectrometry 119
8.5 Microbial Diversity Assessment Through Enzymes Detection 119
8.6 Culture-Dependent Versus Culture-Independent Methods 119
8.7 Conclusions 120
References 120
Further Reading 123

9. Barrel Aging; Types of Wood 125
FERNANDO ZAMORA

9.1 Brief Historical Introduction 125
9.2 The Main Tree Species Used in Cooperage 125
9.3 The Main Forests Providing Wood For Cooperage 126
9.4 The Concept of Wood Grain in Cooperage 126
9.5 Obtaining the Staves: Hand Splitting and Sawing 127
9.6 Drying Systems: Natural Seasoning and Artificial Drying 129
9.7 Assembly and Toasting of the Barrel 129
9.8 Types of Barrels and Barrel Parts 130
9.9 What Happens to a Wine During Barrel Aging 131
9.10 Volatile Substances Released by Oak Wood During Barrel Aging 132
9.11 Phenolic Compounds Released by Oak Wood During Barrel Aging 134
9.12 Oxygen Permeability of Oak Wood 136
9.13 Influence of Wood Grain 136
9.14 Influence of Botanical and Geographic Origin 137
9.15 Influence of Natural Seasoning and Artificial Drying 138
9.16 Influence of Toasting Level 139
9.17 Influence of the Repeated Use of Barrels 141
9.18 Barrel Aging Process 143
Acknowledgments 144
References 144
Further Reading 147

10. Emerging Technologies for Aging Wines: Use of Chips and Micro-Oxygenation 149
ENCARNA GÓMEZ-PLAZA AND ANA B. BAUTISTA-ORTÍN

10.1 Why Aging Wines in Barrels? 149
10.2 The Micro-Oxygenation Technique 150
10.3 Positive Factors of Using Micro-Oxygenation 151
10.4 The Application of the MOX Technique 152
10.5 The Use of Oak Chips 153
10.6 When and How Use Them 153
10.7 Effect of Adding Oak Chips on Wine Characteristics 153
10.8 Comparing the Effect of Chips or MOX With Aging Wine in Barrels 155
10.9 The Combined Used of MOX + Chips 156
10.10 Innovations in MOX and Chips Application 157
References 159
Further Reading 162

11. New Trends in Aging on Lees 163
ANTONIO MORATA, FELIPE PALOMERO, IRIS LOIRA AND JOSE A. SUÁREZ-LEPE

11.1 Introduction 163
11.2 Use of Non-*Saccharomyces* Yeasts 166
11.3 Accelerated Aging on Lees 167
11.4 Lees Aromatization 171
11.5 Conclusions 174
References 174
Further Reading 176

12. Evolution of Proanthocyanidins During Grape Maturation, Winemaking, and Aging Process of Red Wines 177
ANTÓNIO M. JORDÃO AND JORGE M. RICARDO-DA-SILVA

12.1 Proanthocyanidins: Composition, Content, and Evolution During Grape Maturation 177
12.2 Evolution of Proanthocyanidins During Fermentative Maceration of Red Wines 183
12.3 Changes on Proanthocyanidins During Red Wine Aging in Contact with Wood 185
12.4 Final Remarks 189
References 189

13. Wine Color Evolution and Stability 195
MARÍA TERESA ESCRIBANO-BAILÓN, JULIÁN C. RIVAS-GONZALO AND IGNACIO GARCÍA-ESTÉVEZ

13.1 Introduction 195
13.2 Anthocyanin Stability 195
13.3 Copigmentation 198
13.4 Red Wine Color Evolution 199
13.5 Winemaking Practices for Stabilizing Red Wine Color 201
Acknowledgments 202
References 203
Further Reading 205

14. Polymeric Pigments in Red Wines 207
JOANA OLIVEIRA, VICTOR DE FREITAS AND NUNO MATEUS

Abbreviations 207
14.1 Introduction 207
14.2 Polymeric Pigments in Red Wines 209
14.3 Analysis of Polymeric Pigments 212
14.4 Stability in Solution and Influence in Red Wine Color 213
14.5 Conclusion 215
References 215

15. Spoilage Yeasts in Red Wines 219
MANUEL MALFEITO-FERREIRA

15.1 Introduction 219
15.2 Description of the Main Yeast Genera/Species Involved in Wine Spoilage 224
15.3 Yeast Monitoring 227
15.4 Control of Yeast Populations in Wines 230
15.5 Future Trends 233
References 233

16. Red Wine Clarification and Stabilization 237
AUDE VERNHET

16.1 Colloids and Colloidal Instabilities in Red Wines 237
16.2 Wine Clarification 240
16.3 Stabilization With Regards to the Crystallization of Tartaric Salts 245
16.4 Microbiological Stabilization 248
16.5 Conclusion 249
References 249

17. Sensory Analysis of Red Wines for Winemaking Purposes 253
PABLO OSSORIO AND PEDRO BALLESTEROS TORRES

17.1 Tasting of Grapes 253
17.2 Tasting in the Production of Red Wine 255
17.3 Tasting During Malolactic Fermentation 256
17.4 Conclusions 256

18. Management of Astringency in Red Wines 257
ALVARO PEÑA-NEIRA

18.1 Introduction 257
18.2 Astringency in Wines 258
18.3 Influence of Winemaking Technology in Wine Astringency 261
18.4 Future Outlook 267
References 267
Further Reading 272

19. Aromatic Compounds in Red Varieties 273
DORIS RAUHUT AND FLORIAN KIENE

19.1 Introduction 273
19.2 Selection of Aromatic Compounds With Distinct Impact 273
19.3 Conclusion 279
References 279

20. The Instrumental Analysis of Aroma-Active Compounds for Explaining the Flavor of Red Wines 283
LAURA CULLERÉ, RICARDO LÓPEZ AND VICENTE FERREIRA

20.1 Introduction 283
20.2 Analytes and an Analytical Classification 284
20.3 The Analysis of "Easy" Aroma Compounds 285
20.4 The Specific Analysis of Volatile Phenols 290
20.5 The Analysis of "Difficult" Aroma Compounds in Red Wine 291
20.6 Final Considerations 300
References 301

21. SO_2 in Wines: Rational Use and Possible Alternatives 309
SIMONE GIACOSA, SUSANA RÍO SEGADE, ENZO CAGNASSO, ALBERTO CAUDANA, LUCA ROLLE AND VINCENZO GERBI

21.1 Sulfur Dioxide: Use in the Winemaking Process and Legal Limits 309
21.2 Different Forms to Use SO_2 311
21.3 SO_2 Action Mechanisms 314
21.4 SO_2 Replacement Products for Red Wine Production 318
References 320

22. Red Wine Bottling and Packaging 323
MARK STROBL

22.1 Glass Bottles 323
22.2 Targets Today 323
22.3 Bottling Lines 324
22.4 Hazards in Bottling Red Wine 325
22.5 Closures 333
22.6 Preparing the Wine for Market 334
22.7 Packaging 336
22.8 Economy 337
22.9 Ecology 338
References 338
Further Reading 339

23. Red Winemaking in Cool Climates 341
BELINDA KEMP, KARINE PEDNEAULT, GARY PICKERING, KEVIN USHER AND JAMES WILLWERTH

23.1 Introduction 341
23.2 Cool Climate Grape Varieties in the Northern and Southern Hemisphere 342
23.3 Chemical Composition of Grapes in Cool and Warm Climates 342
23.4 Innovations in Cool Climate Winemaking 347
23.5 Making Wine From Red Interspecific Hybrid and Fungus-Resistant Varieties 349
23.6 Yeast Assimilable Nitrogen 350
23.7 Tannins and Anthocyanin 350
References 353

24. Red Winemaking in Cold Regions With Short Maturity Periods 357
MA TENGZHEN, KAI CHEN, HAO YAN, HAN SHUNYU AND BI YANG

24.1 Introduction 357

24.2 Wintering Adaptability and Cold Resistance of Grape Vine	363	24.8 Winemaking Technology	367
24.3 Influence of Low Temperature on Different Tissues of Grape Vine	364	24.9 Final Comments	370
		References	370
24.4 Influence of Low Temperature on Grape Cells	364	Further Reading	372
24.5 The Reasons for Freeze Damage	365	**Author Index**	**373**
24.6 Maturity Analysis	365	**Subject Index**	**383**
24.7 Anthocyanin Accumulation by Viticulture Process	366		

List of Contributors

Alvaro Peña-Neira Department of Agro-Industry and Enology, Faculty of Agronomical Sciences, University of Chile, Santiago, Región Metropolitana, Chile

Pedro Ballesteros Torres Agronomical Engineer, Master of Wine, Av Bourgmestre Herinckx 16, Brussels, Belgium

Jorge Barros-Velázquez Department of Analytical Chemistry, Nutrition and Food Science, School of Veterinary Sciences/College of Biotechnology, University of Santiago de Compostela, Lugo, Spain

Begoña Bartolomé Suáldea Instituto de Investigación en Ciencias de la Alimentación (CIAL), CSIC-UAM, Madrid, Spain

Franco Battistutta Department of Agricultural, Food, Environmental and Animal Sciences (Di4A), University of Udine, Udine, Italy

Ana B. Bautista-Ortín Department of Food Science and Technology, University of Murcia, Murcia, Spain

Karola Böhme Technological Agroalimentary Center of Lugo, Lugo, Spain

Eduardo Boido Enology and Fermentation Biotechnology Area, Food Science and Technology Department, School of Chemistry of the Universidad de la Republica, Montevideo, Uruguay

Enzo Cagnasso Department of Agricultural, Forestry and Food Sciences, University of Turin, Grugliasco (TO), Italy

Pilar Calo-Mata Department of Analytical Chemistry, Nutrition and Food Science, School of Veterinary Sciences/College of Biotechnology, University of Santiago de Compostela, Lugo, Spain

Francisco Carrau Enology and Fermentation Biotechnology Area, Food Science and Technology Department, School of Chemistry of the Universidad de la Republica, Montevideo, Uruguay

Alberto Caudana Department of Agricultural, Forestry and Food Sciences, University of Turin, Grugliasco (TO), Italy

Kai Chen College of Food Science and Nutritional Engineering, China Agricultural University, Beijing, P.R. China

Maurizio Ciani Department of Life and Environmental Sciences, Polytechnic University of Marche, Ancona, Italy

Francesca Comitini Department of Life and Environmental Sciences, Polytechnic University of Marche, Ancona, Italy

Piergiorgio Comuzzo Department of Agricultural, Food, Environmental and Animal Sciences (Di4A), University of Udine, Udine, Italy

Laura Culleré Laboratory for Flavor Analysis and Enology, Instituto Agroalimentario de Aragón (IA2), Department of Analytical Chemistry, Faculty of Sciences, Universidad Zaragoza, Zaragoza, Spain

Victor de Freitas REQUIMTE - LAQV - Department of Chemistry and Biochemistry, Faculty of Science, Oporto University, Oporto, Portugal

María Teresa Escribano-Bailón Analytical Chemistry, Nutrition and Food Science, Faculty of Pharmacy, University of Salamanca, Salamanca, Spain

Vicente Ferreira Laboratory for Flavor Analysis and Enology, Instituto Agroalimentario de Aragón (IA2), Department of Analytical Chemistry, Faculty of Sciences, Universidad Zaragoza, Zaragoza, Spain

Sergi Ferrer ENOLAB, Interdisciplinary Research Structure for Biotechnology and Biomedicine (BIOTECMED) Universitat de València, Burjassot, Spain

Ignacio García-Estévez Analytical Chemistry, Nutrition and Food Science, Faculty of Pharmacy, University of Salamanca, Salamanca, Spain

Vincenzo Gerbi Department of Agricultural, Forestry and Food Sciences, University of Turin, Grugliasco (TO), Italy

Simone Giacosa Department of Agricultural, Forestry and Food Sciences, University of Turin, Grugliasco (TO), Italy

Irene Gil-Sánchez Instituto de Investigación en Ciencias de la Alimentación (CIAL), CSIC-UAM, Madrid, Spain

Encarna Gómez-Plaza Department of Food Science and Technology, University of Murcia, Murcia, Spain

Carmen González Department of Chemistry and Food Technology, Technical University of Madrid, Madrid, Spain

António M. Jordão Department of Food Industries, Polytechnic Institute of Viseu (CI&DETS), Viseu, Portugal; CQ-VR, Chemistry Research Centre, Vila Real, Portugal

Belinda Kemp Cool Climate Oenology and Viticulture Institute (CCOVI), Brock University, St Catharines, ON, Canada; Department of Biological Sciences, Brock University, St Catharines, ON, Canada

Florian Kiene Department of Microbiology and Biochemistry, Hochschule Geisenheim University, Geisenheim, Germany

Iris Loira Department of Chemistry and Food Technology, Technical University of Madrid, Madrid, Spain

Ricardo López Laboratory for Flavor Analysis and Enology, Instituto Agroalimentario de Aragón (IA2), Department of Analytical Chemistry, Faculty of Sciences, Universidad Zaragoza, Zaragoza, Spain

Manuel Malfeito-Ferreira Linking Lanscape, Environment, Agriculture and Food Research Centre (LEAF), University of Lisbon, Lisbon, Portugal

Valentina Martin Enology and Fermentation Biotechnology Area, Food Science and Technology Department, School of Chemistry of the Universidad de la Republica, Montevideo, Uruguay

Nuno Mateus REQUIMTE - LAQV - Department of Chemistry and Biochemistry, Faculty of Science, Oporto University, Oporto, Portugal

Karina Medina Enology and Fermentation Biotechnology Area, Food Science and Technology Department, School of Chemistry of the Universidad de la Republica, Montevideo, Uruguay

Antonio Morata Department of Chemistry and Food Technology, Technical University of Madrid, Madrid, Spain

Joana Oliveira REQUIMTE - LAQV - Department of Chemistry and Biochemistry, Faculty of Science, Oporto University, Oporto, Portugal

Pablo Ossorio Winemaker and Oenology Consultant, Bodegas Hispano-Suizas, Ctra. Nacional 322, Valencia, Spain

Felipe Palomero Department of Chemistry and Food Technology, Technical University of Madrid, Madrid, Spain

Isabel Pardo ENOLAB, Interdisciplinary Research Structure for Biotechnology and Biomedicine (BIOTECMED) Universitat de València, Burjassot, Spain

Karine Pedneault Département des Sciences, Université Sainte-Anne, NS, Canada; Institut de Recherche en Biologie Végétale, Jardin Botanique de Montréal, Montreal, QC, Canada

Gary Pickering Cool Climate Oenology and Viticulture Institute (CCOVI), Brock University, St Catharines, ON, Canada; Department of Biological Sciences, Brock University, St Catharines, ON, Canada; Charles Sturt University, Wagga Wagga, NSW, Australia; Sustainability Research Centre, University of the Sunshine Coast, Sippy Downs, QLD, Australia

Doris Rauhut Department of Microbiology and Biochemistry, Hochschule Geisenheim University, Geisenheim, Germany

Jorge M. Ricardo-da-Silva LEAF, Linking Landscape, Environment, Agriculture and Food, Higher Institute of Agronomy - ISA, University of Lisbon, Lisbon, Portugal

Susana Río Segade Department of Agricultural, Forestry and Food Sciences, University of Turin, Grugliasco (TO), Italy

Julián C. Rivas-Gonzalo Analytical Chemistry, Nutrition and Food Science, Faculty of Pharmacy, University of Salamanca, Salamanca, Spain

Luca Rolle Department of Agricultural, Forestry and Food Sciences, University of Turin, Grugliasco (TO), Italy

Han Shunyu Gansu Key Laboratory of Viticulture and Enology, College of Food Science and Engineering Gansu Agricultural University, Lanzhou, P.R. China

Mark Strobl Hochschule Geisenheim University, Geisenheim, Germany

Jose A. Suárez-Lepe Department of Chemistry and Food Technology, Technical University of Madrid, Madrid, Spain

Ma Tengzhen Gansu Key Laboratory of Viticulture and Enology, College of Food Science and Engineering Gansu Agricultural University, Lanzhou, P.R. China

Wendu Tesfaye Department of Chemistry and Food Technology, Technical University of Madrid, Madrid, Spain

Kevin Usher Agriculture and Agri-food Canada, Summerland Research and Development Centre, Summerland, BC, Canada

Aude Vernhet Montpellier SupAgro, Institute for Higher Education in Vine and Wine Sciences, Joint Research Unit Sciences for Enology, Montpellier, France

M. Victoria Moreno-Arribas Instituto de Investigación en Ciencias de la Alimentación (CIAL), CSIC-UAM, Madrid, Spain

James Willwerth Cool Climate Oenology and Viticulture Institute (CCOVI), Brock University, St Catharines, ON, Canada; Department of Biological Sciences, Brock University, St Catharines, ON, Canada

Hao Yan Institute of Fruit and Floriculture Research, Gansu Academy of Agricultural Science, Lanzhou, P.R. China

Bi Yang Gansu Key Laboratory of Viticulture and Enology, College of Food Science and Engineering Gansu Agricultural University, Lanzhou, P.R. China

Fernando Zamora Department of Biochemistry and Biotechnology, University Rovira i Virgili, Tarragona, Spain

Prologue

This book covers a gap in the specialized bibliography in enology, most of the books are focused in enology, wine microbiology or wine chemistry, but there is not a specific reference on red wine technology at the international level. Red wine technology has enough weight, repercussions, and specificities to be studied and explained as an independent discipline. Red wines are peculiar in terms of grape processing, the extraction of phenolic compounds during maceration, the development of malolactic fermentation, and the use of aging processes especially in high quality wines.

It includes the main aspects that must be considered during red winemaking treated by international experts of prestigious universities, research centers, and wineries from nine countries of traditional and emerging vine growing areas. Most of them are located in warm regions where most of the vine surface at international level is located, but the peculiarities of winemaking in cold regions are also considered by experts from these areas.

Concerning the contents, grape selection and quality is analyzed by Professor Rolle considering traditional and modern analytical techniques, either destructives or nondestructives, such as NIR or spectral image, also including the use of drones to automatize the work and information related to grape tasting. Acidity management in musts and also in wines is profusely described by Professor Comuzzo including the use of conventional acidification and deacidification agents, but also physical technologies like ion exchange resins or electrodialysis. The blend with acidic fractions of nonmatured grapes or musts is considered as well.

The maceration, as a fundamental process to extract phenolic compounds from grape skins into the wine during fermentation, is analyzed by my team, including conventional operations to facilitate the extraction. The use of emerging physical technologies is described with updated information on high hydrostatic pressures, pulsed electric fields, ultrasounds, irradiation, pulsed light, and others, considering how their use can speed the extraction while simultaneously improving the application of new biotechnologies of fermentation and helping to reduce SO_2 levels.

Use of non-*Saccharomyces* in enology is a trending topic in current enology. Professor Ciani has elaborated a deep revision of the main species used nowadays, with the potential applications to improve aromatic profile, color, acidity, structure, ethanol reduction and antimicrobial activities. The fermentation biotechnologies are described in detail by Professor Carrau considering the role of yeast in all the fermentative process and how the main parameters (nutrition, temperature and redox) affect yeasts performance.

Malolactic fermentation is a key process in red wine quality. Lactic acid bacteria species and their requirements are described by Dr. Moreno-Arribas together with technological strategies to improve the performance of this fermentation. The impact in sensory properties is included, but also the formation of off-flavors and the repercussions in wine safety. Complementarily, Professors Pardo and Ferrer made a full description of the cofermentation with yeast and bacteria as a current technology to reduce whole fermentation time, and with positive effects in lowering alcoholic degree and wine acidification.

Conventional molecular tools and high throughput techniques to analyze microbial populations in wines are extensively described by Professors Calo and Barros. The advantages and possibilities of RAPD-PCR, PCR-RFLP, T-RFLP, PCR-DGGE, QPCR, CE-SSCP, ARISA, and MALDI-TOF-MS are described.

Special attention is paid to wine aging with several chapters focused on the conventional barrel aging. A detailed description prepared by Professor Zamora about the management of the oxidative aging, wood types and origins, and the repercussion in wine composition and quality is included. Emerging accelerated aging is described by Professor Gómez Plaza explaining the advantages and management of wood chips and microoxygenation techniques. The suitable doses and aging time to better emulate barrel processes are discussed. Finally, my team includes the possibilities of biological aging on lees of red wines. This technique improves mouthfeel, roundness, and aromatic quality together with a protective effect in fruitiness and wine color. New applications as the use of non-*Saccharomyces* yeasts, physical techniques to speed yeast autolysis and aromatic impregnation of lees are described.

The evolution and management of the proanthocyanidin fraction are exhaustively described by Professors Ricardo da Silva and Jordão. Typical contents in different grape varieties and the evolution according to the winemaking process are studied with technical information about the monomeric, oligomeric and polymeric fractions.

Color is a key factor in red wine quality; the evolution and stability of anthocyanins are reviewed by Professors Escribano and Rivas. The effects of enological parameters like temperature, pH, and sulfites in anthocyanin stability are included, together with the copigmentation effects and color evolution during aging. The formation of derived anthocyanins and polymeric pigments with their influence in color stability is analyzed by Professor Mateus.

Main spoilage yeasts and their physiological, metabolic and growing parameters are described by Professor Malfeito. The production of ethyl phenols by *Brettanomyces/Dekkera* yeasts is analyzed together with the refermentations in bottle by *Zygosaccharomyces bailii*. Control measures and management of yeast populations are also considered.

Fining and wine stabilization are vital processes to keep the wine limpidity in attractive conditions for consumers. The understanding and management of the colloidal processes involved in most of the stabilization techniques are described by Professor Vernhet. Clarification, centrifugation and filtration, cold stabilization, and electrodialysis are considered as tools to improve wine stability.

Sensory analysis during winemaking is an irreplaceable tool used with a prospective intention to reach the sensory profile intended in the final wine. It is necessary to know how the evolution of the polyphenolic fraction, color and aroma will be, deciding previously how to work the maceration, fermentation and aging to reach the desired quality at the end of the process. Tasting techniques and strategies in berry maturation and during winemaking are described by the Master of wines Ballesteros and the wine consultor Ossorio.

Astringency is a typical property of red wines with strong repercussion in sensory quality. The origin of wine astringency, the mechanisms of its perception in mouth and its management is reviewed deeply by Professor Peña-Neira.

Aroma is an essential parameter affecting red wine quality. The aromas can originate in the plant and depend on the grape varieties or they can be developed during the winemaking and aging processes. Professor Rauhut reviewed in detail the different aromatic compounds that can be found in grapes, describing them and reporting information on their chemical structure, descriptors, and sensory thresholds. Complementarily, Professor Ferreira has revised the relation between the instrumental analysis and the aroma-active compounds with influence in wine flavor.

Sulfur dioxide is the most useful additive to control oxidation and microbial spoilage in wines but, at the same time, it is a toxic molecule that the wine industry tries to reduce year by year. Professor Gerbi has analyzed properties and applications of sulfur dioxide together with the most suitable ways to reduce its concentration in wines. Along with the use of sulfur dioxide, the bottle is not only the main way to serve wines to the consumer but also protects wine quality during distribution. Professor Strobl reviews main ways of wine bottling, the equipment used, closures, labeling, and packaging systems.

Finally, the specificities of winemaking in cold areas are included in two chapters focused on northern regions of Canada and China. The special management of the viticulture, winemaking, and the ways to control unsuitable maturity of grapes is reviewed by Professor Kemp. Professor Ma describes the use of buried viticulture in extreme cold regions to get the survival of the vine during winter.

An edited book means a complex work of coordination and management; more than 600 emails are in my mail box concerning this book. I would like to acknowledge all the prestigious and knowledgeable Professors and experts that have contributed because they have made chapters of very high quality and some of them in very stressful situations. The quality of this book is due to their work and experience. Special thanks go to my colleagues at the Food Technology department at the Universidad Politécnica de Madrid for their help and understanding during the editing of this book and also for their useful advice, especially to professors José Antonio Suárez, Karen González, Iris Loira, Wendu Tesfaye, Felipe Palomero, María Jesús Callejo, and my PhD students Carlos Escott and Juan Manuel del Fresno. I am grateful to Patricia Osborn, Senior Editor of Elsevier, for the confidence in me as editor of this book and the support received during the entire project. My gratitude also to Tasha Frank, Senior Editorial Project Manager of Elsevier, and Denny Mansingh, Project Manager of Elsevier, for the full support and kindness, helping me manage the editing of all the chapters. And finally, but especially, to my family: Cari, Jaime, and María, most of the time spent on this book belongs to them.

CHAPTER 1

Grape Maturity and Selection: Automatic Grape Selection

Susana Río Segade, Simone Giacosa, Vincenzo Gerbi and Luca Rolle
Department of Agricultural, Forestry and Food Sciences, University of Turin, Grugliasco (TO), Italy

1.1 PHYSICOCHEMICAL CHARACTERISTICS OF ENOLOGICAL INTEREST

Modern enology bases the premises about wine quality on the synergic "vineyard—winery" alliance that permitted to surpass the dualism that, in recent years, has often characterized an enological production that is particularly quantity oriented. The basis of wine quality is intrinsically confined in that of its raw material: the grape. In particular, grape quality at harvest derives from the compositional balance of several primary and secondary metabolites: sugars, organic acids, colorant and tannic substances, odorous compounds, and their precursors. In particular, the phenolic component in red wines is strategic for wine characteristics, since it influences the color, the gustative perceptions, and overall the longevity. Together with the aroma precursors, the phenolic substances are therefore mainly responsible for the wine "typicity." These aspects are even more relevant for a varietal enology (i.e., winemaking of a single grape variety, which is often the expression of a determined territory), where the emphasizing of the raw material must be at the maximum level. In fact, in this case there is not the possibility of finding the right sensorial balance that can derive from the blend of grapes from different cultivars.

The importance of phenolic compounds contained in the solid parts of the berries (skins and seeds) and their essential role in the sensorial characterization of different red wines are nowadays a known aspect, and have been known in scientific literature since the 1980s (Arnold et al., 1980; Robichaud and Noble, 1990). The chromatic characteristics, astringency, and bitterness of a wine are highly influenced by the content and degree of polymerization and/or condensation of phenolic compounds (Blanco-Vega et al., 2014; Cheynier et al., 2006). Knowing the polyphenolic content and profile of the grapes allows the maceration and winemaking process to be planned, allowing winemakers to fully exploit the potentiality that the grape attains in the vineyard (Zanoni et al., 2010).

Nowadays, thanks to advanced chromatographic techniques, better performing detectors, and specific analytical protocols dedicated to grape extracts and wines (Aleixandre-Tudo et al., 2017; Valls et al., 2009), it is possible to determine and quantify the several molecules belonging to the different classes of the phenolic compounds and their derivatives, which have been classified as: anthocyanins, hydroxybenzoic acids, hydroxycinnamic acids, stilbenes, flavanol, flavonol, and others (Ribèreau-Gayon et al., 2006). In parallel with this new knowledge, numerous recent studies have researched and valued the relationships existing among the mouthfeel properties, the different polyphenolic classes/molecules, and the relative sensory perceptions threshold (Brossaud et al., 2001; Gonzalo-Diago et al., 2014; Laguna et al., 2017a). In particular, regarding grape and wine tannins, the molecular sizes and the monomeric composition of proanthocyanidins have a large influence on the sensation of astringency. More specifically, the greater degree of polymerization and the greater percentage of galloylation will cause a greater sensation of astringency (Laguna et al., 2017b; Kontoudakis et al., 2011; Vidal et al., 2003).

Despite this important new knowledge, because of the high scientific competence requested from the analysts and the cost itself of the analysis, these chromatographic methodologies are still limited by not being very applicable directly to the evaluation of the grape in preharvest, at least when there are a lot of samples to be examined. Therefore, spectrophotometric analysis of phenolic compounds in grapes and wines are still the most used in the viticultural-enology sector (Aleixandre-Tudo et al., 2017). Although characterized by a major spectra of determinable molecules, global indexes such as total anthocyanin, total polyphenol, oligomeric flavanols (flavan reactive to vanillin), or polymeric tannins (proanthocyanindin) are evaluated in commercial and winery laboratories by spectrophometric methods based on their ease of use as a routine analytical technique (Di Stefano and Cravero, 1991; Rolle et al., 2011a).

Even though the use of spectrophotometric indexes permits to best assess the content of anthocyanins and tannins present in the grape skin and seed, in many cases adequate direct correlations were not found with the corresponding analytical parameters in related produced wines.

In fact, in maceration during skin contact, only part of the phenolic compounds is extracted. Consequently, in the last two decades, winemakers have been very interested in a new concept called "Phenolic Maturity," defined by Glories and Augustin (1993) as the concentration of phenolic compounds in grapes, and the ease with which they are released. This definition encompasses the anthocyanin concentration in the skin, their degree of extractability (cell maturity index; EA%), and the flavanol concentration in the seeds and skins and their degree of polymerization (seed maturity index; Mp%). These two indexes, determined with some modifications with respect to the original methods useful only for French varieties, are nowadays quite commonly used in the enology sector in order to select the grape ripeness thanks to high correlation with the color index (CIE $L^*a^*b^*$ parameters, color intensity and hue) and phenolic characteristics of wines (Cagnasso et al., 2008; Romero-Cascales et al., 2005). There have been authors, though, that proposed different methodologies, that are cheaper and less laborious, to determine a correct grape phenolic maturity (Celotti et al., 2007; Kontoudakis et al., 2010).

Grape maturity is associated with physicochemical changes in the skin and pulp cell walls. During the ripening, changes in the composition and structure of the cell wall, as well as in the structure of the tissue, may determine the mechanical resistance and the texture of the berry (Zouid et al., 2013). These structural changes can be assessed in a subjective way by simple berry tasting, an immediate means of grape evaluation. In addition to berry mechanical traits and sugars/acidity ratio estimation, the berry sensory assessment was found to be useful for the evaluation of skin and seed astringency, important parameters related also to overall grape ripeness. However, a high experience of the matter is necessary to perform meaningful sensory assessments, especially for seed phenolic ripeness evaluation, since it is a complicated task and it leads very quickly to sensory fatigue (Le Moigne et al., 2008; Olarte Mantilla et al., 2012). To exploit these changes, a very innovative instrumental approach was proposed by Río Segade et al. (2008), where phenolic ripeness of grape skin was assessed by texture analysis. A significant multiple linear regression was found between EA%, berry skin break force, and thickness. Subsequent studies have shown that the break force and the thickness of berry skin can be considered mechanical properties, adequate for the estimation of the degradability of the skin cell wall and, therefore, of the extractability of anthocyanins from skin to must/wine (Río Segade et al., 2011a; Rolle et al., 2008). More recently, the evolution of mechanical and acoustic texture parameters of grape seed during maturation was studied (Rolle et al., 2012a,b), and important relationships with extractable phenols were found (Rolle et al., 2013). Like this, the characterization of the mechanical properties of grape berries appears to be an important parameter to understand grape ripening. However, published studies have shown that the changes of several grape textural properties during ripening are strongly influenced by the growing conditions and vintage. So, the "terroir" effect on mechanical behavior must be considered in the data interpretation (Le Moigne et al., 2008; Río Segade et al., 2011b).

All of this research activity, to some extent even pioneering, permitted, on one hand, to give useful tools and methodologies to the sector operators in order to objectively define the level of grape ripeness, and, on the other hand, to increase the awareness of the red wine winemakers to find new criteria and indexes in addition to the traditional measurement of technological variables (i.e., sugars, organic acid content, pH), in the choice of the harvest date. Therefore, over the last years there has been a clear inclination of the enterprises toward increased consideration of polyphenolic maturity or aroma. Harvest dates need to be carefully considered since they can be based on objective chemical—physical data interpretation or also only subjective evaluations of optimum berry composition in view of ultimate wine quality, whose definition is exposed to individual interpretation and trends, and may also depend on commercial targets, market constraints, processing capacity, and other factors.

The synthesis and accumulation of primary and secondary metabolites in the skin, pulp, and seed of the grape berry during maturation has been studied in depth. The concentration of sugars, amino acids, phenolic compounds, and potassium increases, while the content in organic acids, particularly malic acid, decreases (Ollat et al., 2002). Regarding phenolic compounds, anthocyanins are gradually accumulated in berry skins from veraison through grape ripening, but the anthocyanin concentration may decrease just before harvest and/or during overripening (Fournand et al., 2006). However, proanthocyanidins are mainly accumulated in berry skins before veraison (Cadot et al., 2006a). The highest concentration of seed proanthocyanidins is achieved at veraison and, from this moment, they decline slowly until close to grape ripeness but thereafter remain relatively constant

(Cadot et al., 2006b). For these behaviors of the grapes during ripening, it is generally thought that insufficiently ripened grapes may produce more astringent and bitter wines because their seeds can release a higher amount of proanthocyanidins, which are highly galloylated (Romeyer et al., 1986). Grape-originated aroma makes an important contribution to the varietal feature of the final wine flavor, in particular for some aromatic varieties used in the production of "special wines" (Torchio et al., 2012, 2016). Therefore, understanding the changes in secondary metabolites during berry ripening may provide predictive information about the link between grape and wine aroma. Free and glycosylated terpenes are accumulated in the grapes from veraison. Normally, the increase of terpene concentration is observed from the first stages of grape ripening until maturity. Some authors observed that terpene concentration starts to decrease before the maximum sugar concentration is reached in grapes (Lasanta et al., 2014), even though studies have reported a continuous accumulation of terpenes even after maturity (Schwab and Wüst, 2015).

However, in spite of the trend toward longer hang times, harvest dates have advanced and grape maturation occurs earlier. A widespread observation is that harvest dates have advanced, especially in the last 10–30 years (Mira de Orduña, 2010). The extremely high sugar concentrations reached at harvest today, especially in warm climates, may rather be associated with the desire to optimize technical or polyphenolic and/or aromatic maturity. Climate change is exerting an increasingly profound influence on vine phenology and grape chemical composition. Furthermore, the relationship between climate and berry mechanical properties has attracted considerable attention as it affects wine quality. In particular, the differences in break force and energy of berry skin among the following years on the berries of the same variety, due to the genotype–environment interaction, are mainly attributable to several bioclimatic indexes (Rolle et al., 2011b). Climatic variables registered during the grape ripening, such as daily minimum temperature, average daily maximum temperature, average daily minimum humidity, average daily maximum humidity, total precipitations, daily maximum precipitations, average daily thermal excursion, leaf wetness duration, daily maximum duration of leaf wetness, absolute maximum temperature, number of rainy days, thermal sum over a 10°C threshold, total thermal excursion, and others, are indexes which are strongly correlated with the berry skin hardness. Softer skins seem to be characterized by a lesser release of red pigments from grape skin into wine during the winemaking process (Rolle et al., 2009). So, climate change ultimately affects winemaking, wine microbiology and chemistry, and sensory aspects (Mira de Orduña, 2010).

In view of these considerations, it can become very difficult for a winemaker to set the optimal harvest date, since there cannot be in every vintage a contemporaneity of the technological, phenolic, aromatic, and "textural" maturity. Depending on the characteristics of the wine planned, it is necessary to make choices and decisions based on objective data, relative to that chemical–physical parameter that majorly impacts on the final wine quality. To further complicate the problem, it is well known that variations in timing and the extent of ripening occur between berries on a cluster, between clusters on a vine, and between different vines in a single cultivar vineyard (Belviso et al., 2017; Letaief et al., 2008; Noguerol-Pato et al., 2012; Tarter and Keuter, 2005).

As already discussed, since grape maturity at harvest strongly affects wine quality, an adequate strategy of berry harvesting and/or a successive selection in winery based on objective quality indicators is a key issue to produce high-quality and premium wines. Therefore, both the possible noncontemporaneity of ripeness and the heterogeneity of ripeness levels inside the vineyard must be managed in order to maximize the enological potentiality of the grape delivered in the winery. In the field, preharvest berry selection makes possible the identification of vineyard sections or even homogeneous groups of vineyards with similar grape composition. Precision Viticulture approaches are highly profitable to manage the harvest but require the use of advanced mapping technologies, involving drones, satellites, global positioning systems, geographic information systems, or local and remote sensors. However, intra- and intercluster variability in berry quality can only be solved in the winery. In this case, postharvest strategies are used, such as manual sorting tables or automatic selection based on berry density and optical measurements [visible/near-infrared (NIR/VIS) spectrometry] combined with multi- and hyperspectral imaging.

1.2 VINEYARD APPROACHES TO GRAPE SELECTION AND HARVEST DATE DETERMINATION

1.2.1 Spatial Variability in Vineyard and Precision Viticulture Tools

The vineyard is the focal point of operation for grape quality. As a complex agro-ecosystem, the vineyard is affected by environmental-orographic conditions such as climate, soil, exposure, altitude, and by the factors

TABLE 1.1 Comparison of air (UAV, helicopter, aircraft) and satellite measurement platforms for vineyard characterization

	Spatial resolution	Field of view	Usability	Payload mass	Cost for data acquisition
UAV	0.5 – 10 cm	50 – 500 m	Very good/easy	Can be limited	Very low
Helicopter	5 – 50 cm	0.2 – 2 km	Pilot mandatory	Almost unlimited	Medium
Airborne	0.1 – 2 m	0.2 – 2 km	Pilot mandatory	Unlimited	High
Satellite	1 – 25 m	10 – 50 km	–	–	Very high, particularly for high-res stereo imagery

Table reported from Candiago et al. (2015).

induced and determined by the growers including the choice of grape variety and rootstock, vine density and spacing, training and pruning system, and the vineyard management. Even when few of the aforementioned parameters change, we may see remarkable differences in grape quality and harvest outcome.

The approach of Precision Viticulture (Bramley and Proffitt, 1999; Bramley, 2010) takes into account the high vineyard variability and tries, with the aid of modern technology, to maximize the results in terms of grape quality, production, or other desired traits with respect to common decision-making choices done in the vineyard, the latter often characterized by simple uniform management. Although the attention is focused on the outcome of grape growing (i.e., harvest and grape final product), the use of these techniques span across the life of the vineyard to accommodate the spatial variability and modulate accordingly the fertilization treatments, pest control, water availability, and other agronomical practices.

Precision Viticulture approaches need an extensive knowledge of the variability found in the vineyard. A model of the vineyard traits must be prepared in advance to be able to differentiate areas inside a vineyard parcel for harvest purposes. This model can be prepared using many sources and different data.

A detailed review by Matese and Di Gennaro (2015) underlined the technologies available to gather vineyard data during the entire vegetative season. Remote and proximal sensing techniques may be used for this purpose, and generally combined to develop a comprehensive vineyard model. Remote sensing tools include satellite measurements and aircraft ground monitoring: these two systems are different in terms of the sensors used, the spatial resolution, and the possibility of setting the time of the data acquisitions. Obviously, also the data acquisition cost may have an impact in evaluating the best choice for vineyard mapping according also to the parcel size. Recently, the use of unmanned aerial vehicles (UAVs) for remote sensing purposes is increasing, as it becomes cheaper than alternative means (such as aircraft deployment), and could allow for higher resolutions depending on the required information. However, the drawbacks in agriculture (and in particular viticulture) are related to the short period of continuous activity, the need to minimize shadowed areas, and weather conditions (Borgogno Mondino and Gajetti, 2017). Some characteristics of airborne and satellite tools are shown in Table 1.1, as reported by Candiago et al. (2015).

Multispectral images obtained using remote sensing sources are required for the calculation of spectral vegetation indices. A common index, widespread in literature for its immediacy and effectiveness, is the normalized different vegetation index (NDVI), first proposed by Rouse et al. (1974) and related to the photosynthetically active biomass present in the field (Bramley, 2010). An extensive list of other spectral vegetation indices used in viticulture is reported in the study by Rey-Caramés et al. (2015). Depending on the desired aim, the use of the most suitable vegetation index permits the building of vigor maps and planning of the vineyard operations, including harvest.

Proximal sensing techniques consist of sensors located directly in the vineyard, which are able to monitor parameters such as soil–water retention, temperature, and sun exposure in different points of the field. In addition to the abovementioned examples, nondestructive portable instruments are a possible solution for the direct evaluation of the grape bunch or of single berries. To improve the measurement density and acquisition speed using proximal sensing techniques, the use of all-terrain vehicles equipped with portable instruments and sensors was also proposed (Diago et al., 2016).

1.2.2 Grape Harvest and Selection in Vineyard

Harvest phase is the most important by a product-oriented view. A selection of grapes during this phase according to several methods is possible, depending on the vineyard characteristics. This selection may be exerted:

(1) by conducting a manual selection based on visual inspection;
(2) by manual picking of parts of the cluster;

(3) by harvesting grapes from some areas of the vineyard separately, thus creating berry groups in the same harvest date;

(4) by conducting the harvest in different times, and in each one of them selectively taking grapes only from parts of the vineyard.

All these approaches may be aided by novel-developing techniques, such as automated grape cluster selection and harvest, which look promising in precise harvest management.

Regardless of the grape selection at harvest strategy, a well-constructed vineyard model could be particularly useful to improve the sampling strategy for the estimation of grape yield, a crucial parameter for production especially when subjected to legal maximum limits. Carrillo et al. (2016) used data gathered by airborne imagery to calculate NDVI in nine different fields, and then sampled at regular space intervals to determine grape yield. With this strategy it was found that a linear correlation existed between NDVI and grape yield, which permitted the improvement of the assessment of the vineyard yield by 10% on average with respect to conventional random sampling.

Proximal sensing techniques may also be useful for yield estimation. Digital image analysis techniques were previously used to evaluate canopy porosity and cluster exposure on cultivar (cv). Cabernet sauvignon vines, where the vine fruit mass was estimated with digital photographs of about 1×1 m area achieving a correlation R^2 of 0.85 in the vineyard, although with the need of white background frames in the background to enhance image recognition capabilities (Dunn and Martin, 2004).

Another practical use of proximal sensors may be related to nondestructive fruit quality determination, in particular regarding phenolic quality. Rolle and Guidoni (2007) tested the use of a portable colorimeter to estimate the red grape anthocyanin content. The authors developed a new grape index (CIRWG) which was found to be correlated with the actual anthocyanin content of the cv. Nebbiolo analyzed grapes ($R^2 = 0.89$). Cerovic et al. (2008) used berry fluorescence to assess grapes phenolic maturity, testing an industrial prototype (Multiplex) presenting multiple excitation sensors directly on grape bunches (Fig. 1.1D,E). The obtained indices were correlated with anthocyanins ($R^2 = 0.96$; Fig. 1.1C) on Pinot noir and Pinot Meunier red varieties. Further studies with the same latter equipment on 12 different cv. Barbera vineyards underlined a correlation between FERARI index (calculated from far-red fluorescence in the red data) and anthocyanins ($R^2 = 0.81$), and a correlation with a new proposed index (FLAV_UV) and flavonols presenting acceptable correlation values ($R^2 > 0.7$; Ferrandino et al., 2016).

The information gathered with these proximal tools may be also used for integration with other sources or existing field models, to assess the vineyard variability, and/or to estimate the harvest date of portions or of whole vineyard parcels.

1.2.2.1 Manual Grape Selection in Vineyard by Visual Inspection

Visual inspection is probably the most ancient selection technique carried out in a vineyard. It is a not-precise, time-consuming but fairly simple activity, characterized by the need of visual assessment and

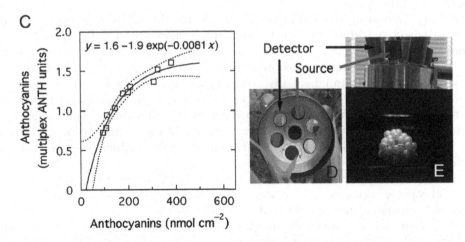

FIGURE 1.1 Validation of the optical sensors for berry fluorescence on grape bunches. In (C), each data point is for skins obtained from 20 representative berries of the bunch measured with the Multiplex. Dotted lines indicate the 95% confidence interval. A front view of the Multiplex sensor is presented in (D). The picture (E) illustrates a Chardonnay grape bunch illuminated by the Multiplex during optical measurements. Figure portion reported from Carrillo et al. (2016).

selective removal of parts of the cluster, or by not picking some clusters at all, based on rapid quality evaluation done by the operator. This technique at some level is involuntarily carried out by manual pickers during harvest, although additional care is needed when harvesting a vineyard with berry damage, dark berries, and/or serious level of infections such as sour rot or black molds, leading to bacteria, *Botrytis* or *Aspergillus* spp., spreading.

An accurate removal of damaged berries may improve the wine quality in terms of ocratoxin A (OTA) concentration reduction. Guzev et al. (2006) evidenced that, at the same bunch level, the parts with VIS decay symptoms infected by *Aspergillus carbonarius* molds contained 0.42–0.72 µg/L OTA, in contrast with bunch parts without symptoms that did not contain this toxin. Given that the maceration process, either lightly conducted during grape transport or in the first phases during cellar winemaking, induces OTA production by *Aspergillus* molds, a selection of grapes could further aid to limit the OTA contents in wines, particularly for red wines production (Gambuti et al., 2005).

1.2.2.2 Selective Harvest of Different Parts of the Cluster

When considering healthy grapes, a selective harvest may be carried out not exclusively based on area or by the time of picking, but also by considering different parts of the cluster. The berries contained in the grape cluster do not ripen homogenously, and some differences may apply in terms of grape sugars, acidity, and secondary metabolites as also induced by berry weight and grape microclimate (Pisciotta et al., 2013).

A study on Brancellao red wine cultivar evidenced that in the last days of ripening important modification occurred in terms of sugars and anthocyanins content according to the bunch position. Twenty days before full ripening the berries belonging to the bunch tips were richer in sugars than the shoulder portions, while no differences were found at full ripening. Regarding anthocyanin accumulation, their concentration in the berries in the shoulder portions remained stable in the given period. However, the bunch tip portion evidenced a lower anthocyanin concentration (−33%) 20 days before ripening, with a rising trend afterwards leading to no differences at full ripening (Figueiredo-González et al., 2012). Therefore, depending on the product and grape cultivar considered the differences induced by berry bunch position may have different impacts according also to the harvest date. An example of this selective harvest approach is used in the production of "Nebbione" sparkling wine from red cv. Nebbiolo grapes by the early harvest of grape bunch tips, achieving higher acidity and lower sugar contents, a positive trait for sparkling wine production (Shah, 2016).

1.2.2.3 Selective Harvest Based on Vineyard Area

Taking into account the vineyard spatial variability, with the tools and methods provided by Precision Viticulture previously described, it is possible to build vigor maps which may be detailed even at single-vine level. Pioneering experiments conducted on grapes, based on airborne imagery systems mounted on a light aircraft, successfully determined three vigor zones and proceeded to separate the winemaking of the determined parcels. The wine produced by grapes harvested from the low vigor zone was found to be different to the other two groups (Johnson et al., 2001).

An extensive study by Gatti et al. (2017) prepared a vigor map by satellite sensing and calculation of the NDVI index of a cv. Barbera red grape vineyard. Then, three vigor zones with equal coverage were determined and separately sampled at harvest. Grape yield and composition was significantly affected by the vigor zone considered: the lowest bunch, berry mass, and yield were registered in the low vigor zone, while the highest leaf area/yield ratio was found here. By contrast, the highest production values (and lowest leaf area/yield ratio) was found in the high vigor zone, therefore setting a clear correlation between remotely-calculated vigor and yield figures. Interestingly, also the bunch compactness parameter (total berry mass/rachis plus shoulder length ratio) was positively correlated resulting in significant differences between the three vigor areas considered. As vigor and yield increased, must reducing sugars and phenolic traits decreased their values.

As previously mentioned, the recent use of UAV or remotely piloted aerial systems is a promising technique for quick and cheaper evaluations of the vineyard. Rey-Caramés et al., (2015) acquired multispectral images using UAV and calculated new normalized indices (NV_1 and NV_2), with the aim to correlate these results with real values gathered by analysis of the grapes from 72 different locations in the vineyard. As one of the outcomes of the work, detailed berry characteristics maps were produced (Fig. 1.2), showing the vineyard spatial variability in terms of the spectral indices determined (NDVI, G, NVI_1, NVI_2) and grape compositional parameters from vineyard sampling (pH, total soluble solids, titratable acidity, anthocyanins, and total phenols). For each sampling point, these results were correlated with multispectral indices. Key vegetative

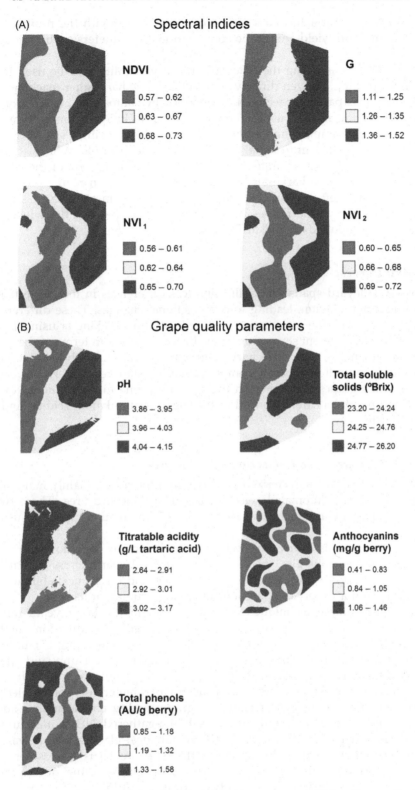

FIGURE 1.2 Maps obtained for spectral indices (A) and grape composition parameters (B) in a Tempranillo vineyard. Maps were represented by terciles. Figures reported from Rey-Caramés et al. (2015).

parameters and mean cluster weight achieved satisfactory correlations with the multispectral indices tested, however the correlation found with yield and grape compositional parameters such as acidity or phenolic content was poor.

Vigor maps or other means of describing the vineyard spatial variability may be used in the organization of manual or machine-assisted picking, or even directly sent to the grape-harvesting machine for conducting a differential harvest "on-the-go." This particular technique, described by Santos et al., (2012), consisted of the preliminary acquisition of multispectral images using remote sensing techniques (aircraft monitoring), and then combining these with in-vineyard sampling and anthocyanin analysis following a spatial grid, culminating in the preparation of a map of the predicted grape quality, with only two possible options (high/low quality). The grape harvester was equipped with two separate grape containers at either side of the machine. At harvest, at each position of the vineyard the grape harvester verified in the electronic map the predicted quality and conveyed the harvested grapes in the left or right container, thus providing in a single pass two separate products according to predicted grape quality. The grapes were separately processed in the winery, obtaining different products in terms of phenolic compounds of finished wines.

1.2.2.4 Time-Differential Harvest

As previously discussed, vineyard spatial variability induces differences in the harvested grapes in terms of grape quality and ripening differentiation, leading to compositional changes. These differences can be exploited to obtain two different products with the same harvest, either by manual picking or using a grape harvester.

Based on these observations, a different separation of the harvest process in terms of the time of harvest is possible. By constructing vineyard maps, a possible harvest may be carried out only on a portion of the vineyard, leaving the other grapes to ripen until satisfactory compositional traits are reached. However, this proposed technique may be expensive, while grapes remaining on the vine are prone to the influence of the weather and to possible pest spreading, and require a careful control of the sugar accumulation, acidity, and secondary metabolites, especially in a hot climate.

1.2.2.5 Toward Automated Grape Cluster Selection and Harvest

Previously described methods were characterized by different approaches, mainly manual handling for visual assessment and selection, or by the automated selection based on vineyard area. These two approaches have either the precise selection capability at bunch level, or the harvest automation capability using grape harvesting machines, but not both.

Robot automation in the vineyard is developing to be able to do similar operations to the human counterpart (Matese and Di Gennaro, 2015). For this aim, the technology required spans across digital imaging, calibration models which may be provided by Precision Viticulture techniques, and mechanical abilities to move in the vineyard autonomously and carry out a variety of operations including pruning or harvest.

To focus on digital imaging, these techniques have improved in the latter years for understanding the agricultural environment for harvest purposes, including eye—hand coordination methods (Zhao et al., 2016). Recent studies underlined the ability for vision devices to detect grape clusters, their form and volume, berry size, and even the ideal cutting point for grape picking (Dunn and Martin, 2004; Cubero et al., 2014; Liu and Whitty, 2015; Luo et al., 2016; Fig. 1.3). To detail an application of digital imaging assessment in the vineyard at harvest, one potentially useful parameter for grape selection is bunch compactness, as it was previously found to be correlated to grape quality characteristics such as variations in sugars, acidity, and phenolic composition, and also to satellite-gathered NDVI measurements (Gatti et al., 2017). Using image analysis and partial least squares (PLS) statistical elaborations it was possible to predict bunch compactness with a predictive capability of 85% with respect to visual assessment carried out according to International Organization of Vine and Wine (OIV) compactness rating descriptor (OIV, 2009). This first prediction result was achieved in lab conditions (Cubero et al., 2015).

Some commercial robots have been proposed in recent years for the vineyard environment. They are mainly used for data gathering (such as multispectral indices or imagery) or for some operations like pruning or grass-cutting, while harvest capabilities are in development (Matese and Di Gennaro, 2015). Although purchase and operational costs may be high in the first years, their improvement will surely represent a milestone in the Precision Viticulture approach toward automation and, if possible, grape quality enhancement, but they will be very difficult to merge with vineyard traditional techniques.

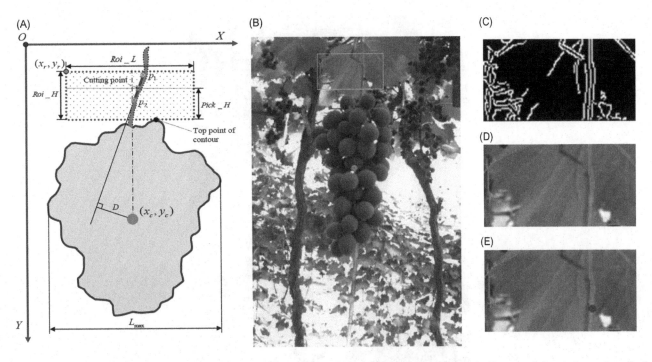

FIGURE 1.3 Solving the process of finding the grape bunch cutting point in vineyard. (A) Schematic diagram of the cutting point calculation model; (B) the detected information on the example image; (C) edge binary image of the stem region of interest (ROI); (D) detected line segment; (E) calculated cutting point (blue point). Figure reported from Luo et al. (2016).

1.3 GRAPE SELECTION IN WINERY

Besides the vineyard, grape selection can be carried out also in cellar. Indeed, the production of high-quality wines may require the postharvest grape selection in the winery with the aim of reducing not only the presence of unripe, damaged, and rotten berries but also all matter other than grape (MOG) from vines, such as petioles, rachis, and leaves, which may affect the sensory perceived quality and chemical composition of the resulting wines (Ward et al., 2015). This selection acquires particular importance in the case of mechanically harvested grapes. In winery, the intra- and intercluster variability in the berry quality could be minimized by homogenizing the berry developmental stage by postharvest sorting strategies. Nevertheless, the grapevine characteristic ripening asynchrony could be exploited to make wines of different qualities and to potentiate specific quality parameters from a single harvest unit, as previously mentioned. Therefore, an adequate berry sorting strategy has to be implemented depending on the quality objectives of each winery. However, berry sorting is a complex operation because the berry sorting based on an attribute will have intrinsic effects on other berry attributes.

1.3.1 Sorting Tables in Winery

Manual sorting tables are usually used to remove undesired grapes and MOG by hand. The winery staff (sorters) working on each side of the sorting table inspect and remove the undesired material. The selection can be carried out directly on grape bunches before destemming (Fig. 1.4A) and/or on individual berries after destemming (Fig. 1.4B). Nevertheless, grape sorting after destemming is becoming increasingly important, particularly in the production of premium wines. On the one hand, belt sorting tables consist of a conveyor belt made of food-grade PVC, on which the harvest material is emptied and driven. It permits the grapes to be spread out making the manual sorting easier. On the other hand, vibrating sorting tables are constructed entirely of stainless steel. The use of vibration as a driving movement facilitates a more homogeneous distribution of the material on the table width, and the presence of perforated areas allows an easier separation of healthy and ripe grapes from free juice, herbaceous materials and undesired berries. In both manual belt and vibrating sorting tables, a speed variator permits control of the speed, which is a critical parameter to ensure a correct balance between productivity and quality. The number of sorters is defined mainly by the length of the sorting table. The main disadvantages of manual sorting are the high personnel requirements and the relatively slow throughput.

FIGURE 1.4 Use of manual sorting table in winery applied on grapes before (A) or after destemming (B). Source: *Photos by Prof. Antonio Morata and Rafael Cuerda, Director técnico de Bodegas Comenge, DO Ribera del Duero, Spain.*

In recent years, automatic sorting tables for destemmed grapes are being increasingly used in the winery. Among mechanical-based devices, vibrating sorting tables can be easily automated to a lesser or greater extent by coupling a feeding hopper. The Mistral sorting line (Vaucher Beguet) and Delta Rflow (Bucher Vaslin) are efficient automatic systems, which combine a first separation step to remove juice and vegetal material with a second sorting step to separate quality berries. A vibrating sorting table is used for grape distribution in a uniform layer by means of vibration while juice and small MOG are removed through perforated areas located in the table. The larger herbaceous material is trapped at the end of the vibrating table via an interchangeable grid and eliminated manually. Afterwards, whole berries are accelerated (only in high-throughput Delta Rflow) and separated with an air jet sorting system of adjustable power. On the one hand, the surface of whole berries offers a low air resistance when exposed to the air stream, and therefore they are deflected very little from their vertical course. On the other hand, damaged berries and other remaining material offering a higher air resistance are blown off course and directly removed. The intact and crushed berries are then better separated using two separator plates placed under the air jet system. Sorted quality berries are then transported usually by a conveyor elevator. Different parameters can be set according to the required grape quality (grid, air jet power, and separator plates). These mechanical sorting devices offer a processing capacity from 3 to 15 t/hour and are made entirely of stainless steel.

1.3.2 Size, Density, and Image Analysis Sorting Equipment

Taking into account that berry diameter is a significant parameter for the wine quality (Friedel et al., 2016), another system has been developed to sort the grape berries according to the diameter, namely Calibaie (AMOS Industrie). The system consists of chain-driven parallel rollers which rotate toward the front of the machine. The berries meeting the minimum size set are driven on the top of roller drum to the end of the roller table, whereas those smaller berries fall into the roller drum and then are driven to the bottom. It is possible to set different grape sizes working at a throughput from 4 to 8 t/hour. However, the grape size sorting is usually carried out after sorting by ripeness.

The latest generation of automatic sorting machines are based on the separation of grape berries by densimetric flotation or image analysis. These systems allow a more accurate berry sorting by setting the operative conditions according to objective quality parameters. Density sorting of berries is a powerful tool for the separation of wine grapes or table grapes (Río Segade et al., 2013; Rolle et al., 2012, 2015; Torchio et al., 2016). In fact, density sorting is already used in the Tribaie system (AMOS Industrie; Parenti et al., 2015), whose separation principle consists of floating grape berries in a grape juice/sugar solution of a well-defined density. Therefore, the sorting operation is based on the sugar concentration, specifically on Brix degree, which is selected according to predefined quality parameters. After a first removal of large green waste by a rotating disc separator and a second gravitational separation of rotten grapes and small herbaceous material using a rotating cylinder, the destemmed whole berries are immersed in the separating solution. Less ripe grapes have a density lower than that corresponding to the solution and they float on the top of the solution, whereas the ripe grapes sink. Then, floating grapes (less ripe) are siphoned off the system, while the ripe sank grapes are separately used for winemaking. Furthermore, the system permits the subsequent recycling of the floating solution by the separation of solid wastes. The Tribaie system is highly versatile because only the density of the sugar solution has to be carefully controlled. Processing flow rates from 5 to 20 t of fruit per hour are achieved.

Optical sorting machines perform the separation of grape berries by analyzing their electronic image via high-speed image recognition. Optical sorters require the use of high-speed cameras and image-processing software for sorting the destemmed grapes according to color and shape parameters (Bucher Vaslin Delta Vistalys, Pellenc Selectiv' Process Vision, Defranceschi-Protec X-Tri, Key VitiSort; Pellenc and Niero, 2009). To get a very accurate sorting, grape berries are spread out and conveyed to be delivered in a uniform way to the camera. In low-throughput machines (up to 5 t/hour), the berries are conveyed by gravity while a roller equipped with spikes and a spreader bar is usually used to facilitate their distribution. In high-throughput machines (up to 12 t/hour), the destemmed berries are driven by a conveyor belt provided with nozzles where they are spread out and then stuck in-between the nozzles at a high speed. Afterwards, the berries are lighted using electroluminescent diodes to ensure a continuous, steady and focused lighting. The images obtained by the high-speed video camera are analyzed, and undesirable matters are removed automatically. The optical sorting machines have the ability to remove MOG and crushed or unripe berries as well as to create berry batches depending on the specific quality criteria by successive sorting.

Image analysis is an important tool for quantitative, objective, and accurate estimation of grape quality as an alternative to the qualitative and subjective visual inspection performed by trained evaluators. In image analysis, the grape berries or seeds are identified in the images, and morphological and colorimetric features can be then extracted from each one. A method was developed to classify table grapes and/or to sort wine grapes before winemaking according to bunch compactness (Cubero et al., 2015). This parameter was already previously discussed as an interesting tool for bunch classification in the vineyard, and it can be useful for winery criteria selection. Acquired images by a digital camera are processed using supervised segmentation, and a predictive PLS model was then developed from the morphological features extracted. A conveyor belt drives the bunches toward the camera equipped with a backlighting illumination system to acquire images with a high contrast between the bunches and background. Other authors proposed a multiperspective simultaneous imaging approach using reflective mirrors with the aim of assessing the morphological quality of grapes. Images are captured from different perspectives of whole grape bunches in-line (Yuan et al., 2016). Furthermore, automated image analysis can be implemented on sorting tables for the accurate determination of the berry size and weight in destemmed wine grapes (Cubero et al., 2014) and to classify berries according to color features (Richter et al., 2015).

1.3.3 New Perspectives for the Direct Grape Quality Evaluation and Selection in Winery

VIS/NIR spectroscopy has often been used in the food industry for quality control, and numerous works have been published on the reliability of this analytical technique for the prediction of quality parameters in grapes in lab conditions by scanning homogenized grape samples or whole berries (Dambergs et al., 2015; Wang et al., 2015). However, until now the real-time applications in the wine industry are still scarce. A NIR/VIS spectrometer (Process Analyzer X—Three industrial spectrometer, NIR-Online) has been successfully integrated into the winery ingress line for on-line quality assessment of crushed grapes (must/mash). This methodology permits the real-time prediction of important quality parameters directly related to the ripeness level, such as relative density, fructose and glucose, pH, tartaric acid and malic acid, titratable acidity, as well as to assess the phytosanitary status of grapes with rot indicators, such as glycerol, gluconic acid, and acetic acid (Porep et al., 2015). Multivariate calibration models were established for the spectral ranges of 450–850 nm (VIS) and 1050–1650 nm (NIR) by PLS regressions. Therefore, despite grape analysis being carried out after destemming/crushing, it is possible to perform an efficient quality control and management as a consequence of rapid and objective grape analysis.

The use of magnetic resonance imaging (MRI) was also proposed for fruits and processed foods. This nondestructive technique is based on the principle of nuclear magnetic resonance and may aid in the assessment of ripeness, quality, and analysis of internal structures and defects of fruits (Kumar Patel et al., 2015). On grapes, Andaur et al. (2004) successful achieved a Brix characterization at berry level, coupled also by berry volume assessment. Furthermore, MRI scans at cluster level showed the internal structure, number of seeds and distribution, and berry volume distribution within the cluster, a group of useful features for possible future sampling purposes.

More advanced optical berry sorting systems are being developed to be implemented on a sorting machine. In this sense, hyperspectral imaging combines imaging techniques with spectral analysis providing a digital image and the spectrum corresponding to each pixel. This is a very promising methodology because of its potential to detect several features simultaneously through a nondestructive and fast analysis. Hyperspectral images in the VIS/NIR wavelength range (from 400 to 1000 nm) have been successfully used for the berry discrimination according to the sugar concentration (Lafontaine et al., 2015). NIR hyperspectral imaging of intact grapes in the wavelength range from 900 to 1700 nm has been also applied for the fast screening of technological maturity of grapes (sugar concentration, titratable acidity, and pH) and grape skin total phenolic concentration in red and white cultivars (Nogales-Bueno et al., 2014), anthocyanins in red intact grapes (Chen et al., 2015; Hernández-Hierro et al., 2013), extractable polyphenols in red grape skins (total phenols, anthocyanins, and flavanols; Nogales-Bueno et al., 2015a), and total and extractable flavanols in grape seeds of white and red cultivars (Rodríguez-Pulido et al., 2014). A discriminant method permitted to isolate the grapes from other parts of image, and the average spectrum of the grape region was then extracted. The prediction models were developed by PLS regression. Nevertheless, hyperspectral imaging is quite restricted for on-line implementation due to its massive data volume and expensive equipment. Multispectral imaging based on the use of characteristic vector analysis has been used firstly to extract the main spectral information from NIR hyperspectral images and then reconstruct the whole NIR spectra from only the most important wavelengths. Particularly, characteristic vector analysis has been applied to predict the technological maturity (sugar concentration, titratable acidity, and pH) and skin total phenolic concentration of red grapes (Nogales-Bueno et al., 2015b). Multi-spectral imaging is a promising tool for on-line implementation in a grape processing line because of important advantages, such as relative little spectral data, low instrument cost, fast image acquisition, and simple algorithms for image processing.

Finally, also seed coat color, texture, and shape have been proposed as indicators of the seed maturity and astringency (Fredes et al., 2010; Ristic and Iland, 2005). Nevertheless, the subjectivity associated with the visual seed inspection should be avoided by using objective methods for the effective discrimination of unripe and ripe seeds. Some authors have developed a sufficiently robust sorting method for its implementation, which is based on the acquisition and analysis of digital seed images (Avila et al., 2014). To address the poor quality and complexity of acquired digital color images, pattern recognition algorithms combining supervised and unsupervised learning were used for the image segmentation, which allows a correct identification of seed images by partitioning the digital image into multiple segments (sets of pixels). The relevance of many descriptors related to color, texture, and shape of seed surface was studied using the Sequential Forward Selection algorithm, and a reduced set of high performance descriptors were selected to describe seed ripeness. Finally, the samples were correctly classified in the two classes (unripe and ripe) by a Simple Perceptron Method from the most representative features. This approach has been improved using a Multi-Layer Perceptron to classify the grape seeds into three classes: unripe, ripe, and overripe (Zuñiga et al., 2014). With the aim of defining more ripeness levels, color scales can be automatically generated from digital seed images using a computer-based method based on a multidimensional regression (Support Vector Regression; Avila et al., 2015; Fig. 1.5). The use of a scanner instead of a camera permits not only the capture of a large number of digital seed images, but also the maintaining of the same lighting conditions regardless of environment conditions.

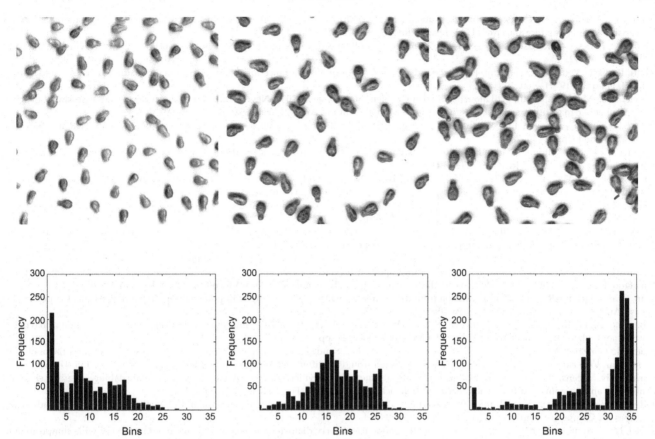

FIGURE 1.5 Association between the color scale and maturity. (A)–(C) Correspond to immature, mature, and overture seeds. (D)–(F) The respective histograms associated with each maturity grade. These histograms were generated by dividing the color scale in 35 bins. Figure reported from Avila et al. (2015).

References

Aleixandre-Tudo, J.L., Buica, A., Nieuwoudt, H., Aleixandre, J.L., du Toit, W., 2017. Spectrophotometric analysis of phenolic compounds in grapes and wines. J. Agric. Food Chem. 65, 4009–4026.

Andaur, J.E., Guesalaga, A.R., Agosin, E.E., Guarini, M.W., Irarrázaval, P., 2004. Magnetic resonance imaging for nondestructive analysis of wine grapes. J. Agric. Food Chem. 52, 165–170.

Arnold, R.A., Noble, A.C., Singleton, V.L., 1980. Bitterness and astringency of phenolics fraction in wine. J. Agric. Food Chem. 28, 675–678.

Avila, F., Mora, M., Fredes, C., 2014. A method to estimate grape phenolic maturity based on seed images. Comput. Electron. Agric. 101, 76–83.

Avila, F., Mora, M., Oyarce, M., Zuñiga, A., Fredes, C., 2015. A method to construct fruit maturity color scales based on support machines for regression: application to olives and grape seeds. J. Food Eng. 162, 9–17.

Belviso, S., Torchio, F., Novello, V., Giacosa, S., de Palma, L., Río Segade, S., et al., 2017. Modelling of the evolution of phenolic compounds in berries of 'Italia' table grape cultivar using response surface methodology. J. Food Compos. Anal. 62, 14–22.

Blanco-Vega, D., Gómez-Alonso, S., Hermosín-Gutiérrez, I., 2014. Identification, content and distribution of anthocyanins and low molecular weight anthocyanin-derived pigments in Spanish commercial red wines. Food Chem. 158, 449–458.

Borgogno Mondino, E., Gajetti, M., 2017. Preliminary considerations about costs and potential market of remote sensing from UAV in the Italian viticulture context. Eur. J. Remote Sens. 50, 310–319.

Bramley, R.G.V., 2010. Precision viticulture: managing vineyard variability for improved quality outcomes. In: Reynolds, A.G. (Ed.), Managing Wine Quality, Volume 1: Viticulture and Wine Quality. Woodhead Publishing Limited, Cambridge, United Kingdom.

Bramley, R.G.V., Proffitt, A.P.B., 1999. Managing variability in viticultural production, Aust. Grapegrower Winemaker, 427. pp. 11–16.

Brossaud, F., Cheynier, V., Noble, A.C., 2001. Bitterness and astringency of grape and wine polyphenols, Aust. J. Grape Wine Res., 7. pp. 33–39.

Cadot, Y., Miñana Castelló, M.T., Chevalier, M., 2006a. Flavan-3-ol compositional changes in grape berries (*Vitis vinifera* L. cv Cabernet Franc) before veraison, using two complementary analytical approaches, HPLC reversed phase and histochemistry. Anal. Chim. Acta 563, 65–75.

Cadot, Y., Miñana Castelló, M.T., Chevalier, M., 2006b. Anatomical, histological, and histochemical changes in grape seeds from *Vitis vinifera* L. cv Cabernet franc during fruit development. J. Agric. Food Chem. 54, 9206–9215.

Cagnasso, E., Rolle, L., Caudana, A., Gerbi, V., 2008. Relationship between grape phenolic maturity and red wine phenolic composition. Ital. J. Food Sci. 20, 365–380.

Candiago, S., Remondino, F., De Giglio, M., Dubbini, M., Gattelli, M., 2015. Evaluating multispectral images and vegetation indices for Precision Farming applications from UAV images. Remote Sens. 2015, 4026–4047.

Carrillo, E., Matese, A., Rousseau, J., Tisseyre, B., 2016. Use of multi-spectral airborne imagery to improve yield sampling in viticulture. Precis. Agric. 17, 74–92.

Celotti, E., Della Vedova, T., Ferrarini, R., Martinand, S., 2007. The use of reflectance for monitoring phenolic maturity curves in red grapes. Ital. J. Food Sci. 19, 91–100.

Cerovic, Z.G., Moise, N., Agati, G., Latouche, G., Ben Ghozlen, N., Meyer, S., 2008. New portable optical sensors for the assessment of winegrape phenolic maturity based on berry fluorescence. J. Food Compos. Anal. 21, 650–654.

Chen, S., Zhang, F., Ning, J., Liu, X., Zhang, Z., Yang, S., 2015. Predicting the anthocyanin content of wine grapes by NIR hyperspectral Imaging. Food Chem. 172, 788–793.

Cheynier, V., Dueñas-Paton, M., Salas, E., Maury, C., Souquet, J.M., Sarni-Manchado, P., et al., 2006. Structure and properties of wine pigments and tannins. Am. J. Enol. Viticult. 57, 298–305.

Cubero, S., Diago, M.P., Blasco, J., Tardaguila, J., Millan, B., Aleixos, N., 2014. A new method for pedicel/peduncle detection and size assessment of grapevine berries and other fruits by image analysis. Biosyst. Eng. 117, 62–72.

Cubero, S., Diago, M.P., Blasco, J., Tardaguila, J., Prats-Montalbán, J.M., Ibáñez, J., et al., 2015. A new method for assessment of bunch compactness using automated image analysis. Aust. J. Grape Wine Res. 21, 101–109.

Dambergs, R., Gishen, M., Cozzolino, D., 2015. A review of the state of the art, limitations, and perspectives of infrared spectroscopy for the analysis of wine grapes, must, and grapevine tissue. Appl. Spectrosc. Rev. 50, 261–278.

Diago, M.P., Rey-Carames, C., Le Moigne, M., Fadaili, E.M., Tardaguila, J., Cerovic, Z.G., 2016. Calibration of non-invasive fluorescence-based sensors for the manual and on-the-go assessment of grapevine vegetative status in the field, Aust. J. Grape Wine Res., 22. pp. 438–449.

Di Stefano, R., Cravero, M.C., 1991. Methods for the study of grape polyphenols. Rivista di Viticoltura ed Enologia 44, 37–45.

Dunn, G.M., Martin, S.R., 2004. Yield prediction from digital image analysis: a technique with potential for vineyard assessments prior to harvest, Aust. J. Grape Wine Res., 10. pp. 196–198.

Ferrandino, A., Pagliarani, C., Carlomagno, A., Novello, V., Schubert, A., Agati, G., 2016. Improved fluorescence-based evaluation of flavonoid in red and white winegrape cultivars., Aust. J. Grape Wine Res., 23. pp. 207–214.

Figueiredo-González, M., Simal-Gándara, J., Boso, S., Martínez, M.C., Santiago, J.L., Cancho-Grande, B., 2012. Anthocyanins and flavonols berries from *Vitis vinifera* L. cv. Brancellao separately collected from two different positions within the cluster. Food Chem. 135, 47–56.

Fournand, D., Vicens, A., Sidhoum, L., Souquet, J.M., Moutounet, M., Cheynier, V., 2006. Accumulation and extractability of grape skin tannins and anthocyanins at different advanced physiological stages. J. Agric. Food Chem. 54, 7331–7338.

Fredes, C., Von Bennewitz, E., Holzapfel, E., Saavedra, F., 2010. Relation between seed appearance and phenolic maturity: a case study using grapes cv. Carmenere. Chilean J. Agric. Res. 70, 381–389.

Friedel, M., Sorrentino, V., Blank, M., Schüttler, A., 2016. Influence of berry diameter and colour on some determinants of wine composition of *Vitis vinifera* L. cv. Riesling. Aust. J. Grape Res. 22, 215–225.

Gambuti, A., Strollo, D., Genovese, A., Ugliano, M., Ritieni, A., Moio, L., 2005. Influence of enological practices on ochratoxin A concentration in wine. Am. J. Enol. Viticult. 56, 155–162.

Gatti, M., Garavani, A., Vercesi, A., Poni, S., 2017. Ground-truthing of remotely sensed within-field variability in a cv. Barbera plot for improving vineyard management. Aust. J. Grape Wine Res. 23, 399–408.

Glories, Y., Augustin, M., 1993. Maturité phénolique du raisin, consèquences technologiques: applications aux millésimes 1991 et 1992, Actes du Colloque "Journée Technique du CIVB", 21. CIVB, Bordeaux, France, p. 56, Janvier.

Gonzalo-Diago, A., Dizy, M., Fernández-Zurbano, P., 2014. Contribution of low molecular weight phenols to bitter taste and mouthfeel properties in red wines. Food Chem. 154, 187–198.

Guzev, L., Danshin, A., Ziv, S., Lichter, A., 2006. Occurrence of ochratoxin A producing fungi in wine and table grapes in Israel. Int. J. Food Microbiol. 111, S67–S71.

Hernández-Hierro, J.M., Nogales-Bueno, J., Rodríguez-Pulido, F.J., Heredia, F.J., 2013. Feasibility study on the use of near-infrared hyperspectral imaging for the screening of anthocyanins in intact grapes during ripening. J. Agric. Food Chem. 61, 9804–9809.

Johnson, L.F., Bosch, D.F., Williams, D.C., Lobitz, B.M., 2001. Remote sensing of vineyard management zones: implications for wine quality. Appl. Eng. Viticult 17, 557–560.

Kontoudakis, N., Esteruelas, M., Fort, F., Canals, J.M., Zamora, F., 2010. Comparison of methods for estimating phenolic maturity in grapes: correlation between predicted and obtained parameters. Anal. Chim. Acta 660, 127–133.

Kontoudakis, N., Esteruelas, M., Fort, F., Canals, J.M., De Freitas, V., Zamora, F., 2011. Influence of the heterogeneity of grape phenolic maturity on wine composition and quality. Food Chem. 124, 767–774.

Lafontaine, M., Bockaj, Z., Freund, M., Vieth, K.-U., Negara, C., 2015. Non-destructive determination of grape berry sugar concentration using visible/near infrared imaging and possible impact on wine quality. Tech. Mess. 82, 633–642.

Laguna, L., Bartolomé, B., Moreno-Arribas, M.V., 2017a. Mouthfeel perception of wine: oral physiology, components and instrumental characterization. Trends Food Sci. Technol. 59, 49–59.

Laguna, L., Sarkar, A., Bryant, M.G., Beadling, A.R., Bartolomé, B., Moreno-Arribas, M.V., 2017b. Exploring mouthfeel in model wines: sensory-to-instrumental approaches. Food Res. Int. 102, 478–486.

Lasanta, C., Caro, I., Gómez, J., Pérez, L., 2014. The influence of ripeness grade on the composition of musts and wines from *Vitis vinifera* cv. Tempranillo grown in a warm climate. Food Res. Int. 64, 432–438.

Le Moigne, M., Maury, C., Bertrand, D., Jourjon, F., 2008. Sensory and instrumental characterization of Cabernet Franc grapes according to ripening stages and growing location. Food Qual. Preference 19, 220–231.

Letaief, H., Rolle, L., Gerbi, V., 2008. Mechanical behavior of winegrapes under compression tests. Am. J. Enol. Viticult. 59, 323–329.

Liu, S., Whitty, M., 2015. Automatic grape bunch detection in vineyards with an SVM classifier. J. Appl. Logic 13, 643–653.

Luo, L., Tang, Y., Zou, X., Ye, M., Feng, W., Li, G., 2016. Vision-based extraction of spatial information in grape clusters for harvesting robots. Biosyst. Eng. 151, 90–104.

REFERENCES

Matese, A., Di Gennaro, S.F., 2015. Technology in precision viticulture: a state of the art review. Int. J. Wine Res. 2015, 69–81.

Mira de Orduña, R., 2010. Climate change associated effects on grape and wine quality and production. Food Res. Int. 43, 1844–1855.

Nogales-Bueno, J., Hernández-Hierro, J.M., Rodríguez-Pulido, F.J., Heredia, F.J., 2014. Determination of technological maturity of grapes and total phenolic compounds of grape skins in red and white cultivars during ripening by near infrared hyperspectral image: a preliminary approach. Food Chem. 152, 586–591.

Nogales-Bueno, J., Baca-Bocanegra, B., Rodríguez-Pulido, F.J., Heredia, F.J., Hernández-Hierro, J.M., 2015a. Use of near infrared hyperspectral tools for the screening of extractable polyphenols in red grape skins. Food Chem. 172, 559–564.

Nogales-Bueno, J., Ayala, F., Hernández-Hierro, J.M., Rodríguez-Pulido, F.J., Echávarri, J.F., Heredia, F.J., 2015b. Simplified method for the screening of technological maturity of red grape and total phenolic compounds of red grape skin: application of the characteristic vector method to near-infrared spectra. J. Agric. Food Chem. 63, 4284–4290.

Noguerol-Pato, R., González-Barreiro, C., Cancho-Grande, B., Santiago, J.L., Martínez, M.C., Simal-Gándara, J., 2012. Aroma potential of Brancellao grapes from different cluster positions. Food Chem. 132, 112–124.

OIV, 2009. OIV Descriptor List for Grape Varieties and Vitis Species, 2nd edition Organisation Internationale de la Vigne et du Vin, Paris, France.

Olarte Mantilla, S.M., Collins, C., Iland, P.G., Johnson, T.E., Bastian, S.E.P., 2012. Review: berry sensory assessment: concepts and practices for assessing winegrapes' sensory attributes. Aust. J. Grape Wine Res. 18, 245–255.

Ollat, N., Diakou-Verdin, P., Carde, J.P., Barrieu, F., Gaudillere, J.P., Moing, A., 2002. Grape berry development: a review. J. Int. des Sciences de la Vigne et du Vin 36, 109–131.

Parenti, A., Spugnoli, P., Masella, P., Guerrini, L., Benedettelli, S., Di Blasi, S., 2015. Comparison of grape harvesting and sorting methods on factors affecting the must quality. J. Agric. Eng. XLVI 456, 19–22.

Patel, K.K., Khan, M.A., Kar, A., 2015. Recent developments in applications of MRI techniques for foods and agricultural produce—an overview. J. Food Sci. Technol. 52, 1–26.

Pellenc, R., Niero, R., 2009. Selective-sorting harvesting machine and sorting chain including one such machine. WO/2009/066020 Patents.

Pisciotta, A., Barbagallo, M.G., Di Lorenzo, R., Hunter, J.J., 2013. Anthocyanin variation in individual 'Shiraz' berries as affected by exposure and position on the rachis. Vitis 52, 111–115.

Porep, J.U., Mattes, A., Pour Nikfardjam, M.S., Kammerer, D.R., Carle, R., 2015. Implementation of an on-line near infrared/visible (NIR/VIS) spectrometer for rapid quality assessment of grapes upon receival at wineries. Aust. J. Grape Wine Res. 21, 69–79.

Rey-Caramés, C., Diago, M.P., Pilar Martín, M., Lobo, A., Tardaguila, J., 2015. Using RPAS multi-spectral imagery to characterise vigour, leaf development, yield components and berry composition variability within a vineyard. Remote Sens. 2015, 14458–14481.

Ribèreau-Gayon, P., Glories, Y., Maujean, A., Dubourdieu, D., 2006. Handbook of enology, 2nd edition The Chemistry of Wine—Stabilization and Treatments, vol. II. Jonh Wiley & Sons, Chichester, United Kingdom.

Richter, M., Langle, T., Beyerer, J., 2015. An approach to color-based sorting of bulk materials with automated estimation of system parameters. Tech. Mess. 82, 135–144.

Río Segade, S., Rolle, L., Gerbi, V., Orriols, I., 2008. Phenolic ripeness assessment of grape skin by texture analysis. J. Food Compos. Anal. 21, 644–649.

Río Segade, S., Giacosa, S., Gerbi, V., Rolle, L., 2011a. Berry skin thickness as main texture parameter to predict anthocyanin extractability in winegrapes. LWT Food Sci. Technol. 44, 392–398.

Río Segade, S., Soto Vazquez, E., Orriols, I., Giacosa, S., Rolle, L., 2011b. Possible use of texture characteristics of winegrapes as markers for zoning and their relationship with anthocyanin extractability index. Int. J. Food Sci. Technol. 46, 386–394.

Río Segade, S., Giacosa, S., de Palma, L., Novello, V., Torchio, F., Gerbi, V., et al., 2013. Effect of the cluster heterogeneity on mechanical properties, chromatic indices and chemical composition of Italia table grape berries (Vitis vinifera L.) sorted by flotation. Int. J. Food Sci. Technol. 48, 103–113.

Ristic, R., Iland, P.G., 2005. Relationships between seed and berry development of Vitis Vinifera L. cv Shiraz: developmental changes in seed morphology and phenolic composition. Aust. J. Grape Wine Res. 11, 43–58.

Robichaud, J.L., Noble, A.C., 1990. Astringency and bitterness of selected phenolics in wine. J. Sci. Food Agric. 53, 343–353.

Rodríguez-Pulido, F.J., Hernández-Hierro, J.M., Nogales-Bueno, J., Gordillo, B., González-Miret, M.L., Heredia, F.J., 2014. A novel method for evaluating flavanols in grape seeds by near infrared Hyperspectral imaging. Talanta 122, 145–150.

Rolle, L., Guidoni, S., 2007. Color and anthocyanin evaluation of red winegrapes by CIE L^*, a^*, b^* parameters. J. Int. des Sciences de la Vigne et du Vin 41, 193–201.

Rolle, L., Torchio, F., Zeppa, G., Gerbi, V., 2008. Anthocyanin extractability assessment of grape skins by texture analysis. J. Int. de la Science de la Vigne et du Vin 42, 157–162.

Rolle, L., Torchio, F., Zeppa, G., Gerbi, V., 2009. Relationship between skin break force and anthocyanin extractability at different ripening stages. Am. J. Enol. Viticult. 60, 93–97.

Rolle, L., Río Segade, S., Torchio, F., Giacosa, S., Cagnasso, E., Marengo, F., et al., 2011a. Influence of grape density and harvest date on the changes in phenolic composition, phenol extractability indices and instrumental texture properties during ripening. J. Agric. Food Chem. 59, 8796–8805.

Rolle, L., Gerbi, V., Schneider, A., Spanna, F., Río Segade, S., 2011b. Varietal relationship between instrumental skin hardness and climate for grapevines (Vitis vinifera L.). J. Agric. Food Chem. 59, 10624–10634.

Rolle, L., Giacosa, S., Torchio, F., Río Segade, S., 2012a. Changes in acoustic and mechanical properties of Cabernet sauvignon seeds during ripening. Am. J. Enol. Viticult. 63, 413–418.

Rolle, L., Torchio, F., Giacosa, S., Río Segade, S., Cagnasso, E., Gerbi, V., 2012b. Assessment of physicochemical differences in Nebbiolo grape berries from different production areas and sorted by flotation. Am. J. Enol. Viticult. 63, 195–204.

Rolle, L., Giacosa, S., Torchio, F., Perenzoni, D., Río Segade, S., Gerbi, V., et al., 2013. Use of instrumental acoustic parameters of winegrape seeds as possible predictors of extractable phenolic compounds. J. Agric. Food Chem. 61, 8752–8764.

Rolle, L., Torchio, F., Giacosa, S., Río Segade, S., 2015. Berry density and size as factors related to the physicochemical characteristics of Muscat Hamburg table grapes (*Vitis vinifera* L.). Food Chem. 173, 105–113.

Romero-Cascales, I., Ortega-Regules, A., López-Roca, J.M., Fernández-Fernández, J.I., Gómez-Plaza, E., 2005. Differences in anthocyanin extractability from grapes to wines according to variety. Am. J. Enol. Viticult. 56, 212–219.

Romeyer, F.M., Macheix, J.J., Sapis, J.C., 1986. Changes and importance of oligomeric procyanidins during maturation of grape seed. Phytochemistry 25, 219–221.

Rouse Jr., J.W., Haas, R.H., Deering, D.W., Schell, J.A., Harlan, J.C., 1974. Monitoring the vernal advancement and retrogradation (green wave effect) of natural vegetation. NASA/GSFC Type III report. NASA, Greenbelt, MD, USA.

Santos, A.O., Wample, R.L., Sachidhanantham, S., Kaye, O., 2012. Grape quality mapping for vineyard differential harvesting. Braz. Arch. Biol. Technol. 55, 193–204.

Schwab, W., Wüst, M., 2015. Understanding the constitutive and induced biosynthesis of mono- and sesquiterpenes in grapes (*Vitis vinifera*): a key to unlocking the biochemical secrets of unique grape aroma profiles. J. Agric. Food Chem. 63, 10591–10603.

Shah, M., 2016. Where tradition meets innovation. Regional Analysis: Piedmont. Meininger's Wine Bus. Int. 11 (2), 56–57.

Tarter, M.E., Keuter, S.E., 2005. Effect of rachis position on size and maturity of Cabernet Sauvignon berries. Am. J. Enol. Viticult. 56, 86–89.

Torchio, F., Río Segade, S., Gerbi, V., Cagnasso, E., Giordano, M., Giacosa, S., et al., 2012. Changes in varietal volatile composition during shelf-life of two types of aromatic red sweet Brachetto sparkling wines. Food Res. Int. 48, 491–498.

Torchio, F., Giacosa, S., Vilanova, M., Río Segade, S., Gerbi, V., Giordano, M., et al., 2016. Use of response surface methodology for the assessment of changes in the volatile composition of Moscato Bianco (*Vitis vinifera* L.) grape berries during ripening. Food Chem. 212, 576–584.

Valls, J., Millán, S., Martí, M.P., Borràs, E., Arola, L., 2009. Advanced separation methods of food anthocyanins, isoflavones and flavanols. J. Chromatogr. A 1216, 7143–7172.

Vidal, S., Francis, L., Guyot, S., Marnet, N., Kwiatkowski, M., Gawel, R., et al., 2003. The mouth-feel properties of grape and apple proanthocyanidins in a winelike medium. J. Sci. Food Agric. 83, 564–573.

Wang, H., Peng, J., Xie, C., Bao, Y., He, Y., 2015. Fruit quality evaluation using spectroscopy technology: a review. Sensors 15, 11889–11927.

Ward, S.C., Petrie, P.R., Johnson, T.E., Boss, P.K., Bastian, S.E.P., 2015. Unripe berries and petioles in *Vitis vinifera* cv. Cabernet sauvignon fermentations affect sensory and chemical profiles. Am. J. Enol. Viticult. 66, 435–443.

Yuan, L.M., Cai, J.R., Sun, L., Ye, C., 2016. A preliminary discrimination of cluster disqualified shape for table grape by mono-camera multi-perspective simultaneously imaging approach. Food Anal. Methods 9, 758–767.

Zanoni, B., Siliani, S., Canuti, V., Rosi, I., Bertuccioli, M., 2010. A kinetic study on extraction and transformation phenomena of phenolic compounds during red wine fermentation. Int. J. Food Sci. Technol. 45, 2080–2088.

Zhao, Y., Gong, L., Huang, Y., Liu, C., 2016. A review of key techniques of vision-based control for harvesting robot. Comput. Electron. Agric. 127, 311–323.

Zouid, I., Siret, R., Jourjon, F., Mehinagic, E., Rolle, L., 2013. Impact of grapes heterogeneity according to sugar level on both physical and mechanical berries properties and their anthocyanins extractability at harvest. J. Texture Stud. 44, 95–103.

Zuñiga, A., Mora, M., Oyarce, M., Fredes, C., 2014. Grape maturity estimation based on seed images and neural networks. Eng. Appl. Artif. Intell. 35, 95–104.

CHAPTER 2

Acidification and pH Control in Red Wines

Piergiorgio Comuzzo and Franco Battistutta

Department of Agricultural, Food, Environmental and Animal Sciences (Di4A), University of Udine, Udine, Italy

2.1 IMPORTANCE OF ACIDIC FRACTION AND pH CONTROL IN RED WINES

The control of acidic fraction and pH is generally considered as a minor technological problem in red winemaking. This is because red wines are generally less acidic with respect to white ones and, in addition, they often undergo malolactic fermentation during aging. Nevertheless, this simplistic approach can lead to underestimating some important technological implications connected with acidity and pH management, potentially compromising the quality of the wine. In recent years, climate changes and the pursuit of a wine style characterized by "sweet" tannins and color stability (to fulfill consumer's expectations), determined the average increase of the pH of red wines with respect to the values normally detected some decades ago.

Acidity and pH are important variables also in red winemaking (Boulton, 1980). The pH of a wine is strictly connected with its microbiological and physicochemical stability (Usseglio-Tomasset, 1985): pH may affect the outset and the behavior of malolactic fermentation (Bousbouras and Kunkee, 1971); it may contribute to the natural selection of microorganisms during winemaking (Dupuy, 1957; Wibowo et al., 1985), and it is directly involved in the definition of the equilibrium of sulfur dioxide in wine, affecting the amounts of free and molecular sulfur dioxide available (Lafon-Lafourcade and Peynaud, 1974). Moreover, pH and acidity (in particular as concerns the presence of tartaric acid) are in close relationship with the solubility of tartaric salts (Berg and Keefer, 1958, 1959), being important factors for tartaric stability. Even the color of red wines may be strongly conditioned by pH, because this variable affects the equilibrium between the different forms of anthocyanins (Berg, 1963; Brouillard and Delaporte, 1977), as well as because certain reactions that lead to the polymerization or condensation of red wine pigments and flavanols may be conditioned by the pH of the medium (Kontoudakis et al., 2012; Sheridan and Elias, 2016). The pH may also affect the behavior of certain technologies: for instance, the effects of microoxygenation on wine color is reported to increase with decreasing pH (Kontoudakis et al., 2012). Finally, total acidity and pH have important repercussions on wine sensory perception and equilibrium (Sowalsky and Noble, 1998; Ribéreau-Gayon et al., 2006a).

This chapter focuses on the main technological implications connected to acidification and deacidification of red wines, considering both traditional and recently introduced techniques.

2.2 MAIN ORGANIC ACIDS IN MUST AND WINE

The main organic acids in grape juice and wine are tartaric, malic, citric, succinic, and lactic, with minor concentrations of other compounds (e.g., acetic and pyruvic acid). The first three molecules derive from grape, while lactic acid, succinic acid, and the others have a microbiological origin, being mainly produced during fermentations (Ribéreau-Gayon et al., 2006a). It is well known that the concentrations of tartaric, malic, and citric acid in grape juice depend on several factors, such as the grape variety, the harvest date, the latitude and climatic conditions, the seasonal behavior, and the vineyard management practices (Amerine, 1956; Kliewer et al., 1967). The relative concentration of organic acids is an important factor, because it directly affects wine pH. Moreover, concentration, pH and anion species may also influence the perception of sourness in wine (Sowalsky and Noble, 1998).

2.2.1 Tartaric Acid

Tartaric acid (H_2T) is the strongest organic acid in must and wine and it is characteristic of grape, where it is present as L-(+)-isomer (Fig. 2.1). It is a diprotic acid, whose pKa (at 25°C) are reported to be 2.98 and 4.34, respectively, for the dissociation of the first and the second acidic function (Lide, 2005). These values of the dissociation constant of the two carboxylic groups, confirm that such compound is mainly present, at wine pH, in form of bitartrate ion (HT^-). In particular, Usseglio-Tomasset (1985) calculated that at pH 3.50, tartaric acid is present at 23.4% in its undissociated form (H_2T), at 67.5% in form of bitartrte (HT^-), and at the remaining 9.1% in the form of tartrate ion (T^{2-}). This percentage of bitartrate ion and the relative high concentration found for tartaric acid in must and wine (2–6 g/L) explain the importance of such acid in the formation of insoluble potassium bitartrate salts in wine. Tartaric acid is relatively stable from the microbiological point of view and its concentration may change due to bitartrate precipitation, or acidification/deacidification treatments.

2.2.2 Malic Acid

Malic acid is a diprotic acid, whose pKa at 25°C are 3.40 and 5.11 (Lide, 2005). Due to these values, at a pH of 3.50, the undissociated form (H_2M) accounts for approximately 47% of the total concentration, while the dissociation of the first carboxylic group (HM^-) represents the 51.7% (Usseglio-Tomasset, 1985). The second acidic function of malic acid is very poorly dissociated at wine pH and the malate form (M^{2-}) at pH 3.50, was estimated to be the 1.2% (Usseglio-Tomasset, 1985). This means that the most important contribution to wine pH, in terms of strength and concentration, is given by the dissociation of tartaric and malic acid (Dartiguenave et al., 2000a) and in particular by their first acidic function. The enantiomeric form of malic acid present in grape is the L-(−)-isomer (Fig. 2.1). Contrary to tartaric acid, in normal conditions malic acid is stable from the physicochemical point of view, but it may be precipitated as calcium tartromalate in specific conditions. Moreover, malic acid may be subjected to microbial transformation, e.g., during malolactic fermentation.

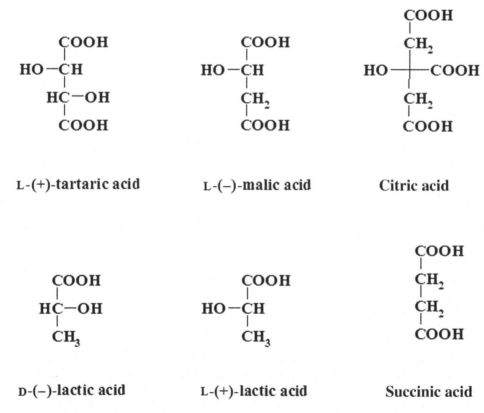

FIGURE 2.1 Main organic acids found in wines.

2.2.3 Citric Acid

Citric acid is a triprotic acid (Fig. 2.1). The pKa reported at 25°C, for the first, second, and third carboxylic group of this acid are 3.13, 4.76, and 6.40, respectively (Lide, 2005). This means that, at a pH close to the pH of red wines (3.50), also for citric acid the most abundant form is the monodissociated one (H_2C^-). According to Usseglio-Tomasset (1985), this form accounts for the 67% of the total, while undissociated citric acid (H_3C) represents the 29% of the total concentration. Due to the high value of its pKa, the dissociation of the third carboxylic function of citric acid practically does not occur at wine pH.

Based on these considerations, citric acid is stronger than malic, but, normally it presents a quite low concentration in wine: in young wines, prior to malolactic fermentation, citric acid amount ranges from 0.5 to 1 g/L (Ribéreau-Gayon et al., 2006a). For this reason, this acid gives generally only a minor contribution to wine pH.

2.2.4 Lactic Acid

Lactic acid can be present in wine in two enantiomeric forms: D-(−)- and L-(+)-lactic acid. The former is generally produced, in small amounts (some hundreds of mg/L), by yeasts during alcoholic fermentation, while the latter is typically originated during malolactic fermentation (Peynaud et al., 1966), from the transformation of malic acid by lactic acid bacteria (LAB). The pKa of lactic acid at 25°C is 3.86 (Lide, 2005); this means that, at wine pH, lactic acid is less dissociated than malic ($pK_1 = 3.40$) and explains the reason for the pH diminution provoked by malolactic fermentation.

The ratio between the two enantiomeric forms of lactic acid, is an indicator of wine quality; in fact, high amounts of the D-(−)-isomer (more than 400 mg/L), is an index of sugar fermentation by LAB. This kind of heterofermentative conversion is described as lactic disease and it may provoke negative modifications in wine, such as the increase of volatile acidity (Wibowo et al., 1985; Ribéreau-Gayon et al., 2006b).

2.2.5 Succinic Acid

Succinic acid accounts for up to the 90% of the nonvolatile acids produced during alcoholic fermentation (Thoukis et al., 1965). The content of this acid in wine ranges normally from 0.5 to 1.5 g/L, but the maximum concentration may reach 3 g/L (de Klerk, 2010). Succinic acid is a diprotic acid. Its pKa at 25°C are 4.21 and 5.64 (Lide, 2005). This means that at pH 3.50, most succinic acid (83.9%) is present in its undissociated form; monodissociated succinate ion accounts only for approximately 16%, while the dissociation of the second carboxylic group is practically negligible (Usseglio-Tomasset, 1985).

2.2.6 Acetic Acid

Acetic acid is normally produced up to 0.1–0.3 g/L, during alcoholic fermentation, but its concentration increases as a consequence of bacterial development (Ribéreau-Gayon et al., 2006b), or in the case of stuck and sluggish fermentations (Zamora, 2009). It is a monoprotic acid, whose pKa at 25°C is 4.76 (Lide, 2005). This means that at pH 3.50 it is mainly present in an undissociated form (Usseglio-Tomasset, 1985). Acetic acid is the major constituent of volatile acidity. In high concentrations, it is an index of low quality in wines, so that, volatile acidity is generally subjected to legal restrictions: as a general provision, European law established, for red wines, a maximum volatile acidity limit of 20 meq/L (Regulation EU 606, 2009).

2.3 TOTAL ACIDITY AND WINE pH

2.3.1 Definition of pH

From a strictly theoretical point of view, pH is defined as the base 10 logarithm of the reciprocal of the relative activity (a_H) of hydrogen ions in solution.

$$pH = \log\left(\frac{1}{a_H}\right) = -\log a_H \tag{2.1}$$

Due to the fact, that a_H is not thermodynamically measurable, Eq. (2.1) is generally written in terms of concentration (Buck et al., 2002; Spedding et al., 1977).

$$pH = -\log c_H \tag{2.2}$$

Hydrogen ions do not occur in the free state in water solution, so that Eq. (2.2) is written in the following form.

$$pH = -\log[H_3O^+] \tag{2.3}$$

The pH of a wine is also described as its "real acidity," because the value of this variable is directly linked with the development of biological reactions (Usseglio-Tomasset, 1985), as well as because it is connected with wine freshness and sensory perception of sourness (Jackson, 2008).

2.3.2 Total Acidity, Titratable Acidity, and Real Acidity

Boulton (1980) described wine total acidity as "the number of protons that the organic acids would contain if they were undissociated" and titratable acidity as "the number of protons recovered during a titration with a strong base to a specified endpoint" (generally pH 7.00).

Practically speaking, these two definitions are generally considered as synonyms in wine chemistry and the O.I.V. (International Organization of Vine and Wine, 2017a) defines the total acidity of a wine as "the sum of its titratable acidities, when it is titrated to pH 7 against a standard alkaline solution," without including carbon dioxide.

Therefore, during the determination of total acidity, as the equivalents of strong base (generally sodium hydroxide) are added to the wine, H_3O^+ ions are neutralized by the base itself and, according to the law of mass action, organic acids are going to increasingly dissociate, until the end point of the titration will be reached (pH 7.00). This means that total acidity reflects the total amount of hydrogen ions releasable by organic acids in the conditions of the titration (Boulton, 1980). Differently, pH is an index of the concentration of hydrogen ions that are actually released by the dissociation of organic acids in a specific condition of chemical equilibrium and any modification of such equilibrium may modify the real acidity of the wine.

Total acidity values in red wines are generally reported to be lower than those detected in white wines (Ribéreau-Gayon et al., 2006a), because red varieties are generally harvested later, because of the occurrence of malolactic fermentation and because of the higher level of salification of organic acids due to the behavior of maceration (Volschenk et al., 2006). However, depending on variety, grapegrowing conditions, and harvest date, in some cases, the total acidities found in red grape varieties may be quite high (Table 2.1).

Despite the relationships between total acidity and wine pH having been studied and predictive models for such relationships having been created (Boulton, 1980; Prenesti et al., 2004), it is difficult to forecast the exact value that pH will reach after a given modification of the acid–base equilibrium of a wine. Briefly, the value of total acidity alone as well as its theoretical variations (e.g., as a consequence of tartaric acid supplementation), do not allow to predict exactly the changes of wine pH these variations will determine (Dartiguenave et al., 2000a; Ribéreau-Gayon et al., 2006a). In addition, further complications come from the fact that the pH of a wine is also affected by tartaric precipitation and malolactic fermentation (Wejnar, 1972). The result is a series of heterogeneous chemical equilibria (Prenesti et al., 2004) that heavily affect wine acidity and pH management practices.

2.3.3 Variations of Acidity During Winemaking

In normal winemaking conditions, even without specific acidification or deacidification treatments, the acidic fraction of wines is subjected to fluctuations. Such variations are connected with different phenomena.

First, after alcoholic fermentation, the formation of ethanol provokes the reduction of the solubility of tartaric salts (Berg and Keefer, 1958, 1959; Thoukis et al., 1965). This phenomena is reported to be responsible for the reduction of titratable acidity after alcoholic fermentation (de Klerk, 2010), together with the consumption of malic acid by the yeasts (de Klerk, 2010; Mayer et al., 1964). However, an increase of total acidity and a change in the organic acid composition after alcoholic fermentation is also described for wine, mainly due to the production of new organic acids during fermentation itself, in particular lactic and succinic acid (Thoukis et al., 1965).

TABLE 2.1 Titratable Acidity (Given in Tartaric Acid), pH Value and Concentration of Tartaric and Malic Acid, Detected in Certain Red Grape Varieties Form Different Locations, Harvest Seasons and Stages of Fruit Maturity

Variety	Location—year	pH	Titratable acidity (g/L)	Malic acid (g/L)	Tartaric acid (g/L)	Ref.
Cabernet Sauvignon EH[a]	U.S.A., California—1966	3.47	6.4	2.8	6.8	Kliewer et al. (1967)
Cabernet Sauvignon LH[b]	U.S.A., California—1966	3.85	4.3	1.9	6.6	
Grenache EH	U.S.A., California—1966	3.06	10.5	5.1	7.4	
Grenache LH	U.S.A., California—1966	3.50	5.7	2.9	6.7	
Grignolino EH	U.S.A., California—1966	3.36	8.7	5.4	7.6	
Grignolino LH	U.S.A., California—1966	3.68	6.5	0.42	6.8	
Malbec EH	U.S.A., California—1966	3.60	8.6	7.0	6.0	
Malbec LH	U.S.A., California—1966	4.01	6.1	5.3	6.0	
Merlot EH	U.S.A., California—1966	3.63	5.4	3.3	6.0	
Merlot LH	U.S.A., California—1966	3.82	4.3	2.4	6.2	
Pinot noir EH	U.S.A., California—1966	3.58	7.7	5.4	6.2	
Pinot noir LH	U.S.A., California—1966	4.02	6.4	5.1	6.3	
Cabernet Sauvignon	France, Medoc—1952	3.38	8.4	3.0	8.1	Boulton et al. (1996)
Cabernet Sauvignon	France, Medoc—1955	3.20	6.6	2.4	9.0	
Merlot	France, Medoc—1952	3.52	5.6	2.1	7.1	
Merlot	France, Medoc—1955	3.40	5.9	1.6	8.9	
Syrah	Australia—Barossa and Renmark—1960	3.43	6.7	2.7	6.5	
Cabernet Sauvignon	South-Africa, Stellenbosch—1964	3.30	8.5	3.5	8.0	
Pinot noir	South-Africa, Stellenbosch—1964	3.10	12.0	6.3	9.0	

[a]*EH: early harvest.*
[b]*LH: late harvest.*

TABLE 2.2 Effect of Fermentation Temperature on the Concentrations of Malic and Succinic Acid in Cabernet Sauvignon Wines from Different Locations

	Malic acid (g/L)			Succinic acid (g/L)		
Location	10°C	21°C	33°C	10°C	21°C	33°C
Davis (CA)	2.9	4.6	3.8	1.6	1.8	1.8
Oakville (CA)	1.3	1.4	1.0	2.0	1.9	1.6

Modified from Ough, C.S., Amerine, M.A., Sparks, T.C., 1969. Studies with controlled fermentations. XI. Fermentation temperature effects on acidity and pH. Am. J. Enol. Vitic. 20, 127-139.

Furthermore, malic acid is also reported to be produced by yeasts in the presence of certain amino acids (Drawert et al., 1965). Such an increase of titratable acidity was found to be higher at middle fermentation temperature (21°C), rather than at extreme temperature values (10 or 33°C) (Ough et al., 1969). In such conditions, the concentration of malic and succinic acid in the wines was higher (Table 2.2).

This situation indicates that it is difficult to hypothesize, on the basis of the acidity of the must, the total acidity that a wine will have after fermentation (Ribéreau-Gayon et al., 2006a). This represents a practical problem for the winemakers, because if the level of acidity is considered to be optimal for a given must, unforeseen variations in titratable acidity after alcoholic fermentation may be detrimental to wine quality (de Klerk, 2010). In addition, for this reason, apart from specific cases of extreme juice pH values, it is generally recommended to perform acidification or deacidification treatments on wines rather than on musts.

2.4 ACID–BASE EQUILIBRIUM AND WINE BUFFER CAPACITY

2.4.1 Acid–Base Equilibrium in Wine

As discussed above, wine pH is connected with the dissociation of the different organic acids, to form their conjugate bases. The pH of a water solution of a weak monoprotic acid (HA) may be calculated as follows.

$$pH = pKa + \log \frac{[A^-]}{[HA]} \tag{2.4}$$

Eq. (2.4) is known as the Henderson–Hasselbach equation and its application to the calculation of wine pH allows some important considerations, related to pH and acidity control in practical winemaking.

The first one is that Henderson–Hasselbach equation is an approximation, because it assumes that each acid contributes to wine acidity separately, in an additive way (Ribéreau-Gayon et al., 2006a) and, moreover, it also assumes that the degree to which each acid is dissociated is independent on the concentration and strength of the others. Actually, musts and wines are not constituted as pure acid solutions (Wejnar, 1972) and the real effect of each acid on wine pH depends on its relative strength and concentration. For instance, if a stronger acid HA (e.g., tartaric acid) is added to a water solution of a weaker acid HB (e.g., malic acid), the dissociation of the former will affect the dissociation of the latter, moving the equilibrium reaction toward its undissociated form. Generally speaking, interactions between organic acids may occur in wine (Dartiguenave et al., 2000a).

The second practical insight of the Henderson–Hasselbach equation applied to wine is that in wine itself, weak acids (HA) and their conjugate bases (A^-) are both present, so that wine is a buffer solution. The buffer capacity of a wine is an important factor to keep in consideration in case of conventional acidification and deacidification treatments.

2.4.2 Buffer Capacity

When an aliquot of a strong base (dB) is added to wine, it provokes an increase of its pH (dpH). The buffer capacity (π) of that wine is defined by the following equation:

$$\pi = \frac{dB}{dpH} \tag{2.5}$$

The same relationship can be written concerning the addition of an acid (dA); in this case, the pH variation will be negative (−dpH) and the pH will decrease.

Eq. (2.5) is obtained from the Henderson–Hasselbach equation (Ribéreau-Gayon et al., 2006a), but more often, π is expressed on the basis of total acidity (T) and alkalinity of ash (A_a), both given in meq/L, as reported in following equation (Usseglio-Tomasset, 1985).

$$\pi = 2.303 \frac{T \times A_a}{T + A_a} \tag{2.6}$$

A_a is defined as the sum of cations, other than ammonium, combined with the organic acids in wine (International Organization of Vine and Wine, 2017b) and represents an index of the concentration of dissociated organic acids, i.e., acids in the form of conjugate base (Usseglio-Tomasset, 1985). Buffer capacity can also be estimated from the titration curve of a wine. In such a case, it is defined as the quantity of acid or base needed to shift wine pH by one unit (Dartiguenave et al., 2000a).

From the practical point of view, the buffer capacity represents the ability of a wine to "react" toward a variation of its acid–base equilibrium and pH and represents a very important practical concept, when conventional acidification/deacidification treatments are carried out.

When an acid (e.g., tartaric acid) is added to wine, it dissociates and the numerator of Eq. (2.6) increases more than the denominator, reflecting an increase of the buffer capacity. This means that the effect of the acidification on the pH may be lower than expected. In the same way, when a base is added to wine (e.g., potassium bicarbonate), it reacts with the acids in solution; moreover potassium ions determine the precipitation of the bitartrate in the form of potassium bitartrate (KHT), reducing the buffer capacity. This may reflect on two opposite phenomena: an increase of the pH, due to the neutralization of tartaric acid by the bicarbonate ion, but also an increase of the HA/A^- ratio, due to bitartrate precipitation. The latter mechanism may provoke a further dissociation of organic acids (according to the law of mass action), which may act in opposition with respect to the former effect, making the increase of pH lower than expected.

Wine buffer capacity depends on several factors. Generally, π value depends on the pH range at which it is measured (Ribéreau-Gayon et al., 2006a), but also on the concentration of organic acids and their conjugate base (Usseglio-Tomasset, 1985; Dartiguenave et al., 2000a), i.e., T and A_a in Eq. (2.6).

This means that buffer capacity tends to decrease after malolactic fermentation (Usseglio-Tomasset, 1985) or tartaric precipitation (Champagnol, 1986), because of the decrease of total acidity (consumption of H_2M) in the first case, and ash alkalinity in the second (precipitation of KHT). In addition, it is reported that the buffer capacity of a given must is higher than the buffer capacity of the corresponding wine, because of the bitartrate precipitation occurring during alcoholic fermentation (Ribéreau-Gayon et al., 2006a) and because the π values measured for single organic acid solutions are lower in 11% v/v hydroalcoholic solution than in water (Dartiguenave et al., 2000a).

The relative concentration of the different acids present in wine may also condition buffer capacity. The π value measured for tartaric acid is conditioned in binary and ternary mixtures by the presence of ethanol and other organic acids. In equimolar concentration, for instance, succinic and citric acid are able to increase the buffer capacity of a hydroalcoholic solution of both pure tartaric acid and a mixture of tartaric and malic acid (Dartiguenave et al., 2000a).

Other factors, such as the presence of —OH groups in the molecular structure of the acid (Dartiguenave et al., 2000a), or the presence in solution of certain amino acids (Dartiguenave et al., 2000b) may also affect the buffer capacity.

Buffer capacity is an important factor, because it directly affects the perception (intensity and duration) of wine acidic sensation (Usseglio-Tomasset and Bosia, 1993), as well as because its knowledge is a fundamental factor for understanding the mechanism of acidification and deacidification treatments in wine.

2.5 TRADITIONAL STRATEGIES FOR CHEMICAL ACIDIFICATION

According to the International Code of Oenological Practices (International Organization of Vine and Wine, 2017c), acidification consists of the increase of the titratable acidity of musts and wines and the consequent decrease of the pH. Different techniques have been recommended for this purpose, based on chemical, microbiological, physical, and physicochemical procedures (Table 2.3). Microbiological acidification practices will be not examined in this chapter: they are based on the use of specific yeast strains producing malic, lactic, or succinic acid (International Organization of Vine and Wine, 2017c). In the present paragraph, chemical acidification will be mainly discussed, while electromembrane techniques and ion exchange resins will be treated later.

The O.I.V. established that all the acidification practices listed in Table 2.3, should be carried out so that the initial acidity content of the wine is not increased by more than 54 meq/L, corresponding to 4 g/L of tartaric acid (International Organization of Vine and Wine, 2017c). This limit is further restricted by the European law, which allows a maximum acidity increase (expressed in tartaric acid) of 1.5 g/L for musts and 2.5 g/L for wines (respectively 20.0 and 33.3 meq/L) (Regulation EU 1308, 2013). Moreover, in Europe, acidification is allowed only

TABLE 2.3 Recommended Techniques for the Acidification of Musts and Wines

Product	Technique	Notes/limitations
Must	Blending with musts with high acidity	
	Microbiological acidification	*Saccharomyces* or non-*Saccharomyces* yeasts producing malic or succinic acid
	Chemical acidification	Addition of lactic acid, L- or D,L-malic acid, L-(+)-tartaric acid
	Use of cation exchange resins	
	Electromembrane treatment	Electrodialysis with bipolar and cationic membranes
Wine	Blending with wines with high acidity	
	Chemical acidification	Addition of lactic acid, L- or D,L-malic acid, L-(+)-tartaric acid, citric acid
	Use of cation exchange resins	Allowed for acidification and tartaric stabilization
	Electromembrane treatment	Electrodialysis with bipolar and cationic membranes

Source: Extracted from: International Organization of Vine and Wine (2017c).

in certain winemaking regions (Regulation EU 1308, 2013). The American Regulation, instead, allows acidification on musts and wines, up to a maximum fixed acid level in the finished wine of 9 g/L of tartaric acid (Electronic Code of Federal Regulations, 2017).

Generally speaking, according to the recommendations reported in the International Code of Oenological Practices, chemical acidification of musts and wines should be only carried out with the products listed in Table 2.3; the addition of mineral acids (e.g., hydrochloric or sulfuric) is not allowed. Moreover, chemical acidification and deacidification are mutually exclusive (International Organization of Vine and Wine, 2017c; Regulation EU 1308, 2013). In the United States fumaric acid may also be used for acidification and stabilization purposes (Electronic Code of Federal Regulations, 2017).

A final remark concerns citric acid. It is interesting to observe that the O.I.V. allows the use of citric acid only for acidifying wines and not for musts (Table 2.3). According to the European and American law (Regulation EU 606, 2009; Electronic Code of Federal Regulations, 2017), such acid is not considered just an acidulant, but a stabilizer. In Europe it is allowed "for wine stabilization purposes," due to its ability to chelate certain metal ions (e.g., iron); its addition is authorized up to a maximum residual level of citric acid in the final wines of 1 g/L (Regulation EU 606, 2009).

2.5.1 Acidification by Blending with Musts or Wines From Low Maturity Grapes

The simplest and probably most ancient technique for increasing wine acidity is the blending of musts or wines with products obtained from low maturity grapes. This practice may be useful when pursuing a low-impact and additive-free winemaking approach, i.e., in organic or biodynamic winemaking. Nevertheless, the increase of acidity by blending with very acidic wines may also present some technological problematics.

The first one is just related with the low maturity of the grapes used for producing the "acidifying wine." If the increase of total acidity to be achieved is high, a nonnegligible amount of the latter must be added. Generally, a low maturity index for grapes is connected with high concentrations of certain C_6 volatile compounds (e.g., hexenols and hexanols), that are connected with vegetal and green sensory notes (Ribéreau-Gayon et al., 2006b). Moreover, in the specific case of red wines, unripe grapes are also characterized by unripe seeds and this might determine the development of harsh and astringent taste, due to the presence of high concentrations of aggressive tannins (Ribéreau-Gayon et al., 2006a, 2006b). In such cases, it would be advisable to use blending, limited to small variations of total acidity, so that only a limited percentage of "acidifying wine" should be added. The same considerations can be done talking about the blending of grapes before red wine maceration.

Furthermore, blending, especially when carried out for modifying acidic fraction, may have other important repercussions on wine stability: it may affect the precipitation of potassium bitartrate and modify wine colloidal equilibrium. For this reason, it would be better to apply blending before tartaric and colloidal stabilization. These problems make it even more difficult to forecast the final effect of the treatment on wine pH. In addition, when blending with acidic wines is chosen, the microbiological aspects must be also taken into account, to avoid the risk of microbial contaminations.

Finally, the blending of wines can be also subjected to legal restrictions (International Organization of Vine and Wine, 2017c; Regulation EU 606, 2009), which may become more severe for the wines with a Denomination of Origin (DOC/AOC).

A final remark concerns the opportunity to increase acidity, by blending with unripe grapes during maceration. Polyphenoloxidase (PPO) activity demonstrated to vary during ripening; in particular, tyrosinase activity was higher at the beginning of ripening, declining at full maturity (Macheix et al., 1991). An increased PPO activity in acidic juice might jeopardize color intensity and stability; for this reason, these practices should be carefully managed also considering these additional aspects.

Recently, an interesting method has been proposed for reducing the problems connected to the use of blending techniques for decreasing wine pH (Kontoudakis et al., 2011). This new approach consists of the use of high acidity/low alcohol products obtained by the fermentation of grapes coming from cluster thinning (carried out at the beginning of veraison). Such products were treated with high amounts of charcoal (5 g/L) and bentonite (1 g/L), to obtain an odorless and colorless wine (total acidity 17.8 g/L of tartaric acid), which was blended with three different red wine typologies, to reduce their pH. Results showed that acidified wines had a similar phenolic composition with respect to their controls, with a higher color intensity due to the pH diminution. In addition, no significant differences were found after sensory evaluation (triangle test), for two of the three cultivars studied (Kontoudakis et al., 2011).

2.5.2 Acidification by Supplementation with Organic Acids

The use of L-(+)-tartaric acid as wine acidulant is the most common practice and, before 2004, it was the only one authorized by the O.I.V. (International Organization of Vine and Wine, 2017d). When added to must or wine, tartaric acid dissociates, mainly forming bitartrate and hydronium ions.

$$H_2T + H_2O \rightleftarrows HT^- + H_3O^+ \tag{2.7}$$

Given that wines are supersaturated potassium bitartrate solutions, the increase of bitartrate concentration may provoke the precipitation of KHT, according to the following equilibrium.

$$HT^- + K^+ \rightleftarrows KHT\downarrow \tag{2.8}$$

This depends on several variables: potassium content, alcoholic strength, pH, amount of protective colloids (Ribéreau-Gayon et al., 2006a), as well as the level of tartaric stability that one wants to achieve with cold treatments.

As mentioned above, since the precipitation of KHT may lead to a decrease in titratable acidity (de Klerk, 2010), the actual yield in terms of wine acidification may be lower than expected and not always easily predictable. Anyway, the removal of bitartrate (conjugate base of H_2T) leads also to decrease ash alkalinity, further provoking pH lowering, supporting the real objective of the acidification process.

Acidification with tartaric acid is generally carried out when small corrections are needed on wine after fermentation, when the acidic structure of wine itself is almost defined. Contrary, on musts, a prefermentative correction may be necessary when the pH of the juice is excessive (e.g., pH >3.50), to avoid potential negative implications connected with the metabolism of LAB (Wibowo et al., 1985; Ribéreau-Gayon et al., 2006b). For instance, the achieving of a must pH value lower than 3.50 is strongly recommended in the case of management of alcoholic and malolactic fermentation by coinoculation practices (Comuzzo and Zironi, 2013).

As mentioned above, acidification with tartaric acid only, is a simple and good practice for minor acidity adjustments. Major additions may lead to harsh or unbalanced wines (Ribéreau-Gayon et al., 2006a) or to a nonequilibrated ratio between tartaric and malic acid. Concerning red wines, these events may be a minor difficulty for young fresh products, but they become a serious problem for the sensory quality of premium aged wines.

The possibility of acidifying musts and wines by malic acid addition is a very interesting proposition, particularly for red wines produced in hot climatic conditions, where the grape metabolism leads to drastic diminutions of such compounds during ripening. Malic acid is not involved in physicochemical equilibria and its addition has a direct and immediate result: increased sourness (Amerine et al., 1965) and freshness for young wines and lower pH variation (Carvalho et al., 2001). However, this compound is a good substrate for LAB development. Concerning this feature, acidification with malic acid should not be considered just as a pH or acidity rebalancing, but this compound may also allow LAB to carry out their own metabolic pathways, i.e., improving aromatic and gustative complexity of the final wine, thanks to the secondary products of malolactic fermentation. Conversely, if malolactic fermentation is not desired in young wines, the use of D,L-malic acid (racemic mixture) is recommended (Table 2.3). The D-isomer is not metabolized by LAB (Boulton et al., 1996; Volschenk et al., 2006) and this may represent an interesting feature for the malolactic stability of bottled wines.

The use of lactic and citric acid completes the overview on conventional acidification treatments. Both of these acids are basically used just for "finishing" the product. The former helps to increase the acidity of the wine by prolonging the sour sensory perception and determining a minor decrease of the pH with respect to tartaric acid (Carvalho et al., 2001). Therefore, it is reported to give a less intense sourness impression during tasting, compared with tartaric and malic (Amerine et al., 1965). A slight "sweet" sensation, characteristic of lactic acid, may also be added to this effect.

Citric acid may contribute to the character of "freshness," with a positive impact in young wines, but this feature is not suitable for aged products. The chelating capacity of this compound is useful to reduce the catalytic effect of certain metal ions (iron and copper), hampering the oxidation mechanisms of wines.

2.6 TRADITIONAL STRATEGIES FOR CHEMICAL DEACIDIFICATION

Deacidification is the decrease of titratable acidity of musts and wines, leading to the increase of their pH (International Organization of Vine and Wine, 2017c). As for acidification, O.I.V. recommends different

TABLE 2.4 Recommended Techniques for the Deacidification of Musts and Wines

Product	Technique	Notes/limitations
Must	Precipitation of potassium bitartrate	Spontaneous or induced by supplementation with KHT and cold treatment
	Blending with musts with low acidity	
	Microbiological degradation of malic acid	*Saccharomyces* or non-*Saccharomyces* yeasts (e.g., *Schizosaccharomyces*), lactic acid bacteria
	Chemical deacidification	Neutral potassium tartrate (K_2T), calcium carbonate[a], potassium hydrogen carbonate (bicarbonate)
	Use of anion exchange resins	Not accepted by O.I.V.
	Electromembrane treatment	Electrodialysis with bipolar and anionic membranes
Wine	Precipitation of potassium bitartrate	Spontaneous or induced by supplementation with KHT and cold treatment
	Blending with wines with low acidity	
	Microbiological degradation of malic acid	Lactic acid bacteria
	Use of anion exchange resins	Not accepted by O.I.V.
	Chemical deacidification	Neutral potassium tartrate (K_2T), calcium carbonate[a], potassium hydrogen carbonate (bicarbonate)
	Electromembrane treatment	Electrodialysis with bipolar and anionic membranes

[a]*Which may contain small amounts of double calcium tartromalate.*
Source: International Organization of Vine and Wine (2017c).

techniques and processing aids for this purpose, which are listed in Table 2.4. Microbiological techniques (e.g., malolactic fermentation or the use of non-*Saccharomyces* yeasts) will be specifically discussed in other chapters of the present book. This paragraph focuses on the main strategies for chemical deacidification.

According to both the O.I.V recommendations and the European regulations, deacidification may be carried out on musts and wines. As a restriction, O.I.V. establishes that the wine produced from a deacidification treatment should contain at least 1 g/L of tartaric acid (International Organization of Vine and Wine, 2017c). In Europe, deacidification is allowed only in certain winemaking regions and, on wines, it may be carried out only up to a limit of 13.3 meq/L (1 g/L) expressed as tartaric acid (Regulation EU 1308, 2013). US Regulation, instead, besides the use of calcium carbonate, potassium carbonate, and potassium bicarbonate, allows the practice of "amelioration," for the wines obtained from juices having a fixed acid level higher than 5 g/L. Fixed acidity may be reduced by adding an ameliorating material, such as water, sugar, or a combination of both. Deacidification treatments as well as the ameliorating material added should not reduce the fixed acidity to less than 5 g/L (Electronic Code of Federal Regulations, 2017).

2.6.1 Deacidification by Using Processing Aids

Deacidification of red wines is used for two main reasons: to facilitate the onset of malolactic fermentation, as the LAB are inhibited by low pH (Wibowo et al., 1985), and to make the wine more pleasant, as wine acids may strengthen the perception of astringency (Sowalsky and Noble, 1998).

The products allowed for deacidification are potassium hydrogen carbonate, calcium carbonate, and potassium tartrate (Table 2.4), with the latter being less used because of its higher cost. All these deacidifying agents act in two phases, involving acid–base and precipitation equilibria, as well as different potential wine instability mechanisms.

The addition of potassium bicarbonate consists of a first acid–base reaction that takes place immediately and leads to a net loss of acidity that occurs by neutralization of H_2T and the release of carbon dioxide.

$$H_2T + K^+ + HCO_3^- \rightleftarrows HT^- + K^+ + H_2O + CO_2\uparrow \tag{2.9}$$

The increase of the pH and the formation of HT^-, together with the increase of potassium concentration, lead to a further supersaturation of the medium. Only after appropriate tartaric stabilization practices and the subsequent precipitation of KHT, will the actual effect of deacidification treatment become evident.

The use of calcium carbonate also involves two reaction steps, but with different consequences. Deacidification, in this case, occurs completely with the first acid—base neutralization reaction, and it is immediate.

$$H_2T + Ca^{2+} + CO_3^{2-} \rightleftarrows T^{2-} + Ca^{2+} + H_2O + CO_2 \uparrow \tag{2.10}$$

However, the increase of the pH, the formation of tartrate (T^{2-}) and the presence of calcium ions, provoke an instability due to the supersaturation of calcium tartrate.

$$Ca^{2+} + T^{2-} \rightleftarrows CaT \cdot 4H_2O \downarrow \tag{2.11}$$

This salt precipitates over longer times with respect to KHT (Ribéreau-Gayon et al., 2006a) and, it does not affect the modification of the wine acidity, as the latter does.

Although $CaT \cdot 4H_2O$ is much more insoluble than KHT (Munyon and Nagel, 1977), it is hardly precipitated by cold stabilization treatments (Ribéreau-Gayon et al., 2006a). This may increase the risk of colloidal instability in bottled wine, for the direct precipitation of calcium tartrate (which may also behave as crystallization nuclei for other precipitations), and for the role of Ca^{2+} ions in wine colloidal instability (Ribéreau-Gayon et al., 2006a).

Contrary to calcium carbonate and potassium bicarbonate, deacidification with potassium tartrate involves a completely different mechanism. A rapid increase of the pH is observed by adding this salt, due to the addition of T^{2-} base, but deacidification actually occurs with the precipitation of KHT, during the following stabilization treatments.

The different practices described above, lead to a decrease of wine acidity, only involving tartaric acid. Since musts and wines with high acidities generally have a high content of malic acid, the risk of imbalances due to a high ratio between H_2M (or lactic after malolactic fermentation) and H_2T is nonnegligible.

To overcome this problem, Münz (1960) proposed a deacidification process with precipitation of double calcium tartromalate (Ca_2TM). The most interesting feature of this salt is its high insolubility at a pH above 4.50. To achieve these extreme pH conditions, an aliquot of wine is treated with an excess of $CaCO_3$. The excess of carbonate keeps the pH high, shifting the dissociation of tartaric and malic acid toward T^{2-} and M^{2-}, thus allowing the precipitation of double calcium tartromalate (Steele and Kunkee, 1978). This precipitation is favored by the presence of crystallization nuclei of Ca_2TM. Commercial calcium carbonate preparations may contain small quantities of double calcium tartromalate to provide this feature (International Organization of Vine and Wine, 2017c, 2017d). Obviously, after precipitation and before the blending of the aliquot treated with the whole lot of wine, crystals must be separated (e.g., filtration), to avoid their fast redissolution at the original pH of wine itself. The excess of added $CaCO_3$ is neutralized according to Eq. (2.10).

2.7 NEW TECHNOLOGIES FOR pH CONTROL

Due to the difficulties in forecast the effects of traditional acidification and deacidification treatments on wine pH, different new technological tools have been developed in the last decades. Besides the conventional practices discussed above, the International Organization of Vine and Wine (2017c) has included, among the recommended practices for the adjustment of the acidity of musts and wines, two main groups of technologies: electromembrane techniques and the use of ion exchangers.

2.7.1 Acidification and Deacidification by Electromembrane Techniques

Acidification and deacidification by electromembrane techniques derive from electrodialysis treatment, used for tartaric stabilization. Classic electrodialysis provides the separation of cations and anions from wine by means of an electric field and a membrane pack, where cationic and anionic membranes are alternatively assembled (Ribéreau-Gayon et al., 2006a; Lasanta and Gómez, 2012). The International Oenological Codex (International Organization of Vine and Wine, 2017d) defines cationic and anionic membranes as thin, dense, insoluble walls composed of polymeric material, permeable to ions; in particular, cationic membranes are permeable only to cations, while anionic membranes allow the passage of anions only (Rayess and Mietton-Peuchot, 2015). Cationic membranes are styrene-divinylbenzene copolymers, which carry sulfonic functional groups ($-SO_3^-$), while anionic membranes are styrene-divinylbenzene copolymers functionalized with quaternary ammonium ($-NR_4^+$), or quaternary ammonium-divinylbenzene copolymers (International Organization of Vine and Wine, 2017d).

While they move toward the opposite poles of the electric field applied, cations (e.g., K^+) and anions (e.g., HT^-) are extracted from wine, and concentrated in the brine, which circulates, inside the membrane pack, in adjacent compartments with respect to those where wine flows (Ribéreau-Gayon et al., 2006a; Rayess and Mietton-Peuchot, 2015). A modification in the assembly of the membrane pack, allows electromembrane technologies to be used for acidification/deacidification purposes.

The first reports about the use of electromembrane techniques for the acidification/deacidification of musts and wines, date back to the 1970–80s (Shpritsman et al., 1972; Wucherpfennig and Keding, 1982; Lopez Leiva, 1988). Nevertheless, the O.I.V. approved these technologies only in 2010–12 (International Organization of Vine and Wine, 2017c) and in Europe, they have been allowed since 2011 (acidification) and 2013 (deacidification) (Regulation EU 53, 2011; Regulation EU 144, 2013).

When electromembrane applications are used for acidification/deacidification, the membrane assembly is modified, introducing bipolar membranes. Bipolar membranes are obtained by laminating together cation- and anion-exchange membranes, through an intermediate junction layer (Rayess and Mietton-Peuchot, 2015). This structure makes these membranes to have both a cationic and an anionic face (International Organization of Vine and Wine, 2017d), so that they do not allow the permeation of either cations nor anions (Rayess and Mietton-Peuchot, 2015). The role of bipolar membranes is fundamental for wine acidification/deacidification treatments, because when the electric field is applied, water molecules are split into hydroxide (OH^-) and hydronium ions (H_3O^+) by a disproportionation reaction, occurring at the membrane junction layer (Wilhelm, 2001; Rayess and Mietton-Peuchot, 2014).

When used for acidification, bipolar membranes are coupled with cationic membranes (Fig. 2.2A). The operating mechanism is well explained by Rayess and Mietton-Peuchot (2015). Briefly, it consists of the circulation of wine inside the membrane pack, so that wine itself flows between cationic membranes and the cationic side of the bipolar ones. In the adjacent compartment, water flows. When the electric field is applied, potassium ions move toward the cathode, cross the cationic membranes and they are extracted from wine, being concentrated in the water, which turns to brine. In the wine compartment, potassium is replaced with the protons (actually hydronium ions) which are formed at the bipolar membrane junction. In the same way, bitartrate ions tend to move toward the anode, but they are forced to remain in wine, because they are unable to cross the cationic layer (−) of the bipolar membrane. The result is that wine preserves bitartrate (and the conjugate bases of the other organic acids), while it is enriched with H_3O^+ ions; consequently, the pH decreases and total acidity increases (Rayess and Mietton-Peuchot, 2014). At the same time, the water used at the beginning of the process becomes gradually more concentrated in K^+ and OH^- ions (brine).

Conversely, for deacidification (Fig. 2.2B), bipolar membranes are coupled with anionic membranes, and wine flows between anionic membranes and the anionic side of the bipolar ones. In this case, when the electric field is applied, bitartrate ions (as well as dissociated malic acid) move to the anode, crossing the anionic membrane; they are extracted from wine and replaced with the OH^- ions, produced at the junction layer of the bipolar membrane. Potassium tends to move to the cathode, but it remains in the wine compartment, because it cannot cross the anionic layer of the bipolar membrane. As the process continues, wine is progressively enriched with OH^- ions (Halama et al., 2015), while organic acids are extracted and concentrated in the brine. The result is an increase of wine pH and a decrease of total acidity (Rayess and Mietton-Peuchot, 2014).

The advantage in the use of electromembrane techniques is that pH variation is independent on bitartrate crystallization and wine buffer capacity. This makes this technology accurate: in acidification treatments, the pH correction has a precision of 0.05 units (Halama et al., 2015; Rayess and Mietton-Peuchot, 2015). The process may be completely automatized and managed continuously, in a single passage, without recirculation. Wine only requires to be filtered before processing to avoid clogging of the modules (Halama et al., 2015). The pH can be measured online for both wine and brine. The cleaning of the installations is simple and it may be managed with the normal cleaning agents used in a winery (Halama et al., 2015). One practical limitation, in acidification treatments, is the alkalinity of the brine (KOH), which needs to be continuously monitored, because if it becomes too high it may damage the membranes. A specific requirement also concerns the water used for the installations: it is recommended to use distilled or osmotized water, because high concentration in carbonates (e.g., in hard water) may provoke precipitations inside the membrane pack. This risk becomes higher as the pH of the brine increases, increasing also the probability to damage the membranes.

Concerning the effects on wine composition, electromembrane techniques are reported to allow a better balance of acidic fraction (e.g., with respect to tartaric acid addition), and a better preservation of wine color and polyphenols (Rayess and Mietton-Peuchot, 2015). Nevertheless, this technique requires nonnegligible water consumption (Halama et al., 2015).

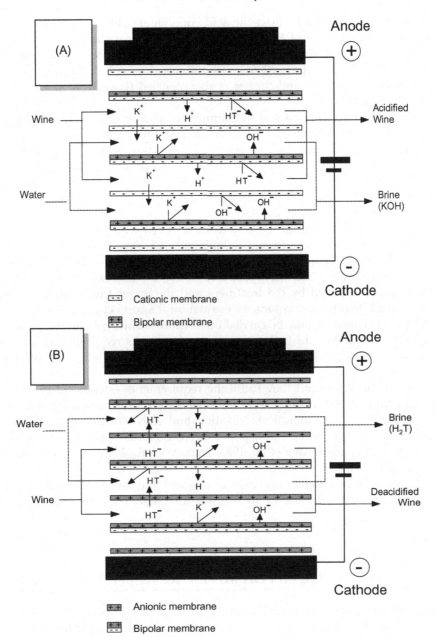

FIGURE 2.2 Scheme of membrane assembly and operating conditions for wine acidification (A) and deacidification (B) by electromembrane techniques.

Specific indications for the treatment, as well as for the characteristics of the membranes are reported in the O.I.V recommendations (International Organization of Vine and Wine, 2017c, 2017d), as well as in the European Regulation (Regulation EU 606, 2009). It is interesting to observe that US legislation does not mention electromembrane techniques for acidification/deacidification of musts and wines (Electronic Code of Federal Regulations, 2017).

2.7.2 Ion Exchange Resins

The European regulation (Regulation EU 606, 2009) and the O.I.V (International Organization of Vine and Wine, 2017c) allow the use of cation exchange resins for the acidification of musts and wines. They are sulfonated styrene-divinylbenzene or divinylbenzene-methacrylic acid copolymers able to extract cations (mainly K^+) from must or wine (International Organization of Vine and Wine, 2017d), providing acidification. Depending on the regeneration method, potassium (and also calcium ion) may be replaced with hydrogen ions, but also with sodium or magnesium (Ribéreau-Gayon et al., 2006a), also reducing the risk of tartaric precipitation. In addition, a diminution of the concentrations of iron and copper is also reported (Lasanta et al., 2013), with potential positive effects on wine oxidative stability.

Ion exchangers may be used in batch or packed in columns (Electronic Code of Federal Regulations, 2017): the former application may be carried out using fluidized bed installations, or simply by introducing a suitable aliquot of resin in the wine and then separating it by appropriate technical methods (International Organization of Vine and Wine, 2017c). Generally, a treatment with cation exchangers does not involve the whole batch of wine, but only one part (10%–20%); O.I.V recommends to carry out the treatments by recirculating wine itself through the resin until the desired pH value is reached, "to avoid the production of fractions" (International Organization of Vine and Wine, 2017c).

The problem in using cation exchangers for the acidification of red wines is connected with the affinity of the styrene-divinylbenzene matrix for anthocyanins and polyphenols; Lasanta et al. (2013) found a slight diminution of anthocyanins and tannins after cation exchange treatment of a red wine. Nevertheless, they also report a decrease of color hue and a higher color intensity, probably due to the decrease of wine pH. Volatile compounds were demonstrated to be poorly affected by the treatment and the wines processed by cation exchangers were perceived with a higher overall quality during sensory evaluation (Lasanta et al., 2013).

The regeneration of cation exchangers may be carried out with concentrated sulfuric acid and sodium chloride solutions (Ribéreau-Gayon et al., 2006a). In both cases, a lot of water is required and effluents shall be treated and recycled as special waste, making this technology quite expensive from the point of view of environmental sustainability.

Cation exchange resins are insoluble and they fulfill the requirements established by the O.I.V. (International Organization of Vine and Wine, 2017d). Conversely, anion exchangers may release traces of quaternary ammonium salts, provoking a slight microbiological stabilization and also modifying wine organoleptic characters (Ribéreau-Gayon et al., 2006a). For this reason, despite their possible use (alone or in combination with cation exchangers) to correct wine acidity (Bonorden et al., 1986; Lasanta et al., 2013), they are not allowed for this purpose by either O.I.V. (International Organization of Vine and Wine, 2017c) or the European Union (Regulation EU 606, 2009). In Europe, anion exchangers are only authorized for the production of concentrated rectified musts; contrary, US regulations allow the use of such resins for must and wine processing (Electronic Code of Federal Regulations, 2017).

2.8 LABORATORY TECHNIQUES FOR MEASURING pH AND ACIDIC FRACTION

The determination of pH at winery scale is one of the simplest analytical parameters used in wine quality control, but at the same time, one of the most important. However, just for the simplicity of the measurements, some critical aspects are sometimes neglected, in particular concerning the choice and the maintenance of the electrodes.

The O.I.V. has established specific recommendations for the characteristics of the electrodes and calibration buffers (International Organization of Vine and Wine, 2017e). In particular, electrodes should be a combination of a glass electrode, and a calomel-saturated potassium chloride reference electrode. Anyway, different typologies of combined glass/calomel electrodes are available commercially. Must and wine have a great ability to dirt the electrodes, due to their high content in phenolic and colloidal substances. This may provoke the passivation of the electrode membrane and diaphragm, resulting in a slow or uncorrected analytical response. These factors are not in agreement with the extreme accuracy required for pH measurements: ± 0.01 pH units, according to O.I.V. recommendations (International Organization of Vine and Wine, 2017e).

Different companies provide practical guidelines for electrode care and maintenance (Barron et al., 2017). Concerning the kind of electrode, for winery use, it is recommended to choose a model in which the diaphragm (that is to say the connection between the outer part of the combined electrode and the reference electrolyte solution) is protected by a plastic sleeve. Another interesting feature is connected with the use of gel-filled electrodes, in which a gel replaces the liquid electrolyte in the bulb; this reduces the formation of air bubbles in the inner buffer, making measurements more reliable.

Besides pH, titratable acidity is another simple but fundamental parameter in wine quality control. It is normally determined by titration, using indicators (bromothymol blue) or by assessing the equivalence point by a pH meter (pH 7.00) (International Organization of Vine and Wine, 2017a). A particular problem, found at winery scale when such determination is carried out on red wines, is the difficulty to visually perceive the color change of the indicator (yellow to green/blue), which may be heavily affected by wine color, especially for varieties very rich in anthocyanins. Different solutions may be available to reduce this problem: the use of a light at the bottom of the flasks used for titration or even the dilution with a small aliquot of distilled water. The latter practice is also suggested in the Compendium of International Methods of Wine and Must Analysis (International Organization of Vine and Wine, 2017a), because it does not affect the number of total acidity equivalents and the final result of the titration. Nowadays, different models of automatic titrators are commercially available, allowing fast and reliable potentiometric assessment of total acidity, on a large number of samples per hour.

Total acidity may be expressed with different units: in most of the European countries and in the United States it is given in grams per liter of tartaric acid, while in other countries (e.g., in France), grams per liter of sulfuric acid is used. The conversion from one expression modality to the other is possible by the following equation:

$$TA_{tartaric} = TA_{sulfuric} \times 1.53 \tag{2.12}$$

Total acidity and pH are fundamental parameters for juice and wine quality control, especially when they are needed to support immediate decisions (i.e., harvest time and grape delivery), because of the ease of their determination. However, they give only a general overview of the acidic fraction; sometimes, more detailed information are needed for supporting correct decisions about specific technological problems.

An important parameter, which is generally poorly considered in the wineries, is the alkalinity of the ash (International Organization of Vine and Wine, 2017b). It represents the aliquot of organic acids in salified form, that is to say in the form of a conjugate base (A^-). As reported in Eq. (2.6), such parameter is fundamental for the calculation of wine buffer capacity and, consequently, to better forecast the effects of conventional acidification/deacidification treatments.

The determination of the concentration of specific organic acids, may be helpful for certain technological needs (e.g., for monitoring malolactic fermentation). One of the most common mistakes made at winery scale, is to compare the sum of the concentrations of the single organic acids, with the total acidity of a wine. Actually, if one sums the concentration of each organic acid (e.g., as determined by high performance liquid chromatography (HPLC) analysis), the value obtained is always higher with respect to the total acidity, because of the fact that HPLC analysis determines total acids, i.e., in free and dissociated form ($[HA] + [A^-]$). For having a correct acidic balance, besides total acidity, pH should also be taken into account, in relationship with ash alkalinity.

Due to its analytical complexity, HPLC determination is hardly used at winery scale and single organic acids may be determined by using enzymatic analyses. Commercial enzymatic kits are available for the evaluation of various organic acids: malic, acetic, citric, succinic, D- and L-lactic. Moreover, enzymatic analyzers allow the automated processing of wine samples. One of the most interesting aspects connected with enzymatic analysis, is the possibility to determine the two enantiomers of lactic acid; in fact, as already mentioned the presence of high concentrations of the D-isomer, is an index of unwanted and uncontrolled fermentations (Ribéreau-Gayon et al., 2006b). Enzymatic analyzers allow a very selective and accurate determination of organic acids, but the cost of the instrument and the commercial kits make this analytical technique quite expensive. An interesting and simple tool for a qualitative determination of the most important organic acids in wine is the use of thin layer chromatography (TLC) (Mato et al., 2005). TLC allows to separate wine organic acids and to effectively monitor the behavior of malolactic fermentation, simply by checking the formation of the spot of lactic acid and the disappearance of that of malic acid. In France, this kind of determination is quite widespread also in small to medium wineries, because of its simplicity and cheapness.

In the last decade, fourier transform infrared spectroscopy (FTIR) analyzers have changed the analytical approach toward wine quality control. The instruments used are interferometers, operating at a wavelength range from 2000 to 10,000 nm, that is to say in the field of near—middle infrared (International Organization of Vine and Wine, 2010). This technique allows the direct and fast multiparametric characterization of must or wine samples, without specific sample preparation procedures: a preliminary filtration/centrifugation and a carbon dioxide elimination (if it exceeds 750 mg/L) are only recommended (Dubernet, 2009). FTIR analyzers have several advantages: they are fast and they can be fully automatized (including self-cleaning), allowing the processing of a large number of samples per day. Nevertheless, FTIR analysis may be subjected to analytical errors if calibration procedures are neglected. The International Organization of Vine and Wine (2010) established detailed guidelines for the correct management of calibration procedures and the selection of calibration samples. The accuracy, repeatability, and the detection limit of FTIR data depend on several factors: the number and typology of the calibration samples, the structure of the compounds to be determined and their concentration, the matrix effect due to interactions between compounds, and the analytical methods used for instrument calibration (Dubernet, 2009). Anyway, FTIR analyzers became very widespread in large-size wine companies and in many private laboratories providing analyses for small-medium wineries. FTIR determinations are accurate and reliable for the most important wine quality control parameters (including the main organic acids, total acidity and pH), provided that calibration is carried out on the basis of a correct approach and that a suitable and effective quality control of the results, by comparison with classical analytical techniques, is applied (International Organization of Vine and Wine, 2010).

Acknowledgments

The authors are grateful to Alessandro Angilella and Marco Marconi from Ju.Cla.S.—Vason Group (S. Pietro in Cariano, VR, Italy), for the practical information supplied about membrane techniques and the use of ion exchangers.

References

Amerine, M.A., 1956. The maturation of wine grapes. Wines Vines 37 (27-32), 53–55.
Amerine, M.A., Roessler, E.B., Ough, C.S., 1965. Acids and the acid taste. I. The effect of pH and titratable acidity. Am. J. Enol. Vitic. 16, 29–37.
Barron, J.J., Ashton, C., & Geary, L., 2017. Care, maintenance and fault diagnosis for pH electrodes. https://reagecon.com/pdf/technicalpapers/Electrode_CM_v5_TSP-02_Issue_4.pdf (accessed 23.12.2017).
Berg, H.W., 1963. Stabilisation des anthocyanes. Comportement de la couleur dans les vins rouges. Ann. Technol. Agricol. 12, 247–259.
Berg, H.W., Keefer, R.M., 1958. Analytical determination of tartrate stability in wine. I. Potassium bitartrate. Am. J. Enol. Vitic. 9, 180–193.
Berg, H.W., Keefer, R.M., 1959. Analytical determination of tartrate stability in wine. II. Calcium tartrate. Am. J. Enol. Vitic. 10, 105–109.
Bonorden, W.R., Nagel, C.W., Powers, J.R., 1986. The adjustment of high pH/high titratable acidity wines by ion exchange. Am. J. Enol. Vitic. 37, 143–148.
Boulton, R., 1980. The relationships between total acidity, titratable acidity and pH in wine. Am. J. Enol. Vitic. 31, 76–80.
Boulton, R.B., Singleton, V.L., Bisson, L.F., Kunkee, R.E., 1996. Principles and Practices of Winemaking. Chapman & Hall, New York, Chapter 15.
Bousbouras, G.E., Kunkee, R.E., 1971. Effect of pH on malo-lactic fermentation in wine. Am. J. Enol. Vitic. 22, 121–126.
Brouillard, R., Delaporte, B., 1977. Chemistry of anthocyanin pigments. 2. Kinetic and thermodynamic study of proton transfer, hydration, and tautomeric reactions of malvidin-3-glucoside. J. Am. Chem. Soc. 99, 8461–8468.
Buck, R.P., Rondinini, S., Covington, A.K., Baucke, F.G.K., Brett, C.M.A., Camões, M.F., et al., 2002. Measurement of pH. Definition, standards, and procedures (IUPAC recommendations 2002). Pure Appl. Chem. 74 (11), 2169–2200.
Carvalho, E.C.P., Costa, S., Franco, C., Curvelo-Garcia, A.S., 2001. Acidification of musts and wines. Application of L-tartaric, L-lactic and DL-malic acids. Bull. O.I.V 849-850, 743–751.
Champagnol, F., 1986. L'acidité des moûts et des vins. Première partie-Facteurs physico-chimiques et technologiques de variation. Rev. Œnol. 104, 26–57.
Commission Regulation (EC) No 606, 2009. Laying down certain detailed rules for implementing Council Regulation (EC) No 479/2008 as regards the categories of grapevine products, oenological practices and the applicable restrictions. Off. J. Eur. Communities L193, 1–59.
Commission Regulation (EC) No 53, 2011. Amending Regulation (EC) No 606/2009 laying down certain detailed rules for implementing CouncilRegulation (EC) No 479/2008 as regards the categories of grapevine products, oenological practices and the applicable restrictions. Off. J. Eur. Communities L19, 1–6.
Commission Regulation (EC) No 144, 2013. Amending Regulation (EC) No 606/2009 as regards certain oenological practices and the applicable restrictions and regulation (EC) No 436/2009 as regards the registering of these practices in the documents accompanying consignments of wine products and the wine sector registers to be kept. Off. J. Eur. Communities L47, 56–62.
Commission Regulation (EU) No 1308, 2013. Establishing a common organisation of the markets in agricultural products and repealing Council Regulations (EEC) No 922/72, (EEC) No 234/79, (EC) No 1037/2001 and (EC) No 1234/2007. Off. J. Eur. Communities L347, 671–854.
Comuzzo, P., Zironi, R., 2013. Biotechnological strategies for controlling wine oxidation. Food Eng. Rev. 5, 217–229.
Dartiguenave, C., Jeandet, P., Maujean, A., 2000a. Study of the contribution of the major organic acids of wine to the buffering capacity of wine in model solutions. Am. J. Enol. Vitic. 51, 352–356.
Dartiguenave, C., Jeandet, P., Maujean, A., 2000b. Changes in the buffering capacity of model solutions of 40 mM tartaric or malic acids in relation to amino acids. Am. J. Enol. Vitic. 51, 347–351.
de Klerk, J.-L., 2010. Succinic acid production by wine yeasts. Master Thesis in Agricultural Sciences. Stellenbosch University, Stellenbosch, South Africa.
Drawert, F., Rapp, A., Ulrich, W., 1965. Über die Bildung von organisden Säuren durd Weinhefen. II. Quantitative Beziehungen zwischen Stickstoffquelle und Weinsäurebildung in Modellgärversuchen. Vitis 5, 199–200.

References

Dubernet, M., 2009. Automatic analysers in oenology. In: Moreno-Arribas, M.V., Polo, M.C. (Eds.), Wine Chemistry and Biochemistry. Springer, New York, pp. 649–676.

Dupuy, P., 1957. Les facteurs du développement de l'acescence dans le vin. Ann. Technol. Agric. 6, 391–407.

Electronic Code of Federal Regulations, 2017. Title 27, Chapter I, Subchapter A, Part 24—Wine. https://www.ecfr.gov/cgi-bin/text-idx?c=ecfr&sid=506cf0c03546efff958847134c5527d3&rgn=div5&view=text&node=27:1.0.1.1.19&idno=27#sp27.1.24.a (accessed 17.12.2017).

Halama, R., Kotala, T., Vrána, J., 2015. Modification of the organoleptic properties of beverages. Desalin. Water Treat. 56, 3181–3190.

International Organization of Vine and Wine, O.I.V, 2010. Resolution OIV/OENO 390/2010. Guidelines on Infrared Analysers in Oenology. O.I.V., Paris.

International Organization of Vine and Wine, O.I.V, 2017a. Total Acidity—OIV-MA-AS313-01: R2015. In: Compendium of International Methods of Wine and Must Analysis, vol. 1. O.I.V., Paris.

International Organization of Vine and Wine, O.I.V, 2017b. Alkalinity of Ash—OIV-MA-AS2-05: R2009. In: Compendium of International Methods of Wine and Must Analysis, vol. 1. O.I.V., Paris.

International Organization of Vine and Wine, O.I.V, 2017c. International Code of Oenological Practices. O.I.V., Paris.

International Organization of Vine and Wine, O.I.V, 2017d. I International Oenological Codex. O.I.V., Paris.

International Organization of Vine and Wine, O.I.V, 2017e. pH—OIV-MA-AS313-15. In: Compendium of International Methods of Wine and Must Analysis, vol. 1. O.I.V., Paris.

Jackson, R.S., 2008. Wine Science Principles and Applications, third ed. Academic Press, London (Chapter 11).

Kliewer, W.K., Howarth, L., Omori, M., 1967. Concentrations of tartaric acid and malic acids and their salts in *Vitis vinifera* grapes. Am. J. Enol. Vitic. 18, 42–54.

Kontoudakis, N., Estruelas, M., Fort, F., Canals, J.M., Zamora, F., 2011. Use of unripe grapes harvested during cluster thinning as a method for reducing alcohol content and pH of wine. Aust. J. Grape Wine Res. 17, 230–238.

Kontoudakis, N., Gonzalez, E., Gil, M., Estruelas, M., Fort, F., et al., 2012. The pH Influence on Micro-Oxygenation. Australian & New Zealand Grapegrower & Winemaker, vol. 578, pp. 54–58.

Lafon-Lafourcade, S., Peynaud, E., 1974. Sur l'action antibacterienne de l'anhydride sulfureux sous forme libre et sous forme combinee. Connaiss. Vigne Vin 10, 187–203.

Lasanta, C., Gomez, J., 2012. Tartrate stabilization of wines. Trends Food Sci. Technol. 28, 52–59.

Lasanta, C., Caro, I., Pérez, L., 2013. The influence of cation exchange treatment on the final characteristics of red wines. Food. Chem. 138, 1072–1078.

Lide, D.R. (Ed.), 2005. CRC Handbook of Chemistry and Physics. eighty fifth ed. CRC Press, Boca Raton, FL (Section 8).

Lopez Leiva, M.H., 1988. The use of electrodialysis in food processing. II. Review of practical applications. Lebensm.-Wiss. Technol. 21, 177–182.

Macheix, J.-J., Sapis, J.-C., Fleuriet, A., 1991. Phenolic compounds and polyphenoloxidase in relation to browning in grapes and wines. Crt. Rev. Food Sci. Nutr. 30, 441–486.

Mato, I., Suárez-Luque, S., Huidobro, J.F., 2005. A review of the analytical methods to determine organic acids in grape juices and wines. Food Res. Int. 38, 1175–1188.

Mayer, K., Busch, I., Pause, G., 1964. Über die Bernsteinsäurebildung während der Weingärung. Z. Lebensm.-Unters. Forsch. 125, 375–381.

Munyon, J.R., Nagel, C.W., 1977. Comparison of methods of deacidification of musts and wines. Am. J. Enol. Vitic. 28, 79–87.

Münz, T., 1960. Die Bildung des Ca-Doppelsalzes der Wein- und Apfelsäure, die Möglichkeiten seiner Fällung durch $CaCO_3$ im Most. Weinberg Keller 7, 239–247.

Ough, C.S., Amerine, M.A., Sparks, T.C., 1969. Studies with controlled fermentations. XI. Fermentation temperature effects on acidity and pH. Am. J. Enol. Vitic. 20, 127–139.

Peynaud, E., Lafon-Lafourcade, S., Guimberteau, G., 1966. L-(+)-lactic acid and D-(−)-lactic acid in wines. Am. J. Enol. Vitic. 17, 302–307.

Prenesti, E., Toso, S., Daniele, P.G., Zelano, V., Ginepro, M., 2004. Acid–base chemistry of red wine: analytical multi-technique characterisation and equilibrium-based chemical modelling. Anal. Chim. Acta 507, 263–273.

Rayess, Y.E., Mietton-Peuchot, M., 2014. Integrated membrane processes in winemaking. In: Cassano, A., Drioli, E. (Eds.), Integrated Membrane Operations in the Food Production. Walter de Gruyter GmbH, Berlin, Germany, pp. 147–162.

Rayess, Y.E., Mietton-Peuchot, M., 2015. Membrane technologies in wine industry: an overview. Crit. Rev. Food. Sci. Nutr. 56, 2005–2020.

Ribéreau-Gayon, P., Glories, Y., Maujean, A., Dubourdieu, D., 2006a. Handbook of enology, second ed. The Chemistry of Wine Stabilization and Treatments, vol. 2. John Wiley & Sons, New York.

Ribéreau-Gayon, P., Dubourdieu, D., Doneche, B., Lonvaud, A., 2006b. Handbook of enology, second ed. The Microbiology of Wine and Vinifications, vol. 1. John Wiley & Sons, New York.

Sheridan, M.K., Elias, R.J., 2016. Reaction of acetaldehyde with wine flavonoids in the presence of sulfur dioxide. J. Agric. Food. Chem. 64, 8615–8624.

Shpritsman, E.M., Shapiro, B.S., Andreev, V.V., 1972. Regulating the ionic composition of wines by electrodialysis. Sadovod., Vinograd. Vinodel. Mold. 27, 26–30.

Sowalsky, R.A., Noble, A.C., 1998. Comparison of the effects of concentration, pH and anion species on astringency and sourness of organic acids. Chem. Sens. 23, 343–349.

Spedding, F.H., Rard, J.A., Habenschuss, A., 1977. Standard state entropies of the aqueous rare earth ions. J. Phys. Chem. 81, 1069–1074.

Steele, J.T., Kunkee, R.E., 1978. Deacidification of musts from the Western United States by the calcium double-salt precipitation process. Am. J. Enol. Vitic. 29, 153–160.

Thoukis, G., Ueda, M., Wright, D., 1965. The Formation of succinic acid during alcoholic fermentation. Am. J. Enol. Vitic. 16, 1–8.

Usseglio-Tomasset, L., 1985. Chimica Enologica, second ed. AEB Edizioni, Brescia.

Usseglio-Tomasset, L., Bosia, P.D., 1993. La sensation acide pour les vins. Bull. O.I.V 753-754, 855–859.

Volschenk, H., van Vuuren, H.J.J., Viljoen-Bloom, M., 2006. Malic acid in wine: origin, function and metabolism during vinification. S. Afr. J. Enol. Vitic. 27, 123–136.

Wejnar, R., 1972. Untersuchungen zur Bedeutung der Weinsäure für die Wasserstoffionenkonzentration des Traubenweines. VIII. Theoretische Erörterungen unter besonderer Berücksichtigung des Äpfelsäureabbaues und der Weinsteinausfällung. Rebe, Wein, Obstbau und Früchteverwertung 22, 19–37.

Wibowo, D., Eschenbruch, R., Davis, C.R., Fleet, G.H., Lee, T.H., 1985. Occurrence and growth of lactic acid bacteria in wine: a review. Am. J. Enol. Vitic. 36, 302–313.

Wilhelm, F.G., 2001. Bipolar Membrane Electrodialysis. University of Twente, Enschede, The Netherlands, PhD Thesis.

Wucherpfennig, K., Keding, K., 1982. Deacidification and acidification of stone, pome and berry fruit juices and grape must by electrodialysis. Flüssiges Obst 49, 590–601.

Zamora, F., 2009. Biochemistry of Alcoholic Fermentation. In: Moreno-Arribas, M.V., Polo, M.C. (Eds.), Wine Chemistry and Biochemistry. Springer, New York, pp. 3–26.

CHAPTER 3

Maceration and Fermentation: New Technologies to Increase Extraction

Antonio Morata, Carmen González, Wendu Tesfaye, Iris Loira and Jose A. Suárez-Lepe

Department of Chemistry and Food Technology, Technical University of Madrid, Madrid, Spain

3.1 INTRODUCTION

The majority of red grape varieties do not contain pigments within their pulp but rather in their skins, so an essential step in the red winemaking process is the extraction of these compounds, along with tannins and aromatic molecules, from skins to the must during maceration. This means that it is necessary to keep skins and seeds in contact with the must in the tank during fermentation to extract and diffuse these phenolic compounds in the must and to obtain the end color of the wine provided by the skin's anthocyanins, as well as the antioxidant capacity and structure provided by tannins. The concentration of tannins (proanthocyanidins) extracted from skins will have a strong influence on the barrel aging capacity of the wine. The simultaneous maceration and fermentation processes make it necessary to balance the optimal conditions for both, which are quite different for many parameters. It is also necessary to optimize tank design to increase extraction during maceration. Typical anthocyanin concentration in full bodied red wines is about 500 mg/L, however for some wines it can be higher than 2 g/L (He et al., 2012).

Maceration and fermentation in red wines depend on several technological factors including the design of fermenters to optimize both processes, with a key factor being the material used to build them. Temperatures and cap management also exert great influence on the extraction of phenolic compounds (anthocyanins and tannins), which is what affects not only the final wine quality in terms of structure, astringency, softness, and color, but also the antioxidant capacity and the wine's aptitude for barrel aging. Other factors may play an important role (SO_2, ethanol content, pH, presence of enzymes...). And finally, a number of new technologies are evolving rapidly and it is likely that in the future, it will be possible to alleviate the traditional discontinuous or batch processes of winemaking, replacing these with a continuous process. Some emerging technologies such as high hydrostatic pressure (HHP), pulsed electric fields (PEFs), ultrasounds (USs), irradiation, pulsed light (PL), ozone and electrolyzed water are opening up new possibilities in wine technology.

3.2 TANK DESIGN FOR RED WINEMAKING

With regard to tanks typically used to make white wines, these are usually tall in height and small in diameter, and have a flat base that can be placed on the ground directly. The tanks are built as such because fermentation occurs to the juice so the most important parameter is temperature control in order to preserve the varietal aromas and fermentative qualities. The absence of skins means the use of special devices to drain the pomace is not necessary. The situation is quite different for red wines because it is necessary to macerate the skins (eventually with seeds or stems) to produce the extraction and the diffusion of pigments and tannins. To facilitate the

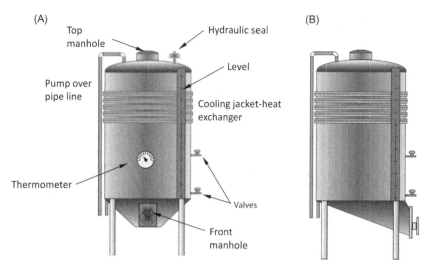

FIGURE 3.1 Metallic tanks and components for red wine maceration and fermentation. (A) Symmetric conic bottom. (B) Asymmetric conic bottom.

mechanical work on skins using either punch downs or pump overs, it is better to form thin caps, and therefore it is necessary to use tanks that are larger in diameter than in height. The use of similar or slightly lower ratios between diameter and height helps to increase extraction.

The inclusion of metallic legs on steel tanks used for red winemaking is quite frequent in order to facilitate the draining of pomace at the end of maceration. Other typical components include several valves to take samples or drain off the liquid, at least two manholes for winemaking operations—one on the upper side for cap management and the other on the bottom to drain the pomace and aid cleaning purposes—a transparent tube to check levels, one or several cooling jackets to control temperature, a thermometer to check temperature, and pipelines to connect the pump over system (Fig. 3.1A). Cone bottoms are quite frequent on fermentation tanks for red wines as they help to remove the pomace after maceration; these can be symmetrically (Fig. 3.1A) or asymmetrically shaped (Fig. 3.1B). In the first figure, the pomace is moved with the assistance of a helix powered by an electric engine, the helix rotates and pushes the macerated skins toward the front manhole. In the second figure, the discharge is carried out through gravity, and the pomace slides onto the inclined cone surface (Fig. 3.1B).

Between 2 and 3 days after the beginning of fermentation, the production of CO_2 as a result of the yeasts causes the skins to float because of its low density and large surface, forming what is known as a skin cap. Within this thick layer is a high concentration of yeasts, not only because of the wild population that occurs naturally in the pruine covering the grape's surface, but also because the layer enables and supports the adsorption of these yeast cells. Moreover, the solids produce a higher concentration of nutrients, especially nitrogen compounds. The higher concentration of yeasts produces a more intense fermentative activity and increases the cap temperature faster than it would in liquid form. If not controlled, the temperature in the cap could easily climb to more than 12°C higher than it would be in liquid form (Schmid et al., 2009) and this helps to increase the extraction of phenolic compounds. Also, typical differences of 8°C between the cap's center and sides can be observed (Schmid et al., 2009). The higher temperatures obtained in the cap in comparison to liquid promotes better extraction and can be used as a strategy to increase extraction rates. Regardless, maximum temperatures must be controlled and sit below 35°C, and if not, yeasts can be stressed and can also suffer damage, which in extreme situations, could lead to the stoppage of the fermentation process.

3.3 VESSEL MATERIALS IN RED WINEMAKING

Materials used on wine tanks have been evolving from the beginning of wine fermentation, in some cases in a cyclical manner. It is likely that the first material used was clay, for the roman amphora, and later, at the end of the 19th century, for the Spanish *Tinaja* (terracotta storage vessels) (Fig. 3.2A and B), which is now making a comeback in winemaking processes because of the preference of some winemakers for tradition, typicity, and the wine quality achieved. Advantages stem from the porosity of the clay, which facilitates more oxygen exposure,

FIGURE 3.2 (A) Traditional Spanish *Tinaja* vessel handmade with clay to ferment red wines. (B) Malolactic fermentation in a *Tinaja*. (C) Oak wood vat. (D) Concrete tank with stainless steel accessories. (E) Stainless steel tank with sloped cone bottom to facilitate pomace extraction. *(A) Courtesy of Cesar Velasco cellar, Villarrobledo, La Mancha, Spain. (B) Courtesy of Cesar Velasco cellar.*

affecting aromatic evolution and development, together with the polishing of tannins and the reduction of seed astringencies. Moreover, their thermal inertia makes them less prone than stainless steel to convention flows, which helps with reaching good colloidal stabilizations within short periods of time. The main drawbacks are the difficulty in extracting the pomace and the problems that exist to clean and eliminate microbial contaminations because of the clay's porosity.

Oak wood barrels and vats have also been used in fermentation processes for several centuries. Moreover, the utility of barrels was a crucial element in the transportation and commercialization of wines. The main advantage of wooden vats is the improvement in aromatic compounds transferred from the oak staves to the wine during the fermentation process. These aromatic compounds are better integrated and less evident when the wine is fermented in wood instead of being aged after stabilization. Generally, the inside of an oak wood vat is charred to increase aromatic compounds. Typical vats are built in a cone shape (Fig. 3.2C) to increase vat resistance as well as to make it more resistant against leaks appearing in joints. In the past, accessories like valves and others were made from bronze but these days most complementary components are made from stainless steel. Oak vats are quite geometrical in order to increase the extraction of phenolic compounds because they are mid-sized and have a similar height and diameter ratio, which facilitates the formation of thin caps and the maceration processes on skins. Good integration between grape tannins and wine can be observed when fermentation takes place in oak vats, which generally produces full-bodied wines with soft astringency. Moreover, the thermal inertia of these vessels makes them really good in facilitating malolactic fermentation and later, faster stabilization. In the past, temperature control could have presented problems, but this is no longer the case today because of the use of cooling plates that can be introduced through the upper hole (see detail in upper side of Fig. 3.2C). In addition, some contribution to microoxygenation through wood porosity can be observed in oak tanks, affecting color stabilization and tannin softening. The main drawback of oak vats is the difficulties encountered in cleaning them—especially when they are several years old—as it is extremely difficult to control microbial developments, and undesired spoilage microorganisms such as some bacteria or *Brettanomyces* yeasts can be problematic.

Concrete tanks are another trend that is returning, and in the past they were used because of their ease of construction and because they were less expensive than alternatives. However, these were later substituted with stainless steel tanks because of the advantages in temperature control and hygiene. Concrete tanks also presented a problem to clean, and one solution was to paint them with an epoxidic covering over the internal surface, making them more inert. Currently, it is possible to find really good concrete tanks, certified with a food safety certificate, in which the migrations of cations from the material are very low and the surface is very dense, smooth, and does not require the use of epoxidic coverings. Besides, refrigeration can be improved by using cooling exchangers that can be introduced through the cap via the upper manhole or which can be placed on the inner surface. These vessels are described as really good for malolactic fermentation and stabilization because of the thermal inertia that helps to stabilize the temperature, making the development of lactic acid bacteria and colloidal precipitations easier. Some authors also describe the advantages of concrete tanks with regard to porosity, which could be useful in improving microoxygenation in some cases and may have some effect on tannin polymerization and the formation of stable pigments. However, this porosity decreases over time with use as a result of the saturation of the pores with colloidal or crystalline depositions, and the porosity-effect is very low in the high density concrete currently used to decrease migrations. They can be made in many shapes; the traditional shape being a cube, which helps to save space in wineries, but other shapes, like a cone or a slightly pyramidal shape (Fig. 3.2D), help with maceration processes because the reduction in section helps to compress the cap by utilizing CO_2 pressure. Moreover, the progressive increase in size of the lower section helps to facilitate punch downs in comparison to vessels whose size is uniform. Other on-trend geometric shapes used today include egg-shaped tanks as some winemakers prefer this form because they favor some of the effects in the maceration process, such as natural continuous stirring processes by CO_2 bubbles, aided by this shape, and their small size helps fermentation rates, while the absence of corners improves cleaning. The shape also helps to keep the cap continuously submerged. Furthermore, some unclear biodynamic connotations mean they are preferred by some winemakers as there are suggestions that this is a naturally maternal structure in which to make wine. The accessories that are required for concrete tanks such as valves, pipelines, connections, manholes, etc., are made from stainless steel to avoid metallic migrations and dirt.

Stainless steel tanks are probably one of the best options for red wine maceration and fermentation. It is easy to build them in any shape, such as cylindrical or cone fermenters, and moreover, it is easy to adapt cone or sloped bottoms to facilitate the draining of pomace (Fig. 3.2E). They can also be placed on metallic legs to make this operation easier. It is easy to make them in whatever dimension is required, especially for red wines where you require a large diameter/height ratio to obtain thin caps. Moreover, it is possible to adapt all the necessary accessories to control fermentation to these tanks (Fig. 3.1). The cleaning of steel tanks is easy and can even be automated. Moreover, the use of chemical solutions or steam makes it possible to obtain high sanitization levels. The main problem with steel tanks is in terms of wine stabilization. They have very good thermal transfer properties so refrigeration during fermentation is very effective, however, it also produces big temperature variations in winter between the day and the night, producing convective currents inside the tank, thus hampering wine settling and colloidal stabilization. This process is much more effectively controlled in concrete or wood vessels. Moreover, the metallic surface of the tank gives them some electrical conductivity, affecting the electrostatic processes that also have some influence on the settling of colloidal particles.

3.4 KINETICS OF EXTRACTION: THE EFFECT OF TEMPERATURE

The most important phenolic compounds in red wines are pigments and tannins. Anthocyanins are the pigments in red wines; they are colored molecules with the flavonoid structure C6—C3—C6 and visible absorbance maximums ranging between 518 and 538 nm in wine grapes (Mazza, 1995). Values depend on the substitution pattern in the B ring and the acylation moiety in sugar. The —C3— structure or pyrilium ring is able to absorb, on average, light of 525 nm and is responsible for the color of red wines. Tannins are also relevant to wine quality and stability. Tannin is a colloquial name used to describe proanthocyanidin compounds with flavonoid structures, which are responsible for the structure, astringency, and antioxidant capacity of red wines. Furthermore, they have some antimicrobial effect, contributing to stability of red wine pH levels and alcohol levels.

As anthocyanins and tannins are mainly found in skins, it is necessary to extract them during maceration to diffuse into the must so that the wine will have a suitable color and structure. This process occurs during

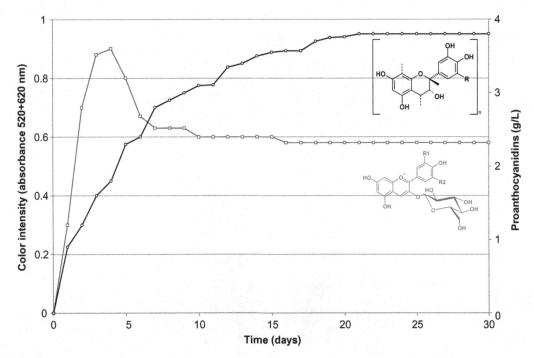

FIGURE 3.3 Kinetics of extraction during maceration of grape anthocyanins expressed as color intensity (line with square marker) and tannins (proanthocyanidins) (line with round marker).

maceration and at the same time as the alcoholic fermentation. However, the extraction kinetics of pigments and tannins are completely different because of the chemical nature of both molecules. Anthocyanins contain a pyrilium ring with a positive charge in oxygen, so they have a certain degree of a cationic nature, making them more soluble in polar solvents. Tannins have a pyrano ring with more apolar properties. So what happens is that at the beginning of fermentation, as must is quite a polar media—basically acidic water with sugars—the solubility of anthocyanins is very high and the extraction is really fast. The height of extraction is normally reached on day 3/4 of fermentation (Fig. 3.3). At the same time as the fermentation is evolving, the ethanol level increases, greatly affecting the polarity. After that time, the ethanol concentration is high enough to reduce anthocyanin solubility, these being precipitated and therefore causing a decrease in color intensity. When all sugars are fermented and the maximum ethanol levels are reached, the concentration of anthocyanins remains stable. With regard to tannins, as they are quite apolar molecules, the concentration increases according to ethanol levels, reaching maximum values at the end of fermentation (Fig. 3.3).

The extraction kinetics can also be improved with enological techniques such as the use of cold soak (cold maceration), dry ice, or enzymes affecting the extraction of anthocyanins and proanthocyanidins. Cold soak is based on the use of low maceration temperatures, 4–15°C (He et al., 2012). Applying refrigeration to the must in order to delay the beginning of fermentation means that maceration is done with the absence of ethanol, so a selective extraction of anthocyanins in regard to tannins is fostered. Moreover, the extraction of varietal aromatic compounds is also improved (Lukić et al., 2017). Similar results can be obtained when cooling is applied using conventional processes like cooling the crushed grapes with external heat exchangers or using the metallic jackets on tanks or by employing alternative techniques like the use of dry ice. In the case of the latter, the effect can be more intense if ice crystals are formed inside the skins, producing their breakage and facilitating the release of phenolic compounds. Especially with dry ice, maximum extraction rates are reached during the first 6 days of fermentation because of the polar media (Busse-Valverde et al., 2011). The cold soak technique and addition of enzymes increase the extraction of proanthocyanidins.

Alternatively, the thermal processing of crushed grapes or skins (40–70°C/15 minutes to more than 1 hour) also accelerates the extraction of phenolic compounds (Fig. 3.4) and eliminates, or strongly reduces, wild yeasts and bacteria. However, wine quality is reduced because of the loss of aromatic compounds and the formation of undesirable odors because of thermal degradation (Geffroy et al., 2015). Moreover, thermovinification produces colloidal instability. Even when the extraction of phenolic compounds is strongly enhanced, it is quite unstable and evolves with precipitations of pigments and tannins. Sometimes, pH levels can also be affected

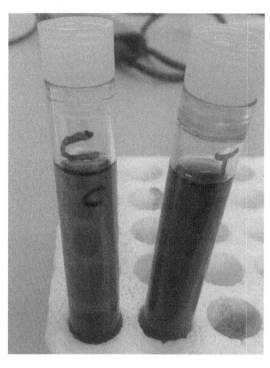

FIGURE 3.4 Color extraction after 1 h of maceration in control must (left) and after thermovinification 1 h at 60°C (right).

because of the higher extraction of cations from skins. In addition, thermovinification produces thermal inactivation of oxidative enzymes (Fischer et al., 2000). This application is especially relevant for grapes affected by fungal developments.

This process can be improved if thermal treatment is applied at a low pressure, a technique known as "flash détente." For this technique, a high temperature (85°C) is applied to the grapes for a short time, increasing their temperature, but this is mainly on the skins and does not affect the pulp. They are later depressurized in a vacuum chamber (40–75 Mbar) resulting in the boiling of the liquid content in the skin cells followed by their explosion. This process creates a fine but extended net of microfissures (depth 0.3 μm, Boulet and Escudier, 2000), facilitating the fast extraction of tannins and pigments. The short duration of the thermal process and the successive cooling produced by the expansion at low pressure reduce the thermal damage of aromatic compounds and phenols. Cold traps are also normally used to recover aromatic compounds released in the vacuum chamber. The result of this technique is the enhanced extraction of anthocyanins and tannins with low maceration times, and later, after pomace separation, fermentation can be developed at low temperatures. There is some effect on pH levels because cation extraction from skins can also be observed, similar to conventional thermovinification.

3.5 MECHANICAL PROCESSES DURING MACERATION

3.5.1 Punch Downs and Pump Overs

Mechanical treatments in cap help to promote extraction and diffusion in must during fermentation. Punch downs and pump overs (*remontage* in French) significantly increase the extraction of phenolic compounds (Fischer et al., 2000). Punch down treatments are typically used during maceration to increase the extraction of skin phenols, to diffuse them in liquid and homogenize skins and liquid, and to balance the temperature in the tank. Punch downs are carried out using a metallic or wooden stick with a flat plate end to increase contact surface (Fig. 3.5A). Pump overs are used in the same way to increase extraction through diffusing wine phenols in liquid. This is normally done by taking the liquid from the bottom of the tank using gravitational force and pumping it to the upper side, thus irrigating the cap surface (Fig. 3.5B). There is a possibility of dropping the liquid into a metallic box before pumping to break up the liquid and increase aeration; this process can also be enhanced using an intermediate grid, breaking up the liquid into fine droplets. In this case, it is easy to saturate the must in oxygen (8–9 mg/L O_2).

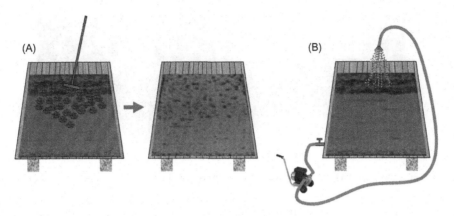

FIGURE 3.5 (A) Punch down treatment and influence on cap disaggregation. (B) Pump over.

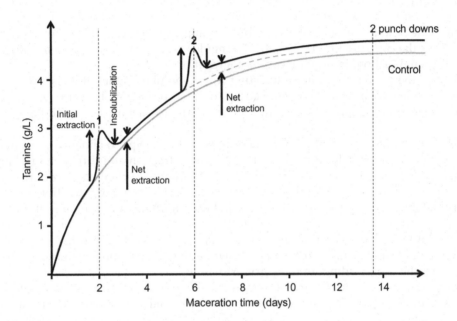

FIGURE 3.6 Effect of mechanical treatments on skins cap in the extraction of tannins. Use of two punch downs on days 2 and 6 of fermentation.

Punch down is a difficult technique because it requires a lot of energy to both completely disaggregate the cap and to homogenize the content. The effect on the extraction is quite significant, and results in an increase in pigment and tannin concentration. It can be carried out daily during the maceration stage, and in some cases, it may be appropriate to do so several times per day. When the tannin or pigment content is analyzed shortly after the punch down treatment, the values of this compound can increase significantly depending on both the fermentation–maceration time and the polarity conditions conditioned by ethanol levels. However, this extraction is quite unstable and after a few hours, the levels decrease until they reach a stable value in accordance with the solubility of each molecule, as well as other parameters including polarity, pH levels, temperature, cations, etc. (Fig. 3.6). So the final stable concentration (net extraction) is reached a few hours after the treatment when the conditions are balanced and the unstable anthocyanins and tannins are precipitated or reabsorbed into seeds, skins, or the tank's surface. When the efficiency of punch downs and pump overs is evaluated through analyses of anthocyanins and tannins, it is better to undertake these tests at least a few hours after the treatment to see the final increment in concentration.

Because punch down is a difficult operation, especially in large tanks, even when the ratio diameter/height is suitable, several mechanical devices have been developed to perform this processing automatically. The most typical is a pedal activated hydraulic piston that gently pushes the cap, thus facilitating extraction; sometimes, several pistons work together (Fig. 3.7A). In this case, pistons can work simultaneously, or even better,

FIGURE 3.7 (A) Automatic system for making mechanical punch downs. (B) Top view detail.

alternately, producing a more effective mechanical work on the cap. The frequency and timing of punching can be established and controlled by software to manage the process more comfortably. These systems are expensive and lighter, and can normally be moved from one tank to another, as the structure is on wheels, and travels over rails or is transported by air using a crane bridge or even by hand if the upper holes of the tanks are on a walkable floor. The pedals can be found in several shapes but a triangular section is most typical, sometimes enlarged by wings to better break the cap structure (Fig. 3.7B). Triangular sections not only help to press the cap, but also to rotate it around the pedal section, improving cap disaggregation.

Should a winery have the capacity, another interesting way to emulate punch downs in large tanks is to use a big steel weight with a cone section that can be moved up to the upper manhole and released over the cap using a crane bridge. This treatment is really effective and increases the extraction, producing a good disaggregation of the cap.

The main objectives of pump overs are the aeration of must, which helps to enhance yeast development at the beginning of fermentation in the initial cell replication stage; the extraction and diffusion of phenolic compounds from skins; and thermal homogenization. It can also help to facilitate the final stage of fermentation, especially when the alcoholic degree is very high. Pump overs can be applied with strong aeration by releasing the must from the bottom of the tank and dropping it into a vessel, then pumping and sprinkling it on the top of the cap, or without aeration by connecting the pump directly to the hose at the bottom of the tank (Fig. 3.5B). The first application can be useful to facilitate yeast development at the beginning, to control sluggish fermentations, and to manage reductive odors formed during fermentation, while the second option is used when it is only necessary to wash and extract the cap avoiding unnecessary oxidations.

With regard to pump overs, a point of discussion is to decide the number of times it should be performed per day and the volume that should be pumped. Excessive treatments only increase costs and are often ineffective. Moreover, they can sometimes increase the extraction of herbal or astringent compounds. When the objective is to enhance the extraction, a good option is to undertake this process between one and three times per day, spaced out evenly to allow the phenolic compounds in the solution to balance. Concerning the volume, the best results are generally obtained when between 40% and 60% of tank volume is moved in each pump over. Of course, these parameters may vary according to variety, phenolic ripeness, skin hardness and thickness, vine age, pH levels of the must, and alcoholic degree.

3.5.2 Rack and Return

Most wineries devoted to wine quality are using a technique to improve phenolic extraction that is called "rack and return" or *délestage*. The most frequently used way to carry out this technique is racking the juice in another tank leaving only the skins in the bottom of the tank, where the fermentation takes place. Normally, the skin cap is floating on the must-wine surface because of the CO_2 pressure of fermentation, so it is very delicate, just like on a waterbed. However, when the juice is removed, the cap descends, coming into contact with the bottom of the tank and producing an increase in pressure, which generates a very intense but gentle extraction. It is common to leave the skins at the bottom for some time, a few hours, and later return the liquid to the upper side of the cap. Dropping the liquid on the cap also produces some extraction effects because of the liquid pressure, but it mainly helps to disaggregate the cap, facilitating the diffusion of pigments and tannins into the liquid. Later, the cap will form once again on the upper side but only following a strong reorganization of the cap particles. This process also produces great aeration and good homogenization of the tank contents; chemical and thermal homogenization being good tools to regulate temperature. The pressure effect of the liquid when returned to the skins may be normal if it is carried out through a pumping process, but if there is the possibility of releasing all the liquid in a few seconds, the effect is much more intense. It is possible to do it when we have another tank above the tank that is being treated, for example, in gravity designed wineries or when we put the liquid in flying vessels that can also be available in gravity wineries equipped with a crane bridge. Before returning the liquid, some winemakers also enhance the mechanical effect on skins by using a big balloon inside the tank that gently increases the pressure on the skins like in a pneumatic press.

Another interesting utility in rack-and-return is the possibility of separating seeds during the racking process. This helps to remove harsh tannins from seeds (Canals et al., 2008). This is of great interest, especially when we have asymmetric maturities between skins and seeds in some varieties. Sometimes, the optimal phenolic maturity in skins happens before suitable seed maturity, affecting the global tannin quality of wine.

The typical effect of this process is a large extraction of phenolic compounds in a smooth form, increasing body and structure but reducing astringency, bitterness, and dryness. In the most extreme situations, total polyphenol index (TPI) can be augmented from 40 to more than 80 with just one rack and return. Normally rack and return requires extra resources in terms of labor, staff, external tanks, pumps, etc., and therefore, is utilized only for better wines or for those in which the enologist considers that wine quality can be improved with this technique. One key question lies in understanding the best time to perform this operation. It is difficult to do it at the beginning of fermentation, because the cap is still not formed, or it may be formed without good consistency. Another consideration is that skins are just at the beginning of the maceration process, so their resistance is strong and the effect of the rack and return will be low. Moreover, the low levels of ethanol at the beginning of fermentation do not help to obtain good tannin extraction. Right at the end is not the best time either, because it is difficult to form the cap again as the fermentation rate is lower. Perhaps the best window of time is sometime during the second third of the fermentation process, when the conditions are optimal to obtain maximum profit from this operation (Fig. 3.8).

3.5.3 Submerged Cap

The cap surface is exposed to air during most of the maceration process, so is sensitive to oxidation. To control this, a useful technique is to keep the cap submerged under the juice level by using some grid or metallic net that holds the cap structure below the liquid surface. Some mechanical effects also take place with the pressure of CO_2, which presses the cap against the metallic grid. The cap will constantly remain wet throughout and this helps to reduce temperature and improve diffusion processes. This technique helps to obtain a higher content of tannins (Bosso et al., 2011) and also improves color intensity in wines (Machado de Castilhos et al., 2015). The main drawback is that a specific tank design with a metallic screen able to resist the CO_2 pressure on the cap is required.

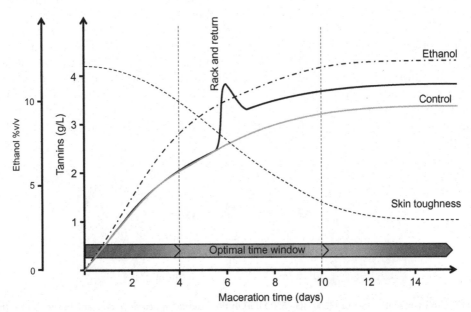

FIGURE 3.8 Effect of rack and return in phenol extraction. Repercussion of alcoholic degree and skin resistance.

3.5.4 Extended Maceration

Maceration time can be extended after the end of fermentation from a few days to, in some cases, more than 1 month. The advantage of delaying the pomace removal is maceration is occurring in the presence of a high ethanol concentration at the end of fermentation, thus improving tannin solubility. Moreover, long macerations promote the extraction processes because of the degradation of skin cell walls. With this technique, higher levels of anthocyanins have been reported in the medium-term, 4 months of bottle aging (Puertas et al., 2008). Sometimes, the main problem can be the development of humid odors or excessive tannin extraction if the process is not well controlled. Another possibility is the development of spoilage microorganisms if the wines are insufficiently attended to, especially with high pH values and low SO_2 levels.

3.6 NEW EXTRACTION TECHNOLOGIES

Due to the difficulties of managing the must with solids during fermentation, it is likely that the future of red winemaking will be to separate the maceration and fermentation processes. If efficient skin extraction can be performed in a short time obtaining suitable concentrations of anthocyanins and tannins, then it would be possible to ferment only the must in better conditions at a lower temperature, and with better management of the yeasts. In addition, most of these new technologies are able to reduce the wild yeast and bacteria populations, promoting a better development of the inoculated yeast starter and reducing inoculum size (Suarez-Lepe and Morata, 2015) and furthermore, they facilitate the use of new fermentation biotechnologies such as the use of non-*Saccharomyces* yeasts and yeast−bacteria coinoculations (Morata et al., 2017).

3.6.1 High Hydrostatic Pressure

HHP is the application of big pressures 300−600 MPa by means of a fluid, commonly water, to transfer the pressure in a hydrostatic way. This means that pressures are not applied in a single direction and opposite components of force can be annulled, therefore external shape and appearance remain unaffected (Fig. 3.9), even at these extremely high pressures. It must be noted that when using gentle pneumatic presses, the grape is completely crushed and most of the juice is released at 0.5 bar (0.05 MPa). Moreover, pressure levels of 400 MPa applied for 10 minutes are enough to eliminate wild yeast populations from skins (Fig. 3.9), but lactic acid bacteria are more resistant and more than 500 MPa is needed to control wild populations on grape skins (Morata et al., 2015a). And more importantly, it works at low temperatures because the adiabatic heating produced by HHP is 2−3°C/100 MPa, so pressurization at 500 MPa increases temperatures less than 15°C, but this temperature can be controlled using cooling systems. Therefore, it is possible to apply HHP in refrigeration conditions. Moreover, HHP do not have enough energy to break covalent bonds, so most of the molecules responsible for wine sensory properties (pigments, aroma, and flavor molecules) remain unaffected by this gentle technology. Another interesting fact is that even when there are no changes in external appearance, cell wall structures in skins and pulp are degraded at a molecular level and the migration of anthocyanins and tannins to pulp and even seeds is promoted (Morata et al., 2015a). HHP produces improvements in phenol extraction and color (Tao et al., 2016), which is important because while it is only applied for a few minutes, usually 3−10, this time is enough to affect final wine color and tannin contents (Corrales et al., 2008, 2009; Morata et al., 2015a).

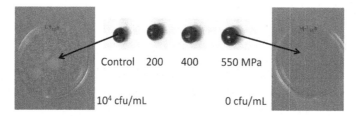

FIGURE 3.9 Effect of high hydrostatic pressure (200, 400, 550 MPa) in the external appearance of grape berries and influence in wild yeast populations on skins.

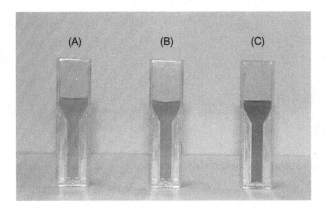

FIGURE 3.10 Effect on color of the PEF processing of red grapes (Garnacha variety) after 1 h of maceration: (A) control; (B) 1 kV/cm; (C) 3 kV/cm. *PEF,* pulsed electric field. *Courtesy of Prof. J. Raso, University of Zaragoza.*

Another interesting advantage is the control of wild microbial populations that has been described as a tool to implement new fermentation biotechnologies such as the use of non-*Saccharomyces*, since some of these are less competitive in must than wild *Saccharomyces* yeasts. HHP also helps to use yeast—bacteria coinoculations to perform simultaneous alcoholic and malolactic fermentations with less risk than in grapes with the wild populations of yeast and bacteria, and lastly, it is also a means of reducing SO_2 levels (Bañuelos et al., 2016). Moreover, HHP has been described as a tool to control spoilage yeasts such as *Brettanomyces* in wines (Morata et al., 2012).

Finally, the application of HHP has also been used to hasten the aging of young wines together with the use of oak chips (Tao et al., 2016). The pressure affects the extraction of aromatic compounds from oak wood when wine is processed with chips by HHP.

3.6.2 Pulsed Electric Fields

PEFs increase the rate and yield in the extraction of phenolic compounds in red grape processing, and can also be used to control microbial populations in grapes and musts, and can be considered as an alternative to reduce SO_2 levels (Garde-Cerdán et al., 2016). The mechanism of action is the proration of the cell membrane when a critical transmembrane potential is overcome. By using PEFs, the extraction of total phenolic compounds can be increased by 10%–50% (Yang et al., 2016). Maceration time can be reduced by 40%–50% using PEF intensities of 5–10 kV/cm (López et al., 2008a, b; Fig. 3.10). Under these conditions, the temperature increment is very low, so it can be considered a nonthermal technology. To control wild yeast and bacteria in crushed grapes, it is necessary to apply higher intensities, which are in the range of 30 kV/cm (Morata et al., 2017).

3.6.3 Ultrasounds

USs are sonic waves in the range of 20–100 kHz producing cavitation, which is the formation of gas bubbles that are expanded and compressed according to the variation of the wave intensity. Through this phenomenon, local pressure can reach values of 50 MPa and temperatures at specific points can also exceed 1000°C (Chemat and Khan, 2011). After several minutes of treatment, sample temperatures can easily increase to more than 60°C. This technique has been frequently used to disaggregate components, to clean materials, and to extract molecules from vegetal tissues. As for grapes, USs help to destroy cell walls of skins, facilitating the extraction of phenolic compounds. It has been verified that by using USs, the extraction and retention of anthocyanins in juices can range between 50% and 90% (Tiwari et al., 2010). This technology can be used in enology to carry out a fast and continuous extraction, extremely reducing maceration times, and the extracted juice can later be fermented at low temperature in the absence of solids (skins and seeds). The best extraction results can be obtained at low frequencies (≈ 30 kHz) inside the USs range.

The prototype Ultrawine from Agrovin SA company has been developed and cofunded by the European program H2020 (Fig. 3.11A). It has been developed across more than 5 years, improving the maceration tubes using hexagonal sections and lenticular flat sonotrodes (Fig. 3.11B) that help to increase the USs effectivity and the cavitation processes. Using this technology, it is possible to reduce maceration time from 7 to 2–3 days with the

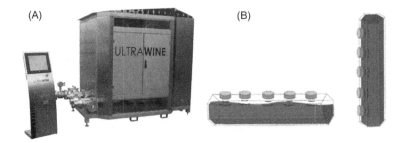

FIGURE 3.11 (A) Ultrawine system to macerate continuously by ultrasounds. (B) Detail of the maceration tubes with the sonotrodes. *(A) Courtesy of Ricardo Jurado, AGROVIN SA.*

FIGURE 3.12 Compared extraction of grapes after crushing (left) and continuous maceration by ultrasounds in Ultrawine (right). *Courtesy of AGROVINSA.*

same degree of extraction in terms of tannins and pigments (Fig. 3.12). In Cabernet sauvignon grapes with a flow of 12 t/hour using an ultrawine system of 50 kW, tannin extraction reaches similar values in 3 days of maceration compared to the control wine using conventional tank maceration for 7 days. In both cases, proanthocyanidins were 600 mg/L (using the methodology based on an acid-catalyzed depolymerization in the presence of phloroglucinol followed by high pressure liquid chromatography (HPLC) analysis) and the total polyphenol index was 45 (Jurado, 2017, AGROVIN SA, personal communication).

3.6.4 Irradiation

β-irradiation or e-beam irradiation is the use of accelerated electrons at a maximum energy of 10 MeV in Europe (12 MeV in the United States) generated in an electrical accelerator device (several brands/technologies can be found commercially: Rhodotron, Dynamitron, Iotron, Linac...). e-Beam irradiation produces the formation of free radicals from water molecules able to oxidize DNA, thus affecting the microorganisms viability. Irradiation doses can be measured using the kGray unit (kGy); 1 Gy is the application of 1 J of energy per kg of mass treated. Doses of 1–3 kGy are able to control many microorganisms in foods, but doses of 20–50 kGy are necessary for sterilization and for controlling spore-forming bacteria. It can be considered a cold technology because grapes can be processed up to 10 kGy, increasing the temperature by less than 5°C. Doses below 10 kGy are effective for controlling grape wild yeasts and bacteria (Fig. 3.13, Morata et al., 2015b), but at the same time do not affect external shape or appearance. e-Beam irradiation produces mechanical degradation of skin cell walls and a higher extraction of anthocyanins and tannins can be observed, especially when grapes are processed at 10 kGy (Morata et al., 2015b). However, aromatic compounds can be degraded at some level by irradiation, both grapes and juice show a less fruity character after being treated, but these differences are less important at the fermentation stage compared to controls. Another advantage of e-beam irradiation is that it is a continuous and fast process and grapes can be processed at a rate of several tons per hour. The thickness that is easily affected by e-beam irradiation is about 3 cm, so several layers of grapes can be treated at the same time with good results.

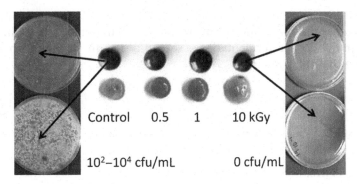

FIGURE 3.13 Effect of e-beam irradiation at 0.5, 1, and 10 kGy in the external appearance of grape berries and influence in wild yeast populations on skins.

3.6.5 Pulsed Light

PL is a continuous irradiation technology using white light in the range 170–2600 nm including ultraviolet, visible, and infrared radiation. The light is produced using xenon flashing lamps capable of creating very high potency, usually >30 MW, in a very short time (1 μs–0.1 second) (Rowan et al., 1999). The pulse energy is lower than 50 J/cm^2, however, when applied in a very short time it may produce potencies at the level of 10^7 W. Flashes have an intensity higher than 100,000 times that of the sunlight at sea level in summer. Treatments are applied as a sequence of pulses, normally in the range of between 1 and 20 pulses/second (Morata, 2010). PL produces very low increments of temperature in foods so it can be considered a nonthermal and gentle technology that helps to control microorganisms preserving the sensory quality of fresh food (Oms-Oliu et al., 2010). A reduction in green and vegetal notes in wines has been reported when grapes are processed using PL (Escott et al., 2017).

After PL processing, external appearance remains unaffected without changes in color or morphology. However, even when global warming of grapes is very low, local heating or overpressure produced in cells by the high potency released in each pulse produce some effects on the extraction. It is probable that the energy is sufficient to degrade or increase the porosity of cell walls facilitating the extraction of anthocyanins and tannins (Escott et al., 2017).

Interestingly, the devices that apply PL are quite developed at an industrial level, especially those produced for the packaging industry, and the cost is quite affordable since the devices can be found for €50–100k, depending on features and processing rate. The power consumption is also quite moderate. This technology can be applied easily in a continuous way, especially when sorting tables are used to select grape berries after destemming. In order to facilitate full exposure to PL and to reduce shadowed areas, it is advisable to use roller belts to transport the berries below the flashing lamps. This design will help to increase antimicrobial effectivity and premaceration effects.

3.6.6 Ozone and Electrolyzed Water

The use of chemical additives such as ozone or electrolyzed water also have some effects on the extraction of phenolic compounds from grape skins (Artés-Hernández et al., 2007; Bellincontro et al., 2017). These additives are quite safe and healthy because of low residual chemicals remaining after treatments, in the case of ozone by the fast degradation to oxygen and in electrolyzed water because of the low amount of salts needed to produce this bioactive solution.

3.7 CONCLUSIONS

The traditional way of extracting pigments and tannins in grape processes for winemaking is the use of conventional mechanical treatments, in many cases, the use of hand-applied techniques such as punch downs, and the development of new technologies, either at pilot or at an industrial level, open new possibilities in the fast extraction of phenolic and aromatic compounds. It is likely that in the future, it will be possible to separate maceration and fermentation processes, making the fermentation of red wines easier and more controlled. Many of these emerging technologies also have an antimicrobial effect that helps to reduce both wild yeasts and bacteria, producing safer and healthier fermentations, sometimes with positive economic and environmental advantages, and contributing to reductions in SO_2 levels.

References

Artés-Hernández, F., Aguayo, E., Artés, F., Tomás-Barberán, F.A., 2007. Enriched ozone atmosphere enhances bioactive phenolics in seedless table grapes after prolonged shelf life. J. Sci. Food Agric. 87 (5), 824—831.
Bañuelos, M.A., Loira, I., Escott, C., D Fresno, J.M., Morata, A., Sanz, P.D., et al., 2016. Grape processing by high hydrostatic pressure: effect on use of non-*Saccharomyces* in must fermentation. Food Bioprocess Technol. 9, 1769—1778.
Bellincontro, A., Catelli, C., Cotarella, R., Mencarelli, F., 2017. Postharvest ozone fumigation of Petit Verdot grapes to prevent the use of sulfites and to increase anthocyanin in wine. Aust. J. Grape Wine Res. 23, 200—206.
Bosso, A., Panero, L., Petrozziello, M., Follis, R., Motta, S., Guaita, M., 2011. Influence of submerged-cap vinification on polyphenolic composition and volatile compounds of Barbera wines. Am. J. Enol. Vitic. 62, 503—511.
Boulet, J.C., Escudier, J.L., 2000. Flash-expansión. In: Flanzy, C. (Ed.), Enología: Fundamentos científicos y tecnológicos. Ed AMV-Mundiprensa, Madrid, p. 490. (In Spanish).
Busse-Valverde, N., Gómez-Plaza, E., López-Roca, J.M., Gil-Muñoz, R., Bautista-Ortín, A.B., 2011. The extraction of anthocyanins and proanthocyanidins from grapes to wine during fermentative maceration is affected by the enological technique. J. Agric. Food. Chem. 59 (10), 5450—5455.
Canals, R., Llaudy, M.C., Canals, J.M., Zamora, F., 2008. Influence of the elimination and addition of seeds on the colour, phenolic composition and astringency of red wine. Eur. Food Res. Technol. 226 (5), 1183—1190.
Chemat, F., Khan, M.K., 2011. Applications of ultrasound in food technology: processing, preservation and extraction. Ultrason. Sonochem. 18 (4), 813—835.
Corrales, M., Toepfl, S., Butz, P., Knorr, D., Tauscher, B., 2008. Extraction of anthocyanins from grape by-products assisted by ultrasonics, high hydrostatic pressure or pulsed electric fields: a comparison. Innov. Food Sci. Emerg. Technol. 9, 85—91.
Corrales, M., Fernández García, A., Butz, P., Tauscher, B., 2009. Extraction of anthocyanins from grape skins assisted by high hydrostatic pressure. J. Food Eng. 90, 415—421.
Escott, C., Vaquero, C., del Fresno, J.M., Bañuelos, M.A., Loira, I., Han, S.-Y., et al., 2017. Pulsed light effect in red grape quality and fermentation. Food Bioprocess Technol. 10 (8), 1540—1547.
Fischer, U., Strasser, M., Gutzler, K., 2000. Impact of fermentation technology on the phenolic and volatile composition of German red wines. Int. J. Food Sci. Technol. 35 (1), 81—94.
Garde-Cerdán, T., Arias, M., Martín-Belloso, O., Ancín-Azpilicueta, C., 2016. Pulsed electric field and fermentation. In: Ojha, K.S., Tiwari, B.K. (Eds.), Novel Food Fermentation Technologies, first ed. In: Food Engineering Series Book Series (FSES). Springer International Publishing Switzerland, pp. 85—123. Available from: https://doi.org/10.1007/978-3-319-42457-6.
Geffroy, O., Lopez, R., Serrano, E., Dufourcq, T., Gracia-Moreno, E., Cacho, J., et al., 2015. Changes in analytical and volatile compositions of red wines induced by pre-fermentation heat treatment of grapes. Food Chem. 187, 243—253.
He, F., Liang, N.-N., Mu, L., Pan, Q.-H., Wang, J., Reeves, M.J., et al., 2012. Anthocyanins and their variation in red wines i. monomeric anthocyanins and their color expression. Molecules 17, 1571—1601. Available from: https://doi.org/10.3390/molecules17021571.
López, N., Puértolas, E., Condón, S., Alvarez, I., Raso, J., 2008a. Application of pulsed electric fields for improving the maceration process during vinification of red wine: influence of grape variety. Eur. Food Res. Technol. 227, 1099—1107.
López, N., Puértolas, E., Condón, S., Alvarez, I., Raso, J., 2008b. Effects of pulsed electric fields on the extraction of phenolic compounds during the fermentation of must of Tempranillo grapes. Innov. Food Sci. Emerg. Technol. 9, 477—482.
Lukić, I., Budić-Leto, I., Bubola, M., Damijanić, K., Staver, M., 2017. Pre-fermentative cold maceration, saignée, and various thermal treatments as options for modulating volatile aroma and phenol profiles of red wine. Food Chem. 224 (1), 251—261.
Machado de Castilhos, M.B., dos Santos Corrêa, O.L., Zanus, M.C., Garcia Maia, J.D., Gómez-Alonso, S., García-Romero, E., et al., 2015. Pre-drying and submerged cap winemaking: effects on polyphenolic compounds and sensory descriptors. Part I: BRS Rúbea and BRS Cora. Food Res. Int. 75, 374—384.
Mazza, G., 1995. Anthocyanins in grape and grape products. Crit. Rev. Food. Sci. Nutr. 35, 341—371.
Morata, A., Benito, S., González, M.C., Palomero, F., Tesfaye, W., Suárez-Lepe, J.A., 2012. Cold pasteurization of red wines with high hydrostatic pressure to control *Dekkera/Brettanomyces*: effect on both aromatic and chromatic quality of wine. Eur. Food Res. Technol. 235, 147—154.
Morata, A., Loira, I., Vejarano, R., Bañuelos, M.A., Sanz, P.D., Otero, L., et al., 2015a. Grape processing by high hydrostatic pressure: effect on microbial populations, phenol extraction and wine quality. Food Bioprocess Technol. 8, 277—286.
Morata, A., Bañuelos, M.A., Tesfaye, W., Loira, I., Palomero, F., Benito, S., et al., 2015b. Electron beam irradiation of wine grapes: effect on microbial populations, phenol extraction and wine quality. Food Bioprocess Technol. 8, 1845—1853.
Morata, A., Loira, I., Vejarano, R., González, C., Callejo, M.J., Suárez-Lepe, J.A., 2017. Emerging preservation technologies in grapes for winemaking. Trends Food Sci. Technol. 67, 36—43.
Oms-Oliu, G., Martín-Belloso, O., Soliva-Fortuny, R., 2010. Pulsed light treatments for food preservation: a review. Food Bioprocess Technol. 3, 13—23.
Puertas, B., Guerrero, R.F., Jurado, M.S., Jimenez, M.J., Cantos-Villar, E., 2008. Evaluation of alternative winemaking processes for red wine color enhancement. Food Sci. Technol. Int. 14 (5), 21—27.
Rowan, N.J., MacGregor, S.J., Anderson, J.G., Fouracre, R.A., McIlvaney, L., Farish, O., 1999. Pulsed-light inactivation of food-related microorganisms. Appl. Environ. Microbiol. 65, 1312—1315.

Schmid, F., Schadt, J., Jiranek, V., Block, D.E., 2009. Formation of temperature gradients in large- and small-scale red wine fermentations during cap management. Aust. J. Grape Wine Res. 15 (3), 249–255. Available from: https://doi.org/10.1111/j.1755-0238.2009.00053.x.

Suarez-Lepe, J.A., Morata, A., 2015. Levaduras para vinificación en tinto. AMV Ediciones, Madrid, Spain, pp. 277–291 (Chapter 8). (In Spanish).

Tao, Y., Sun, D.-W., Górecki, A., Błaszczak, W., Lamparski, G., Amarowicz, R., et al., 2016. Preliminary study about the influence of high hydrostatic pressure processing in parallel with oak chip maceration on the physicochemical and sensory properties of a young red wine. Food Chem. 194, 545–555.

Tiwari, B.K., Patras, A., Brunton, N., Cullen, P.J., O'Donnell, C.P., 2010. Effect of ultrasound processing on anthocyanins and color of red grape juice. Ultrason. Sonochem. 17, 598–604.

Yang, N., Huang, K., Lyu, C., Wang, J., 2016. Pulsed electric field technology in the manufacturing processes of wine, beer, and rice wine: a review. Food Control 61, 28–38.

CHAPTER

4

Use of Non-*Saccharomyces* Yeasts in Red Winemaking

Maurizio Ciani and Francesca Comitini

Department of Life and Environmental Sciences, Polytechnic University of Marche, Ancona, Italy

4.1 INTRODUCTION

The red wine production in the vast majority of cases involves the maceration process during fermentation to promote the extraction of coloring substances from the grape skins. This winemaking practice favors the colonization of "wild" yeasts widely present on the grape surface even if a selected *Saccharomyces cerevisiae* starter culture is inoculated. These yeasts can influence both the analytical composition and sensorial profile of the final wine and for this reason a brief description of yeast ecology of grape berries is given below.

In the past the non-*Saccharomyces* wine yeasts were negatively considered because of reduced fermentation power and high production of undesired products that affect the aromatic profile of wines. However, recently, several studies have revaluated the role of non-*Saccharomyces* yeasts during alcoholic fermentation and their metabolic impact in both white and red wines. Indeed, demand for new and improved wine-yeast strains able to produce different types and styles of wines is growing (Pretorius, 2000). In this context, to improve the chemical composition and sensory properties of wine, the inclusion of some selected non-*Saccharomyces* wine yeasts, together with *S. cerevisiae* strains has been proposed as a tool to take advantage of spontaneous fermentation, avoiding the risks of stuck fermentation. In this regard, the production of wines with peculiar flavor profiles has been one of the main reasons for including selected non-*Saccharomyces* yeasts in the winemaking market. However, promising approaches to lower alcohol content of wines, to modify the acidification level, to control wine spoilage, or to improve some specific enological properties are being explored, and undoubtedly, they represent new biotechnological natural opportunities to exploit in wine production. In particular, in red wine production the use of non-*Sacccharomyces* wine yeasts could add further specific beneficial effects such as color evolution and astringency perception.

4.2 YEAST ECOLOGY OF GRAPE BERRY

For a long time, the characteristics and management of grapes used to produce wine have been studied and a vast amount of information has been collected on yeasts occurrence in winemaking environments, even if ecological relationships among them have yet to be fully studied (Barata et al., 2012).

Over the years, it has been clearly demonstrated that the microbial population of grape berries is a complex microbial consortium that includes filamentous fungi, yeasts, and bacteria with different physiological characteristics and different impacts on the grape metabolome and final wine quality (Pinto et al., 2015; Verginer et al., 2010). Indeed, the first population encountered by grape must prior to the fermentation can crucially affect the metabolic profile of wine and its quality, even when commercial starters are used (Bokulich et al., 2016).

Kurtzman and Fell (1998) already many years ago, ascribed yeasts associated with grape/wine ecosystem into 15 different yeast genera: *Dekkera/Brettanomyces, Candida, Cryptococcus, Debaryomyces, Hanseniaspora/Kloeckera,*

Kluyveromyces, Metschnikowia, Pichia, Rhodotorula, Saccharomyces, Saccharomycodes, Schizosaccharomyces, and *Zygosaccharomyces*. The yeasts belonging to *Koeckera* and *Hanseniaspora* genera are predominant, on the surface of the grape, accounting for more than 50% of the total yeast population (Fleet and Heard, 1993). To a smaller extent, it is also possible to detect the species of *Candida, Cryptococcus, Rhodotorula, Pichia, Metschnikowia*, and *Kluyveromyces* (Heard and Fleet, 1988, Mills et al., 2002; Rosini et al., 1982). However, this diversity may be reduced to a few groups of similar physiological characteristics. For instance, the ubiquitous *Candida* spp. and *Pichia* spp. are highly heterogeneous, and new species are likely to be found in each new survey because the accuracy of molecular identifications is constantly increasing. Barata et al. (2012) proposed a division of yeast microbiota of grape berries into three main groups, characterized by similar characteristics: (1) oligotrophic oxidative basidiomycetous yeasts, such as *A. pullulans*; (2) copiothrophic oxidative ascomycetes (several *Candida* spp, *Hanseniaspora* spp., *Pichia* spp, *Candida zemplinina*, *Metschnikowia* spp.) yeasts; (3) copiotrophic strongly fermentative yeasts (*Saccharomyces* spp., *Torulaspora* spp., *Zygosaccharomyces* spp., *Lachancea* spp., and *Pichia* spp.). This grouping is particularly dependent on nutrient availability on the berry surface. In Table 4.1 are summarized the main yeast species colonizing the surface of grape berries.

From a quantitative point of view, the yeast populations of grapes are reported in many works, and all results agree showing an occurrence roughly comprised between 10^2 and 10^4 cells/g (Fleet et al., 2002). Since the grape is a natural ecological niche, this range could be considered very wide, but this may be partially explained by sampling carried out without accurate separation between damaged and undamaged berries. When this is done, smaller variations are found (Barata et al., 2008). Indeed, damaged grapes, with a consequent higher sugar availability on the surface of berries, induced at least an increase of one log cycle of epiphytic yeast population. Many other factors can influence the occurrence and complexity of the yeast community on grape berry surface: biotic factors, including microbial interactions such as competition and symbiosis or antimicrobial and enzymatic activity; abiotic factors such as climatic conditions, geographic location, and age, size and variety of vineyard, and also phytochemical treatments. In this regard, Pretorius et al. (1999) described the influence of each of these variables and also considered that many factors influenced indirectly the grape microbiota affecting primarily the skin integrity. This is the case for rainfall, wind, temperature, diseases, pests, and viticultural practices.

More recently, Clavijo et al. (2010) carried out an ecological survey of wine yeasts present in grapes growing in two vineyards located in the southern Spain (Serranía de Ronda region). They found that, although *Kluyveromyces thermotolerans, Hanseniaspora guilliermondii, Hanseniaspora uvarum*, and *Issatchenkia orientalis* are the most frequent species, a specific distribution of strains was found in the three grape varieties studied. Studying the influence of grape varieties on the indigenous yeast community of grape berries, Raspor et al. (2006) evaluated the impact of the geographical locations in the Dolenjska vine-growing region. The frequency of occurrence of particular yeast species showed their preferences for certain grape varieties. The white grape variety mostly attracted pigmented basidiomycetous yeasts belonging to the genera *Rhodotorula, Sporobolomyces*, and *Cryptococcus* that dominated on all sampling locations. Differently, yeast populations isolated from the red grape surfaces belonged both to ascomycetous and basidiomycetous yeasts in the ratio of 1:1. Hierro et al. (2005) evaluating the effect of grape maturity and cold maceration prior to fermentation on the yeast ecology during wine fermentation found that cold maceration favors the presence of *Hanseniaspora osmophila, Candida tropicalis*, and *Zygosaccharomyces bisporus*, that are the species only isolated from the unripe and overripe fermentations after cold maceration.

The presence of yeasts on the surface of grape berries can vary at the quantitative and qualitative levels also as a function of agrochemical treatments (Cordero-Bueso et al., 2011; Milanović et al., 2013; Pretorius, 2000). However, the main vineyard factors studied are related to the use of chemical pesticide treatments, and the studies are based on analyzing grapes after vineyard treatment, which do not exclude the influence of other factors, such as climatic conditions or grape variety, which cannot be correctly extrapolated to evaluate the single effect on berry microbiota. However, Ganga and Martínez (2004) detected less diversity of non-*Saccharomyces* species, which was explained by the use of fungicides against *B. cinerea*, while Regueiro et al. (1993) and Valero et al. (2007) recovered lower numbers of this species. Milanović et al. (2013) found that *C. zemplinina* and *Hanseniaspora* species colonized the surface of grapes treated with both organic and conventional treatment, while *Metschnikowia pulcherrima* was widely found in conventional samples and only occasionally in organic grapes.

Recently, Salvetti et al. (2016) used a metagenome sequencing approach to analyze the microbial consortium associated with Corvina berries at the end of the withering process performed in two different conditions, such as traditional withering and accelerated withering. Within the eukaryotic community, they found that microorganisms belonging to *Saccharomycetales* order, including *Saccharomyces* and non-*Saccharomyces* yeasts, represented a minority of the Ascomycota fraction of the samples. In particular, the detection of *Saccharomyces* species,

TABLE 4.1 Principal non-*Saccharomyces* yeasts found in grape berry surface and their description

Non-*Saccharomyces*		Fermentation	Morphology	Type strain	Ecology
Candida pyralidae		Glucose, trealose	Budding cells with pseudohyphae	CBS 5035	From insect–fungus–tree interface
Debaryomyces carsonii		absent	Spherical-elongated single cells, in pairs or chains	CBS 2285	Variable habitats including spoiled wine, agricultural processing wasted, and insects
Hanseniaspora vinae		Glucose	Apiculate cells lemon-shaped	CBS 2171	Grape, orange juice
Hanseniaspora guillermondii		Glucose	Apiculate cells lemon-shaped	CBS 465	Widespread, associated mainly with fruits
Hanseniaspora uvarum		Glucose	Apiculate cells lemon-shaped	CBS 314	Wide spread, most frequently isolated from soil, fruits
Kloeckera apiculata		Glucose	Apiculate cells lemon-shaped	CBS 2171	Widespread, fruits
Lachancea thermotolerans		Glucose	Spherical or cylindrical cells	CBS 2745	Tree exudates, associated with insect
Metscnikowia pulcherrima		Fructose Glucose	Spherical cells	CBS 5833	Ripe fruit, flowers, nectar
Pichia kudriavzevii		Glucose	Ovoid elongate cells	CBS 573	Widely distributed in nature often occurring in soil, on fruits, and in various natural fermentations
Pichia membranifaciens		Glucose (weak)	Elongated cells	CBS 209	Natural fermentations, including rotting plant material

(Continued)

TABLE 4.1 (Continued)

Non-Saccharomyces	Fermentation	Morphology	Type strain	Ecology
Saccharomycodes	Glucose sucrose	Apiculate cells lemon-shaped	CBS 821	Isolated from soil, fruits, and fruit juices, fermented beverages and oak trees
Schizosaccharomyces pombe	Glucose Sucrose Maltose (v)	Fisson yeast, globose, ellipsoidal, or cylindrical cells	CBS 10391	High sugar habitats, beer
Starmerella bacillaris (synonymus Candida zemplinina)	Glucose, fructose	Spherical to ovoidal	CBS 9494	Grape surfaces, botrytized grape must
Starmerella bombicola	Glucose; Sucrose; Raffinose	Spherical to ovoid, usually found as single cells but may be arranged in a star-like configuration of cells	CBS 6009	Floricolous insects, grape surfaces
Torulaspora delbrueckii	Glucose Sucrose (v) Maltose (v)	Spherical to ellipsoidal	CBS 1146	Soil, fermenting grapes and other berry juices, agave juice
Wickerhamomyces anomalus	Glucose, Sucrose, Galactose (v) Maltose (v)	Spheroidal to elongate	CBS 6986	Fermentation contaminant, soil, grain, ensilage, water, plants (especially fruits and fermenting matter), sewage
Zygosaccharomyces bailii	Glucose, Sucrose (v)	Spherical to ellipsoidal	CBS 680	Dehydrated, "mummified" fruit, fruit trees or gummy exudates from trees
Zygotorulaspora florentina	Glucose, galactose, maltose, sucrose, raffinose	Spherical to ovoidal	CBS 3020	Fruit and fruit products

especially *S. cerevisiae*, has been found very rarely, and only after enrichment procedures. This trend is in agreement with a lot of studies that indicated a scarce presence of *S. cerevisiae* in sound berries (about 0.05%–0.1%) with a slightly increased presence only in damaged berries (Barata et al., 2012; Cordero-Bueso et al., 2011).

The relation between microbial community and regional wine producing area was recently investigated. Currently, studies on autochthonous yeasts strongly adapted to specific grape musts are growing, both to study the biodiversity associated with different areas and to select new indigenous strains associated with "*terroir*" (Capozzi et al., 2016; Zarraonaindia et al., 2015). These studies showed that microbiomes (bacteria and fungi) associated with grapes and then with the earlier stage of fermentation, showed defined biogeography, illustrating that different wine growing regions stably maintain different microbial communities. Bokulich et al. (2014) using a high-throughput sequencing approach, demonstrated that regional, site-specific, and grape variety factors shape the fungal consortia inhabiting wine-grape surfaces. These microbial communities are correlated to specific climatic features, suggesting a link between vineyard environmental conditions and microbial inhabitation patterns. Overall these factors shape the unique microbial inputs to regional wine fermentations, posing the existence of nonrandom *"microbial terroir"* as a determining factor in regional variation among wine grapes.

4.3 CONTROLLED FERMENTATION: THE ROLE OF SACCHAROMYCES CEREVISIAE

S. cerevisiae is isolated with extreme difficulty from natural wine related habitats, such as vineyard soil or the surface of ripe grapes, while it is easily found on the surfaces of a winery and its equipment (Martini, 1993). In general, *S. cerevisiae* is predominantly found in association with human activities, linked to the production of alcoholic beverages. This led to the idea that *S. cerevisiae* is a "domesticated" species that is strongly specialized for fermenting high sugar substrates while the isolation of *S. cerevisiae* strains from sources not correlated with alcoholic beverages is simply due to a migration from these fermentation processes (Fay and Benavides, 2005). This observation is supported by the evidence that *S. cerevisiae* strains are generally clustered in separated genotyping for use in the lab and the production of wine, beer, and bread. Indeed, all *sensu strictu* species of the genus *Saccharomyces* are yeast specialized for growth on sugary substrates. Indeed, numerous strains of *S. cerevisiae* have been isolated, the majority of which have been found associated with the alcoholic fermentations (Mortimer and Polsinelli, 1999; Naumova et al., 2003; Teresa et al., 2003). Cavalieri et al. (2003) with a study based on extraction of DNA from ancient wine containers demonstrated that *S. cerevisiae* has been unknowingly associated with winemaking since 3150 BC and the earliest evidence for winemaking is from 7000 BC from the molecular analysis of pottery jars found in China (McGovern et al., 2004).

Spontaneous must fermentation has been employed for centuries in winemaking and only in the 1960s did scientific and technological improvements allow starter cultures of *S. cerevisiae* to develop with the diffusion of active dry yeasts commercial preparation. Thereafter, commercial starters have been extensively used to control the effectiveness of the fermentative process and standardize the final product. Indeed, fermentations inoculated with selected commercially available strains of *S. cerevisiae* are useful to manage the fermentations. After the wide diffusion of the use of selected *S. cerevisiae* starter cultures many studies were conducted with the aim to select, from various habitats, strains of yeast with physiological characteristics useful to be used as commercial starters. However, recently their utilization for their effect on mitigation of distinctive characteristics that can affect the specificity and peculiarity of wine has begun to be discussed.

The scientific controversy on the use of commercial *S. cerevisiae* strains is based on a dual aspect. First, the maintenance of biological patrimony including autochthonous non-*Saccharomyces* yeasts is essential to exalt the typical flavor and aroma of wines originating from different grapevine cultivars (Pretorius, 2000). On the other hand, commercial starters guarantee a correct trend of fermentation avoiding negative deviations of the analytical and sensory profile of the wine. Moreover, the use of indigenous selected *S. cerevisiae* strains in the winemaking process requires the development of molecular techniques able to distinguish commercial from indigenous strains. It is generally assumed that indigenous yeasts are suppressed by the starter, however, studies show that indigenous yeasts can still participate in inoculated fermentation (Capece et al., 2010).

A compromise proposed by the scientific community and accepted in the winemaking market is the selection of new yeast strains with all the characteristics of a good starter but isolated from the grapes or winery to the nearest vineyard (Schuller et al., 2005; Valero et al., 2005, 2007). This feature allows the control of the undesired microflora and, at the same time, guarantees the control of the fermentation process. However, there is little information about the fitness of these yeast strains and about their possible dissemination in the vineyard as the vintage advances during a given campaign. Another chance to enhance the complex peculiarities of wine is the controlled use of non-*Saccharomyces* wine yeasts (Ciani et al., 2010). Most of the non-*Saccharomyces* species coming from wine-related environments have limited fermentation potential, such as low fermentation power and rates, as well as low SO_2 resistance (Ciani and Maccarelli, 1997; Jolly et al., 2006) and some of these may produce by-products such as acetic acid, ethyl acetate, acetaldehyde, and acetoin at high concentrations. However, in the last decades, several studies have revaluated the role of non-*Saccharomyces* yeasts revealing their positive impact on the composition and aroma complexity of the final product (Jolly et al., 2003; Domizio et al., 2011; Jolly et al., 2014; Varela et al., 2017). In the next section we analyze and discuss the influence of non-*Saccharomyces* wine yeasts in winemaking and in the red winemaking process that present some specific technological and biological features during the production and maturation/aging process.

4.4 NON-SACCHAROMYCES YEASTS FEATURES IN RED WINE

In the following paragraphs the main characteristics of non-*Saccharomyces* yeasts and the modalities of their use will be reported. The contribution of selected non-*Saccharomyces* yeasts during red wine production will be provided, focusing the attention on the aromatic profile, color stability, polysaccharides production, total acidity, ethanol reduction, and antimicrobial activity (Fig. 4.1; Table 4.2).

4.4.1 The Enzymatic Activities

Extracellular enzymatic activities by yeasts have a relevant role in winemaking, releasing volatile aromatic components from their precursors and stabilizing the wine. In this regard, non-*Saccharomyces* showed relevant extracellular enzymatic activities that positively affect the analytical composition and sensorial profile of wine. A screening of yeasts belonging to genera *Kloeckera*, *Candida*, *Debaryomyces*, *Rhodotorula*, *Pichia*, *Zygosaccharomyces*, *Hanseniaspora*, and *Kluyveromyces* to produce extracellular pectinases, proteases β-glucanases, lichenases, β-glucosidases, cellulases, xylanases, amylases, and sulfite reductase activity revealed the potential of non-*Saccharomyces* wine yeasts to produce a wide range of useful extracellular enzymes during the initial phase of wine fermentation. Indeed, enzymes produced by indigenous non-*Saccharomyces* yeasts associated with grapes might be harnessed to catalyze desired biotransformations during wine fermentation (Strauss et al., 2001). The β-glucosidase activity is involved in the aroma enhancement in winemaking. Among several non-*Saccharomyces* yeast species tested *H. uvarum* showed the highest activity in winemaking conditions and exhibited catalytic specificity for aromatic glycosides of C_{13}-norisoprenoids and some terpenes, enhancing fresh floral, sweet, berry, and nutty aroma characteristics in wine (Hu et al., 2016). β-xylosidase can also contribute to the production of

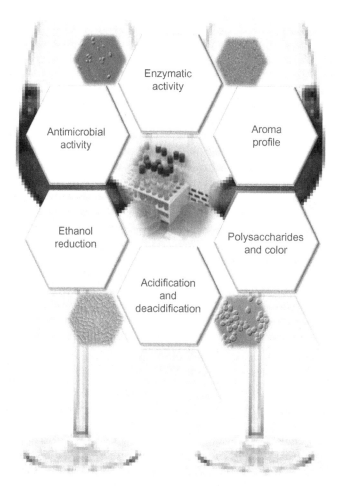

FIGURE 4.1 CONTRIBUTIONS OF SELECTED NON-SACCHAROMYVCES YEASTS IN RED WINE PRODUCTION.

TABLE 4.2 Principal non-*Saccharomyces* wine yeasts and their activities in red wine production

Non-*Saccharomyces* species	Role during fermentation	References
Schizosaccharomyces pombe	Color stabilization through pyruvic acid mediation and urea removal	Escott et al. (2016)
	Increase of pigments	Palomero et al. (2009)
	Deacidification agent	Suárez-Lepe et al. (2012)
Pichia kudriavzevii	Deacidification	Moreno et al. (2014)
Lachancea thermotolerans	Increase of total acidity: lactic acid production	Gobbi et al. (2013)
	Color stabilization (polymeric pigments)	Escott et al. (2018)
Hanseniaspora uvarum	β-Glucosidase, β-xylosidase activities	Hu et al. (2016)
	Production of isoamyl acetate, ethyl acetate	Moreira et al. (2008)
	Color stabilization (polymeric pigments)	Del Fresno et al. (2017)
	Quorum sensing tyrosol and tryptophol mediated	Zupan et al. (2013)
Pichia membranifaciens	β-Glucosidase, β-xylosidase activities	Lopez et al. (2015)
Wickerhamomyces anomalus	Enhancement isoamyl acetate and ethyl esters	Kurita (2008)
	Enzymatic activities	Lopez et al. (2015)
	Increase of aromatic quality (floral fruity notes)	Cañas et al. (2014)
	Killer activity toward *Brettanomyces/Dekkera*	Comitini et al. (2004)
Debaryomyces carsonii	β-Glucosidase activity	Hernandez-Orte et al. (2008)
Kloeckera apiculata	Acetate formation	Rojas et al. (2003)
Hanseniaspora guillermondii	High producers of 2-phenylethyl acetate, 1-propanol, 3-(methylthio) propionic acid	Rojas et al. (2003)
	Low producers of ethylexanoate, pentanoic acid, free fatty acids	Moreira et al. (2011)
Hanseniaspopra vinae	Positive aromatic characters (floral and fruity)	Tristezza et al. (2016)
Torulaspora delbrueckii	Increase of aroma, increase acetate ester, thiols, terpens, 2-phenylethanol	Cus and Jenko (2013)
	Production of zymocin with antimicrobial activity vs *Brettanomyces* spp. and *Pichia* spp.	Villalba et al. (2016)
	Quorum sensing tyrosol and tryptophol and 2-phenylethanol mediated	Avbelj et al. (2016)
Metscnikowia pulcherrima	Positive aromatic characters, citrus aroma	González-Royo et al. (2015)
	Increment of higher alcohols, esters, terpenoids	Varela et al. (2017)
	Color stabilization polymeric(pigments)	Medina et al. (2017)
	Pulcherrimin secretion with inhibitory effect vs. *Brettanomyces* growth	Oro et al. (2014)
	Ethanol reduction	Contreras et al. (2015)
		Varela et al. (2017)
Zygotorulaspora florentina	Increase of fruity/floral notes, aroma complexity	Lencioni et al. (2016)
	Low astringency	
Sacharomycodes	Color stabilization and preservation (release of mannoproteins)	Domizio et al. (2011)
Candida pyralidae	Antimicrobial activity toward *Brettanomyces* spoilage yeasts	Mehlomakulu et al. (2014)
Zygosaccharomyces bailii	Quorum sensing tyrosol and tryptophol, 2-phenylethanol mediated	Zupan et al. (2013)

(Continued)

TABLE 4.2 (Continued)

Non-*Saccharomyces* species	Role during fermentation	References
Starmerella bacillaris (synonymus Candida zemplinina)	Quorum sensing tyrosol and tryptophol,2-phenylethanol mediated	Zupan et al. (2013)
	Ethanol reduction	Englezos et al. (2016)
Saccharomyces uvarum	Ethanol reduction	Varela et al. (2016)
	Increase of 2 phenylethanol and 2 phenyl acetate	
	Increase of higher alcohols	

quality wines. *Pichia membranifaciens, Hanseniaspora vineae, H. uvarum*, and *Wickerhamomyces anomalus*, in addition to β-glucosidase showed also β-xylosidase activities, showing an overall terpenes and volatile phenols (terpineol, 4-vinyl-phenol and 2-methoxy-4-vinylphenol) increase (López et al., 2015). Among enzymatic activity of non-*Saccharomyces*, proteolytic activity could lead to a reduction in protein levels with a consequent increase in protein stability of wine. Differently, Dizy and Bisson (2000) reported that increased yeast proteolytic activity did not lead to a reduction in haze formation.

4.4.2 The Influence on the Aroma Profile

Certainly, the aromatic profile is one of most important traits that contribute to the quality of wine. As in many foods, wine aroma is composed by hundreds of different compounds with concentrations that can vary between 10^{-1} and 10^{-10} mg/mL. The balance and interaction of all of them determine the wine aromatic quality (Padilla et al., 2016).

The maceration process during fermentation is a standard practice in red wine production (Bautista-Ortín et al., 2007). The main purpose of maceration is the extraction of color compounds (anthocyanins and phenolic substances) from the solid components of the grape. However, this process also affects the sensory profile of the resulting wines because other compounds, such as aromatic substances and precursors, nitrogen compounds, polysaccharides, and minerals, are released during the maceration process, whose effects influence the quality of the wine (De Beer et al., 2017). The transfer of these compounds from grape skins also implies a modification of initial yeasts composition in must. During this winemaking phase the yeast species present on the grape surface survive and give their contribution during the first step of fermentation. For this purpose, Francesca et al. (2014) investigated the microbiological, chemical, and sensory characteristics of red wine subjected to postfermentation maceration using "Aglianico di Taurasi" red grape, a typical variety of the South Italy. The main objective of this work was the evaluation of the influence of long postfermentation maceration on the evolution of yeast populations of red wine. A high initial concentration of both yeast and Lactic Acid Bacteria (LAB) population and a large variation during the process was seen. During the entire maceration process, *S. cerevisiae* was the main species found together with *Debaryomyces carsonii*, a yeast species that is not commonly associated with the wine environment that possesses high β-glucosidase activity (Hernandez-Orte et al., 2008) and contributed to the characteristic aroma of grapevine varieties (Zott et al., 2008). However, the most representative yeast group during the first stages of fermentation (in the presence of grape skins) are apiculate yeasts. They have become an object of increasing interest, as their presence and proliferation in competition with *S. cerevisiae* influence the sensory quality of wine due also to the enzymatic activities on aromatic varietal precursors. Consumer expectations regarding the greater complexity of wines with complex profiles led to the importance of apiculate yeasts in mixed starter cultures with *S. cerevisiae*, and toward innovative biotechnological approaches. In this regard, several works investigated the production of esters by different non-*Saccharomyces* yeast species that contribute to enhance the aroma profile of wines (Moreira et al., 2008; Rojas et al.,2003; Viana et al., 2008). Plata et al. (2003) examined the production of ethyl acetate and isoamyl acetate by seven wine yeast strains in natural grape juice during fermentation. In this case, as expected the amounts in which the two esters are produced depended on the yeast strain, but also on its specific production rate and survival rate. In particular, *Kloeckera apiculata* exhibited the highest ability for acetate formation. *Hansenula subpelliculosa, Kluyveromyces marxianus, Torulaspora delbrueckii*, and *S. cerevisiae* produced intermediate levels and *Pichia membranaefaciens* and *Candida guilliermondii* very low levels of the two esters measured in this study. Other non-*Saccharomyces* yeast strains belonging to the genera

Candida, Hanseniaspora, Pichia, Torulaspora, and *Zygosaccharomyces* were screened for ester formation on synthetic microbiological medium. *H. osmophila* was found to be a strong producer of 2-phenylethyl acetate suggesting it is a good candidate for mixed starters during natural wine fermentation (Viana et al., 2008). Further studies confirmed this behavior (Medina et al., 2013; Moreira et al., 2011; Tristezza et al.,2016). The high esters production by apiculate yeast was confirmed by Rojas et al. (2003). Indeed, these authors found that ethyl acetate was the main ester produced, and isoamyl acetate and 2-phenylethyl acetate made up the next largest group of ester compounds. *H. guilliermondii* was found to be a strong producer of 2-phenylethyl acetate in both pure and mixed cultures. This behavior was confirmed in selected *H. uvarum* and *H. guilliermondii* strains used in pure or mixed fermentation with *S. cerevisiae*. The high production of esters did not negatively influence the aromatic profile of wines Moreira et al. (2008). In particular, the highest levels of 2-phenylethyl acetate were obtained with *H. guilliermondii* whereas *H. uvarum* increased the isoamyl acetate content of wines. On the other hand, both these apiculate yeasts produced high amounts of ethyl acetate that can negatively affect the sensorial properties of wine. However, the production of this ester was reduced in mixed cultures with *S. cerevisiae*.

Also *H. vineae* was proposed in mixed fermentation to enhance the aromatic profile of wines. Indeed, this species showed an increase, other than in acetate esters, also in terpens with an improvement of positive aromatic characters such as "floral" and "fruity" (Hu et al., 2016; Lleixà et al., 2016). Red grape must pilot fermentations were performed and an inoculated fermentation with *H. guilliermondii* was compared to a spontaneous fermentation. The grape must inoculated with *H. guilliermondii* led to the production of wine with higher concentrations of 1-propanol, 2-phenylethyl acetate, and 3-(methylthio) propionic acid, and lower amounts of ethyl hexanoate, pentanoic acid, free fatty acids, 2-methyltetrahydrothiophen-3-one, and acetic acid-3-(methylthio)propyl ester (Moreira et al., 2011).

The use of non-*Saccharomyces* in mixed fermentation to enhance flavor and aroma complexity of wine is not limited to apiculate yeasts. Several investigations focused the attention on *T. delbrueckii*, a species infrequently isolated on the grape surface but one of the most studied species to increase flavor and aroma complexity in alcoholic beverages. Indeed, *T. delbrueckii* possesses several positive features that could be profitably used. Several investigations agree that *T. delbrueckii* impacts on aromatic composition and sensory attributes of wines in both simultaneous and sequential fermentation. Indeed, these investigations found an increase in acetate ester (Cordero-Bueso et al., 2013), thiols (3-sulfanylhexan-1-ol and 3-sulfanylhexyl acetate (Renault et al., 2015; Zott et al., 2011), terpenes (α terpineol and linalool) (Cus and Jenko, 2013), and β phenyl-ethanol (Comitini et al., 2011). In red wine production the impact of *T. delbrueckii* in mixed fermentation (simultaneous or sequential) on aroma profile was also evaluated at pilot or winery scale. In this regard, *T. delbrueckii* and *S. cerevisiae* mixed cultures were evaluated on fermentation and aroma of Amarone wine (a red wine made with dry grapes) (Azzolini et al., 2012). The most significant changes caused by the presence of *T. delbrueckii* were observed among alcohols, fermentative esters, fatty acids, and lactones, which are important in the Amarone wine flavor. Renault et al. (2015) studied ester formation and the aromatic impact of *T. delbrueckii* when used in association with *S. cerevisiae* during the alcoholic fermentation of must. Fermentation trials carried out at winery scale using Merlot must from the Bordeaux area, showed that sequential inoculations increased total esters concentration such as isoamyl acetate ethyl propanoate, ethyl isobutanoate, and to a lesser extent ethyl dihydrocinnamate and isobutyl acetate.

Another non-*Saccharomyces* yeast is *M. pulcherrima*, frequently present on the grape surface and recovered during the initial stages of alcoholic fermentation. *M. pulcherrima* is a high producer of β-glucosidase (Rodríguez et al., 2010), and its presence in mixed cultures can provide significant enhancements in the wine of higher alcohols, esters, and terpenoids. Its aromatic profile in mixed fermentation was characterized by "citrus/grape fruits" some smoky and flowery attributes in Reisling and Macabeo grape varieties, respectively (Benito et al., 2015a; González-Royo et al., 2015). Mixed fermentation with *M. pulcherrima* at pilot scale using red must (Merlot) was carried out to reduce the ethanol content in final wines. The sensorial evaluation of resulting wines indicated that mixed fermentations received relatively high scores for sensory descriptors such as "red fruit" and "fruit flavor" and overall exhibited a sensory profile similar to that of wine made with *S. cerevisiae* (Varela et al., 2017). Also *W. anomalus* (formerly *Pichia anomala*) resulted in positive contribution to aroma profile of wines in mixed fermentation determining an enhancement of isoamyl acetate and ethyl esters (Kurita, 2008). Inquiero-Cañas et al. (2014) studied the influence of mixed fermentation of *W. anomalus* and *S. cerevisiae* on the aroma of red musts from Mazuela variety. The red wines elaborated by sequential inoculation showed higher levels of acetates and ethyl esters, compounds that supply a fruity note, higher levels of linear alcohols, which are responsible for herbaceous notes, and lower concentrations of organic acids, that contribute to increase the aromatic quality. Moreover, sensory analysis showed that red wines obtained by mixed fermentations were largely preferred and were particularly appreciated for their floral and/or fruity notes.

Finally, an interesting non-*Saccharomyces* wine yeast able to enhance complexity and overall aroma profile is *Zygotorulaspora florentina*. At winery scale (10 hL), sequential fermentation (*Z. florentina*/*S. cerevisiae*) of Sangiovese grape juice (a red variety used to produce Chianti and Brunello di Montalcino wines) showed an increase in fruity and floral notes as well as a lower perception of astringency (Lencioni et al., 2016).

4.4.3 The Polysaccharides Production and Color Stability

In general, during the last 10 years several studies have been carried out to characterize non-*Saccharomyces* yeasts for their ability to produce desirable metabolites for wine flavor, such as acetate esters, or enzymes to enhance wine aroma and complexity (Fia et al., 2005; Rojas et al., 2003; Swiegers et al., 2005).

In the past, little attention has been paid to the ability of non-*Saccharomyces* yeast strains to release cell wall polysaccharides, particularly mannoproteins, in fermentation media. Polysaccharides improve wine mouthfullness and body, have positive effects on aroma persistence, and contribute to protein and tartrate stability. This feature is important in red wines and particularly in those with long maturation and aging and without filtration where stability, mouthfullness, and body of wine are particularly desired. The ability to release polysaccharides by 13 non-*Saccharomyces* strains belonging to different species coming from wine and grape was investigated (Giovani et al., 2012). Results obtained showed that non-*Saccharomyces* strains released polysaccharides into the fermentation medium and that the amount was strictly dependent on the yeast species, the number of cells formed, and physiological conditions. Domizio et al. (2011), analyzing the main enological characteristics of other non-*Saccharomyces*, found a wide biodiversity within each genus. In particular, polysaccharides production was higher in the non-*Saccharomyces* strains than in the *S. cerevisiae* strains and those belonging to the genera *Pichia* and *Saccharomycodes* showed the highest polysaccharides production. Wide biodiversity for this feature was seen within the genera *Hanseniaspora* and *Zygosaccharomyces*. One of the major polysaccharides found in wine is mannoproteins, released from the yeast cell wall during the alcoholic fermentation and wine aging. The mannoproteins exhibit positive enological properties improving mouth-feel and fullness, decreasing astringency, adding complexity and aromatic persistence, increasing sweetness and roundness, and reducing protein and tartrate instability (Carvalho et al., 2006; Charpentier et al., 2004; Gonzalez-Ramos et al., 2008; Guadalupe et al., 2010; Vidal et al., 2004). To increase the amount of mannoproteins released during the alcoholic fermentation, some researchers developed autolytic thermosensitive mutants (Giovani and Rosi, 2007) or genetically engineered wine yeast strains belonging to *S. cerevisiae* species (Brown et al., 2007; Gonzalez-Ramos et al., 2008). However, the possibility to increase the content of mannoproteins naturally by the use of non-*Saccharomyces* yeasts could represent an added value to the already heightened interest towards these wine yeasts for flavor modification (Ciani et al., 2010).

Anthocyanins and polymers of athocyanins with tannins are the main pigments in red wine. These compounds give red wine its color and organoleptic character; all the differences between white and red wines depend on these compounds. The chemical changes in the coloring materials constitute the basic process of aging in red wines; these changes include the disappearance of red anthocyanins from grapes and the chemical transformation of tannins which gives them the yellow—red color typical of old wines. In this regard, prefermentative addition of enological tannin could be useful for the wines fermented with some selected non-*Saccharomyces* strains, to improve both wine color and anthocyanin composition. Indeed, Chen et al. (2018) found that mixed fermentations with *Schizosaccharomyces pombe* can be effective to increase the level of vinylphenolic pyranoanthocyanin, which is a positive for wine color stability.

Many winemaking variables and techniques, such as maceration temperature and clarification method, storage temperature, and length of storage, are known to affect the phenolic composition and color of red wines (Sacchi et al., 2005). In this regard, mannoproteins can influence the color parameters in red wine. With regards to this last aspect, Guadalupe et al. (2010) tested some commercially available mannoprotein preparations produced by yeasts in Tempranillo winemaking to determine their influence on color composition, and found that mannoprotein addition did not modify the content and composition of anthocyanins or phenols, and it did not affect monomeric anthocyanin color.

The use of mannoprotein overproducing yeast strains strongly increased high-molecular-weight mannoprotein content during the maceration. No differences were observed in terms of grape polysaccharides, monomeric phenolics, or color parameters, but proanthocyanidin content was significantly reduced when using the overproducing strain. Except for wine color, mannoproteins released from yeast strains produced the same effects as commercial mannoproteins. Winemakers could use both alternatives to reduce red wine astringency and increase wine smoothness and body, although commercial mannoproteins may cause loss of red wine stable color

(Guadalupe et al., 2010). Palomero et al. (2009) studying the osmophilic yeast genera *Schizosaccharomyces* and *Saccharomycodes* found that the aging of red wines is associated with a degradation of the pigments present and their development into other compounds, which in both cases can lead to lower color indices and more orange hues. In aging over-lees, the release of biopolymers such as mannoproteins and cell wall polysaccharides exerts a protective effect on anthocyanin monomers. These anthocyanin monomers are responsible for the lasting blue—red colors. Indeed, aging overlees is now considered to be a technique that helps preserve wine color (Escott et al., 2016; Vivas et al., 2001).

Recently, Benito et al. (2014) reported benefits in addition to the demalication activity deriving from using *S. pombe* yeast in wine fermentation, such as the production of pyruvic acid and the possibility to reduce ethyl carbamate in wine, through the removal of urea (its precursor), as a consequence of urease activity. Pyruvic acid seems to be of interest for the color stability of the wine. Indeed, a strong correlation between the amount of pyruvic acid released and the formation of vitisin A (a pyranoanthocyanin, a natural polyphenol found in grapes) has been observed (Morata et al., 2003). In this regard, the influence of non-*Saccharomyces* on color of red wine was recently investigated. More recently works showed that *Schizosaccharomyces* species in mixed fermentation determined and increased the production of pigments and large amounts of polysaccharides (Domizio et al., 2017; Escott et al., 2018).

During winemaking, colorless phenolics increased during alcoholic fermentation, reached maximum values at pressing, and remained stable during malolactic fermentation and subsequent storage. Aged red wines possess significantly different polyphenolic composition compared with young ones, mainly due not only to the formation of polymeric compounds but also because of oxidation, hydrolysis, and other transformations that may occur in native grape phenolics during aging.

Non-*Saccharomyces* could be effects on the production of more stable derived pigments. In this regard, *Metschnikowia* and *Hanseniaspora*, were screened for their effect on red wine color in artificial red grape must containing phenolic extract from Tannat grapes in cofermentation with *S. cerevisiae*. The key role of acetaldehyde in the enhancement of vitisin B was suggested (Medina et al., 2016). Moreover, the formation of vitisin A, vitisin B, malvidine-3-glucoside-4-vinylphenol, and malvidin-3-glucoside-4-vinylguaiacol are reported for species of genera *Hanseniaspora* and *Metschnikowia* (Medina et al., 2017). In addition to these genera *S. pombe* can increase the amounts of vitisins having an important role in color stability (Del-Fresno et al., 2017). The involvement of some non-*Saccharomyces* (*Lachancea thermotolerans*, *M. pulcherrima*, and *T. delbrueckii*) in sequential fermentation with S. *cerevisiae* and *S. pombe* was investigated. Sequential fermentations led to larger amounts of polymeric pigments, such as catechin, which is the flavanol predisposed to forming such pigments (Escott et al., 2018).

4.4.4 Acidification and Deacidification Activities

One of the first applications and a valuable use of non-*Saccharomyces* wine yeasts in winemaking is the deacidification of grape juice and wine. Indeed, the biological deacidification through maloalcoholic fermentation is limited in *S. cerevisiae* populations varying between 0% and 40%. On the contrary, within the *Schizosaccharomyces* genus several strains are able to completely degrade the malic acid present. *S. pombe* has been proposed for a long time as biological deacidification agent (Ciani, 1995; Munyon and Nagel, 1977; Rankine, 1966; Silva et al., 2003; Snow and Gallander, 1979; Suárez-Lepe et al., 2012) and could be profitably used since it possesses other characteristics that are beneficial for winemaking (see below). Moreover, the consumption of malic acid by *Schizosaccharomyces* yeasts permits the avoidance of bacterial biological deacidification (malolactic fermentation) and averting the production of amines.

The ability of grape juice/wine deacidification was also found in *Pichia kudriavzevii*, another non-*Saccharomyces* wine yeast isolated from Patagonian (Moreno et al., 2014). On the other hand, the characteristic to produce organic acids during the fermentation may be a desired feature in some winemaking environments, process conditions, and consumer's preferences. Indeed, consumers required well-structured, full body wines, and optimal phenolic maturity of grapes, characteristics obtained with late harvests. This practice results in a noticeable increase in the sugar content of the berries (Mira de Orduña, 2010) and consequent reduction of total acidity. On the other hand, global climate change has deeply influenced the vine phenology and the grape composition, resulting in grapes with lower acidity, altered phenolic maturation, and tannin content (Jones et al., 2005). Both these features have been determined in extensive areas of vine cultivation problems concerning low acidity and high pH values. In this regard, *L. thermotolerans* in simultaneous and sequential fermentation exhibited the ability to produce lactic acid determining an increase in total acidity of wine, a desired feature in grape juices deficient in acidity generally coming from the wines of warm climates (Gobbi et al., 2013; Kapsopoulou et al., 2007). An alternative to the traditional malolactic fermentation in red wine production using non-*Saccharomyces* yeasts was proposed (Benito et al., 2015b). This red winemaking methodology consists of combined use of *S. pombe* and *L. thermotolerans* where the former consumes malic acid stabilizing the product and the latter produces lactic acid, thus avoiding the reduction or even determining the increase of acidity.

4.4.5 Reduction of Ethanol Content

Nowadays, the progressive increase in alcohol levels in wine is a growing problem affecting the winemaking industry. Indeed, over the last two decades, there has been a progressive increase in the ethanol content in wines of c.a. two degrees over the viticulture areas (Alston et al., 2011; Gonzalez et al., 2013; MacAvoy, 2010). This increase is mainly due to two main concerns: global climate change and the new wine styles often associated with increased grape maturity. For example, in red wine harvesting the grapes at complete phenolic maturation may determine overripe grapes and consequently the production of wines with high ethanol content. On the other hand, global climate change has deeply influenced the vine phenology and the grape composition, resulting in grapes with lower acidity, altered phenolic maturation and tannin content, modifying other wine sensory attributes (MacAvoy, 2010).

In order to overcome these issues, the market focus is directed to wines with a moderate alcohol content. In addition, lowering ethanol content has an economic interest due to the high taxes imposed in some countries (Gil et al., 2013). In this context, there is rising interest in an ethanol reduction in wine. A microbiological approach for decreasing ethanol concentrations appears a promising way and there is a growing interest to evaluate the use of non-*Saccharomyces* wine yeasts. There are several features possessed by non-*Saccharomyces* wine yeast that are a potential tool for the reduction of alcohol content in wine: a wide variability in ethanol yield (Contreras et al., 2014; Contreras et al., 2015; Gobbi et al., 2014; Magyar and Toth, 2011) and the differences in regulatory respirofermentative metabolism with *S. cerevisiae* (Gonzalez et al., 2013). Indeed, among non-*Saccharomyces* wine yeasts some strains/species showed sugar consumption by respiration (Crabtree negative). Therefore, both features of non-*Saccharomyces* yeasts have indicated a promising way to limit ethanol production. The approach using non-*Saccharomyces* wine yeasts to limit the production of ethanol is the mixed culture (simultaneous or sequential) due to the inability of these yeasts of completing alcoholic fermentation (Ciani et al., 2016). In this way, the production of quality wines with decreased ethanol concentration is an objective of wine producers. After a setup of mixed fermentations using different non-*Saccharomyces* wine yeasts some investigations in red wine grape varieties were conducted. The aim was to obtain ethanol reduction and final wines comparable or better to the control wines.

Production of quality wines with decreased alcohol concentration continues to be one of the major challenges facing wine producers. Therefore, there is considerable interest in the isolation or generation of wine yeasts less efficient at transforming grape sugars into ethanol. Shiraz grape juice trials fermented with *Saccharomyces uvarum* and *M. pulcherrima* showed a reduction in ethanol content. Wines fermented with *S. uvarum* and with a combination of *M. pulcherrima* and *S. uvarum* were characterized by increased concentrations of β-phenyl ethanol and 2-phenylethyl acetate, both associated with positive sensory attributes, while *M. pulcherrima* alone showed concentrations of ethyl acetate likely to negatively affect wine aroma (Contreras et al., 2015; Varela et al., 2016). The ethanol reduction in these sequential fermentations using Shiraz wines were approximately 0.9%–1.7% v/v lower than wines produced with *S. cerevisiae* alone. Merlot wines produced at pilot scale with *M. pulcherrima* and *S. uvarum* confirmed the ethanol reduction from 1 to 1.7 vol.% less than *S. cerevisiae* wines. The volatile composition and sensory profiles of reduced-alcohol wines indicated that *M. pulcherrima* showed higher concentrations of ethyl acetate, total esters, total higher alcohols, and total sulfur compounds, while wines fermented with *S. uvarum* were characterized by the highest total concentration of higher alcohols. *M. pulcherrima* wines received relatively high scores for sensory descriptors, whereas the main sensory descriptors associated with wines fermented with *S. uvarum* were "barnyard" and "meat" (Varela et al., 2017). Another non-*Saccharomyces* yeast species investigated to reduce ethanol content in wine is *Starmerella bacillaris* (synonym *C. zemplinina*) in combination with *S. cerevisiae*, in mixed coinoculated and sequential fermentation (Englezos et al., 2015; Englezos et al., 2016). In Barbera grape juice, a red grape variety, with skin maceration, *M. pulcherrima* sequential fermentation with *S. cerevisiae* showed wines with a slightly lower alcohol concentration (0.2%–0.3% v/v), while facilitating the release of anthocyanin and some esters of fatty acids that could contribute positively to wine aroma (Rolle et al., 2017). In Table 4.3 are summarized the non-*Saccharomyces* mixed fermentations carried out in red grape juice with the aim to reduce the ethanol content.

TABLE 4.3 Non-*Saccharomyces* wine yeasts in reduced-ethanol wines production using red grape varieties

Mixed fermentation	Grape variety	Ethanol reduction (v/v%)	References
M. pulcherrima/S. cerevisiae	Shiraz	1.6	Contreras et al. (2014)
L. thermotolerans/S. cerevisiae	Sangiovese	0.7	Gobbi et al. (2013)
M. pulcherrima—S. uvarum/S. cerevisiae	Shiraz	0.9	Contreras et al. (2015)
M. pulcherrima/S. uvarum		1.7	Varela et al. (2016)
S. bacillaris/S. cerevisiae	Barbera	0.5	Englezos et al. (2016)
M. pulcherrima/S. cerevisiae	Merlot	1.0–1.7	Varela et al. (2017)
M. pulcherrima/S. uvarum			
S. bacillaris/S. cerevisiae	Barbera	0.2–0.3	Rolle et al. (2017)

4.4.6 Antimicrobial Activities

Another possible applicative use of non-*Saccharomyces* yeasts in winemaking regards the control of spoilage microorganisms. During different stages of fermentation, a punctual and timely control of potential spoilage microorganisms is needed. In particular, during fermentation and aging stages of red wine, the spoilage yeast is *Brettanomyces bruxellensis* is most responsible for undesired odors and considered the current major concern for winemakers, since an effective method to control their growth has not yet been developed (Schumaker et al., 2017).

Dekkera/Brettanomyces are described in the literature as part of the microbiota of many fermented beverages including cider, some types of beer, kombucha, and kefyr, etc. (Morrissey et al., 2004). *Dekkera/Brettanomyces* can grow during the red wine aging and even after their bottling; conversely, these yeasts are rarely found during the alcoholic fermentation of grape must. A few studies have reported their presence on grapes due to the difficult cultivation while in the winery, in vats, pumps, or equipment that are difficult to sanitizes, *Brettanomyces* yeasts are more easily found (Fugelsang and Zoecklein, 2003; Pretorius, 2000; Renouf and Lonvaud-Funel, 2007a,b).

Different strains of *Brettanomyces* can show great differences in their production of volatile phenols. The variety of grape used also affects the sensorial perception of ethylphenols. Phister and Mills (2003) indicated detection thresholds to be high in monovarietal Cabernet sauvignon wines, and lower in Tempranillo wines. Recently, it was determined the concentrations at which consumers from two different geographic areas were able to detect *Brettanomyces* volatile compounds present in a red wine containing different concentrations of 4-ethylphenol (4-EP), 4-ethylguiacol, and 4-ethylcatechol in a 5:1:1 ratio, respectively (Schumaker et al., 2017). Results demonstrated a high ability by consumers to detect differences in red wines due to volatile compounds. Indeed, volatile phenols greatly influence the aroma of wine and elevated concentrations of 4-EP in red wine are associated with disagreeable aromas often described as "phenolic," "leather," "horse sweat," "stable," or "varnish," etc. (Chatonnet et al., 1993; Rodrigues et al., 2001). In addition to these compounds, *Dekkera/Brettanomyces* produce a large variety of other volatile compounds with significant olfactory repercussions, such as tetrahydropyridines synthesized from lysine (Heresztyn, 1986).

The treatments to reduce the negative effects caused by *Dekkera/Brettanomyces* are based on both preventive and curative actions. Certain additives can inhibit the growth of *Brettanomyces*, including sulfur dioxide (SO_2). The recommended molecular dose of SO_2 is highly variable, from 0.3 to 0.8 mg/L. But these doses do not consider differences of strain resistance to sulfites or yeast population levels. Moreover, SO_2 is known as a chemical stressor inducing a viable but nonculturable state of *B. bruxellensis* that is nondetectable by plate counting, and can lead to new contamination when the amount of sulfite decreases over time (Capozzi et al., 2016). Moreover, the SO_2 preservative agent has been largely demonstrated to have negative effects in wine consumers, including allergic reactions, asthma, and headaches. This led to the establishment of strict regulations governing its use in the wine industry (Guerrero and Cantos-Villar, 2015) with a direct consequence that the industry is interested in new ways to reduce sulfur dioxide levels, without changing the sensory quality of the wine. Based on this, a valid and natural alternative could be represented by bioactive compounds produced by yeasts (Muccilli and Restuccia, 2015). Biopreservation or biocontrol refers to the use of natural or controlled microorganisms, or their antimicrobial products, to extend the shelf life and to enhance the safety of food and beverages. This can be achieved by the addition of antimicrobial metabolites, such as killer toxins, or the direct application of protechnological killer strain. A number of microorganisms and other biological agents have been regarded to be crucial in the biopreservation of food and beverages. In this context, a large group of non-*Saccharomyces* killer yeasts, able to produce killer toxins, can counteract *Dekkera/Brettanomyces* spoilage yeasts in red wine.

Yeast killer toxins, also named mycocins or zymocins were initially defined as extracellular proteins, glycoproteins or glycolipids that disrupt the cell membrane function in susceptible yeast bearing receptors for the compound, whose activity is directed primarily against yeast closely related to the producer strain, which has a protective factor. The first mycocins were identified in association with *S. cerevisiae* in the brewing industry, but several others have since been isolated, frequently where yeast populations occur in high density and in highly competitive conditions, as for example fermented olive brine and fermenting grape must. Biological control could have an important application during the maturation and wine aging of red wines. In this regard, killer toxins secreted by *W. anomalus* (Pikt) and *Kluyveromyces wickerhamii* (Kwkt) were tested to control *Dekkera/Brettanomyces* spoilage yeasts. The stability in red wine and the fungicidal effect of these two zymocins were demonstrated (Comitini et al., 2004). Thus, a potential application for the two toxins as antimicrobial agents active on *Dekkera/Brettanomyces* during wine aging and storage can be hypothesized. Also, Santos et al. (2011) studied the killer activity of *Ustilago maydis* to control *B. bruxellensis*, in mixed cultures under winemaking conditions, showing that the killer yeast can inhibit *B. bruxellensis*, while the *S. cerevisiae* inoculated strain is fully resistant to its killer activity, indicating that it could be used in wine fermentation to avoid the development of *B. bruxellensis* without undesirable effects on the fermentative yeast.

Recently, two new killer toxins produced by *Candida pyralidae* with an antimicrobial effect against *B. bruxellensis*, were tested in wine (Mehlomakulu et al., 2014). The killer toxins were stable under winemaking conditions and the activity was not affected by the ethanol and sugar concentrations typically found in grape juice and wine. Also, Belda et al. (2017) studied and widely characterized the mode of action of two killer toxins from *P. membranifaciens* (PMKT1 and PMKT2) able to inhibit *B. bruxellensis*, without any negative effect on *S. cerevisiae* that was fully resistant to their fungicide effect. Another new killer toxin from *T. delbrueckii* was identified and partially characterized (Villalba et al., 2016). This zymocin showed also a potential biocontrol effect on *B. bruxellensis* and other spoilage non-*Saccharomyces* yeasts, such as *Pichia guilliermondii*, *Pichia manshurica*, and *P. membranifaciens*. However, other biological methods besides killer yeasts, were evaluated to control *B. bruxellensis* using non-*Saccharomyces* specific strains. For example, Oro et al. (2014) showed that *M. pulcherrima* secretes pulcherriminic acid that exhibits an effective inhibitory effect on the growth of *B. bruxellensis*. In this case, the antimicrobial activity of *M. pulcherrima* does not seem due to proteinaceous compounds such as killer phenomenon, but to the precursor of pulcherrimin pigment that depletes iron present in the medium, making it not available to the other yeasts. Moreover, cell-to-cell contact and quorum sensing have been investigated as mechanisms involved in non-*Saccharomyces*-mixed fermentation. Quorum sensing was recently examined in *H. uvarum*, *Torulaspora pretoriensis*, *Zygosaccharomyces bailii*, *C. zemplinina*, and *B. bruxellensis*. Results indicated species-specific kinetics to produce 2-phenylethanol, tryptophol, and tyrosol, considered the main molecules involved in the quorum sensing mechanism (Zupan et al., 2013; Avbelj et al., 2016).

References

Alston, J.M., Fuller, K.B., James, T., Lapsley, J.T., Soleas, G., 2011. Too much of a good thing? Causes and consequences of increases in sugar content of California wine grapes. J. Wine Econ. 6, 135–159.

Avbelj, M., Zupan, J., Raspor, P., 2016. Quorum-sensing in yeast and its potential in wine making. Appl. Microbiol. Biotechnol. 100 (18), 7841–7852.

Azzolini, M., Fedrizzi, B., Tosi, E., Finato, F., Vagnoli, P., Scrinzi, C., 2012. Effects of *Torulaspora delbrueckii* and *Saccharomyces cerevisiae* mixed cultures on fermentation and aroma of Amarone wine. Eur. Food Res. Technol. 235, 303–313.

Barata, A., González, S., Malfeito-Ferreira, M., Querol, A., Loureiro, V., 2008. Sour rot-damaged grapes are sources of wine spoilage yeasts. FEMS Yeast Res. 8 (7), 1008–1017.

Barata, A., Malfeito-Ferreira, M., Loureiro, V., 2012. The microbial ecology of wine grape berries. Int. J. Food Microbiol. 153 (3), 243–259.

Bautista-Ortín, A.B., Fernández-Fernández, J.I., López-Roca, J.M., Gómez-Plaza, E., 2007. The effects of enological practices in anthocyanins, phenolic compounds and wine color and their dependence on grape characteristics. J. Food Compos. Anal. 20 (7), 546–552.

Belda, I., Ruiz, J., Alonso, A., Marquina, D., Santos, A., 2017. The biology of Pichia membranifaciens killer toxins. Toxins 9 (4), 112.

Benito, S., Hofmann, T., Laier, M., Lochbühler, B., Schüttler, A., Ebert, K., et al., 2015a. Effect on quality and composition of Riesling wines fermented by sequential inoculation with non-*Saccharomyces* and *Saccharomyces cerevisiae*. Eur. Food Res. Technol. 241 (5), 707–717.

Benito, Á., Calderón, F., Palomero, F., Benito, S., 2015b. Combine use of selected *Schizosaccharomyces pombe* and *Lachancea thermotolerans* yeast strains as an alternative to the traditional malolactic Fermentation in red wine production. Molecules 20, 9510–9523.

Benito, S., Palomero, F., Gálvez, L., Morata, A., Calderón, F., Palmero, D., et al., 2014. Quality and composition of red wine fermented with *Schizosaccharomyces pombe* as sole fermentative yeast, and in mixed and sequential fermentations with *Saccharomyces cerevisiae*. Food Technol. Biotechnol. 52 (3), 376–382.

Bokulich, N.A., Thorngate, J.H., Richardson, P.M., Mills, D.A., 2014. Microbial biogeography of wine grapes is conditioned by cultivar, vintage, and climate. Proc. Natl. Acad. Sci. U.S.A. 111 (1), E139–E148.

Bokulich, N.A., Collins, T.S., Masarweh, C., Allen, G., Heymann, H., Ebeler, S.E., et al., 2016. Associations among wine grape microbiome, metabolome, and fermentation behavior suggest microbial contribution to regional wine characteristics. MBio 7 (3), e00631-16.

Brown, S.L., Stockdale, V.J., Pettolino, F., Pocock, K.F., de Barros Lopes, M., Williams, P.J., et al., 2007. Reducing haziness in white wine by overexpression of *Saccharomyces cerevisiae* genes YOL155c and YDR055w. Appl. Microbiol. Biotechnol. 73 (6), 1363.

Cañas, P.M.I., García-Romero, E., Manso, J.M.H., Fernández-González, M., 2014. Influence of sequential inoculation of *Wickerhamomyces anomalus* and *Saccharomyces cerevisiae* in the quality of red wines. Eur. Food Res. Technol. 239 (2), 279–286.

Capece, A., Romaniello, R., Siesto, G., Pietrafesa, R., Massari, C., Poeta, C., et al., 2010. Selection of indigenous *Saccharomyces cerevisiae* strains for Nero d'Avola wine and evaluation of selected starter implantation in pilot fermentation. Int. J. Food Microbiol. 144 (1), 187–192.

Capozzi, V., Di Toro, M.R., Grieco, F., Michelotti, V., Salma, M., Lamontanara, A., et al., 2016. Viable but not culturable (VBNC) state of *Brettanomyces bruxellensis* in wine: new insights on molecular basis of VBNC behaviour using a transcriptomic approach. Food Microbiol. 59, 196–204.

Carvalho, E., Mateus, N., Plet, B., Pianet, I., Dufourc, E., De Freitas, V., 2006. Influence of wine pectic polysaccharides on the interactions between condensed tannins and salivary proteins. J. Agric. Food. Chem. 54 (23), 8936–8944.

Cavalieri, D., McGovern, P.E., Hartl, D.L., Mortimer, R., Polsinelli, M., 2003. Evidence for *S. cerevisiae* fermentation in ancient wine. J. Mol. Evol. 57 (Suppl 1), S226–S232.

Charpentier, C., Dos Santos, A.M., Feuillat, M., 2004. Release of macromolecules by *Saccharomyces cerevisiae* during ageing of French flor sherry wine "Vin jaune". Int. J. Food Microbiol. 96 (3), 253–262.

Chatonnet, P., Dubourdieu, D., Boidron, J.N., Lavigne, V., 1993. Synthesis of volatile phenols by *Saccharomyces cerevisiae* in wines. J. Sci. Food Agric. 62 (2), 191–202.

Chen, K., Escott, C., Loira, I., del Fresno, J.M., Morata, A., Tesfaye, W., et al., 2018. Use of non-*Saccharomyces* yeasts and oenological tannin in red winemaking: Influence on colour, aroma and sensorial properties of young wines. Food Microbiol. 69, 51–63.

Ciani, M., 1995. Continuous deacidification of wine by immobilized *Schizosaccharomyces pombe* cells: evaluation of malic acid degradation rate and analytical profiles. J. Appl. Microbiol. 79 (6), 631–634.

Ciani, M., Maccarelli, F., 1997. Oenological properties of non-*Saccharomyces* yeasts associated with wine-making. World J. Microbiol. Biotechnol. 14 (2), 199–203.

Ciani, M., Comitini, F., Mannazzu, I., Domizio, P., 2010. Controlled mixed culture fermentation: a new perspective on the use of non-*Saccharomyces* yeasts in winemaking. FEMS Yeast Res. 10 (2), 123–133. 9.

Ciani, M., Morales, P., Comitini, F., Tronchoni, J., Canonico, L., Curiel, J.A., et al., 2016. Non-conventional yeast species for lowering ethanol content of wines. Front. Microbiol. 7, 642.

Clavijo, A., Calderón, I.L., Paneque, P., 2010. Diversity of *Saccharomyces* and non-*Saccharomyces* yeasts in three red grape varieties cultured in the Serranía de Ronda (Spain) vine-growing region. Int. J. Food Microbiol. 143 (3), 241–245.

Comitini, F., Ingeniis, De, J., Pepe, L., Mannazzu, I., Ciani, M., 2004. *Pichia anomala* and *Kluyveromyces wickerhamii* killer toxins as new tools against *Dekkera/Brettanomyces* spoilage yeasts. FEMS Microbiol. Lett. 238 (1), 235–240.

Comitini, F., Gobbi, M., Domizio, P., Romani, C., Lencioni, L., Mannazzu, I., et al., 2011. Selected non-*Saccharomyces* wine yeasts in controlled multistarter fermentations with *Saccharomyces cerevisiae*. Food Microbiol. 28 (5), 873–882.

Contreras, A., Hidalgo, C., Schmidt, S., Henschke, P.A., Curtin, C., Varela, C., 2014. Evaluation of non-*Saccharomyces* yeasts for the reduction of alcohol content in wine. Appl. Environ. Microbiol. 80, 1670–1678.

Contreras, A., Curtin, C., Varela, C., 2015. Yeast population dynamics reveal a potential 'collaboration' between *Metschnikowia pulcherrima* and *Saccharomyces uvarum* for the production of reduced alcohol wines during Shiraz fermentation. Appl. Microbiol. Biotechnol. 99, 1885–1895.

Cordero-Bueso, G., Arroyo, T., Serrano, A., Tello, J., Aporta, I., Vélez, M.D., et al., 2011. Influence of the farming system and vine variety on yeast communities associated with grape berries. Int. J. Food Microbiol. 145 (1), 132–139.

Cordero-Bueso, G., Esteve-Zarzoso, B., Cabellos, J.M., Gil-Díaz, M., Arroyo, T., 2013. Biotechnological potential of non-*Saccharomyces* yeasts isolated during spontaneous fermentations of Malvar (*Vitis vinifera* cv. L.). Eur. Food Res. Technol. 236 (1), 193–207.

Cus, F., Jenko, M., 2013. The influence of yeast strains on the composition and sensory quality of Gewürztraminer wine. Food Technol. Biotechnol. 51 (4), 547.

De Beer, D., Joubert, E., Marais, J., Manley, M., 2017. Maceration before and during fermentation: Effect on Pinotage wine phenolic composition, total antioxidant capacity and objective color parameters. S. Afr. J. Enol. Vitic. 27 (2), 137–150.

Del Fresno, J.M., Morata, A., Loira, I., Bañuelos, M.A., Escott, C., Benito, S., et al., 2017. Use of non-*Saccharomyces* in single-culture, mixed and sequential fermentation to improve red wine quality. Eur. Food Res. Technol. 1–11.

Dizy, M., Bisson, L.F., 2000. Proteolytic activity of yeast strains during grape juice fermentation. Am. J. Enol. Vitic. 51 (2), 155–167.

Domizio, P., Romani, C., Lencioni, L., Comitini, F., Gobbi, M., Mannazzu, I., et al., 2011. Outlining a future for non-*Saccharomyces* yeasts: selection of putative spoilage wine strains to be used in association with *Saccharomyces cerevisiae* for grape juice fermentation. Int. J. Food Microbiol. 147 (3), 170–180.

Domizio, P., Liu, Y., Bisson, L.F., Barile, D., 2017. Cell wall polysaccharides released during the alcoholic fermentation by *Schizosaccharomyces pombe* and *S. japonicus*: quantification and characterization. Food Microbiol. 61, 136–149.

Englezos, V., Kalliopi, R., Torchio, F., Rolle, L., Gerbi, V., Cocolin, L., 2015. Exploitation of the non-*Saccharomyces* yeast *Starmerella bacillaris* (synonym *Candida zemplinina*) in wine fermentation: physiological and molecular characterizations. Int. J. Food Microbiol. 199, 33–40.

Englezos, V., Rantsiou, K., Cravero, F., Torchio, F., Ortiz-Julien, A., Gerbi, V., et al., 2016. *Starmerella bacillaris* and *Saccharomyces cerevisiae*. Appl. Microbiol. Biotechnol. 100 (12), 5515–5526.

Escott, C., Morata, A., Loira, I., Tesfaye, W., Suarez-Lepe, J.A., 2016. Characterization of polymeric pigments and pyranoanthocyanins formed in microfermentations of non-*Saccharomyces* yeasts. J. Appl. Microbiol. 121 (5), 1346–1356.

Escott, C., Del Fresno, J.M., Loira, I., Morata, A., Tesfaye, W., González, M.C., et al., 2018. Formation of polymeric pigments in red wines through sequential fermentation of flavanol-enriched musts with non-*Saccharomyces* yeasts. Food Chem. 239, 975–983.

Fay, J.C., Benavides, J.A., 2005. Evidence for domesticated and wild populations of *Saccharomyces cerevisiae*. PLoS Genet. 1 (1), e5.

Fia, G., Giovani, G., Rosi, I., 2005. Study of β-glucosidase production by wine-related yeasts during alcoholic fermentation. A new rapid fluorimetric method to determine enzymatic activity. J. Appl. Microbiol. 99 (3), 509–517.

Fleet, G.H., Heard, G.M., 1993. Yeasts-growth during fermentation. In: Fleet, G.H. (Ed.), Wine Microbiology and Biotechnology. Harwood Academic Publishers, Chur, Switzerland, pp. 27–54.

Fleet, G.H., Prakitchaiwattana, C., Beh, A.L., Heard, G., 2002. The yeast ecology of wine grapes. In: Ciani, M. (Ed.), Biodiversity and Biotechnology of Wine Yeast. Research Signpost, Trivandrum, Kerala, India, pp. 1–17.

Francesca, N., Romano, R., Sannino, C., Le Grottaglie, L., Settanni, L., Moschetti, G., 2014. Evolution of microbiological and chemical parameters during red wine making with extended post-fermentation maceration. Int. J. Food Microbiol. 171, 84–93.

Fugelsang, K.C., Zoecklein, B.W., 2003. Population dynamics and effects of *Brettanomyces bruxellensis* strains on Pinot noir (*Vitis vinifera* L.) wines. Am. J. Enol. Vitic. 54 (4), 294–300.

Ganga, M.A., Martínez, C., 2004. Effect of wine yeast monoculture practice on the biodiversity of non-*Saccharomyces* yeasts. J. Appl. Microbiol. 96 (1), 76–83.

Gil, M., Estevez, S., Kontoudakis, N., Fort, F., Canals, J.M., Zamora, F., 2013. Influence of partial dealcoholization by reverse osmosis on red wine composition and sensory characteristics. Eur. Food Res. Technol. 237, 481–488.

Giovani, G., Rosi, I., 2007. Release of cell wall polysaccharides from *Saccharomyces cerevisiae* thermosensitive autolytic mutants during alcoholic fermentation. Int. J. Food Microbiol. 116 (1), 19–24.

Giovani, G., Rosi, I., Bertuccioli, M., 2012. Quantification and characterization of cell wall polysaccharides released by non-*Saccharomyces* yeast strains during alcoholic fermentation. Int. J. Food Microbiol. 160 (2), 113–118.

Gobbi, M., Comitini, F., Domizio, P., Romani, C., Lencioni, L., Mannazzu, I., et al., 2013. *Lachancea thermotolerans* and *Saccharomyces cerevisiae* in simultaneous and sequential co-fermentation: a strategy to enhance acidity and improve the overall quality of wine. Food Microbiol. 33 (2), 271–281.

Gobbi, M., De Vero, L., Solieri, L., Comitini, F., Oro, L., Giudici, et al., 2014. Fermentative aptitude of non *Saccharomyces* wine yeast for reduction in the ethanol content in wine. Eur. Food Res. Technol. 239, 41–48.

Gonzalez-Ramos, D., Cebollero, E., Gonzalez, R., 2008. A recombinant *Saccharomyces cerevisiae* strain overproducing mannoproteins stabilizes wine against protein haze. Appl. Environ. Microbiol. 74 (17), 5533–5540.

Gonzalez, R., Quirós, M., Morales, P., 2013. Yeast respiration of sugars by non-*Saccharomyces* yeast species: a promising and barely explored approach to lowering alcohol content of wines. Trends Food Sci. Technol. 29, 55–61.

González-Royo, E., Pascual, O., Kontoudakis, N., Esteruelas, M., Esteve-Zarzoso, B., Mas, A., et al., 2015. Oenological consequences of sequential inoculation with non-*Saccharomyces* yeasts (*Torulaspora delbrueckii* or *Metschnikowia pulcherrima*) and *Saccharomyces cerevisiae* in base wine for sparkling wine production. Eur. Food Res. Technol. 240 (5), 999–1012.

Guadalupe, Z., Martínez, L., Ayestarán, B., 2010. Yeast mannoproteins in red winemaking: effect on polysaccharide, polyphenolic, and color composition. Am. J. Enol. Vitic. 61 (2), 191–200.

Guerrero, R.F., Cantos-Villar, E., 2015. Demonstrating the efficiency of sulphur dioxide replacements in wine: A parameter review. Trends in Food Sci. Technol 42 (1), 27–43.

Heard, G.M., Fleet, G.H., 1988. The effects of temperature and pH on the growth of yeast species during the fermentation of grape juice. J. Appl. Bacteriol. 65, 23–28.

Heresztyn, T., 1986. Formation of substituted tetrahydropyridines by species of *Brettanomyces* and *Lactobacillus* isolated from mousy wines. Am. J. Enol. Vitic. 37 (2), 127–132.

Hernandez-Orte, P., Cersosimo, M., Loscos, N., Cacho, J., Garcia-Moruno, E., Ferreira, V., 2008. The development of varietal aroma from non-floral grapes by yeasts of different genera. Food Chem. 107 (3), 1064–1077.

Hierro, N., González, Á., Mas, A., Guillamón, J.M., 2005. Diversity and evolution of non-*Saccharomyces* yeast populations during wine fermentation: effect of grape ripeness and cold maceration. FEMS Yeast Res. 6 (1), 102–111.

Hu, K., Qin, Y., Tao, Y.S., Zhu, X.L., Peng, C.T., Ullah, N., 2016. Potential of glycosidase from non-*Saccharomyces* isolates for enhancement of wine aroma. J. Food Sci. 81 (4), 935–943.

Jolly, N.P., Augustyn, O.P.H., Pretorius, I.S., 2003. The occurrence of non-*Saccharomyces cerevisiae* yeast species over three vintages in four vineyards and grape musts from four production regions of the Western Cape, South Africa. S. Afr. J. Enol. Vitic. 24 (2), 35–42.

Jolly, N.P., Augustyn, O.P.H., Pretorius, I.S., 2006. The role and use of non-*Saccharomyces* yeasts in wine production. S. Afr. J. Enol. Vitic. 27 (1), 2006.

Jolly, N.P., Varela, C., Pretorius, I.S., 2014. Not your ordinary yeast: non-*Saccharomyces* yeasts in wine production uncovered. FEMS Yeast Res. 14 (2), 215–237.

Jones, G.V., White, M.A., Cooper, O.R., Storchmann, K., 2005. Climate change and global wine quality. Clim. Change 73, 319–343.

Kapsopoulou, K., Mourtzini, A., Anthoulas, M., Nerantzis, E., 2007. Biological acidification during grape must fermentation using mixed cultures of *Kluyveromyces thermotolerans* and *Saccharomyces cerevisiae*. World J. Microbiol. Biotechnol. 23 (5), 735–739.

Kurita, O., 2008. Increase of acetate ester-hydrolysing esterase activity in mixed cultures of *Saccharomyces cerevisiae* and *Pichia anomala*. J. Appl. Microbiol. 104 (4), 1051–1058.

Kurtzman, C., Fell, J.W. (Eds.), 1998. The Yeasts-A Taxonomic Study. Elsevier.

Kurtzman, C.P., Fell, J.W., 2006. Yeast systematics and phylogeny—implications of molecular identification methods for studies in ecology. In Biodiversity and ecophysiology of yeasts, Springer, Berlin, Heidelberg 11–30.

Lencioni, L., Romani, C., Gobbi, M., Comitini, F., Ciani, M., Domizio, P., 2016. Controlled mixed fermentation at winery scale using *Zygotorulaspora florentina* and *Saccharomyces cerevisiae*. Int. J. Food Microbiol. 234, 36–44.

Lleixà, J., Martín, V., Portillo, M.D.C., Carrau, F., Beltran, G., Mas, A., 2016. Comparison of fermentation and wines produced by inoculation of *Hanseniaspora vineae* and *Saccharomyces cerevisiae*. Front. Microbiol. 7.

López, M.C., Mateo, J.J., Maicas, S., 2015. Screening of β-glucosidase and β-xylosidase activities in four non-*Saccharomyces* yeast isolates. J. Food Sci. 80 (8), 455–481.

MacAvoy, M.G., 2010. Wine—harmful or healthy? What is being considered in Australia and New Zealand?. In: Proceedings of Fourteenth Australian Wine Industry Technical Conference. AWITC Inc, Glen Osmond, Adelaide, South Australia, Australia.

Magyar, I., Toth, T., 2011. Comparative evaluation of some oenological properties in wine strains of *Candida stellata*, *Candida zemplinina*, *Saccharomyces uvarum* and *Saccharomyces cerevisiae*. Food Microbiol. 28, 94–100.

Martini, G., 1993. Origin and domestication of the wine yeast *Saccharomyces cerevisiae*. J. Wine Res. 4, 165–176.

McGovern, P.E., Zhang, J., Tang, J., Zhang, Z., Hall, G.R., 2004. Fermented beverages of pre- and proto-historic China. Proc. Natl. Acad. Sci. U. S.A. 101, 17593–17598. A1011759317598.

Medina, K., Boido, E., Fariña, L., Gioia, O., Gomez, M.E., Barquet, M., et al., 2013. Increased flavour diversity of Chardonnay wines by spontaneous fermentation and co-fermentation with *Hanseniaspora vineae*. Food Chem. 141 (3), 2513–2521.

Medina, K., Boido, E., Fariña, L., Dellacassa, E., Carrau, F., 2016. Non-*Saccharomyces* and *Saccharomyces* strains co-fermentation increases acetaldehyde accumulation: effect on anthocyanin-derived pigments in Tannat red wines. Yeast 33 (7), 339–343.

Medina, K., Boido, E., Dellacassa, E., Carrau, F., 2017. Effects of non-*Saccharomyces* yeasts on color, anthocyanin and anthocyanin-derived pigments of Tannat grapes during fermentation. Am. J. Enol. Vitic. 5, 752–763.

Mehlomakulu, N.N., Setati, M.E., Divol, B., 2014. Characterization of novel killer toxins secreted by wine-related non-*Saccharomyces* yeasts and their action on *Brettanomyces* spp. Int. J. Food Microbiol. 188, 83–91.

Milanović, V., Comitini, F., Ciani, M., 2013. Grape berry yeast communities: influence of fungicide treatments. Int. J. Food Microbiol. 161 (3), 240–246.

Mills, D.A., Johannsen, E.A., Cocolin, L., 2002. Yeast diversity and persistence in Botrytis-affected wine fermentations. Appl. Environ. Microbiol. 68 (10), 4884–4893.

Mira de Orduña, R., 2010. Climate change associated effects on grape and wine quality and production. Food Res. Int. 43, 1844–1855.

Morata, A., Gómez-Cordovés, M.C., Colomo, B., Suárez, J.A., 2003. Pyruvic acid and acetaldehyde production by different strains of *Saccharomyces cerevisiae*: relationship with vitisin A and B formation in red wines. J. Agric. Food. Chem. 51, 6475–6481.

Moreira, N., Mendes, F., De Pinho, P.G., Hogg, T., Vasconcelos, I., 2008. Heavy sulphur compounds, higher alcohols and esters production profile of *Hanseniaspora uvarum* and *Hanseniaspora guilliermondii* grown as pure and mixed cultures in grape must. Int. J. Food Microbiol. 124 (3), 231–238.

Moreira, N., Pina, C., Mendes, F., Couto, J.A., Hogg, T., Vasconcelos, I., 2011. Volatile compounds contribution of *Hanseniaspora guilliermondii* and *Hanseniaspora uvarum* during red wine vinifications. Food Control 22 (5), 662–667.

Moreno, P.I., Vilanova, I., Villa-Martínez, R., Garreaud, R.D., Rojas, M., De Pol-Holz, R., 2014. Southern Annular Mode-like changes in southwestern Patagonia at centennial timescales over the last three millennia. Nat. Commun. 5, 4375.

Morrissey, W.F., Davenport, B., Querol, A., Dobson, A.D.W., 2004. The role of indigenous yeasts in traditional Irish cider fermentations. J. Appl. Microbiol. 97 (3), 647–655.

Mortimer, R., Polsinelli, M., 1999. On the origins of wine yeast. Res. Microbiol. 150, 199–204.

Muccilli, S., Restuccia, C., 2015. Bioprotective role of yeasts. Microorganisms 3 (4), 588–611.

Munyon, J.R., Nagel, C.W., 1977. Comparison of methods of deacidification of musts and wines. Am. J. Enol. Vitic. 28 (2), 79–87.

Naumova, E.S., Bulat, S.A., Mironenko, N.V., Naumov, G.I., 2003. Differentiation of six sibling species in the *Saccharomyces* sensu stricto complex by multilocus enzyme electrophoresis and UP-PCR analysis. Antonie Van Leeuwenhoek 83, 155–166.

Oro, L., Ciani, M., Comitini, F., 2014. Antimicrobial activity of *Metschnikowia pulcherrima* on wine yeasts. J. Appl. Microbiol. 116 (5), 1209–1217.

Padilla, B., Gil, J.V., Manzanares, P., 2016. Past and future of non-*Saccharomyces* yeasts: from spoilage microorganisms to biotechnological tools for improving wine aroma complexity. Front. Microbiol. 7, 411.

Palomero, F., Morata, A., Benito, S., Calderón, F., Suárez-Lepe, J.A., 2009. New genera of yeasts for over-lees aging of red wine. Food Chem. 112 (2), 432–441.

Phister, T.G., Mills, D.A., 2003. Real-time PCR assay for detection and enumeration *of Dekkera bruxellensis* in wine. Appl. Environ. Microbiol. 69 (12), 7430–7434.

Pinto, M., Coelho, E., Nunes, A., Brandão, T., Coimbra, M.A., 2015. Valuation of brewers spent yeast polysaccharides: a structural characterization approach. Carbohydr. Polym. 116, 215–222.

Plata, C., Millan, C., Mauricio, J.C., Ortega, J.M., 2003. Formation of ethyl acetate and isoamyl acetate by various species of wine yeasts. Food Microbiol. 20 (2), 217–224.

Pretorius, I.S., 2000. Tailoring wine yeast for the new millennium: novel approaches to the ancient art of winemaking. Yeast 16 (8), 675–729. 15.

Pretorius, I.S., Van der Westhuizen, T.J., Augustyn, O.P.H., 1999. Yeast biodiversity in vineyards and wineries and its importance to the South African wine industry. A review. S. Afr. J. Enol. Vitic. 20 (2), 61–70.

Rankine, B.C., 1966. Decomposition of L-malic acid by wine yeasts. J. Sci. Food Agric. 17 (7), 312–316.

Raspor, P., Milek, D.M., Polanc, J., Možina, S.S., Čadež, N., 2006. Yeasts isolated from three varieties of grapes cultivated in different locations of the Dolenjska vine-growing region, Slovenia. Int. J. Food Microbiol. 109 (1), 97–102.

Regueiro, L.A., Costas, C.L., Rubio, J.E.L., 1993. Influence of viticultural and enological practices on the development of yeast populations during winemaking. Am. J. Enol. Vitic. 44 (4), 405–408.

Renault, P., Coulon, J., de Revel, G., Barbe, J.C., Bely, M., 2015. Increase of fruity aroma during mixed *Torulaspora delbrueckii/Saccharomyces cerevisiae* wine fermentation is linked to specific esters enhancement. Int. J. Food Microbiol. 207, 40–48.

Renouf, V., Lonvaud-Funel, A., 2007a. Development of an enrichment medium to detect *Dekkera/Brettanomyces bruxellensis*, a spoilage wine yeast, on the surface of grape berries. Microbiol. Res. 162 (2), 154–167.

Renouf, V., Claisse, O., Lonvaud-Funel, A., 2007b. Inventory and monitoring of wine microbial consortia. Appl. Microbiol. Biotechnol. 75 (1), 149−164.

Rodrigues, N., Gonçalves, G., Pereira-da-Silva, S., Malfeito-Ferreira, M., Loureiro, V., 2001. Development and use of a new medium to detect yeasts of the genera Dekkera/Brettanomyces. J. Appl. Microbiol. 90 (4), 588−599.

Rodríguez, M.E., Lopes, C.A., Barbagelata, R.J., Barda, N.B., Caballero, A.C., 2010. Influence of Candida pulcherrima Patagonian strain on alcoholic fermentation behaviour and wine aroma. Int. J. Food Microbiol. 138 (1), 19−25.

Rojas, V., Gil, J.V., Piñaga, F., Manzanares, P., 2003. Acetate ester formation in wine by mixed cultures in laboratory fermentations. Int. J. Food Microbiol. 86 (1), 181−188.

Rolle, L., Englezos, V., Torchio, F., Cravero, F., Río Segade, S., Rantsiou, K., et al., 2017. Alcohol reduction in red wines by technological and microbiological approaches: a comparative study. Aust. J. Grape Wine Res. 1, 127−136.

Rosini, G., Federici, F., Martini, A., 1982. Yeast flora of grape berries during ripening. Microb. Ecol. 8, 83−89.

Sacchi, K.L., Bisson, L.F., Adams, D.O., 2005. A review of the effect of winemaking techniques on phenolic extraction in red wines. Am. J. Enol. Vitic. 56 (3), 197−206.

Salvetti, E., Campanaro, S., Campedelli, I., Fracchetti, F., Gobbi, A., Tornielli, G.B., et al., 2016. Whole-metagenome-sequencing-based community profiles of Vitis vinifera L. cv. Corvina berries withered in two post-harvest conditions. Front. Microbiol. 7.

Santos, A., Navascués, E., Bravo, E., Marquina, D., 2011. Ustilago maydis killer toxin as a new tool for the biocontrol of the wine spoilage yeast Brettanomyces bruxellensis. Int. J. Food Microbiol. 145 (1), 147−154.

Schuller, D., Alves, H., Dequin, S., Casal, M., 2005. Ecological survey of Saccharomyces cerevisiae strains from vineyards in the Vinho Verde Region of Portugal. FEMS Microbiol. Ecol. 51 (2), 167−177.

Schumaker, M.R., Chandra, M., Malfeito-Ferreira, M., Ross, C.F., 2017. Influence of Brettanomyces ethylphenols on red wine aroma evaluated by consumers in the United States and Portugal. Food Res. Int. 100, 161−167.

Silva, S., Ramón-Portugal, F., Andrade, P., Abreu, S., de Fatima Teixeira, M., Strehaiano, P., 2003. Malic acid consumption by dry immobilized cells of Schizosaccharomyces pombe. Am. J. Enol. Vitic. 54 (1), 50−55.

Snow, P.G., Gallander, J.F., 1979. Deacidification of white table wines through partial fermentation with Schizosaccharomyces pombe. Am. J. Enol. Vitic. 30 (1), 45−48.

Strauss, M.L.A., Jolly, N.P., Lambrechts, M.G., Van Rensburg, P., 2001. Screening for the production of extracellular hydrolytic enzymes by non-Saccharomyces wine yeasts. J. Appl. Microbiol. 91 (1), 182−190.

Suárez-Lepe, J.A., Palomero, F., Benito, S., Calderón, F., Morata, A., 2012. Oenological versatility of Schizosaccharomyces spp. Eur. Food Res. Technol. 235, 375−383.

Swiegers, J.H., Bartowsky, E.J., Henschke, P.A., Pretorius, I.S., 2005. Yeast and bacterial modulation of wine aroma and flavour. Aust. J. Grape Wine Res. 11 (2), 139−173.

Teresa, F.-E.M., Barrio, E., Querol, A., 2003. Analysis of the genetic variability in the species of the Saccharomyces sensu stricto complex. Yeast 20, 1213−1226.

Tristezza, M., Tufariello, M., Capozzi, V., Spano, G., Mita, G., Grieco, F., 2016. The oenological potential of Hanseniaspora uvarum in simultaneous and sequential co-fermentation with Saccharomyces cerevisiae for industrial wine production. Front. Microbiol. 7, 670.

Valero, E., Schuller, D., Cambon, B., Casal, M., Dequin, S., 2005. Dissemination and survival of commercial wine yeast in the vineyard: a large-scale, three-years study. FEMS Yeast Res. 5 (10), 959−996.

Valero, A., Begum, M., Leong, S.L., Hocking, A.D., Ramos, A.J., Sanchis, V., et al., 2007. Effect of germicidal UVC light on fungi isolated from grapes and raisins. Lett. Appl. Microbiol. 45 (3), 238−243.

Varela, C., Sengler, F., Solomon, M., Curtin, C., 2016. Volatile flavour profile of reduced alcohol wines fermented with the non-conventional yeast species Metschnikowia pulcherrima and Saccharomyces uvarum. Food Chem. 209, 57−64.

Varela, C., Barker, A., Tran, T., Borneman, A., Curtin, C., 2017. Sensory profile and volatile aroma composition of reduced alcohol Merlot wines fermented with Metschnikowia pulcherrima and Saccharomyces uvarum. Int. J. Food Microbiol. 252, 1−9.

Verginer, M., Leitner, E., Berg, G., 2010. Production of volatile metabolites by grape-associated microorganisms. J. Agric. Food. Chem. 58 (14), 8344−8350.

Viana, F., Gil, J.V., Genovés, S., Vallés, S., Manzanares, P., 2008. Rational selection of non-Saccharomyces wine yeasts for mixed starters based on ester formation and enological traits. Food Microbiol. 25 (6), 778−785.

Vidal, S., Francis, L., Williams, P., Kwiatkowski, M., Gawel, R., Cheynier, V., et al., 2004. The mouth-feel properties of polysaccharides and anthocyanins in a wine like medium. Food Chem. 85 (4), 519−525.

Villalba, M.L., Sáez, J.S., del Monaco, S., Lopes, C.A., Sangorrín, M.P., 2016. TdKT, a new killer toxin produced by Torulaspora delbrueckii effective against wine spoilage yeasts. Int. J. Food Microbiol. 217, 94−100.

Vivas, N.G., Nonier, M.-F., Guerra, C., Vivas, N., 2001. Anthocyanin in grape skins during maturation of Vitis vinifera L. cv. Cabernet Sauvignon and Merlot Noir from different bordeaux terroirs. J. Int. Sci. Vigne Vin 35, 149−156.

Zarraonaindia, I., Owens, S.M., Weisenhorn, P., West, K., Hampton-Marcell, J., Lax, S., et al., 2015. The soil microbiome influences grapevine-associated microbiota. MBio 6 (2), e02527-14.

Zott, K., Miot-Sertier, C., Claisse, O., Lonvaud-Funel, A., Masneuf-Pomarede, I., 2008. Dynamics and diversity of non-Saccharomyces yeasts during the early stages in winemaking. Int. J. Food Microbiol. 125 (2), 197−203.

Zott, K., Thibon, C., Bely, M., Lonvaud-Funel, A., Dubourdieu, D., Masneuf-Pomarede, I., 2011. The grape must non-Saccharomyces microbial community: impact on volatile thiol release. Int. J. Food Microbiol. 151 (2), 210−215.

Zupan, J., Avbelj, M., Butinar, B., Kosel, J., Šergan, M., Raspor, P., 2013. Monitoring of quorum-sensing molecules during minifermentation studies in wine yeast. J. Agric. Food. Chem. 61 (10), 2496−2505.

CHAPTER

5

Yeast Biotechnology for Red Winemaking

Karina Medina, Valentina Martin, Eduardo Boido and Francisco Carrau

Enology and Fermentation Biotechnology Area, Food Science and Technology Department, School of Chemistry of the Universidad de la Republica, Montevideo, Uruguay

5.1 INTRODUCTION

Yeasts are the key microbial cells that transform grape juice into wine. As grape juice is affected by many variables such as soil, climate, harvest, and winemaking procedures, etc., research on this topic was always considered a complex challenge for microbiologists and winemakers. However, this challenge of controlling and understanding wine fermentation has been the ideal workshop system to learn about fermentation biotechnology. Louis Pasteur is the traditional example of how grape and wine knowledge can contribute to opening up new areas of fermentation technology and creating cutting edge microbial biotechnologies.

In red wine fermentation, yeast species are subject to more complex effects because the maceration process introduces less reductive conditions compared with white wine fermentation. The effect of higher polyphenol content due to skin and seed contact are also expected to increase the stress pressure for yeast cells. These compounds, together with high sugar content, low pH, high temperature conditions, and increased alcohol concentration during the process, result in a very stressful and selective pressure condition.

Within yeast species, *Saccharomyces* strains are the main group that can survive and contribute to wine fermentation. It is known that this genus uses the wine fermentation niche as an ideal environment to compete against other species that are less efficient fermenters. Alcohol and CO_2 are the main compounds that can change the original environment of grape juice to a more aggressive and selective medium.

The origin of *Saccharomyces cerevisiae* in grape and wine was a topic of discussion in the 1980s, when it was demonstrated that winery and cellar equipment was the main source of strains of this species. However, it is also known that there are natural *Saccharomyces* flora in vineyards, which are dependent on the year, climate, region, and vineyard management (Bokulich et al., 2014). We are also going to show here that they are dependent on red grape variety, with our studies on Tannat red grape, and the initial steps of fermentation.

The third important compound produced by *S. cerevisiae* in terms of concentration is glycerol. Red wines usually always have a higher concentration of glycerol than white wines, even doubling the level up to 12 g/L (Nieuwoudt et al., 2002). It is well known that this is caused by various physicochemical factors in the red wine maceration process and *Saccharomyces* cell metabolism. In red wine fermentation, glycerol synthesis in *Saccharomyces* is stimulated by traditional practices such as increasing aeration, temperature of fermentation, and osmotic pressure by sugars and a higher pH level at harvest (Oreglia, 1978). Consumption of malic acid by *Saccharomyces* strains during red winemaking is another attractive yeast characteristic for red wine production (Delcourt et al., 1995). This might reduce the malolactic fermentation (MLF) process and result in a lower risk of adverse outcomes such as losses of color and varietal flavors, or histamine production (Lonvaud-Funel, 1999; Bartowsky and Henschke, 2004).

From a sensory point of view, studies on flavor compound thresholds in red wines are very limited (Swiegers et al., 2005). However, it is known that threshold values in red wines are usually significantly higher than in white wines (Guth and Sies, 2002). One of the first studies on this topic in red wines concerned the presence of diacetyl (Rankine et al., 1969). The diacetyl flavor threshold was found to be 0.2 mg/L in white Chardonnay, but 2.8 mg/L in red Cabernet sauvignon (Martineau et al., 1995). Dimethyl sulfide, for example, is a nice and complex flavor that appears after a few years in bottled wines and has been described as corn and cooked notes; it is noticed in white wine above 25 μg/L compared with 60 μg/L in red wine.

5.2 YEAST DIVERSITY IN RED GRAPES AND MUSTS

There are two distinct ecosystems within the yeast habitat where winemaking processes occur. The vineyard (surface of the grapes), where there is a certain community of yeasts, and the winery (equipment, grape must, and wine), where there is another community of yeasts. The natural population of vineyards may vary according to the degree of maturity and/or sanity of the grape, e.g., the more mature the fruit, the higher the level of sugars and other nutrients that increase yeast population. Many other factors will affect yeast population, such as grape variety (thickness of berry skin, tight or loose bunches), climate (temperature, rainfall, humidity), and vineyard management (canopy management, fertilization, irrigation, use of fungicides, etc.). On the other hand, the community within the winery ecosystem will be affected by winery hygiene procedures, grape must characteristics (pH, concentration of sugars), and winemakers' influences (addition of SO_2 and enzymes, racking, cooling, maceration management, etc.).

It was once thought that the fermentation process, by which grape juice becomes wine, was spontaneously caused by the action of the yeasts present on the surface of the grape. In 1970, it was theorized that *S. cerevisiae* wine yeast cells are ubiquitous in nature, with a tangible propensity for vineyard soils where they thrive on grapes fallen from vines (Martini et al., 1996). Finally, it was determined that the yeast of the wine "par excellence" is isolated with extreme difficulty from conventional habitats, such as the vineyard floor or the surface of mature grapes, while it is considered the main species that colonizes the surfaces of the cellar. When *S. cerevisiae* is isolated directly from grapes, preculture in a rich liquid medium was needed to recover its very low population (Martini, 1993).

It is thought that *S. cerevisiae* strains that start the first fermentation of grape must in a particular new winery undergo radical modifications of their enobiotechnological characteristics over time. Some data indicate that a genetic selection occurs in each winery, leading to progressive adaptations of the resident population of *S. cerevisiae* to extreme grape must conditions (Martini, 2003). This natural population will be affected by the application of commercial strains, even into the vineyard surroundings (Valero et al., 2005).

We have made isolations from different varieties of grapes, during successive harvests in different Uruguayan regions. Among the most commonly planted red grape varieties are Tannat, the flagship wine of the country, and Merlot, as the second in quantity. In Fig. 5.1 we can see the diversity of yeasts obtained from 10 different vineyards of each variety. In both cases, 13 different species were identified. In the case of Merlot, all the yeasts identified were non-*Saccharomyces* (first *Hanseniaspora uvarum*, followed by *Issatchenkia terricola*). In contrast, Tannat grapes were dominated by *Metschnikowia pulcherrima*, followed by *Hanseniaspora uvarum*. Interestingly, within the yeast isolates of Tannat variety, 3% were *S. cerevisiae* native strains, although these usually represent less than 1% of the population present on grape surfaces (Martini, 1993). The effects of variety on yeast natural populations will require further research, as this is obviously another "terroir" factor that it is necessary to better understand in wine regions (Bokulich et al., 2014).

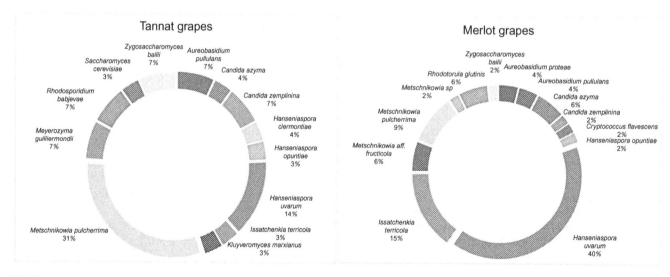

FIGURE 5.1 Yeast diversity isolated from grape berries at harvest time, average percentages of species found in two consecutive vintages. Results for Tannat and Merlot red, the main varieties of Uruguayan vineyards. *Saccharomyces cerevisiae* was isolated in Tannat grapes at a very high percentage compared to other reports that are usually under 1%.

5.3 INFLUENCE OF RED WINE TECHNOLOGY ON SACCHAROMYCES STRAINS

5.3.1 *Saccharomyces* Strains Dominate in the Wine Ecosystem; *Saccharomyces* Specific Niche

Saccharomyces strains are known to build the ideal ecosystem in a winery to compete with other yeasts that are less efficient in fermentation. The production of higher rates of CO_2 and alcohol, together with the heat produced by fermentation and a rapid removal of yeast-assimilable nitrogen (YAN), favors the formation of a selective environment ideal for *Saccharomyces* strains (Goddard, 2008). However, the presence of skins and seeds significantly affects the population dynamics of the yeast cells, and makes it more difficult for *Saccharomyces* to dominate the natural flora of grapes. These solids delay the formation of an anaerobic environment, and as will be discussed, the size and style of tank for grape maceration significantly affect this situation.

5.3.2 Nitrogen Competition During Winemaking

It is well known that the YAN can significantly affect fermentation rate (Jiranek et al., 1991) and the final flavor profile of wine fermentation (Carrau et al., 2008). There are some particular situations that happen with *Saccharomyces* industrial strains caused by the appearance of hydrogen sulfide notes at the end of fermentation in red winemaking processes; this is due to low YAN levels at this step (Henschke, 1996; Park et al., 2000; Sweigers et al., 2005). The presence of a higher diversity of yeast species at the beginning of fermentation of red grape musts due to the combination of whole berries and a lack of settling processes such as in white wines, promotes cell competition for YAN in the juice. Some of the more active strains at the beginning of fermentation usually are not the most efficient in finishing the fermentation and they remove YAN and retain it inside their cells for many days (Medina et al., 2012). This is particular important in the competition of the native or inoculated *Saccharomyces* strains against the natural non-*Saccharomyces* flora from the grapes (Medina et al., 2012).

Paradoxically, adding higher levels of di-ammonium phosphate (DAP) to red wines is usually considered not ideal for red-color stability in red wines. Furthermore, the residual nitrogen excess could affect microbial stability during the following steps of barrel aging and bottling. Moreover, key compounds that are related to stable derived anthocyanin formation in red wine, such as acetaldehyde and vinyl phenols, are negatively sensitive to DAP addition (Carrau et al., 2008). So the use of DAP to prevent the formation of hydrogen sulfide will be a delicate practice in red winemaking. In a *Saccharomyces* strain screening program for red winemaking, low nitrogen-demand strains should be a key characteristic to evaluate. Interestingly, in practice, formation of higher levels of hydrogen sulfide are particularly found at the end of fermentation in some red grape varieties such as Merlot. This might be due to a particular amino acid profile in these varieties. These statements also show that a low YAN level in the selection medium for screening new yeast strains is not only important for aroma compounds, as discussed by Carrau et al. (2015), but also for pigment and red wine aging potential.

5.3.3 Redox and Temperature Effects in Red Winemaking

It was demonstrated that under microaerobic conditions the synthesis of esters is decreased and alcohols are increased, so esters will have a more moderate effect on flavor profile in red winemaking compared with white wines. Furthermore, smaller fermentation tanks or barrels create more extreme microaerobic conditions during vinification. Microaerobic conditions offer a better environment for non-*Saccharomyces* species, so they might compete with *Saccharomyces* strains for a longer period in fermentation (Fleet, 2003).

On the other hand, the higher temperatures (above 20°C) usually reached in red wine batch fermentations do not stimulate ester formation and retention in the medium (Ramey and Ough, 1980). These conditions also favor premature cell death in the majority of non-*Saccharomyces* strains, leaving a more comfortable environment for *Saccharomyces* performance (Fleet, 2003). However, the development of non-*Saccharomyces* strains together with *Saccharomyces* at the beginning of vinification results in a slower rate of fermentation; this has the advantage of decreasing heat production and thus demanding less energy for cooling the system (Medina et al., 2013).

5.3.4 Alcohol and Polyphenol Contents in Red Winemaking

Although during the whole-berry grapes maceration in red winemaking there is increased presence of yeast cells of many species, red grapes usually contain higher levels of sugars that increase osmotic pressure and the potential ethanol level in the vinification process. These factors might accelerate cell death of many non-*Saccharomyces* cells

(Henschke, 1997). Polyphenol compounds of red grape musts are extracted from skin and seeds, and they are modified by yeast and fermentation conditions. They also act as active antimicrobial compounds against many bacteria and non-*Saccharomyces* species (Swiegers et al., 2005). In addition, it is known that tannins are strong inhibitors of the synthesis of volatile phenols (Chatonnet et al., 1997), including *Saccharomyces* carboxymethyl decarboxylases (Smit et al., 2003).

5.3.5 *Saccharomyces cerevisiae* and Red Wine Color

Wine color depends on grape pigments and how those pigments are modified during vinification, storage, and aging (Boulton, 2001; Schwarz et al., 2003). It has been demonstrated that yeast interacts with anthocyanins in different ways during fermentation. The interaction of yeast cell walls results in anthocyanin adsorption (Morata et al., 2003b; Vasserot et al., 1997). This is a very important consideration for the wine industry, because some yeast strains might remove and decrease the color of red wines due to high color-adsorption events. This phenomenon is now well understood and yeast selection programs for red wine should include tests to measure the adsorption characteristics of each strain.

5.3.6 Cell Wall Adsorption and Cell Lysis Effects on Anthocyanins

This topic started to be studied in 1976, when a group of French researchers observed color differences between red wines fermented with different strains of yeast (Bourzeix and Heredia, 1976). Two explanations were considered at that moment. One of them referred to the existence of different anthocyanin extraction potentials from the grapes, depending on the strain of yeast used. This idea was discarded, since it was observed that there is no specific enzymatic activity in yeast that will differentially extract polyphenolic compounds from the grape berry skin. The other possible explanation of color differences in red wines was the different absorption abilities among yeast cell walls. Dynamics of yeast—anthocyanin interaction is explained by the following three mechanisms:

5.3.6.1 Cell Wall Anthocyanins Adsorption

This topic was taken up by other authors, with microvinifications of other varieties (Castino, 1982; Cuinier, 1988). The results demonstrated that wines obtained after alcoholic fermentation, made with the same grape must but with different yeast, have different colors and also different lees. Subsequent assays confirmed anthocyanin removal from an extensive variety of red wines, attributing this to mechanisms not yet fully elucidated (Loiseau et al., 1994; Vasserot et al., 1997; Conterno et al., 1997; Cuinier, 1997; Morata et al., 2003b, 2005, 2016; Caridi et al., 2004; Medina et al., 2005).

In some of the previous studies, the yeast cell wall was posited as the responsible cellular structure. It has been shown that yeast cell envelopes possess the ability to retain many of the compounds present in wine, such as volatile compounds (Lubbers et al., 1994a), fatty acids (Larue et al., 1984), and polyphenolic compounds (Vasserot et al., 1997). The consequence of this retention is the loss of these substances once the lees are removed from the wines. Another factor related to anthocyanins adsorption is cell wall porosity. It is maximal in the exponential growth phase and tends to fall rapidly in later phases (Boivin et al., 1998). During the stage in which the porosity is greater, the formation of disulfide bridges (De Nobel et al., 1990) and hydrogen bridges (Vasserot et al., 1997) is facilitated, so the exponential stage is where there is the greatest loss of anthocyanins in wines. It is known that large inoculum sizes are not ideal for flavor synthesis in white winemaking (Carrau et al., 2010). Further studies on how inoculum size affects anthocyanin red wine composition would be interesting.

The surface provided by the interstitial spaces of the cell wall also favors adsorption. The area of yeast cells during alcoholic fermentation is greater than $10\,m^2/L$ of must and therefore the amount of anthocyanins adsorbed during fermentation can become very large (Morata et al., 2003b). According to some authors (Chassagne et al., 2005; Morata et al., 2003b), acylated anthocyanin monomers, acetates, and coumarates are strongly adsorbed by the yeast cell wall. Straightforward experimental procedures have allowed strains with the low anthocyanin adsorption trait ideal for red wine yeasts to be selected (Caridi et al., 2015). It has also been shown that yeast strain has an effect on reducing the tannin content in wines (Ribereau-Gayon et al., 2006). It has been observed that with some strains it is possible to obtain wines higher in catechins, epicatechins, and gallic acid. It is postulated that the mechanism involved would be the same as for anthocyanin adsorption.

5.3.6.2 β-Glycosidase Activity

Red wine anthocyanins bind to glucose as β-D-glucopyranosides, and the glucose unit can also be acylated with acetic, coumaric, or caffeic acids (Ribereau-Gayon et al., 2006). Nevertheless, literature reports showed that the action of β-glucosidase could be applied also on anthocyanins composed of a glycosylated flavylium ion, the main component responsible for the color of red wine. Specific β-D-glucosidases are reported to release the corresponding anthocyanidin, which spontaneously converts to undesirable brown or colorless compounds (Vernocchi et al., 2011). These authors evaluate the suitability of four strains of *S. cerevisiae* endowed with in vitro β-glucosidase activity to improve Sangiovese wine aroma profiles. Two strains used resulted in wine characterized by high levels of nerolidol and citronellol, without detrimental effects on wine color. Conversely, the wine obtained using two other strains showed high concentrations of linalool and nerolidol; this had a strong sensorial impact but with decreased color, indicating the involvement of anthocyanin-β-glucosidases with these strains.

5.3.6.3 Polysaccharide Release

During alcoholic fermentation and wine aging over lees, *S. cerevisiae* release many of their cellular compounds, including the cell wall polysaccharides, into the medium (Escot et al., 2001). Selecting yeast with good potential for production and release of polysaccharides is an important enological tool for increasing polyphenolic stability and color in red wines. The formation of these new complexes results in a decrease in astringency (Saucier et al., 1996) and an increase in color stability (Dupin et al., 2000; Escot et al., 2001).

Mannoproteins from yeasts also have other enological interests since they improve protein stability (Lubbers et al., 1993; Ledoux et al., 1992) and tartaric stability (Moutonnet et al., 1999), increase aromatic persistence in the nose (Lubbers et al., 1994a, 1994b), increase mouth volume (Feuillat et al., 1988), promote MLF (Rosi et al., 1999), and increase microbiological stability (Dupin et al., 2000). The effect of mannoproteins on phenolic red wine compounds should receive more attention in the near future (Ghanem et al., 2017).

5.3.7 Formation of Derived Anthocyanin Compounds by Yeast Fermentation Improves Color

During winemaking, anthocyanins may be modified to create stable oligomeric pigments. Some of these pigments have been identified in wine (Fulcrand et al., 1996, 1998; Bakker and Timberlake, 1997; Bakker et al., 1997; Asenstorfer et al., 2001); their relevance is related to their resistance to bisulfite bleaching, pH-induced color loss, and to oxidation (Bakker and Timberlake, 1997).

In the last decade, the release of yeast secondary metabolic products such as pyruvic acid, vinylphenols, and acetaldehyde into the medium was identified as a potential precursor in the formation of new pigments. These compounds were demonstrated for *Saccharomyces* strains to react with anthocyanins-producing derivatives like vitisin A, vitisin B, and ethyl-linked anthocyanin-flavanol pigments (Medina et al., 2005; Asenstorfer et al., 2003; Lee et al., 2004; Morata et al., 2003a, 2007a, 2007b; Eglinton et al., 2004; Benito et al., 2009). This suggests that yeast strain selection strongly affects color intensity and the final concentration of anthocyanins (Monagas et al., 2007; Morata et al., 2006) and other phenolics (Monagas et al., 2007) (Fig. 5.2).

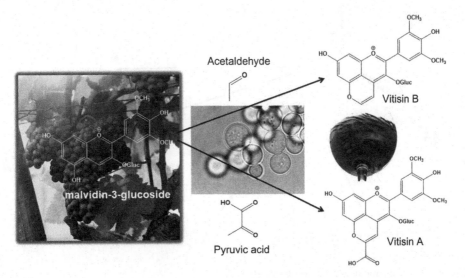

FIGURE 5.2 Yeast anthocyanin derived compounds result in increased wine red color stability.

We have developed a Tannat grape juice model medium that allows for screening of *S. cerevisiae* strains' capacity to synthetize anthocyanin derived compounds while avoiding the interference of grape solids, such as the skin and seeds (Medina et al., 2005). Tannat is a widely grown cultivar in Uruguay and one of the richest varieties in polyphenolic compounds (Da Silva et al., 2013). In this work, the authors showed that differences in wine composition were observed depending on the particular yeast strain used in the vinification experiments.

According to several authors (Romano et al., 1994; Vasserot et al., 2010; Mateos et al., 2006; Longo et al., 1992), some strains of *S. cerevisiae* have the ability to metabolize acetic acid during alcoholic fermentation and produce different amounts of acetaldehyde. The study of Romano et al. (1994) allowed to assign the strains to different phenotypes: Low, medium, and high acetaldehyde producers. According to Medina et al. (2016), when *S. cerevisiae* was fermented in pure cultures, it produced higher concentrations of acetaldehyde and vitisin B (acetaldehyde reaction-dependent) compared with some non-*Saccharomyces*. However, cofermentation of non-*Saccharomyces* with *S. cerevisiae* resulted in a significantly higher concentration of acetaldehyde compared with the pure *S. cerevisiae* control. For these reasons, acetaldehyde production would be a desirable selection characteristic for *Saccharomyces* red wine yeasts.

5.4 SACCHAROMYCES CEREVISIAE AND FLAVOR COMPOUNDS

5.4.1 *Saccharomyces cerevisiae* Synthesis of Flavor Compounds

The chemical identification of aroma compounds derived from the metabolic activity of yeasts has been a research objective in recent decades (Lambrechts and Pretorius, 2000; Rapp and Versini, 1996; Swiegers et al., 2005; Carrau et al., 2008). Even though *S. cerevisiae* is responsible for most of the ethanol in wine, it also has a significant effect on the production of aroma compounds, including ethyl and acetate esters, higher alcohols, fatty acids, lactones, sulfur compounds, monoterpenes, and benzenoids (Fig. 5.3). These compounds are especially important for the sensory perception of different wine types (Cozzolino et al., 2006; Ferreira et al., 1996; Guth and Sies, 2002; Smyth et al., 2005; Carrau et al., 2005).

The main higher alcohols compounds produced by *Saccharomyces* are 2- and 3-methylbutanol, propanol, 2-methylpropanol, butanol, pentanol, 2-phenylethanol, and 3-methylthio-1-propanol. The nitrogen-dependent common trend to produce higher alcohols and isoacids suggests that their metabolism and production may be coordinated with and dependent on the nicotinamide adenine dinucleotide (NAD/NADH) redox balance of the

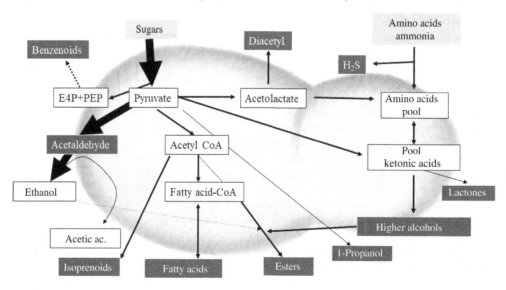

FIGURE 5.3 Fermentation flavor compounds produced by yeasts.

cell (Nordstrom, 1964; Reazin et al., 1970; Fariña et al., 2012). On the other hand, it was shown that significantly increased production of higher alcohols by some strains might be stimulated by microaerobic and low-YAN conditions (Fariña et al., 2012). In contrast, 1-propanol is known to be formed by the condensation of pyruvic acid and acetyl-CoA (Nykanen, 1986), and assimilable nitrogen has an opposite behavior compared with other higher alcohols (Carrau et al., 2008).

Yeasts also produce fatty acids of different chain lengths, with a chain length of less than 12 carbons being volatile. Acetic acid is present in the highest concentration and, having a relatively low threshold of perception, it can be perceived in many cases. In yeasts, the production of acetic acid has important metabolic roles and its biosynthesis results from the enzyme acetaldehyde decarboxylase. Long-chain fatty acids with 4–12 carbons are produced as by-products of the synthesis of saturated fatty acids by yeasts.

Esters are the volatile compounds of fermentative origin that have the greatest impact on wine flavor profiles. However, they have a moderate contribution in red winemaking compared with white wines. Esters can be divided into ethyl esters of fatty acids and acetates of higher alcohols. In the case of ethyl esters, their formation typically occurs through the esterification of activated fatty acids in the form of acetyl-CoA. The limiting factor during their biosynthesis is the concentration of the fatty acid. Acetate esters are formed through the condensation of alcohol with acetyl-CoA catalyzed in the cell by alcohol acetyltransferase enzymes. The expression of alcohol acetyl transferases is the most important factor in determining acetate ester levels during fermentation. These acetate esters might be synthesized more in red grape wine conditions due to the increased formation of higher alcohols.

Knowledge about the formation of terpenoids has been limited to the production of trace concentrations by *S. cerevisiae* (Carrau et al., 2005). When *S. cerevisiae* strain M522 was used to evaluate the factors affecting terpene production, it was demonstrated that, strikingly, the YAN and oxygen content of the fermentation medium positively influence monoterpene formation (Carrau et al., 2005). High YAN and oxygen concentrations in the medium stimulate the formation of monoterpenes but not sesquiterpenes (nerolidol and farnesol) (Carrau et al., 2005). It is expected that microaerobic fermentation conditions such as found in red wine vinification increase this family of compounds.

5.4.2 *Saccharomyces* Enzymes Effects on Flavor

In *Saccharomyces* strains, the enzymes that appear and can affect flavor and color are β-glycosidases. In several grape varieties, the dominating aroma compounds are bound to sugars, such as the volatile group of benzenoid/phenylpropanoid, monoterpenes, and norisoprenoids, which substantially contribute to wine aroma during fermentation or aging (Hjelmeland and Ebeler, 2015). These bound compounds can represent 10%–90% of the total volatile fraction after hydrolysis in final wines of grapes such as Chardonnay (Sefton et al., 1993), Cabernet sauvignon, Merlot (Francis et al., 1998), Tannat (Boido et al., 2003), Pinot noir (Fang and Qian, 2006), and aromatic varieties such as Muscats.

In Tannat grapes, phenylpropanoid and norisoprenoids are quantitatively the most important group of compounds, representing almost 80% of the total glycosidic components (Boido et al., 2013). Recently, the glycosidic form of ethyl-β-(4-hydroxy-3-methoxy-phenyl)-propionate was identified for the first time in grapes of this variety (Boido et al., 2013). The bound-compound hydrolysis can occur during wine aging, under mild acidic conditions. But unlike acid hydrolysis, enzymatic hydrolysis is efficient and does not result in modification of the aromatic composition of the bound fraction (Günata et al., 1990). The hydrolysis of monoglucosides requires the action of a β-glucosidase, but many of these compounds are bound to disaccharides and require the sequential activity of an appropriate exoglycosidase to remove the outermost sugar residue, followed by a β-glucosidase to remove the remaining glucose (Fig. 5.4) (Günata et al., 1988).

Duerksen and Halvorson (1958) reported the production and subsequent purification of a β-glucosidase in *S. cerevisiae*. Since then, the information present in the literature regarding *Saccharomyces* and the activity of this enzyme is contradictory. Some studies emphasize that the enzymes of *S. cerevisiae* have very low β-glucosidase activity, and that it is severely inhibited in the conditions of winemaking (low pH, and high glucose and alcohol in the medium) (Rosi et al., 1994; Winterhalter and Skouroumounis, 1997). However, Delcroix et al. (1994) and Grossmann et al. (1987) report the hydrolysis of monoterpenol glycosides using *S. cerevisiae* possessing broad spectrum β-glucosidase activity. Grossmann et al. (1987) argue that it is much more effective to add these yeasts to a wine that has finished fermentation than to add them to a must that is in the process of fermentation, since in a must with high fermentative activity the enzyme β-glucosidase would be subject to denaturation by proteases and to activity inhibition by glucose. Conversely, some authors suggest that in the early stages of alcoholic fermentation,

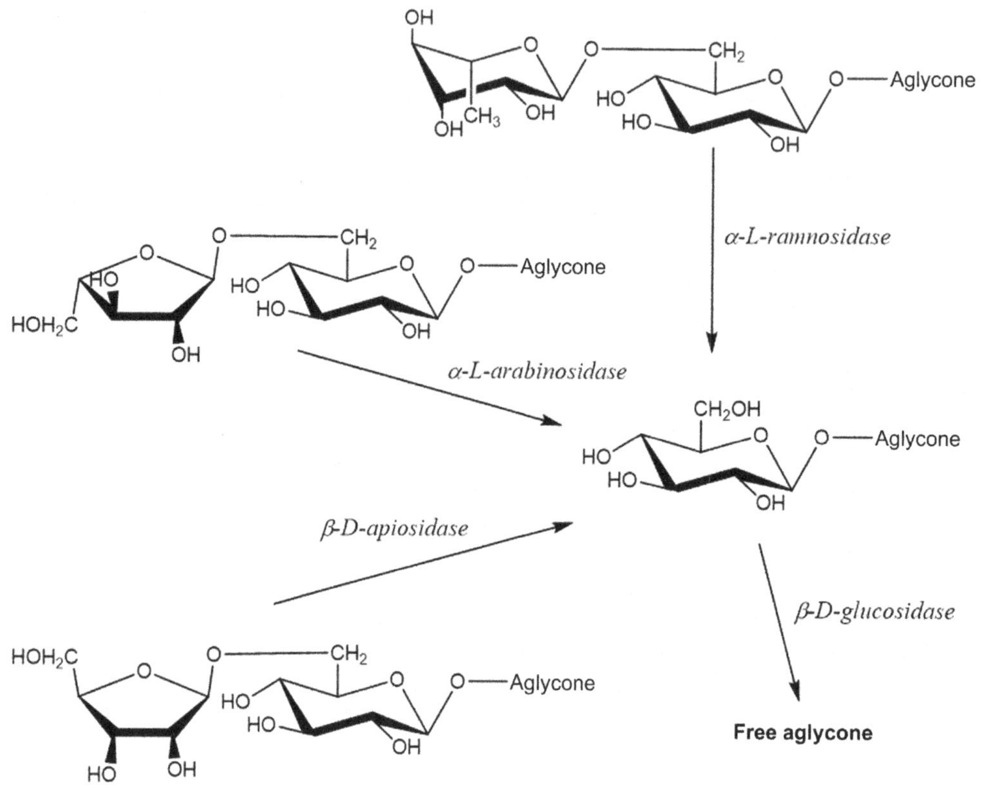

FIGURE 5.4 Hydrolysis of disaccharide glycosides by sequential action of an appropriate exoglycosidase, followed by a β-glucosidase to remove the remaining glucose. The different aglycones found in grapes and wines are terpenes, norisoprenoids, phenylpropanoids, anthocyanidins, etc.

S. cerevisiae could actively contribute to the release of volatile compounds from Moscatel grapes (Darriet et al., 1988; Mateo and Di Stefano, 1997). These authors maintain that the inhibitory effect of ethanol would restrict the enzymatic activity of this β-glucosidase on the glycosylated compounds. In contrast, Delcroix et al. (1994) show that the enzyme β-glucosidase originating from *Saccharomyces* yeasts is not significantly affected by the ethanol present in the medium. For a concentration of 15% (v/v) of ethanol, the activity decreased only 10%, and in some cases it was observed that the enzymatic activity was favored (Delcroix et al., 1994; Rosi et al., 1994).

Nevertheless, in recent work, it has been demonstrated that permeabilized cells of *Saccharomyces* species can exhibit β-glycosidase activity as high as non-*Saccharomyces* species, suggesting that the transport of glycosidic precursors is a limiting factor in the aromatic release and could guide breeding programs for the construction of new interspecific wine yeasts (Bisotto et al., 2015). Spagna et al. (2002), working with β-glucosidase enzymes of *S. cerevisiae* isolated from musts and wines from Sicily, observed that the content of free terpenes increased by at least 30%. This enzyme was not inhibited by ethanol in concentrations between 12% and 14% (v/v) but, on the contrary, was activated by it. The activity of this enzyme was also not inhibited by fructose and/or glucose concentrations of 15%—20% in the medium. Similarly, Hernández et al. (2003), working with a strain of *S. cerevisiae* isolated from a must from Valdepeña, Spain, and Ugliano et al. (2005) with two strains of *S. cerevisiae* and one strain of *Saccharomyces bayanus*, observed that in the fermentations by these yeasts there was a significant increase in the concentration of several volatile compounds. Pérez et al. (2011) selected two *S. cerevisiae* strains, isolated from grapes and musts, with β-glucosidase activity on Muscat Miel grape must fermentation, rich in bound monoterpenes. The levels of free linalool, hodiol I, and geraniol increased significantly compared with fermentation with a commercial wine yeast strain.

5.5 PRACTICAL RED WINEMAKING AND YEAST PERFORMANCE

Here we discuss some topics that we consider key points for working in red winemaking and taking advantage of yeast technology. There are many very good references for red winemaking that should be considered

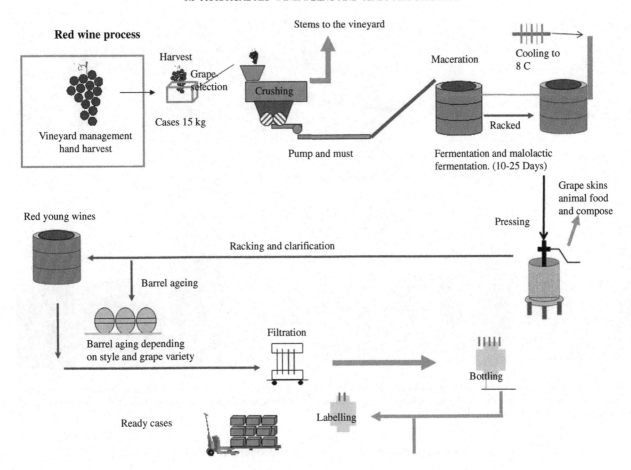

FIGURE 5.5 Red wine making production scheme.

by the reader. We assume the reader has knowledge of works such as Peynaud (1984), Flanzy (1998), Ruiz-Hernández (2004), Ribereau-Gayon et al. (2006), and Boulton et al. (2013).

Fig. 5.5 shows the key points of the process flow for the production of quality red wines. In the vineyard, many parameters should be controlled to avoid complicating fermentations even by strong strains of *S. cerevisiae*. *Fungi* and *acetic* bacteria could be a main source of contaminations that affect the development of *S. cerevisiae* and also might contribute flavor taints, such as acetic acid, hydrogen sulfide, or gluconic acid, all of which are compounds that will negatively affect the final wine.

At a second step, the style of harvest and transport of the grapes to the crushing facility should avoid breaking berries during the process. Hand-harvested grapes in small cases are known to be the ideal method; this procedure prevents the development of aerobic non-*Saccharomyces* strains that compete with *S. cerevisiae* for YAN, and also could be antagonistic to it (Medina et al., 2012). Broken berries, a few hours of delay in crushing, and temperatures above 15°C will promote the development of natural flora that is not good for flavor quality and for a healthy skin vinification process. If a mechanical harvest is used, the winemaker should take care to do this work at the lowest temperature during night or early morning. Adding sulfites or an adequate yeast pre-inoculum into the mechanical harvest machine might improve the conservation of the grape must formed during harvest through berries broken by the machine. At the winery, again depending on the grape variety, temperature, aeration, pH control, gentle crushing, and avoiding breaking seeds when pumping the must should be controlled in order to obtain a good grape must to inoculate and start maceration.

Various maceration technologies are shown in Fig. 5.6. It is known that yeast are affected by differences in the tanks' shapes, sizes, or maceration styles, and that these physical chemical variables will result in different flavors and aroma characteristics by changing cell metabolism. For example, open tank macerations result in microaerobic redox situations that increase higher alcohols and monoterpene synthesis, and decrease esters synthesis (Fariña et al., 2012).

FIGURE 5.6 Different styles of fermentation tanks for red wine will affect *S. cerevisiae* performance resulting in flavor and color diversity of the final wines due to redox, temperature and solids/juice conditions. Open bins, submerged cap tanks, truncated-cone shaped design in oak or stainless steel are different examples.

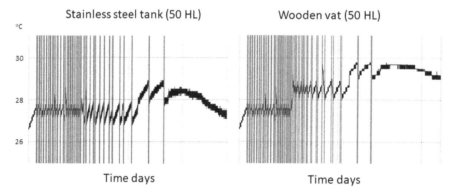

FIGURE 5.7 Fermentation temperature differences between stainless steel and wooden vats of the same size and design. A more rapid decrease of temperature control is obtained in the steel tanks. Red lines indicate the automatic open/close of cold water valves. Both tanks are truncated cone-shaped designs of 5000 L.

The contemporary winemaking strategy of increasing flavor complexity in wines demands different maceration and vinification processes to create diverse situations for yeast metabolism. Fermentation tank design and size (barrel, bins, vats, etc.) can be used to increase the diversity of vinification styles, and the same block of vines can be fermented separately and then blended by sensory analysis. Stainless steel and wooden vats, although with the same shape design also affects the temperature fermentation profiles of the process and this situation give diverse flavors to the final wine (Fig. 5.7).

5.5.1 Use of Commercial Yeasts

The use of native *Saccharomyces* yeasts also contributes to increased diversity of flavors, and helps to differentiate our wines from the more massive commercial market. Using the same *S. cerevisiae* strain for Malbec red grapes in Argentina and for Tannat grapes in Uruguay will increase the similarities between these two wines.

This is why the concept of using specific yeasts for different varieties was developed in yeast selection and application at the industrial level. Some yeast strains that were developed in different wine regions were later produced for industrial applications with specific red grape varieties, such as *Tempranillo* in Rioja, *Pinotage* in South Africa, or *Syrah* in Australia.

5.5.2 Saccharomyces-Lactic Acid Bacteria Interactions During Winemaking

MLF is a key step in red winemaking to obtain softer and less acidic wines. The interaction between bacteria and yeast during alcohol fermentation (AF) and/or MLF has a direct effect on lactic acid bacteria (LAB) growth and malolactic activity. Various studies have attempted to elucidate the interaction between yeast and bacteria (Henick-Kling and Park, 1994; Rosi et al., 2003; Arnink and Henick-Kling, 2005; Guilloux-Benatier et al., 2006; Jussier et al., 2006; Osborne and Edwards, 2006). Alexandre et al. (2004) proposed that the nature of these interactions rely on three factors. First, the combination of yeast and bacteria strain; second, the uptake and release of nutrients by the yeast, which will in turn affect the nutrients available for the LAB; and third, the ability of the yeast to produce metabolites that can have either a stimulating or inhibiting/toxic effect on LAB.

The combined effect of yeast and bacteria strain is complex and has been studied with different results by different authors. Nehme et al. (2008) studied this interaction during the winemaking process and found that inhibition between these microorganisms is largely dependent on the selected strains of yeast and bacteria. It was demonstrated that this inhibition correlated to a decrease in bacterial growth, rather than a decline in the malolactic activity of the bacteria. In contrast, Arnink and Henick-Kling (2005), in a study of commercial pairings of microorganisms, found the differences between vintages and grape varieties to be more influential on LAB and MLF than the effect of a yeast/bacteria strain combination. Costello et al. (2003) proposed a method for testing the compatibility between yeast and bacteria, which is an effective tool for screening yeast/LAB combinations to be used in vinification.

Oenococcus oeni is inhibited by *S. cerevisiae* at the start of AF, due to the rapid uptake of certain grape metabolites from the must. These metabolites include sterols, amino acids, and vitamins (Larsen et al., 2003), which result in a low nutrient environment for the bacteria. During AF, the amino acids and vitamins that are essential for bacteria are depleted by yeast metabolism, and bacterial growth is delayed until yeast cells lyse (Nygaard and Prahl, 1997; Alexandre et al., 2004; Arnink and Henick-Kling, 2005). Yeast autolytic activity can release mannoproteins, which is important because they can stimulate bacterial growth by adsorbing medium-chain fatty acids and thus detoxifying the wine medium. Lonvaud-Funel et al. (1988) identified medium-chain fatty acids as one of the main inhibitory products to bacterial growth. Mannoproteins can also be enzymatically hydrolyzed by bacterial enzyme activity, which will enhance the nutritional content of the wine, and some strains will behave as friendly yeasts that stimulate bacterial growth (Guilloux-Benatier et al., 2006; Alexandre et al., 2004). Comitini et al. (2005) related part of the inhibitory effect of *S. cerevisiae* to the production of extracellular compounds via metabolic activity of the yeast, rather than to a competition for nutrients. The yeast-derived inhibitory compounds include ethanol, SO_2, medium-chain fatty acids, and proteins. The choice of a yeast strain with low production of medium-chain fatty acids and SO_2 is important for the natural development of LAB fermentation. In a more recent study, Nehme et al. (2008) reported the inhibition of an *O. oeni* strain, resulting in a decrease of the malic acid consumption rate mainly due to ethanol (75%) and to a peptidic fraction (25%) having a molecular weight between 5 and 10 kDa.

5.5.3 Aging and Microbial Stability

The challenge today is to mature wines to soften the astringent characteristics of young polyphenols, preventing over-oak flavors so as to not lose varietal character and to reach microbial stability. Racking, blending, and lees contact are the typical operations at the winery that could affect the final flavor of wines. The periodic control of tank or barrel wines during aging by SO_2 analysis and wine tasting is still the ideal way to control the aging process at the cellar. At the beginning of the barrel aging, free SO_2 levels always decrease, and we need to monitor this initially every 15 days and then on a monthly base.

Racking or blending can change the redox situation of the wine and obviously the SO_2 levels will decrease, demanding adjustments for a period of approximately 45–90 days. The best way to measure microbial stability of a wine being prepared for bottling is to blend the barrels or batches and keep the final wine in a tank with a stable temperature for a certain number of weeks that will be determined when the free SO_2 levels are stable. If free SO_2 continues decreasing every month, the wine is considered not stable. Usually after about 2 months this stability is obtained and the wine can be bottled even without a deep filtration, just with the right level of free SO_2. Some strains of *S. cerevisiae, Zygosaccharomyces,* or *Brettanomyces* could be active in dry red wine with an alcohol content of 12%–15% by volume. Some lactic acid bacteria could still be active if malic acid is still present in the wine, and if residual assimilable nitrogen is also present.

Acknowledgments

We wish to thank Comisión Sectorial de Investigación Científica (CSIC), CSIC Group Project No. 656 and CSIC Productive Sector Project No. 602 of UdelaR, Uruguay (Grant No. ANII Postgraduate POS_NAC_2012_1_9099), and Agencia Nacional de Investigación e Innovación (ANII) for financial support of our wine projects.

References

Alexandre, H., Costello, P.J., Remize, F., Guzzo, J., Guilloux-Benatier, M., 2004. *Saccharomyces cerevisiae–Oenococcus oeni* interactions in wine: current knowledge and perspectives. Int. J. Food Microbiol. 93, 141–154.

Arnink, K., Henick-Kling, T., 2005. Influence of *Saccharomyces cerevisiae* and *Oenococcus oeni* strains on successful malolactic conversion in wine. Am. J. Enol. Vitic. 56, 228–237.

Asenstorfer, R.E., Hayasaka, Y., Jones, G.P., 2001. Isolation and structures of oligomeric wine pigments by bisulfite-mediated ion-exchange chromatography. J. Agric. Food Chem. 49, 5957–5963.

Asenstorfer, R.E., Markides, A.J., Iland, P.G., Jones, G.P., 2003. Formation of vitisin A during red wine fermentation and maturation. Aust. J. Grape Wine Res. 9, 40–46.

Bakker, J., Timberlake, C.F., 1997. Isolation, identification, and characterization of new color-stable anthocyanins occurring in some red wines. J. Agric. Food Chem. 45, 35–43.

Bakker, J., Bridle, P., Honda, T., Kuwano, H., Saito, N., Terahara, N., et al., 1997. Identification of an anthocyanin occurring in some red wines. Phytochemistry 44, 1375–1382.

Bartowsky, E.J., Henschke, P.A., 2004. The 'buttery' attribute of wine—diacetyl—desirability, spoilage and beyond. Int. J. Food Microbiol. 96 (3), 235–252.

Benito, S., Palomero, F., Morata, A., Uthurry, C., Suárez-Lepe, J.A., 2009. Minimization of ethylphenol precursors in red wines via the formation of pyranoanthocyanins by selected yeasts. Int. J. Food Microbiol. 132, 145–152.

Bisotto, A., Julien, A., Rigou, P., Schneider, R., Salmon, J., 2015. Evaluation of the inherent capacity of commercial yeast strains to release glycosidic aroma precursors from Muscat grape must. Aust. J. Grape Wine Res. 21, 194–199.

Boido, E., Lloret, A., Medina, K., Fariña, L., Carrau, F., Versini, G., et al., 2003. Aroma composition of *Vitis vinifera* cv. Tannat: the typical red wine from Uruguay. J. Agric. Food Chem. 51, 5408–5413.

Boido, E., Fariña, L., Carrau, F., Dellacassa, E., Cozzolino, D., 2013. Characterization of glycosylated aroma compounds in Tannat grapes and feasibility of the Near Infrared Spectroscopy application for their prediction. Food Anal. Methods 6 (1), 100–111.

Boivin, S., Feuillat, M., Alexyre, H., Charpentier, C., 1998. Effect of must turbity on cell wall porosity and macromolecule excretion of *Saccharomyces cerevisiae* cultivated on grape juice. Am. J. Enol. Vitic. 49, 325–332.

Bokulich, N.A., Thorngated, J.H., Richardsone, P.M., Millset, D.A., 2014. Microbial biogeography of wine grapes is conditioned by cultivar, vintage, and climate. Proc. Natl. Acad. Sci. U.S.A. 111 (1), 139–148.

Boulton, R., 2001. The copigmentation of anthocyanins and its role in the color of red wine: a critical review. Am. J. Enol. Vitic. 52, 67–87.

Boulton, R.B., Singleton, V.L., Bisson, L.F., Kunkee, R.E., 2013. Principles and Practices of Winemaking. Springer Science and Business Media, New York, USA.

Bourzeix, M., Heredia, N., 1976. Etude des colorants anthocyaniques fixes par les levures de vinification. C. R. Acad. Agric. 62, 750–753.

Caridi, A., Cufari, A., Lovino, R., Palumbo, R., Tedesco, I., 2004. Influence of yeast on polyphenol composition of wine. Food Technol. Biotechnol. 42, 37–40.

Caridi, A., Sidari, R., Kraková, L., Kuchta, T., Pangallo, D., 2015. Assessment of color adsorption by yeast using grape skin agar and impact on red wine color. OENO ONE 49 (3), 195–203.

Carrau, F., Medina, K., Boido, E., Fariña, L., Gaggero, C., Dellacassa, E., et al., 2005. De novo synthesis of monoterpenes by *Saccharomyces cerevisiae* wine yeasts. FEMS Microbiol. Lett. 243, 107–115.

Carrau, F., Medina, K., Fariña, L., Boido, E., Henschke, P.A., Dellacassa, E., 2008. Production of fermentation aroma compounds by *Saccharomyces cerevisiae* wine yeasts: Effects of yeast assimilable nitrogen on two model strains. FEMS Yeast Res. 8, 1196–1207.

Carrau, F., Medina, K., Fariña, L., Boido, E., Dellacassa, E., 2010. Effect of inoculums size of *Saccharomyces cerevisiae* on wine fermentation aroma compounds and its relation with the nitrogen content. Int. J. Food Microbiol. 143 (1-2), 81–85.

Carrau, F., Gaggero, C., Aguilar, P.S., 2015. Yeast diversity and native vigor for flavor phenotypes. Trends Biotechnol. 33 (3), 148–154.

Castino, M.R., 1982. Lieviti e polifenoli. Rivista di Viticoltura e di Enologia 34, 333–348.

Chassagne, D., Vernizeau, S., Nedjmac, M., Alexandre, H., 2005. Hydrolysis and sorption by *Saccharomyces cerevisiae* strains of Chardonnay grape must glycosides during fermentation. Enzyme Microb. Technol. 37, 212–217.

Chatonnet, P., Viala, C., Dubourdieu, D., 1997. Influence of polyphenolic components of red wines on the microbial synthesis of volatile phenols. Am. J. Enol. Vitic. 48, 443–448.

Comitini, F., Ferretti, R., Clementi, F., Mannazzu, I., Ciani, M., 2005. Interactions between *Saccharomyces cerevisiae* and malolactic bacteria: preliminary characterization of a yeast proteinaceous compound(s) active against *Oenococcus oeni*. J. Appl. Microbiol. 99, 105–111.

Conterno, L., Tortia, C., Minati, J., Trioli, G., 1997. La selezione di un ceppo di lievito per il Barolo. Vignevini 24, 30–33.

Costello, P.J., Henschke, P.A., Markides, A.J., 2003. Standardized methodology for testing malolactic bacteria and wine yeast compatibility. Aust. J. Grape Wine Res. 9, 127–137.

Cozzolino, D., Smyth, H.E., Lattey, K.A., Cynkar, W., Janik, L., Dambergs, R., et al., 2006. Combining mass spectrometry based electronic nose, visible-near infrared spectroscopy and chemometrics to assess the sensory properties of Australian Riesling wines. Anal. Chim. Acta 563, 319–324.

Cuinier, C., 1988. Influence of yeasts on the phenolic compounds of wine, Bull. O.I.V, 61. pp. 596–601.

Cuinier, C., 1997. Ceppi di lievito e composizione fenolica dei vini rossi. Vignevini 7, 39–42.

Darriet, P., Boidron, J.N., Dubourdieu, D., 1988. L'hydrolyse des hétérosides terpéniques du Muscat a Petits Grains par les enzymes périplasmiques de Saccharomyces cerevisiae. Connaissance de la Vigne et du Vin 22, 189–195.

Da Silva, C., Zamperin, G., Ferrarini, A., Minio, A., Dal Molin, A., Venturini, L., et al., 2013. The high polyphenol content of grapevine cultivar Tannat berries is conferred primarily by genes that are not shared with the reference genome. Plant Cell 25, 4777–4788.

Delcroix, A., Günata, Z., Sapis, J.C., Salmon, J.M., Bayonove, C., 1994. Glycosidase activities of three enological yeast strains during winemaking: effect on the terpenol content of Muscat wine. Am. J. Enol. Vitic. 45, 291–296.

Delcourt, F., Taillandier, P., Vidal, F., Strehaiano, P., 1995. Influence of pH, malic acid and glucose concentrations on malic acid consumption by Saccharomyces cerevisiae. Appl. Microbiol. Biotechnol. 43 (2), 321–324.

De Nobel, J., Klis, G.F.M., Priem, J., Munnik, T., Van Den Ende, H., 1990. The glucanase-soluble mannoproteins limit cell wall porosity in Saccharomyces cerevisiae. Yeast 6, 491–499.

Duerksen, J.D., Halvorson, H., 1958. Purification and properties of an inducible β-glucosidase of yeast. J. Biol. Chem. 233, 1113–1120.

Dupin, I.V.S., Mckinnon, V.M., Ryan, C., Boulay, M., Markides, A.J., Jones, G., et al., 2000. Saccharomyces cerevisiae mannoproteins that protect wine from protein haze: their release during fermentation and lees contact and a proposal for their mechanism of action. J. Agric. Food Chem. 48, 3098–3105.

Eglinton, J., Griesser, M., Henschke, P., Kwiatkowski, M., Parker, M., Herderich, Y.M., 2004. Yeast-mediated formation of pigmented polymers in red wine. In: Waterhouse, A.L., Kennedy, J.A. (Eds.), Red Wine Color: Exploring the Mysteries. American Chemical Society, Washington, DC, pp. 7–21.

Escot, S., Feuillat, M., Dulau, L., Charpentier, C., 2001. Release of polysaccharides by yeasts and the influence of released polysaccharides on colour stability and wine astringency. Aust. J. Grape Wine Res. 7, 153–159.

Fang, Y., Qian, M.C., 2006. Quantification of selected aroma-active compounds in Pinot noir wines from different grape maturities. J. Agric. Food Chem. 54, 8567–8573.

Fariña, L., Medina, K., Urruty, M., Boido, E., Dellacassa, E., Carrau, F., 2012. Redox effect on volatile compound formation in wine during fermentation by Saccharomyces cerevisiae. Food Chem. 134, 933–939.

Ferreira, V., Fernández, P., Cacho, J.F., 1996. A study of factors affecting wine volatile composition and its application in discriminant analysis. Food Sci. Technol. 29, 251–259.

Feuillat, M., Charpentier, C., Picca, C., Et Bernard, P., 1988. Production de colloïdes par les levures dans les vins mousseux élaborés selon la méthode champenoise. Revue Française des Oenologues 111, 36–45.

Flanzy, C., 1998. "Oenologie", Fondements scientifiques et technologiques. Tec & Doc Lavoisier, Paris.

Fleet, G.H., 2003. Yeast interactions and wine flavour. Int. J. Food Microbiol. 86 (1-2), 11–22.

Francis, I.L., Kassara, S., Noble, A.C., Williams, P.J., 1998. The contribution of glycoside precursors to Cabernet Sauvignon and Merlot aroma: sensory and compositional studies. In: Waterhouse, A., Ebeler, S. (Eds.), Chemistry of Wine Flavor, ACS Symposium Series. American Chemical Society, Washington, DC, pp. 13–30.

Fulcrand, H., Cameira Dos Santos, P.J., Sarni-Manchado, P., Cheynier, V., Favre-Bonvin, J., 1996. Structure of new anthocyanin-derived wine pigments. J. Chem. Soc., Perkin Trans. 1, 735–739.

Fulcrand, H., Benabdeljalil, C., Rigaud, J., Cheynier, V., Moutounet, M., 1998. A new class of wine pigments generated by reaction between pyruvic acid and grape anthocyanins. Phytochemistry 47, 1401–1407.

Ghanem, C., Taillandier, P., Rizk, M., Rizk, Z., Nehme, N., Souchard, J.P., et al., 2017. Analysis of the impact of fining agents types, oenological tannins and mannoproteins and their concentrations on the phenolic composition of red wine. LWT Food Sci. Technol. 83, 101–109.

Goddard, M.R., 2008. Quantifying the complexities of Saccharomyces cerevisiae's ecosystem engineering via fermentation. Ecology 89, 2077–2082.

Grossmann, M., Rapp, A., Rieth, W., 1987. Enzymatische freisetzung gebundener aromastoffe in wein. Dtsch. Lebensm. Rundsch. 83, 7–12.

Guilloux-Benatier, M., Remize, F., Gal, L., Guzzo, J., Alexandre, H., 2006. Effects of yeast proteolytic activity on Oenococcus oeni and malolactic fermentation. FEMS Microbiol. Lett. 263, 183–188.

Günata, Z., Bayonove, C., Tapiro, C., Cordonnier, R., 1990. Hydrolysis of grape monoterpenyl β-glucosides by various β-glucosidases. J. Agric. Food Chem. 38, 1232–1236.

Günata, Z.Y., Bitteur, S., Brillouet, J.M., Bayonove, C.L., Cordonnier, R.E., 1988. Sequential enzymic hydrolysis of potentially aromatic glycosides from grapes. Carbohydr. Res. 184, 139–149.

Guth, H., Sies, A., 2002. Flavour of wines: towards an understanding by reconstitution experiments and an analysis of ethanol's effect on odour activity of key compounds. In: Blair, R.J., Williams, P.J., Høj, P.B. (Eds.), Proceedings of the Eleventh Australian Wine Industry Technical Conference. Adelaide, Australia, pp. 128–139.

Henick-Kling, T., Park, Y.H., 1994. Considerations for the use of yeast and bacterial starter cultures: SO_2 and timing of inoculation. Am. J. Enol. Vitic. 45, 464–469.

Henschke, P.A., 1996. Hydrogen sulfide production by yeast during fermentation. Proceedings Eleventh International Oenological Symposium, Sopron, Hungary. International Association for Winery Technology and Management, Breisach, Germany, pp. 83–102.

Henschke, P.A., 1997. Wine Yeast. In: Zimmermann, F.K., Entian, K.D. (Eds.), Yeast Sugar Metabolism. Technomic Publishing Co, Lancaster, PA, pp. 527–560.

Hernández, L.F., Espinosa, J.C., Fernández-González, M., Briones, A., 2003. β-Glucosidase activity in a Saccharomyces cerevisiae wine strain. Int. J. Food Microbiol. 80 (2), 171–176.

Hjelmeland, A.K., Ebeler, S.E., 2015. Glycosidically bound volatile aroma compounds in grapes and wine: a review. Am. J. Enol. Vitic. 66, 1–11.

Jiranek, V., Langridge, P., Henschke, P.A., 1991. Yeast nitrogen demand: selection criterion for wine yeast for fermenting low nitrogen musts. In: Rantz, J.M. (Ed.), Proceedings of the International Symposium on Nitrogen in Grapes and Wine. American Society for Enology and Viticulture, Davis, CA, pp. 266–269.

Jussier, D., Morneau, A.D., Mira de Orduña, R.M., 2006. Effect of simultaneous inoculation with yeast and bacteria on fermentation kinetics and key wine parameters of cool-climate Chardonnay. Appl. Environ. Microbiol. 72, 221–227.

Lambrechts, M.G., Pretorius, I.S., 2000. Yeast and its importance to wine aroma a review. S. Afr. J. Enol. Vitic. 21, 97–129.

Larsen, J.T., Nielsen, J.C., Kramp, B., Richelieu, M., Bjerring, P., Riisager, N.A., et al., 2003. Impact of different strains of *Saccharomyces cerevisiae* on malolactic fermentation by *Oenococcus oeni*. Am. J. Enol. Vitic. 54, 246–251.

Larue, F., Geneix, C., Lafon-Lafourcade, S., Bertry, A., Ribereeau-Gayon, P., 1984. Premiers observtions sur le mode d' action des écores de levures. Connaissance de la Vigne et du Vin 18, 155–163.

Ledoux, V., Dulau, L., Dubourdieu, D., 1992. Interpretation de l'amélioration de la stabilité proteique des vins au cours de l'élevage sur lies. J. Int. Sci. Vigne Vin 26, 239–251.

Lee, D.F., Swinny, E.E., Jones, G.P., 2004. NMR identification of ethyl-linked anthocyanin-flavanol pigments formed in model wine ferments. Tetrahedron Lett. 45, 1671–1674.

Loiseau, G., Cuinier, C., Berger, J.L., Peyron. D., 1994. Fixation des anthocyanes par les levures œnologiques au cours de la vinification en rouge. In: Microbiologie, S.F.D. (Ed.), Proceedings for Société Francaise de Microbiologie. Dijon, France, pp. 175–178.

Longo, E., Velázquez, J.B., Sieiro, C., Cansado, J., Calo, P., Villa, T.G., 1992. Production of higher alcohols, ethyl acetate, acetaldehyde and other compounds by 14 *Saccharomyces cerevisiae* wine strains isolated from the same region (Salnés, N.W. Spain). World J. Microbiol. Biotechnol. 8, 539–541.

Lonvaud-Funel, A., 1999. Lactic acid bacteria in the quality improvement and depreciation of wine. In: Konings, W.N., Kuipers, O.P., In'tVeld, J.H. (Eds.), Lactic Acid Bacteria: Genetics, Metabolism and Applications. Springer, Dordrecht.

Lonvaud-Funel, A., Joyeux, A., Desens, C., 1988. The inhibition of malolactic fermentation of wines by products of yeast metabolism. J. Food Sci. Technol. 44, 183–191.

Lubbers, S., Leger, B., Charpentier, C., 1993. Essai colloides protecteurs d'extraits de parois de levures sur la stabilité tartrique d'un vin modêle. J. Int. Sci. Vigne Vin 27, 13–22.

Lubbers, S., Charpentier, C., Feuillat, M., Voilley, A., 1994a. Influence of yeast walls on the behavior of aroma compounds in a model wine. Am. J. Enol. Vitic. 45, 29–33.

Lubbers, S., Voilley, A., Feuillat, M., Charpentier, C., 1994b. Influence of mannoproteins from yeast on the aroma intensity of a model wine. Lebensm. Wiss. Technol. 27, 108–114.

Martineau, B., Acree, T.E., Henick-Kling, T., 1995. Effect of wine type on the detection threshold for diacetyl. Food Res. Int. 28, 139–143.

Martini, A., 1993. Origin and domestication of the wine yeast *Saccharomyces cerevisiae*. J. Wine Res. 4 (3), 165–176.

Martini, A., 2003. Biotechnology of natural and winery-associated strains of *Saccharomyces cerevisiae*. Int. Microbiol. 6 (3), 207–209.

Martini, A., Ciani, M., Scorzetti, G., 1996. Direct enumeration and isolation of wine yeasts from grape surfaces. Am. J. Enol. Vitic. 47 (4), 435–440.

Mateo, J.J., Di Stefano, R., 1997. Description of the β-glucosidase activity of wine yeasts. Food Microbiol. 14 (6), 583–591.

Mateos, J.A., Pérez-Nevado, F., Ramírez Fernández, M., 2006. Influence of *Saccharomyces cerevisiae* yeast strain on the major volatile compounds of wine. Enzyme Microb. Technol. 40, 151–157.

Medina, K., Boido, E., Dellacassa, E., Carrau, F., 2005. Yeast interactions with anthocyanins during red wine fermentation. Am. J. Enol. Vitic. 56, 104–109.

Medina, K., Boido, E., Dellacassa, E., Carrau, F., 2012. Growth of non-*Saccharomyces* grape must yeasts affects nutrient availability for *Saccharomyces cerevisiae* during wine fermentation. Int. J. Food Microbiol. 157, 245–250.

Medina, K., Boido, E., Fariña, L., Gioia, O., Gómez, M.E., Barquet, M., et al., 2013. Increased flavour diversity of Chardonnay wines by spontaneous fermentation and co-fermentation with *Hanseniaspora vineae*. Food Chem. 141, 2513–2521.

Medina, K., Boido, E., Fariña, L., Dellacassa, E., Carrau, F., 2016. Non-*Saccharomyces* and *Saccharomyces* strains co-fermentation increases acetaldehyde accumulation: effect on anthocyanin-derived pigments in Tannat red wines. Yeast 33, 339–343.

Monagas, M., Gómez-Cordovés, C., Bartolomé, B., 2007. Evaluation of different *Saccharomyces cerevisiae* strains for red winemaking. Influence on the anthocyanin, pyranoanthocyanin and non-anthocyanin phenolic content and colour characteristics of wines. Food Chem. 104, 814–823.

Morata, A., Gómez-Cordovés, M.C., Suberviola, J., Bartolome, B., Colomo, B., Suárez, J.A., 2003a. Adsorption of anthocyanins by yeast cell walls during the fermentation of red wines. J. Agric. Food Chem. 51, 4084–4088.

Morata, A., Gómez-Cordovés, M.C., Colomo, B., Suárez-Lepe, J.A., 2003b. Pyruvic acid and acetaldehyde production by different strains of *Saccharomyces cerevisiae*: relationship with vitisin A and B formation in red wines. J. Agric. Food Chem. 51, 7402–7409.

Morata, A., Gómez-Cordovés, M.C., Colomo, B., Suárez, J.A., 2005. Cell wall anthocyanin adsorption by different *Saccharomyces* strains during the fermentation of *Vitis vinifera* L. cv Graciano grapes. Eur. Food Res. Technol. 220 (3-4), 341–346.

Morata, A., Gómez-Cordoves, M.C., Calderon, F., Suárez, J.A., 2006. Effects of pH, temperature and SO_2 on the formation of pyranoanthocyanins during red wine fermentation with two species of *Saccharomyces*. Int. J. Food Microbiol. 106, 123–129.

Morata, A., Calderón, F., González, M.C., Gómez-Cordovés, M.C., Suárez-Lepe, J.A., 2007a. Formation of the highly stable pyranoanthocyanins (vitisins A and B) in red wines by the addition of pyruvic acid and acetaldehyde. Food Chem. 100, 1144–1152.

Morata, A., González, C., Suárez-Lepe, J.A., 2007b. Formation of vinylphenolic pyranoanthocyanins by selected yeasts fermenting red grape musts supplemented with hydroxycinnamic acids. Int. J. Food Microbiol. 116, 144–152.

Morata, A., Loira, I., Heras, J.M., Callejo, M.J., Tesfaye, W., González, C., et al., 2016. Yeast influence on the formation of stable pigments in red winemaking. Food Chem. 197, 686–691.

Moutonnet, M., Battle, J.L., Saint Pierre, B., Escudier, J.L., 1999. Stabilisation tartrique. Détermination du degree d'instabilité des vins. Mesure de l'efficacité des inhibiteurs de cristallisation. In: Aline Lonvaud-Funel (Ed.), 6e Symposium International d'Oenologie, Lavoisier Tec-Doc, Paris, France, pp. 531–534.

Nehme, N., Mathieu, F., Taillandier, P., 2008. Quantitative study of interactions between *Saccharomyces cerevisiae* and *Oenococcus oeni* strains. J. Ind. Microbiol. Biotechnol. 35, 685–693.

Nieuwoudt, H., Prior, B., Pretorius, S., Bauer, F., 2002. Glycerol and wine quality: fact and fiction. Wynboer . Available from: www.wynboer.co.za/recentarticles/1102glycerol.php3

Nordstrom, K., 1964. Studies on the formation of volatile esters in fermentation with brewer's yeast. Sven. Kem Tidskr. 76, 510–543.

Nygaard, M., Prahl, C., 1997. Compatibility between strains of *Saccharomyces cerevisiae* and *Leuconostoc oenos* as an important factor for successful malolactic fermentation. Am. J. Enol. Vitic. 48, 270.

Nykanen, L., 1986. Formation and occurrence of flavour compounds in wine and distilled alcoholic beverages. Am. J. Enol. Vitic. 37, 84–96.

Oreglia, F., 1978. Enología teórico práctica 2ª Edición. Ed. Instituto Salesiano de Artes Gráficas, Buenos Aires.

Osborne, J.P., Edwards, C.G., 2006. Inhibition of malolactic fermentation by *Saccharomyces* during alcoholic fermentation under low- and high nitrogen conditions: a study in synthetic media. Aust. J. Grape Wine Res. 12, 69–78.

Park, S.K., Boulton, R.B., Noble, A.C., 2000. Automated HPLC analysis of glutathione and thiol-containing compounds in grape juice and wine using pre-column derivatization with fluorescence detection. Food Chem. 68 (4), 475–480.

Pérez, G., Fariña, L., Barquet, M., Boido, E., Gaggero, C., Dellacassa, E., et al., 2011. A quick screening method to identify β-glucosidase activity in native wine yeast strains: application of Esculin Glycerol Agar (EGA) medium. World J. Microbiol. Biotechnol. 27, 47–55.

Peynaud, E., 1984. Knowing and Making Wine. John Wiley, New York, NY.

Ramey, D.D., Ough, C.S., 1980. Volatile ester hydrolysis or formation during storage of model solutions and wines. J. Agric. Food Chem. 28 (5), 928–934.

Rankine, B.C., Fornachon, J.C.M., Bridson, D.A., 1969. Diacetyl in Australian dry red wines and its significance in wine quality. Vitis 8, 129–134.

Rapp, A., Versini, G., 1996. Influence of nitrogen compounds in grapes on aroma compounds of wines. Vitic. Enol. Sci. 51, 193–203.

Reazin, G., Scales, H., Andreasen, A., 1970. Mechanism of major congener formation in alcoholic grain fermentations. J. Agric. Food Chem. 18, 585–589.

Handbook of Enology. In: Ribereau-Gayon, P., Dubourdieu, D., Doneche, B., Lanvaud, A. (Eds.), The Microbiology of Wine and Vinifications, vol. 1. Wiley and Sons, Chichester, UK.

Romano, P., Suzzi, G., Turbanti, L., Polsinelli, M., 1994. Acetaldehyde production in *Saccharomyces cerevisiae* wine yeasts. FEMS Microbiol. Lett. 118, 213–218.

Rosi, I., Vinella, M., Domizio, P., 1994. Characterization of β-glucosidase activity in yeasts of oenological origin. J. Appl. Bacteriol. 77, 51119–51217.

Rosi, I., Gheri, A., Domizio, P., Fia, G., 1999. Production of parietal macromolecules by *Saccharomyces cerevisiae* and their influence on malolactic fermentation. Colloids and mouthfeel in wines. In: Lallemand (Ed.), Lallemand Technical Meeting. Montreal, Canada.

Rosi, I., Fia, G., Canuti, V., 2003. Influence of different pH values and inoculation time on the growth and malolactic activity of a strain of *Oenococcus oeni*. Aust. J. Grape Wine Res. 9, 194–199.

Ruiz-Hernández, M., 2004. Los compuestos fenólicos y el color del vino tinto. Tratado de Vinificación en Tinto, first ed Mundi-Prensa, Madrid.

Saucier, C., Roux, D., Glories, Y. 1996. Stabilité colloidale polymers catéchiques. Influence des polysaccharides. In: Aline Lonvoud-Funel (Ed.), 5e Symposium International d'Oenologie. Lavoisier Tec-Doc, Paris, France, pp. 395–400.

Schwarz, M., Wabnitz, T.C., Winterhalter, P., 2003. Pathway leading to the formation of anthocyanin-vinylphenol adducts and related pigments in red wines. J. Agric. Food Chem. 51, 3682–3687.

Sefton, M., Francis, I., Williams, P., 1993. The volatile composition of Chardonnay juices: a study by flavor precursor analysis. Am. J. Enol. Vitic. 44, 359–370.

Smit, A., Cordero Otero, R.R.C., Lambrechts, M.G., Pretorius, I.S., Van Rensburg, P., 2003. Enhancing volatile phenol concentrations in wine by expressing various phenolic acid decarboxylase genes in *Saccharomyces cerevisiae*. J. Agric. Food Chem. 51, 4909–4915.

Smyth, H.E., Cozzolino, D., Herderich, M.J., Sefton, M.A., Francis, I.L., 2005. Relating volatile composition to wine aroma: identification of key aroma compounds in Australian white wines. In: Blair, R., Williams, P., Pretorius, S. (Eds.), Proceedings of the Twelfth Australian Wine Industry Technical Conference. Australian Wine Industry Technical Conference. Melbourne, Adelaide, Australia, pp. 31–33.

Spagna, G., Barbagallo, R.N., Palmeri, R., Restuccia, C., Giudici, P., 2002. Properties of endogenous β-glucosidase of a *Saccharomyces cerevisiae* strain isolated from Sicilian musts and wines. Enzyme Microb. Technol. 31 (7), 1030–1035.

Swiegers, J.H., Bartowsky, P.A., Henschke, P.A., Pretorius, I.S., 2005. Yeast and bacterial modulation of wine aroma and flavour. Aust. J. Grape Wine Res. 11, 139–173.

Ugliano, M., Moio, L., 2005. Changes in the concentration of yeast-derived volatile compounds of red wine during malolactic fermentation with four commercial starter cultures of *Oenococcus oeni*. J. Agric. Food Chem. 53 (26), 10134–10139.

Valero, E., Schuller, D., Cambon, B., Casal, M., Dequinet, S., 2005. Dissemination and survival of commercial wine yeast in the vineyard: a large-scale, three-years study. FEMS Yeast Res. 5, 959–969.

Vasserot, Y., Caillet, S., Maujean, A., 1997. Study of anthocyanin adsorption by yeast lees. Effect of some physicochemical parameters. Am. J. Enol. Vitic. 48, 433–437.

Vasserot, Y., Mornet, F., Jeandet, P., 2010. Acetic acid removal by *Saccharomyces cerevisiae* during fermentation in oenological conditions. Metabolic consequences. Food Chem. 119, 1220–1223.

Vernocchi, P., Ndagijimana, M., Serrazanetti, D.I., Chaves López, C., Fabiani, A., Gardini, F., et al., 2011. Use of *Saccharomyces cerevisiae* strains endowed with β-glucosidase activity for the production of Sangiovese wine. World J. Microbiol. Biotechnol. 27, 1423–1433.

Winterhalter, P., Skouroumounis, G.K., 1997. Glycoconjugated aroma compounds: ooccurrence, role and biotechnological transformation. In: Scheper, T. (Ed.), Advances in Biochemical Engineering Biotechnology. Springer, Germany, pp. 73–105.

Further Reading

Carrau, F., Boido, E., Gaggero, C., Medina, K., Fariña, L., Disegna, E., et al., 2011. *Vitis vinifera* Tannat, chemical characterization and functional properties. Ten years of research. In: Filip, R. (Ed.), Multidisciplinary Approaches on Food Science and Nutrition for the XXI Century. Transworld Research Network, Kerala, India.

CHAPTER 6

Malolactic Fermentation

Irene Gil-Sánchez, Begoña Bartolomé Suáldea and M. Victoria Moreno-Arribas

Instituto de Investigación en Ciencias de la Alimentación (CIAL), CSIC-UAM, Madrid, Spain

6.1 INTRODUCTION

Malolactic fermentation (MLF), also known as secondary fermentation or malolactic deacidification, is technically not a fermentation but the enzymatic decarboxylation of the dicarboxylic L-malic acid to the monocarboxylic L-lactic acid) in a reaction requiring $NAD+$ and Mn^{2+} as cofactors and devoid of free intermediates (Naouri et al., 1990). This process usually follows primary alcoholic fermentation (AF) of wine but may also occur concurrently. This conversion may occur spontaneously by indigenous lactic acid bacteria (LAB), or be induced by a commercial starter culture. *Oenococcus oeni* is the main bacterial species responsible for conducting this biochemical stage, due to its ability to tolerate the harsh physiochemical properties of wine after completion of AF (Bartowsky, 2005). Nevertheless, certain strains belonging to the genus *Lactobacillus* and *Pediococcus* can also induce and/or contribute to this process. In particular, the species *Lactobacillus plantarum* has also been shown to be suitable to drive this process (du Toit et al., 2011; Berbegal et al., 2016) and, in fact, there are some commercial malolactic starters of these species.

MLF is a process required for most red wines but also for some white wines and base sparkling wines; it makes wines more palatable by reducing the tart taste associated with malic acid, and provides additional improvements, like microbial stability and enhanced aroma and flavor complexity. Apart from its main sensory effect (depleting malic acid from wine) this secondary fermentation could modify the aromatic properties of wines by releasing notable concentrations of diacetyl (2,3-butanedione) and other carbonyl compounds, which contributes to the buttery aroma of wines. Many other biochemical reactions which occur at the same time also enhance wine aroma and quality. Esterase activity, methionine metabolism, and some transformations involving glycosidases have also been demonstrated to be carried out by LAB (Moreno-Arribas and Polo, 2005; Muñoz et al., 2011).

Despite the importance of MLF in wines, its occurrence is often unpredictable and it is difficult to control or manipulate. Uncontrolled MLF may risk wine acceptability due to the formation of off-flavors compounds (including acetic acid, volatile phenols, and mousiness, among others), and even, it may risk wine safety due to the generation of compounds hazardous to human health [such as ethyl carbamate (EC) and biogenic amines]. Therefore, it is essential to control this biochemical process during the winemaking in order to ensure the final quality of wines.

This chapter describes the main wine LAB involved in MLF as well as the influence of some wine-related physicochemical factors on their growth and metabolism. After discussing different technological strategies for MLF management at the winery, the most important bacterial metabolic pathways produced during this stage and their impact on wine organoleptic properties and human health are also detailed.

6.2 LACTIC ACID BACTERIA IN WINEMAKING

A complex ecological niche is involved in the winemaking process, including yeast, bacteria, fungi, and virus (du Toit and Pretorius, 2000). LAB can both be naturally present on grape skins and in the cellular environment (including barrels, tank, pipelines…), and/or be inoculated by winemakers through the addition of starters. A large amount of research has focused on the description and ecology of LAB in wine; and their involvement in winemaking, their distribution, and their succession in musts have been extensively studied in wine and during fermentation (Liu, 2002; Sumby et al., 2014).

LAB are Gram positive, microaerophilic, and characterized by the formation of lactic acid as a primary metabolite of sugar (glucose) (Dicks and Endo, 2009). They are divided into two groups according to their glucose metabolic activity: homofermentative (which ferment glucose with lactic acid as the primary by-product), and heterofermentative (which ferment glucose producing lactic acid, ethanol/acetic acid and carbon dioxide as by-products). The most common isolates from wine are in the genera *Oenococcus*, *Lactobacillus*, and *Pediococcus* (see below), and *Leuconostoc* to a lower extent.

6.2.1 *Oenococcus oeni*

Of the three *Oenococcus* species, *O. oeni* is associated with wine. *O. oeni* species is nonmotile and asporogenous with ellipsoidal-to-spherical cells usually arranged in pairs or short chains, and has an optimal growth range between 20°C and 30°C and pH 4.8–5.5. While lactobacilli predominate on grape skins, the *O. oeni* population increases throughout alcoholic (yeast) fermentation to typically become the only species found in wine at the completion of MLF. For this reason and because of its desirable flavor effects, *O. oeni* is the preferred species for this process, which is applied to most red, aged white, and sparkling wine styles (Henick-Kling et al., 1993; Lafon-Lafourcade et al., 1983b). Real-time quantitative Polymerase chain reaction (PCR) (q-PCR) methods are being developed to enable the rapid detection and quantification of these bacteria in wine samples during fermentation. The main advantage of q-PCR methods is that they enable rapid corrective action to be taken in order to control bacterial growth (Pinzani et al., 2004). This genus comprises a compact genome, forming around 1.8 Mb (Borneman et al., 2012). The three genes responsible for MLF are present in a single cluster, with mleA (encoding malolactic enzyme) and mleP (encoding malate permease) on the same operon and mleR encoding the regulatory protein transcribed in the opposite direction. Maximal activity of mleA is seen at pH 5.0 and 37°C and it is noncompetitively inhibited by ethanol, underscoring the less-than-ideal nature of the wine environment.

6.2.2 *Lactobacillus* sp.

Wine *Lactobacillus* sp. are facultative heterofermentative and have regular elongated shapes, often with long rod-like forms. To date, several species of *Lactobacillus* genus have been isolated from grape, must, and wine, including *L. plantarum*, *L. casei*, *L. brevis*, *L. fermentum*, *L. bobalius*, *L. buchneri*, *L. collinoides*, *L. fermentum*, *L. fructivorans*, *L. hilgardii*, *L. kunkeei*, *L. lindneri*, *L. mali*, *L. nagelii*, *L. oeni*, *L. paracasei*, *L. paraplantarum*, *L. uvarum*, and *L. vini*, some of which have been sequenced (Mtshali et al., 2012; Zhao et al., 2016; Lamontanara et al., 2015). Wine lactobacilli possess a large number of enzymes encoding important genes for the production of wine aroma compounds. In 1988, the potential of *L. plantarum* as a malolactic starter culture was realized by Prahl (1988) with the first freeze-dried culture being released. Today, there are a few *L. plantarum* strains commercially available as MLF starter cultures (Lerm et al., 2011; Fumi et al., 2010). Some relevant characteristics of *L. plantarum*, such as the ability to function well in high pH conditions, the tolerance of ethanol up to 14%, tolerance of SO_2 similar to *O. oeni*, and more diverse array of enzymes that could lead to more aroma compounds being produced, make *L. plantarum* as the up-to-date generation wine MLF starter cultures (Lerm et al., 2011; du Toit et al., 2011; Spano et al., 2002).

6.2.3 *Pediococcus* sp.

The genus *Pediococcus* is homofermentative and has ellipsoidal or spherical shape. Within this genus, there are only four species that play an important role in the MLA: *Pediococcus damnosus*, *Pediococcus parvulus*, *Pediococcus pentosaceus*, and *Pediococcus inopinatus* (Gonzalez-Centeno et al., 2017). They are commonly considered as spoilage bacteria in wine as some strains have the capacity to synthesize exopolysaccharides, consequently providing a viscous and thick texture to the wine (Walling et al., 2005), and also producing high acetic acid concentrations and biogenic amines (Lafon-Lafourcade et al., 1983a; Landete et al., 2005). Nonetheless, as in the case of *Lactobacillus*, *Pediococcus* species produce a great number of enzymes that generate desirable wine aroma compounds (Juega et al., 2014).

6.3 FACTORS IMPACTING LAB AT WINERY

In normal conditions, once AF is complete, there is a lag phase lasting between 10 and 15 days during which the population of LAB remains unchanged as their growth is inhibited by the presence of live yeasts and inhibitory substances secreted by these. Once this phase is complete, the bacteria begin to multiply until they reach a density of approximately 10^6 CFU/mL, and MLF begins. Therefore, at the winery it is essential to control the underlying conditions under which LAB carry out MLF, thereby assuring that the fermentation process runs properly. There are several wine physicochemical parameters that influence the growth and metabolism of LAB. Growth is favored by a relatively high pH level (>3.5), a sulfur dioxide concentration of no more than 50 mg/L, an ethanol content of 13% (vol/vol), and a temperature of between 19°C and 26°C. However, as shown Table 6.1, the typical wine conditions differ from these optimal conditions. The stressors have various cellular targets and mechanisms (see below), which often work in combination to produce a more severe impact on growth or the enzymes involved in this biochemical process (Betteridge et al., 2015).

6.3.1 Ethanol

Ethanol interferes with the growth and metabolism of LAB. The most wine LAB are tolerant to ethanol levels of up to 14% (v/v), so wine's alcohol concentration does not inhibit the growth of these microorganisms as it is expected to be 11%–14% (v/v). However, ethanol tolerance is influenced by temperature and pH. As temperature increases, ethanol toxicity also increases (Asmundson and Kelly, 1990). This means that at 25°C (optimal growth temperature) ethanol levels of 14% (v/v) inhibit the growth of most malolactic bacteria; this is why the MLF takes place between 18°C and 22°C (Henick-Kling, 1993). Furthermore, ethanol increases the passive proton flux into bacteria, which, in turn, affects cell mechanisms that maintain pH homeostasis. Therefore, high ethanol levels and low pH hinder the bacteria to maintain an equilibrium in their hydrogen ion gradients. Although the degree of ethanol tolerance is strain-dependent, species of *Lactobacillus* and *Pediococcus* are in general more resistant to a high ethanol concentration than *O. oeni* (Davis et al., 1988).

6.3.2 pH

pH is a key parameter as it determinates which LAB survive during this biochemical process. In general, MLF takes place most of the time in acidic conditions. Thus, wines of pH 3.2–3.6 exhibit less problems in undergoing MLF (Kunkee, 1967). In particular, *O. oeni* strains are able to decarboxylate malic acid in wines with pH values closed to 3.5. Low pH (2.9–3.2) slows down or inhibits (<2.8) the growth and metabolic activities of the main LAB, and makes them more sensible to SO_2 and ethanol effects. High pH values (3.6–4.2) encourage the growth of strains belonging to lactobacilli and *Pediococcus* spp., which can produce unfavorable taste and aroma components (Bauer and Dicks, 2004).

TABLE 6.1 Main inhibitors of malolactic fermentation in wines

Inhibitors	Description	Optimal condition for MLF	Wine conditions	Action mechanisms of inhibitors
Ethanol	Primary metabolite produced by alcoholic fermentation	Up to 5% favors growth	12%–15% (v/v)	Affect physical structure of cell membranes
Low pH	Acidity from organic acids extracted from grapes and intervention by enologist during winemaking	4.8–5.5	2.5–3.5	Slow down or inhibit the bacterial growth and metabolic activities
Low temperature	Temperature controlled by enologist	25°C	12–20°C	Increase lag phase affecting growth rate
SO_2	Intermediate metabolite produced by yeast and antimicrobial agent used in winemaking	0 mg/L	10–210 mg/L	Decrease the ATPase specific activity and produces a loss of cell viability

Adapted from Betteridge, A., Grbin, P., Jiranek, V., 2015. Improving Oenococcus oeni *to overcome challenges of wine malolactic fermentation. Trends Biotechnol. 33, 547–553.*

6.3.3 Sulfur Dioxide

Adding sulfur dioxide (SO_2) is a common practice during the vinification. This additive is used to preserve the quality of wine due to its antioxidant and selective antimicrobial effects (especially against spoilage LAB). Two classes of sulfites are found in wine: bound and free. The bound sulfites are those that have reacted with other molecules into wine matrix, such as acetaldehyde, anthocyanins, or acids, and thus cannot exert their germicidal effects, while the free sulfites are those available to protect the wine against undesirable microorganisms. At wine pH 3.5, the predominant species are the free sulfites, the more active form of sulfites. Despite its beneficial use in the winemaking, the presence of a large amount of SO_2 can interfere with the ability of LAB to start with MLF. For this reason, the concentration of this additive must be closely monitored and regulated. Although the bacterial adaptation phenomenon to SO_2 is still unknown, it seems to be related to pH tolerance (Guzzo et al., 1998).

6.3.4 Temperature

Temperature is a factor that affects directly the growth rate and lag phase length of LAB (Bauer and Dicks, 2004). The optimal growth of most LAB strains, especially for *O. oeni*, is achieved at temperature close to 25°C. To date, some studies have reported that higher temperatures accelerate the growth of bacteria, leading to deterioration of the wine, while lower temperatures inhibit or slow it down, avoiding the fermentation process (Henick-Kling, 1993; Van der Westhuizen and Loos, 1981). In general, MLF takes place at temperatures around 18–20°C for white wines and 18–22°C for red wines.

6.4 TECHNOLOGICAL STRATEGIES FOR MANAGING THE MLF PERFORMANCE

Industrial malolactic strains, mainly strains of *O. oeni* and *L. plantarum*, have been selected on the basis of their performance for malic acid conversion, no biogenic amine production, and a positive contribution to wine sensory properties. MLF usually takes place after AF, therefore meaning that LAB must face a harsh environment in which high alcohol content, low concentration of nutrients, low pH, and high SO_2 content are the main but not the only hurdles (G-Alegría et al., 2004).

Stress-inducing factors such as ethanol, acidic pH, phenolic compounds, sulfur dioxide, and fatty acids in wine have an inhibitory effect on MLF performance that has been linked to inhibition of ATPase activity (Carreté et al., 2002). The expression of the malolactic operon in *O. oeni* appears to be regulated by another factor linked to metabolic energy (Galland et al., 2003). Genes previously implicated in the stress response in *O. oeni* (clpX, clpLP, trxA, hsp18, ftsH, ormA, and the operons groESL and dnaK) were also found in the fully sequenced PSU-1 strain (Mills et al., 2005). As far as oxidative stress is concerned, like all LAB, *O. oeni* is microaerophilic and does not possess catalase activity. It does, however, have the genes trxA and trxB and systems to eliminate reactive oxygen species (ROS) such as NADH-oxidase. In order to identify novel genes involved in adaptation to wine, new approaches using genome-wide analysis based on stress-related genes was performed in strain *O. oeni* PSU-1, and 106 annotated stress genes were identified. The in silico analysis revealed the high similarity of all those genes through 57 *O. oeni* genomes; however, seven variable regions of genomic plasticity could be determined for their different presence observed among these strains (Margalef-Català et al., 2017).

Slow or incomplete MLF due to failure of *O. oeni* to successfully implant, or complete metabolism of malic acid can have undesirable consequences in wine quality and winery economic affairs. Wine nutrient status is a crucial parameter for ensuring the success of MLF. LAB are strictly fermentative and have complex nutritional needs, requiring the presence of carbohydrates, amino acids, peptides, fatty acid esters, salts, nucleic acid derivatives, and vitamins (Hébert et al., 2004). Yeasts usually consume the most wine essential nutrients during the AF. Once this stage is complete, some of these nutrients may be returned to wine by the lysis of yeast cells, however, in most cases these amounts are insufficient. Consequently, it is often necessary to supplement wine with nutrients to ensure MLF performance. Furthermore, in an effort to avoid protracted or "stuck" MLF or encourage spontaneous MLF, the addition of protective amounts of SO_2 may be delayed, thereby increasing the risk of spoilage by yeast or bacteria or oxidation of the juice/wine. Because of their more precarious state, such batches demand closer monitoring by the winemaker. Since wines are rarely sterile filtered, packaging wine with residual malic acid carries a risk of spoilage organisms growing to produce haze, off-odors, and/or dissolved CO_2 in the bottle. Solutions include one or more reinoculations with fresh bacterial starter cultures, addition of nutrients, removal of inhibitors (e.g., SO_2), warming of the wine, or abandonment of the MLF with stability by greater SO_2 addition, which itself compromises quality (Betteridge et al., 2015).

Nutrients or activators formed by inactive yeasts and substances such as casein and cellulose are often used to activate MLF. These cultures contain amino acids and vitamins that function as growth factors for LAB and also absorb inhibitory substances such as sulfites and medium-chain fatty acids (Lonvaud-Funel et al., 1988). A possible alternative to these activators is the use of *O. oeni* cells immobilized in different matrices. This strategy can increase the productivity of fermentation because of the higher packing density and the greater protection afforded to cells. Examples of different materials used as immobilization matrices in studies analyzing the use of immobilized forms of *O. oeni* to deacidify wine include alginates, polyacrylamide, wood shavings, and cellulose sponges (Maicas et al., 2001). Not all of these agents, however, have been accepted by winemakers as they imply the use of additional chemical compounds.

Different investigations have reported the presence of compounds capable of inhibiting or slowing down the MLF in wines. Some polyphenols including hydroxybenzoic acids and their derivatives, hydroxycinnamic acids, phenolic alcohols as well as other related compounds as stilbenes, flavan-3-ols, and flavonols have shown inhibitory effects on the growth of enological LAB (García-Ruiz et al., 2011; Campos et al., 2009). Other compounds such as residues of plant-protection products or some yeast metabolites (lysozyme and short-chain fatty acids) have shown to negatively affect to the development of MLF (Cabras et al., 1994).

On the other hand, taking into account that MLF can be a source of concern to winemakers due to the unpredictability of the spontaneous process, including difficulties often encountered by commercial MLF starters to develop under industrial conditions, wine yeast strains capable of performing MLF have been developed in last years. In fact, the first recombinant yeast strain to get official approval by appropriate food safety authorities (in the United States and Canada) was a malolactic wine yeast. Several variants of malolactic wine yeast were engineered including different sources of malolactic enzyme and malate permease, before the commercial strain ML01 was developed. This strain carries the *Schizosaccharomyces pombe* malate permease gene (*mae1*) and the *O. oeni* malolactic gene (*mleA*). The strain was shown to fully decarboxylate 5.5 g/L of malate in Chardonnay grape must during AF (Coulon et al., 2006). However, despite the ongoing development of these cutting-edge techniques and their potential enormous application, strict genetically modified organism regulation and consumer demands and preferences, raising issues related to food and environmental safety, have limited their application in the wine industry (Gonzalez et al., 2016).

In relation to sulfur dioxide (SO_2), while only the free form of sulfur dioxide has antiseptic properties against yeast in wine, all forms of sulfur dioxide have activity against bacteria. The antibacterial effect of sulfur dioxide depends mainly on the pH of the wine. The levels of free sulfur dioxide required to inhibit the activity of LAB range from 10 to 20 mg/L for wines with a low pH and from 20 to 50 mg/L for wines with a high pH. Coccoid species (*Pediococcus*, *Oenococcus*, and *Leuconostoc*) are less resistant than *Lactobacillus* species to the effect of sulfur dioxide. Sulfites may lead to adverse health effects in a small but significant proportion of the population, therefore the search for alternatives is a new challenge in the enology field (Guerrero and Cantos-Villar, 2015; Santos et al., 2012; Monge and Moreno-Arribas, 2016; Vally et al., 2009).

6.5 IMPACT OF MLF ON WINE ORGANOLEPTIC PROPERTIES

Metabolism of LAB responsible for MLF has been extensively studied over the last years (Liu, 2002; Lonvaud-Funel, 1999; Swiegers et al., 2005; Belda et al., 2017). It reduces the total acidity of wine in terms of 1–3 g/L (the decline of pH is proportional to initial concentration of malic acid) (Costantini et al., 2009). Besides lactic acid, the main substrate of MLF, there are a large number of metabolic end products, produced by specific bacterial species/strains that are responsible for modifying aroma and flavor perception of wine. Below, it is given a brief overview of the main metabolic pathways related to MLF that contribute to modify the wine organoleptic properties.

6.5.1 Carbonyl Compounds

Diacetyl (2,3-butanedione) is the main flavor compound derived from MLF. It confers "buttery," "nutty," and "butterscotch" characters to wines (Bartowsky, 2002). Diacetyl is an intermediate compound mainly produced during citric acid metabolism by LAB. Citric acid, one of the acids present in both grapes and must, is generally found at lower concentrations (0.1–1 g/L) than major organic acids such as tartaric (2–8 g/L) and malic acids (1–7 g/L). The citric acid is converted by citrate lyase and oxaloacetate decarboxylase into pyruvate, which is mostly reduced to lactate in the presence of NADH. However, pyruvate is sometimes converted by acetolactate decarboxylase to acetolactic acid, giving rise to acetoin and 2,3-butanediol following decarboxylation. The chemical oxidation of acetoin, in turn, yields diacetyl. The degradation of citric acid by LAB automatically leads to an increase in volatile acidity in wine (as an average, 1.2 molecules of acetic acid are produced from each molecule of citric acid). However, due to the small quantities concerned, this phenomenon is not detrimental to wine quality.

Wines that undergo MLF generally have a greater concentration of diacetyl than those that do not (Martineau et al., 1995). Moreover, this transformation is promoted by the prolonged contact with bacterial biomass or yeast lees. While moderate levels of diacetyl have a positive effect on aroma, high levels cause an unpleasant aroma, leading to spoilage (Nielsen and Richelieu, 1999). Consequently, winemakers try to control diacetyl concentrations to enhance aroma by eliminating the microorganisms earlier in the process or, in contrast, by maintaining the wine with yeast lees. The final concentration of diacetyl in wine also depends on various factors, including bacterial strain, wine type, and sulfur dioxide and oxygen concentrations.

Analyses of the *O. oeni* genome showed the presence of the typical *cit* gene group, which includes genes that encode citrate lyase (*cit-DEF*), citrate lyase ligase (*citC*), oxaloacetate decarboxylase (*mae*), and the citrate transporter (*maeP* o *citP*) (Mills et al., 2005). The genome also contains genes involved in the butandiol pathway (*ilvB*, *alsD*, *butA*).

6.5.2 Esters

The majority of wine esters are produced as secondary products by yeast during the AF, however once this process is finished, the esters profile can be modified by the esterase activity of LAB (Matthews et al., 2004). In this vein, MLF is associated with increases in the concentration of ethyl fatty acid esters (ethyl lactate, ethyl acetate, ethyl hexanoate, and ethyl octanoate) rather than acetate esters (Cappello et al., 2017; Lerm et al., 2010; Ugliano and Moio, 2005; Liu, 2002). As in the case of diacetyl, it is important to note that all these volatile compounds are responsible for the desirable fruity aroma of wines at appropriate concentrations. By contrast, at high concentration, they can also provide a detrimental effect on wine aroma. Table 6.2 contains the aroma descriptors of the main esters, as well as the concentrations and odor thresholds detected in wines.

Different studies demonstrated that wine LAB exhibit enzyme activities that can augment the ethyl ester content of wine (Pozo-Bayón et al., 2005; Antalick et al., 2012). However, while the esterases from yeasts have been widely studied, esterase activity from wine-related LAB is not well documented (Sumby et al., 2014; Pérez-Martín et al., 2013). For example, *O. oeni* produced significant levels of ethyl hexanoate and ethyl octanoate following growth in an ethanolic test medium, and also esterified 1-propanol to produce propyl octanoate (Costello et al., 2013). However, the concentrations of some of these compounds appear to be influenced by the LAB species and the strain used, reflecting a degree of diversity among strains of the same species (Pozo-Bayón et al., 2005). Intracellular esterases from *O. oeni* and *L. hilgardii* were characterized under wine-like conditions. Both

TABLE 6.2 Aroma descriptors, wine concentrations, and odor thresholds of most esters that contribute to aroma during MLF

Ester	Aroma descriptors	Wine concentrations (µg/L)	Odor threshold (µg/L)	References
Ethyl acetate	Fuit, nail polish	22,500–208,000	12,264	Francis and Newton (2005)
Isoamyl acetate	Banana, sweet	118–7900	30	Francis and Newton (2005)
Phenylethyl acetate	Flowery, honey, tobacco	0.54–960	250	Francis and Newton (2005)
Isobutyl acetate	Fruit	10–1600	1600	Swiegers et al. (2005)
Hexyl acetate	Sweet, perfume	0–4800	700	Swiegers et al. (2005)
Ethyl butyrate	Fruit	20–1118	20	Francis and Newton (2005)
Ethyl hexanoate	Fruit, brandy	153–2556	5–14	Francis and Newton (2005)
Ethyl octanoate	Floral, fruit, sweet	138–783	2–5	Francis and Newton (2005)
Ethyl decanoate	Floral	0–2100	200	Swiegers et al. (2005)

esterases were stable and retained activity under conditions that would be encountered in wine. They have the potential to reduce short-chain ethyl esters such as ethyl acetate (Sumby et al., 2013). Among wine LAB, besides *O. oeni*, *L. plantarum* strains are also used as malolactic starters. *L. plantarum* is a good source of esterase enzymes; in fact, some esterase proteins have been purified and characterized in this species (Esteban-Torres et al., 2013).

6.5.3 Monoterpenes

The aromatic profile of many wines depends on the varietal compounds of the grapes that have been employed in their production. These varietal compounds can be present in grapes as free volatile compounds and, in much higher concentrations, as aroma precursors. Among them, nonvolatile sugar-bound conjugates are odorless molecules which represent a natural reservoir of odorant compounds in wines, which can be naturally and slowly released during wine aging or intentionally released by using enological enzymes during winemaking. A large number of glycosidic conjugates (including α-D-glucosides, α-L-arabinofuranosyl-β-D-glucosides, α-L-rhamnopyranosyl-β-D-glucosides, or β-D-apiofuranosyl-β-D-glucosides) originating from the grape, yeast metabolism, or by the oak used have been reported in wine (Swiegers et al., 2005). The hydrolysis of these molecules releases compounds such as C13 norisoprenoids, volatile phenols, benzene derivatives, or aliphatic compounds, which contribute to the aroma wine (Gastón Orrillo et al., 2007). For this reason, glycosidic conjugates are considered an important aromatic reservoir in wine. The cleavage glycosidic bounds required the action of bacterial glucosidases. In the case of monoglycosides, only the action of β-glucosidase is needed, while for disaccharides, glycosides are required together with specific exo-glycosydases (Günata et al., 1988). Nevertheless, the degree of enzyme hydrolysis of odorless nonvolatile glycosides depends on other factors, such as chemical structure of substrates, pH, temperature, and/or ethanol (Grimaldi et al., 2000, 2005b).

Some transformations involving glycosidases have been demonstrated in *O. oeni* (Grimaldi et al., 2005b). For example, Saguir et al. (2009) showed that different *O. oeni* cultures at the end of their exponential growth possessed a detectable and variable level of β-glucosidase activity. In accordance with these findings, Grimaldi et al. (2000) also demonstrated a significant β-glucosidase activity in several *O. oeni* strains. In addition, this enzymatic activity has been also found in other wine LAB strains belong to genus *Lactobacillus* and *Pediococcus* (Grimaldi et al., 2005a). For instance, *L. brevis* and *L. casei* strains were able to increase the concentration of C13-norisoprenoides and monoterpenes after performing MLF (Hernandez-Orte et al., 2009). The release of variety-specific volatile compounds has been observed for Tannat, Chardonnay, and Muscat wines comparing the impact of several malolactic cultures (Boido et al., 2002; Ugliano et al., 2003). The grape variety aromas released by bacteria vary widely, depending on the strains and terpene substrates involved. Some authors have evidenced that *L. plantarum* shows a different enzymatic profile compared to other LAB species, which suggests that this species plays an important role in the wine aroma profile (Lerm et al., 2011; Iorizzo et al., 2016).

6.6 PRODUCTION OF OFF-FLAVORS BY LACTIC ACID BACTERIA

LAB can be responsible for the off-flavors in wine, including volatile phenols. Wine contains a great variety of phenolic compounds including phenolic acids, flavonols, proanthocyanidins, flavanols, and anthocyanins. Hydroxycinnamic acids (mainly ferulic, *p*-coumaric and caffeic acids) are phenolic acids very susceptible to the bacterial metabolism. These compounds are transported into bacterial cells to be converted into volatile phenols. In wines, hydroxycinnamic acids can be present in their free form or esterified with tartaric acid (fertaric, cutaric, and caftaric acids). These latter can be released into matrix by cinnamoyl esterase activities from fungal, commercial, and bacterial enzymes (Dugelay et al., 1993; Cabrita et al., 2008). The concentrations of volatile phenols in wine are usually low and depend on the amount of their corresponding precursors.

As shown in Fig. 6.1, phenolic acids are decarboxylated into 4-vinyl derivatives and then are eventually reduced to 4-ethyl derivatives. Some of these compounds, particularly vinyl and ethyl guaiacol (generated from ferulic acid) are associated with animal and medical aromas such as horse sweat, horse stable, barnyard, and elastoplast (Lonvaud-Funel, 1999). Although an increasing number of authors have reported that *Brettanomyces* yeasts (teleomorph, *Dekkera*), with *B. bruxellensis* as the most frequent representative, are considered the main producer of volatile phenols, different studies have reported the ability of several LAB strains to produce these undesirable volatile phenols under winemaking conditions. It seems that the genera *Lactobacillus* and *Pediococcus* had higher capacity for producing these compounds than *O. oeni* (Cavin et al., 1993). De las Rivas et al. (2009)

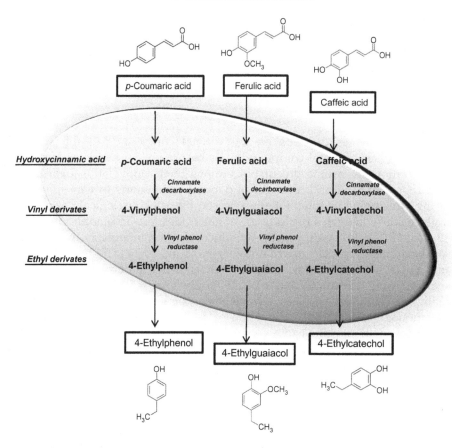

FIGURE 6.1 Formation of most phenol volatile compounds by LAB during MLF.

analyzed the capacity of LAB to produce volatile phenols in wine and described a PCR method for detecting bacteria with this potential. *L. plantarum*, *L. brevis*, and *P. pentosaceus* strains produced vinyl derivatives from hydroxycinnamic acids, but only *L. plantarum* strains produced the corresponding ethyl derivatives. *O. oeni*, *L. hilgardii*, and *Lc. mesenteroides* strains, in contrast, did not decarboxylate the hydroxycinnamic, p-coumaric, and ferulic acids, suggesting that they are not responsible for the production of volatile phenols (De las Rivas et al., 2009).

On the other hand, the production of undesirable aromas and flavors in wine described as "mousy" or "acetamide" has been associated with several LAB (Costello et al., 2001). A mousy odor or flavor is specifically attributed to the production of three volatile heterocyclic compounds: 2-ethyltetrahydropyridine, 2-acetyltetrahydopyridine, and 2-acetylpyrroline. Certain winemaking conditions such as high pH (>3.5) or low sulfur dioxide levels can favor the growth of the bacterial strains involved in the production of these bases. This flaw has been associated with heterofermentative strains of *Lactobacillus*, in particular *L. hilgardii*, followed by *O. oeni* and *Pediococcus* strains, as well as some homofermentative *Lactobacillus* species. A mousy taint can render an unpalatable wine and cannot be eliminated. Very few studies have analyzed the origin of this flaw and little is known about its repercussion on wine quality due to the complexity of the process, but also because it occurs in conjunction with other defects (Costello and Henschke, 2002).

6.6.1 Volatile Sulfur Compounds

Volatile sulfur compounds contribute to the complexity of wine by adding aromatic notes, in particular those arising from the typical grape variety. During the MLF, these compounds are produced by the metabolism of the sulfur-containing amino acids cysteine and methionine (these latter being commonly found in wine). It still remains unclear what is the precise biochemical pathway of wine LAB that makes up sulfur metabolism, but it is assumed that it shares a similar route to the dairy lactic bacteria industry (Liu et al., 2008). In recent years, some authors have investigated the methionine catabolism of malolactic bacteria at the winemaking condition, as well as the involved potential enzymes and codified their genes (Pripis-Nicolau et al., 2004; Vallet et al., 2008). As other metabolic products derived from MLF, volatile sulfur compounds (methanethiol, dimethyl disulfide, methionol, and 3-(methylsulfanyl)propionic acid) can positively contribute to the wine bouquet or decrease its quality depending on their concentrations (Swiegers et al., 2005; Pripis-Nicolau et al., 2004; Landaud et al., 2008).

6.7 IMPLICATIONS OF LAB AND MLF IN WINE SAFETY

Wine, like other fermented foods, is not without risk for human health. During MLF, two hazardous compounds, namely biogenic amines and EC, can be synthesized by LAB. Both compounds derive from microbial metabolism of amino acids (see below).

6.7.1 Biogenic Amines

Biogenic amines are low molecular weight, organic bases, frequently occurring in fermented food and beverages (Silla Santos, 1996; Ten Brink et al., 1990). Foods likely to contain high levels of biogenic amines include fish, fish products, and fermented foodstuffs (meat, dairy, some vegetables, beers, and wines). In the case of wine, up to 25 different amines have been described. The total content of amines in wine varies from trace up to 130 mg/L (Soufleros et al., 1998). Although these amounts are very low in comparison to other fermented foods, they are quite significant since wine ethanol can intensify their toxic effects by inhibiting the amine-oxidases responsible for their catabolism (Silla Santos, 1996).

The biogenic amines most commonly found in wine are putrescine, histamine, and tyramine. These compounds are mainly formed from their respective amino acid precursors (ornithine, histidine, and tyramine) via the substrate-specific enzymatic decarboxylation (Ten Brink et al., 1990). Consumption of high concentrations biogenic amines can have adverse effects for sensitive individuals such as headaches, gastrointestinal disorders, shortness of breath, fall in blood pressure, unconsciousness, and even cardiac arrhythmia in severe cases (Moreno-Arribas et al., 2010; Alvarez and Moreno-Arribas, 2014). Histamine and tyramine are considered as the most toxic and particularly relevant for food safety. Putrescine and cadaverine are known to potentiate these effects.

At present, a shared regulation limiting the amounts of biogenic amines in foods is still lacking, although their presence beyond the limits recommended by scientific literature may have negative commercial implications. For example, to minimize histamine toxicological effects, it is suggested that its concentration should not exceed 2 mg/L in fermented beverages, such as wine (Ten Brink et al., 1990). The only country with a limit for histamine in wine (10 mg/L) was Switzerland until 2008, but currently there is no legal or regulatory limit for histamine content in wine in any country in the world. In 2011, the International Organization of Vine and Wine (OIV) published the "Code OIV of good enological practices to minimize the presence of biogenic amines in wine," which highlights the importance of this issue for the food sector. One of the recommended strategies to prevent the formation of biogenic amines is the inoculation of safe malolactic starter cultures. However, despite a great deal of research, including toxicological studies, it has not been scientifically proven that biogenic amines are responsible for serious health problems (EFSA, 2011).

The role of LAB in the biosynthesis of biogenic amines has been widely reported (Smit et al., 2008; Moreno-Arribas and Lonvaud-Funel, 2001; Moreno-Arribas et al., 2003). Biogenic amine production in wines requires the availability of precursors (i.e., amino acids), the presence of bacteria synthesizing amino acids decarboxylases, and favorable conditions for their growth and decarboxylating activity (Costantini et al., 2009). The capacity to produce biogenic amines seems to be strain-dependent, and not a species-specific characteristic (Smit et al., 2008). Although certain *O. oeni* strains have demonstrated their ability to produce histidine, putrescine, and cadaverine (Guerrini et al., 2002; Izquierdo cañas et al., 2009), the species of the *Pediococcus* and *Lactobacillus* genera are primarily responsible for formation of biogenic amines in wines (Moreno-Arribas and Polo, 2008). The production of these compounds is one of the key criteria to take into account for the selection of MLF starters. Recently, a study demonstrated the capacity of an indigenous (noncommercial) *O. oeni* strain used as a malolactic starter to lower histamine formation in winery (Berbegal et al., 2017).

Detection of amino acid decarboxylase-positive microorganisms, both involving in vitro differential growth media and sensitive and specific PCR protocols based on the detection of gene-encoding decarboxylases, have been shown by several authors (Coton and Coton, 2005; de las Rivas et al., 2006; Landete et al., 2007), including multiplex PCR methods for the simultaneous detection of wine LAB with the potential to produce histamine, tyramine, and putrescine (Marcobal et al., 2005).

The preponderance of LAB microbiota involved in the fermentation process is mainly approached by adopting microbial starters lacking the pathways to degrade precursor amino acids (Moreno-Arribas et al., 2003; Landete et al., 2007; Del Prete et al., 2009). However, in an attempt to control the production of biogenic amines, microorganisms intended to be used as starter cultures in any fermented food should be confirmed not only as not producing biogenic amines but also as being able to outgrow autochthonous microbiota under conditions of production and storage (Moreno-Arribas and Polo, 2008; Gardini et al., 2002; Marcobal et al., 2006).

During wine production, surveillance of parameters that influence bacterial growth, such as pH, T^a, presence of organic acids, and/or some typical enological practices such as maceration or prolonged contact with yeast lees, have been proposed to prevent LAB proteolytic activity and decarboxylase activity (Martín-ÁLvarez et al., 2006).

At present, in the market there are no effective procedures or treatments used to limit the content of biogenic amines in wine. Enzymatic removal of amines may be a safe and economic way to eliminate these troublesome compounds from wines and other fermented foods. García-Ruiz et al. (2011) reported for the first time the ability of LAB of food origin (i.e., wine) to degrade putrescine. Lately, a procedure based on the use of an enzymatic extract from *Penicillium citrinum* CIAL-274,760, isolated from vineyards, which added to the wine reduces or even completely eliminates BA in synthetic wines has been reported (Cueva et al., 2012). Further research is needed to provide conclusive evidence of the applicability of these strategies in real wine systems.

6.7.2 Ethyl Carbamate

EC, also known as urethane, is a carcinogenic compound found in fermented food and beverages. The EC formation is due to a nonenzymatic reaction between ethanol and a compound containing a carbamyl group. The main carbamyl group involved in EC production is urea produced by the yeast catabolism of arginine, but there are also other compounds such as citrulline (derived from bacterial metabolism) (Vincenzini et al., 2009) and carbamyl phosphate that, to a lesser extent, participate in the formation of EC. Citrulline is produced from the catabolism of residual arginine amounts, via the arginine-deiminase pathway, by wine LAB. Arginine-degrading strains belong mainly to heterofermentative *lactobacilli*, although some studies have reported that some *O. oeni* strains do so (Lonvaud-Funel, 2016).

The content of EC can be reduced by decreasing the generation of its precursors. Therefore, it is important to select LAB that produce low concentrations of EC precursors (Jiao et al., 2014). Thanks to the knowledge that has been generated in the area, today's winemakers are better equipped than ever to take steps to prevent or reduce the formation of EC during the winemaking process. They can now implement in-process controls to monitor levels and also inoculate selected strains of yeasts and LAB that do not produce EC during alcoholic or MLF (Araque et al., 2009). There are limits for EC concentration in an alcoholic beverage, but in wines, the content of these compounds are generally below this threshold.

6.8 CONCLUSION

MLF is a traditional issue in enology that has continuing relevance in modern winemaking. As it has been discussing, MLF plays a substantially beneficial role by reducing the wine total acidity (making wines more palatable as result of the loss of strong green taste), contributing to the stabilization and enrichment of aroma and flavor complexity (releasing of great amount of volatile compounds) and imparting microbiology stability to wines. Nowadays, changes in wine flavor and aroma during MLF, which have been detected by sensory analyses, have now been identified at a molecular level. However, more research should be made on this topic, considering also LAB species different to *O. oeni*. It is essential to consider two facts for ensuring the success of MLF: first, the growth and metabolism of selected LAB; and second, the control of some physicochemical parameters. Methods for strain selection that take advantage of the diversity of existing microbiota at the winery, and for strain improvement by nonrecombinant techniques, are being pursued for industrial applications. Finally, in our current state of knowledge, biogenic amines and EC are the only undesirable compounds of microbial origin in wine that raise potential health concerns. The concentrations of these compounds in the finished wine depend mainly on the LAB species/strains present and the availability of amino acids. However, the concentration and variety of nitrogen compounds also have a significant impact on fermentation rates and completion. It is possible to issue recommendations to winemakers to avoid the potential risks of these undesirable compounds. Enological practices, including nitrogen addition, contact with grape solids and lees, the use of starter cultures, must take into account the interactions among all the supplies in the winery.

Acknowledgments

We thank the Spanish Ministry of Economy, Industry and Competitiveness (AGL2015-64522-C2-1), and the Comunidad de Madrid Program (ALIBIRD-CM S2013/ABI-2728). I.G.-S. is the recipient of fellowships from the Spanish FPU-MECD (FPU14/05760) programme.

References

Alvarez, M.A., Moreno-Arribas, M.V., 2014. The problem of biogenic amines in fermented foods and the use of potential biogenic amine-degrading microorganisms as a solution. Trends Food Sci. Technol. 39, 146–155.

Antalick, G., Perello, M.-C., De revel, G., 2012. Characterization of fruity aroma modifications in red wines during malolactic fermentation. J. Agric. Food. Chem. 60, 12371–12383.

Araque, I., Gil, J., Carreté, R., Bordons, A., Reguant, C., 2009. Detection of arc genes related with the ethyl carbamate precursors in wine lactic acid bacteria. J. Agric. Food. Chem. 57, 1841–1847.

Asmundson, R.V., Kelly, W.J., 1990. Temperature and ethanol effects on growth of *Leuconostoc oenos*. In: Yu, P.-L. (Ed.), Fermentation Technologies: Industrial Applications. Elsevier, London.

Bartowsky, E.J., 2005. *Oenococcus oeni* and malolactic fermentation—moving into the molecular arena. Aust. J. Grape Wine Res. 11, 174–187.

Bartowsky, E.J., Francis, I.L., Bellon, J.R., Henschke, P.A., 2002. Is buttery aroma perception in wines predictable from the diacetyl concentration? Aust J Grape Wine Res 8, 180–185.

Bauer, R., Dicks, L.M.T., 2004. Control of malolactic fermentation in wine. A review. S. Afr. J. Enol. Vitic. 25, 74–88.

Belda, I., Ruiz, J., Esteban-Fernández, A., Navascués, E., Marquina, D., Santos, A., et al., 2017. Microbial contribution to wine aroma and its intended use for wine quality improvement. Molecules 22.

Berbegal, C., Peña, N., Russo, P., Grieco, F., Pardo, I., Ferrer, S., et al., 2016. Technological properties of *Lactobacillus plantarum* strains isolated from grape must fermentation. Food Microbiol. 57, 187–194.

Berbegal, C., Benavent-Gil, Y., Navascués, E., Calvo, A., Albors, C., Pardo, I., et al., 2017. Lowering histamine formation in a red Ribera del Duero wine (Spain) by using an indigenous *O. oeni* strain as a malolactic starter. Int. J. Food Microbiol. 244, 11–18.

Betteridge, A., Grbin, P., Jiranek, V., 2015. Improving *Oenococcus oeni* to overcome challenges of wine malolactic fermentation. Trends Biotechnol. 33, 547–553.

Boido, E., Lloret, A., Medina, K., Carrau, F., Dellacassa, E., 2002. Effect of β-glycosidase activity of *Oenococcus oeni* on the glycosylated flavor precursors of Tannat wine during malolactic fermentation. J. Agric. Food. Chem. 50, 2344–2349.

Borneman, A.R., Mccarthy, J.M., Chambers, P.J., Bartowsky, E.J., 2012. Comparative analysis of the *Oenococcus oeni* pan genome reveals genetic diversity in industrially-relevant pathways. BMC Genomics 13, 373.

Cabras, P., Meloni, M., Melis, M., Farris, G.A., Budroni, M., Satta, T., 1994. Interactions between lactic bacteria and fungicides during lactic fermentation. J. Wine Res. 5, 53–59.

Cabrita, M.J., Torres, M., Palma, V., Alves, E., Patão, R., Costa freitas, A.M., 2008. Impact of malolactic fermentation on low molecular weight phenolic compounds. Talanta 74, 1281–1286.

Campos, F.M., Couto, J.A., Figueiredo, A.R., Tóth, I.V., Rangel, A.O.S.S., Hogg, T.A., 2009. Cell membrane damage induced by phenolic acids on wine lactic acid bacteria. Int. J. Food Microbiol. 135, 144–151.

Cappello, M.S., Zapparoli, G., Logrieco, A., Bartowsky, E.J., 2017. Linking wine lactic acid bacteria diversity with wine aroma and flavour. Int. J. Food Microbiol. 243, 16–27.

Carreté, R., Vidal, M.T., Bordons, A., Constantí, M., 2002. Inhibitory effect of sulfur dioxide and other stress compounds in wine on the ATPase activity of *Oenococcus oeni*. FEMS Microbiol. Lett. 211, 155–159.

Cavin, J.F., Andioc, V., Etievant, P.X., Divies, C., 1993. Ability of wine lactic acid bacteria to metabolize phenol carboxylic acids. Am. J. Enol. Vitic. 44, 76–80.

Costantini, A., García-Moruno, E., Moreno-Arribas, M.V., 2009. Biochemical transformations produced by malolactic fermentation. In: Moreno-Arribas, M.V., Polo, M.C. (Eds.), Wine Chemistry and Biochemistry. Springer, New York, NY.

Costello, P.J., Henschke, P.A., 2002. Mousy off-flavor of wine: precursors and biosynthesis of the causative N-heterocycles 2-ethyltetrahydropyridine, 2-acetyltetrahydropyridine, and 2-acetyl-1-pyrroline by *Lactobacillus hilgardii* DSM 20176. J. Agric. Food Chem. 50, 7079–7087.

Costello, P.J., Lee, T.H., Henschke, P., 2001. Ability of lactic acid bacteria to produce N-heterocycles causing mousy off-flavour in wine. Aust. J. Grape Wine Res. 7, 160–167.

Costello, P.J., Siebert, T.E., Solomon, M.R., Bartowsky, E.J., 2013. Synthesis of fruity ethyl esters by acyl coenzyme A: alcohol acyltransferase and reverse esterase activities in *Oenococcus oeni* and *Lactobacillus plantarum*. J. Appl. Microbiol. 114, 797–806.

Coton, E., Coton, M., 2005. Multiplex PCR for colony direct detection of Gram-positive histamine- and tyramine-producing bacteria. J. Microbiol. Methods 63, 296–304.

Coulon, J., Husnik, J.I., Inglis, D.L., Van Der merwe, G.K., Lonvaud, A., Erasmus, D.J., et al., 2006. Metabolic engineering of *Saccharomyces cerevisiae* to minimize the production of ethyl carbamate in wine. Am. J. Enol. Vitic. 57, 113.

Cueva, C., García-Ruiz, A., González-Rompinelli, E., Bartolome, B., Martín-ÁLvarez, P.J., Salazar, O., et al., 2012. Degradation of biogenic amines by vineyard ecosystem fungi. Potential use in winemaking. J. Appl. Microbiol. 112, 672–682.

de las Rivas, B., Marcobal, A., Carrascosa, A.V., Muñoz, R., 2006. PCR detection of foodborne bacteria producing the biogenic amines histamine, tyramine, putrescine, and cadaverine. J. Food Prot. 69, 2509–2514.

De las Rivas, B., Rodríguez, H., Curiel, J.A., Landete, J.M., Muñoz, R., 2009. Molecular screening of wine lactic acid bacteria degrading hydroxycinnamic acids. J. Agric. Food. Chem. 57, 490–494.

Davis, C.R., Wibowo, D., Fleet, G.H., Lee, T.H., 1988. Properties of wine lactic acid bacteria: their potential enological significance. Am. J. Enol. Vitic. 39, 137–142.

Del Prete, V., Costantini, A., Cecchini, F., Morassut, M., Garcia-Moruno, E., 2009. Occurrence of biogenic amines in wine: the role of grapes. Food Chem. 112, 474–481.

Dicks, L.M.T., Endo, A., 2009. Taxonomic status of lactic acid bacteria in wine and key characteristics to differentiate species. S. Afr. J. Enol. Vitic. 30, 72–90.

du Toit, M., Pretorius, I.S., 2000. Microbial spoilage and preservation of wine: using weapons from Nature's own arsenal—a review. S. Afr. J. Enol. Vitic. 21, 74–96.

du Toit, M., Engelbrecht, L., Lerm, E., Krieger-Weber, S., 2011. Lactobacillus: the next generation of malolactic fermentation starter cultures—an overview. Food Bioprocess Technol. 4, 876–906.

Dugelay, I., Gunata, Z., Sapis, J.C., Baumes, R., Bayonove, C., 1993. Role of cinnamoyl esterase activities from enzyme preparations on the formation of volatile phenols during winemaking. J. Agric. Food. Chem. 41, 2092–2096.

Esteban-Torres, M., Reverón, I., Mancheño, J.M., DE Las rivas, B., Muñoz, R., 2013. Characterization of a feruloyl esterase from *Lactobacillus plantarum*. Appl. Environ. Microbiol. 79, 5130–5136.

EFSA, 2011. Scientific opinion on risk based control of biogenic amine formation in fermented foods. EFSA J. 9 (10), 2393. 1–93.

Francis, I.L., Newton, J.L., 2005. Determining wine aroma from compositional data. Aust. J. Grape Wine Res. 11, 114–126.

Fumi, M.D., Krieger-Weber, S., Déléris-Bou, M., Silva, A., Du toit, M., 2010. A new generation of malolactic starter cultures for high pH wines. In: Proceedings International IVIF Congress 2010, WB3 Microorganisms-Malolactic-Fermentation.

G-Alegría, E., López, I., Ruiz, J.I., Sáenz, J., Fernández, E., Zarazaga, M., et al., 2004. High tolerance of wild *Lactobacillus plantarum* and *Oenococcus oeni* strains to lyophilisation and stress environmental conditions of acid pH and ethanol. FEMS Microbiol. Lett. 230, 53–61.

Galland, D., Tourdot-Maréchal, R., Abraham, M., Chu, K.S., Guzzo, J., 2003. Absence of malolactic activity is a characteristic of H(+)-ATPase-deficient mutants of the lactic acid bacterium *Oenococcus oeni*. Appl. Environ. Microbiol. 69, 1973–1979.

García-Ruiz, A., Moreno-Arribas, M.V., Martín-Álvarez, P.J., Bartolomé, B., 2011. Comparative study of the inhibitory effects of wine polyphenols on the growth of enological lactic acid bacteria. Int. J. Food Microbiol. 145, 426–431.

Gardini, F., Martuscelli, M., Crudele, M.A., Paparella, A., Suzzi, G., 2002. Use of *Staphylococcus xylosus* as a starter culture in dried sausages: effect on the biogenic amine content. Meat Sci. 61, 275–283.

Gastón Orrillo, A., Ledesma, P., Delgado, O.D., Spagna, G., Breccia, J.D., 2007. Cold-active α-L-rhamnosidase from psychrotolerant bacteria isolated from a sub-Antarctic ecosystem. Enzyme Microb. Technol. 40, 236–241.

Gonzalez, R., Tronchoni, J., Quirós, M., Morales, P., 2016. Genetic improvement and genetically modified microorganisms. In: Moreno-Arribas, M.V., Bartolomé suáldea, B. (Eds.), Wine Safety, Consumer Preference, and Human Health. Springer International Publishing, Cham.

Gonzalez-Centeno, M.R., Chira, K.A.-O.H.O.O., Teissedre, P.L., 2017. Comparison between malolactic fermentation container and barrel toasting effects on phenolic, volatile, and sensory profiles of red wines. J. Agric. Food. Chem. 65, 3320–3329.

Grimaldi, A., Mclean, H., Jiranek, V., 2000. Identification and partial characterization of glycosidic activities of commercial strains of the lactic acid bacterium, *Oenococcus oeni*. Am. J. Enol. Vitic. 51, 362.

Grimaldi, A., Bartowsky, E., Jiranek, V., 2005a. Screening of *Lactobacillus* spp. and *Pediococcus* spp. for glycosidase activities that are important in oenology. J. Appl. Microbiol. 99, 1061–1069.

Grimaldi, A., Bartowsky, E., Jiranek, V., 2005b. A survey of glycosidase activities of commercial wine strains of *Oenococcus oeni*. Int. J. Food Microbiol. 105, 233–244.

Guerrero, R.F., Cantos-Villar, E., 2015. Demonstrating the efficiency of sulphur dioxide replacements in wine: a parameter review. Trends Food Sci. Technol. 42, 27–43.

Guerrini, S., Mangani, S., Granchi, L., Vincenzini, M., 2002. Biogenic amine production by *Oenococcus oeni*. Curr. Microbiol. 44, 374–378.

Günata, Z., Bitteur, S., Brillouet, J.-M., Bayonove, C., Cordonnier, R., 1988. Sequential enzymic hydrolysis of potentially aromatic glycosides from grape. Carbohydr. Res. 184, 139–149.

Guzzo, J., Jobin, M.-P., Diviès, C., 1998. Increase of sulfite tolerance in *Oenococcus oeni* by means of acidic adaptation. FEMS Microbiol. Lett. 160, 43–47.

Hébert, E.M., Raya, R.R., Savoy De Giori, G., 2004. Evaluation of minimal nutritional requirements of lactic acid bacteria used in functional foods. In: Walker, J.M., Spencer, J.F.T., Ragout De spencer, A.L. (Eds.), Environmental Microbiology: Methods and Protocols. Humana Press, Totowa, NJ.

Henick-Kling, T., 1993. Malolactic fermentation. In: Fleet, G.H. (Ed.), Wine Microbiology and Biotechnology. Harwood Academic Publishers, Switzerland.

Henick-Kling, T., Acree, T., Gavitt, B., Krieger, S.A., Laurent, M.H., 1993. Sensory aspects of malolactic fermentation. In: Stockley, C.S., Johnston, R.S., Leske, P.A., Lee, T.H. (Eds.), Proceedings of the Eighth Australian Wine Industry Technical Conference. Winetitles, Adelaide, Australia.

Hernandez-Orte, P., Cersosimo, M., Loscos, N., Cacho, J., Garcia-Moruno, E., Ferreira, V., 2009. Aroma development from non-floral grape precursors by wine lactic acid bacteria. Food Res. Int. 42, 773–781.

Iorizzo, M., Testa, B., Lombardi, S.J., García-Ruiz, A., Muñoz-González, C., Bartolomé, B., et al., 2016. Selection and technological potential of *Lactobacillus plantarum* bacteria suitable for wine malolactic fermentation and grape aroma release. LWT Food Sci. Technol. 73, 557–566.

Izquierdo cañas, P.M., Gómez alonso, S., Ruiz pérez, P., Seseña prieto, S., García romero, E., Palop herreros, M.L., 2009. Biogenic amine production by *Oenococcus oeni* isolates from malolactic fermentation of Tempranillo wine. J. Food Prot. 72, 907–910.

Jiao, Z., Dong, Y., Chen, Q., 2014. Ethyl carbamate in fermented beverages: presence, analytical chemistry, formation mechanism, and mitigation proposals. Compr. Rev. Food Sci. Food Saf. 13, 611–626.

Juega, M., Costantini, A., Bonello, F., Cravero, M.C., Martinez-Rodriguez, A.J., Carrascosa, A.V., et al., 2014. Effect of malolactic fermentation by *Pediococcus damnosus* on the composition and sensory profile of Albariño and Caiño white wines. J. Appl. Microbiol. 116, 586–595.

Kunkee, R.E., 1967. Control of malo-lactic fermentation induced by *Leuconostoc citrovorum*. Am. J. Enol. Vitic. 18, 71–77.

Lafon-Lafourcade, S., Carre, E., Ribéreau-Gayon, P., 1983a. Occurrence of lactic acid bacteria during the different stages of vinification and conservation of wines. Appl. Environ. Microbiol. 46, 874–880.

Lafon-Lafourcade, S., Lonvaud-Funel, A., Carre, E., 1983b. Lactic acid bacteria of wines: stimulation of growth and malolactic fermentation. Antonie Van Leeuwenhoek 49, 349–352.

Lamontanara, A., Caggianiello, G., Orrù, L., Capozzi, V., Michelotti, V., Bayjanov, J., et al., 2015. Draft genome sequence of *Lactobacillus plantarum* Lp90 isolated from wine. Genome Announce. 3, e00097-15.

Landaud, S., Helinck, S., Bonnarme, P., 2008. Formation of volatile sulfur compounds and metabolism of methionine and other sulfur compounds in fermented food. Appl. Microbiol. Biotechnol. 77, 1191–1205.

Landete, J.M., Ferrer, S., Pardo, I., 2005. Which lactic acid bacteria are responsible for histamine production in wine? J. Appl. Microbiol. 99, 580–586.

Landete, J.M., DE Las rivas, B., Marcobal, A., Muñoz, R., 2007. Molecular methods for the detection of biogenic amine-producing bacteria on foods. Int. J. Food Microbiol. 117, 258–269.

Lerm, E., Engelbrecht, L., Du toit, M., 2010. Malolactic fermentation: the ABCs of MLF. S. Afr. J. Enol. Vitic. 31, 186–212.

Lerm, E., Engelbrecht, L., Du toit, M., 2011. Selection and characterisation of *Oenococcus oeni* and *Lactobacillus plantarum* South African wine isolates for use as malolactic fermentation starter cultures. S. Afr. J. Enol. Vitic. 32, 280–295.

Liu, M., Nauta, A., Francke, C., Siezen, R.J., 2008. Comparative genomics of enzymes in flavor-forming pathways from amino acids in lactic acid bacteria. Appl. Environ. Microbiol. 74, 4590–4600.

Liu, S.Q., 2002. Malolactic fermentation in wine—beyond deacidification. J. Appl. Microbiol. 92, 589–601.

Lonvaud-Funel, A., 1999. Lactic acid bacteria in the quality improvement and depreciation of wine. Antonie Van Leeuwenhoek 76, 317–331.

Lonvaud-Funel, A., 2016. Undesirable compounds and spoilage microorganisms in wine. In: Moreno-Arribas, M.V., Bartolomé suáldea, B. (Eds.), Wine Safety, Consumer Preference, and Human Health. Springer International Publishing, Cham.

Lonvaud-Funel, A., Joyeux, A., Desens, C., 1988. Inhibition of malolactic fermentation of wines by products of yeast metabolism. J. Sci. Food Agric. 44, 183–191.

Maicas, S., Pardo, I., Ferrer, S., 2001. The potential of positively-charged cellulose sponge for malolactic fermentation of wine, using *Oenococcus oeni*. Enzyme Microb. Technol. 28, 415–419.

Marcobal, A., DE Las rivas, B., Moreno-Arribas, M.V., Muñoz, R., 2005. Multiplex PCR method for the simultaneous detection of histamine-, tyramine-, and putrescine-producing lactic acid bacteria in foods. J. Food Prot. 68, 874–878.

Marcobal, A., Martin-Alvarez, P.J., Polo, M.C., Muñoz, R., Moreno-Arribas, M.V., 2006. Formation of biogenic amines throughout the industrial manufacture of red wine. J. Food Prot. 69, 397–404.

Margalef-Català, M., Felis, G.E., Reguant, C., Stefanelli, E., Torriani, S., Bordons, A., 2017. Identification of variable genomic regions related to stress response in *Oenococcus oeni*. Food Res. Int. 102, 625–638.

Martín-ÁLvarez, P.J., Marcobal, A., Polo, C., Moreno-Arribas, M.V., 2006. Technological factors influencing biogenic amine production during red wine manufacture. Eur. Food Res. Technol. 222, 420–424.

Martineau, B., Henick-Kling, T., Acree, T., 1995. Reassessment of the influence of malolactic fermentation on the concentration of diacetyl in wines. Am. J. Enol. Vitic. 46, 385.

Matthews, A., Grimaldi, A., Walker, M., Bartowsky, E., Grbin, P., Jiranek, V., 2004. Lactic acid bacteria as a potential source of enzymes for use in vinification. Appl. Microbiol. Biotechnol. 70, 5715–5731.

Mills, D.A., Rawsthorne, H., Parker, C., Tamir, D., Makarova, K., 2005. Genomic analysis of *Oenococcus oeni* PSU-1 and its relevance to winemaking. FEMS Microbiol. Rev. 29, 465–475.

Monge, M., Moreno-Arribas, M.V., 2016. Applications of nanotechnology in wine production and quality and safety control. In: Moreno-Arribas, M.V., Bartolomé suáldea, B. (Eds.), Wine Safety, Consumer Preference, and Human Health. Springer International Publishing, Switzerland.

Moreno-Arribas, M.V., Polo, M.C., 2005. Winemaking biochemistry and microbiology: current knowledge and future trends. Crit. Rev. Food Sci. Nutr. 45, 265–286.

Moreno-Arribas, M.V., Polo, M.C., 2008. Occurrence of lactic acid bacteria and biogenic amines in biologically aged wines. Food Microbiol. 25, 875–881.

Moreno-Arribas, M.V., Polo, M.C., Jorganes, F., Muñoz, R., 2003. Screening of biogenic amine production by lactic acid bacteria isolated from grape must and wine. Int. J. Food Microbiol. 84, 117–123.

Moreno-Arribas, M.V., Smit, A.Y., Du toit, M., 2010. Biogenic amines and the winemaking process. In: Reynolds, A.G. (Ed.), Managing Wine Quality: Oenology and Wine Quality. Woodhead Publishing Limited, Cambridge, UK.

Moreno-Arribas, V., Lonvaud-Funel, A., 2001. Purification and characterization of tyrosine decarboxylase of *Lactobacillus brevis* IOEB 9809 isolated from wine. FEMS Microbiol. Lett. 195, 103–107.

Mtshali, P.S., Divol, B., Du toit, M., 2012. Identification and characterization of *Lactobacillus florum* strains isolated from South African grape and wine samples. Int. J. Food Microbiol. 153, 106–113.

Muñoz, R., Moreno-Arribas, M.V., DE Las rivas, B., 2011. Lactic acid bacteria. In: Carrascosa santiago, A.V., Muñoz, R., González, R. (Eds.), Molecular Wine Microbiology. Academic Press, San Diego, CA.

Naouri, P., Chagnaud, P., Arnaud, A., Galzy, P., 1990. Purification and properties of a malolactic enzyme from *Leuconostoc oenos* ATCC 23278. J. Basic Microbiol. 30, 577–585.

Nielsen, J.C., Richelieu, M., 1999. Control of flavor development in wine during and after malolactic fermentation by *Oenococcus oeni*. Appl. Environ. Microbiol. 65, 740–745.

Pérez-Martín, F., Seseña, S., Izquierdo, P.M., Palop, M.L., 2013. Esterase activity of lactic acid bacteria isolated from malolactic fermentation of red wines. Int. J. Food Microbiol. 163, 153–158.

Pinzani, P., Bonciani, L., Pazzagli, M., Orlando, C., Guerrini, S., Granchi, L., 2004. Rapid detection of *Oenococcus oeni* in wine by real-time quantitative PCR. Lett. Appl. Microbiol. 38, 118–124.

Pozo-Bayón, M.A., G-Alegría, E., Polo, M.C., Tenorio, C., Martín-ÁLvarez, P.J., Calvo De La banda, M.T., et al., 2005. Wine volatile and amino acid composition after malolactic fermentation: effect of *Oenococcus oeni* and *Lactobacillus plantarum* starter cultures. J. Agric. Food. Chem. 53, 8729–8735.

Prahl, C. 1988. Method of inducing the decarboxylation of malic acid in must or fruit juice. European patent filed 24.01.1989, priority 25.01.1988, International application number PCT/DK89/00009.

Pripis-Nicolau, L., De revel, G., Bertrand, A., Lonvaud-Funel, A., 2004. Methionine catabolism and production of volatile sulphur compounds by *Oenococcus oeni*. J. Appl. Microbiol. 96, 1176–1184.

Saguir, F.M., Loto Campos, I.E., Maturano, C., Manca De Nadra, M.C., 2009. Identification of dominant lactic acid bacteria isolated from grape juices. Assessment of its biochemical activities relevant to flavor development in wine. Int. J. Wine Res. 1, 175–185.

Santos, M.C., Nunes, C., Saraiva, J.A., Coimbra, M.A., 2012. Chemical and physical methodologies for the replacement/reduction of sulfur dioxide use during winemaking: review of their potentialities and limitations. Eur. Food Res. Technol. 234, 1–12.

Silla Santos, M.H., 1996. Biogenic amines: their importance in foods. Int. J. Food Microbiol. 29, 213–231.

Smit, A.Y., Du toit, W.J., Du toit, M., 2008. Biogenic amines in wine: understanding the headache. S. Afr. J. Enol. Vitic. 29, 109–127.

Soufleros, E., Barrios, M.-L., Bertrand, A., 1998. Correlation between the content of biogenic amines and other wine compounds. Am. J. Enol. Vitic. 49, 266.

Spano, G., Beneduce, L., Tarantino, D., Zapparoli, G., Massa, S., 2002. Characterization of *Lactobacillus plantarum* from wine must by PCR species-specific and RAPD-PCR. Lett. Appl. Microbiol. 35, 370–374.

Sumby, K.M., Grbin, P.R., Jiranek, V., 2013. Characterization of EstCOo8 and EstC34, intracellular esterases, from the wine-associated lactic acid bacteria *Oenococcus oeni* and *Lactobacillus hilgardii*. J. Appl. Microbiol. 114, 413–422.

Sumby, K.M., Grbin, P.R., Jiranek, V., 2014. Implications of new research and technologies for malolactic fermentation in wine. Appl. Microbiol. Biotechnol. 98, 8111–8132.

Swiegers, J.H., Bartowsky, E.J., Henschke, P.A., Pretorius, I.S., 2005. Yeast and bacterial modulation of wine aroma and flavour. Aust. J. Grape Wine Res. 11, 139–173.

Ten Brink, B., Damink, C., Joosten, H.M., Huis in 't Veld, J.H., 1990. Occurrence and formation of biologically active amines in foods. Int. J. Food Microbiol. 11, 73–84.

Ugliano, M., Moio, L., 2005. Changes in the concentration of yeast-derived volatile compounds of red wine during malolactic fermentation with four commercial starter cultures of *Oenococcus oeni*. J. Agric. Food. Chem. 53, 10134–10139.

Ugliano, M., Genovese, A., Moio, L., 2003. Hydrolysis of wine aroma precursors during malolactic fermentation with four commercial starter cultures of *Oenococcus oeni*. J. Agric. Food. Chem. 51, 5073–5078.

Vallet, A., Lucas, P., Lonvaud-Funel, A., De revel, G., 2008. Pathways that produce volatile sulphur compounds from methionine in *Oenococcus oeni*. J. Appl. Microbiol. 104, 1833–1840.

Vally, H., Misso, N.L., Madan, V., 2009. Clinical effects of sulphite additives. Clin. Exp. Allergy 39, 1643–1651.

Van der Westhuizen, L.M., Loos, M.A., 1981. Effect of pH, temperature and SO_2 concentration on the malo-lactic fermentation abilities of selected bacteria and on wine colour. S. Afr. J. Enol. Vitic. 2, 61–65.

Vincenzini, M., Guerrini, S., Mangani, S., Granchi, L., 2009. Amino acid metabolisms and production of biogenic amines and ethyl carbamate. In: König, H., Unden, G., Fröhlich, J. (Eds.), Biology of Microorganisms on Grapes, in Must and in Wine. Springer Berlin Heidelberg, Berlin, Heidelberg.

Walling, E., Gindreau, E., Lonvaud-Funel, A., 2005. A putative glucan synthase gene dps detected in exopolysaccharide-producing *Pediococcus damnosus* and *Oenococcus oeni* strains isolated from wine and cider. Int. J. Food Microbiol. 98, 53–62.

Zhao, M., Liu, S., He, L., Tian, Y., 2016. Draft genome sequence of *Lactobacillus plantarum* XJ25 isolated from Chinese red wine. Genome Announce. 4, e01216-16.

CHAPTER

7

Yeast-Bacteria Coinoculation

Isabel Pardo and Sergi Ferrer

ENOLAB, Interdisciplinary Research Structure for Biotechnology and Biomedicine (BIOTECMED)
Universitat de València, Burjassot, Spain

7.1 INTRODUCTION

Winemaking involves two biochemical processes: alcoholic fermentation (AF) and malolactic fermentation (MLF). *Saccharomyces cerevisiae* is responsible mainly for AF and contributes chiefly to sugar transformation into ethanol. *Oenococcus oeni* is the microorganism responsible mainly for MLF, and transforms naturally-occurring L-malic acid in grape must into L-lactic acid (Ribéreau-Gayon et al., 2006a). In addition to these main microorganisms, a plethora of strains belonging to different yeast and bacterial species contribute differently to the final characteristics of wine (Esteve-Zarzoso et al., 1998; Mateo and Maicas, 2016; Viana et al., 2008). Yeast strains belonging to the non-*Saccharomyces* group improve the organoleptic complexity of wine in two ways: producing compounds of their own metabolism (alcohols and esters); transforming the precursors present in grape must, or which derive from *S. cerevisiae* metabolism (Andorrà et al., 2010; Maturano et al., 2015; Viana et al., 2008). These transformations are possible because non-*Saccharomyces* strains are able to produce exoenzymes (Padilla et al., 2016; Claus, 2017).

MLF is desirable for several reasons: it diminishes excessive acidity, transforms the malic herbaceous taste into the milder lactic taste of lactic acid wines, improves wine's sensory characteristics (de Cort et al., 1994; Bordons, 1997; de Revel et al., 1999; Delaquis et al., 2000; Grimaldi et al., 2000; Liu, 2002), and reduces the microbial spoilage risk (Osborne and Edwards, 2005; du Toit et al., 2011; Vendrame, 2013; Lucio et al., 2014; Piao et al., 2015). Although MLF is promoted mainly in red wine production, white wines have been seen to also benefit from this process (Davis et al., 1985; Lonvaud-Funel, 1999; Liu, 2002).

Currently, MLF takes place once AF is completed. This sequentiality is the consequence of competition between yeasts and lactic acid bacteria (LAB). After grape crushing, the yeast concentration is around 10^5 cfu/mL, whereas that of LAB is 10^2–10^3 cfu/mL (Ribéreau-Gayon et al., 2006a). Yeasts, mainly *S. cerevisiae*, rapidly develop and exhaust some LAB essential nutrients of grape musts. Yeasts also produce ethanol and other products that inhibit LAB growth (Lucio, 2014; Lucio et al., 2014). When yeasts die, lysis promotes the release of intracellular nutrients to wine at the end of AF, and ethanol-resistant LAB use them to grow (Fugelsang and Edwards, 2007). At this time, the concentration of these bacteria dramatically increases. When commercial yeasts and LAB are added to perform AF and MLF, they are generally used sequentially, which reproduces the natural process. Malolactic starter cultures can be inoculated before AF, coinoculated with yeast at the beginning or in the last phase of AF, and also sequentially (Table 7.1) after AF and running-off (Edwards and Beelman, 1989; Kunkee et al., 1964; Lasik, 2013). Among the variety of possibilities available for performing yeasts and LAB coinoculation to conduct AF and MLF (almost) simultaneously, the three most widely used combinations are (decreasingly): (1) *S. cerevisiae* + *O. oeni*; (2) *S. cerevisiae* + *Lactobacillus plantarum*; (3) non-*Saccharomyces* + *S. cerevisiae* + LAB (*O. oeni* or *L. plantarum*). The *S. cerevisiae* + *O. oeni* couple is the classic combination, but has only been used intermittently given the fear of a dramatic increase in acetic acid due to the heterofermentative metabolism of this bacterium (Antalick et al., 2013; Edwards et al., 1999; Muñoz et al., 2014). However, this fear is unreal for many authors as they have never observed such an increase (Lucio et al., 2016; Saerens et al., 2015; Onetto and Bordeu, 2015). In any case, the combination (2) (*S. cerevisiae* + *L. plantarum*) should do away with

TABLE 7.1 Timing of Inoculation of LAB (*L. Plantarum* and *O. Oeni*), Showing the Main and Additional Research Objectives and the Yeasts Employed, Indicating Some References as Examples

LAB inoculation time	Main research objective	Microorganisms' mixture	Additional research objectives	References
Before AF	MLF	*L. plantarum*/*S. cerevisiae*		Lucio et al. (2017)
	Acidification	*L. plantarum*/*S. cerevisiae*		Lucio (2014) and Onetto and Bordeu (2015)
	Ethanol reduction	*L. plantarum*/*S. cerevisiae*		Saerens et al. (2015)
Beginning AF	MLF	*L. plantarum*/*S. cerevisiae*		Brizuela et al. (2017), Berbegal et al. (2016), Lerena et al. (2016), Krieger-Weber et al. (2016), Krieger-Weber (2017), Henick-Kling and Park (1994), Lucio et al. (2017), Miller et al. (2011), Onetto and Bordeu (2015), and Prahl et al. (1988)
	MLF	*O. oeni*/*S. cerevisiae*	Aromatic impact	Knoll et al. (2012)
	MLF	*O. oeni*/*S. cerevisiae*		Rosi et al. (2003)
	MLF	*O. oeni*/*S. cerevisiae*		Zapparoli et al. (2009)
	MLF	*O. oeni*/*S. cerevisiae*	Aromatic impact, risk spoilage	Massera et al. (2009)
	MLF	*L. plantarum*/*S. cerevisiae*		Onetto and Bordeu (2015)
	MLF	*O. oeni*/*S. cerevisiae*		Onetto and Bordeu (2015)
	MLF	*O. oeni*/*S. cerevisiae*	Aromatic impact	Pan et al. (2011)
	Acidification	*L. plantarum*/*S. cerevisiae*		Onetto and Bordeu (2015)
	Microbial interactions	*O. oeni*/*S. cerevisiae*		Edwards et al (1990), Henick-Kling and Park (1994), and Tristezza (2016)
	Microbial interactions	non-*Saccharomyces*/*S. cerevisiae*/*O. oeni*	MLF, aromatic impact	du Plessis et al. (2017)
	Ethanol reduction	*L. plantarum*/*S. cerevisiae*		Saerens et al. (2015)
	Aromatic impact	*O. oeni*/*S. cerevisiae*		Antalick et al. (2013) and Abrahamsen and Bartowsky (2012a)
	Aromatic impact	*O. oeni*/*S. cerevisiae*		Garofalo et al. (2015)
	Overall vinification time reduction	*O. oeni*/*S. cerevisiae*	Aromatic impact	Abrahamsen and Bartowsky (2012b), Jussier et al. (2006), Izquierdo Cañas et al. (2014), and Azzolini et al. (2010)
	Overall vinification time reduction	*L. plantarum*/*O. oeni*/*S. cerevisiae*	Aromatic impact, risk spoilage	Lerm (2010)
	Overall vinification time reduction	*L. plantarum*/*S. cerevisiae*	Aromatic impact, risk spoilage	Lerm (2010)
	Overall vinification time reduction	*O. oeni*/*S. cerevisiae*	Aromatic impact, risk spoilage	Lerm (2010)
	BA reduction	*O. oeni*/*S. cerevisiae*		Smit and du Toit (2013), Massera et al. (2009), and Izquierdo Cañas et al. (2012)
During AF	Microbial interactions	*O. oeni*/*S. cerevisiae*		Tristezza (2016)
	MLF	*O. oeni*/*S. cerevisiae*	Aromatic impact	Knoll et al. (2012)
	MLF	*O. oeni*/*S. cerevisiae*		Rosi et al. (2003)
	Aromatic impact	*O. oeni*/*S. cerevisiae*		Antalick et al. (2013)

(Continued)

TABLE 7.1 (Continued)

LAB inoculation time	Main research objective	Microorganisms' mixture	Additional research objectives	References
After AF	MLF	*L. plantarum*		Lucio et al. (2017)
	MLF	*O. oeni*		Henick-Kling and Park (1994)
	MLF	*O. oeni*	Aromatic impact	Lerm (2010)
	MLF	*L. plantarum*	Aromatic impact	Lerm (2010)
	MLF	*O. oeni*	Aromatic impact	Knoll et al. (2012)
	MLF	*O. oeni*		Rosi et al. (2003)
	MLF	*O. oeni*		Zapparoli et al. (2009)
	MLF	*O. oeni*	Aromatic impact	Massera et al. (2009)
	MLF	*O. oeni*	Aromatic impact	Pan et al. (2011)
	Microbial interactions	*O. oeni*	MLF, aromatic impact	du Plessis et al. (2017)
	Aromatic impact	*O. oeni*		Garofalo et al. (2015)
	OTA reduction	*O. oeni*	MLF	Mateo et al. (2010)
	BA reduction	*O. oeni*		Smit and du Toit (2013), Massera et al. (2009), and Izquierdo Cañas et al. (2012)

LAB, lactic acid bacteria; *MLF*, malolactic fermentation; *AF*, alcoholic fermentation; *BA*, biogenic amines; *OTA*, ochratoxin A.

these fears as this is a homofermentative bacterium (De Vos et al., 2011). This strategy has continued to gain followers in recent times, and several commercial products are already available (Berbegal et al., 2016; Brizuela et al., 2017, 2018; Iorizzo et al., 2016; Krieger-Weber et al., 2016, 2017; Lerena et al., 2016; Lerm et al., 2011; Lerm, 2010; Lucio et al., 2017; Onetto and Bordeu, 2015; Pilatte and Prahl, 1997; Pozo-Bayón et al., 2005; Sun et al., 2013). The third strategy, which employs non-*Saccharomyces* + *S. cerevisiae* + LAB (*O. oeni* and/or *L. plantarum*), is also very interesting and some trials have already been run (du Plessis et al., 2017; Petruzzi et al., 2017).

Nowadays, yeasts and bacteria are inoculated simultaneously, or closely in time, to meet different enological objectives, with a major goal of performing MLF. Other main research objectives are acidification, lowering volatile acidity and ethanol yields, and reduction of overall vinification time. Additional research objectives that sometimes are aromatic profile modification, microbial spoilage control, wine toxin reduction, and selection of the appropriate microbial couples.

7.2 OBJECTIVES

7.2.1 Controlling Wine Acidity

Acidity is one of the most important parameters in wines, and it depends on the type and quantity of organic acids present in it. These acids come from fruit [L(+)-tartaric, L(−)-malic, citric, coumaric, and caffeic acids], from microbial metabolism [pyruvic acid, L(+)-lactic, D(−)-lactic, succinic, acetic, citramalic, oxalacetic, and fumaric acids], from the transformation of biological products, such as CO_2 (carbonic acid), or from enological practices, such as addition of SO_2 (sulfurous acids). These acids greatly contribute to the composition, stability, organoleptic qualities, and evolution of wines. Their preservative properties also enhance wines' microbiological and physicochemical stability (Ribéreau-Gayon et al., 2006b). Sometimes acidity is not well-balanced and makes wines excessively acidic or, conversely, poorly acidic. Excess acidity can be corrected by physicochemical biological methods. Chemical deacidification can be achieved using calcium or potassium carbonate ($CaCO_3$, K_2CO_3), potassium bicarbonate ($KHCO_3$), potassium hydroxide (KOH), or calcium sulfate ($CaSO_4$). These compounds promote the formation and precipitation of the insoluble salts that derive from tartaric and malic acids, which lead to lower acidity (Dziezak, 2003). Another chemical method used to deacidify wine is ion exchange; in this case, tartrate or malate ions are exchanged with hydroxyl ions thereby removing them from must/wine. Physical deacidification can be achieved by cooling wine; cooling promotes the precipitation of naturally-produced potassium tartrate salts because solubility of salt drastically reduces at low temperatures. For acidifying musts or wines, chemical methods are the most widely used and are based on the addition of authorized organic acids, such as malic, tartaric, citric, or lactic acids (OIV, 2015). As an alternative, physical methods like electrodialysis and ion exchange resins can also be used (Lucio et al., 2017; OIV, 2015). Besides these physicochemical methods, biological (microbial) methods can be followed to manage wine acidity (Dequin et al., 1999; Kapsopoulou et al., 2007; Yeramian, 2003; Lucio et al., 2016; Onetto and Bordeu, 2015). Biological deacidification methods are based mainly on the elimination of L(−)-malic acid, which can be transformed into L(+)-lactic acid by most winemaking-related LAB in a reaction called MLF (Ribéreau-Gayon et al., 2006a). MLF is catalyzed by the malolactic enzyme (MLE), and was purified and characterized for the first time in *L. plantarum* (Caspritz and Radler, 1983). However, the presence of this enzyme has been previously demonstrated in lots of wine bacteria (Henick-Kling, 1993). Although many species have the ability to perform MLF, the most frequently associated species with this process is *O. oeni* (Garofalo et al., 2015). Cells of this species have been used in a free or an immobilized form to carry out MLF (Maicas, 2001; Kourkoutas et al., 2004; Simó et al., 2017; Bauer and Dicks, 2004). In addition, MLE bioreactors, instead of cell bioreactors, have been used to accomplish MLF (Formisyn et al., 1997; Vaillant and Formisyn, 1996), and different yeast species have been employed to eliminate L(+)-malic acid from must, e.g., *Schizosaccharomyces pombe* (Mylona et al., 2016; Ciani, 1995; Charpentier, 1985; Charpentier et al., 1985). Fig. 7.1 shows cells of *S. cerevisiae* and *O. oeni* coimmobilized on oak chips prepared as described by Ferrer et al. (2017).

One of the main advantages of yeasts and LAB coinoculation is the advance in MLF, which reduces winemaking overall duration (Krieger-Weber, 2017; Bartowsky et al., 2015; Jussier et al., 2006). The coinoculation of *S. cerevisiae* and *O. oeni* or *L. plantarum* has promoted early MLF and reduced vinification times (Abrahamse and Bartowsky, 2012a; Azzolini et al., 2010; Krieger-Weber et al., 2010; Izquierdo Cañas et al., 2014; Jussier et al., 2006; Knoll et al., 2012). Izquierdo Cañas et al. (2014) report shortening vinification time by four days by following a coinoculation procedure of bacteria and yeasts. Massera et al. (2009) indicate having shortened vinification times by between 12 and 14 days by adding malolactic starters to Malbec grape must 12 hours after yeast inoculation, compared to its addition at the end of AF. This strategy allows to complete MLF quickly and to save between 20 and 25 days compared to the sequential inoculation of yeasts first, and of LAB afterwards, after accomplishing AF (Garofalo et al., 2015; Azzolini et al., 2010). This reduction in time enables wines to be stabilized more rapidly than sequential systems and, consequently, the risk of microbial spoilage is reduced (Abrahamse and Bartowsky, 2012b; Knoll et al., 2012; Lerm et al., 2010; Azzolini et al., 2010). Besides, MLF can be more efficiently performed in yeast coinoculation than after AF because of low ethanol levels and higher nutrient concentrations (Lerm et al., 2010). Even though MLF is not immediately started, an early inoculation of LAB in grape must allow bacteria to gradually adapt to ethanol before it becomes toxic (Azzolini et al., 2010).

To perform MLF in yeast coinoculation, two different LAB species are mainly employed: *O. oeni* and *L. plantarum* (Bartowsky et al., 2015). *O. oeni* is the most regularly used LAB in yeast coinoculation, although *L. plantarum* has been recovered as a renewed putative MLF starter in recent years (Berbegal et al., 2016; Iorizzo et al., 2016; Miller et al., 2011; Sun et al., 2013; Lucio et al., 2017; Onetto and Bordeu, 2015). *L. plantarum* was one of the first

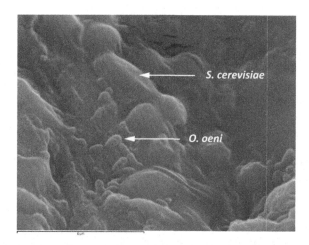

FIGURE 7.1 Electron microscopy photograph showing *S. cerevisiae* and *O. oeni* cells immobilized on oak chips by lyophilization.

bacteria to be used as a freeze-dried malolactic starter (Prahl et al., 1988; Prahl, 1989a,b) before implementing *O. oeni* (syn. *Leuconostoc oenos*) as a direct freeze-dried inoculum (Nielsen et al., 1996). Nowadays, *L. plantarum* is regaining its protagonist status among MLF starters, mostly because it is used early in simultaneous fermentations with yeasts (Lerena et al., 2016; Krieger-Weber et al., 2016, 2017; Lucio et al., 2017; Lerm et al., 2011; Brizuela et al., 2017; Berbegal et al., 2016, 2017b; Iorizzo et al., 2016), especially for high-ethanol wines (Miller et al., 2011). One of the advantages of coinoculation is that no ethanol is synthesized in early winemaking stages. Thus, more sensitive LAB, such as *L. plantarum*, can be used instead of *O. oeni* (Lerm et al., 2011). It is even possible to sometimes schedule the coinoculation of a mixture of different LAB (*L. plantarum* and *O. oeni*) with yeasts to complete MLF more quickly and safely (Lerm, 2010). Another advantage of using *L. plantarum* is that this organism is homofermentative for glucose (De Vos et al., 2011), which avoids the risk of increasing volatile acidity. Recent evidence suggests the further use of selected *L. plantarum* strains for the prealcoholic acidification of grape must (Berbegal et al., 2016; Lucio, 2014; Lucio et al., 2014, 2016, 2017; Onetto and Bordeu, 2015; Pilatte and Prahl, 1997).

As for timing, different MLF inoculation possibilities exist (Table 7.1) (Knoll et al., 2012; Rosi et al., 2003; Lucio, 2014; Lucio et al., 2014). Among them, the most inhibitory ones for MLF are LAB inoculation halfway through AF, whereas ethanol can inhibit or delay MLF when MLF is carried out at the end of AF. Time is gained when LAB are coinoculated as early as possible, less ethanol is present, and bacteria can better adapt to harsh conditions. For high-alcohol wines, Zapparoli et al. (2009) propose using cells that have been adapted for 48 hours to be inoculated simultaneously with yeast to conduct AF and MLF.

Climate change modifies vine development and fruit maturation patterns. Increasing temperature affects the total acidity and pH of grape musts, which implies low acidity and high pH wines due to overipening (Mira de Orduña, 2010). In this case, grape musts or wines need to be corrected by following different approaches to increase their acidity. Microorganisms (yeasts and bacteria) are employed to increase wine acidity (Dequin et al., 1999; Kapsopoulou et al., 2007; Yeramian, 2003; Lucio et al., 2016; Onetto and Bordeu, 2015). Val et al. (2002) were the first to report the coinoculation of strains *L. plantarum* and *S. cerevisiae* to acidify Pedro Ximénez low-acidity wines, whose titratable acidity (expressed as tartaric acid) was 4.5 g/L. These authors report increases in titratable acidity of 0.7–1.6 g/L when grape musts are coinoculated with *L. plantarum* C51 and different *S. cerevisiae* strains. Later Lucio (2014), Lucio et al. (2014), and Onetto and Bordeu (2015) described the production of 2.5 g/L of lactic from sugars when they coinoculated *S. cerevisiae* and *L. plantarum* simultaneously at the beginning of the vinification of low-acidity grape musts. These researchers achieved higher acidification levels when they inoculated *L. plantarum* some days before inoculating *S. cerevisiae* (7 days for Lucio, 2014 and 10 days for Onetto and Bordeu, 2015). This inoculation strategy allowed Lucio (2014) to obtain more than 11 g/L of lactic acid, whereas Onetto and Bordeu (2015) achieved 6 g/L. These data demonstrate that applying yeast inoculation some days later to bacteria ensures higher acidification levels than simultaneous inoculation. This high level is possible given the temporary lack of competition exerted by yeast on bacteria growth on the first few days (Henick-Kling and Park, 1994; Carreté et al., 2002; Osborne and Edwards, 2007; Edwards and Beelman, 1987; Lonvaud-Funel et al., 1988; Capucho and San Romão, 1994). When the delay between *L. plantarum* and *S. cerevisiae* inoculation was reduced to 2 days (a more realistic enological situation), lactic acid production from the bacterium is 5.5 g/L (Lucio, 2014; Lucio et al., 2014).

7.2.2 Reducing Ethanol Yields and Volatile Acidity

Another consequence of climate change is the high sugar concentration of grape musts (Santisi, 2011). Higher sugar triggers shifts in alcohol that alter flavors and mouthfeel, which consumers do not desire. There are several ways to minimize this problem, and some can be applied to vineyards and others at the vinification level (Mozell and Thach, 2014). A high sugar grape must results in high-ethanol wines that consumers well appreciate. To avoid high sugar levels in grapes, one possibility is to advance harvesting since veraison comes earlier. However, the anthocyanins concentration is too low in many cases when the sugar concentration is optimal (Mozell and Thach, 2014). Other physicochemical measures can be employed to reduce sugar content in grape musts, such as ultrafiltration and reverse osmosis (Mozell and Thach, 2014; E-VitiClimate, 2012; Pickering, 2000; Schmidtke et al., 2012). Biological procedures, e.g., yeast and LAB coinoculation, result in wines having lower ethanol and acetic acid contents. Reduced ethanol is the result of the partial transformation of grape must sugars into lactic acid by LAB via the homofermentative fermentation of hexoses (Ribéreau-Gayon et al., 2006a). This transformation has two effects: on the one hand, less ethanol is produced by yeast; thus, less acetic acid is generated. The selection of proper LAB to transform sugars into lactic acid is of crucial interest to increase total acidity, but not the acetic acid concentration. In line with this, Lucio et al. (2016) have published a paper that records the criteria adopted to select the most active strains to transform sugars into L-lactic acid among 31 strains that belong to *Lactobacillus vini*, *L. plantarum*, *Lactobacillus mali*, *Lactobacillus paracasei*, *Lactobacillus pantheris*, and *Lactobacillus satsumensis*. These species have a homofermentative or facultative heterofermentative metabolism of sugars (Dellaglio et al., 1994; Endo and Okada, 2005; Liu and Dong, 2002; Rodas et al., 2006). Saerens et al. (2015) indicate reductions of 0.5%–3% in ethanol (v/v) in the fermented beverages obtained by coinoculating *L. plantarum* and *S. cerevisiae* in different fruit juices (grape, apple, peach, etc.). The strategies that result in higher ethanol reduction consist of inoculating *L. plantarum* in fresh juice and 12–60 hours later *S. cerevisiae*. Up to 20 g/L of sugars are converted into lactic acid. Lucio et al. (2016) and Onetto and Bordeu (2015) report lactic acid production from sugars of up to 14 and 6.35 g/L for *L. plantarum*. Taking this into account, the theoretical yields of lactic acid production from glucose and fructose represent a sugar consumption of 28 and 12.7 g/L, respectively, for Lucio et al. (2016) and for Onetto and Bordeu (2015). Hence, there is a theoretical ethanol reduction of 1.65% and 0.75% (v/v) ethanol when considering that 17 g/L of sugar produces 1% alcohol in volume.

Neither Lucio et al. (2016) nor Saerens et al. (2015) have reported acetic acid production in the fruit juices fermented with different LAB strains, as expected from their homofermentative hexose metabolism. However, Onetto and Bordeu (2015) indicate an acetic acid production of 0.35 g/L in the grape juices fermented with *L. plantarum* alone; the synthesis of this acid might be due to either citric acid degradation or pentose metabolism (Desmazeaud and de Rosissart, 1994). For some researchers, the AF rate and volatile acidity levels do not present significant differences between sequential and coinoculated wines (Abrahamse and Bartowsky, 2012b; Izquierdo Cañas et al., 2014; Pan et al., 2011). *S. cerevisiae* and LAB coinoculation with homofermentative hexose metabolism guarantees lower acetic acid contents in wine than coinoculation with heterofermentative bacteria (*O. oeni*) (Onetto and Bordeu, 2015) or single inoculation with low ethanol-producing non-*Saccharomyces* yeasts. In addition, the partial degradation of sugars by LAB decreases the osmotic stress that *S. cerevisiae* faces to a high sugar grape must. Mira de Orduña (2010) warns about grape musts with high sugar concentrations causing a stress response in yeast, which leads to an increased formation of fermentation coproducts, such as acetic acid (Erasmus et al., 2003; Nurgel et al., 2004; Pigeau and Inglis, 2005). High sugar stress has been found to upregulate glycolytic and pentose phosphate pathway genes (Erasmus et al., 2003).

7.2.3 Controlling Microbial Spoilage

Many authors have studied the influence of the timing malolactic bacteria inoculation (Knoll et al., 2012; Barrajón-Simancas et al., 2011). Among the four methodologies used to promote MLF described in Table 7.1, the most appropriate is early LAB coinoculation (at the beginning of that AF) because it ensures better and more complete MLF. When LAB are inoculated in grape must instead of in wine, bacteria can easily develop in the absence of the inhibitors produced by yeasts (ethanol, SO_2, acetic acid, medium-chain fatty acids, β-phenylethanol, succinic acid, and certain peptides) (Caridi and Corte, 1997; Capucho and San Romão, 1994; Lerm et al., 2010; Lerm, 2010; Nehme et al., 2010; Rizk et al., 2016; Osborne and Edwards, 2007; Lafon-Lafourcade et al., 1983). However, if bacteria are inoculated in grape must before yeasts, vigorous bacterial growth can compromise ulterior yeast development and lead to stuck fermentations (King and Beelman, 1986), in which dangerous acetic acid and LAB bacteria, and spoilage yeasts, can develop (Drysdale and Fleet, 1988; Krieger and Arnink, 2003). Yeast and LAB coinoculation before AF ensures that MLF takes place before or on the first days of AF without compromising yeast development (Davis et al., 1985). This strategy permits the rapid completion of both alcoholic and MLFs and, when they finish, wines can be stabilized and remain more protected from the microbial development of unwanted bacteria (acetic acid bacteria, LAB that produce amines) or yeasts, e.g., *Brettanomyces/Dekkera* and *Zygosaccharomyces* (Lasik, 2013; Renouf et al., 2009; Bartowsky et al., 2015; Ribéreau-Gayon et al., 2006a; OIV, 2014; Smit et al., 2013; Suriano et al., 2015; Masqué et al., 2011; Abrahamse and Bartowsky, 2012a,b; Bartowsky and Pretorius, 2009; Henick-Kling and Park, 1994; Massera et al., 2009; Rosi et al., 2003; Ciani and Comitini, 2015; Azzolini et al., 2010; Gerbaux et al., 2009; Alexandre et al., 2004; Lallemand, 2013; Laffort, 2008; Berbegal et al., 2017a). When malolactic bacteria are sequentially inoculated, the delay between the end of AF and the beginning of MLF implies a risk period as the abovementioned spoilage microorganisms can develop in the absence of stabilization treatments. The early coinoculation with *S. cerevisiae* and of some *L. plantarum* strains with antibacterial activities can protect wine from contamination by unwanted lactobacilli and cocci (Ribéreau-Gayon et al., 2006a).

Knowledge of yeast interactions will contribute to the management of wine fermentation by controlling undesirable or spoilage microbiota by using controlled mixed fermentations (Ciani and Comitini, 2015).

7.2.4 Reducing Wine Toxins: Ochratoxin, Biogenic Amines, Ethyl Carbamate

The collateral consequences of the coinoculation that derives from the above-described main applications may include a reduction in the toxins produced by microorganisms associated with vineyards or winemaking.

As Russo et al. (2016) state, "An outbreak in wine safety generally results in the recovery of a wide variety of harmful compounds with a strong biological activity on human health (carbamate, amines, mycotoxins, heavy metals and residues from wrong production practices, methanol etc.)." Here we consider two of the most important ones of biological origin: OTA and BA.

OTA is produced by some fungi like *Penicillium* and *Aspergillus* species, which have been described on surfaces of grapevine clusters and berries (Kassemeyer and Berkelmann-Löhnertz, 2009). OTA has harmful effects on human health because it causes severe nephrotoxic effects, and has mutagenic, teratogenic, and carcinogenic activities (Mateo et al., 2010). The presence of OTA in grape juice and wine was first reported by Zimmerli and Dick (1996). OTA concentrations ranging from 0.01 to 3.5 µg/L have been recorded in grape products, and this toxin is present at higher concentrations in products from southern European regions (Mateo et al., 2010).

Although physicochemical treatments have been proposed to remove mycotoxins on the laboratory scale, not all of them are feasible in practical terms (Mateo et al., 2010; Berbegal et al., 2017b; Quintela et al., 2012). New biological methods constitute an interesting alternative to these methods. Indeed some strains of winemaking-related species, such as *L. plantarum, O. oeni, Saccharomyces bayanus,* and *S. cerevisiae*, have been described as being able to partially remove OTA (Dalié et al., 2010; Bejaoui et al., 2004; Fuchs et al., 2008; Piotrowska and Zakowska, 2005; Mateo et al., 2010; Petruzzi et al., 2014). The main OTA removing mechanism is adsorption on the cell surface. However, Piotrowska and Zakowska (2005) point out that another form of elimination must exist, while Del Prete et al. (2007) demonstrate that no intracellular enzymatic activity is able to degrade OTA and conclude that mycotoxin removal is a cell-binding phenomenon in the wine LAB they tested.

The factors that affect the percentage of removed OTA are strain, the pH of the media in which experiments are performed, cell concentration, and the cell's physiological state (alive or inactivated). Bejaoui et al. (2004) indicate that 90% of OTA is rapidly bound within the first 5 minutes and up to 72 hours of incubation with the heat-treated cells of either *S. cerevisiae* or *S. bayanus*. Fuchs et al. (2008) report an 8.6% reduction with the most active *L. plantarum* strain, whereas Piotrowska and Zakowska (2005) indicate a 56% reduction when one of the three tested *L. plantarum* strains is used. Mateo et al. (2010) demonstrate an OTA reduction above 60% by using the *O. oeni* living cells of strain 124 M. The majority of the above-recorded results have been obtained by incubating cells in either buffers or synthetic media, and only Bejaoui et al. (2004) have tested the ability of heat-treated *S. cerevisiae* LALVIN Rhône 2056 cells to remove OTA in red grape juice. In that experiment, yeast cells completely removed 10 µg/L of the mycotoxin in spiked grape juice. These findings offer the possibility of using cocultures of pairs *L. plantarum* or *O. oeni* and *S. cerevisiae* which, in addition to performing AF and MLF, are able to remove the OTA present in grape musts.

Twenty-four BA have been described in wine (Lehtonen, 1996). Some are naturally present in grape musts, while others are synthesized mainly by LAB, but also by some yeasts. These microorganisms decarboxylate the amino acidic precursors present in fruit juice (Landete et al., 2005; Caruso et al., 2002; Torrea and Ancín, 2001; Sahu et al., 2015). The most abundant BA in these products are histamine, tyramine, and putrescine and, to a lesser extent, phenylethylamine and cadaverine, but also agmatine, methylamine, and ethylamine (Nuñez et al., 2016). The most toxic BA for human health are histamine and tyramine, and tryptamine and phenylethylamine have important physiological effects (García-Moruno and Muñoz, 2012; Sahu et al., 2015). When ingested at high concentrations, BA may induce headache, respiratory distress, heart palpitation, hypo- or hypertension, enteric histaminosis, several allergic disorders, anaphylactic shock, and even death (Galgano et al., 2009; Sahu et al., 2015). The wine LAB described as histamine-producers are especially *Lactobacillus hilgardii, L. mali, O. oeni,* and *Pediococcus parvulus* (Landete et al., 2007; Coton et al., 2010). However, other bacteria recently found in winemaking processes, such as *Enterococcus faecium* and *Staphylococcus epidermidis*, are histamine-producers (Benavent-Gil et al., 2016; Pérez-Martín et al., 2014).

The negative consequences that BA have on health are a driving factor to develop methodologies which lead to foods with these toxins at low levels. Preventive and corrective strategies have been applied to produce low BA wines: not adding amino acid to help yeast development; short skin maceration and lees contact avoid BA precursors from accumulating. The use of SO_2 to prevent undesirable putative LAB BA-producers growth, and the inoculation of safe malolactic starters, prevent the synthesis of these toxins (Berbegal et al., 2017b). Recently, some LAB strains belonging to the species Lactobacillus *brevis*, Lactobacillus *casei*, Lactobacillus *collinoides*, Lactobacillus *farciminis*; *L. hilgardii*, Lactobacillus *delbrueckii*, *L. paracasei*, Lactobacillus *pentosus*, *L. plantarum*, *L. vini*, *O. oeni*, *Pediococcus acidilactici*, *P. parvulus*, and *Pediococcuspentosaceus* have been described as BA-degraders (Callejón et al., 2014; Capozzi et al., 2012; García-Ruiz et al., 2011). BA degradation positive results have been achieved in synthetic media or wine-like media, but have been demonstrated only in wine by García-Ruiz et al. with a *L. casei* strain (17.5%, 15.3%, and 7.6% reduction in histamine, tyramine, and putrescine, respectively), and by Callejón et al. (2016, 2017) with six strains belonging to *L. farciminis*, *L. plantarum*, and *P. acidilactici* (13.4%–27.8%, 16.2%–28.4%, and 26.5%–44% reduction in histamine, tyramine, and putrescine, respectively). The enzymes that catalyze amine degradation have been determined in *L. plantarum* and *P. acidilactici* and identified as laccases (Callejón et al., 2016, 2017).

Based on these findings, another innovative biological strategy to reduce wine BA content can be envisaged. New cocultures consisting of strains of *O. oeni* or *L. plantarum*, and in *S. cerevisiae* or *Torulaspora delbrueckii* unable to produce amines, can perform alcoholic and MLFs, and reduce the risk of BA formation in a preventive manner (Smit and du Toit, 2013; Massera et al., 2009; Izquierdo Cañas et al., 2012). LAB coinoculation with yeast starters is a promising approach to enhance both the quality and safety of wine (Tristezza et al., 2016). In addition, if the LAB that are able to degrade BA are used to perform sequential MLF, a reduction in BA, already produced by autochthonous microorganisms, can be achieved (in a corrective manner). Several authors point out the importance of taking into account the inability to produce BA among the selection criteria of malolactic and AF starters (Torrea and Ancín, 2001).

Another health concern in wine is ethyl carbamate (EC) or urethane. This compound is classified as a carcinogen by the International Agency for Research on Cancer, which reclassifies EC in Group 2A, which means that it is probably carcinogenic to humans (Patrignani et al., 2011). The origin of this compound in wine is urea (yeasts) and citrulline and carbamoyl phosphate (LAB), which are all produced from arginine metabolism (Vincenzini et al., 2009). Masqué et al. (2011) performed a study about the influence of LAB inoculation time, strain performing MLF, and type of wine (white Chardonnay and red Tempranillo) on EC content. The inoculation strategies that they tested were coinoculation of LAB and early and late yeast (12/24 hours from the beginning of AF and in the last moments of AF), and sequential inoculation of bacteria (once AF had finished). As malolactic starters, these authors used one strain of *L. plantarum*, one of *O. oeni*, or a combination of both. They did not find any evidence for the influence of the LAB inoculation point on EC content in either wine type.

7.2.5 Modification of the Organoleptic Characteristics

The sensorial characteristics of wines depend on grape variety characteristics (Pozo-Bayón et al., 2010), the metabolism of the microorganisms associated with vinification, mainly those responsible for AF and MLF (Torrens et al., 2008; Boido et al., 2009), and also on enological practices (Kyraleou et al., 2016; Boido et al., 2009; Pozo-Bayón et al., 2003). Apart from malic acid or sugars, LAB can also metabolize other precursors present in fermenting grape must and can affect, therefore, the chemical composition of wine and result in increased wine aroma and flavor complexity (Petruzzi et al., 2017). Differences in volatile composition are described in wines in which MLF has been performed with *O. oeni* or *L. plantarum* strains (Pozo-Bayón et al., 2005; Maicas et al., 1999). Brizuela et al. (2018) demonstrates that the *O. oeni* UNQOe 73.2 strain exhibits a superior capacity to improve wine flavor than the *L. plantarum* UNQLp 11 strain. These authors relate the better performance of the *O. oeni* strain to the presence of esterase activity in it, which increases fruity and creamy odorants. Pozo-Bayón et al. (2005) tested the influence of two *O. oeni* and two *L. plantarum* strains used to perform MLF on the volatile composition of Tempranillo red wines, and their data revealed significant metabolic differences between both species. Namely, *O. oeni* strains produced more γ-butyrolactone and decanoic acid and lower ethyl lactate than *L. plantarum*. These and others authors report significant intraspecific differences: Maicas et al. (1999) reported ranges of total alcohol concentrations of 55–70 mg/L in wines fermented by four different *O. oeni* strains and, similarly, total esters and total acids vary between 17–23 and 28–42 mg/L, respectively. Izquierdo Cañas et al. (2013) report different aldehydes, esters, and alcohols in Tempranillo wines fermented with different *O. oeni* strains. Chescheir et al. (2015) state that several *O. oeni* strains with distinct abilities to degrade tartaric and hydroxycinnamic acids can modify the free hydroxycinnamic concentration, which can be used by *Brettanomyces bruxellensis* to synthesize ethyl phenol compounds. Thus, LAB strains can directly or indirectly modify the aromatic profile of wine.

In addition to the abovedescribed factors, some authors have studied the influence of LAB inoculation timing on volatile wine composition, although no clear trends can be deduced from these studies. Frequently, the overall quality of final wines is not altered when coinoculation of microorganisms is used, and most metabolites synthesized by yeasts and bacteria are similar to those in conventional winemaking sequential strategies (Lasik-Kurdyś et al., 2017; Izquierdo Cañas et al., 2014; Pan et al., 2011; Abrahamse and Bartowsky, 2012b; Jussier et al., 2006). Bleve et al. (2016) report that coimmobilization produces a wine with comparable organoleptic profiles to that produced by sequential inoculation. Krieger and Arnink (2003) and by Massera et al. (2009) indicate minor (nonsignificant) differences, although coinoculation offers the best results. Simultaneous inoculation favors differences in fruitiness flavor, color intensity, and color type. Devi et al. (2017) and Antalick et al. (2013) tested the impact of timing inoculation with *O. oeni* on a Merlot red wine and found that the inoculation regime impacts the metabolic profile of wines. Variations in ester concentrations depend on the considered group of esters: coinoculation tends to increase the levels of ethyl acetate and ethyl succinates, mainly with diethyl succinate, but no trend can be emphasized on the impact on individual higher alcohol acetate contents or fatty acid ethyl esters and fatty acid. Knoll et al. (2012) describe how *O. oeni* and *S. cerevisiae* coinoculation tends to have higher concentrations of ethyl and acetate esters in Riesling white wine, which confer wines a fruitier character.

In sensory terms, Antalick et al. (2013) explain how coinoculated wines differ significantly for sequentially inoculated wines as far as fruity attributes are concerned. However, these authors conclude that coinoculation can increase or decrease the intensity of fruit and lactic descriptors by the production and degradation of metabolites, or by an aromatic mask developing in the short and long terms. du Plessis et al. (2017) explain how berry, astringency, and acid balance descriptors are significantly affected by the bacterial inoculating time, but other fruity, fresh vegetative, cook vegetative, spicy, floral, body, bitterness, and overall quality descriptors are not. Tristezza et al. (2016) describe how the Negroamaro wines produced according to a coinoculation strategy show a profile dominated by red and ripe fruits notes in association with esters and buttery and creamy notes.

Abrahamse and Bartowsky (2012b) indicate no influence of *O. oeni* and *S. cerevisiae* coinoculation on the phenolic composition of Shiraz and Cabernet Sauvignon. They studied the influence of the effect of different inoculation regimes in Shiraz wine on overall wine chemical composition, including wine color and the volatile profile, using the *O. oeni* Viniflora oenos (Christian Hansen) and *S. cerevisiae* strain AWRI 1490 couple. Coinoculation of bacteria with yeast produces a most distinct wine; at mid-AF, pressing, and post-AF inoculation, the produced wines have a similar chemical composition, but they all differ from wines not subjected to MLF. These authors found no significant differences in spectral wine color properties due to MLF timing, but found instead lower tannin, catechin, and epicatechin concentrations in coinoculated Shiraz. Coinoculated wines also display the highest concentration of total ethyl esters, which are compounds that confer wines a fruity sensation (Francis and Newton, 2005).

Although some authors hint at the existence of numerous aroma-related genes in *L. plantarum*, the correlation with the final aroma changes is poor in the wine produced by this species (Cappello et al., 2017; Costello et al., 2013; Grimaldi et al., 2005a,b; Matthews et al., 2004; Sestelo et al., 2004; Spano et al., 2005; Rodriguez et al., 2008; Esteban-Torres et al., 2015). Schöltz (2013) and Maarman (2014) describe some aroma changes, and indicate that the impact of bacterial strains on the modification of aroma compounds is stronger in coinoculation than in sequential inoculation. A general increase in total esters, which contributes to the fruity character of wines, especially ethyl lactate and ethyl acetate, has been reported (Maarman, 2014).

7.3 INTERACTIONS BETWEEN WINE MICROORGANISMS

Many interactions (both synergies and antagonisms) occur among microorganisms throughout winemaking, including both inoculated and non-inoculated autochthonous ones. These interactions are of utmost importance when planning coinoculation. The inhibition properties of some organisms on others seem species-specific (Fornachon, 1968), or even strain-specific (Arnink and Henick-Kling, 2005; Nehme et al., 2008; Costello et al., 2003). Although the biochemical bases for the antagonistic interactions between wine yeast and bacteria may be unclear, several factors, or their combination, can be involved (Alexandre et al., 2004). These factors include the production of bioactive yeast metabolites, such as ethanol (Capucho and San Romão, 1994), SO_2 (Fornachon, 1968; Carreté et al., 2002; Henick-Kling and Park, 1994), medium-chain fatty acids (Edwards and Beelman, 1987; Lonvaud-Funel et al., 1988; Edwards et al., 1990), antibacterial yeast metabolites of a protein nature (Dick et al., 1992; Comitini et al., 2005; Mendoza et al., 2010; Rizk et al., 2016), etc. Other factors depend on competition for nutrients since fermenting yeasts often deplete these nutrients, such as essential vitamins or amino acids for fastidious LAB (Fornachon, 1968; Beelman et al., 1982). Depletion of amino acids during AF delays the onset of bacterial growth until amino acid availability increases once again as yeast cells lyse or leak (Arnink and Henick-Kling, 2005).

While the commonest kind of interaction described by several researchers is bacterial inhibition by yeasts, stimulation and neutralism appear less frequently. Nehme et al. (2008) quantitatively studied the interactions between *S. cerevisiae* and *O. oeni* strains. They observed that some yeast strains lead to the inhibition of the bacterial component, which correlates to weak bacterial growth rather than to reduced bacterial malolactic activity. Some other yeasts display total neutrality toward bacteria, and the third group of yeast strains stimulates the growth and malolactic activity of bacterial strains. These interactions can also vary depending on the winemaking moment: Rosi et al. (2003) inoculated a bacterial culture at the beginning, halfway through, and at the end of AF. They observed that the midpoint of AF is when the antagonistic effect of yeasts toward bacteria can be stronger, and there is no guarantee that bacteria can overcome this antagonism. They indicate that this effect can be explained by the removal of nutrients, and, also, by accumulation of SO_2, ethanol, and other toxic metabolites. When inoculation is carried out at the end of AF, the presence of yeasts favors bacterial viability and malolactic activity (Rosi et al., 2003): the release of polysaccharides, especially mannoproteins during AF, can stimulate malolactic bacteria growth (Guilloux-Benatier et al., 1995, Rosi, 2000). At times some negative effects of non-*Saccharomyces* yeasts, such as *Candida zemplinina* and *Lachancea thermotolerans*, have been reported for LAB growth and the progress of MLF; for example, when LAB are used in simultaneous inoculation, but the same effect is not observed for sequential MLF (du Plessis et al., 2017).

Occasionally LAB can have negative effects against yeasts, particularly *S. cerevisiae*, or can change its metabolism and the aroma compound profiles of the resulting wines. Krieger and Arnink (2003) indicate the putative effect of acetic acid formed by some lactobacilli on *Saccharomyces* growth. Muñoz et al. (2014) report that bacterial growth and malolactic activity can negatively affect AF performance by leaving residual sugar. This effect is stronger for some yeast strains. In this case, the entailed risk is the higher production of D-lactic and acetic acids from hexose degradation by bacteria, which is higher than in sequential inoculations (Muñoz et al., 2014; Mendoza et al., 2011). Although the D-lactic acid produced from hexose degradation by bacteria is higher than in sequential inoculations, volatile acidity shows a higher level in wines that derive from simultaneous inoculation (Muñoz et al., 2014). Rossouw et al. (2012) point out that a significant number of genes are differentially expressed in *S. cerevisiae* in yeast-only and in coinoculated fermentations. Some differentially expressed genes respond to chemical changes in fermenting must that are linked to bacterial metabolic activities, whereas others might represent a direct response of yeast to the presence of a competing organism (Rossouw et al., 2012).

In many cases, no negative or positive interactions between yeasts and bacteria are reported; e.g. the same AF rate (Abrahamse and Bartowsky, 2012a,b), lack of inhibition, and simultaneous AF and MLF (Carew et al., 2015). Briefly, choosing the appropriate yeast-bacteria couple it is crucial to guarantee a vinification without problems.

Acknowledgments

The present work has been partially financed by Project AGL2015-71227-R (ERDF and the Spanish Ministry of Economy and Competitiveness). Helen Warburton, an English language reviewer, has revised the English.

References

Abrahamse, C., Bartowsky, E., 2012a. Inoculation for MLF reduces overall vinification time. Aust. N.Z. Grapegrower Winemaker 578, 41–46.

Abrahamse, C.E., Bartowsky, E.J., 2012b. Timing of malolactic fermentation inoculation in Shiraz grape must and wine: influence on chemical composition. World J. Microbiol. Biotechnol. 28, 255–265.

Alexandre, H., Costello, P.J., Remize, F., Guzzo, J., Guilloux-Benatier, M., 2004. *Saccharomyces cerevisiae–Oenococcus oeni* interactions in wine: current knowledge and perspectives. Int. J. Food Microbiol. 93, 141–154.

Andorrà, I., Berradre, M., Rozès, N., Mas, A., Guillamón, J.M., Esteve-Zarzoso, B., 2010. Effect of pure and mixed cultures of the main wine yeast species on grape must fermentations. Eur. Food Res. Technol. 231, 215–224.

Antalick, G., Perello, M.C., de Revel, G., 2013. Co-inoculation with yeast and LAB under winery conditions: modification of the aromatic profile of Merlot wines. S. Afr. J. Enol. Vitic. 34, 223–232.

Arnink, K., Henick-Kling, T., 2005. Influence of *Saccharomyces cerevisiae* and *Oenococcus oeni* strains on successful malolactic conversion in wine. Am. J. Enol. Vitic. 56, 228–237.

Azzolini, M., Tosi, E., Vagnoli, P., Krieger, S., Zapparoli, G., 2010. Evaluation of technological effects of yeast-bacterial co-inoculation in red table wine production. Ital. J. Food Sci. 22, 257–263.

Barrajón-Simancas, N., Giese, E., Arévalo-Villena, M., Úbeda, J., Briones, A., 2011. Amino acid uptake by wild and commercial yeasts in single fermentations and co-fermentations. Food Chem. 127, 441–446.

Bartowsky, E., Pretorius, I.S., 2009. Microbial formation and modification of flavor and off-flavor compounds in wine. In: Köning, H., Unden, G., Fröhlich, J. (Eds.), Biology of Microorganisms on Grapes, in Must and in Wine. Berlin, Gemany, Springer-Verlag.

Bartowsky, E.J., Costello, P.J., Chambers, P.J., 2015. Emerging trends in the application of malolactic fermentation. Aust. J. Grape Wine Res. 21, 663–669.

Bauer, R., Dicks, L., 2004. Control of malolactic fermentation in wine: a review. S. Afr. J. Enol. Vitic. 25, 74–88.

Beelman, R.B., Keen, R.M., Banner, M.J., King, S.W., 1982. Interactions between wine yeast and malolactic bacteria under wine conditions. Dev. Ind. Microbiol. 23, 107–121.

Bejaoui, H., Mathieu, F., Taillandier, P., Lebrihi, A., 2004. Ochratoxin A removal in synthetic and natural grape juices by selected oenological *Saccharomyces* strains. J. Appl. Microbiol. 97, 1038–1044.

Benavent-Gil, Y., Berbegal, C., Lucio, O., Pardo, I., Ferrer, S., 2016. A new fear in wine: isolation of *Staphylococcus epidermidis* histamine producer. Food Control 62, 142–149.

Berbegal, C., Peña, N., Russo, P., Grieco, F., Pardo, I., Ferrer, S., et al., 2016. Technological properties of *Lactobacillus plantarum* strains isolated from grape must fermentation. Food Microbiol. 57, 187–194.

Berbegal, C., Spano, G., Fragasso, M., Grieco, F., Russo, P., Capozzi, V., 2017a. Starter cultures as biocontrol strategy to prevent *Brettanomyces bruxellensis* proliferation in wine. Appl. Microbiol. Biotechnol. 102, 569–576.

Berbegal, C., Spano, G., Tristezza, M., Grieco, F., Capozzi, V., 2017b. Microbial resources and innovation in the wine production sector. S. Afr. J. Enol. Vitic. 38, 156–166.

Bleve, G., Tufariello, M., Vetrano, C., Mita, G., Grieco, F., 2016. Simultaneous alcoholic and malolactic fermentations by *Saccharomyces cerevisiae* and *Oenococcus oeni* cells co-immobilized in alginate beads. Front. Microbiol. 7, 943.

Boido, E., Medina, K., Fariña, L., Carrau, F., Versini, G., Dellacassa, E., 2009. The effect of bacterial strain and aging on the secondary volatile metabolites produced during malolactic fermentation of Tannat red wine. J. Agric. Food. Chem. 57, 6271–6278.

Bordons, A., 1997. Las bacterias lácticas del vino y la fermentación maloláctica. In Ayuntamiento de Haro (Ed.), Criterios de valoración de la calidad de la uva: XII Cursos Rioja '97. Haro, Spain. pp. 9–31.

Brizuela, N.S., Bravo-Ferrada, B.M., La Hens, D.V., Hollmann, A., Delfederico, L., Caballero, A., et al., 2017. Comparative vinification assays with selected Patagonian strains of *Oenococcus oeni* and *Lactobacillus plantarum*. LWT Food Sci. Technol.

Brizuela, N.S., Bravo-Ferrada, B.M., Pozo-Bayón, M.Á., Semorile, L., Elizabeth Tymczyszyn, E., 2018. Changes in the volatile profile of Pinot noir wines caused by Patagonian *Lactobacillus plantarum* and *Oenococcus oeni* strains. Food Res. Int. 106, 22–28.

Callejón, S., Sendra, R., Ferrer, S., Pardo, I., 2014. Identification of a novel enzymatic activity from lactic acid bacteria able to degrade biogenic amines in wine. Appl. Microbiol. Biotechnol. 98, 185–198.

Callejón, S., Sendra, R., Ferrer, S., Pardo, I., 2016. Cloning and characterization of a new laccase from *Lactobacillus plantarum* J16 CECT 8944 catalyzing biogenic amines degradation. Appl. Microbiol. Biotechnol. 100, 3113–3124.

Callejón, S., Sendra, R., Ferrer, S., Pardo, I., 2017. Recombinant laccase from *Pediococcus acidilactici* CECT 5930 with ability to degrade tyramine. PLoS ONE 12, e0186019.

Capozzi, V., Russo, P., Ladero, V., Fernandez, M., Fiocco, D., Alvarez, M.A., et al., 2012. Biogenic amines degradation by *Lactobacillus plantarum*: toward a potential application in wine. Front. Microbiol. 3, 122.

Cappello, M.S., Zapparoli, G., Logrieco, A., Bartowsky, E.J., 2017. Linking wine lactic acid bacteria diversity with wine aroma and flavour. Int. J. Food Microbiol. 243, 16–27.

Capucho, I., San Romão, M.V., 1994. Effect of ethanol and fatty acids on malolactic activity of *Leuconostoc oenos*. Appl. Microbiol. Biotechnol. 42, 391–395.

Carew, A.L., Close, D.C., Dambergs, R.G., 2015. Yeast strain affects phenolic concentration in Pinot noir wines made by microwave maceration with early pressing. J. Appl. Microbiol. 118, 1385–1394.

Caridi, A., Corte, V., 1997. Inhibition of malolactic fermentation by cryotolerant yeasts. Biotechnol. Lett. 19, 723–726.

Carreté, R., Vidal, M., Bordons, A., Constantí, M., 2002. Inhibitory effect of sulfur dioxide and other stress compounds in wine on the ATPase activity of *Oenococcus oeni*. FEMS Microbiol. Lett. 211, 155–159.

Caruso, M., Fiore, C., Contursi, M., Salzano, G., Paparella, A., Romano, P., 2002. Formation of biogenic amines as criteria for the selection of wine yeasts. World J. Microbiol. Biotechnol. 18, 159–163.

Caspritz, G., Radler, F., 1983. Malolactic enzyme of *Lactobacillus plantarum*. Purification, properties, and distribution among bacteria. J. Biol. Chem. 258, 4907–4910.

Charpentier, C. 1985. La desacidification biologique des vins par les Schizosaccharomyces. In: Communications a les Journèe de Recontres Oenologiques Montpellier, 1985.

Charpentier, C., Feuillat, M., Gervaux, V., Auther, R., 1985. La désacidification biologique des vins blancs par les Schizosaccharomyces. C.R. Acad. Agric. Fr. 71, 425–432.

Chescheir, S., Philbin, D., Osborne, J.P., 2015. Impact of *Oenococcus oeni* on wine hydroxycinnamic acids and volatile phenol production by *Brettanomyces bruxellensis*. Am. J. Enol. Vitic. 66, 357–362.

Ciani, M., 1995. Continuous deacidification of wine by immobilized *Schizosaccharomyces pombe* cells: evaluation of malic acid degradation rate and analytical profiles. J. Appl. Bacteriol. 79, 631–634.

Ciani, M., Comitini, F., 2015. Yeast interactions in multi-starter wine fermentation. Curr. Opin. Food Sci. 1, 1–6.

Claus, H., 2017. Microbial enzymes: relevance for winemaking. In: König, H., Unden, G., Fröhlich, J. (Eds.), Biology of Microorganisms on Grapes, in Must and in Wine. Springer International Publishing, Cham.

Comitini, F., Ferretti, R., Clementi, F., Mannazzu, I., Ciani, M., 2005. Interactions between *Saccharomyces cerevisiae* and malolactic bacteria: preliminary characterization of a yeast proteinaceous compounds active against *Oenococcus oeni*. J. Appl. Microbiol. 99, 105–111.

Costello, P.J., Henschke, P.A., Markides, A.J., 2003. Standardised methodology for testing malolactic bacteria and wine yeast compatibility. Aust. J. Grape Wine Res. 9, 127–137.

Costello, P.J., Siebert, T.E., Solomon, M.R., Bartowsky, E.J., 2013. Synthesis of fruity ethyl esters by acyl coenzyme A: alcohol acyltransferase and reverse esterase activities in *Oenococcus oeni* and *Lactobacillus plantarum*. J. Appl. Microbiol. 114, 797–806.

Coton, M., Romano, A., Spano, G., Ziegler, K., Vetrana, C., Desmarais, C., et al., 2010. Occurrence of biogenic amine-forming lactic acid bacteria in wine and cider. Food Microbiol. 27, 1078–1085.

Dalié, D.K.D., Deschamps, A.M., Richard-Forget, F., 2010. Lactic acid bacteria—potential for control of mould growth and mycotoxins: a review. Food Control 21, 370–380.

Davis, C.R., Wibowo, D., Eschenbruch, R., Lee, T.H., Fleet, G.H., 1985. Practical implications of malolactic fermentation: a review. Am. J. Enol. Vitic. 36, 290–301.

de Cort, S., Shantha Kumara, H.M.C., Verachert, H., 1994. Localization and characterization of alpha-glucosidase activity in *Lactobacillus brevis*. Appl. Environ. Microbiol. 60, 3074–3078.

Dellaglio, F., de Roissart, H., Torriani, S., Curk, M.C., Janssens, D., 1994. Caractéristiques générales des bactéries lactiques. In: de Roissart, H., Luquet, F.M. (Eds.), Bactéries Lactiques, vol. I. Lorica, Uriage, France, pp. 25–116.

Delaquis, P., Cliff, M., King, M., Giraud, B., Hall, J., Reynolds, A., 2000. Effect of two commercial malolactic cultures on the chemical and sensory properties of Chancellar wines vinified with differrent yeast and fermentation temperatures. Am. J. Enol. Vitic. 51, 42–48.

Del Prete, V., Rodríguez, H., Carrascosa, A.V., de las Rivas, B., García-Moruno, E., Muñoz, R., 2007. *In vitro* removal of ochratoxin a by wine lactic acid bacteria. J. Food Prot. 70, 2155–2160.

de Revel, G., Marin, N., Pripis-Nicolau, L., Lonvaud-Funel, A., Bertand, A., 1999. Contribution to the knowledge of malolactic fermentation influence of wine aroma. J. Agric. Food. Chem. 47, 4003–4008.

Dequin, S., Baptista, E., Barre, P., 1999. Acidification of grape musts by *Saccharomyces cerevisiae* wine yeast strains genetically engineered to produce lactic acid. Am. J. Enol. Vitic. 50, 45–50.

Desmazeaud, M.J., de Rosissart, H., 1994. Métabolime général des bactéries lactiques. In: de Roissart, H., Luquet, F.M. (Eds.), Bactéries Lactiques, vol. I, Lorica, Uriage, France, pp. 169–207.

Devi, A., Archana, K.M., Bhavya, P.K., Anu-Appaiah, K.A., 2017. Non-anthocyanin polyphenolic transformation by native yeast and bacteria co-inoculation strategy during vinification. J. Sci. Food Agric. 98, 1162–1170.

De Vos, P., Garrity, G., Jones, D., Krieg, N.R., Ludwig, W., Rainey, F.A., et al., 2011. Bergey's Manual of Systematic Bacteriology: The Firmicutes. Springer Dordrecht Heidelberg London, New York, NY.

Dick, K.J., Molan, P.C., Eschenbruch, R., 1992. The isolation from *Saccharomyces cerevisiae* of two antibacterial cationic proteins that inhibit malolactic bacteria. Vitis 31, 105–116.

Drysdale, G.S., Fleet, G.H., 1988. Acetic acid bacteria in winemaking: a review. Am. J. Enol. Vitic. 39, 143–154.

du Plessis, H., du Toit, M., Nieuwoudt, H., van der Rijst, M., Kidd, M., Jolly, N., 2017. Effect of *Saccharomyces*, non-*Saccharomyces* yeasts and malolactic fermentation strategies on fermentation kinetics and flavor of Shiraz wines. Fermentation 3, 64.

du Toit, M., Engelbrecht, L., Lerm, E., Krieger-Weber, S., 2011. *Lactobacillus*: the next generation of malolactic fermentation starter cultures—an overview. Food Bioprocess Technol. 4, 876–906.

Dziezak, J.D., 2003. ACIDS natural acids and acidulants A2—Caballero, Benjamin, Encyclopedia of Food Sciences and Nutrition, second ed. Academic Press, Oxford.

E-VitiClimate, 2012. Lifelong learning project, E-VitiClimate [Online]. EuroProject Lifelong Learning Programme. Available: http://www.e-viticlimate.eu/ (accessed 05.01.18).

Edwards, C.G., A.G, R., Rodriguez, A.V., M.J., S., J.M., M., 1999. Implication of acetic acid in the induction of slow/stuck grape juice fermentations and inhibition of yeasts by *Lactobacillus* sp. Am. J. Enol. Vitic. 50, 204–210.

Edwards, C.G., Beelman, R.B., 1987. Inhibition of the malolactic bacterium *Leuconostoc oenos* (PSU-1), by decanoic acid and subsequent removal of the inhibition by yeast ghosts. Am. J. Enol. Vitic. 38, 239–242.

Edwards, C.G., Beelman, R.B., Bartley, C.E., McConnell, A.L., 1990. Production of decanoic acid and other volatile compounds and the growth of yeast and malolactic bacteria during vinification. Am J. Enol. Vitic. 41, 48–56.

Edwards, C.G., Beelman, R.B., 1989. Inducing malolactic fermentation in wines. Biotechnol. Adv. 7, 333–360.

Endo, A., Okada, S., 2005. *Lactobacillus satsumensis* sp. nov., isolated from mashes of shochu, a traditional Japanese distilled spirit made from fermented rice and other starchy materials. Int. J. Syst. Evol. Microbiol. 55, 83–85.

Erasmus, D.J., van der Merwe, G.K., van Vuuren, H.J.J., 2003. Genome-wide expression analyses: metabolic adaptation of *Saccharomyces cerevisiae* to high sugar stress. FEMS Yeast Res. 3, 375–399.

Esteban-Torres, M., Landete, J.M., Reveron, I., Santamaría, L., de las Rivas, B., Muñoz, R., 2015. A *Lactobacillus plantarum* esterase active on a broad range of phenolic esters. Appl. Environ. Microbiol. 81, 3235–3242.

Esteve-Zarzoso, B., Manzanares, P., Ramón, D., Querol, A., 1998. The role of non-*Saccharomyces* yeasts in industrial winemaking. Int. Microbiol. 1, 143–148.

Ferrer, S., Pardo, I., Berbegal, C., Lucio, O., Polo, L., 2017. Wood shavings with microorganisms, preparation thereof and use thereof. WO/2017/114999.

Formisyn, P., Vaillant, H., Lantreibecq, F., Bourgois, J., 1997. Development of an enzymatic reactor for initiating malolactic fermentation in wine. Am. J. Enol. Vitic. 48, 345–351.

Fornachon, J.C.M., 1968. Influence of different yeasts on the growth of lactic acid bacteria in wine. J. Sci. Food Agric. 19, 374–378.

Francis, I.L., Newton, J.L., 2005. Determining wine aroma from compositional data. Aust. J. Grape Wine Res. 11, 114–126.

Fuchs, S., Sontag, G., Stidl, R., Ehrlich, V., Kundi, M., Knasmüller, S., 2008. Detoxification of patulin and ochratoxin A, two abundant mycotoxins, by lactic acid bacteria. Food Chem. Toxicol. 46, 1398–1407.

Fugelsang, K.C., Edwards, C.G., 2007. Wine Microbiology: Practical Applications and Procedures. Springer Science + Business Media LLC, New York, NY, USA.

Galgano, F., Caruso, M., Favati, F., 2009. Biogenic amines in wines: a review. In: O'Byrne, P. (Ed.), Red Wine and Health. Nova Science Publishers Inc, New York, NY.

García-Moruno, E., Muñoz, R., 2012. Does *Oenococcus oeni* produce histamine? Int. J. Food Microbiol. 157, 121–129.

García-Ruiz, A., González-Rompinelli, E.M., Bartolomé, B., Moreno-Arribas, M.V., 2011. Potential of wine-associated lactic acid bacteria to degrade biogenic amines. Int. J. Food Microbiol. 148, 115–120.

Garofalo, C., El-Khoury, M., Lucas, P., Bely, M., Russo, P., Spano, G., et al., 2015. Autochthonous starter cultures and indigenous grape variety for regional wine production. J. Appl. Microbiol. 118, 1395–1408.

Gerbaux, V., Briffox, C., Dumont, A., Krieger, S., 2009. Influence of inoculation with malolactic bacteria on volatile phenols in wines. Am. J. Enol. Vitic. 60, 233–235.

Grimaldi, A., McLean, H., Jiranek, V., 2000. Identification and partial characterization of glycosidic activities of commercial strains of the lactic acid bacterium, *Oenococcus oeni*. Am. J. Enol. Vitic. 51, 362–364.

Grimaldi, A., Bartowsky, E., Jiranek, V., 2005a. Screening of *Lactobacillus* spp. and *Pediococcus* spp. for glycosidase activities that are important in oenology. J. Appl. Microbiol. 99, 1061–1069.

Grimaldi, A., Bartowsky, E., Jiranek, V., 2005b. A survey of glycosidase activities of commercial wine strains of *Oenococcus oeni*. Int. J. Food Microbiol. 105, 233–244.

Guilloux-Benatier, M., Guerreau, J., Feuillat, M., 1995. Influence of initial colloid content on yeast macromolecule production and on the metabolism of wine microorganisms. Am. J. Enol. Vitic. 46, 486–492.

Henick-Kling, T., 1993. Malolactic fermentation. In: Fleet, G.H. (Ed.), Wine Microbiology and Biotechnology. Harwood Academic Publishers, Chur, Switzerland.

Henick-Kling, T., Park, Y.H., 1994. Considerations for the use of yeast and bacterial starter cultures: SO_2 and timing of inoculation. Am. J. Enol. Vitic. 45, 464–469.

Iorizzo, M., Testa, B., Lombardi, S.J., García-Ruiz, A., Muñoz-González, C., Bartolomé, B., et al., 2016. Selection and technological potential of *Lactobacillus plantarum* bacteria suitable for wine malolactic fermentation and grape aroma release. LWT Food Sci. Technol. 73, 557–566.

Izquierdo Cañas, P.M., Pérez-Martín, F., García Romero, E., Seseña Prieto, S., Palop Herreros, M.L.L., 2012. Influence of inoculation time of an autochthonous selected malolactic bacterium on volatile and sensory profile of Tempranillo and Merlot wines. Int. J. Food Microbiol. 156, 245–254.

Izquierdo Cañas, P.M., García Romero, E., Pérez Martín, F., Seseña Prieto, S., Heras Manso, J.M., Palop Herreros, M.L.L., 2013. Behavior during malolactic fermentation of three strains of *Oenococcus oeni* used as direct inoculation and acclimatisation cultures. S. Afr. J. Enol. Vitic. 34, 1–9.

Izquierdo Cañas, P.M., García Romero, E., Pérez-Martín, F., Seseña, S., Palop, M.L., 2014. Sequential inoculation versus co-inoculation in Cabernet Franc wine fermentation. Food Sci. Technol. Int. 21, 203–212.

Jussier, D., Dubé Morneau, A., Mira de Orduña, R., 2006. Effect of simultaneous inoculation with yeast and bacteria on fermentation kinetics and key wine parameters of cool-climate Chardonnay. App. Environ. Microbiol. 72, 221–227.

Kapsopoulou, K., Mourtzini, A., Anthoulas, M., Nerantzis, E., 2007. Biological acidification during grape must fermentation using mixed cultures of *Kluyveromyces thermotolerans* and *Saccharomyces cerevisiae*. World J. Microbiol. Biotechnol. 23, 735–739.

Kassemeyer, H.-H., Berkelmann-Löhnertz, B., 2009. Fungi of grapes. In: König, H., Unden, G., Fröhlich, J. (Eds.), Biology of Microorganisms on Grapes, in Must and in Wine. Springer, Berlin, Germany.

King, S.W., Beelman, R.B., 1986. Metabolic interactions between *Saccharomyces cerevisiae* and *Leuconostoc oenos* in a model grape juice/wine system. Am. J. Enol. Vitic. 37, 53–60.

Knoll, C., Fritsch, S., Schnell, S., Grossmann, M., Krieger-Weber, S., du Toit, M., et al., 2012. Impact of different malolactic fermentation inoculation scenarios on Riesling wine aroma. World J. Microbiol. Biotechnol. 28, 1143–1153.

Kourkoutas, Y., Bekatorou, A., Banat, I.M., Marchant, R., Koutinas, A.A., 2004. Immobilization technologies and support materials suitable in alcohol beverages production: a review. Food Microbiol. 21, 377–397.

Krieger, S., Arnink, K., 2003. Malolactic fermentation—a review of recent research on timing of inoculation and possible yeast-bacteria combinations. In: 32nd Annual New York Wine Industry Workshop.

Krieger-Weber, S., 2017. Application of yeast and bacteria as starter cultures. In: König, H., Unden, G., Fröhlich, J. (Eds.), Biology of Microorganisms on Grapes, in Must and in Wine, second ed. Springer International Publishing, Cham.

Krieger-Weber, S., Déléris-Bou, M., Heras, J.M., 2010. Cultivos iniciadores malolácticos en forma seca activa. Un enfoque rentable y eficiente para el control de las fermentaciones maloláctica. EnoReports 1–7. Available from: http://www.enoreports.com/enoreports/pdf/lallemand_sep10.pdf.

Krieger-Weber, S., Déléris-Bou, M., Dumont, A., Vidal, F., Suárez, C., Heras, J.M., 2016. Un nuevo concepto de *Lactobacillus plantarum* seleccionada para vinos de pH elevado. La Semana Vitivinícola 3473, 2294–2297.

Krieger-Weber, S., Silvano, A., Déléris-Bou, M., Vidal, F., Dumont, A., Lo Paro, J.F., 2017. Uno specifico ceppo di *Lb. plantarum* ad elevata efficienza per malolattiche in coinoculo: gestione e vantaggi. L'Enologo 7-8, 93–97.

Kunkee, R.E., Ough, C.S., Amerine, M.A., 1964. Induction of malo-lactic fermentation by inoculation of must and wine with bacteria. Am. J. Enol. Viti 15, 178–183.

Kyraleou, M., Tzanakouli, E., Kotseridis, Y., Chira, K., Ligas, I., Kallithraka, S., et al., 2016. Addition of wood chips in red wine during and after alcoholic fermentation: differences in color parameters, phenolic content and volatile composition. 2016, 50.

Laffort, 2008. Fermentation management—specific case: yeast/bacteria co-inoculation. Technical Booklet [Online]. Available from: http://vinestovintages.ca/LaffortHelpfulHints/Co-Inoculation%20(yeast%20&%20bacteria).pdf.

Lafon-Lafourcade, S., Carre, E., Ribéreau-Gayon, P., 1983. Occurrence of lactic acid bacteria during the different stages of vinification and conservation of wines. App. Environ. Microbiol. 46, 874–880.

Lallemand, 2013. Co-inoculation of selected wine bacteria. The Wine Expert [Online], 7, April 5th. Available from: http://www.lallemand-wine.com/wp-content/uploads/2013/07/WE4-Australia.pdf.

Landete, J.M., Ferrer, S., Pardo, I., 2005. Which lactic acid bacteria are responsible of histamine production in wine? J. Appl. Microbiol. 99, 580–586.

Landete, J.M., Ferrer, S., Pardo, I., 2007. Biogenic amine production by lactic acid bacteria, acetic bacteria and yeast isolated from wine. Food Control 18, 1569–1574.

Lasik, M., 2013. The application of malolactic fermentation process to create good-quality grape wine produced in cool-climate countries: a review. Eur. Food Res. Technol. 237, 843–850.

Lasik-Kurdyś, M., Gumienna, M., Nowak, J., 2017. Influence of malolactic bacteria inoculation scenarios on the efficiency of the vinification process and the quality of grape wine from the Central European region. Eur. Food Res. Technol.

Lehtonen, P., 1996. Determination of amines and amino acids in wine: a review. Am. J. Enol. Vitic. 47, 127–133.

Lerena, M.C., Rojo, M.C., Sari, S., Mercado, L.A., Krieger-Weber, S., Combina, M., 2016. Malolactic fermentation induced by *Lactobacillus plantarum* in Malbec wines from Argentina. S. Afr. J. Enol. Vitic. 37, 115–123.

Lerm, E., 2010. The Selection and Characterisation of Lactic Acid Bacteria to be Used as a Mixed Starter Culture for Malolactic Fermentation. Master of Agricultural Science. Stellenbosch University.

Lerm, E., Engelbrecht, L., du Toit, M., 2010. Malolactic fermentation: the ABC's of MLF. S. Afr. J. Enol. Vitic. 31, 186–212.

Lerm, E., Engelbrecht, L., du Toit, M., 2011. Selection and characterisation of *Oenococcus oeni* and *Lactobacillus plantarum* South African wine isolates for use as malolactic fermentation starter cultures. S. Afr. J. Enol. Vitic. 32, 280–295.

Liu, B., Dong, X., 2002. *Lactobacillus pantheris* sp. nov., isolated from faeces of a jaguar. Int J. Syst. Evol. Microbiol. 52, 1745–1748.

Liu, S.-Q., 2002. Malolactic fermentation in wine—beyond deacidification. J. Appl. Microbiol. 92, 589–601.

Lonvaud-Funel, A., 1999. Lactic acid bacteria in the quality improvement and depreciation of wine. Antonie Van Leeuwenhoek 76, 317–331.

Lonvaud-Funel, A., Joyeux, A., Desens, C., 1988. Inhibition of malolactic fermentation of wines by products of yeast metabolism. J. Sci. Food Agric. 44, 183–191.

Lucio, O., 2014. Acidificación biológica de vinos de pH elevado mediante la utilización de bacterias lácticas. Tesis Doctoral Ph.D. Thesis. Universidad de Valencia.

Lucio, O., Pardo, I., Heras, J.M., Krieger, S., Ferrer, S., 2014. Effect of yeasts/bacteria co-inoculation on malolactic fermentation of Tempranillo wines. Infowine [Online]. http://www.infowine.com/default.asp?scheda = 13630.

Lucio, O., Pardo, I., Krieger-Weber, S., Heras, J.M., Ferrer, S., 2016. Selection of *Lactobacillus* strains to induce biological acidification in low acidity wines. LWT Food Sci. Technol. 73, 334–341.

Lucio, O., Pardo, I., Heras, J.M., Krieger-Weber, S., Ferrer, S., 2017. Use of starter cultures of *Lactobacillus* to induce malolactic fermentation in wine. Aust. J. Grape Wine Res. 23, 15–21.

Maarman, B.C., 2014. Interaction between wine yeast and malolactic bacteria and the impact on wine aroma and flavour. Department of Viticulture and Oenology, Institute for Wine Biotechnology, Stellenbosch University.

Maicas, S., 2001. The use of alternative technologies to develop malolactic fermentation in wine. Appl. Microbiol. Biotechnol. 56, 35–39.

Maicas, S., Gil, J.V., Pardo, I., Ferrer, S., 1999. Improvement of volatile composition of wines by controlled addition of malolactic bacteria. Food Res. Int. 32, 491–496.

Masqué, M.C., Soler, M., Zaplana, B., Franquet, R., Rico, S., Elorduy, X., et al., 2011. Ethyl carbamate content in wines with malolactic fermentation induced at different points in the vinification process. Ann. Microbiol. 61, 199–206.

Massera, A., Soria, A., Catania, C., Krieger, S., Combina, M., 2009. Simultaneous inoculation of Malbec (*Vitis vinifera*) musts with yeast and bacteria: effects on fermentation performance, sensory and sanitary attributes of wines. Food Technol. Biotechnol. 47.

Mateo, E.M., Medina, Á., Mateo, F., Valle-Algarra, F.M., Pardo, I., Jiménez, M., 2010. Ochratoxin A removal in synthetic media by living and heat-inactivated cells of *Oenococcus oeni* isolated from wines. Food Control 21, 23–28.

Mateo, J., Maicas, S., 2016. Application of non-*Saccharomyces* yeasts to wine-making process. Fermentation 2, 14.

Matthews, A., Grimaldi, A., Walker, M., Bartowsky, E., Grbin, P., Jiranek, V., 2004. Lactic Acid Bacteria as a potential source of enzymes for use in vinification. Appl. Environ. Microbiol. 70, 5715–5731.

Maturano, Y.P., Mestre, M.V., Esteve-Zarzoso, B., Nally, M.C., Lerena, M.C., Toro, M.E., et al., 2015. Yeast population dynamics during prefermentative cold soak of Cabernet Sauvignon and Malbec wines. Int. J. Food Microbiol. 199, 23–32.

Mendoza, L., Merín, M., Morata, V., Farías, M., 2011. Characterization of wines produced by mixed culture of autochthonous yeasts and *Oenococcus oeni* from the northwest region of Argentina. J. Ind. Microbiol. Biotechnol. 38, 1777–1785.

Mendoza, L.M., Manca de Nadra, M.C., Farías, M.E., 2010. Antagonistic interaction between yeasts and lactic acid bacteria of oenological relevance: partial characterization of inhibitory compounds produced by yeasts. Food Res. Int. 43, 1990–1998.

Miller, B.J., Franz, C.M.A.P., Cho, G.-S., du Toit, M., 2011. Expression of the malolactic enzyme gene (*mle*) from *Lactobacillus plantarum* under winemaking conditions. Curr. Microbiol. 62, 1682–1688.

Mira de Orduña, R., 2010. Climate change associated effects on grape and wine quality and production. Food Res. Int. 43, 1844–1855.

Mozell, M.R., Thach, L., 2014. The impact of climate change on the global wine industry: challenges & solutions. Wine Econ. Policy 3, 81–89.

Muñoz, V., Beccaria, B., Abreo, E., 2014. Simultaneous and successive inoculations of yeasts and lactic acid bacteria on the fermentation of an unsulfited Tannat grape must. Braz. J. Microbiol. 45, 59–66.

Mylona, A.E., Del Fresno, J.M., Palomero, F., Loira, I., Bañuelos, M.A., Morata, A., et al., 2016. Use of *Schizosaccharomyces* strains for wine fermentation—effect on the wine composition and food safety. Int. J. Food Microbiol. 232, 63–72.

Nehme, N., Mathieu, F., Taillandier, P., 2008. Quantitative study of interactions between *Saccharomyces cerevisiae* and *Oenococcus oeni* strains. J. Ind. Microbiol. Biotechnol. 35, 685–693.

Nehme, N., Mathieu, F., Taillandier, P., 2010. Impact of the co-culture of *Saccharomyces cerevisiae–Oenococcus oeni* on malolactic fermentation and partial characterization of a yeast-derived inhibitory peptidic fraction. Food Microbiol. 27, 150–157.

Nielsen, J.C., Prahl, C., Lonvaud-Funel, A., 1996. Malolactic fermentation in wine by direct inoculation with freeze-dried *Leuconostoc oenos* cultures. Am. J. Enol. Vitic. 47, 42–48.

Nuñez, M., del Olmo, A., Calzada, J., 2016. Biogenic amines. Encyclopedia of Food and Health. Academic Press, Oxford.

Nurgel, C., Pickering, G.J., Inglis, D.L., 2004. Sensory and chemical characteristics of Canadian ice wines. J. Sci. Food Agricult. 84, 1675–1684.

OIV, 2014. Code of good vitivinicultural practices in order to avoid or limit contamination by *Brettanomyces*. OIV, Resolution OIV-CST 462-2014, 1–7.

OIV, 2015. International Code of Oenological Practices Paris, France.

Onetto, C., Bordeu, E., 2015. Pre-alcoholic fermentation acidification of red grape must using *Lactobacillus plantarum*. Antonie Van Leeuwenhoek 108, 1469–1475.

Osborne, J.P., Edwards, C.G., 2005. Bacteria important during winemaking. Advances in Food and Nutrition Research. Academic Press, San Diego, USA.

Osborne, J.P., Edwards, C.G., 2007. Inhibition of malolactic fermentation by a peptide produced by *Saccharomyces cerevisiae* during alcoholic fermentation. Int. J. Food Microbiol. 118, 27–34.

Padilla, B., Gil, J.V., Manzanares, P., 2016. Past and future of non-*Saccharomyces* yeasts: from spoilage microorganisms to biotechnological tools for improving wine aroma complexity. Front. Microbiol. 7, 411.

Pan, W., Jussier, D., Terrade, N., Yada, R.Y., Mira de Orduña, R., 2011. Kinetics of sugars, organic acids and acetaldehyde during simultaneous yeast-bacterial fermentations of white wine at different pH values. Food Res. Int. 44, 660–666.

Patrignani, F., Ndagijimana, M., Belletti, N., Gardini, F., Vernocchi, P., Lanciotti, R., 2011. Biogenic amines and ethyl carbamate in Primitivo wine: survey of their concentrations in commercial products and relationship with the use of malolactic starter. J. Food Prot. 75, 591–596.

Pérez-Martín, F., Seseña, S., Izquierdo, P.M., Palop, M.L., 2014. Are *Enterococcus* populations present during malolactic fermentation of red wine safe? Food Microbiol. 42, 95–101.

Petruzzi, L., Corbo, M.R., Sinigaglia, M., Bevilacqua, A., 2014. Yeast cells as adsorbing tools to remove ochratoxin A in a model wine. Int. J. Food Sci. Technol. 49, 936–940.

Petruzzi, L., Capozzi, V., Berbegal, C., Corbo, M.R., Bevilacqua, A., Spano, G., et al., 2017. Microbial resources and enological significance: opportunities and benefits. Front. Microbiol. 8, 995.

Piao, H., Hawley, E., Kopf, S., DeScenzo, R., Sealock, S., Henick-Kling, T., et al., 2015. Insights into the bacterial community and its temporal succession during the fermentation of wine grapes. Front. Microbiol. 6, 809.

Pickering, G.J., 2000. Low- and reduced-alcohol wine: a review. J. Wine Res. 11, 129–144.

Pigeau, G.M., Inglis, D.L., 2005. Upregulation of ALD3 and GPD1 in *Saccharomyces cerevisiae* during icewine fermentation. J. Appl. Microbiol. 99, 112–125.

Pilatte, E., Prahl, C., 1997. Biological deacidification of acide grape varieties by inoculation on must with a freeze-dried culture of *Lactobacillus plantarum*. In: Rautz, J. (Ed.), Annual Meeting ASEV, San Diego, California, 1–2 de Julio 1997.

Piotrowska, M., Zakowska, Z., 2005. The elimination of ochratoxin A by lactic acid bacteria strains. Pol. J. Microbiol. 54, 279–286.

Pozo-Bayón, M.A., Pueyo, E., Martín-Álvarez, P.J., Martínez-Rodríguez, A.J., Polo, M.C., 2003. Influence of yeast strain, bentonite addition, and aging time on volatile compounds of sparkling wines. Am. J. Enol. Vitic. 54, 273–278.

Pozo-Bayón, M.A., G-Alegría, E., Polo, M.C., Tenorio, C., Martín-Álvarez, P.J., Calvo de la Banda, M.T., et al., 2005. Wine volatile and amino acid composition after malolactic fermentation: effect of *Oenococcus oeni* and *Lactobacillus plantarum* starter cultures. J. Agric. Food. Chem. 53, 8729–8735.

Pozo-Bayón, M.A., Martín-Álvarez, P.J., Moreno-Arribas, M.V., Andujar-Ortiz, I., Pueyo, E., 2010. Impact of using Trepat and Monastrell red grape varieties on the volatile and nitrogen composition during the manufacture of rosé Cava sparkling wines. LWT Food Sci. Technol. 43, 1526–1532.

Prahl, C., 1989a. La décarboxylation de l'acide L-malique dans le moût par l'ensemecement de lactobacilles homofermentatives. Revue des œnologues et des techniques vitivinicoles et œnologiques 54, 13–17.

Prahl, C., 1989b. A method of inducing the decarboxylation of malic acid in must or fruit juice. Patent WO 89/06685.

Prahl, C., Lonvaud-Funel, A., Korsgaard, E.M., Joyeux, A., 1988. Étude d'un nouveau procédé de déclenchement de la fermentation malolactique. Connaissance de la Vigne et du Vin 22, 197–207.

Quintela, S., Villarán, M.C., De Armentia, I.L., Elejalde, E., 2012. Ochratoxin A removal from red wine by several oenological fining agents: bentonite, egg albumin, allergen-free adsorbents, chitin and chitosan. Food Addit. Contam., A 29, 1168–1174.

Renouf, V., Murat, M.-L., Guerche, S.L., Moine, V., 2009. Incidence de différentes techniques d'utilisation des levains malolactiques sur la qualité des vins rouges. Revue Française d'Oenologie 237, 2–7.

Ribéreau-Gayon, P., Dubourdieu, D., Donèche, B., Lonvaud, A., 2006a. Handbook of enology. The Microbiology of Wine and Vinifications. John Wiley & Sons, Chippenham, Wiltshire.

Ribéreau-Gayon, P., Glories, Y., Maujean, A., Dubourdieu, D., 2006b. Handbook of enology. The Chemistry of Wine Stabilization and Treatments. John wiley & Sons Ltd, England.

Rizk, Z., El Rayess, Y., Ghanem, C., Mathieu, F., Taillandier, P., Nehme, N., 2016. Impact of inhibitory peptides released by *Saccharomyces cerevisiae* BDX on the malolactic fermentation performed by *Oenococcus oeni* Vitilactic F. Int. J. Food. Microbiol. 233, 90–96.

Rodas, A.M., Chenoll, E., Macian, M.C., Ferrer, S., Pardo, I., Aznar, R., 2006. *Lactobacillus vini* sp. nov., a wine lactic acid bacterium homofermentative for pentoses. Int. J. Sys. Evol. Microbiol. 56, 513–517.

Rodriguez, H., de las Rivas, B., Gomez-Cordoves, C., Munoz, R., 2008. Characterization of tannase activity in cell-free extracts of *Lactobacillus plantarum* CECT 748^T. Int. J. Food Microbiol. 121, 92–98.

Rosi, I., 2000. Production de macromolécules pariétales de *Saccharomyces cerevisiae* au cours de la fermentation et leur influence sur la fermentation malolactique. Revue des oenologues et des techniques vitivinicoles et oenologicques: magazine trimestriel d'information professionnelle 27, 18–20.

Rosi, I., Fia, G., Canuti, V., 2003. Influence of different pH values and inoculation time on the growth and malolactic activity of a strain of *Oenococcus oeni*. Aust. J. Grape Wine Res. 9, 194–199.

Rossouw, D., Du Toit, M., Bauer, F.F., 2012. The impact of co-inoculation with *Oenococcus oeni* on the trancriptome of *Saccharomyces cerevisiae* and on the flavour-active metabolite profiles during fermentation in synthetic must. Food Microbiol. 29, 121–131.

Russo, P., Capozzi, V., Spano, G., Corbo, M.R., Sinigaglia, M., Bevilacqua, A., 2016. Metabolites of microbial origin with an impact on health: ochratoxin A and biogenic amines. Front. Microbiol. 7, 482.

Saerens, S., Edwards, N., Soerensen, K.I., Badaki, M., Swiegers, J.H. 2015. Production of a low-alcohol fruit beverage. WO/2015/110484.

Sahu, L., Panda, S.K., Paramithiotis, S., Zdolec, N., Ray, R.C. 2015. Biogenic amines in fermented foods: overview. Fermented Foods—Part I: Biochemistry and Biotechnology. CRC Press, Boca Raton, USA.

Santisi, J., 2011. Warming up the wine industry. E–Environ. Mag. 22, 12.

Schmidtke, L.M., Blackman, J.W., Agboola, S.O., 2012. Production technologies for reduced alcoholic wines. J. Food Sci. 77, R25–R41.

Schöltz, M., 2013. Assessing the compatibility and aroma production of NT 202 Co-inoculant with different wine yeasts and additives. MSc Thesis. Stellenbosch University Stellenbosch. South Africa.

Sestelo, A.B.F., Poza, M., Villa, T.G., 2004. ß-Glucosidase activity in a *Lactobacillus plantarum* wine strain. World J. Microbiol. Biotechnol. 20, 633–637.

Simó, G., Fernández-Fernández, E., Vila-Crespo, J., Ruipérez, V., Rodríguez-Nogales, J.M., 2017. Silica–alginate-encapsulated bacteria to enhance malolactic fermentation performance in a stressful environment. Aust. J. Grape Wine Res. 23, 342–349.

Smit, A.Y., du Toit, M., 2013. Evaluating the influence of malolactic fermentation inoculation practices and ageing on lees on biogenic amine production in wine. Food Bioprocess Technol. 6, 198–206.

Smit, A.Y., du Toit, W.J., Stander, M., du Toit, M., 2013. Evaluating the influence of maceration practices on biogenic amine formation in wine. LWT Food Sci. Technol. 53, 297–307.

Spano, G., Rinaldi, A., Ugliano, M., Moio, L., Beneduce, L., Massa, S., 2005. A β-glucosidase gene isolated from wine *Lactobacillus plantarum* is regulated by abiotic stresses. J. Appl. Microbiol. 98, 855–861.

Sun, S.Y., Gong, H.S., Zhao, K., Wang, X.L., Wang, X., Zhao, X.H., et al., 2013. Co-inoculation of yeast and lactic acid bacteria to improve cherry wines sensory quality. Int. J. Food Sci. Technol. 48, 1783–1790.

Suriano, S., Savino, M., Basile, T., Tarricone, L., Di Gennaro, D., 2015. Management of malolactic fermentation and influence on chemical composition of Aglianico red wines. Ital. J. Food Sci. 27, 310–319.

Torrea, D., Ancín, C., 2001. Influence of yeast strain on biogenic amines content in wines: relationship with the utilization of amino acids during fermentation. Am. J. Enol. Vitic. 52, 185–190.

Torrens, J., Urpí, P., Riu-Aumatell, M., Vichi, S., López-Tamames, E., Buxaderas, S., 2008. Different commercial yeast strains affecting the volatile and sensory profile of cava base wine. Int. J. Food Microbiol. 124, 48–57.

Tristezza, M., di Feo, L., Tufariello, M., Grieco, F., Capozzi, V., Spano, G., et al., 2016. Simultaneous inoculation of yeasts and lactic acid bacteria: effects on fermentation dynamics and chemical composition of Negroamaro wine. LWT Food Sci. Technol. 66, 406–412.

Vaillant, H., Formisyn, P., 1996. Purification of the malolactic enzyme from a *Leuconostoc oenos* strain and use in a membrane reactor for achieving the malolactic fermentation of wine. Biotechnol. Appl. Biochem. 24, 217–223.

Val, P., Ferrer, S., Pardo, I., 2002. Acidificación biológica de vinos con bajo contenido en ácidos. Enólogos 20, 35–42.

Vendrame, M. 2013. Study of *Oenococcus oeni* to Improve Wine Quality. Doctorate Ph.D. Thesis. Università degli Studi di Udine.

Viana, F., Gil, J.V., Genovés, S., Vallés, S., Manzanares, P., 2008. Rational selection of non-*Saccharomyces* wine yeasts for mixed starters based on ester formation and enological traits. Food Microbiol. 25, 778–785.

Vincenzini, M., Guerrini, S., Mangani, S., Granchi, L., 2009. Amino acid metabolism and production of biogenic amines and ethyl carbamate. Biology of Microorganims on Grapes, in Musts and in Wine. Springer, Berlin, Heidelberg.

Yeramian, N., 2003. Acidificación Biológica de mostos en zonas cálidas. Tesis Doctoral. Universidad Politécnica de Madrid.

Zapparoli, E., Tosi, E., Azzolini, M., Vagnoli, P., Krieger, S., 2009. Bacterial inoculation strategies for the achievement of malolactic fermentation in high alcohol wines. S. Afr. J. Enol. Vitic. 30, 49–55.

Zimmerli, B., Dick, R., 1996. Ochratoxin A in table wine and grape-juice: occurrence and risk assessment. Food Addit. Contam. 13, 655–668.

CHAPTER

8

Molecular Tools to Analyze Microbial Populations in Red Wines

Karola Böhme[1], Jorge Barros-Velázquez[2] and Pilar Calo-Mata[2]

[1]Technological Agroalimentary Center of Lugo, Lugo, Spain [2]Department of Analytical Chemistry, Nutrition and Food Science, School of Veterinary Sciences/College of Biotechnology, University of Santiago de Compostela, Lugo, Spain

8.1 INTRODUCTION

Grape must fermentation is a process involving alcoholic fermentation and malolactic fermentation (MLF), with the latter being usually encouraged in red winemaking. These processes involve several genera and species of yeasts and bacteria which produce several metabolites that contribute to the final wine composition and quality. The transformation of grape must into wine can be either spontaneous or induced. The microorganisms involved in spontaneous fermentations originate from different niches around the vineyard and the winery (Fleet et al., 2002). The microbial population in grape berries may vary from 10^3 cfu/g berry to 10^8 cfu/g, depending on the maturity and/or damage of the grapes (Fleet and Heard, 1993; Barata et al., 2012a; Fleet, 2003; Marzano et al., 2016). The kind of yeast may vary depending on the state of the grape. Thus, damaged grapes carry more osmotolerant kinds of yeast, (mainly ascomycetes such as *Torulaspora* and *Zygosaccharomyces*) than undamaged grapes (carrying more basidiomycetous yeast such as *Rhodotorula* spp. and *Cryptococcus* spp., some ascomycetous yeasts such as *Hanseniaspora uvarum*, oxidative yeasts such as *Candida* spp. and the film forming yeast *Pichia* spp.) (Barata et al., 2012b). The yeast community associated with grape must can be divided into three main categories: (1) oxidative yeasts such as *Aureobasidium pullulans*, *Cryptococcus* spp. and *Rhodotorula* spp. having no fermentation ability; (2) weakly fermentative yeasts such as *Candida* spp. and *Pichia* spp.; (3) strongly fermentative yeasts such as *Torulaspora delbrueckii*, *Lachancea thermotolerans*, and *Saccharomyces* spp. At this stage, *Saccharomyces cerevisiae*, the main wine fermentation yeast, is usually present below the detection level. Alcoholic fermentation involves several stages at which the activity, composition, and interaction of the different microbiota are gradually evolving. Thus, in the early stages of spontaneous alcoholic fermentation non-*Saccharomyces* yeasts are dominating (10^7 cfu/mL) (Jolly et al., 2014; Grangeteau et al., 2017). At this stage, *S. cerevisiae* is still at a very low level (around 50 cfu/mL) (Fleet and Heard, 1993; Fleet et al., 2002). After this stage, the non-*Saccharomyces* yeast population declines and the population of *S. cerevisiae* increases (10^7–10^8 cfu/mL) mainly due to the presence of ethanol and organic acids, decrease in the pH values, low oxygen, lack of some nutrients, killer toxins, and antimicrobial peptides (Fleet, 2003).

MLF, carried out by lactic acid bacteria (LAB), contributes to the reduction of acidity and may enhance the sensory properties and microbial stability of wine (Davis et al., 1985; Marzano et al., 2016). LAB are classified into two groups based on their metabolic end products: (1) the homofermentative LAB producing lactic acid as the end product of sugar metabolism; (2) the heterofermentative LAB producing lactic acid, CO_2 and acetate (König and Fröhlich, 2010). The main LAB species involved in MLF comprise *Lactobacillus* spp., *Oenococcus oeni*, *Leuconostoc mesenteroides*, and *Pediococcus* spp. (König and Fröhlich, 2010).

Besides alcoholic and MLF microbiota, other species may be part of the microbial community in the wine fermentation environment, such as acetic acid bacteria and spoilage microorganisms. Acetic acid bacteria are Gram-negative, aerobic, catalase-positive, that have been recently classified into four groups: *Acetobacter*, *Acidomonas*,

Gluconobacter, and *Gluconacetobacter* (Du Toit and Pretorius, 2002). Otherwise, in the maturation stage, wine spoilage yeasts such as *Brettanomyces bruxellensis* and *Zygosaccharomyces bailii* may grow producing undesirable off-flavors (Barata et al., 2008). Other grape must microbiota such as *Chryseobacterium* (Bacteroidetes), *Methylobacterium*, *Sphingomonas* (Alphaproteobacteria), *Arcobacter* (Eplisoniproteobacteria), *Naxibacter*, *Ralstonia* (Betaproteobacteria), *Frigoribacterium* (Actinobacteria), *Pseudomonas*, *Zymobacter*, and *Acinetobacter* (Gammaproteobacteria) have been found through metagenomic studies, although those do not seem to play an important role in wine fermentation (Bokulich et al., 2012).

Considering the great diversity of microbiota found in the process of wine production, many studies have focused on the dynamics and interaction of such microbiota. A wide variety of methods have been applied for the identification of the microbiota throughout all stages of wine production. These methods may be classified as cultivation-dependent and cultivation-independent techniques (Ghosh, 2015). However, some of them may be applied with and without cultivation, such as polymerase chain reaction (PCR)-restriction fragment length polymorphism (PCR-RFLP).

8.2 CLASSICAL AND PHENOTYPIC METHODS

The classical methods of microbial culturing serve for the determination of the total bacterial content or the detection of the presence of a particular species or group of microorganisms. The identification of wine microbiota species is usually accomplished through macroscopic criteria such as the examination of morphological characteristics of colonies isolated on specific culture media by microscopy, and finally, with physiological and biochemical tests for both yeast and bacteria (Marzano et al., 2016). A first classification of isolated strains takes into account the morphology of the colonies, Gram staining, the presence of catalase and oxidase, and growth conditions, such as the lack of oxygen and the ability to grow in selective media. For a better classification, further tests are necessary, including type of hemolysis, presence of aminopeptidases, ureases, coagulases, indole production, and resistance tests. There are commercial biochemical tests available, such as the API and VITEK identification systems, with which several tests can be performed at the same time in a minimized space. However, these techniques require long times and the identification is limited to biochemical characterization tests, which do not allow a precise determination of the identity of the microbiota. In addition, phenotypic tests are designed for specific genera and groups of bacteria, but cannot be applied to the identification of a wide range of different bacterial genera.

8.3 DNA-BASED METHODS

The methods based on the analysis of DNA are being widely used in the field of microbial identification, reaching remarkable importance in food analysis laboratories. DNA has the advantage that it does not degrade when food is subjected to physical, chemical, thermal, or high pressure treatments, thus DNA can be analyzed even in highly processed and/or degraded samples. Molecular methods are rapid and reliable. Microbial identification techniques based on the analysis of nucleic acids dramatically improved with the discovery of the PCR (Mullis and Faloona, 1989). Thus, genes amplified by PCR can be sequenced and identified by comparing with sequences in reference databases, such as the International Nucleotide Sequence Database Collaboration which includes the European Nucleotide Archive at the EBI, GenBank at the NCBI, and the DNA Database of Japan, allowing the inter- and intraspecific differentiation of isolates. In yeast, the DNA regions more studied include the 5.8S, 18S, and 26S ribosomal genes which are grouped in tandem to form transcription units that are repeated about 100–200 times through the genome. The transcription units are separated by regions called internal transcribed spacer (ITS) and are separated from other genes by intergenic spacers called (IGSs). These regions were used for identifying phylogenetic relationships between yeasts (Kurtzman and Robnett, 1998). The bacterial gene most studied and on which most of the methods of bacterial identification are based is the 16S rRNA gene (about 1500 nucleotides of length), because it is a highly conserved gene that allows the design of universal primers for the simultaneous amplification of most of the bacterial species. At the same time, the 16S rRNA gene presents variable regions, unique to each species, that reflect the genetic evolution. These conserved and variable regions allow differentiation at the genus and species level.

The analysis of DNA is also commonly applied to phylogenetic studies. By this way, the relationship of isolated strains to reference strains or other isolates can be determined. Based on PCR, several molecular methods have been developed including culture-independent methods that allow a faster identification and the monitorization of populations that are in low numbers, as well as those in the viable but nonculturable (VBNC) state (Andorrà et al., 2008; Cocolin et al., 2013).

8.3.1 Randomly Amplified Polymorphic DNA PCR Fingerprints (RAPD-PCR)

Random amplified polymorphic DNA-PCR (RAPD-PCR) fingerprints is a method using short and well defined primers that hybridize at a variety of random locations on the microbial genome, giving a species-specific fingerprint. This method was evaluated by several authors for its use in the taxonomic identification of wine. Thus, it has been applied to the typification of wine yeast (Quesada and Cenis, 1995), as well as of *O. oeni* strains and study of population dynamics during MLF (Zapparoli et al., 2000; Reguant and Bordons, 2003). Otherwise, RAPD-PCR and phylogenetic clustering have also been used to study the microbiota isolated from spoiled red wine such as *Acetobacter pasteurianus* (Bartowsky et al., 2003). This method has been recently applied to the detection and identification of *O. oeni* bacteriophages based on endolysin gene (Doria et al., 2013).

8.3.2 PCR-Restriction Fragment Length Polymorphism

PCR-RFLP has been applied to both, inter- and intraspecies differentiation of wine yeast and bacteria (Dlauchy et al., 1999; Hierro et al., 2004; Guillamón and Barrio, 2017). PCR-RFLP requires the extraction and amplification of DNA, using well defined phylogenetic marker genes, such as 16S rRNA gene in the case of bacteria or the ITS-5.8S rRNA-ITS2 gene (Guillamon et al., 1998) and the D1–D2 domains of the 26S rRNA gene (Kurtzman and Robnett, 1998; Leaw et al., 2006) in the case of fungi. Then, the DNA is digested by well defined restriction enzymes and the resulting restriction fragments are separated according to their lengths by gel electrophoresis, obtaining different banding profiles that are species-specific and that can be compared to a known species allowing the identification of the isolate (Esteve-Zarzoso et al., 1999). Using this method, Espinosa et al. (2002) identified wine yeasts without previous isolation on plate. This method has also been applied to the identification of wine bacteria. Thus, LAB *O. oeni* isolated from red wine have been characterized by PCR-RFLP analysis (Sato et al., 2000).

8.3.3 Terminal Restriction Fragment Length Polymorphism

Terminal restriction fragment length polymorphism (T-RFLP) uses PCR amplification of DNA using well defined primer pairs that have been labeled with fluorescent tags. The amplified PCR products are then digested using well defined restriction enzymes. The resulting patterns are then visualized using a DNA sequencer and analyzed either by comparing bands or peaks from the T-RFLP runs to a database of known species. This method has been applied for characterizing bacterial communities in mixed-species samples such as to profile the wine microbial diversity (Dunbar et al., 2001; Blackwood et al., 2003; Martins et al., 2012; Sun and Liu, 2014).

8.3.4 Gradient Gel Electrophoresis

According to Ercolini (2004) culture-independent techniques are used due to the difficulty to develop culture media for microbial species resembling their natural habitat. Another difficulty is when species enter a VBNC state. Among these techniques, PCR-denaturing gradient gel electrophoresis (PCR-DGGE) has been applied to investigate microbial communities in different ecosystems such as water, grapevine, grapes, soil, wine equipment, and environment, as well as to analyze the wine microbiota and microbial diversity throughout the grape ripening process and fermentation (Morgan et al., 2017).

PCR-DGGE is based on the separation of PCR amplified DNA of the same size but of different sequences that are partially melted double-stranded DNA, affecting the mobility through polyacrylamide gels consisting either of a linear gradient of DNA denaturants (mixture of urea and formamide) (DGGE) or temperature [temperature gradient gel electrophoresis (TGGE) and temporal temperature gradient electrophoresis (TTGE)]. Several studies on wine microbiota species identification are based on DGGE (Prat et al., 2009; Lopez et al., 2003;

Cocolin et al., 2000; Prakitchaiwattana et al., 2004; De Vero et al., 2006; Siebrits, 2007; Cameron et al., 2013; De Vero and Giudici, 2008; David et al., 2014; Shankar et al., 2016; Bester et al., 2016), TGGE (Muyzer, 1999; Gadanho and Sampaio, 2004), and TTGE (Fernández-González et al., 2001). The ITS-rRNA and other genomic DNA regions such as tye *rpo*B gene have been used for PCR-DGGE (Renouf et al., 2005, 2007; Ruiz et al., 2010). PCR-DGGE has been used to monitor the diversity and dynamics of yeast and bacteria at the different stages of winemaking, either red, white, or botrytized wines (Mills et al., 2002; Prakitchaiwattana et al., 2004; Cocolin et al., 2001). This method has also been applied for the detection of LAB in Rioja red wine (González-Arenzana et al., 2013).

8.3.5 Quantitative Real-Time PCR (QPCR) and Reverse Transcription Quantitative Real-Time PCR (RT-qPCR)

Quantitative real-time PCR was developed to rapidly detect and quantify the total number of microorganisms without culturing in a certain environment. This method has been applied to the identification of several wine yeast species using universal yeast primers from the variable D1/D2 domains of the 26S rRNA gene (Martorell et al., 2005; Hierro et al., 2006, 2007; Phister et al., 2007). This method has also been applied to the quantification of *B. bruxellensis* and *Pediococcus damnosus* (Delaherche et al., 2004). In the same sense, reverse transcription quantitative real-time PCR has also been applied to the quantification of wine yeast species (Hierro et al., 2006). Recently, assays were developed for the quantification of the fungal species *Saccharomycopsis fibuligera*, *Rhizopus oryzae*, and *Monascus purpureus* during the traditional brewing of Hong Qu glutinous rice wine using species-specific primers of the ITS-5.8S rRNA gene.

8.3.6 Capillary Electrophoresis Single-Strand Conformation Polymorphism

Capillary electrophoresis single-strand conformation polymorphism (CE-SSCP) is a technique in which single-stranded DNA fragments of the same length are separated based on their sequence. The migration position of each single-stranded DNA is compared with an internal standard. This technique is used for population monitoring and microbial identification. Studies on yeast using CE-SSCP revealed changes in the yeast community throughout the ripening process, as well as due the impact of chemical compounds, such as fungicides (Grube et al., 2011; Schmid et al., 2011; Martins et al., 2014).

8.3.7 Automated Ribosomal Intergenic Spacer Analysis

Automated ribosomal IGS analysis is a culture independent technique used for determining the microbial diversity and estimating the microbial population. This technique has been applied to wine microbial diversity profiling (Ranjard et al., 2001; Kitts, 2001; Brezna et al., 2010; Chovanová et al., 2011; Kraková et al., 2012; Pancher et al., 2012; Setati et al., 2012; Ženišová et al., 2014; Jami et al., 2014; Ghosh et al., 2015). Using this technique some authors monitored the yeast population dynamics at different fermentation stages using a PCR-based amplification of the ITS (IGS) region (Brezna et al., 2010).

8.3.8 Next Generation Sequencing

In recent decades, DNA sequencing techniques have undergone great advances and the capacities have improved significantly for high-throughput next generation sequencing (NGS) and third generation sequencing techniques. The simultaneous analysis of millions of sequences allows the characterization of the microbiomes of complex matrices, such as food and environmental samples.

In a number of studies based on high-throughput sequencing, the microbial communities of wine has been characterized (Bokulich et al., 2012, 2014; Bokulich and Mills, 2013; Taylor et al., 2014). Likewise, the biogeographical distribution of the wine associated microorganisms has been investigated by NGS, revealing that the location of vineyards may contribute to distinctive metabolites and introduce an authenticity terroir to the region (Jolly et al., 2006; Fleet, 2008; Setati et al., 2012; Taylor et al., 2014). These studies also demonstrated a direct influence of the vineyard environment, such as climate and soil on the grape vine/must microbial consortium.

When comparing to other microbiome profiling techniques, high-throughput sequencing has the advantage to allow quantification of the biodiversity (David et al., 2014). In further studies of microbial populations during alcoholic fermentation of grapes, more species could be detected by high-throughput sequencing fermentation that were undetectable with the other techniques (Pinto et al., 2015).

Besides high-throughput sequencing that targets just specific genes with the aim to profile the microbiome diversity of a sample, whole metagenome sequencing approaches provide much more information allowing the detection of changes in the microbial structure, as well as in the gene expression patterns at different processing and storage stages (Bokulich et al., 2012).

8.4 MATRIX-ASSISTED LASER DESORPTION/IONIZATION–TIME OF FLIGHT MASS SPECTROMETRY

Matrix-assisted laser desorption/ionization–time of flight mass spectrometry (MALDI–TOF MS) emerged as a new tool for bacterial identification due to its rapidness, simplicity, and cost-effectiveness. Whole bacterial cells can be analyzed directly without any sample pretreatment and the resulting spectral profiles are highly specific, representing a fingerprint for the corresponding organism (Clark et al., 2013).

Bacterial identification by MALDI–TOF MS is carried out by comparing the spectral profile to a previously created library of reference spectra. The MALDI Biotyper system from Bruker Daltonics (Bremen, Germany) includes an ample database of bacterial strains, mycobacteria, and fungi. Another microbial identification system based on MALDI–TOF MS, including a spectral archive, is the Vitek MS platform from bioMérieux (Marcyl' Etoile, France). Both systems have been validated for routine microbial identification, achieving 92%–98% of correct species identification. That is a significantly better result than obtained with commonly applied microbial identification tools. MALDI–TOF MS has been applied to the identification and typing of wine yeasts (Blättel et al., 2013; Usbeck et al., 2014; Gutiérrez et al., 2017).

8.5 MICROBIAL DIVERSITY ASSESSMENT THROUGH ENZYMES DETECTION

Proteomic techniques such as shotgun proteomics in which a mixture of proteins extracted from microbial cells are digested with proteases such as trypsin and then the mixture of peptides are analyzed by liquid chromatography-electrospray ionization-tandem mass spectrometry (LC–ESI–MS/MS) obtaining spectra that allow the identification of the peptides and, thus, the identification of the extracted proteins (Carrera et al., 2017). Many of these proteins include extracellular hydrolytic enzymes mainly related with the cell wall but also some intracellualar enzymes. It is well known that *Saccharomyces* and non-*Saccharomyces* yeasts secrete a wide range of enzymes that lead to the production of certain metabolites that contribute to the wine sensorial properties. Enzymes such as proteases, pectinases, glucanases, glucosidases, xylanases, and amylases can be detected by LC–ESI–MS/MS, allowing both the identification of the microbial species at certain stages and the potential metabolite profile. Otherwise, using this technique several virulence factors can be detected (Carrera et al., 2017), allowing the identification of enzymes leading to undesirable metabolites such as biogenic amines.

8.6 CULTURE-DEPENDENT VERSUS CULTURE-INDEPENDENT METHODS

The principal limitation of culture-dependent methods is that they only allow the identification of cultivable and dominant microbiota. Nevertheless, in some stages of the fermentation of wine, microbiota may overgrow less active microbiota. Thus, species that are less abundant at a certain stage become difficult to recover through cultivation. Adaptation to the culture medium may also hinder the growth of certain cells since the transfer from their specific environment to a rich cultivation medium constitutes a shock that some cells may not be able to overcome. Unculturable and slow-growing fungal and bacterial species detection might lead to the description of new taxons (Divol and Lonvaud-Funel, 2005; Millet and Lonvaud-Funel, 2000; Serpaggi et al., 2012; Agnolucci et al., 2010; Du Toit et al., 2005; David et al., 2014).

Also, changes in the growing environment of the wine microbiota or sudden adverse conditions such as high concentrations of SO_2, changes of temperature, pH, and sugar concentration, the deprivation of certain nutrients, or the presence of certain metabolites, may cause some microorganisms to enter into a VBNC state (Divol and Lonvaud-Funel, 2005; Salma et al., 2013). These microorganisms are sublethally injured or viable but weakly

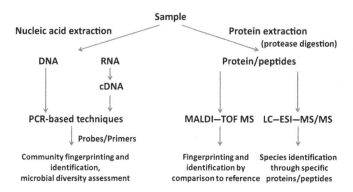

FIGURE 8.1 Scheme of identification and fingerprinting methods used to study the microbiome of wine.

metabolically active, losing their ability to grow on culture media (Charoenchai et al., 1998). Thus, wine yeast species may enter into VBNC physiological state, and lactic acid and acetic acid bacteria enter the VBNC state in wine when they are deprived of O_2 and in the presence of sulfites (Millet and Lonvaud-Funel, 2000). Although in the past years knowledge on vine and wine microbiota has increased greatly, much remains to be explored because the vast majority of these microbes cannot be cultured by standard techniques (Marzano et al., 2016).

Culture-independent methods circumvent the culturing step. When comparing culture-dependent and molecular culture-independent methods, some authors found that culture-independent methods were better than culture-depending methods for detecting relevant species (Boase et al., 2013; Pandya et al., 2017).

The combined use of genomic methods with the study of the protein profile and the volatile compounds allow the simultaneous analysis of multiple samples enabling the comparison of the genetic diversity of microbial communities either from different habitats or at different stages of wine fermentation, as well as to study the behavior and evolution of individual communities over time (Fig. 8.1).

8.7 CONCLUSIONS

The winemaking process involves a wide range of filamentous fungi, yeasts, and bacterial genera, species, and strains that may interact leading to a microbial population dynamics that confer to wine specific quality characteristics and attributes. Microbial species detection, identification, and characterization using new genomic and proteomic techniques will lead to a better understanding of the interaction and behavior of the microbiota in winemaking.

References

Agnolucci, M., Rea, F., Sbrana, C., Cristani, C., Fracassetti, D., Tirelli, A., et al., 2010. Sulphur dioxide affects culturability and volatile phenol production by *Brettanomyces/Dekkera bruxellensis*. Int. J. Food Microbiol. 143 (1–2), 76–80.

Andorrà, I., Landi, S., Mas, A., Guillamón, J.M., Esteve-Zarzoso, B., 2008. Effect of oenological practices on microbial populations using culture-independent techniques. Food Microbiol. 25, 849–856.

Barata, A., Gonzalez, S., Malfeito-Ferreira, M., Querol, A., Loureiro, V., 2008. Sour rot-damaged grapes are sources of wine spoilage yeasts. FEMS Yeast Res. 8, 1008–1017.

Barata, A., Malfeito-Ferreira, M., Loureiro, V., 2012a. The microbial ecology of wine grape berries. Int. J. Food Microbiol. 153, 243–259.

Barata, A., Malfeito-Ferreira, M., Loureiro, V., Malfeito-Ferreira, M., Loureiro, V., 2012b. Changes in sour rotten grape berry microbiota during ripening and wine fermentation. Int. J. Food Microbiol. 154, 152–161.

Bartowsky, E.J., Xia, D., Gibson, R.L., Fleet, G.H., Henschke, P.A., 2003. Spoilage of bottled red wine by acetic acid bacteria. Lett. Appl. Microbiol. 36 (5), 307–314.

Bester, L., Cameron, M., Du Toit, M., Witthuhn, R.C., 2016. PCR and DGGE detection limits for wine spoilage microbes. S. Afr. J. Enol. Vitic. 31 (1), 26–33.

Blackwood, C.B., Marsh, T., Kim, S.H., Paul, E.A., 2003. Terminal restriction fragment length polymorphism data analysis for quantitative comparison of microbial communities. Appl. Environ. Microbiol. 69 (2), 926–932.

Blättel, V., Petri, A., Rabenstein, A., Kuever, J., König, H., 2013. Differentiation of species of the genus *Saccharomyces* using biomolecular fingerprinting methods. Appl. Microbiol. Biotechnol. 97 (10), 4597–4606.

Boase, S., Foreman, A., Cleland, E., Tan, L., Melton-Kreft, R., Pant, H., et al., 2013. The microbiome of chronic rhinosinusitis: culture, molecular diagnostics and biofilm detection. BMC Infect. Dis. 13 (1), 210.

References

Bokulich, N.A., Mills, D.A., 2013. Improved selection of internal transcribed spacer-specific primers enables quantitative, ultra-high-throughput profiling of fungal communities. Appl. Environ. Microbiol. 79 (8), 2519–2526.

Bokulich, N.A., Joseph, C.L., Allen, G., Benson, A.K., Mills, D.A., 2012. Next-generation sequencing reveals significant bacterial diversity of botrytized wine. PLoS One 7 (5), e36357.

Bokulich, N.A., Thorngate, J.H., Richardson, P.M., Mills, D.A., 2014. Microbial biogeography of wine grapes is conditioned by cultivar, vintage, and climate. Proc. Natl. Acad. Sci. U.S.A. 111 (1), E139–E148.

Brezna, B., Zenisova, K., Chovanova, K., Chebenova, V., Krakova, L., Kuchta, T., et al., 2010. Evaluation of fungal and yeast diversity in Slovakian wine-related microbial communities. Antonie Van Leeuwenhoek 98, 519–529.

Cameron, M., Siebrits, L., Du Toit, M., Witthuhn, R.C., 2013. PCR-based DGGE fingerprinting and identification of the microbial population in South African red grape must and wine. J. Int. Sci. Vigne Vin 47 (1), 47–54.

Carrera, M., Böhme, K., Gallardo, J.M., Barros-Velázquez, J., Cañas, B., Calo-Mata, P., 2017. Characterization of foodborne strains of *Staphylococcus aureus* by shotgun proteomics: functional networks, virulence factors and species-specific peptide biomarkers. Front. Microbiol. 8, 2458.

Charoenchai, C., Fleet, G.H., Henschke, P.A., 1998. Effects of temperature, pH, and sugar concentration on the growth rates and cell biomass of wine yeasts. Am. J. Enol. Viticult. 49 (3), 283–288.

Chovanová, K., Kraková, L., Ženišová, K., Turcovská, V., Brežná, B., Kuchta, T., et al., 2011. Selection and identification of autochthonous yeasts in Slovakian wine samples using a rapid and reliable three-step approach. Lett. Appl. Microbiol. 53 (2), 231–237.

Clark, A.E., et al., 2013. Matrix-assisted laser desorption ionization-time of flight mass spectrometry: a fundamental shift in the routine practice of clinical microbiology. Clin. Microbiol. Rev. 26 (3), 547–603.

Cocolin, L., Bisson, L.F., Mills, D.A., 2000. Direct profiling of the yeast dynamics in wine fermentations. FEMS Microbiol. Lett. 189 (1), 81–87.

Cocolin, L., Heisey, A., Mills, D.A., 2001. Direct identification of the indigenous yeasts in commercial wine fermentations. Am. J. Enol. Vitic. 52 (1), 49–53.

Cocolin, L., Alessandria, V., Dolci, P., Gorra, R., Rantsiou, K., 2013. Culture independent methods to assess the diversity and dynamics of microbiota during food fermentation. Int. J. Food Microbiol. 167, 29–43.

David, V., Terrat, S., Herzine, K., Claisse, O., Rousseaux, S., Tourdot-Maréchal, R., et al., 2014. High-throughput sequencing of amplicons for monitoring yeast biodiversity in must and during alcoholic fermentation. J. Ind. Microbiol. Biotechnol. 41 (5), 811–821.

De Vero, L., Giudici, P., 2008. Genus-specific profile of acetic acid bacteria by 16S rRNA gene PCR-DGGE. Int. J. Food Microbiol. 125 (1), 96–101.

Davis, C.R., Wibowo, D., Eschenbruch, R., Lee, T.H., Fleet, G.H., 1985. Practical implications of malolactic fermentation: a review. Am. J. Enol. Viticult. 36 (4), 290–301.

De Vero, L., Gala, E., Gullo, M., Solieri, L., Landi, S., Giudici, P., 2006. Application of denaturing gradient gel electrophoresis (DGGE) analysis to evaluate acetic acid bacteria in traditional balsamic vinegar. Food Microbiol. 23, 809–813.

Divol, B., Lonvaud-Funel, A., 2005. Evidence for viable but nonculturable yeasts in botrytis-affected wine. J. Appl. Microbiol. 99 (1), 85–93.

Delaherche, A., Claisse, O., Lonvaud-Funel, A., 2004. Detection and quantification of *Brettanomyces bruxellensis* and 'ropy' *Pediococcus damnosus* strains in wine by real-time polymerase chain reaction. J. Appl. Microbiol 97 (5), 910–915.

Dlauchy, D., Tornai-Lehoczki, J., Péter, G., 1999. Restriction enzyme analysis of PCR amplified rDNA as a taxonomic tool in yeast identification. Syst. Appl. Microbiol. 22 (3), 445–453.

Doria, F., Napoli, C., Costantini, A., Berta, G., Saiz, J.C., Garcia-Moruno, E., 2013. Development of a new method for detection and identification of *Oenococcus oeni* bacteriophages based on endolysin gene sequence and randomly amplified polymorphic DNA. Appl. Environ. Microbiol. 79 (16), 4799–4805.

Du Toit, W.J., Pretorius, I.S., 2002. The occurrence, control and esoteric effect of acetic acid bacteria in winemaking. Ann. Microbiol. 52, 155–179.

Du Toit, W.J., Pretorius, I.S., Lonvaud-Funel, A., 2005. The effect of sulphur dioxide and oxygen on the viability and culturability of a strain of *Acetobacter pasteurianus* and a strain of *Brettanomyces bruxellensis* isolated from wine. J. Appl. Microbiol. 98 (4), 862–871.

Dunbar, J., Ticknor, L.O., Kuske, C.R., 2001. Phylogenetic specificity and reproducibility and new method for analysis of terminal restriction fragment profiles of 16S rRNA genes from bacterial communities. Appl. Environ. Microbiol. 67 (1), 190–197.

Ercolini, D., 2004. PCR-DGGE fingerprinting: novel strategies for detection of microbes in food. J. Microbiol. Methods 56 (3), 297–314.

Espinosa, J.C., Fernandez-Gonzalez, M., Ubeda, J., Briones, A., 2002. Identification of wine yeasts by PCR-RFLP without previous isolation on plate. Food Technol. Biotech. 40 (2), 157–160.

Esteve-Zarzoso, B., Belloch, C., Uruburu, F., Querol, A., 1999. Identification of yeasts by RFLP analysis of the 5.8S rRNA gene and the two ribosomal internal transcribed spacers. Int. J. Syst. Bacteriol. 49 (Pt 1), 329–337.

Fernández-González, M., Espinosa, J.C., Úbeda, J.F., Briones, A.I., 2001. Yeasts present during wine fermentation: comparative analysis of conventional plating and PCR-TTGE. Syst. Appl. Microbiol. 24 (4), 634–638.

Fleet, G.H., 2003. Yeast interactions and wine flavour. Int. J. Food Microbiol. 86, 11–22.

Fleet, G.H., 2008. Wine yeasts for the future. FEMS Yeast Res. 8, 979–995.

Fleet, G.H., Heard, G.M., 1993. Yeast-growth during fermentation. In: Fleet, G.H. (Ed.), Wine, Microbiology and Biotechnology. Harwood Academic, Lausanne, pp. 27–54.

Fleet, G.H., Prakitchaiwattana, C., Beh, A.L., Heard, G., 2002. The yeast ecology of wine grapes, Biodiversity and Biotechnology of Wine Yeasts, 95. Research Signpost, Kerala, India, pp. 1–17.

Gadanho, M., Sampaio, J.P., 2004. Application of temperature gradient gel electrophoresis to the study of yeast diversity in the estuary of the Tagus river, Portugal. FEMS Yeast Res. 5 (3), 253–261.

González-Arenzana, L., López, R., Santamaría, P., López-Alfaro, I., 2013. Dynamics of lactic acid bacteria populations in Rioja wines by PCR-DGGE, comparison with culture-dependent methods. Appl. Microbiol. Biotechnol. 97 (15), 6931–6941.

Ghosh, S., 2015. Metagenomic Screening of Cell Wall Hydrolases, Their Anti-fungal Activities and Potential Role in Wine Fermentation. Stellenbosch University, Stellenbosch, Doctoral Dissertation.

Ghosh, S., Bagheri, B., Morgan, H.H., Divol, B., Setati, M.E., 2015. Assessment of wine microbial diversity using ARISA and cultivation-based methods. Ann. Microbiol. 65 (4), 1833—1840.

Grangeteau, C., Roullier-Gall, C., Rousseaux, S., Gougeon, R.D., Schmitt-Kopplin, P., Alexandre, H., et al., 2017. Wine microbiology is driven by vineyard and winery anthropogenic factors. Microb. Biotechnol. 10 (2), 354—370.

Grube, M., Schmid, F., Berg, G., 2011. Black fungi and associated bacterial communities in the phyllosphere of grapevine. Fungal Biol. 115, 978—986.

Guillamón, J.M., Barrio, E., 2017. Genetic polymorphism in wine yeasts: mechanisms and methods for its detection. Front. Microbiol. 8, 806.

Guillamon, J.M., Sabate, J., Barrio, E., Cano, J., Querol, A., 1998. Rapid identification of wine yeast species based on RFLP analysis of the ribosomal internal transcribed spacer (ITS) region. Arch. Microbiol. 169, 387—392.

Gutiérrez, C., et al., 2017. Wine yeasts identification by MALDI—TOF MS: optimization of the preanalytical steps and development of an extensible open-source platform for processing and analysis of an in-house MS database. Int. J. Food Microbiol. 254, 1—10.

Hierro, N., Gonzalez, A., Mas, A., Guillamon, J.M., 2004. New PCR-based methods for yeast identification. Appl. Microbiol. Biotechnol. 97 (4), 792—801.

Hierro, N., Esteve-Zarzoso, B., González, Á., Mas, A., Guillamón, J.M., 2006. Real-time quantitative PCR (QPCR) and reverse transcription-QPCR for detection and enumeration of total yeasts in wine. Appl. Environ. Microbiol. 72 (11), 7148—7155.

Hierro, N., Esteve-Zarzoso, B., Mas, A., Guillamón, J.M., 2007. Monitoring of *Saccharomyces* and *Hanseniaspora* populations during alcoholic fermentation by real-time quantitative PCR. FEMS Yeast Res. 7 (8), 1340—1349.

Jami, E., Shterzer, N., Mizrahi, I., 2014. Evaluation of automated ribosomal intergenic spacer analysis for bacterial fingerprinting of rumen microbiome compared to pyrosequencing technology. Pathogens 3 (1), 109—120.

Jolly, N., Augustyn, O., Pretorius, I.S., 2006. The role and use of non-*Saccharomyces* yeasts in wine production. S. Afr. J. Enol. Vitic. 27, 15—39.

Jolly, N.P., Varela, C., Pretorius, I.S., 2014. Not your ordinary yeast: non-*Saccharomyces* yeasts in wine production uncovered. FEMS Yeast Res. 14, 215—237.

Kitts, C.L., 2001. Terminal restriction fragment patterns: a tool for comparing microbial communities and assessing community dynamics. Curr. Issues Intest. Microbiol. 2, 17—25.

Konig, H., Frohlich, J., 2009. Lactic acid bacteria. In: Konig, H., Unden, G., Frohlich, J. (Eds.), Biology of Microorganisms on Grapes, in Must and in Wine. Springer, Verlag, Berlin Heidelberg, pp. 3—29.

Kraková, L., Chovanová, K., Ženišová, K., Turcovská, V., Brežná, B., Kuchta, T., et al., 2012. Yeast diversity investigation of wine-related samples from two different Slovakian wine-producing areas through a multistep procedure. LWT-Food Sci. Technol. 46 (2), 406—411.

Kurtzman, C.P., Robnett, C.J., 1998. Identification and phylogeny of ascomycetous yeast from analysis of nuclear large subunit 26S ribosomal DNA partial sequences. Antonie Van Leeuwenhoek 73, 331—371.

Leaw, S.N., Chang, H.C., Sun, H.F., Barton, R., Bouchara, J.P., Chang, T.C., 2006. Identification of medically important yeast species by sequence analysis of the internal transcribed spacer regions. J. Clin. Microbiol. 44 (3), 693—699.

Lopez, I., Ruiz-Larrea, F., Cocolin, L., Orr, E., Phister, T., Marshall, M., et al., 2003. Design and evaluation of PCR primers for analysis of bacterial populations in wine by denaturing gradient gel electrophoresis. Appl. Environ. Microbiol. 69 (11), 6801—6807.

Martins, G., Miot-Sertier, C., Lauga, B., Claisse, O., Lonvaud-Funel, A., Soulas, G., et al., 2012. Grape berry bacterial microbiota: impact of the ripening process and the farming system. Int. J. Food Microbiol. 158, 93—100.

Martins, G., Vallance, J., Mercier, A., Albertin, W., Stamatopoulos, P., Rey, P., et al., 2014. Influence of the farming system on the epiphytic yeasts and yeast-like fungi colonizing grape berries during the ripening process. Int. J. Food Microbiol. 177, 21—28.

Martorell, P., Querol, A., Fernández-Espinar, M.T., 2005. Rapid identification and enumeration of *Saccharomyces cerevisiae* cells in wine by real-time PCR. Appl. Environ. Microbiol. 71 (11), 6823—6830.

Marzano, M., Fosso, B., Manzari, C., Grieco, F., Intranuovo, M., Cozzi, G., et al., 2016. Complexity and dynamics of the winemaking bacterial communities in berries, musts, and wines from Apulian grape cultivars through time and space. PLoS One 11 (6), e0157383.

Millet, V., Lonvaud-Funel, A., 2000. The viable but non-culturable state of wine micro-organisms during storage. Lett. Appl. Microbiol. 30 (2), 136—141.

Mills, D.A., Johannsen, E.A., Cocolin, L., 2002. Yeast diversity and persistence in botrytis-affected wine fermentations. Appl. Environ. Microbiol. 68, 4884—4893.

Morgan, H.H., du Toit, M., Setati, M.E., 2017. The grapevine and wine microbiome: insights from high-throughput amplicon sequencing. Front. Microbiol. 8, 820.

Mullis, K.B., Faloona, F.A., 1989. Specific synthesis of DNA in vitro via a polymerase-catalyzed chain reaction. In Recombinant DNA Methodology, pp. 189—204.

Muyzer, G., 1999. DGGE/TGGE a method for identifying genes from natural ecosystems. Curr. Opin. Microbiol. 2 (3), 317—322.

Pancher, M., Ceol, M., Corneo, P.E., Longa, C.M.O., Yousaf, S., Pertot, I., et al., 2012. Fungal endophytic communities in grapevines (*Vitis vinifera* L.) respond to crop management. Appl. Environ. Microbiol. 78 (12), 4308—4317.

Pandya, S., Ravi, K., Srinivas, V., Jadhav, S., Khan, A., Arun, A., et al., 2017. Comparison of culture-dependent and culture-independent molecular methods for characterization of vaginal microflora. J. Med. Microbiol. 66 (2), 149—153.

Prat, C., Ruiz-Rueda, O., Trias, R., Anticó, E., Capone, D., Sefton, M., et al., 2009. Molecular fingerprinting by PCR-denaturing gradient gel electrophoresis reveals differences in the levels of microbial diversity for musty-earthy tainted corks. Appl. Environ. Microbiol. 75 (7), 1922—1931.

Phister, T.G., Rawsthorne, H., Joseph, C.L., Mills, D.A., 2007. Real-time PCR assay for detection and enumeration of *Hanseniaspora* species from wine and juice. Am. J. Enol. Vitic. 58 (2), 229—233.

Pinto, C., Pinho, D., Cardoso, R., Custódio, V., Fernandes, J., Sousa, S., et al., 2015. Wine fermentation microbiome: a landscape from different Portuguese wine appellations. Front. Microbiol. 6, 905.

Prakitchaiwattana, C.J., Fleet, G.H., Heard, G.M., 2004. Application and evaluation of denaturing gradient gel electrophoresis to analyse the yeast ecology of wine grapes. FEMS Yeast Res. 4 (8), 865—877.

Prat, C., Ruiz-Rueda, O., Trias, R., Anticó, E., Capone, D., Sefton, M., et al., 2009. Molecular fingerprinting by PCR-denaturing gradient gel electrophoresis reveals differences in the levels of microbial diversity for musty-earthy tainted corks. Appl. Environ. Microbiol. 75 (7), 1922–1931.

Quesada, M.P., Cenis, J.L., 1995. Use of random amplified polymorphic DNA (RAPD-PCR) in the characterization of wine yeasts. Am. J. Enol. Vitic. 46 (2), 204–208.

Ranjard, L., Poly, F., Lata, J.C., Mougel, C., Thioulouse, J., Nazaret, S., 2001. Characterization of bacterial and fungal soil communities by automated ribosomal intergenic spacer analysis fingerprints: biological and methodological variability. Appl. Environ. Microbiol. 67 (10), 4479–4487.

Reguant, C., Bordons, A., 2003. Typification of Oenococcus oeni strains by multiplex RAPD-PCR and study of population dynamics during malolactic fermentation. J. Appl. Microbiol 95 (2), 344–353.

Renouf, V., Claisse, O., Lonvaud-Funel, A., 2005. Understanding the microbial ecosystem on the grape berry surface through numeration and identification of yeast and bacteria. Aust. J. Grape Wine Res. 11, 316–327.

Renouf, V., Strehaiano, P., Lonvaud-Funel, A., 2007. Yeast and bacterial analysis of grape, wine and cellar equipments by PCR-DGGE. J. Int. Sci. Vigne Vin 41, 51–61.

Ruiz, P., Sesena, S., Izquierdo, P.M., Palop, M.L., 2010. Bacterial biodiversity and dynamics during malolactic fermentation of Tempranillo wines as determined by a culture-independent method (PCR-DGGE). Appl. Microbiol. Biotechnol. 86, 1555–1562.

Salma, M., Rousseaux, S., Sequeira-Le Grand, A., Divol, B., Alexandre, H., 2013. Characterization of the viable but nonculturable (VBNC) state in Saccharomyces cerevisiae. PLoS One 8 (10), e77600.

Sato, H., et al., 2000. Restriction fragment length polymorphism analysis of 16S rRNA genes in lactic acid bacteria isolated from red wine. J. Biosci. Bioeng. 90 (3), 335–337.

Schmid, F., Moser, G., Müller, H., Berg, G., 2011. Functional and structural microbial diversity in organic and conventional viticulture: organic farming benefits natural biocontrol agents. Appl. Environ. Microbiol. 77, 2188–2191.

Serpaggi, V., Remize, F., Recorbet, G., Gaudot-Dumas, E., Sequeira-Le Grand, A., Alexandre, H., 2012. Characterization of the "viable but nonculturable" (VBNC) state in the wine spoilage yeast Brettanomyces. Food Microbiol. 30 (2), 438–447.

Setati, M.E., Jacobson, D., Andong, U.C., Bauer, F., 2012. The vineyard yeast microbiome, a mixed model microbial map. PLoS One 7 (12), e52609.

Shankar, P.S., Anupama, A., Pradhan, P., Prasad, G.S., Tamang, J.P., 2016. Identification of yeasts by polymerase-chain-reaction-mediated denaturing gradient gel electrophoresis in marcha, an ethnic amylolytic starter of India. J. Ethnic Foods 3 (4), 292–296.

Siebrits, L., 2007. PCR-Based DGGE Identification of Bacteria and Yeasts Present in South African Grape Must and Wine. Stellenbosch University, Stellenbosch, Doctoral Dissertation.

Sun, Y., Liu, Y., 2014. Investigating of yeast species in wine fermentation using terminal restriction fragment length polymorphism method. Food Microbiol. 38, 201–207.

Taylor, M.W., Tsai, P., Anfang, N., Ross, H.A., Goddard, M.R., 2014. Pyrosequencing reveals regional differences in fruit-associated fungal communities. Environ. Microbiol. 16, 2848–2858.

Usbeck, J.C., et al., 2014. Wine yeast typing by MALDI–TOF MS2010 Appl. Microbiol. Biotechnol. 98 (8), 3737–3752.

Zapparoli, G., Reguant, C., Bordons, A., Torriani, S., Dellaglio, F., 2000. Genomic DNA fingerprinting of Oenococcus oeni strains by pulsed-field gel electrophoresis and randomly amplified polymorphic DNA-PCR. Curr. Microbiol. 40 (6), 351–355.

Ženišová, K., Chovanová, K., Chebeňová-Turcovská, V., Godálová, Z., Kraková, L., Kuchta, et al., 2014. Mapping of wine yeast and fungal diversity in the Small Carpathian wine-producing region (Slovakia): evaluation of phenotypic, genotypic and culture-independent approaches. Ann. Microbiol. 64 (4), 1819–1828.

Further Reading

Chovanova, K., Krakova, L., Zenisova, K., Turcovska, V., Brezna, B., Kuchta, T., et al., 2011. Selection and identification of autochthonous yeasts in Slovakian wine samples using a rapid and reliable three-step approach. Lett. Appl. Microbiol. 53, 231–237.

Cocolin, L., Campolongo, S., Alessandria, V., Dolci, O., Rantsiou, K., 2010. Culture independent analyses and wine fermentation: an overview of achievements 10 years after first application. Ann. Microbiol. 61, 17–23.

Radler, F., Schmitt, M., Meyer, B., 1990. Killer toxin of Hanseniaspora uvarum. Arch. Microbiol. 154, 175–178.

Verginer, M., Leitner, E., Berg, G., 2010. Production of volatile metabolites by grape-associated microorganisms. J. Agric. Food Chem. 58 (14), 8344–8350.

CHAPTER 9

Barrel Aging; Types of Wood

Fernando Zamora
Department of Biochemistry and Biotechnology, University Rovira i Virgili, Tarragona, Spain

9.1 BRIEF HISTORICAL INTRODUCTION

Wine and barrels have a long common history that goes back to very early times (Taransaud, 1976; Vivas, 2002). The first references to the use of barrels are by Pliny the Elder, who attributed their origin to the Celts who used them to store and transport beer and other goods. The Romans originally stored and transported wine in amphorae, but after they arrived in Gaul, they quickly learned to build barrels. The greater capacity and resistance of the casks meant that they gradually replaced amphorae throughout the Mediterranean.

For a long time, barrels were used for the maritime transport of wine, oil, and other products. In fact, the transport capacity of a ship is still defined as tonnage. Contrary to what one might believe, the term "tonnage" does not come from the word "tonne" because the metric system of units was not adopted until the General Conference on Weights and Measures in 1889). Instead, the term "tonnage" originally referred to an English tax. In England (17th century), the King's tax collectors controlled the number of "tuns" of wine that were imported and imposed a tax (called "tonnage") which later became the standard unit for measuring the carrying capacity of a ship (Jones, 2016). Hence, the term "tonnage" comes from the amount of "barrels" that a boat could carry. To this day, a number of similar terms are still in use; for example the production and price of oil is still measured in barrels. Until the mid-20th century, a great number of products such as wine, olive oil, salted fish, whale oil, salt, sugar, etc., were stored and transported in wooden barrels. Over the last 50–60 years, other materials—especially stainless steel and plastics—have replaced wood, which nowadays is only used for aging wines.

9.2 THE MAIN TREE SPECIES USED IN COOPERAGE

Historically, barrels were made of the most abundant wood in the area where they were produced and their manufacture was very much conditioned by the development and needs of the shipping industry. For this reason, many woods from different species of trees have been used in cooperage but nowadays barrels are almost exclusively made of oak wood (Vivas, 2002; Zamora, 2003a). Other trees such as chestnut (*Castanea sativa*) (Fernández de Simón et al., 2009; Castro-Vázquez et al., 2013), acacia (*Robinia pseudoacacia*) (Fernández de Simón et al., 2009; Sanz et al., 2011), cherry (*Prunus avium*) (Fernández de Simón et al., 2009), ash (*Fraxinus excelsior*) (Sanz et al., 2012), beech (*Fagus sylvatica*) (Singleton, 1974), Raulí (*Nothofagus alpina*) (Lacoste et al., 2015), and even canary pine (*Pinus canariensis*) (Climent et al., 1998) are also used but to a very small extent.

The genus *Quercus* contains over 500 species mainly growing in the northern hemisphere (Camus, 1934–1955; Manos et al., 1999; Oldfield and Eastwood, 2007), although only a few of these are used in cooperage. In fact, nowadays three species of oak practically monopolize the market, two French (or European) oaks, *Quercus petraea* and *Quercus robur*, and one American oak, *Quercus alba* (Vivas, 2002; Zamora, 2003a). Other species of oak such as *Quercus pyrenaica* (Fernández de Simón et al., 2008), *Quercus mongolica* (YanLong et al., 2010), or *Quercus macrocarpa* (Chatonnet and Dubourdieu, 1998) have also been used in cooperage but to a much lesser degree (Mosedale et al., 1999).

FIGURE 9.1 Main forest for oak production in France and USA.

9.3 THE MAIN FORESTS PROVIDING WOOD FOR COOPERAGE

The genus *Quercus* is widely distributed throughout Europe, North and Central America, Southeast Asia, and to a lesser extent North Africa and the northern part of South America (Keller, 1987; Nixon, 2006). However, the two major areas supplying oak wood for cooperage are France and the United States (Chatonnet and Dubourdieu, 1998; Mosedale et al., 1999), hence the widespread use of the terms "French oak" and "American oak" in cooperage. Fig. 9.1 shows the main forests for oak production in France and the USA.

The main oak producing areas in France are Limousin, which is the principal producer of *Q. robur*, also called *Quercus pedunculata* or pedunculate oak, and Nevers, Allier, Tronçais, Centre, Vosges, Bertrange, and Jupilles, which produce *Q. petraea*, also called *Quercus sessile* or sessile oak (Camus, 1934–1955).

In the United States, the main oak production areas are the forests of the states of Virginia, Missouri, Minnesota, Pennsylvania, North Carolina, California, and Oregon. The primary species produced for use in cooperage is *Q. alba* (Fernández de Simón et al., 2009); however, *Q. macrocarpa* and *Quercus muehlenbergii* are also produced for this purpose, albeit to a lesser extent (Chatonnet and Dubourdieu, 1998).

In recent years, forests in other countries such as Hungary, Romania, Russia, Ukraine, etc., have also started to be exploited, which has led to even more types of oak on the market. Even so, most oak barrels are currently French or American in origin.

Consequently, it can be asserted that the term "American oak" refers exclusively to *Q. alba*, while the term "French oak" includes two species, *Q. robur* and *Q. petraea*, although it is important to state that if the origin of French oak is not specified as Limousin, then the term refers to *Q. petraea*.

9.4 THE CONCEPT OF WOOD GRAIN IN COOPERAGE

One of the most important characteristics of the wood used in cooperage is its grain, which depends to a great extent on the botanical and geographical origin of the trees (Vivas, 1995; Doussot et al., 2002). The Oxford English Dictionary (2009) defines grain as the longitudinal arrangement of wood fibers. However, in cooperage, grain refers to the size and regularity of the tree's annual growth circles (Feuillat et al., 1992; Vivas, 1995). Fig. 9.2 shows the structure of the cross-section of an oak trunk.

The figure clearly shows the rings or annual growth circles and spinal radii or medullary rays in the wood cut.

The heartwood is composed of the succession of spring wood and summer wood layers which make up the annual growth rings (Fig. 9.3).

The spring wood is very rich in wide vessels because normally in spring the rainfalls abound and the tree needs to pump a lot of sap up its trunk to form the buds and leaves. In contrast summer wood has fewer and smaller vessels because less water is available and less sap is needed. Moreover, during summer, through photosynthesis the leaves synthesize sugars and polysaccharides which are then stocked in the tree's fibers, thus making summer wood denser than spring wood. Fig. 9.3 also shows how the width of the annual growth rings defines what in cooperage is known as grain.

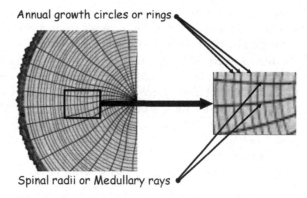

FIGURE 9.2 Structure of the cross-section of an oak trunk.

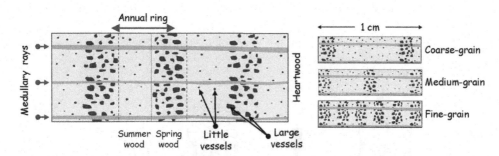

FIGURE 9.3 Scheme of the structure of heartwood; relationship with the grain.

Normally, oak is classified as fine-grain when there are five or more rings per cm of wood, as medium-grain oak when there are between 3 and 4 rings per cm, and finally as coarse-grain oak is when there are two or fewer rings per cm (Feuillat et al., 1992). The size and regularity of the grain obviously depend on the species of oak, but also on the edaphoclimatic conditions of the terrain in which it is planted (Vivas, 1995; Feuillat et al., 1992). In any case, it is generally accepted that *Q. petraea* has a fine to medium grain (1–3 mm of annual growing/year), *Q. robur* has a medium to coarse grain (3–10 mm annual growing/year), and finally American Oak (*Q. alba*) has a fine to medium grain although it tends to be more variable (1–5 mm annual growing/year) (Chatonnet and Dubourdieu, 1998).

However, the size of the grain can vary significantly within the same species depending on where the tree was grown. For example, *Q. petraea* grown in the Allier region is regarded has having a particularly fine grain (average of 1.2 mm/year), while *Q. petraea* from the Vosges region has a medium grain (average of 2.6 mm/year) (Vivas et al., 2001). Moreover, grain size varies significantly among trees within the same forest (Polgue and Keller, 1973; Ribéreau-Gayon et al., 1999a). For these reasons some cooperages have begun to classify their wood not on the basis of its geographical origin but on the actual size of the grain. Thus, the names Allier type, Vosges type, or Limousin type applied to oak casks do not necessarily indicate the geographical origin of the wood used for the cask, but rather are a guarantee that the wood used belongs to the main species in these forests and has the grain size characteristic of these areas.

9.5 OBTAINING THE STAVES: HAND SPLITTING AND SAWING

At present, most cooperages are supplied with oak from controlled forests, where only trees over 100 years old are exploited. Given the fact that the forest is a limited resource, felled trees need to be replaced with new specimens for exploitation in the future. Oak silviculture for the purpose of making barrels requires rigorous control of the growth process and tree selection (Remy, 1991). In fact only when the trees reach the appropriate height (i.e., when they are between 120 and 160 years old) are the best specimens ready for commercial exploitation.

The process of making barrels begins, obviously, by selecting the most suitable trunks for use in cooperage. The oak barrels will be made only with the finest wood, which comes from those parts of the trunk with the finest grain and devoid of branches (Taransaud, 1976; Vivas, 2002).

The selected parts of the logs are then sawed into shorter pieces. The length of these shorter logs will define the size and volume of the future barrel. These short trunks will then be cut to obtain staves using one of two techniques: hand-splitting or sawing. Fig. 9.4 illustrates how staves are obtained by each of these techniques.

Hand-splitting consists of breaking the trunk by means of a hydraulic wedge. When the wedge is driven into the wood, it naturally splits lengthwise along the planes whose edges are the medullary rays. These log fragments, called quarters, are then used to obtain the staves. In contrast, sawing consists simply of cutting the log lengthwise with an electric band or circular saw.

Fig. 9.5 illustrates the different impacts that hand-splitting and sawing have on the structure of the staves and the advantages and drawbacks of both techniques (Chatonnet, 1995a; Vivas, 2002; Zamora, 2003a). Hand-splitting has the advantage of obtaining staves whose medullary rays are parallel to the contact surface with the wine (see Fig. 9.5). This reduces the risk of leaks but has the drawback of yielding fewer barrels; that is, 5 m^3 of timber are required to make 1 m^3 of staves which in turn produce only 2.2 bbl/m^3 of wood. In contrast sawing has a higher yield; only 3 m^3 of timber are needed to make 1 m^3 of staves, which in turn produce 3.7 bbl/m^3 of wood.

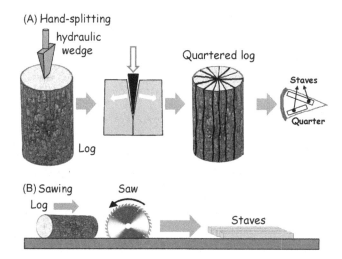

FIGURE 9.4 Wood cutting techniques for obtaining staves.

Technique	Advantages	Drawbacks
Hand splitting	The medullary rays of the staves are parallel to the contact surface with the wine. Low risk of wine leaking through the staves	Low yield ↓ 5 m^3 of logs ↓ 1 m^3 of staves ↓ 11 barrels (225 L) ↓ 2.2 barrel/m^3 of logs
Sawing	High yield ↓ 3 m^3 of logs ↓ 1 m^3 of staves ↓ 11 barrels (225 L) ↓ 3.7 barrel/m^3 of logs	The medullary rays of the staves are more or less oblique to the contact surface with the High risk of wine leaking through the staves

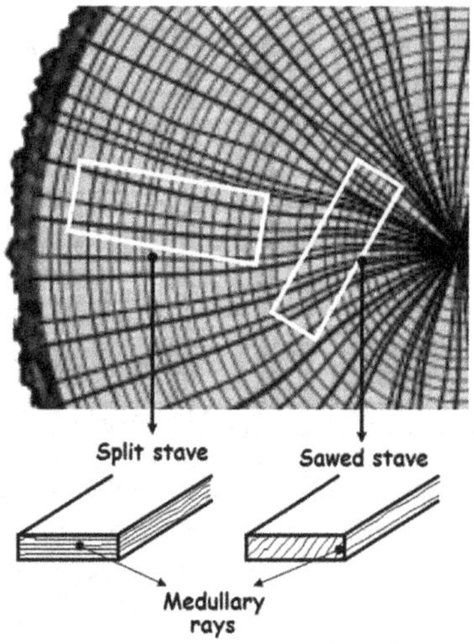

FIGURE 9.5 Impact of hand-splitting and sawing on the structure of the staves: advantages and drawbacks.

However, the drawback of sawing is that it produces staves whose medullary rays are more or less oblique to the contact surface with the wine, thus significantly increasing the risk of leaks.

The decision to use either of these techniques depends basically on the type of oak (Chatonnet, 1995a; Vivas, 2002; Zamora, 2003a). Basically, French oak timber (*Q. petraea* and *Q. robur*) must be hand-split while American oak timber can be sawed. The reason why French and American oak require different techniques is due to the different structures of their tyloses. Tyloses are outgrowths of parenchyma cells adjacent to a vessel which, under certain conditions, will block the vessel. The spring vessels of a tree may be somewhat permeable to liquids if they are not sufficiently sealed by tyloses, and this in turn can cause barrels to leak. Tyloses are thin and scarce in European oaks, thus making spring vessels more permeable to liquids and increasing the risk of leakage in the barrels made from this wood. However, splitting the wood, as opposed to sawing it, produces staves whose medullary radii are parallel to the contact surface with the wine parallel (see Fig. 9.5), thus minimizing the risk of leaks (Moutounet et al., 1999; Zamora, 2003b). In contrast, American oak (*Q. alba*) can be sawed because its vessels are more occluded by large numbers of tyloses, thus making the risk of leakage negligible. The different yields obtained through these two techniques are the main reason why French oak barrels are more expensive than American oak barrels.

9.6 DRYING SYSTEMS: NATURAL SEASONING AND ARTIFICIAL DRYING

Once the staves have been obtained, they must be dried before they can be used to manufacture barrels (Zamora, 2003b; Vivas et al., 1991). Traditionally, staves are dried simply by leaving them stacked in the open for two or three years (Taransaud, 1976; Vivas, 2002) in a process known as natural seasoning. During natural seasoning the staves are washed by rainwater and dried by the sun until they have the right level of moisture needed to mold them and give the barrel its characteristic shape.

The continuous washing by rain during natural seasoning extracts and eliminates certain compounds that would otherwise be released into the wine during aging (Vivas, 2002). It has also been shown that during natural seasoning certain fungi grow on the surface of the wood producing enzymatic transformations that degrade some substances and transform others (Vivas et al., 1991). Other studies have shown that natural seasoning decreases the phenolic compounds content (Cadahía et al., 2001; Doussot et al., 2002) and increases the concentration of volatile compounds released by the wood to the wine (Chatonnet et al., 1994; Cadahía et al., 2003).

Obviously, natural seasoning can be accelerated by artificial drying which simply consists of dehydrating the wood in drying chambers. Nevertheless, the chemical components that artificially dried barrels release into the wine differ significantly from those released by naturally seasoned barrels and produce wines that are poorer, more bitter, and less complex (Vivas and Glories, 1996a). Consequently, the vast majority of the cooperages use natural seasoning to dry their staves.

9.7 ASSEMBLY AND TOASTING OF THE BARREL

Assembling the barrel is a very complex process requiring great skill. Fig. 9.6 illustrates the process. First, the staves must be shaped by hollowing, tapering, and beveling to give them the necessary form. The staves have different widths so they need to be selected carefully to ensure that the barrel has the required diameter. Normally around 30 staves are needed and one of them, the master stave, is wider than the others because it will be where the bung hole is drilled to allow the barrel to be filled.

Once the staves have been selected, the cooper places them side by side in a circle with the help of the toasting hoops. This operation, called rosing (from the French expression "Mise en rose"), consists of placing the staves vertically inside hoops to form the shape of a truncated cone. The cooper then places a steel cable, set with a winch, in a half rigging key knot around the bottom of the barrel. The barrel is then heated over an open fire, in a charcoal pan or brazier. The fire is normally fed from scraps left over from the cutting of the barrel heads. This operation, called toasting, consists of two stages. During the first stage, called tightening or bending or *cintrage*, the heat of the open fire increases the flexibility of the staves which allows the winch to bend the staves until they form the characteristic shape of the barrel. The second stage, properly named toasting or *bousinage*, changes the wood's physicochemical composition and physical structure and thus determines which substances the barrel will release into the wine during aging.

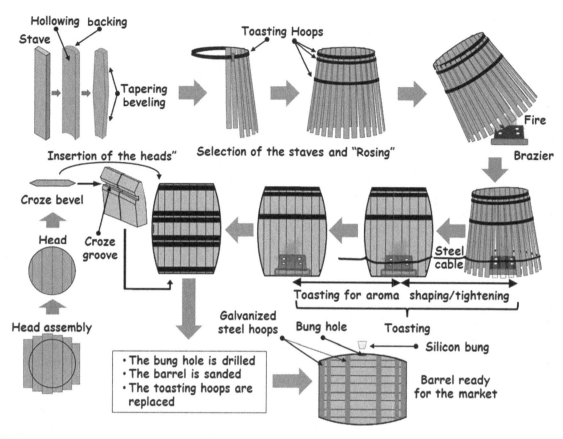

FIGURE 9.6 Barrel manufacture.

The bending usually lasts about 15 minutes, whereas the duration of the *bousinage* will depend on which characteristics are wanted to be imparted to the wine. The *bousinage* is an essential phase in the manufacture of high quality barrels because it mellows the wood's harsh ellagitannins, mitigates raw oak flavors, and generates several new aromas that increase the intensity and complexity of wine.

Toasting is classified as light, medium, or heavy depending on the length of the *bousinage*. Each toasting length has a different effect on the wood's chemical composition and, therefore, on the substances that are released into the wine during aging, which in turn influence the wine's sensory profile. The time required for light toasting is usually considered to be 20 minutes, for medium toasting it is 30 minutes and for heavy toasting it is 40 minutes. However, these criteria can differ from one cooperage to another as each has its own toasting protocol. Moreover, some cooperages specify different levels of medium toasting, differentiating between "medium +" or "medium −." Some cooperages even offer extra-heavy toasting and also complete toasting in which the barrel heads are also toasted.

Once the barrel has been toasted, it is ready for the final stage of assembly. A croze groove is cut at each end of the barrel to allow the insertion of the heads. The heads are made from shorter staves that are rounded and beveled to form an edge that can be inserted into the croze groove. The head is then inserted and the toasting hoops are hammered and pressed to ensure tightness. After that the bung hole is drilled into the master stave and the toasting hoops are removed and replaced with galvanized steel hoops. Pressurized water is then used to check the barrel is watertight and to detect any possible leaks. The surface of the barrel is then polished and the barrel is ready for the market.

9.8 TYPES OF BARRELS AND BARREL PARTS

The market offers different types of barrel that can be classified according to different criteria. Thus, barrels can be classified according to the botanical/geographical origin of the oak, the degree of toasting, capacity, and

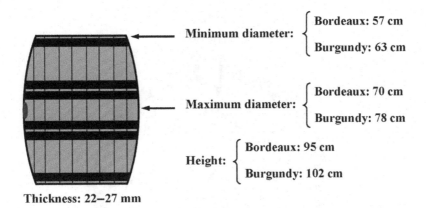

FIGURE 9.7 Dimensions of Burgundy (300 L) and Bordeaux barrels (225 L).

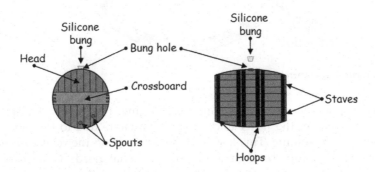

FIGURE 9.8 The structure of a barrel and its various component parts.

purpose. The first two criteria have already been discussed extensively in the preceding sections so this section will focus on the remaining two criteria.

In terms of capacity, barrels are usually classified as follows: Bordeaux barrel (225 L), Burgundy barrel (300 L), Sherry pipe (500 L), Porto pipe (650 L), and, for the highest volumes, Foudres or Vats. There are also lower-volume barrels known usually as miniatures, but they are almost never used in wineries. At present, however, most of the market is focused on 300 and (in particular) 225 L barrels. These barrels have the physical dimensions shown in Fig. 9.7.

Barrels can also be classified according to how they are used. Thus, barrels used for aging were originally called "Château barrels," while barrels used to transport the wine were called "Transport barrels." The differences between them was that Transport barrels were thicker (27 mm) than Château barrels (22 mm) in order to make them stronger and that Château barrels were equipped with a crossboard and complementary holes in the bottom, called spouts, to facilitate sampling and racking.

In recent years, the understanding of how wine ages has increased and most wineries no longer regard the crossboard and the spouts as necessary, hence the steady decline in demand for Château barrels. At the same time, the classic Transport barrel has also practically disappeared from the market. The reason for this is that wine is no longer traded and transported in barrels. In this context, the greater weight and thickness of Transport barrels make them more costly and difficult to handle. For this reason, at present, the most commonly used barrel is the "Light Transport" barrel, which has the thickness of the Château barrel and the structure of the Transport barrel (Fig. 9.8).

9.9 WHAT HAPPENS TO A WINE DURING BARREL AGING

The aging of wine in oak barrels is a complex process in which various phenomena are involved (Ribéreau-Gayon et al., 1999a). Fig. 9.9 shows the processes that take place during aging (Feuillat et al., 1998; Zamora, 2003a).

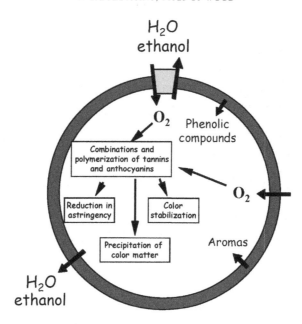

FIGURE 9.9 Phenomena that occur during barrel aging.

As Fig. 9.9 shows, different phenomena occur during barrel aging. First, the wood absorbs some of the wine, especially in new barrels, and some of the water and ethanol evaporate due to the porosity of the wood, which causes a nonnegligible reduction in volume. This loss of wine depends on the relative humidity and temperature (3%–10% of the wine volume) and represents an economic cost that needs to be taken into account (Stamp, 2015). Second, the oak wood releases several volatile substances, making the wine aroma more intense and complex (Chatonnet et al., 1994; Chatonnet, 1995a; Cadahía et al., 2003; Navarro et al., 2018). Third, the oak wood also releases ellagitannins, which influence texture sensations such as astringency and mouthfeel (Cadahía et al., 2001; Chira and Teissedre, 2013a; Navarro et al., 2016a). Fourth, the moderate diffusion of oxygen through the wood causes different reactions between anthocyanins and proanthocyanidins which stabilize the color and reduce the astringency (Ribéreau-Gayon et al., 1999a; Zamora, 2003a). Finally, the protracted period that wine spends in the barrel allows the natural sedimentation of unstable colloidal matter, thus leaving the wine limpid and stable.

All these phenomena are influenced by several factors including the botanical and geographic origin of the wood, the seasoning technique, the degree of toasting, and the number of times the barrel has been used previously.

9.10 VOLATILE SUBSTANCES RELEASED BY OAK WOOD DURING BARREL AGING

The volatile compounds released by oak into the wine during barrel aging may be naturally present in the original oak wood or may derive from other wood's substances during the toasting process (Chira and Teissedre, 2013a). Many volatile compounds from a wide range of chemical families have been described (Cadahía et al., 2003), but few of them are significant with regard to their impact on the sensory characteristics of wines. In terms of sensory impact, the main volatile substances released by oak wood are furans, phenolic aldehydes and ketones, volatile phenols, and β-methyl-γ-octalactones (Garde-Cerdán et al., 2004; Prida and Chatonnet, 2010). Table 9.1 shows the main volatile substances released from oak into wine, along with their origin, their estimated thresholds, and the aroma descriptor for each molecule.

The family of furans, which includes furfural, methylfurfural, hydroxymethylfurfural, and furfurilic alcohol, is produced from wood polysaccharides by means of the Maillard reaction during the toasting of the staves needed for barrel assembly (Fazzalari and Stahl, 1978; Boidron et al., 1988; Fors, 1998). It is usually accepted that furans contribute smoked and toasted nut notes (Boidron et al., 1988; Navarro et al., 2018). However, these furanic aldehydes are present in wine at concentrations well below their olfactory perception thresholds, so they have little impact on the aroma (Ribéreau-Gayon et al., 1999a).

TABLE 9.1 Main volatile substances released from oak into wine

Origin	Family	Compound	Indicative threeshold	References	Aroma
Polysaccharides	Furans	Furfural	3–20 mg/L	Fazzalari and Stahl, 1978; Boidron et al., 1988; Fors, 1988	Toasted almonds
		Methyfurfural	45 mg/L		
		Hydroxymethylfurfural	45 mg/L		
		Furfurilic alcohol	50 mg/L		
	Other volatile heteocycles	Maltol	35 mg/L	Pittet et al., 1970; Seifert et al., 1970; Fors, 1988	Caramel, Cocoa
		Dimethylpyrazines	0.2–35 mg/L		
		Acetic acid	0.6–1.0 g/L	Guth, 1997; Macías et al., 2013	Vinegar
Lignin	Phenolic aldehydes	Vanillin	20–200 µg/L	Pittet et al., 1970; Fazzalari and Stahl, 1978; Ohloff, 1978; Boidron et al., 1988; Spillman et al., 2004;	Vanilla
		Syringaldehyde	15–50 mg/L		
		Sinapaldehyde	~50 mg/L		
		Coniferaldehyde			
	Phenolic ketones	Acetophenone	65 µg/L	Paterson and Piggott, 1994; Pino and Queris, 2011	
		Acetovanillone	1–9 mg/L		
		Propiovanillone	Unknown		
		Butyrivanillone			
	Volatile phenols	Guaiacol	75 µg/L	Boidron et al., 1988; Chatonnet et al., 1992; Sterckx et al., 2011	Smoked toasted
		Methylguaicol	65 µg/L		
		Ethyguaiacol	140–150 µg/L		
		Vinylguaiacol	380 µg/L		
		Vinylphenol	180–725 µg/L	Chatonnet et al., 1993; Lapez et al., 2002	Medicinal band-aids
		Ethylphenol	400–620 µg/L	Chatonnet et al., 1992; Romano et al., 2009	Animal notes horse
		Eugenol	100–500 µg/L	Fazzalari and Stahl, 1978; Boidron et al., 1988; Garde-Cerdán and Ancín-Azpilicueta, 2006; Prida and Chatonnet, 2010	Clove
		t-Isoeugenol			
Lipids	β-methyl-γ-octalactones	cis-β-methyl-γ-octalactone	46–83 µg/L	Chatonnet, 1992; Abbott et al., 1995; Wilkinson et al., 2004	Coconut Brazil nuts
		trans-β-methyl-γ-octalactone	390–460 µg/L		

The Maillard reaction also produces other volatile substances from polysaccharides such as maltol and dimethylpyrazines, which have the aromas of caramel and cocoa, respectively (Pittet et al., 1970; Seifert et al., 1970). Finally, the toasting process also produces acetic acid from wood polysaccharides. Some xylose units of the hemicelluloses have acetyl groups in positions 2 and/or 3, which can lead to the formation of free acetic acid by hydrolysis during the toasting of the staves (Vivas, 2002; Zamora, 2003a). This phenomenon leads to a nonnegligible increase in the volatile acidity of barrel-aged wines (around 0.1–0.15 g/L), especially when the barrels are new. This increase is below the sensory threshold of acetic acid (Guth, 1997; Macías et al., 2013) but it must be taken into account.

The family of phenolic aldehydes and ketones, which includes vanillin, syringaldehyde, acetovanillone, and propiovanillone, contributes to the characteristic vanilla aroma of aged wines (Fazzalari and Stahl, 1978; Ohloff, 1978; Boidron et al., 1988; Paterson and Piggott, 1994; Spillman et al., 2004; Pino and Queris, 2011), with vanillin being the major contributor (Fors, 1998; Prida and Chatonnet, 2010). Phenolic aldehydes and ketones result from the thermal degradation of lignin during barrel toasting (Fors, 1998; Cadahía et al., 2003).

The family of volatile phenols includes ethylphenol, vinylphenol, guaiacol, methylguaiacol, ethylguaiacol, vinylguaiacol, eugenol, and trans-isoeugenol. Some of these volatile phenols have a positive effect on wine aroma, some are clearly negative, and others can be positive or negative depending on their concentration. For example, eugenol and *t*-isoeugenol contribute pleasant spicy notes of clove (Fazzalari and Stahl, 1978; Boidron et al., 1988; Prida and Chatonnet, 2010), whereas ethylphenol and vinylphenol, which are mainly produced by spoilage microorganisms (essentially by *Brettanomyces*) resulting from the decarboxylation of coumaric acids (Chatonnet et al., 1992), contribute disagreeable odors of horse sweat and medicinal notes, respectively (Chatonnet et al., 1993; López et al., 2002; Romano, et al., 2009; Sterckx et al., 2011). Finally, other volatile phenols such as guaiacol, methylguaiacol, ethylguaiacol, and vinylguaiacol are mainly produced from lignin during barrel toasting (Vivas, 2002; Zamora, 2003a; Chira and Teissedre, 2013a). Depending on their concentration, their effect can be negligible if they are below their sensory threshold, pleasant when their presence provides slight notes of grilled bread, or unpleasant when their concentration is so high that their notes of burnt bread dominate the rest of the olfactory matrix (Navarro et al., 2018).

Finally, β-methyl-γ-octalactones (also known as whiskey lactones), which are present in the form of two isomers (*cis* and *trans*), are responsible for the coconut and Brazil nut flavors (Chatonnet, 1992; Garde-Cerdán and Ancín-Azpilicueta, 2006). These lactones are produced by dehydration of certain fatty acid derivatives present in the wood (Wilkinson et al., 2004). It should be noted that the *cis* isomer has a much lower perception threshold than the *trans* isomer, making its contribution to coconut perception much more intense (Abbott et al., 1995).

All these volatile substances are released from oak wood into wine during barrel aging and the amount and proportion of them will depend on multiple factors such as the botanical and geographical origin of the oak, the seasoning technique, the degree of toasting, and the number of times the barrel was used previously. All these aspects will be discussed in the following points.

9.11 PHENOLIC COMPOUNDS RELEASED BY OAK WOOD DURING BARREL AGING

Oak wood also releases certain nonvolatile substances such as phenolic acids, coumarins, and ellagitannins into the wine (Zhang et al., 2015) which contribute to wine taste and texture sensations (Ribéreau-Gayon et al., 1999b).

Particularly important among the group of phenolic acids released by oak wood are gallic acid and its dimer, ellagic acid. Their direct organoleptic contribution is not important but they contribute to red wine color by acting as copigments (Mazza and Brouillard, 1990) and they also protect anthocyanins from oxidation (Vivas and Glories, 1996a).

Coumarins are derivatives of cinnamic acids and are formed by the intramolecular esterification of a phenol hydroxyl group onto the α position of the carbon chain. They can be present in glycosylated form (heterosides) and in aglycone form (Salagoity-Auguste et al., 1987). The main coumarins are esculin/esculetin and scopoline/scopoletin depending on whether the molecule is an heteroside or an aglycone (Puech and Moutounet, 1988. Coumarins can contribute to wine bitterness, especially when they are present as heterosides (Ribéreau-Gayon et al., 1999b).

Ellagitannins are hydrolyzable tannins with a specific structure consisting of open-chain glucose esterified at positions 4 and 6 by a hexahydroxydiphenyl unit and a nonahydroxyterphenyl unit esterified at positions 2, 3, and 5 with a C-glycosidic bond between the carbon of the glucose and position 2 of trihydroxyphenyl unit (Quideau et al., 2004; Okuda et al., 2009). Fig. 9.10 shows the structures of gallic acid, ellagic acid, and the main ellagitannins.

Ellagitannins account for around 10% of the dry weight of oak heartwood, and contribute to the wood's high durability (Scalbert et al., 1988). The most abundant ellagitannins in oak wood are castalagin and vescalagin, which represent between 40% and 60% of the total (Fernández de Simón et al., 1999).

The solubilization of ellagitannins in wine takes place during barrel aging or when wood alternatives (wood chips, blocks, staves, …) are used and is facilitated by the hydroalcoholic nature of wine (Moutounet et al., 1989; Jordão et al., 2005. Nevertheless, the high reactivity of ellagitannins means that their levels in wine are much

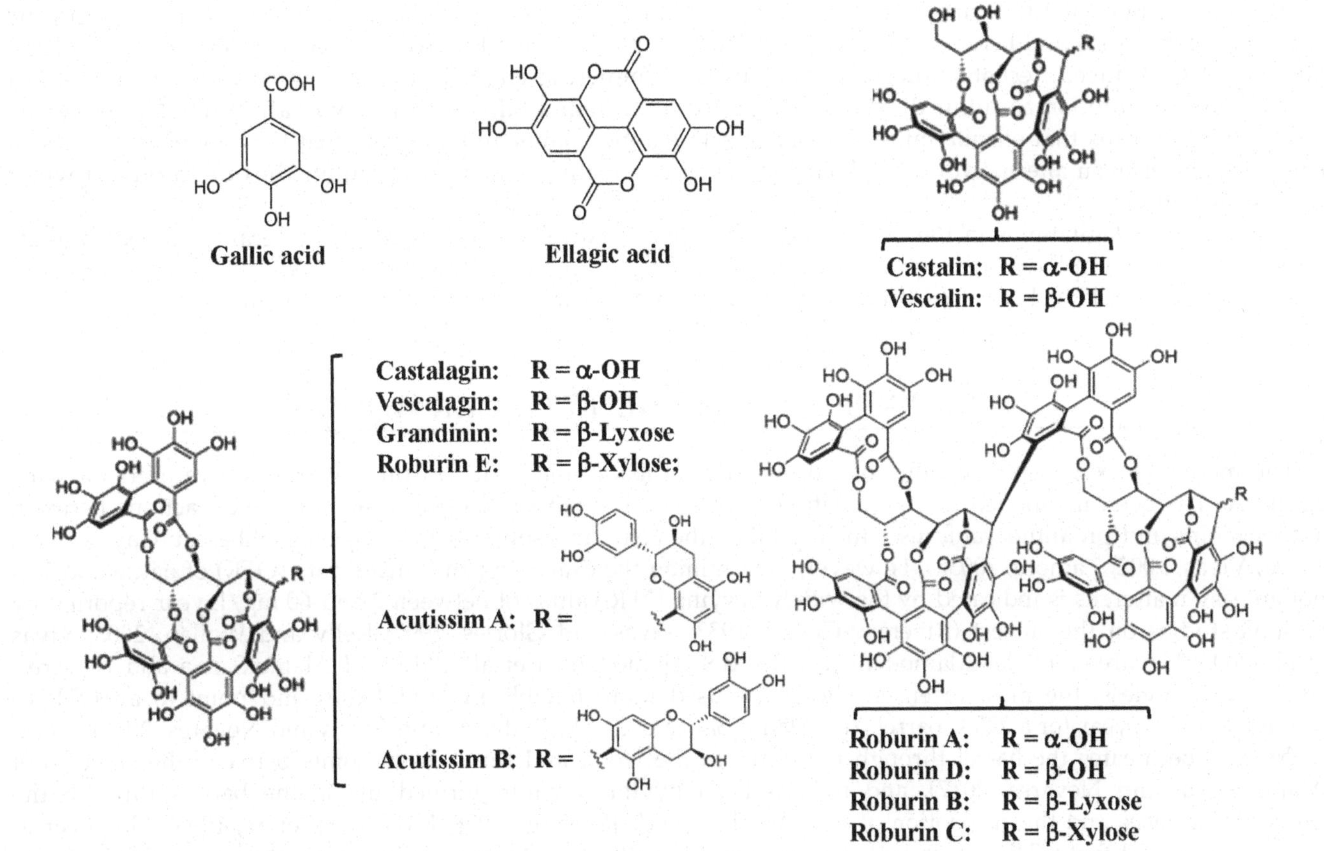

FIGURE 9.10 Structures of gallic acid, ellagic acid, and the main ellagitannins.

TABLE 9.2 Astringency and bitterness thresholds of the different ellagitannins

Ellagitannin	Astringency threshold (mg/L)	Bitterness threshold (mg/L)
Castalagin	1.03	1578
Vecalagin	1.03	1578
Grandinin	0.21	657
Roburin A	5.37	1535
Roburin B	12.09	1160
Roburin C	12.49	1200
Roburin D	5.55	1373
Roburin E	0.21	437

lower than might be expected (Michel et al., 2011). In addition, new ellagitannins can be formed in wine when vescalagin reacts with (+)-catechins, ethanol, or even malvidin-3-glucoside to form new adducts (acutissimins A and B) (Jourdes et al., 2009; Chassaing et al., 2010).

Ellagitannins mainly affect astringency. Table 9.2 shows the astringency and bitterness thresholds of the different ellagitannins reported by Glabasnia and Hofmann (2006).

These data show that ellagitannins have a greater contribution to astringency than bitterness because their astringency thresholds are significantly lower than their bitterness thresholds. In sufficient quantity, ellagitannins can reinforce the wine structure, especially during its first years of life (Michel et al., 2011). However, an excess of ellagitannins can be negative because the associated wood taste can overpower the wine. In any case, this sensation of structure imparted by the ellagitannins will disappear as the wine ages because the acidic medium leads to their hydrolysis.

It has been suggested that ellagitannins can improve the color of red wine because they can act as copigments (Gombau et al., 2016) and because they exert a protective effect against the oxidation of anthocyanins (Vivas and Glories, 1996a). Indeed, recent studies have shown that commercial enological ellagitannins have a similar effect on oxygen consumption as sulfur dioxide at usual doses for both additives (Navarro et al., 2016b; Pascual et al., 2017). This confirms that ellagitannins are very effective antioxidants and suggests that they could be used as a complement (or even alternative) for reducing (or even eliminating) the need for sulfur dioxide to protect wines against oxidation.

The level of ellagitannins in the oak wood used to make barrels also depends on several factors such as its botanical and geographical origin, grain size, the seasoning technique, the degree of toasting, and the number of times that the barrel has been used previously. All these aspects are discussed in the following sections.

9.12 OXYGEN PERMEABILITY OF OAK WOOD

The moderate oxygen permeability of oak wood is probably the main reason why barrels are used for wine aging. As has been mentioned, the permeation of oxygen through the wood causes different reactions between anthocyanins and proanthocyanidins which stabilize the color and soften the astringency (Ribéreau-Gayon et al., 1999a; Vivas, 2002; Zamora, 2003a). However, determining the exact oxygen transfer rate (OTR) of oak wood is it not an easy matter, as is indicated by the widely varying OTR values of between 2 and 60 mg/L year reported by various studies on the subject (Ribéreau-Gayon, 1933; Vivas and Glories, 1997; Kelly and Wollan, 2003; Vivas et al., 2003; Nevares and del Alamo-Sanza, 2015; del Alamo-Sanza et al., 2017; del Alamo-Sanza and Nevares, 2017). Nevertheless, the most recent studies have used more reliable methodologies and point to an OTR of around 12 mg/L year for a 225 L barrel (del Alamo-Sanza et al., 2017; del Alamo-Sanza and Nevares, 2017).

Oxygen permeates the barrel through the pores in the wood and through the joints between the staves (del Alamo-Sanza and Nevares, 2017), and it is assisted by the vacuum formed inside the barrels through the exchange of gases and the absorption of wine by the wood (Peterson, 1976; Moutounet et al., 1998). However, it has been reported that OTR of barrels decreases over time after they have been filled (Nevares et al., 2014). These data suggest that OTR is determined by wood moisture because oxygen diffuses better through dry wood (del Alamo-Sanza and Nevares, 2014). The influence of wood grain and botanical origin on the OTR is discussed in the following sections.

9.13 INFLUENCE OF WOOD GRAIN

Historically the most valued oak wood for cooperage has been that with the finest grain (Polgue and Keller, 1973; Feuillat et al., 1992). This is because the size of the grain has a great influence on the substances that the oak wood will release into the wine during barrel aging (Vivas, 1995, 2002). Fig. 9.11 shows how the grain size

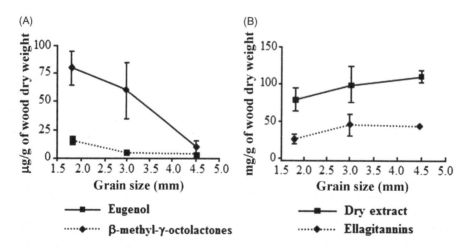

FIGURE 9.11 Influence of grain size on the aromas (A) and phenolic compounds (B) released by oak wood.

affects the release of volatile (A) (eugenol and β-methyl-γ-octolactones) and nonvolatile substances (B) (dry extract and ellagitannins).

The figure indicates that as grain size increases, so too does the release of dry extract and ellagitannins. In contrast, the release of eugenol (clove) and β-methyl-γ-octolactone (coconut) increases as the grain size reduces (Vivas, 1995).

The grain size also exerts an effect on the OTR because more oxygen enters when the grain is smaller (del Alamo-Sanza and Nevares, 2014, 2017). Wines therefore mature more quickly when they are aged in fine-grain oak barrels.

These data confirm something that enologists have known empirically for a long time, which is that fine-grain oaks provide wine with greater aromatic richness but have only a slight effect on its body, whereas coarse-grained oaks improve the wine's body but have only a slight effect on its aromas. It is for these reasons that fine-grain oaks are considered to be the best for cooperage (Chatonnet, 1995b).

9.14 INFLUENCE OF BOTANICAL AND GEOGRAPHIC ORIGIN

By now it is clear that the species of oak and its forest of origin are the main factors to take into account when choosing a barrel. Table 9.3 shows the composition of hydroalcoholic extracts released by wood from the three most common species, specifically *Q. petraea* from Allier, *Q. robur* from Limousin, and *Q. alba* from the United States (Chatonnet and Dubourdieu, 1998; Masson et al., 1997).

These data can be summarized as follows:

- Limousin oak (*Q. robur*) provides the highest levels of dry extract, A280, and ellagic tannins and the lowest concentrations of β-methyl-γ-octolactone, eugenol, and vanillin.
- American oak (*Q. alba*) provides the lowest levels of dry extract, A280, and ellagic tannins and the highest levels of vanillin and β-methyl-γ-octolactone. Moreover, American oak has levels of more than 90% for the most aromatic *cis* isomer of β-methyl-γ-octolactone, which increases the impact of coconut notes.
- Allier oak provides intermediate levels of dry extract, A280, ellagic tannins, β-methyl-γ-octolactone, and vanillin. In contrast it provides the highest levels of eugenol.

In recent years, other studies have reported about volatile substances (Chatonnet, 1992; Chatonnet and Dubourdieu, 1998; Mosedale et al., 1999; Cadahía et al., 2003; Spillman et al., 2004; Chira and Teissedre, 2015; Navarro et al., 2018) and ellagitannins (Masson et al., 1995; Fernández de Simón et al., 1999; Prida and Puech, 2002; Cadahía et al., 2003, 2007; Navarro et al., 2016a), particularly with regard to the two most commonly used oak tree species in cooperage (*Q. petraea* and *Q. alba*). All these investigations confirm the previous data. Fig. 9.12 summarizes the contribution of French (*Q. petraea*) and American (*Q. alba*) oaks to the composition of wine and its sensory qualities (Navarro et al., 2016a, 2018).

This figure shows the concentration of volatile compounds (A) (Navarro et al., 2018) and ellagitannins (B) (Navarro et al., 2016a) in a red wine aged for 12 months in 225 L, medium-toasted barrels of French (*Q. petraea*) and American (*Q. alba*) oak. This figure also shows the theoretical sensory impact of the main aromas (C) and astringency (D) in the different samples. These sensory impacts were calculated by dividing the individual

TABLE 9.3 Influence of botanical and geographical origin on the composition of wood hydroalcoholic extracts

		Quercus petraea (Allier)	*Quercus robur* (Limousin)	*Quercus alba* (United States)
Dry extract		90 ± 15	140 ± 7	57 ± 24
A280 (nm)		22.4 ± 2.9	30.0 ± 1.8	17.0 ± 5.6
Ellagitannins (mg/g)		8.0 ± 1.4	15.0 ± 1.5	6.0 ± 2.4
Vanillin (μg/g)		580 ± 217	435 ± 181	797 ± 398
Eugenol (μg/g)		8.0 ± 1.0	2.0 ± 1.5	4.0 ± 0.15
β-Methyl-γ-octolactones (μg/g)	Total	87.2 ± 27.3	16.0 ± 1.5	158.1 ± 27.0
	cis Isomer (%)	58 ± 21	51 ± 26	93 ± 3

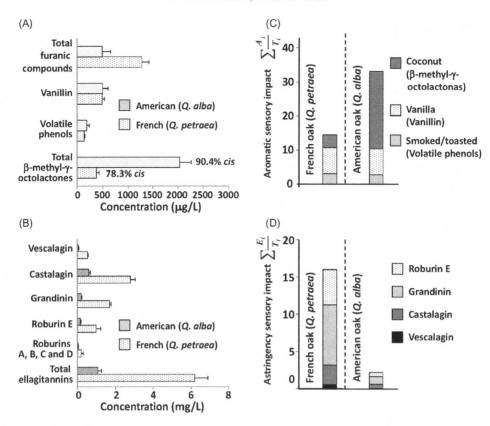

FIGURE 9.12 Influence of French and American oak on the concentration of volatile compounds (A) and ellagitannins (B) and on their aromatic (C) and astringency sensory impact (D).

concentration of each volatile substance or ellagitannin by their corresponding thresholds (Navarro et al., 2018; Glabasnia and Hofmann, 2006).

These data and the findings reported in the literature show that the American oak (*Q. alba*) has the highest aromatic impact, marked in particular by the aromas of coconut (β-methil-γ-octolactone) and to a lesser extent vanilla (vanillin). In contrast American oak's contribution to wine body/structure is very slight because of its poor release of ellagitannins. The French oak from Limousin (*Q. robur*) will slightly aromatize the wine because it releases small amounts of volatile substances but its contribution to wine structure is considerable since it releases high amounts of ellagitannins. Finally, the French oak from Allier (*Q. petraea*) contributes an aroma that is slightly less intense than the American oak but which is nevertheless more complex, balanced, and elegant because it is less marked by vanilla and coconut and introduces a distinctive nuance of clove spice. This oak also contributes to wine structure, but to a lesser extent than *Q. robur* because it releases intermediate levels of ellagitannins.

In relation to the OTR, it has been reported that of the two French oaks, *Q. petraea* is more permeable to oxygen than *Q. robur* because its grain is thinner. It has been also reported that of the two most commonly used oaks in cooperage, the American oak (*Q. alba*) is less permeable than the French oak (*Q. petraea*) (del Alamo-Sanza and Nevares, 2014, 2017).

For all these reasons, *Q. petraea* and *Q. alba* are considered the best oak trees for aging wine, whereas *Q. robur*, due to its high ellagitannin content and its poor aromatic contribution, seems more apt for aging spirit beverages (Chatonnet and Dubourdieu, 1998).

9.15 INFLUENCE OF NATURAL SEASONING AND ARTIFICIAL DRYING

As previously mentioned, it is essential to dry the oak staves in order to be able to shape the barrel. However, drying staves by natural seasoning means setting aside large amounts of wood stock, which in turn makes the process more expensive and hinders the production planning. For this reason, artificial drying of staves with

TABLE 9.4 Comparison between natural seasoning and artificial drying

		Natural seasoning	Artificial drying
Dry extract		90 ± 15	113 ± 4
A280 (nm)		22.4 ± 2.9	27.2 ± 1.9
Ellagitannins (mg/g)		8.0 ± 1.4	11.9 ± 1.2
Cumarins (µg/g)	Total	39.5 ± 2.3	52.8 ± 5.3
	Heterosides	4.3 ± 0.9	30.0 ± 3.5
	Aglycones	35.2 ± 2.7	22.8 ± 3.0
Vanillin (µg/g)		580 ± 45	282 ± 23
Eugenol (µg/g)		8.0 ± 1.0	4.0 ± 0.9
β-Methyl-γ-octolactones (µg/g)	Total	87.2 ± 27.3	149.0 ± 17.1
	cis Isomer	50.2 ± 24.1	25.0 ± 7.1
	trans Isomer	37.0 ± 4.5	124.2 ± 21.4

warm air has been tested in order to reduce the time and cost of the process. However, the drying process has a great impact on the quality of the wine because it determines to a large extent what the wood releases during aging (Polgue and Keller, 1973; Vivas, 2002; Cadahía et al., 2001, 2007; Doussot et al., 2002).

Table 9.4 analyzes the products released into a model wine synthetic solution by naturally seasoned and artificially dried pieces of Q. petraea from the Allier forest (Chatonnet, 1991; Chatonnet et al., 1994; Vivas, 1997, 2002).

These data clearly show that naturally seasoned wood releases lower levels of dry extract, A280, ellagitannins, and coumarins concentration than artificially dried wood. The lower presence of these molecules is probably due to rainfall repeatedly washing the staves during the natural seasoning process, thus solubilizing and eliminating them.

Another important aspect is that natural seasoning enables the hydrolysis of coumarin heterosides due to the development of certain fungi, (mainly *Trichoderma harzianum* and *Aureobasidium pullulans*) (Vivas et al., 1991). Since coumarin heterosides are more bitter than their corresponding aglycones (Ribéreau-Gayon et al., 1999b), this transformation is clearly beneficial to wine quality.

It must be also highlighted that natural seasoning results in higher levels of certain volatile molecules that affect wine aroma, such as vanillin and eugenol (Vivas and Glories, 1996b; Vivas et al., 1997). Natural seasoning means that the overall percentage of β-methyl-γ-octolactones is lower; however, the most aromatic of the two isomers (cis), is present at a higher concentration than it is after artificial drying, which in turn means that the final contribution of coconut notes will also be higher in naturally dried wood (Chatonnet et al., 1994).

In summary, artificially dried oak provides fewer aromas and more substances with a bitter and astringent taste. For this reason the cooperage industry generally prefers natural seasoning.

9.16 INFLUENCE OF TOASTING LEVEL

As has been mentioned, a large number of the molecules that are transferred from the oak barrel to the wine and that have a direct and positive effect on the wine's quality are produced or degraded during the process of toasting the staves. This makes toasting level a key point when selecting barrels. Moreover, the expertise of the toasting process is one of the criteria that sets apart the most renowned cooperages (Zamora, 2003a).

Several studies have reported the influence of toasting level on volatile substances (Chatonnet, 1999; Cadahía et al., 2003; Fernández de Simón et al., 2010; Chira and Teissedre, 2013a; Collins et al., 2015; Navarro et al., 2018) and ellagitannins (Cadahía et al., 2001; Doussot et al., 2002; Fernández de Simón et al., 2010; Sanz et al., 2011, 2012; Chira and Teissedre, 2013b; Navarro et al., 2016a). Fig. 9.13 shows the influence of toasting level on volatile compound concentration (A) and the theoretical sensory impact on the main aromas (B) (Navarro et al., 2018) of a red wine aged for 12 months in American (Q. alba) and French (Q. petraea) oak barrels with different toasting

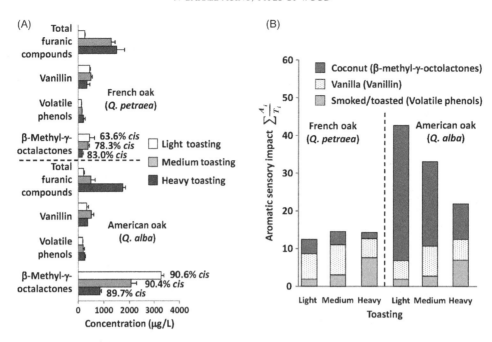

FIGURE 9.13 Influence of toasting level on volatile compounds concentration (A) and aromatic impact (B).

levels (light, medium, and heavy). Again, these sensory impacts were calculated by dividing the individual concentration of each volatile substance by its corresponding threshold.

Toasting level clearly affected all volatile substances in both species in a similar way. In short, β-methyl-γ-octalactones decreased when the toasting level was higher, whereas furanic compounds and volatile phenols did the opposite. Vanillin also increased between light and medium toasting but decreased with heavy toasting. These changes exerted by toasting level on volatile substances have a clear effect on the sensory impact of both oak species. Basically the coconut notes decrease as the toasting level increases whereas the smoked/toasted and vanilla notes increase, although in the case of heavy-toasting the vanilla aroma diminishes.

Fig. 9.14 shows the influence of toasting level on ellagitannin concentration (A) and its theoretical sensory impact on the astringency (B) (Navarro et al., 2018) of a red wine aged for 12 months in American (*Q. alba*) and French (*Q. petraea*) oak barrels with different toasting levels (light, medium, and heavy). The astringency impact was calculated on the basis of the astringency thresholds of the different ellagitannins (Glabasnia and Hofmann, 2006; Navarro et al., 2016a).

These results confirm that toasting level also exerts a significant influence on total ellagitannin concentration. In short, the data clearly indicate that the higher the toasting level the lower the total ellagitannin concentration in both oak species (this behavior was observed in all individual ellagitannins), which in turn means that the higher the toasting level, the lower the impact on wine astringency. This effect has also been reported elsewhere (Cadahía et al., 2001; Chira and Teissedre, 2015) and confirms that ellagitannins from oak heartwood are degraded or transformed by high temperatures, thus reducing their extractability (Peng et al., 1991; Chira and Teissedre, 2013a).

If these data are translated to the sensory level, the following conclusions can be drawn.

- Light-toasting releases high amounts of ellagitannins and therefore can indicate an excess of coconut notes, especially in the case of American oak (*Q. alba*), and an excess of astringency/structuration, especially in the case of French oak (*Q. petraea*).
- Medium-toasting provides a good balance between coconut, vanilla, and smoked/toasted notes. In the case of French oak (*Q. petraea*), aroma complexity is increased by spice notes and the structure of the wine is complemented by the release of ellagitannin.
- Heavy-toasting can provide too many smoked/toasted notes which can excessively mark the wine aroma and significantly diminish the impact of ellagitannins.

For all these reasons, most winemakers usually opt for medium-toasting because it confers the best sensorial balance to the wine.

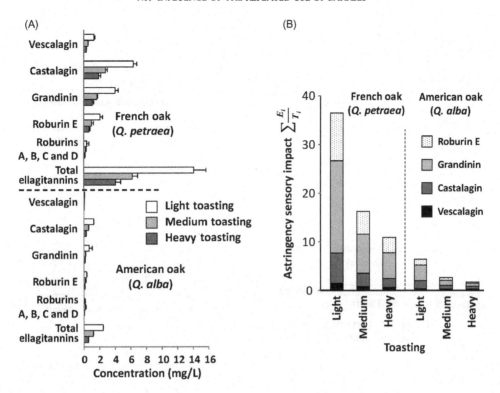

FIGURE 9.14 Influence of toasting level on ellagitannin concentration (A) and astringency impact (B).

9.17 INFLUENCE OF THE REPEATED USE OF BARRELS

Repeatedly using a barrel for successive batches of wine depletes the substances that it can release into the wine during aging and therefore diminishes its capacity to enrich the wine with aromas and ellagitannins (Ribéreau-Gayon et al., 1999a; Zamora, 2003a). The number of times a barrel has been used previously is, therefore, another key point considered by winemakers. However, almost all research into the volatile and nonvolatile substances contributed by oak wood barrels has been conducted with new barrels or with wood chip macerations and only a few studies have investigated the repeated use of barrels and the effects that this has on these substances (Vivas and Saint-Cricq de Gaulejac, 1998; Garde-Cerdán et al., 2002; Navarro et al., 2016a, 2018).

Figs. 9.15 and 9.16 show the volatile compound (15A) and ellagitannins (16A) concentrations and their theoretical sensory impact of the main aromas (15B) and astringency (16B) on a red wine aged for 12 months in new and 1-year used American (*Q. alba*) and French (*Q. petraea*) oak barrels with a medium-toasting level. Again, these sensory impacts were calculated by dividing the individual concentration of each substance by its corresponding threshold (Glabasnia and Hofmann, 2006; Navarro et al., 2018).

The results are clear and confirm that, for both French (*Q. petraea*) and American (*Q. alba*) oak, all volatile substances and ellagitannins and their corresponding sensory impact on the wine decrease drastically after one year of use due to their progressive depletion. In fact the total sensory aromatic impact decreases to less than half and the astringency impact decreases even more after the barrel has been used for 1 year.

These data agree with the limited existing information (Vivas and Saint-Cricq de Gaulejac, 1998; Garde-Cerdán et al., 2002) and confirm that after one year of use the capacity of the barrels to aromatize and give structure to wine significantly diminishes.

Another aspect to keep in mind is that the repeated use of barrels means that their pores become clogged by the gradual precipitation of tartrates, color matter, and other colloids, which in turn gradually diminishes their capacity for oxygenation (Vivas et al., 1992) and consequently their capacity to stabilize wine color and reduce wine astringency.

Since a barrel's capacity for oxygenation and to contribute substances to the wine decreases with repeated use, wineries need to ensure the regular turnover of their barrels. The turnover ratio will depend basically on the characteristics of the wine. In general, the more tannic the wine, the more new barrels are needed. However, excess use of new wood can be negative because it can mask other sensory attributes such as fruit notes and

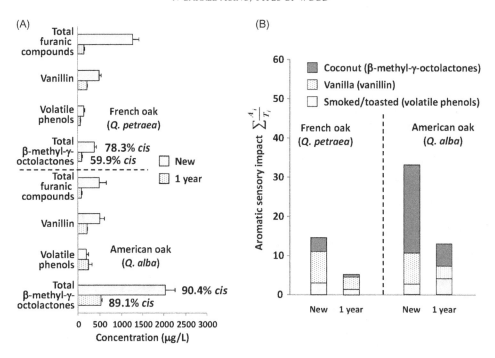

FIGURE 9.15 Influence of the repeated use of barrels on volatile compounds concentration (A) and aromatic impact (B).

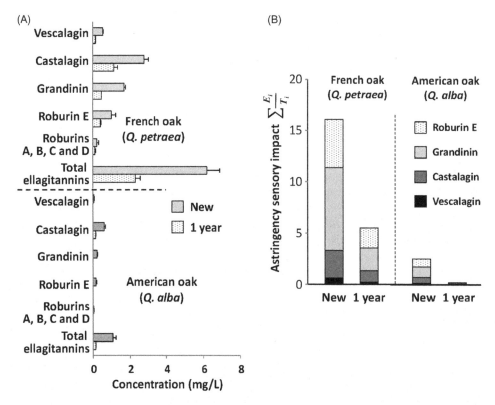

FIGURE 9.16 Influence of the repeated use of barrels on ellagitannins concentration (A) and astringency impact (B).

other varietal characteristics such as freshness, spicy notes, etc. Moreover, always using new barrels to age a wine can sometimes lead it to become too uniform and lose the characteristics of its terroir.

To summarize, every year wineries need to obtain a certain number of new barrels in order to preserve the desired characteristics of their wines but they must avoid excess use of new barrels to prevent wood notes from overpowering the wine's specific attributes. Moreover, the barrel turnover ratio will have a clear impact on production costs which will have a knock-on effect on the price of the wine.

9.18 BARREL AGING PROCESS

It is extremely difficult to establish a standard protocol for aging wine in barrels since it is conditioned by so many possibilities and variables. Even so, Fig. 9.17 shows a general protocol that could be applied to traditional barrel aging.

The barrel is an extremely difficult container to fill. Its shape hinders this operation by facilitating the formation of bubbles. For this reason it needs to be tapped with a rubber hammer to facilitate the escape of air bubbles. Moreover, as has been mentioned, wood absorbs wine, thus making it necessary to refill the barrels over the following days until the level of the wine remains stable. This operation is known by the French term of *ouillage*.

Traditionally, once they had been completely filled, barrels were turned in order to protect the wine against an excessive oxidation. This operation was obligatory when barrels were still sealed with wooden bungs wrapped in burlap but it is not necessary nowadays because modern silicon bungs ensure that the barrel remains airtight. Since turning barrels severely complicates several processes such as sampling and racking, nowadays almost all wineries work with silicon bungs which eliminate the need to turn the barrels.

Traditionally, wines were racked from barrel to barrel every three months. This operation eliminates the lees and causes an important exchange of gases between the wine and the atmosphere. Since lees can be a source of reduction fault and its presence favors the development of spoilage microorganisms, many wineries still systematically rack their wines. However, other wineries prefer to maintain the wines in contact with lees and only rack them if reduction or other problems appear. Racking barrel to barrel is not a simple operation and requires highly trained workers. Consequently, nowadays most wineries rack directly from the barrels to a tank and then return the wine to the barrels if further aging is necessary.

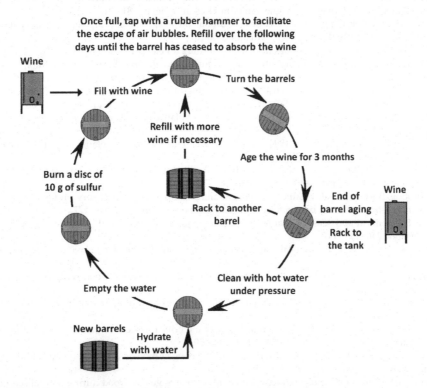

FIGURE 9.17 Traditional barrel aging process for wines.

The barrels, once emptied of wine, must be cleaned with hot water under pressure, and after removing the cleaning water, a disc of sulfur (10 g for 225 L barrels) should be burned to aseptise the wood. After that barrels are ready to be filled again with wine.

The number and frequency of rackings depends on several factors, but as a rule it is usually done every three months, which in theory would mean four times a year, although racking is usually omitted during the summer.

Finally, when the enologist considers that the wine has aged sufficiently, the wine is racked from the barrels to a tank in which the wine is homogenized and prepared for bottling.

Acknowledgments

I would like to thank the CICYT (Projects AGL2011-29708-C02-01, AGL2011-29708-C02-02, AGL2014-56594-C2-1-R, and AGL2014- 56594-C2-2-R) for its financial support. I am also grateful to "Boteria Torner" cooperage for providing the barrels used in some of the assays.

References

Abbott, N., Puech, J.L., Bayonove, C., Baumes, R., 1995. Determination of the aroma threshold of the *cis* and *trans* racemic forms of β-methyl-γ-octalactone by gas chromatography-sniffing analysis. Am. J. Enol. Vitic. 46, 292–294.
Boidron, J.N., Chatonnet, P., Pons, M., 1988. Influence du bois sur certaines substances odorantes des vins. Conn. Vigne Vin 22, 275–294.
Cadahía, E., Varea, S., Muñoz, L., Fernández de Simón, B., García-Vallejo, M.C., 2001. Evolution of ellagitannins in Spanish, French, and American oak woods during natural seasoning and toasting. J. Agric. Food Chem. 49, 3677–3684.
Cadahía, E., Fernández de Simón, B., Jalocha, J., 2003. Volatile compounds in Spanish, French, and American oak woods after natural seasoning and toasting. J. Agric. Food Chem. 51, 5923–5932.
Cadahía, E., Fernández de Simón, B., Vallejo, R., Sanz, M., Broto, M., 2007. Volatile compound evolution in Spanish oak wood (*Quercus petraea* and *Quercus pyrenaica*) during natural seasoning. Am. J. Enol. Vitic. 58, 163–172.
Camus, A., 1936–1954. Les Chênes. Monographie du Genre Quercus [et Lithocarpus]. Paul Lechevalier, París.
Castro-Vázquez, L., Alañón, M.E., Ricardo-da-Silva, J.M., Pérez-Coello, M.S., Laureano, O., 2013. Study of phenolic potential of seasoned and toasted Portuguese wood species (*Quercus pyrenaica* and *Castanea sativa*). J. Int. Sci. Vigne Vin 47, 311–319.
Chassaing, S., Lefeuvre, D., Jacquet, R., Jourdes, M., Ducasse, L., Galland, S., et al., 2010. Physicochemical studies of new anthocyano-ellagitannin hybrid pigments: about the origin of the influence of oak c-glycosidic ellagitannins on wine color. Eur. J. Org. Chem. 1, 55–63.
Chatonnet, P., 1991. Incidences du bois de chêne sur la composition chimique et les qualités organoleptiques des vins – Applications Technologiques. Thesis DER. Université de Bordeaux II, Bordeaux.
Chatonnet, P., 1992. Les composés aromatiques du bois de chêne cédés aux vins. Influence des opérations de chauffe en tonnellerie. In: Le bois et la qualité des vins et des eaux-de-vie. J. Inter. Sci. Vigne Vin núm Hors de série, pp. 81–91.
Chatonnet, P., 1995a. Influence des procédés de tonnellerie et des conditions d'élevage sur la composition et la qualité des vins élevés en fûts de chêne. Thèse de doctorat. Université de Bordeaux II, Bordeaux (France).
Chatonnet, P., 1995b. Principales origines et caracteristiques des chênes destinés à l'élevage des vins. Rev. Oenol 75, 15–18.
Chatonnet, P., 1999. Discrimination and control of toasting intensity and quality of oak wood barrels. Am. J. Enol. Vitic. 50, 479–494.
Chatonnet, P., Dubourdieu, D., 1998. Comparative study of the characteristics of American white oak (*Quercus alba*) and European oak (*Quercus petraea* and *Q. robur*) for production of barrels used in barrel aging of wines. Am. J. Enol. Vitic. 49, 79–85.
Chatonnet, P., Dubourdieu, D., Boidron, J.N., Pons, M., 1992. The origin of ethylphenols in wines. J. Sci. Food Agric. 60, 165–178.
Chatonnet, P., Dubourdieu, D., Boidron, J.N., Lavigne, V., 1993. Synthesis of volatile phenols by *Saccharomyces cerevisiae* in wines. J. Sci. Food Agric. 62, 191–202.
Chatonnet, P., Boidron, J.N., Dubourdieu, D., Pons, M., 1994. Evolution de certains composés volatils du bois de chêne au cours de son séchage premiers résultats. J. Int. Sci. Vigne Vin 28 (4), 359–380.
Chira, K., Teissedre, P.L., 2013a. Relation between volatile composition, ellagitannin content and sensory perception of oak wood chips representing different toasting processes. Eur. Food Res. Technol. 236, 735–746.
Chira, K., Teissedre, P.L., 2013b. Extraction of oak volatiles and ellagitannins compounds and sensory profile of wine aged with French winewoods subjected to different toasting methods: behaviour during storage. Food Chem. 140, 168–177.
Chira, K., Teissedre, P.L., 2015. Chemical and sensory evaluation of wine matured in oak barrel: effect of oak species involved and toasting process. Eur. Food Res. Technol. 240, 533–547.
Climent, J., Gil, L., Pardos, J.A., 1998. Xylem anatomical traits related to resinous heartwood formation in *Pinus canariensis* Sm. Trees 12, 139–145.
Collins, T.S., Miles, J.L., Boulton, R.B., Ebeler, S.E., 2015. Targeted volatile composition of oak wood samples taken during toasting at a commercial cooperage. Tetrahedron 71, 2971–2982.
del Alamo-Sanza, M., Nevares, I., 2014. Recent advances in the evaluation of the oxygen transfer rate in oak barrels. J. Agric. Food Chem. 62, 8892–8899.
del Alamo-Sanza, M., Nevares, I., 2017. Oak wine barrel as an active vessel: a critical review of past and current knowledge. Crit. Rev. Food Sci. Nutr. Available from: https://doi.org/10.1080/10408398.2017.1330250.
del Alamo-Sanza, M., Cárcel, L.M., Nevares, I., 2017. Characterization of the oxygen transmission rate of oak wood species used in cooperage. J. Agric. Food Chem. 65, 648–655.
Doussot, F., de Jéso, B., Quideau, S., Pardon, P., 2002. Extractives content in cooperage oak wood during natural seasoning and toasting; influence of tree species, geographic location, and single-tree effects. J. Agric. Food Chem. 50, 5955–5961.

REFERENCES

Fazzalari, F.A., Stahl, W.H., 1978. Compilation of Odor and Taste Threshold Data, ASTM Data Series. American Society for Testing and Materials, Philadelphia, PA, USA.

Fernández de Simón, B., Cadahía, E., Conde, E., García-Vallejo, M.C., 1999. Ellagitannins in woods of Spanish, French and American oaks. Holzforschung 53, 147–150.

Fernández de Simón, B., Cadahía, E., Sanz, M., Poveda, P., Pérez-Magariño, S., Ortega-Heras, M., et al., 2008. Volatile compounds and sensorial characterization of wines from four Spanish denominations of origin aged in Spanish rebollo (Quercus pyrenaica Wild.) oak wood barrels. J. Agric. Food Chem. 56, 9046–9055.

Fernández de Simón, B., Esteruelas, E., Muñoz, A., Cadahía, E., Sanz, M., 2009. Volatile Compounds in acacia, chestnut, cherry, ash, and oak woods with a view to their use in cooperage. J. Agric. Food Chem. 57, 3217–3227.

Fernández de Simón, B., Cadahía, E., Muiño, I., del Álamo, M., Nevares, I., 2010. Volatile composition of toasted oak chips and staves and of red wine aged with them. Am. J. Enol. Vitic. 61, 157–165.

Feuillat, F., Huber, F., Keller, R., 1992. Mise au point sur: "La notion du grain utilisée pour le classement des merrains de chêne". Rev. Fran. Œnol 32, 65–69.

Feuillat, F., Keller, R., Masson, G., Puech, J.L., 1998. Bois de chêne. In: Flancy, C. (Ed.), Œnologie: Fondements scientifiques et technologiques. Lavoisier, París, pp. 1002–1027.

Fors, S., 1998. Sensory properties of volatile Maillard reaction products and related compounds. In: Waller, G.R., Feather, M.S. (Eds.), The Maillard Reaction in Foods and Nutrition. ACS Symposium Series, 215. American Chemical Society, Washington, DC, USA, pp. 185–286.

Garde-Cerdán, T., Ancín-Azpilicueta, C., 2006. Review of quality factors on wine ageing in oak barrels. Trends Food Sci. Technol. 17, 438–447.

Garde-Cerdán, T., Rodríguez, S., Ancín-Azpilicueta, C., 2002. Volatile composition of aged wine in used barrels of French oak and American oak. Food Res. Int. 35, 603–610.

Garde-Cerdán, T., Torrea-Goñi, D., Ancín-Azpilicueta, C., 2004. Accumulation of volatile compounds during ageing of two red wines with different composition. J. Food Eng. 65, 349–356.

Glabasnia, A., Hofmann, T., 2006. Sensory-directed identification of taste-active ellagitannins in American (Quercus alba L.) and European oak wood (Quercus robur L.) and quantitative analysis in bourbon whiskey and oak-matured red wines. J. Agric. Food Chem. 54, 3380–3390.

Gombau, J., Vignault, A., Pascual, O., Canals, J.M., Teissedre, P.L., Zamora, F., 2016. Influence of supplementation with different oenological tannins on malvidin-3-monoglucoside copigmentation. In: BIO Web of Conferences. vol. 7, pp. 02033.

Guth, H., 1997. Quantitation and sensory studies of character impact odorants of different white wine varieties. J. Agric. Food Chem. 45, 3027–3032.

Jones, E.T., 2016. Inside the Illicit Economy; Reconstructing the Smuggler's Trade of Sixteenth Century Bristol. Routledge Taylor and Francis Groups, New York, NY.

Jordão, A.M., Ricardo-da-Silva, J.M., Laureano, O., 2005. Extraction and evolution of some ellagic tannins and ellagic acid of oak wood chips (Quercus pyrenaica) in model wine solutions: effect of time, pH, temperature and alcoholic content. S. Afr. J. Enol. Vitic. 26, 83–89.

Jourdes, M., Lefeuvre, D., Quideau, S., 2009. C-Glycosidic ellagitannins and their influence on wine chemistry. In: Quideau, S. (Ed.), Chemistry and Biology of Ellagitannins; An Underestimated Class of Bioactive Plant Polyphenols. World Scientific Publishing Co, Singapore, pp. 320–365.

Keller, R., 1987. Differentes varietes de chêne et leur repartition dans le monde. Connaiss. Vigne Vin 21, 191–229.

Kelly, M., Wollan, D., 2003. Micro-oxygenation of wine in barrels. Aust. N. Z. Grapegrower Winemaker 473, 29–32.

Lacoste, P., Castro, A., Briones, F., Mujica, F., 2015. The Pipeño: story of a typical wine Southern Central Valley of Chile. Idesia 33, 87–96.

López, R., Aznar, M., Cacho, J., Ferreira, V., 2002. Determination of minor and trace volatile compounds in wine by solid-phase extraction and gas chromatography with mass spectrometric detection. J. Chromatogr. A 966, 167–177.

Macías, M.M., García Manso, A., García Orellana, C.J., González Velasco, H.M., Gallardo Caballero, R., Peguero Chamizo, J.C., 2013. Acetic acid detection threshold in synthetic wine samples of a portable electronic nose. Sensors (Basel) 13, 208–220.

Manos, P.S., Doyle, J.J., Nixon, K.C., 1999. Phylogeny, biogeography, and processes of molecular differentiation in Quercus subgenus Quercus (Fagaceae). Mol. Phylogenet. Evol. 12, 333–349.

Masson, G., Moutounet, M., Puech, J.L., 1995. Ellagitannin content of oak wood as a function of species and of sampling position in the tree. Am. J. Enol. Vitic. 46, 262–268.

Masson, G., Guichard, E., Fournier, N., Puech, J.L., 1997. Teneurs en stéréo-isomeres de la β-metil-γ-octalactone des bois de chêne européens et américains. Application aux vins et aux eaux-de-vie. J. Sci. Tech. Tonnellerie 3, 1–8.

Mazza, G., Brouillard, R., 1990. The mechanism of co-pigmentation of anthocyanins in aqueous solutions. Phytochemistry 29, 1097–1102.

Michel, J., Jourdes, M., Silva, M.A., Giordanengo, T., Mourey, N., Teissedre, P.L., 2011. Impact of concentration of ellagitannins in oak wood on their levels and organoleptic influence in red wine. J. Agric. Food Chem. 59, 5677–5683.

Mosedale, J.R., Puech, J.L., Feuillat, F., 1999. The influence on wine flavor of the oak species and natural variation of heartwood components. Am. J. Enol. Vitic. 50, 503–512.

Moutounet, M., Rabier, P.H., Puech, J.L., Verette, E., Barillere, J.M., 1989. Analysis by HPLC of extractable substances in oak wood. Application to a Chardonnay wine. Sci. Aliment. 9, 35–51.

Moutounet, M., Mazauric, J.P., Saint-Pierre, B., Hanocq, J.F., 1998. Gaseous exchange in wines stored in barrels. J. Sci. Technol. Tonnellerie 4, 131–145.

Moutounet, M., Puech, J.L., Keller, R., Feuillat, F., 1999. Les caractéristiques du bois de chêne en relation avec son utilisation en oenologie; Le phénomène de duramisation et ses conséquences. Rev. Fr. Oenol. 174, 12–17.

Navarro, M., Kontoudakis, N., Gómez-Alonso, S., García-Romero, E., Canals, J.M., Hermosín-Gutíerrez, I., et al., 2016a. Influence of the botanical origin and toasting level on the ellagitannin content of wines aged in new and used oak barrels. Food Res. Int. 87, 197–203.

Navarro, M., Kontoudakis, N., Giordanengo, T., Gómez-Alonso, S., García-Romero, E., Fort, F., et al., 2016b. Oxygen consumption by oak chips in a model wine solution; influence of the botanical origin, toast level and ellagitannin content. Food Chem. 199, 822–827.

Navarro, M., Kontoudakis, N., Gómez-Alonso, S., García-Romero, E., Canals, J.M., Hermosín-Gutíerrez, I., et al., 2018. Influence of the volatile substances released by oak barrels into a Cabernet Sauvignon red wine and a discolored Macabeo white wine on sensory appreciation by a trained panel. Eur. Food Res. Technol. 244, 245–258.

Nevares, I., del Alamo-Sanza, M., 2015. Oak stave oxygen permeation: a new tool to make barrels with different wine oxygenation potentials. J. Agric. Food Chem. 63, 1268–1275.

Nevares, I., Crespo, R., González, C., del Alamo-Sanza, M., 2014. Imaging of oxygen permeation in the oak wood of wine barrels using optical sensors and a colour camera. Aust. J. Grape Wine Res. 20, 353–360.

Nixon, K.C., 2006. Global and neotropical distribution and diversity of oak (genus *Quercus*) and oak forests. Ecological Studies. In: Kappelle, M. (Ed.), Ecology and Conservation of Neotropical Montane Oak Forests, Vol. 185. Springer-Verlag Berlin Heidelberg, New York, NY, pp. 4–13.

Ohloff, G., 1978. Recent developments in the field of naturally occurring aroma components. Prog. Chem. Org. Nat. Prod. 35, 431–527.

Okuda, T., Yoshida, T., Hatano, T., Ito, H., 2009. Ellagitannins renewed the concept of tannins. In: Quideau, S. (Ed.), Chemistry and Biology of Ellagitannins; An Underestimated Class of Bioactive Plant Polyphenols. World Scientific Publishing Co, Singapore, pp. 1–54.

Oldfield, S., Eastwood, A., 2007. The Red List of Oaks. Fauna & Flora International, Cambridge.

Oxford English Dictionary, 2009. Oxford University Press, Oxford.

Pascual, O., Vignault, A., Gombau, J., Navarro, M., Gómez-Alonso, S., García-Romero, E., et al., 2017. Oxygen consumption rates by different oenological tannins in a model wine solution. Food Chem. 234, 26–32.

Paterson, A., Piggott, J., 1994. Understanding Natural Flavors. Springer US, New York, NY, USA.

Peng, S., Scalbert, A., Monties, B., 1991. Insoluble ellagitannins in *Castanea sativa* and *Quercus petraea* woods. Phytochemistry 30, 775–778.

Peterson, R.G., 1976. Formation of reduced pressure in barrels during wine aging. Am. J. Enol. Vitic. 27, 80–81.

Pino, J.A., Queris, O., 2011. Characterization of odor-active compounds in guava wine. J. Agric. Food Chem. 59, 4885–4890.

Pittet, A., Rittersbacher, P., Muralidhara, R., 1970. Flavor properties of compounds related to maltol and isomaltol. J. Agric. Food Chem. 18, 929–933.

Polgue, H., Keller, R., 1973. Qualité du bois et longueur d'acroissement en forêt de Tronçais. Ann. Sci. For. 30, 91–125.

Prida, A., Puech, J.L., 2002. Ellagitannins in differents oak species. Vinodel Vinograd 5, 24–25.

Prida, A., Chatonnet, P., 2010. Impact of oak-derived compounds on the olfactory perception of barrel-aged wines. Am. J. Enol. Vitic. 61, 408–413.

Puech, J.L., Moutounet, M., 1988. Liquid chromatographic determination of scopoletin in hydroalcoholic extract of oak wood and in distilled alcoholic beverages. J. Assoc. Anal. Chem. 71, 512–514.

Quideau, S., Varadinova, T., Karagiozova, D., Jourdes, M., Pardon, P., Baudry, C., et al., 2004. Main structural and stereochemical aspects of the antiherpetic activity of nonahydroxyterphenoyl-containing C-glycosidic ellagitannins. Chem. Biodivers. 1, 247–258.

Ribéreau-Gayon, J., 1933. Contribution à l'étude des oxydations et reductions dans les vins. Ph.D. Thesis. Université de Bordeaux, Bordeaux.

Ribéreau-Gayon, P., Glories, Y., Maujean, A., Dubourdieu, D., 1999a. Aging red wines in vat and barrel: phenomena occurring during aging, Handbook of Enology, Vol 2. John Wiley & Sons, Ltd, Chichester, pp. 387–428.

Ribéreau-Gayon, P., Glories, Y., Maujean, A., Dubourdieu, D., 1999b. Phenolic Compounds. Hand Book of Enology, Vol 2, The Chemistry of Wine, Stabilization and Treatmets. John Wiley & Sons, Ltd, Chichester, pp. 129–186.

Remy, B., 1991. Silviculture, cooperage and oenology: good cask wood for good grapes. Rev. For. Fr. 43, 290–300.

Romano, A., Perello, M.C., Lonvaud-Funel, A., Sicard, G., de Revel, G., 2009. Sensory and analytical re-evaluation of ''Brett character''. Food Chem. 114, 15–19.

Salagoity-Auguste, M.H., Tricard, C., Sudraud, P., 1987. Dosage simultané des aldéhydes aromatiques et des coumarines par chromatographie liquide haute performance. J. Chromatogr. 392, 379–387.

Sanz, M., Fernández de Simón, B., Esteruelas, E., Muñoz, A., Cadahía, E., Hernández, T., et al., 2011. Effect of toasting intensity at cooperage on phenolic compounds in acacia (*Robinia pseudoacacia*) heartwood. J. Agric. Food Chem. 59, 3135–3145.

Sanz, M., Fernández de Simón, B., Cadahía, E., Esteruelas, E., Muñoz, A., Hernández, T., et al., 2012. LC-DAD/ESI-MS/MS study of phenolic compounds in ash (*Fraxinus excelsior* L. and *F. americana* L) heartwood. Effect of toasting intensity at cooperage. J. Mass Spectrom. 47, 905–918.

Scalbert, A., Monties, B., Favre, J.M., 1988. Polyphenols of *Quercus robur*: adult tree and in vitro grown calli and shoots. Phytochemistry 27, 3483–3488.

Seifert, R.M., Buttery, R.G., Guadagni, D.G., Black, D.R., Harris, J.G., 1970. Synthesis of some 2-methoxy-3-alkylpyrazines with strong bell pepper-like odors. J. Agric. Food Chem. 18, 246–249.

Singleton, V.L., 1974. Some aspects of the wooden container as a factor in wine maturation, Chemistry of Winemaking. Advances in Chemistry, Volume 137. American Chemical Society, Whasington DC, pp. 254–277.

Spillman, P.J., Sefton, M.A., Gawel, R., 2004. The effect of oak wood source, location of seasoning and coopering on the composition of volatile compounds in oak-matured wines. Aust. J. Grape Wine Res. 10, 216–226.

Stamp, C., August 2015. The Economics of Wine Barrels; how to determine the effect of barrel choices on profits. Wines & Vines 96, 54–57.

Sterckx, F.L., Missiaen, J., Saison, D., Delvaux, F.R., 2011. Contribution of monophenols to beer flavour based on flavour thresholds, interactions and recombination experiments. Food Chem. 126, 1679–1685.

Taransaud, J., 1976. Le Livre de la tonnellerie. La roue à livres diffusión, Paris.

Vivas, N., 1995. Sur la notion de grain en tonnellerie. J. Sci. Technol. Tonnellerie 1, 17–32.

Vivas, N., 1997. Recherches sur la qualité de chêne français de tonnellerie et sur les méchanismes d'oxidoréducction des vins rouges au cours de leur élevage en barriques. PhD Thesis. Université de Bordeaux II, Bordeaux.

Vivas, N., 2002. Manuel de tonnellerie à l'usage des utilisateurs de futaille. Féret Éditions, Bordeaux.

Vivas, N., Glories, Y., 1996a. Role of oak wood ellagitannins in the oxidation process of red wines during aging. Am. J. Enol. Vitic. 47, 103–107.

Vivas, N., Glories, Y., 1996b. Étude et optimisation des phénomènes impliqués dans le séchage natural du bois de chêne. Rev. Fran. Œnol. 158, 28–35.

Vivas, N., Glories, Y., 1997. Modélisation et calcul du bilan des apports d'oxygène au cours de l'élevage des vins rouges. II. Les apports liés au passage d'oxygène au travers de la barrique. Prog. Agric. Vitic 114, 315–316.

Vivas, N., Glories, Y., Doneche, B., Gueho, E., 1991. Observation sur la flore fongique du bois de chêne au cours de son sechage naturel. Ann. Sci. Nat. Bot. 13, 149–153.

Vivas, N., Zamora, F., Glories, Y., 1992. Etude des phénomenes d'oxydorreduction dans les vins. Mise au point d'une méthode rapide de mesure du potentiel d'oxydorreduction. J. Int. Sci. Vigne Vin 26, 271–285.

Vivas, N., Saint Cricq de Gaulejac, N., Doneche, B., Glories, Y., 1997. The duration effect of natural seasoning of *Quercus petraea* Liebl. and *Quercus robur* L. on the diversity of existing fungi flora and some aspects of its ecology. J. Sci. Technol. Tonnellerie 3, 27–35.

Vivas, N., Saint-Cricq de Gaulejac, N., 1998. Influence de la durée d'utilisation des barriques sur leurs apports aux vins. Actes du Colloque Sciences et Techniques de la Tonnellerie. Vigne et Vin publications Internationales, Bordeaux, pp. 65–74.

Vivas, N., Absalon, C., Benoist, F., Vitry, C., Grazillier, S., De Revel, G., et al., 2001. Les chênes européens *Q. Robur* et *Q. Petraea*: Analyse des potentialités œnologiques des différents massifs forestiers. In: Connaissances actuelles & Avenir de l'élevage en barriques. Bordeaux, pp 31–37.

Vivas, N., Debeda, H., Menil, F., Vivas de Gaulejac, N., Nonier, M.F., 2003. Mise en evidence du passage de l'oxygène au travers des douelles constituant les barriques par l'utilisation d'un dispositif original de mesure de la porosité du bois. Premiers résultats. Sci. Aliment. 23, 655–678.

Wilkinson, K.L., Elsey, G.M., Prager, R.H., Tanaka, T., Sefton, M.A., 2004. Precursors to oak lactone. Part 2. Synthesis, separation and cleavage of several β-D-glucopyranosides of 3-methyl-4-hydroxyoctanoic acid. Tetrahedron 60, 6091–6100.

YanLong, L., XueYan, L., JiuYin, P., YanJie, L., 2010. Production process and performances of oak barrel made of *Quercus mongolica* wood from Changbai Mountains. J. NE For. Univ. 38, 123–127.

Zamora, F., 2003a. Elaboración y crianza del vino tinto; aspectos científicos y prácticos. Editorial Mundi-Prensa; AMV Ediciones, Madrid.

Zamora, F., 2003b. El concepto de grano en tonelería; un criterio para clasificar el roble. Enólogos 24, 24–28.

Zhang, B., Cai, J., Duan, C.Q., Reeves, M.J., He, F., 2015. A review of polyphenolics in oak woods. Int. J. Mol. Sci. 16, 6978–7014.

Further Reading

Prida, A., Puech, J.L., 2006. Influence of geographical origin and botanical species on the content of extractives in American, French, and east European oak woods. J. Agric. Food Chem. 54, 8115–8126.

CHAPTER 10

Emerging Technologies for Aging Wines: Use of Chips and Micro-Oxygenation

Encarna Gómez-Plaza and Ana B. Bautista-Ortín

Department of Food Science and Technology, University of Murcia, Murcia, Spain

10.1 WHY AGING WINES IN BARRELS?

In the past, barrels and vats with a volume of 5–10 hL were commonly used. They acted as fermentor vessels as well as storage cooperage following fermentation. However, nowadays, these wooden tanks have been replaced by more durable and easily cleaned tanks such as those made from stainless steel, epoxy-lined carbon steel, fiberglass, and concrete. The wooden tanks have now, as the main objective, the maturation of wines.

Experience has shown that aging wines, especially red wines, in wood barrels improves their quality and organoleptic characteristics. After a period of oak barrel maturation, the wines are enriched in aromatic substances, the color is more stable, and mouthfeel complexity is improved.

The increased color stability is due to the small amounts of air that penetrate through the wood pores and bunghole (Pérez Prieto et al., 2002). Moutounet et al. (1998) explained that the walls of wine barrels act as semipermeable "membranes," allowing oxygen gas in the atmosphere outside the barrel to permeate through the barrel walls and diffuse into the wine. This osmotic exchange is driven by the partial pressure difference of the atmospheric oxygen outside the barrel and the low partial pressure of oxygen inside, a low pressure originated by the continuous consumption of oxygen by the wine. Del Alamo-Sanza and Nevares (2014) calculated that the total dose of oxygen that wine would receive in a fine-grain American oak barrel was 11.62 mg/L per year, values lower than the results that were published by other authors, such as Ribéreau-Gayon (1933) who observed an oxygen entry rate of 15 – 20 mL/L per year, and Vivas and Glories (1997) who observed an oxygen entry rate of 20 mg/L per year in two new Limousin barrels from central France. Differences can be due to the methodology used by these authors, which considered oxygen transfer rate (OTR) as a constant rate with time and extrapolated values obtained after controlling the OTR for only 6 months or less.

How does this limited oxygenation improve wine color stability? It does so by participating in reactions with polyphenols. Anthocyanins are the most significant components responsible for the purple-red color of young wines. However, they are unstable and participate in reactions during winemaking and aging to form more complex and stable pigments, which mainly arise from the interaction between anthocyanins and other phenolic compounds, especially flavan-3-ols. These processes are favored by the presence of oxygen, especially the reactions where acetaldehyde is involved, such as the formation of adducts linked by an ethyl bridge (Cano-López et al., 2006; Saucier, Little, and Glories, 1997) or the formation of pyranoanthocyanins through the reaction between anthocyanins and other compounds, such as flavanols, vinylphenols, and piruvic acid (Cano-López et al., 2006; Fulcrand et al., 2004; Romero and Bakker, 2000). Moreover, these changes in phenolic compounds, especially when complex flavan-3-ols are involved, also positively affect mouthfeel and astringency.

It is well known that not only color and mouthfeel are improved after wood barrel aging. The complexity of the aroma is increased due to the extraction of compounds present in the wood, which are transferred to wine during the aging period (Pérez Prieto et al., 2002). Oak wood is mainly composed of large insoluble polymers such as cellulose, hemicellulose, and lignin but also contains other compounds of lower molecular weight, such as acids, sugars, terpenes, volatile phenols, and lactones, which can be extracted into wine during storage (Arapitsas et al., 2004; Maga, 1984; Nykänen, 1986).

From a sensory point of view, among the most important compounds released from the wood are the oak lactones, cis and trans-β-methyl-γ-octalactones (Masson et al., 1996; Moutounet et al., 1999), with a woody aroma also described as coconut and present in the green wood; guaiacol and 4-methylguaiacol, with their smoky aroma (Pollnitz, 2000; Weeks and Sefton, 1999), and formed during the toasting process of the wood from lignin degradation; vanillin, a compound normally present in green wood (Pollnitz, 2000); the furfuryl compounds (furfural and 5-methylfurfural), whose almond aroma is formed through the degradation of cellulose and hemicellulose during the toasting process (Chatonnet et al., 1989, 1990; Cutzach et al., 1999); and eugenol, which possesses a spicy, clove-like aroma and is present in raw oak, although it is reported to increase during open-air wood seasoning and with toasting level (Pérez-Coello and Diaz-Maroto, 2009).

Ellagitannins are also important compounds extracted from wood. These are polymeric compounds derived from ellagic acid (ellagitannins) or gallic acid (gallotannins). They are potent antioxidants and protect wine from oxidation (Navarro et al., 2016). These molecules are soluble in an hydroalcoholic environment; that is why they are gradually extracted by wine during its aging in oak barrels. Although present in raw wood, the processing of wood in cooperage, i.e., the type and length of drying and toasting, also has an influence (Michel et al., 2011).

However, we cannot forget that there is a negative side to the use of barrels for wine aging. Barrels are expensive and with a limited time of use. Moreover, in wines aged in barrels, mainly if the barrels have been reused, ethylphenols could be formed and these are undesirable compounds for wine quality since they confer unpleasant odors to wine (Garde-Cerdan and Ancin-Azpilicueta, 2006). These compounds have a microbiological origin; some yeasts capable of contaminating wood (*Brettanomyces/Dekkera* genera) decarboxylate cinnamic acids and form these phenols in wines (Chatonnet et al., 1992; Suarez et al., 2007).

All these facts give some idea of the lights and shadows regarding wine aging in wood and the large number of factors that may have an influence on the final quality of those wines. Therefore, and mainly for economic reasons, alternatives to the oak barrel are being looked at to carry out the wine-aging process.

10.2 THE MICRO-OXYGENATION TECHNIQUE

One of these alternatives is the use of the technique called micro-oxygenation (MOX), a technique that tries to mimic the small oxygenation that occurs during wine oak barrel aging, although that can also be used to improve wine quality. Nowadays, it has become a powerful tool for winemakers to produce color-stable wines and it may reduce the length of the winemaking process.

As stated before, the effect of oxygen on the sensorial quality of red wines has been recognized to be positive at low levels, improving the sensorial characteristics of the wine, and negative at high levels, since it is responsible for the oxidation of phenolic and volatile compounds (Arapitsas et al., 2012).

For the introduction of these small and controlled amounts of oxygen into wines MOX needs the use of specialized equipment for a fine control of the oxygen doses applied (Cano-López and Gómez-Plaza, 2011; Parish et al., 2000; Paul, 2002). It normally consists of a ceramic or stainless steel sparger that produces small bubbles, which can dissolve in the wine (Cano-López and Gómez-Plaza, 2011; Du Toit et al., 2006b). This is achieved by filling a camera with a known volume with gas at a high pressure. The volume is then transferred via a low-pressure circuit to the diffuser and into the wine. Something very important in the MOX technique is that the oxygen should be introduced at a rate equal to or less than the oxygen uptake rate of the wine (Paul, 2002) so there is no accumulation of dissolved oxygen. It is for this reason that the success of MOX depends strongly on controlling the rate of oxygen exposure (small doses are normally used, ranging from 2 to 90 mg of O_2/L of wine/month, depending on the type of wine and the stage of the winemaking process (Cano-López and Gómez-Plaza, 2011; Dykes, 2007). Also, the dimensions of the tanks where the wine is being microoxygenated are very important since the height should be enough to allow the dissolution of the oxygen bubbles (Fig. 10.1).

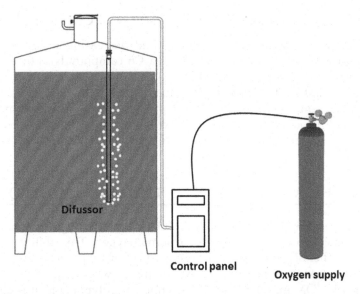

FIGURE 10.1 Scheme of a micro-oxygenation system.

10.3 POSITIVE FACTORS OF USING MICRO-OXYGENATION

10.3.1 Incidence on Yeast Development During Alcoholic Fermentation

Oxidative stress is one of the fundamental causes of early death of yeasts during fermentation. A correct addition of oxygen may confer upon yeasts a higher tolerance to ethanol and a higher fermentative activity, decreasing the formation of sulfur compounds and promoting a faster end to fermentation (Du Toit et al., 2006b).

Yeasts, in general, need 8–10 mg/L of oxygen during the initial phase of yeast growth for optimal development. At the beginning of fermentation, must is saturated with oxygen and no oxygen additions are needed. However, at the end of the growing phase, yeasts must/wine can be depleted of oxygen and this could be a good moment for oxygen addition. However, with an ethanol content around 10%, oxygen should not be added since yeasts are no longer taking nutrients nor oxygen and the presence of oxygen could impact wine quality. A correct combination of oxygen and ammonium salts could avoid stuck fermentations (Du Toit et al., 2006b).

10.3.2 Wine Chromatic Characteristics and Stability

As stated before, the phenolic compounds are one of the most important wine constituents due to their high concentration and their role in the organoleptic characteristics of wines (Cheynier et al., 1997; Kennedy, 2008; Ribéreau-Gayon et al., 1998; Ribéreau-Gayon and Glories, 1987). Oxygen favors the reactions involving phenolic compounds that leads to the formation of new anthocyanin-derived and polymeric compounds that stabilize wine color since they are partially resistant to SO_2 bleaching and more stable at wine pH. The formation, relative importance, and final structure of these new pigments, not only depend on the initial phenolic composition of wine but also to the presence of some yeast metabolites (aldehydes, pyruvic acid, etc.) and oxygen exposure so their formation is favored by MOX (Cano-López and Gómez-Plaza, 2011).

10.3.3 Improvement of Astringency and Mouthfeel

Different phenolic compounds are involved in the red wine bitterness, astringency, and complexity, with the flavan-3-ols being the main compounds (Vidal and Aagaard, 2008). They are very reactive compounds. Low-molecular-weight flavan-3-ols are mainly bitter and high-molecular-weight polymers (tannins) are mainly astringents (Vidal et al., 2004; Vidal and Aagaard, 2008). These tannins suffer depolymerization and repolymerization reactions during aging, due to the low pH of wine. These reactions may occur either in the presence or the absence of oxygen resulting in different structures and those formed in the presence of oxygen are less reactive towards proteins and thus, less astringent (Tanaka et al., 1994).

10.3.4 Improvement of Wine Aroma and Reduction of Vegetal Characteristics

MOX may improve wine aroma, increasing fruity notes and decreasing vegetal notes. This decrease is not related with the typical herbaceous compounds (the so-called C6 compounds), however, the reduction of other compounds such as pyrazines and thiols are oxidized by oxygen and could contribute to the herbaceous character of wines (Parpinello et al., 2011). MOX also could help to solve reduction odors, attributed to sulfur compounds (Jackson, 2000). Oxygen could help to control this problem, since during fermentation it could limit the impact of sulfur compounds by protecting yeast health and increasing wine redox potential. Thus, compounds such as thiols will be oxidized to compounds with a much higher detection threshold, such as disulfites.

10.4 THE APPLICATION OF THE MOX TECHNIQUE

One important point when deciding the application of MOX is when to apply it and the correct doses to apply. Wines present marked differences in their abilities to consume oxygen. This ability is generally related to their relative concentrations of polyphenols, since phenolic compounds are the main consumers of oxygen (Devatine et al., 2007), thus a red wine can consume much more oxygen than a white wine.

Since most of the benefits of MOX are related to the reactions of polyphenols, it is clear that the initial phenol content of the wine is one of the most important conditions at the time of designing a MOX treatment. Those wines with high phenolic content will be easily improved by the MOX treatment and much more care should be taken when micro-oxygenating a wine with a low phenolic content since overtreatment can occur fairly quickly with low-colored wines (Du Toit, 2007). The results of Cano-López et al. (2008b) showed how different wines were differently affected by MOX. The formation of anthocyanin-derived pigments was much more favored in wines with high total phenol content, these compounds being responsible for the higher color intensity of the treated wines. Moreover, among wines with high polyphenol content, those with the highest quantities of free anthocyanins were those that led to the best results. The wine with the lowest phenolic content was less influenced by MOX (Cano-López et al., 2008b), moreover, the MOX of a wine with few free anthocyanins will favor a quite large polymerization of tannins. This large polymerization of tannins in these wines was also detected in the work of Arapitsas et al. (2012) who could demonstrate that the application of MOX after malolactic fermentation (MLF), due to low quantities of free anthocyanins at this moment of winemaking, led to an easier polymerization of tannins, which, in turn, could lead to dryness and browning.

As regard the doses of oxygen to be applied, it clearly will depend on the moment of application and on the capacity of the wine (its polyphenol content) to uptake oxygen. The dose of oxygen recommended for improving wine's phenolic structure before MLF is around 10–30 mL/L/month, always depending on the wine phenolic content (Cano-López et al., 2006; Cejudo-Bastante et al., 2011a, 2011b), and post-MLF, the doses may range from 1 to 5 mL/L/month, during 1–4 months (Cano-López et al., 2006; Ertan Anli and Cavuldak, 2012; Lesica and Kosmerl, 2009).

We should not forget the role of temperature, since temperature also affects oxygen solubility. If MOX is applied when the temperature is too low, below 15°C, the oxygen solubility increases but the speed of the different chemical reactions slow down, thus oxygen may accumulate in the wine. Too high temperatures may lead to poor oxygen solubility (Cano-López and Gómez-Plaza, 2011).

Important care should be taken to prevent an excess of oxygen, which can lead to the formation of large molecules of high molecular weight that are unable to stay in solution, causing the precipitation of polymeric material and leaving the wines with a reduced color intensity as Cano-López et al. (2006) demonstrated.

Nowadays, it is easier to determine the key compounds that could help to monitor the application of MOX. The advances in the use of metabolomics allow the determination of how MOX affects the wine composition and to find biomarkers of micro-oxygenated wines. As it was expected, the results of Arapitsas et al. (2012) determined that different phenolic compounds were candidate biomarkers. They could demonstrate that MOX had an influence not only on the grape anthocyanins but also on the pigments formed during wine aging, such as vitisins, vinylphenol-pyranoanthocyanins (products of the reaction between anthocyanins and cinnamic acid derivatives, e.g., malvidin 3-glucoside 4-vinylcatechol), and the products of direct or indirect reactions between anthocyanins and flavanols. In total the authors identified 58 pigments as candidate biomarkers. Previously, it was stated that pre-MLF application of MOX had a higher influence on phenolic compounds (Cano-López et al., 2008b) and this was coincident with these results, the number was higher when MOX was applied in pre-MLF, and most of the pigments belonged to the group of flavanol-anthocyanin (direct and ethyl-bridged) derivatives, the vitisin group was the second highest, and the pinotin group was the third highest. They also stated that tannins are another class of compounds found among candidate biomarkers, with a total of 38 condensed tannins being found among the candidate biomarkers. They could also demonstrate that for the application of MOX after MLF, due to lower quantities of free anthocyanins, the polymerization of tannins became easier. In the same study, other compounds different from polyphenols were found to be candidate biomarkers, including organic acids, such as succinic acid, lactic acid, malic acid, abscisic acid, indolic acid, and gluconic acid; some amino acids; fatty acids, such as linolenic acid, arachidonic acid, hexenoic acid, and oleic acid; oligosaccharides and raffinose; sterols; and the purine bases xanthine and uridine.

10.5 THE USE OF OAK CHIPS

The MOX system does not provide either the flavors or the ellagic tannins that the contact with the oak wood provides. Therefore, the practice of macerating wine with small fragments of toasted oak pieces has become another practice to mimic barrel aging. These pieces provide similar tastes, aromas, and wooden notes to the wine as those obtained with traditional barrel aging, but much faster and cheaper (Hernandez-Orte et al., 2014). This practice has been largely used in the United States, Australia, and other countries such as Chile for several years and became authorized in the EU countries in 2006 [(CE) No 1507/2006]. The EU regulation approved the use of pieces of oak wood in winemaking. The regulation states that "the pieces of oak wood must come exclusively from the *Quercus* genus. They may be left in their natural state, or heated to a low, medium or high temperature, but they may not have undergone combustion, including surface combustion, nor be carbonaceous or friable to the touch. They may not have undergone any chemical, enzymatic or physical processes other than heating. No product may be added for the purpose of increasing their natural flavor or the amount of their extractible phenolic compounds. The dimensions of the particles of wood must be such that at least 95% in weight are retained by a 2 mm mesh filter (9 mesh)."

10.6 WHEN AND HOW USE THEM

The oak chips can be added at different moments during winemaking and at different doses (Bautista-Ortin et al., 2008; Dumitriu et al., 2016; González-Saiz et al., 2014):

- During alcoholic fermentation, chips may be added to the tank, affecting the wine aroma and color. Some of the phenolic compounds extracted from wood could help to stabilize wine color. However, a high volatilization of the volatile compounds, accompanying the CO_2 formed during alcoholic fermentation could occur. At this moment, doses may vary from 1 to 10 g/L, depending on the final objective.
- After the end of alcoholic fermentation and before MLF, the addition of chips is very effective for improving aroma and wine structure, especially if it is combined with MOX, increasing its stability. Doses may vary between 2 and 15 g/L.
- During aging, doses should be more controlled to avoid an excess of wood aroma compounds that could hinder other wine aroma components and may harden the wine. Doses around 2 g/L are the most common used. The equivalent extraction of compounds obtained after 1 year in barrels could be obtained with a total of 10–15 g/L of chips, added to the wine at different moments. The cost could be around 10 times lower than the cost of barrel aging (Karvela et al., 2008).

10.7 EFFECT OF ADDING OAK CHIPS ON WINE CHARACTERISTICS

All the studies have shown that the use of chips increased the oak derived volatile compounds. As regards the liberation of these volatile compounds, several factors are important (wood origin, toast level, size), in a similar way that occurs in barrel aging (Rodríguez-Rodríguez et al., 2012). Among these factors, the size of the pieces has been proved as a very important factor since the wine will penetrate and soak the chips, facilitating the process of aroma compounds diffusing from wood to wine (Fernandez de Simon et al., 2010; Rodríguez-Rodríguez et al., 2012). They can be large pieces, similar to the barrel staves or smaller pieces, such as cubes or shavings. Some studies have stated that the smaller fragments release more volatile compounds and are more combustible so that more toasting-related compounds are formed compared with large staves (Fernandez de Simon et al., 2010; Rous and Alderson, 1983). Bautista-Ortin et al. (2008) compared the effect of the addition of small wood fragments (cubes, shavings, and powder) to a red wine during 3, 6, or 9 months of contact time, and they found that the cube format led to a higher enrichment of the wine in oak-related volatile compounds and that longer contact times could lead to a decrease in the concentration of these compounds (Fig. 10.2).

The time of contact of wine with oak chips is usually shorter than in the case of barrel aging. The studies of Rous and Alderson (1983) and Kadim and Mannheim (1999) revealed that it took around 70 days for a model wine solution to reach a total phenolic content of 200 mg/L during aging in new French oak barrels. Meanwhile, the time required for a model wine solution to reach the same phenolic content was approximately 35 days

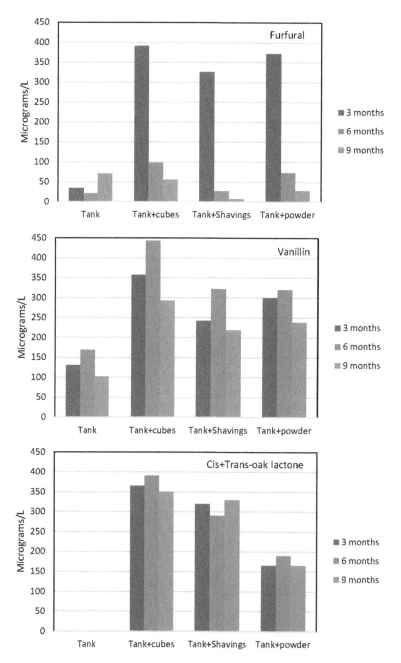

FIGURE 10.2 Changes in oak-related volatile compounds as a function of the contact time (3, 6, or 9 months) and size of the chips (powder, shavings or cubes). *Adapted from Bautista-Ortin, A.B., Lencina, A., Cano-López, M., Pardo-Minguez, F., López-Roca, J.M., Gómez-Plaza, E., 2008. The use of oak chips during the ageing of a red wine in stainless steel tanks or used barrels: effect of the contact time and size of the oak chips on aroma compounds. Aust. J. Grape Wine Res. 14, 63–70.*

during aging with French oak chips (Karvela et al., 2008). However, it has been reported that the oak-related volatile compound content in wines treated with oak chips appears to increase even after oak chips are removed (Tao et al., 2014), probably because soluble precursors extracted from oak chips, are transformed into additional amounts of oak-derived volatiles. This hypothesis has been suggested by several observations. For example, guaiacol and vanillin concentrations were seen to increase up to 25% of the total after the shavings were removed (Rous and Alderson, 1983).

One important thing regarding the studies on the accumulation of oak-derived volatile compounds in wine is the effect of different factors on the final concentration of these compounds. The study of Rodríguez-Rodríguez and Gómez-Plaza (2011) was designed to test the importance that the physicochemical properties (acidity and alcohol level) may have in the extraction processes. The concentration of compounds such as lactones could increase in wine as the percentage of ethanol and titratable acidity decreases, while the concentration of compounds such as vanillin increased at higher alcohol percentage and acidity values. This fact may play an important role in wine flavor because the alcohol content and acidity of the wines will modulate the extraction of lactones and vanillin, two of the most powerful compounds in the aroma of oak-aged wines.

This study also showed that the behavior of compounds such as furfuryl derivates and vanillin depended on the matrix where the oak chips were soaked. The concentration of furfuryl compounds and vanillin gradually increased when the study was done in model wines, but it substantially decreased during the second part of the experiment if the study was done with wine samples, a decrease attributed to the enzymatic activity that is always present in wines and/or to adsorption to molecules suspended in the wines. Winemakers should be aware of these findings, with lower concentrations of most oak-derived volatile compounds being expected when high microbial activity or suspension material is present in their wines, especially if oak chips are going to be in contact with wine more than 3 months (Rodríguez-Rodríguez and Gómez-Plaza, 2011).

It is also important to indicate that the use of oak chips also affects wine color. During the time that a wine is in contact with oak chips, phenolic compounds such as phenolic acids, flavanols, and ellagitannins are extracted from the wood, in a similar way as they are from barrels (Del Alamo-Sanza et al., 2004; Fernández de Simón et al., 1996; Frangipane et al., 2007) and they contribute to changes in the color of wines aged with chips, these changes being very dependent to the length of contact time. Thus, the studies of Ortega-Heras et al. (2010) showed that with short aging times, wines aged with chips present a higher color intensity than those aged in barrels, while color intensity decreases were observed when the wines were aged with chips for longer periods of time. It seems to be a loss of phenolic compounds probably due to a precipitation or absorption of these compounds in the oak chips (Vivar-Quintana et al., 1999).

The changes in color and the improvement of its stability are also related with the liberation of ellagitannins, especially the C-glucosidic ellagitannins, that can represent up to 10% of dry weight of all of the extractives from oak chips (Moutounet et al., 1994). The studies of Jourdes et al. (2011) showed that French oak chips released significantly higher amounts of ellagitannins than American oak chips at any toast level. They are important compounds because they can protect against oxidation. It is well known that wine, especially red wine, consumes oxygen and, although oak chips consumed oxygen more slowly than wine, the oxygen consumed by ellagitannins and the other phenolic compounds released from oak chips may compete with the oxygen consumed by the wine itself, and this can play a major role in protecting the wine against oxidation (Navarro et al., 2016).

10.8 COMPARING THE EFFECT OF CHIPS OR MOX WITH AGING WINE IN BARRELS

Several studies have been accomplished to compare the effect of aging in barrels and with chips. The results observed by Bautista-Ortin et al. (2008) showed that chips release aroma compounds into wine very rapidly, an effect that was clearly seen when they were added to the wines stored in tanks. However, wines aged in new barrels continued to extract aroma compounds for a longer time, and higher concentrations were reached in these wines for most aroma compounds. If wines are stored in used barrels and oak chips were added, the extraction behaved in an intermediate manner.

These wines were also sensorially evaluated (Cano-López et al., 2008a). The results showed that when the wines were aged in tanks, the effect of adding oak chips was readily appreciated by the panelists, however, the panelists were able to clearly differentiate between wines aged in new barrels and those aged in tanks and in contact with wood chips. The wood pieces were also added to used barrels and the results indicated that the differences between the resulting wine and that aged in new barrels, although smaller, were still perceived, the latter reaching higher scores in the sensory analysis.

Also, the comparison of adding chips or barrel aging on the wine phenolic composition was studied. The results of the study of Ortega-Heras et al. (2010) pointed to significant differences in phenolic composition depending on the aging system assayed and showed that it was not possible to obtain wines aged with chips with sensory characteristics similar to those aged for a long time in new barrels, although the use of oak chips could be a good alternative for elaborating young wines with olfactory and gustative wood notes quite similar to wines aged in new barrels and for shorter periods of time (about 3 months).

Other modern methodologies have been used for comparing the effect of chips and oak barrels. The combination of the data obtained by means of an e-nose (based on resistive MOX sensors), an e-tongue (based on voltammetric sensors), and an e-eye (based on CIE Lab coordinates) has been used to monitor the aging of a red wine (Apetrei et al., 2012), without the need for separation techniques to evaluate simple components. The study showed that after 10 months of aging it was possible to discriminate between the wine aged in a French oak barrel and the same wine soaked with oak chips of the same origin and toasting level, even when treated with MOX. These electronic systems could be of interest to monitor periodically the changes of wines aged by different methods and to evaluate the rate of maturing in each case.

One fear of the enologists and consumers is that the use of chips could lead to fraud if their use is not clearly indicated in the label. Therefore, some researchers have been working to try to find compounds or relation between compounds for the differentiation of wines aged with chips or barrels. Hernandez-Orte et al. (2014) used volatile compounds as discriminant factors. It is very complicated since the compounds that can be extracted are exactly the same and the extraction of wood-derived compounds is affected by many factors such as the age of the barrel, the application of fermentation or maceration in wines, the dose, etc. and the same occurs for chips. It was found that those compounds directly related to the wood have greater discriminative power for differentiating wines aged in barrels from those macerated with oak fragments, but they could not find a single compound that permitted flawless classification.

This study of Hernandez-Orte et al. (2014) was very deep, studying the effect of the addition of oak fragments of different origins, different oak types, different formats, and subjected to different toasting processes on a set of 231 samples from six Spanish Denominations of Origin wines, and compared them to those same wines aged in oak barrels. The principal component analysis of the volatile compounds showed that vanillin, acetovanillone, and syringaldehyde were the compounds that explain the variance of wines fermented or macerated with wood fragment wines; they are present in higher concentrations than in wines aged in barrels. The vanillin + acetovanillone/eugenol ratio was found to be essential for discrimination. It was observed that when wines were aged in barrels, the ratio was lower than 20, whereas when the wines were fermented or macerated with wood fragments, the ratio was higher than 20.

Also, the effect of MOX and barrel aging was compared. Most of the studies on MOX claim that the effect of this technique on wine color can be compared to that which occurs during maturation in oak barrels (Atanasova et al., 2002; Castellari et al., 2000; Du Toit et al., 2006a; Pérez-Magariño et al., 2007). To prove this, Cano-López et al. (2010) compared both techniques studying the evolution of a wine's chromatic parameters during 3 months of MO, simultaneously aging the same wine in American oak barrels for three and 6 months. Since the age of the barrel influences the natural MOX of the wines, both new and 3-year-old barrels were used in this study. Wines were analyzed at the end of the MOX and barrel aging period (3 and 6 months) and again 6 months after bottling. The application of MOX for 3 months produced wines with a lower concentration of monomeric anthocyanins and a higher concentration of vitisin-related pigments than the control wine, the oak mature wines showing similar results than MOX wines when aged for the same period of time (3 months). The effect of MOX and barrel aging was also observed in the chromatic characteristics, the micro-oxygenated and the oak matured wines showing a higher color intensity than the control wine. However, after 6 months in the bottle differences were found between the micro-oxygenated wines and oak matured wines, the latter showing a more stable color, probably due to the beneficial effects of compounds extracted from the wood (e.g., ellagitannins or wood aldehydes). That is, MOX alone cannot totally mimic the results obtained after barrel aging, not only not obtaining the aroma that barrel aging will give to the wine but also resulting in a wine stabilization lower than that seen with barrels.

10.9 THE COMBINED USED OF MOX + CHIPS

The closest way to mimic barrel aging is the combination of chips and the MOX technique. The studies of Cejudo-Bastante et al. (2011a, 2011b) with Petit Verdot and Merlot wines presented the results of applying MOX before MLF and oak chip treatments, evaluating the color characteristics, the phenolic compounds related to the color of red wine, the volatile compounds, and the sensory characteristics of the wines. The MOX treatment promoted the stabilization of red wine color by increasing the formation of color-related phenolic compounds (higher concentrations of pyranoanthocyanins and anthocyanin-ethyl-flavan-3-ol adducts) and red wine aroma quality was improved with the addition of oak chips, although the typical oak chip aromas (vanilla and woody) were observed to a lesser extent in wines that were also treated by MOX.

However, the studies of Ortega-Heras et al. (2008) showed different results, since from a sensorial point of view, the MOX treatment promoted an increase of fruity flavors and a better integration of the aroma of the wood, reducing green and herbaceous aromas. Llaudy et al. (2006), Ortega-Heras et al. (2008), and Rodríguez-Bencomo et al. (2008) also observed an intensification of wood attributes when wine were also micro-oxygenated.

Pizarro et al. (2014) studied the influence of chip maceration and microoxygenation-related factors (oxygen doses, chip doses, wood origin, toasting degree, and maceration time) on the volatile profile of red wines during the accelerated aging process. The results obtained indicated that the volatile profile of wines could be modulated by applying different combinations of factor conditions and these results would be used to obtain wines with specific volatile profiles that would lead to particular olfactory attributes according to consumers' preferences.

10.10 INNOVATIONS IN MOX AND CHIPS APPLICATION

10.10.1 Innovations in MOX

The basic technology of active MOX described above is still the most common approach to MOX. However, some information regarding alternative methodologies can be found, including diffusion methods, oxygen-permeable tanks, and electrochemical micro-oxygenation (ELMOX).

One of the newer proposed technologies is based on the diffusion of oxygen across a permeable membrane instead of using the usual bubbling method (Kelly and Wollan, 2003; Paul and Kelly, 2005). The technique uses polytetrafluoroethylene (Teflon) or polydimethylsiloxane (Silastic) tubing, permeable to oxygen. The application of these technologies for wine oxygenation is a continuous process as permeation across the membrane and diffusion into the wine occurs while gas pressure is maintained within the tube. Oxygen is dissolved into the wine without the formation of bubbles. The wall thickness of the tube is in the range from 0.3 to 0.5 mm and preferably about 0.3 mm. Oxygen is absorbed into the polymer on the gas side and transported by diffusion to the liquid side where desorption of the oxygen into the liquid occurs. Conversely the liquid phase penetrates the pores of hydrophilic microporous polymers and, consequently, no bubbles are formed.

In the basic MOX system, an acute control is necessary in order to produce suitable fine bubbles that need to pass through the wine for a predetermined distance so that sufficient oxygen is absorbed into the wine. Typically, this distance is of the order of about 2 m. This makes the technique unsuitable for use in small or shallow tanks since oxygen may not be dissolved in the wine and accumulates in the headspace. Because headspace accumulation of oxygen is not a problem with diffusion technology, one advantage of the system is that it can be used in small tanks (Blaauw, 2009) and even barrels.

Another solution is the use of oxygen-permeable tanks (Flecknoe-Brown, 2004, 2005). Moutounet et al. (1998) showed that a typical new oak barrique allows oxygen permeation through its walls, in the range of 20–30 mg/L per year. Kelly and Wollan (2003) report an estimated oxygen permeation rate into a typical oak cask of 2.2 mL/L/month or 26.4 mL/L/year.

These new wine maturation vessels are made from high density polyethylene (HDPE) which is permeable to oxygen so that the vessels actually allow oxygen through the walls in a similar way to normal oak barrels. The amount of oxygen delivered depends on the size of the tank but the oxygen dose from the standard 1000 L tank is equivalent to MOX in stainless steel tanks (Flecknoe-Brown, 2004).

As described in a patent (Flecknoe-Brown, 2002), the container is self-supporting and the walls of the container comprise a rigid plastic material which allows oxygen (typically at atmospheric partial pressure of 0.18 atm) to permeate the walls directly from the atmosphere into the liquid in contact with the walls, at a rate of 13–65 mg of oxygen per square meter of the wall area for each 1 mm of the walls thickness per 24 hours period at room temperature.

If oxygen accumulates on the surface, aerobic bacteria may propagate on this oxygen-rich layer, generating volatile acidity and acetaldehyde. Therefore, the lower the free surface area, the longer that wine can be safely kept in the tank. In order to limit the transfer of oxygen into the surface of the wine, a barrier system may be floated on the surface. Suitably, the barrier member has a peripheral portion which is in sliding contact with the container walls to separate the liquid surface from the head space in the container.

Typically, the wine may be stored in the container for a period ranging between 4 and 36 months with the total rate of oxygen transmission into the wine being maintained at less than 55 mg/L of wine/year.

Also, Del Alamo-Sanza et al. (2015) studied the use of HDPE tanks. They calculated the OTR and it was 28 mg/L of oxygen per year, in the upper range of values previously suggested for new oak barrels (10–28 mg/L of oxygen per year) and within the range of MOX dosage used for red wines in contact with oak wood, after MLF (Cano-López et al., 2006).

These authors used these tanks combined with oak. They found that the chemical composition of the wines stored in barrels differ significantly from those aged in HDPE tanks containing oak barrel alternatives, the later showing lower levels of free anthocyanins and more color due to polymeric pigments than those aged in barrels, suggesting that the typical reactions associated with oxidation that result in a loss of monomeric anthocyanins and the formation of polymerized colored species were more acute in the wines aged in HDPE tanks with chips; probably, as a result of a lower oxygen incorporation within the barrels. A greater intensity of sweet and cinnamon aromas was observed in the barrel-aged wines, as compared to the HDPE-aged wines.

Another variation to classic MOX systems is the ELMOX, that uses electrolysis to produce oxidation reactions in wine (Fell et al., 2007). The chemistry of reduction and oxidation, at a basic level, is simply the transfer of electrons. The oxidation of polyphenols can be achieved by the application of an electrical potential across two appropriate conducting surfaces contacting the wine. One advantage of this approach is that it oxidizes phenols and oxidizes ethanol to acetaldehyde without producing hydrogen peroxide (H_2O_2) or other reactive oxygen species (Blaauw, 2009). Currents of between 0.1 and 100 μA/L were assayed, and an initial trial produced effects similar to traditional MOX methods (Fell et al., 2007).

A potential advantage of using ELMOX methods is that the control is more precise. This is particularly useful when very small dosage rates are required (Dykes, 2007). Traditional micro-oxygenating, with bubbles, has the disadvantage of the residual dissolved oxygen remaining in the treatment vessel after the dosage has been stopped. With ELMOX, however, once the current is switched off the oxidation process finishes (Dykes, 2007).

10.10.2 Innovations in the Treatment With Chips

Some techniques are being tested to accelerate the transfer of oak-related compounds from oak wood into wine. Tao et al. (2015) tested high hydrostatic pressure (HHP) to process wines macerated with oak chips, to accelerate the release of oak-related compounds and modify wine composition within a short processing time. The results indicated that the release of compounds from oak chips into wine could be faster with this technique, probably due to HHP causing modifications in the raw wood that increased the rate of extraction (Jun, 2009). The larger extraction of phenolic compounds increased wine antioxidant activity and wine color, although, as a negative part, an artificial taste arose after HHP treatment. The authors concluded that more studies need to be done.

Also, ultrasound has been tested (Tao et al., 2014). Ultrasound technique is a nonhazardous, environmentally friendly, and relatively cost-effective system for enhancing various food processes, and is considered to be suitable to intensify the mass transfer process occurred in wine (Bautista-Ortin et al., 2017). The basis of ultrasonic enhancement of the extraction process is the ultrasonic cavitation and the generated physical and chemical phenomena, including local extremely high temperature and pressure, microstreaming, and microturbulence (Chemat et al., 2011; Cheung et al., 2013; Tao and Sun, 2013). These authors studied the release kinetics of total phenolics from oak chips into a model wine at various conditions. Generally, a significant increase of total phenolic content in model wine was observed under ultrasound treatment for 150 min, although the results depended on the acoustic energy density and temperature, especially the latter. However, due to wine complexity and the dependence of oak compounds extraction with wine composition, more studies should be carried out since, as mentioned previously, the ultrasound treatment at proper conditions may modify wine composition positively, thus shortening wine aging time, whereas sonication in an improper way may result in the decrease of wine quality (Pingret et al., 2013). Future innovations may lead to systems that could work continuously, in a similar manner to that observed in the work of Bautista-Ortin et al. (2017) (Fig. 10.3).

As a conclusion, it is reported that only a small portion of wines have the potential to improve significantly during aging. For these wines with high aging potential, the wine aging in oak barrels is still the most reliable method to enhance their quality at present. However, most studies have indicated that the use of oak chips

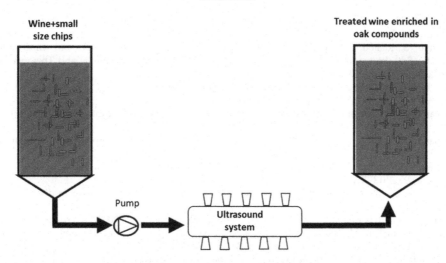

FIGURE 10.3 Scheme of a continuous ultrasound system for accelerating the extraction of oak-related compounds in wines.

and/or MOX can be considered a good choice for short-aged wines and to extend the lifetime of good sanitary used barrels. The most pronounced improvement would appear in mid-range wines, which show a decrease in vegetal characters, improved complexity, increased volume, and mid-palate tannin intensity.

References

Apetrei, I., Rodríguez-Méndez, M.L., Apetrei, C., Nevares, I., Del Alamo-Sanza, M., 2012. Monitoring of evolution during red wine aging in oak barrels and alternative method by means of an electronic panel test. Food Res. Int. 45, 244–249.

Arapitsas, A., Antonopoulos, A., Stefanou, E., Dourtoglou, V., 2004. Artificial aging of wines using oak chips. Food Chem. 86, 563–570.

Arapitsas, A., Scholz, M., Vrhovsek, U., Di Blasi, S., Biondi, A., Masuero, D., et al., 2012. A metabolomic approach to the study of wine microoxygenation. PLoS ONE 7, e37783.

Atanasova, V., Fulcrand, H., Cheynier, F.V., Moutounet, M., 2002. Effect of oxygenation on polyphenol changes occurring in the course of wine-making. Anal. Chim. Acta 458, 15–27.

Bautista-Ortin, A.B., Lencina, A., Cano-López, M., Pardo-Minguez, F., López-Roca, J.M., Gómez-Plaza, E., 2008. The use of oak chips during the ageing of a red wine in stainless steel tanks or used barrels: effect of the contact time and size of the oak chips on aroma compounds. Aust. J. Grape Wine Res. 14, 63–70.

Bautista-Ortin, A.B., Jiménez-Martínez, M.D., Jurado, R., Iniesta, J.A., Terrades, S., Andres, A., et al., 2017. Application of high power ultrasounds during red wine vinification. Int. J. Food Sci. Technol. 52, 1314–1323.

Blaauw, D., 2009. Micro-Oxygenation in Contemporary Winemaking. Ph.D. Thesis. Cape Wine Academy, Stellenbosch, South Africa.

Cano-López, M., Gómez-Plaza, E., 2011. A review on micro-oxygenation of red wines: claims, benefits and the underlying chemistry. Food Chem. 125, 1131–1140.

Cano-López, M., Pardo-Minguez, F., López-Roca, J.M., Gómez-Plaza, E., 2006. Effect of micro-oxygenation on anthocyanin and derived pigments content and chromatic characteristics of red wines. Am. J. Enol. Vitic. 57, 331–336.

Cano-López, M., Bautista-Ortin, A.B., Pardo-Minguez, F., López-Roca, J.M., Gómez-Plaza, E., 2008a. Sensory descriptive analysis of a red wine aged with oak chips in stainless steel tanks or used barrels: effect of the contact time and size of the oak chips. J. Food Qual. 31, 645–660.

Cano-López, M., Pardo-Minguez, F., Schmauch, G., Saucier, C., Teissedre, J., López-Roca, J.M., et al., 2008b. Effect of micro-oxygenation on color and anthocyanin-related compounds of wines with different phenolic contents. J. Agric. Food Chem. 56, 5932–5941.

Cano-López, M., López-Roca, J.M., Pardo-Mínguez, F., Gómez-Plaza, E., 2010. Oak barrel maturation vs. micro-oxygenation: effect on the formation of anthocyanin-derived pigments and wine color. Food Chem. 119, 191–195.

Castellari, M., Matricardi, L., Arfelli, G., Amati, A., 2000. Level of single bioactive phenolics in red wine as a function of the oxygen supplied during storage. Food Chem. 69, 61–67.

Cejudo-Bastante, M.J., Hermosin-Gutierrez, I., Pérez-Coello, M.S., 2011a. Micro-oxygenation and oak chip treatments of red wines: effects on colour-related phenolics, volatile composition and sensory characteristics. Part I: Petit Verdot wines. Food Chem. 124, 727–737.

Cejudo-Bastante, M.J., Hermosin-Gutierrez, I., Pérez-Coello, M.S., 2011b. Micro-oxygenation and oak chip treatments of red wines: effects on colour-related phenolics, volatile composition and sensory characteristics. Part II: Merlot wines. Food Chem. 124, 738–748.

Chatonnet, P., Boidron, J.N., Pons, M., 1989. Incidence du traitement thermique du bois de chêne sur sa composition chimique. 2e Partie: Évolution de certains composés en fonction de l'intensite de brûlage. Conn. Vigne Vin 23, 223–250.

Chatonnet, P., Boidron, J.N., Pons, M., 1990. Élevages des vins rouges en fûts de chéne: évolution de certains composés volatils et de leur impact aromatique. Sci. Aliments 10, 565–587.

Chatonnet, P., Dubourdie, D., Boidron, J., Pons, M., 1992. The origin of ethylphenols in wines. J. Sci. Food Agric. 60, 165–178.

Chemat, F., Huma, Z., Khan, M., 2011. Applications of ultrasound in food technology: processing, preservation and extraction. Ultrason. Sonochem. 18, 813–835.

Cheung, K., Siu, K., Wu, J., 2013. Kinetic models for ultrasound assisted extraction of water soluble components and polysaccharides from medical fungi. Food Bioprocess Technol. 6, 2659–2665.

Cheynier, V., Prieur, C., Guyot, S., Rigaud, J., Moutounet, M., 1997. The structures of tannins in grapes and wines and their interaction with proteins. In: Watkins, T.R. (Ed.), Proccedings of the ACS Symposium Series 661, Wine: Nutritional and Therapeutical benefits. American Chemical Society, Nueva York, NY, pp. 81–93.

Cutzach, I., Chatonnet, P., Henry, R., Dubourdieu, D., 1999. Identifying new volatile compounds in toasted oak. J. Agric. Food Chem. 47, 1663–1667.

Del Alamo-Sanza, M., Nevares, I., 2014. Recent advances in the evaluation of the oxygen transfer rate in oak barrels. J. Agric. Food Chem. 62, 8892–8899.

Del Alamo-Sanza, M., Fernández-Escudero, J.A., De Castro-Torio, J., 2004. Changes in phenolic compounds and colour parameters of red wine aged with oak chips and in oak. Food Sci. Technol. Int. 10, 233–241.

Del Alamo-Sanza, M., Laurie, F., Nevares, I., 2015. Wine evolution and partial distribution of oxygen during storage in high-density polyethylene tanks. J. Sci. Food Agric. 95, 1313–1321.

Devatine, A., Chiciuc, I., Poupot, C., Mietton-Peuchot, M., 2007. Micro-oxygenation of wine in presence of dissolved carbon dioxide. Chem. Eng. Sci. 62, 4579–4588.

Du Toit, M., 2007. Micro-oxygenation in South Africa wine. www.wynboer.co.za/recenarticles/200707micro.php3.

Du Toit, W., Lisjak, K., Marais, J., du Toit, M., 2006a. The effect of micro-oxygenation on the phenolic composition, quality and aerobic wine-spoilage microorganisms of different South African red wines. S. Afr. J. Enol. Vitic. 27, 57–67.

Du Toit, W., Marais, J., Pretorius, I.S., du Toit, M., 2006b. Oxygen in must and wine: aA review. S. Afr. J. Enol. Vitic. 57, 76–94.

Dumitriu, G., López de Lerma, N., Cotea, V., Zamfir, C., Peinado, R., 2016. Effect of aging time, dosage and toasting level of oak chips on the color parameters, phenolic compounds and antioxidant activity of red wines (var. Feteasca neagra). Eur. Food Res. Technol. 242, 2171–2180.

Dykes, S., 2007. The Effect of Oxygen Dosage Rate on the Chemical and Sensory Changes Occurring During Micro-Oxygenation of New Zealand Red Wine. Ph.D. Thesis. University of Auckland, Auckland, New Zealand.

Ertan Anli, R., Cavuldak, O., 2012. A review of microoxygenation application in wine. J. Inst. Brew. 118, 368–385.

Fell, A., Dykes, S., Nicolau, L., Kilmartin, P., 2007. Electrochemical microoxygenation of red wine. Am. J. Enol. Vitic. 58, 443–450.

Fernández de Simón, B., Cadahía, E., Conde, E., García Vallejo, M.C., 1996. Low molecular weight phenolic compounds in Spanish oak woods. J. Agric. Food Chem. 44, 1507–1511.

Fernandez de Simon, B., Cadahia, E., Del Alamo-Sanza, M., Nevares, I., 2010. Effect of size, seasoning and toasting in the volatile compounds in toasted oak wood and in a red wine treated with them. Anal. Chim. Acta 660, 211–220.

Flecknoe-Brown, A., 2002. Liquid food and wine storage bladder within a container (patent).

Flecknoe-Brown, A., 2004. Controlled Permeability Moulded Wine Tanks. New Development in Polymer Catalyst Technology. The Australian & New Zealand Grapegrower & Winemaker, vol. 480 pp. 59–61.

Flecknoe-Brown, A., 2005. Oxygen Permeable Polyethylene Vessels: A New Approach to Wine Maturation. The Australian & New Zealand Grapegrower & Winemaker, vol. 494 pp. 53–57.

Frangipane, M.T., Santis, D.D., Ceccarelli, A., 2007. Influence of oak woods of different geographical origins on quality of wines aged in barriques and using oak chips. Food Chem. 103, 46–54.

Fulcrand, H., Atanasova, V., Salas, E., Cheynier, V., 2004. The fate of anthocyanins in wine. Are they determining factors? In: Waterhouse, A., Kennedy, J.A. (Eds.), Red Wine Color. Revealing the Mysteries. ACS Symposium Series 886. American Chemical Society, Washington, DC, pp. 68–88.

Garde-Cerdan, T., Ancin-Azpilicueta, C., 2006. Review of quality factors on wine ageing in oak barrels. Trends Food Sci. Technol. 7, 438–447.

González-Saiz, J.M., Esteban-Diez, N., Rodríguez-Tecedor, S., Pérez del Notario, N., Arenzana-Ramila, I., Pizarro, C., 2014. Modulation of the phenolic composition and colour of red wines subjected to accelerated ageing by controlling process variables. Food Chem. 165, 271–281.

Hernandez-Orte, P., Franco, E., González-Huerta, C., Martínez-García, J., Cabellos, M., Suberviola, J., et al., 2014. Criteria to discriminate between wines aged in oak barrels and macerated with oak fragments. Food Res. Int. 57, 237–241.

Jackson, R., 2000. Wine Science: Principles, Practice and Perception. Academic Press, San Francisco, CA.

Jourdes, M., Michel, J., Saucier, C., Quideau, S., Teissedre, J., 2011. Identification, amounts, and kinetics of extraction of C-glucosidic ellagitannins during wine aging in oak barrels or in stainless steel tanks with oak chips. Anal. Bioanal. Chem. 401, 1531–1539.

Jun, X., 2009. Caffeine extraction from green tea leaves assisted by high pressure processing. J. Food Eng. 94, 105–109.

Kadim, D., Mannheim, C., 1999. Kinetics of phenolic extraction during aging of model wine solution and white wine in oak barrels. Am. J. Enol. Vitic. 50, 33–39.

Karvela, E., Makris, D., Kefalas, P., Moutounet, M., 2008. Extraction of phenolics in liquid model matrices containing oak chips: kinetics, liquid chromatography-mass spectroscopy characterization and association with in vitro antiradical activity. Food Chem. 110, 263–272.

Kelly, M., Wollan, D., 2003. Micro-Oxygenation of Wines in Barrels, 447. The Australian & New Zealand Grapegrower & Winemaker, 45.

Kennedy, J., 2008. Grape and wine phenolics: observations and recent findings. Cienc. Tecnol. Aliment. 35, 107–120.

Lesica, M., Kosmerl, T., 2009. Microoxygenation of red wines. Acta Agric. Slov. 93, 327–336.

Llaudy, M.C., Canals, R., González-Manzano, S., Canals, J., Santos-Buelga, C., Zamora, F., 2006. Influence of micro-oxygenation treatment before oak aging on phenolic compounds composition, astringency and color of red wine. J. Agric. Food Chem. 54, 4246–4252.

Maga, J.A., 1984. Flavour contribution of wood to alcoholic beverages. In: Adda, J. (Ed.), Progress in Flavour Research. Dourdan, France, pp. 409–416.

Masson, G., Puech, J.L., Moutounet, M., 1996. Composition chimique du bois de chêne de tonnellerie. Bull O. I. V. 785-786, 635–657.

Michel, J., Jourdes, M., Silva, M., Giordanengo, T., Mourey, N., Teissedre, P., 2011. Impact of concentration of ellagitannins in oak wood on their levels and organoleptic influence in red wine. J. Agric. Food Chem. 59, 5677–5683.

Moutounet, M., Barbier, P., Sarni, F., Scalbert, A., 1994. Les tanins du bois de chêne. Les conditions de leur presence dans les vins. Conn. Vigne Vin. Hors Série 75–81.

Moutounet, M., Mazauric, J.P., Saint-Pierre, B., Hanocq, J.F., 1998. Gaseous exchange in wines stored in barrels. J. Sci. Tech. Tonnellerie 4, 131–145.

Moutounet, M., Puech, J.L., Keller, R., Feuillat, F., 1999. Les caractéristiques du bois de chêne en relation avec son utilisation en œnologie. Le phénomène de duramisation et ses conséquences. Revue Française d'Œnologie 174, 12–17.

Navarro, M., Kontoudakis, N., Giordanengo, T., Gómez-Alonso, S., García-Romero, E., Fort, F., et al., 2016. Oxygen consumption by oak chips in a model solution: influence of the botanical origin, toast level and ellagitannin content. Food Chem. 199, 822–829.

Nykänen, L., 1986. Formation and occurrence of flavor compounds in wine and distilled alcoholic beverages. Am. J. Enol. Vitic. 37, 84–96.

Ortega-Heras, M., Rivero-Perez, M., Pérez-Magariño, S., González-Huerta, C., González-San José, M.L., 2008. Changes in volatile composition of red wines during aging in oak barrels due to microoxygenation treatment applied before malolactic fermentation. Eur. Food Res. Technol. 226, 1485–1493.

Ortega-Heras, M., Pérez-Magariño, S., Cano-Mozo, E., González-San José, M.L., 2010. Differences in the phenolic composition and sensory profile between red wines aged in oak barrels and wines aged with oak chips. LWT 43, 1533–1541.

Parish, M., Wollan, D., Paul, R., 2000. Micro-Oxygenation. A Review. Australian and New Zealand Grapegrower and Winemaker, vol. 438a pp. 47–50.

Parpinello, G., Plumejeau, F., Maury, C., Versari, A., 2011. Effect of micro-oxygenation on sensory characteristics and consumer preference of Cabernet Sauvignon wine. J. Sci. Food Agric. 92, 1238–1244.

Paul, R., 2002. Micro-oxygenation. Where now? In: Allen, M., Bell, S., Rowe, N., Wall, G (Eds.), Use of Gases in Winemaking. ASVO, Adelaide, Australia, pp. 18–31.

Paul, R., Kelly, M., 2005. Diffusion. A new approach to micro-oxygenation. In: Blair, R., Williams, P., Pretorius, I. (Eds.), Proceedings of the 12th Australian Wine Industry Technical Conference. AWITC Ins, Adelaide, Australia, pp. 121–122.

Pérez-Coello, M.S., Diaz-Maroto, M., 2009. Volatile compounds and wine aging. In: Moreno, M.V., Polo, C. (Eds.), Wine Chemistry and Biochemistry. Springer, New York, NY, USA, pp. 295–311.

Pérez Prieto, L.J., López Roca, J.M., Martínez Cutillas, A., Pardo Minguez, F., Gómez Plaza, E., 2002. Maturing wine in oak barrels. Effects of origin, volume, and age of the barrel on the wine volatile composition. J. Agric. Food Chem. 50, 3272–3276.

Pérez-Magariño, S., Sanchez-Iglesias, M., Ortega-Heras, M., Gónzalez-Huerta, C., González-San José, M.L., 2007. Colour stabilization of red wines by microoxygenation treatment before malolactic fermentation. Food Chem. 101, 881–893.

Pingret, A., Fabiano-Tixier, F., Chemat, F., 2013. Degradation during the application of ultrasound in food processing: a review. Food Control 31, 593–606.

Pizarro, C., Rodríguez-Tecedor, S., Esteban-Diez, N., Pérez-del Notario, J., Gonzalez-Saiz, J.M., 2014. Experimental design approach to evaluate the impact of oak chips and micro-oxygenation on the volatile profile of red wines. Food Chem. 148, 357–366.

Pollnitz, A.P., 2000. The Analysis of Volatile Wine Components Derived from Oak Products During Winemaking and Storage. Ph.D. Thesis. University of Adelaide, Australia.

Ribéreau-Gayon, P., 1933. Contribution a l'etude des oxydations et reductions dans les vins. Ph.D. Thesis. University of Bordeaux. Bordeaux, France.

Ribéreau-Gayon, P., Glories, Y., 1987. Phenolics in grapes and wines. In: Lee, T.H. (Ed.), The Sixth Australian Wine Industry Technical Conference. Australian Industrial Publishers, Adelaide, pp. 247–256.

Ribéreau-Gayon, P., Glories, Y., Maujean, A., Dubourdieu, D., 1998. Traité d'Oenologie.2. Chimie du vin. Stabilisation et traitements. Dunod, Paris.

Rodríguez-Bencomo, J.J., Ortega-Heras, M., Pérez-Magariño, S., Gonzaléz-Huerta, C., González-San José, M.L., 2008. Importance of chip selection and elaboration process on the aromatic composition of finished wines. J. Agric. Food Chem. 56, 5102–5111.

Rodríguez-Rodríguez, P., Gómez-Plaza, E., 2011. Differences in the extraction of volatile compounds from oak chips in wine and model solutions. Am. J. Enol. Vitic. 62, 127–132.

Rodríguez-Rodríguez, P., Bautista-Ortin, A.B., Gómez-Plaza, E., 2012. Increasing wine quality through the use of oak barrels: factors that will influence aged wine color and aroma. In: Peeters, A. (Ed.), Wines: Types, Production and Health. Nova Science Publishers Inc, New York, NY, USA, pp. 301–317.

Romero, C., Bakker, J., 2000. Effect of acetaldehyde and several acids on the formation of vitisin A in model wine anthocyanin and colour evolution. Int. J. Food Sci. Technol. 35, 129–140.

Rous, C., Alderson, B., 1983. Phenolic extraction curves for white wine aged in French and American oak barrels. Am. J. Enol. Vitic. 34, 211–215.

Saucier, C., Little, D., Glories, Y., 1997. First evidence of acetaldehyde-flavanol condensation products in red wine. Am. J. Enol. Vitic. 48, 370–372.

Suarez, R., Suarez-Lepe, J.A., Morata, A., Calderon, F., 2007. The production of ethylphenols in wine by yeasts of the genera *Brettanomyces* and *Dekkera*: a review. Food Chem. 102, 10–21.

Tanaka, T., Takahashi, R., Kuono, I., Nonaka, K., 1994. Chemical evidence for the de-astringency (insolubilization of tannins) of persimon fruit. J. Chem. Soc. Perkin Trans. 301, 3113–3122.

Tao, Y., Sun, D., 2013. Enhancement of food processes by ultrasound: a review. Crit. Rev. Food Sci. Nutr. 55, 570–594.

Tao, Y., Zhang, Z., Sun, D., 2014. Experimental and modeling studies of ultrasound-assisted release of phenolics from oak chips into model wine. Ultrason. Sonochem. 21, 1839–1848.

Tao, Y., Sun, D., Gorecki, A., Blaszczak, W., Laparski, G., Amarowicz, R., et al., 2015. A preliminary study about the influence of high hydrostatic pressure processing in parallel with oak chip maceration on the physicochemical and sensory properties of a young red wine. Food Chem. 194, 545–554.

Vidal, S., Aagaard, O., 2008. Oxygen management during vinification and storage of Shiraz wine. Wine Ind. J. 5, 56–63.

Vidal, S., Francis, L., Noble, A., Kwiatkowski, M., Cheynier, V., Waters, E., 2004. Taste and mouth-feel properties of different types of tannin-like polyphenolic compounds and anthocyanins in wine. Anal. Chim. Acta 513, 57–65.

Vivar-Quintana, A.M., Santos-Buelga, C., Francia-Aricha, E.M., Rivas-Gonzalo, C., 1999. Formation of anthocyanin-derived pigments in experimental red wines. Food Sci. Technol. Int. 5, 347–352.

Vivas, N., Glories, Y., 1997. Mode lisation et calcul du bilan des apports d'oxygene au cours de l'elevage des vins rouges. II. Les aportes lie sau passage de l'oxygene au travers de la barrique. Prog. Agric. Vitic. 114, 315–316.

Weeks, S., Sefton, M.A., 1999. Analysis of oak-derived wine flavours. Wine Ind. J. 14, 42–43.

Further Reading

Jordao, A., Ricardo da Silva, J.M., Laureano, O., 2005. Extraction of some ellagic tannins and ellagic acid from oak wood chips in model wine solutions: effect of pH, temperature and alcoholic content. S. Afr. J. Enol. Vitic. 26, 83–89.

CHAPTER 11

New Trends in Aging on Lees

Antonio Morata, Felipe Palomero, Iris Loira and Jose A. Suárez-Lepe
Department of Chemistry and Food Technology, Technical University of Madrid, Madrid, Spain

11.1 INTRODUCTION

The current wine market is characterized by a certain homogeneity and a saturation of competing products. Therefore, new techniques and technologies not only reduce costs but aid to obtain highly distinguishable and quality products that are much sought after. Many research groups in enological microbiology and winemaking have focused their efforts on achieving these overall objectives, motivated by an industry that understands that the differentiation of its products at a moderate cost can significantly increase a brand's competitiveness.

In this context, it is important to note that there is a particular component subject to hedonistic tendencies, fashions, and popular trends in the consumption of wine. Wooded wines with great extraction, high alcohol content, and long periods of aging in barrels have been replaced by others in which primary or varietal fruit aromas are further respected through better integration and balance with wood volatiles.

Traditionally, aging on lees (AOL) has been used to improve the sensory profile and mouthfeel of white wines, especially for those that are barrel fermented. But it is also a helpful technique for red wines, improving the softness of tannins as a result of the interaction between mannoproteins and cell wall polysaccharides with wine phenols (Escot et al., 2001). This parameter can be modulated depending on the yeast strain used for AOL (Loira et al., 2013). The reductive effect of lees and the release of some antioxidant components of cell structures like glutathione (GSH) protect the aromatic compounds, thus preserving the fruitiness and freshness, even after long periods of AOL. Antioxidant activity of lees surface, have shown a protective role on thiols during the aging (Gallardo-Chacón et al., 2010). The high ability of yeast lees to scavenge oxygen has also been reported (Salmon et al., 2000); and the antioxidant role of yeast lees can also improve the color stability of wines (Escot et al., 2001; Palomero et al., 2007). Moreover, it has been observed that red wines aged on lees show a high limpidity, even when cell wall polysaccharides increase the colloidal load in wines. Lees polysaccharides has the ability to protect wines from protein haze (Dupin et al., 2000). In addition, wine lees have been described as good binding agents for certain volatile compounds such as ethyl esters and fusel alcohol acetates (Rodríguez-Bencomo et al., 2010). Although at first this feature may seem something negative for wine quality, it was found that, in general, the nature of this linking is transitory and the union is reversed with the passage of time; that is, the volatile compounds temporarily bound to the yeast cell wall can be released back into the wine.

During yeast growth, cell wall porosity depends on the nutritional composition of the media (De Nobel et al., 1990). Then, autolysis occurs after the death of the yeast cell affecting the degradation of the cell wall by auto-enzymatic activity and the release of cytoplasmic contents and coverings fragments. The release of cell wall polysaccharides during fermentation and the dependence of that process on yeast strains has been reported (Escot et al., 2001). The released polysaccharides can reach 100 mg/L. *Saccharomyces cerevisiae* cell wall is formed by a net of fibrillary polysaccharides containing globular mannoproteins ($\approx 40\%$). Fiber polysaccharides are mainly β-glucans ($\approx 60\%$) and chitin ($\approx 2\%$) formed by N-acetylglucosamine units (Magnelli et al., 2002). Glucans can be β1,3- or β1,6- branched structures (Fig. 11.1).

It has been reported that AOL and yeast autolysis is a slow process requiring 7–9 months before there are any repercussions in the sensory profile of wine. Not all yeasts behave in the same way with regard to autolysis and it is possible to use yeast selection to obtain optimal strains of *S. cerevisiae* with short autolysis periods, or in other terms (Palomero et al., 2007), with a greater release of cell wall components. The release of yeast cell wall polysaccharides into the media can be monitored by using liquid chromatography coupled with refractive index detection (LC-RID) in model media or wines. Using this technique, it is possible to study the strains that are able to release higher amounts in a shorter time, and therefore, to select those with shorter autolysis periods (Figs. 11.1A and B and 11.2).

A typical problem when using traditional AOL for red wines involves microbiological instability, because generally after red wine fermentation, together with the maceration of solids (skins and seeds), the lees are quite dirty, containing not only the yeast cells that have been fermenting the must, but also colloidal particles and cell wall fragments from grapes as well as a broad microbial population formed by yeast and bacteria species. Hence, this gross lees aging is prone to produce microbial developments and sensory deviations with associated reductive off-flavors. Reductive off-smells can include unpleasant molecules like diethyl sulfide (garlic) or dimethyl sulfide (stewed cabbage). To control this, an initial improvement is to simply use fine lees cleaning the colloidal particles settling the wine. However, this is just a partial solution because some contaminant load will still remain and at the same time, a large fraction of the yeast cells is removed. A better solution is the external production of a biomass with the yeast that is required to perform the AOL; this technique has several advantages: (1) the yeast

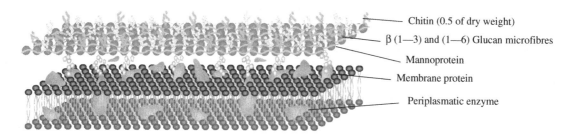

FIGURE 11.1 Plasmatic membrane and cell wall structure of *Saccharomyces cerevisiae*. Adapted from Palomero, F., Benito, S., Morata, S., Calderón, F., Suárez-Lepe, J.A., 2008. New yeast genera for over lees ageing in red wines. In: XXX World Congress of the International Organization of Grape and Wine (OIV). 16−19 June, Verona, Italy.

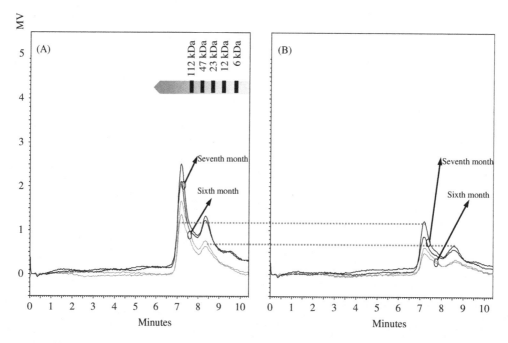

FIGURE 11.2 LC-RID chromatograms of polysaccharides releasing for 2 yeast strains 5CV (A) and 2EV (B) of *Saccharomyces cerevisiae* during the autolysis analyzed after 6 and 7 months of aging on lees. Chromatograms from duplicate processes. Pullulanes were used as molecular size markers in kDa. *LC-RID*, Liquid chromatography coupled with refractive index detection.

biomass can be produced using a yeast species or strain with improved performance for AOL being selected by a fast autolysis and positive sensory impact; (2) even non-*Saccharomyces* yeasts can be used; (3) the addition will be just pure yeasts cells so there will be no collateral contaminants—no bacteria, undesired yeasts or colloidal particles; and (4) the wine can be partially cleaned before the addition of lees to obtain an even safer process, including settling and filtration procedures. The lees can be produced in an external fermenter under aerobic conditions to enhance biomass production and the yeast cells can later be purified by rinsing with water and subsequent centrifugation to get a pure biomass (Suárez-Lepe and Morata, 2009). Moreover, it is possible to apply it as not only a fresh biomass but it can also be applied after a drying process (e.g., lyophilization) to facilitate storage and dosage.

Red wines aged on lees frequently show an intense fruity smell even when they are barrel matured for long periods. Glutathione, a cell wall peptide component formed by three amino acids and currently used as antioxidant in enology, is one of the factors responsible for this effect (Comuzzo et al., 2015a,b). The direct release of GSH during AOL or the application as an active yeast by-products affects the protective effect on fruity and varietal aromas (Rodríguez-Bencomo et al., 2014). The selection of yeasts that are able to release high contents of GSH during AOL is one way to enhance this protection (Suárez-Lepe and Morata, 2012).

AOL can be used synergistically with barrel aging or oak chips, improving the aromatic richness of wines (Loira et al., 2013). Moreover, the reductive effect of AOL by increasing the GSH content helps to balance the oxidative effect of barrel aging, reducing the aggressive repercussions of the use of new barrels.

It is interesting to note the high degree of chance and empiricism underlying the link between wine and wood. Do you ever wonder why a barrel of wine is shaped the way it is? The answer is that originally, they were only used as a container for trade of goods. At the time, wood was robust enough, cheap, and an abundant material, and such geometry allowed its heavy weight (volume) to be effortlessly moved by rolling it, upright if on rails or on its side on smooth surfaces. It was some time later that the positive influence of the container over the content, was perceived and recognized. This influence is due to the migration of certain chemicals, whether or not volatile, and mostly known and fully identified molecules that enrich the wine's sensory profile (Sanz et al., 2012).

Conventional wine aging in French or American oak barrels is a slow and costly process, where a large volume of wine must be immobilized in storage for varying periods before being marketed. The aromatic potential of the barrels is not unlimited, and therefore, their purchase, maintenance, and replacement, constitute a significant proportion of the variable costs for wineries. Even though these drawbacks are even more important if AOL is performed, this novel technique in red winemaking is gaining popularity since it allows the production of wines with distinctive sensory attributes.

In order to solve some of the previously mentioned inconveniences, oak chips, staves, or fragments of variable sizes and shapes are usually used by winemakers globally in stainless steel tanks (Spillman, 1999). Its direct addition to the wine coupled with micro-oxygenation techniques, increases color stability and enhances the diffusion and integration of oak aromas into the wine. This procedure significantly reduces aging periods and allows wineries to release their wines into the market sooner. On the other hand, these techniques are not free of problems. The migration of oak constituents primarily depends—among other factors involved—on the exposed surface area. The quantity, shape, and size of chips can sometimes produce an excessive wood aroma extraction when the dosage is incorrect. The repeatability of the processes is not always good since a great variation between batches has been a matter of concern for some winemakers. Besides the traditional woods commonly employed to age wines irrespective of the presence of lees, woods other than oak and chestnut have been studied (see Section 11.4).

The simultaneous use of chips or barrel aging with AOL helps to improve the aromatic profile with a more softening impact than just single barrel aging (Loira et al., 2013). Furthermore, cell wall polysaccharides released from the yeast during AOL soften the tannin dryness contributed by the barrel, promoting a faster and better integration of barrel tannins (Rodrigues et al., 2012; Loira et al., 2013).

Another application of lees concerns the yeast cell wall adsorption capacity, which has been studied in order to remove some specific and undesirable wine compounds such as ochratoxin A and 4-ethylphenols (Caridi et al., 2012; Palomero et al., 2011). In fact, AOL has been proposed as a palliative treatment to mitigate excessive oak aromas from wines (Chatonnet et al., 1992). In contrast, yeast lees adsorb anthocyanins and other grape phenols during fermentation that can be partially released during aging and autolysis stages if they are not removed (Morata et al., 2003).

11.2 USE OF NON-SACCHAROMYCES YEASTS

Non-*Saccharomyces* yeasts are a trend in current enology (Morata and Suárez Lepe, 2016), with applications in grape sanitization, fermentation, stabilization, and aging. Traditional *Saccharomyces* yeasts commonly used in AOL (Fig. 11.3A) can also be substituted with non-*Saccharomyces* yeast and this has some advantages. *Saccharomycodes ludwigii* (Fig. 11.3B), *Schizosaccharomyces pombe* (Fig. 11.3C), and even *Brettanomyces bruxellensis* (Fig. 11.3D) have been described for of their optimal release of polysaccharides during AOL (Palomero et al., 2009; Kulkarni et al., 2015). *S. pombe* also releases 3–7 times more polysaccharides than commercial *S. cerevisiae* during fermentation (Domizio et al., 2017).

The use of osmophilic non-*Saccharomyces* yeasts can accelerate the release of cell wall polysaccharides and mannoproteins (Palomero et al., 2009). This behavior is as a result of their special structure and the composition of their cell walls. In some osmophilic yeasts like *S. pombe*, a multilayer cell wall formed by α-galactomannose with α-1,3 glucan fibers has been described (Kopeckà et al., 1995) (Fig. 11.4). The average composition is 9%–14% α-galactomannan, 18%–28% α-1,3-glucan, 42% β-1,3-glucan, and 2% β-1,6-glucan (Manners and Meyer, 1977; Kopeckà et al., 1995). A three-layer structure with an external and internal electron dense structure and a less dense layer in the middle can be observed by transmission electron microscopy (Humbel et al., 2001).

The thick multilayer structure of *S. pombe* contains more polysaccharides and cell wall globular proteins than *S. cerevisiae* cells, allowing a higher release of polymers during the AOL processes. When the release of cell wall polysaccharides has been recorded by LC-RID during the AOL across different non-*Saccharomyces* yeast species, it was observed that the contents were much greater when *S. pombe* was used (Palomero et al., 2009; Kulkarni et al., 2015). Moreover, polysaccharides were released in a shorter time (Fig. 11.5) and with higher molecular sizes (Palomero et al., 2009). This last peculiarity can influence the tactile sensation of polysaccharides in the mouth

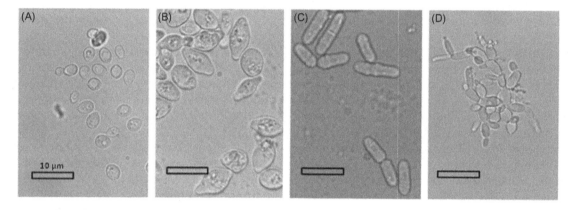

FIGURE 11.3 Optical microscopy of yeasts. (A) *Saccharomyces cerevisiae*. (B) *Saccharomycodes ludwigii*. (C) *Schizosaccharomyces pombe*. (D) *Brettanomyces bruxellensis*.

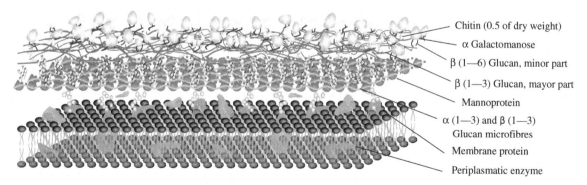

FIGURE 11.4 Plasmatic membrane and cell wall structure in *Schizosaccharomyces pombe*. Adapted from Palomero, F., Benito, S., Morata, S., Calderón, F., Suárez-Lepe, J.A., 2008. New yeast genera for over lees ageing in red wines. In: XXX World Congress of the International Organization of Grape and Wine (OIV). 16–19 June, Verona, Italy.

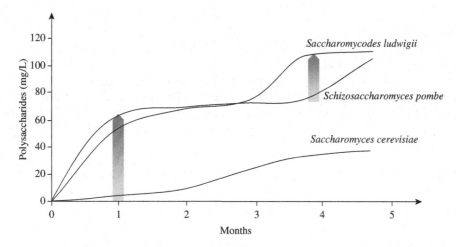

FIGURE 11.5 Cell wall polysaccharides released by several yeast species during AOL. *AOL*, Aging on lees.

and, in fact, wines were positively perceived by tasters (Palomero et al., 2009). Another interesting fact is a two-stage kinetics in the release of cell wall polysaccharides with a maximum peak and a first plateau after 1 month, but later a second maximum and plateau at 4 and 5 months for *S. ludwigii* and *S. pombe*, respectively (Fig. 11.5; Palomero et al., 2009). Moreover, the maximum value in *S. cerevisiae* is lower than the first plateau for *S. ludwigii* and *S. pombe*. Therefore, the use of some non-*Saccharomyces* is a powerful tool to enhance mouthfeel by increasing polysaccharide content.

Several wine related species of non-*Saccharomyces* yeasts, such as, for example, *Kloeckera apiculata* and *Candida stellata*, have been reported to produce high amounts of extracellular β-glucanase enzymes (Strauss et al., 2001), and therefore, could be considered a means to hasten the breakage of the yeast cell wall.

11.3 ACCELERATED AGING ON LEES

Several techniques have been described as useful to accelerate the breakage of the yeast and the fragmentation of cell walls. Among them, β-glucanases enzymes (Palomero et al., 2007), HHP (high hydrostatic pressure) (Nida, 2010), HPH (high pressure homogenization) (Comuzzo et al., 2015a,b), US (ultrasound) (Kulkarni et al., 2015; Liu et al., 2016), MW (microwaves) (Liu et al., 2016), and PEF (pulsed electric fields) (Martínez et al., 2016) have all reported advantages for a faster release of polysaccharides and/or cytoplasmic proteins when yeast cells are processed using them.

β-glucanases enzymes produce the hydrolysis of β-glucans, thus facilitating the cell wall disruption and consequently increasing the release of polysaccharides. Yeast lysis time can be reduced from several months to a few weeks in comparison to conventional natural autolysis (Palomero et al., 2007). The presence of some collateral enzymatic activities such as β-glycosidases can affect anthocyanin stability (Palomero et al., 2007). However, some authors reported no effect on the release of polysaccharides when commercial β-glucanases enzymes (4 g/HL) were added to the lees (3% v/v of fine lees) (Del Barrio-Galán et al., 2011). It is likely that there may be some influence of the doses of enzyme versus lees concentration or of the conditions in which the aging was carried out. Regarding the influence of this enzyme addition to the aromatic quality of the wine, it was found that using β-glucanases during AOL at a dose equal to or higher than 30 mg/L allows the obtaining of wines with a higher content of certain volatile compounds, mainly ethyl esters and 2-phenylethanol, but also interestingly hexanol and trans-3-hexenol, even though these last two compounds are not involved in yeast metabolism (Masino et al., 2008).

Lees pressurization at 100 MPa affects cell wall structure, promoting depolymerization and facilitating yeast lysis and polysaccharides release (Nida, 2010). The use of HPH at 150 MPa produces similar levels of colloids polysaccharides, proteins, and free amino acids than thermal autolysis but increases ethyl esters and reduces fatty acids (Comuzzo et al., 2015a,b).

The use of US is an effective technology to disrupt the cells facilitating the depolymerization of cell coverings and releasing mannoproteins and cell wall polysaccharides. When US is employed for a few minutes to a yeast biomass at a power of 400 W and a frequency of 24 KHz, with a sonotrode S24D14D (Ø14 mm, length 100 mm) the breakage of cell coverings and the release of cytosolic contents facilitating the autolysis process can be observed by optical microscopy (Fig. 11.6). The effect of US can be enhanced when inert abrasives such as sand or glass beads are used together (Liu et al., 2016).

The application of US to non-*Saccharomyces* produces a fast lysis of cells (Fig. 11.7) and it can be observed that US treated non-*Saccharomyces* produces 2—4 times more polysaccharides than conventional *Saccharomyces* yeasts (Kulkarni et al., 2015). Maximum concentrations after US treatments are observed after 2—3 weeks in all yeast species, so the use of US is a powerful tool to accelerate yeast lysis and to speed up AOL especially when used together with non-*Saccharomyces* yeasts.

Thermal treatments are traditionally used to produce yeast derivatives used in enology such as inactivated yeasts or cell wall mannoproteins. The use of emerging heating systems including MW facilitates yeast cell disruption in a short time, thus increasing the release of polysaccharides (Liu et al., 2016).

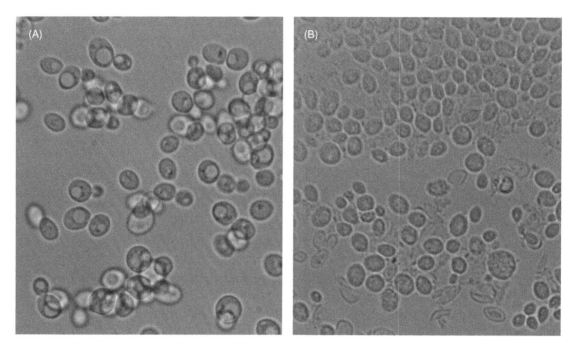

FIGURE 11.6 (A) Control. (B) Ultrasound treated lees.

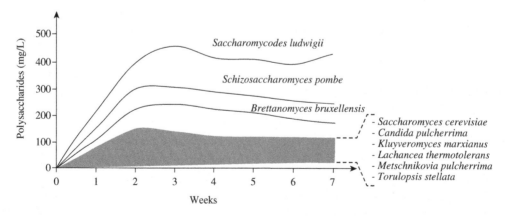

FIGURE 11.7 Release of polysaccharides by both *Saccharomyces* and non-*Saccharomyces* yeasts during aging on lees accelerated by ultrasound. *Adapted from Kulkarni, P., Loira, I., Morata, A., Tesfaye, W., González, M.C., Suárez-Lepe, J.A., 2015. Use of non-Saccharomyces yeast strains coupled with ultrasound treatment as a novel technique to accelerate ageing on lees of red wines and its repercussion in sensorial parameters. LWT Food Sci. Technol. 64, 1255–1262.*

TABLE 11.1 Main emerging technologies available to accelerate and/or improve wine aging on lees: features and observed results

	Mechanism of action	Advantages	Drawbacks	References
β-Glucanases enzymes	Hydrolysis of βglucans forming part of the yeast cell wall	Shorter duration of the autolytic process (2–3 weeks). Conventional autolysis (nonaccelerated AOL): >5 months	Slight loss of color (undesired β-glucosidase activity)	Palomero et al. (2007), Masino et al. (2008), and Fernández et al. (2011)
		Faster and higher release of polysaccharides from the cell wall	Increased risk of contamination by altering microorganisms (e.g., *Brettanomyces*)	
		Enrichment of the aromatic profile	Homogeneity in the size of the released polysaccharides	
HHP	Inflicts damage on cytoplasmic membrane favoring depolymerization	Higher release of polysaccharides	High investment costs in equipment acquisition	Nida (2010)
			Industrial application in continuous processes	
HPH	Cavitation, turbulence and shear, which occur when the yeast suspension is forced through the homogenization valve	Increase of ethyl esters and reduction of fatty acids		Comuzzo et al. (2015) and Popper and Knorr (1990)
US	Sound waves used to cause thinning of cell membranes, localized heating and production of free radicals	Faster and higher release of polysaccharides from the cell wall	Possible slight oxidation of the wine	Kulkarni et al. (2015) and Liu et al. (2016)
		No negative effects on the sensory quality of the wine		
		Possible influence in wine color stability (higher amounts of acetaldehyde)		
PEF	Electroporation and electric breakdown	Acceleration of autolysis	Increase in metal ions (Fe, Cr, Zn, and Mn) concentrations	Martínez et al. (2016) and Yang et al. (2016)
	Destabilization of the lipid bilayer and proteins of the cell membranes	Faster and higher release of mannoproteins		
		No negative effect on wine quality has been reported		
Micro-oxygenation	Controlled addition of oxygen into the wine in small doses leads to wine components transformation	Increased color stability	Decrease in total anthocyanin content	Del Barrio-Galán et al. (2011)
		Prolonged aging on lees	Residual dissolved oxygen in the medium (problems of overoxygenation)	
Use of nontoasted oak chips	Transfer of wood compounds to wine	Enrich wine in polysaccharides from wood (not from the yeast cell wall)	Difficulty in controlling the proper transfer of compounds (dose and contact time dependent)	Del Barrio-Galán et al. (2011)
		Artificially accelerate the aging process in barrel		

HHP, High hydrostatic pressure; *HPH*, High pressure homogenization; *US*, Ultrasounds. *AOL*, Aging on lees; *PEF*, Pulsed electric fields.

The PEF inactivation mechanism is mainly based on the electroporation of the cytoplasmic membrane that leads to disorders in the membrane's electrical potential, osmotic imbalance, and, finally, cell lysis (Barba et al., 2018, Page 115). This nonthermal and efficient method in the use of energy technology has little effect on wine composition (Yang et al., 2016). These days, its main drawback is the release of metal ions from the electrode that may significantly increase Fe, Cr, Zn, and Mn concentrations. However, further research in this field is needed to properly understand all the implications surrounding electrode degradation.

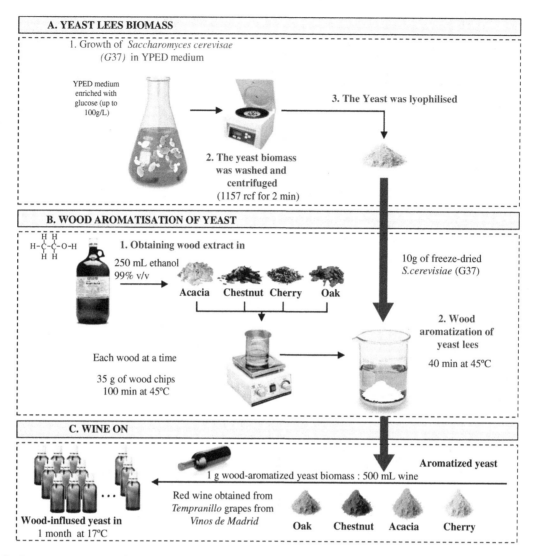

FIGURE 11.8 Lees impregnation with wood-extract procedure.

Another suggested alternative to increase the content of polysaccharides in wine is the use of nontoasted oak chips (for example: 4 g/L) with the aim of extracting these polymers from the wood (Del Barrio-Galán et al., 2011). Despite these polysaccharides having a different origin, they can act in the same manner as polysaccharides released from yeast cell walls. That is, increasing the wine's mouthfeel and body, as well as reducing astringency and bitter flavors.

The yeast cell disruption and the nonthermal permeabilization of cell membrane provoked by the aforementioned emerging technologies have been shown to be effective for wine microbial inactivation (García et al., 2013; Ganeva et al., 2014; Abca and Akdemir Evrendilek, 2014). But, at the same time, they are also useful to accelerate the lees yeast lysis with the subsequent release of protective colloids and biological material (e.g., mannoproteins, polysaccharides, peptides, amino acids, fatty acids, nucleotides, etc.) from the cytoplasm or the coating structures (cell wall and cytoplasmic membrane) to the wine (Martínez et al., 2016).

Regarding complementary techniques to AOL, micro-oxygenation (5 mL/L/month of O_2) applied to AOL improves the color stability of the wine through the formation of new pigments that may increase or remain constant until the end of the barrel aging process (Del Barrio-Galán et al., 2011). These micro-oxygenated + AOL wines are characterized by higher color intensities and blue tonalities.

Similarly, the use of nontoasted oak chips is an effective technique to accelerate the transfer of wood components to the wine in order to improve aromatic complexity and mouthfeel without the need of aging in barrels.

By adjusting the appropriate dose, it is possible to obtain the desired wood characteristics in the wine in shorter periods of time thanks to the greater transfer surface on the chips.

Table 11.1 summarizes the main novel techniques that can be coupled with AOL to improve results or speed up the AOL process.

11.4 LEES AROMATIZATION

A new technique proposed to incorporate wood aromatic compounds from several tree species is the use of yeast lees as a carrier for the wood extracts during AOL (Palomero et al., 2015). The woods studied were oak, chestnut, cherry, and acacia. A biomass of a yeast strain previously selected according to its good autolytic behavior was used to conduct the experiments. Once the yeast was soaked with an ethanol wood-extract and dried, the wood-aromatized yeast biomasses for each type of wood were poured into a red wine made from *Vitis Vinifera* L.cv. Tempranillo grapes from D.O. Vinos de Madrid. One gram of these biomasses was added to 500 mL of wine. Untreated yeasts were used as a control. Wines were left in contact with the lees for 1 month. All the treatments were performed in triplicate. Fig. 11.8 illustrates the processes carried out during the experimental stage.

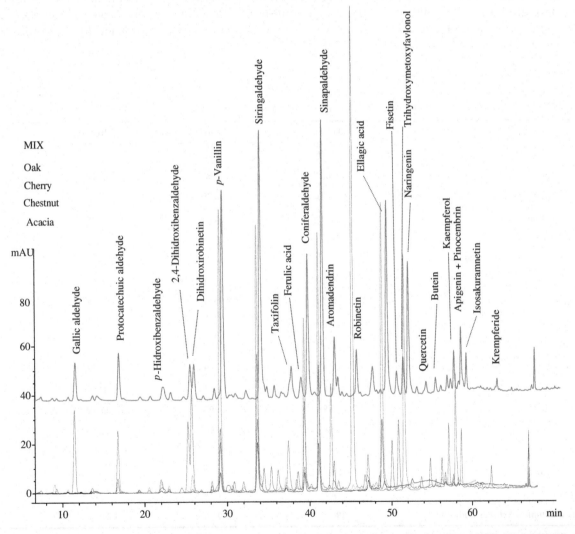

FIGURE 11.9 Phenolic compounds (HPLC/DAD chromatograms) detected on wood-aromatized yeast, and for an equal part mixture of all wood types (MIX).

The analysis of polyphenols and volatile compounds (by HPLC/DAD and GC-MS, respectively) performed following the methodology described in Palomero et al. (2015) on both the wines and the woody yeast biomasses confirmed that the adsorption/diffusion of these compounds from the wood to the yeast, and then to the wines occurs. Finally, a sensory assessment of all wines was undertaken by 10 expert tasters.

Some of the 34 wood phenols adsorbed by yeast cell walls, and then detected by HPLC/DAD, are graphically presented in Fig. 11.9. All of these compounds have been found and described previously (Sanz et al., 2012). The data confirmed sorption phenomena occurred and that the phenolic composition of the studied woods were significantly different.

This technique confirms its potential usefulness because of the great concentration of phenolics aldehydes observed in the chestnut samples. Compounds such as vanillin and syringaldehyde are some of the most distinctive wood odorants in barrel-aged wines. The concentration of vanillin, the phenolic aldehyde with the lowest sensory threshold, could be enhanced significantly through the use of chestnut instead of oak (Palomero et al., 2015).

As reported previously by other authors, there are some specific compounds only found in one type of wood and therefore, could potentially be used as identification markers (Fernández de Simón et al., 2014; Sanz et al., 2012). In this sense, one of the clearest examples can be seen in Fig. 11.9 for the highest peak corresponding to robinetin, only found in aromatized acacia samples (and in the mixture).

Table 11.2 summarizes the main volatile compounds (of the 71 volatile molecules identified that are mainly formed during wood toasting) found in the wood-aromatized yeasts. Both lactones and carbohydrate derivatives (such as furfural and 5-hydroxymethylfurfural), lignin-derived volatile phenols, and phenolic aldehydes (such as vanillin) concentrations are significantly higher in chestnut-infused yeast. Generally, large differences have been observed between the quantitative values for the kinds of wood used in our experiments. Our data correlate well with the findings of Fernández de Simón et al. (2014).

TABLE 11.2 Main volatile compounds (µg/g) in aromatized yeast biomasses

Compound	Acacia	Cherry	Chestnut	Oak
Furfural	0.56	0.75	3.05	0.82
5-Methylfurfural	0.23	0.37	1.18	0.21
5-Hydroxymethylfurfural	4.51	2.68	6.01	1.89
Trans-whiskylactone	nd	nd	0.25	0.34
Cis-whiskylactone	nd	nd	0.14	0.13
$\sum$ lactones and carbohydrate derivatives[a]	**84.74**	**94.04**	**102.95**	**66.41**
Phenol	0.17	0.21	0.20	0.15
2-Phenylethanol	4.29	5.31	6.28	3.21
4-Methyl guaiacol	0.59	0.71	4.74	0.16
4-Ethyl guaiacol	0.09	0.09	0.21	0.17
4-Vinylguaiacol	0.45	0.54	0.69	0.58
Eugenol	0.19	0.31	0.87	0.65
$\sum$ volatile phenols[a]	**11.99**	**16.03**	**26.39**	**15.87**
Vanillin	35.9	19.5	238	179
Syringaldehyde	98.2	124	252	214
2,4-Dihydroxybenzaldehyde	38.5	nd	nd	nd
p-Anisaldehyde	nd	1.12	nd	nd
$\sum$ phenolic aldehydes, ketones and other[a]	**197**	**200**	**624**	**488**

[a] Summations include other compounds not shown.
Note: Bold indicates to summary of a family of compounds.
Results in wines with wood-aromatized yeast lees.

TABLE 11.3 Phenolic composition (μg/L) of red wines enriched with wood-aromatized yeast lees[a]

Compound	Acacia	Cherry	Chestnut	Oak
Gallic acid	23.8 ± 0.5a	24.4 ± 1.0a	28.9 ± 3.5b	26.4 ± 2.3ab
Vanillic acid	1.8 ± 0.0a	1.9 ± 0.7a	3.2 ± 1.0b	2.4 ± 0.7ab
Ellagic acid	4.2 ± 0.3a	3.8 ± 0.7a	11.8 ± 0.6c	7.7 ± 1.7b
(+)-Catechin	59.7 ± 1.8a	53.6 ± 7.2a	57.9 ± 1.0a	55.4 ± 6.2a
Procyanidin B2	11.4 ± 0.7b	8.4 ± 0.9a	11.3 ± 1.1b	8.6 ± 2.8ab
(−)-Epicatechin	5.7 ± 0.2a	5.2 ± 0.5a	5.7 ± 0.3a	5.2 ± 0.6a
Myricetin-3-glucuronide	11.6 ± 0.5bc	12.9 ± 0.8c	10.3 ± 0.9ab	9.5 ± 1.2a
Quercetin-3-galactoside	9.2 ± 0.6bc	8.6 ± 1.7abc	7.2 ± 0.7ab	6.5 ± 2.0a
Quercetin-3-glucoside	5.6 ± 0.2b	5.2 ± 1.2ab	4.5 ± 0.3ab	3.9 ± 1.0a
Quercetin	1.5 ± 0.1a	2.0 ± 0.1b	1.8 ± 0.1ab	1.6 ± 0.1a
Dihydrorobinetin	0.9 ± 0.0a	nd	nd	nd
Robinetin	10.6 ± 0.0a	nd	nd	nd
Aromadendrin	nd	1.7 ± 0.0a	nd	nd

[a]Values are mean ± standard deviation (n = 3). Values in the same row with different letter denotes a statistical difference ($p < 0.05$).
nd: not detected.

TABLE 11.4 Main volatile compounds (μg/L) in red wines aged for 1 month with wood-aromatized yeast lees[a]

Compounds	Acacia	Cherry	Chestnut	Oak
Furfural	65.3 ± 7.4b	55.3 ± 3.0ab	66.4 ± 19.2b	68.2 ± 5.9b
5-Methylfurfural	5.83 ± 1.22a	5.40 ± 1.61a	6.93 ± 0.68a	6.23 ± 1.18a
5-Hydroxymethylfurfural	224 ± 19.4ab	249 ± 158ab	301 ± 83.9ab	307 ± 84.9b
Trans-β-methyl-γ-octalactone	nd	nd	18.8 ± 4.3b	19.7 ± 1.1b
Cis-β-methyl-γ-octalactone	nd	nd	19.4 ± 12.4b	24.2 ± 10.7b
4-Methylguaiacol	2.9 ± 0.3abc	2.3 ± 0.7ab	4 ± 0.5c	3.3 ± 1.1bc
4-Ethylguaiacol	0.47 ± 0.05b	0.51 ± 0.14b	0.55 ± 0.07b	0.44 ± 0.28b
4-Vinylguaiacol	2.99 ± 0.22b	3.18 ± 0.12b	4.04 ± 0.82b	3.91 ± 0.77b
Phenol	0.96 ± 0.08a	1.06 ± 0.16a	0.96 ± 0.03a	1.03 ± 0.24a
4-Ethylphenol	0.47 ± 0.04a	0.52 ± 0.18a	0.68 ± 0.03a	0.69 ± 0.24a
4-Vinylphenol	500 ± 43.7b	506 ± 118b	532 ± 90.3b	643 ± 114b
Eugenol	1.6 ± 0.1a	2.2 ± 1a	1.9 ± 0.3a	1.9 ± 0.0a
Vanillin	1132.7 ± 18.6c	864.5 ± 201.9b	3840.3 ± 167.6e	704.3 ± 87.6b
Vanillyl ethyl ether	1721 ± 303a	1783 ± 361a	3569 ± 79.9b	2786 ± 410ab
Syringaldehyde	2009.7 ± 476.3a	2938.4 ± 863.8c	4812.7 ± 478.4d	1940.6 ± 378.2b
Acetosiringone	338 ± 34.3ab	403 ± 39.3ab	654 ± 118c	283 ± 25.7a
p-Anisaldehyde	nd	28.1 ± 12.9b	nd	nd

[a]Values are mean ± standard deviation (n = 3). Values in the same row with different letter denotes a statistical difference ($p < 0.05$).
nd: not detected.

The phenolic composition of *Tempranillo* wines was significantly different between treatments. Table 11.3 shows some of the phenolics found in the wines. As previously observed during dried yeast phenolic analysis, robinetin and dihydrorobinetin were only found in wine samples aged with acacia-infused yeast. A similar conclusion can be established for aromadendrin in the case of cherry wood.

A great ratio of gallic and ellagic acid was seen in chestnut samples compared with oak ones. Some authors have claimed that these compounds could contribute to the chemical recognition of chestnut wood (Fernández de Simón et al., 2014). It is probable that these compounds will be easily perceived in wines aged in chestnut vessels.

Generally, the volatile composition of the initial wine was significantly modified before treatments. As previously seen, there are no methyl-octolactone isomers in acacia and cherry (Tables 11.2 and 11.4), and no statistical differences were observed between chestnut and oak for these compounds.

Phenolic aldehydes and phenyl ketones were detected to be at higher levels in wines aged with chestnut-infused yeast lees (Table 11.4). These compounds impart vanilla-like aromas into the wines; and woods with higher concentrations would be useful in order to reduce aging costs (Palomero et al., 2015).

No differences were observed with regard to color intensity and chromatic parameters, a fact that is reasonable when taking into account that wines were left in contact with lees only for 1 month. All the kinds of wines with wood-infused yeast were reported to be more complex than the control wine, so wine aroma was improved with the aging treatment.

Chestnut aromatized biomasses released the highest amounts of wood-related compounds like phenolic aldehydes, and it is perhaps for this reason that they were valued with the highest scores for plum, currant, spicy, roasted, and nutty aromas.

11.5 CONCLUSIONS

AOL is an interesting technique for the biological aging of red wines, softening phenolic fraction and decreasing astringency, with a protective effect on color and varietal smells. The use of selected *Saccharomyces* yeasts or some non-*Saccharomyces* species accelerates this slow process, improving industrial application. In the same way, lees processing using emerging technologies, such as HHP, HPH, US, MW, PEF, etc., accelerates the process, increasing industrial feasibility. Moreover, this is a technique that can be synergistically used with barrel aging or wood chips to increase aromatic quality.

References

Abca, E.E., Akdemir Evrendilek, G., 2014. Processing of red wine by pulsed electric fields with respect to quality parameters. J. Food Process. Preserv. 39 (6), 758–767.

Barba, F.J., Ahrné, L., Xanthakis, E., Landerslev, M.G., Orlien, V., 2018. Innovative Technologies for Food Preservation. Inactivation of Spoilage and Pathogenic Microorganisms, first ed Academic Press, Elsevier Ed, London, UK.

Caridi, A., Sidari, R., Pulvirenti, A., Meca, G., Ritieni, A., 2012. Ochratoxin A adsorption phenotype: an inheritable yeast train. J. Gen. Appl. Microbiol. 58 (3), 225–233.

Chatonnet, D., Dubourdieu, D., Boidron, J.N., 1992. Incidence des conditions de fermentation et d'élevage des vins blancs secs en barriques sur leur composition en substances cédées par le bois de chêne. Sci. Aliments 12, 665–685.

Comuzzo, P., Battistutta, F., Vendrame, M., Páez, M.S., Luisi, G., Zironi, R., 2015a. Antioxidant properties of different products and additives in white wine. Food Chem. 168, 107–114.

Comuzzo, P., Calligaris, S., Iacumin, L., Ginaldi, F., Palacios Paz, A.E., Zironi, R., 2015b. Potential of high pressure homogenization to induce autolysis of wine yeasts. Food Chem. 185 (15), 340–348.

Del Barrio-Galán, R., Pérez-Magariño, S., Ortega-Heras, M., 2011. Techniques for improving or replacing ageing on lees of oak aged red wines: the effects on polysaccharides and the phenolic composition. Food Chem. 127 (2), 528–540.

De Nobel, J.G., Klis, F.M., Munnik, T., Priem, J., Van Den Ende, H., 1990. An assay of relative cell wall porosity in *Saccharomyces cerevisiae*, *Kluyveromyces lactis* and *Schizosaccharomyces pombe*. Yeast 6, 483–490.

Domizio, P., Liu, Y., Bisson, L.F., Barile, D., 2017. Cell wall polysaccharides released during the alcoholic fermentation by *Schizosaccharomyces pombe* and *S. japonicus*: quantification and characterization. Food Microbiol. 61, 136–149.

Dupin, I.V.S., McKinnon, B.M., Ryan, C., Boulay, M., Markides, A.J., Jones, G.P., et al., 2000. *Saccharomyces cerevisiae* mannoproteins that protect wine from protein haze: Their release during fermentation and lees contact and a proposal for their mechanism of action. J. Agric. Food Chem. 48, 3098–3105.

REFERENCES

Escot, S., Feuillat, M., Dulau, L., Charpentier, C., 2001. Release of polysaccharides by yeasts and the influence of released polysaccharides on colour stability and wine astringency. Aust. J. Grape Wine Res. 7, 153–159.

Fernández, O., Martínez, O., Hernández, Z., Guadalupe, Z., Ayestarán, B., 2011. Effect of the presence of lysated lees on polysaccharides, color and main phenolic compounds of red wine during barrel ageing. Food Res. Int. 44 (1), 84–91.

Fernández de Simón, B., Sanz, M., Cadahía, E., Martínez, J., Esteruelas, E., Muñoz, A.M., 2014. Polyphenolic compounds as chemicals markers of wine ageing in contact with cherry, chestnut, false acacia, ash and oak wood. Food Chem. 143, 66–76.

Gallardo-Chacón, J.J., Vichi, S., Urpí, P., López-Tamames, E., Buxaderas, S., 2010. Antioxidant activity of lees cell surface during sparkling wine sur lie aging. Int. J. Food Microbiol. 143, 48–53.

Ganeva, V., Galutzov, B., Teissie, J., 2014. Evidence that pulsed electric field treatment enhances the cell wall porosity of yeast cells. Appl. Biochem. Biotechnol. 172 (3), 1540–1552.

García, J.F., Guillemet, L., Feng, C., Sun, D.W., 2013. Cell viability and proteins release during ultrasound-assisted yeast lysis of light lees in model wine. Food Chem. 141 (2), 934–939.

Humbel, B.M., Konomi, M., Takagi, T., Kamasawa, N., Ishijima, S.A., Osumi, M., 2001. In situ localization of β-glucans in the cell wall of Schizosaccharomyces pombe. Yeast 18, 433–444.

Kopeckà, M., Fleet, G.H., Phaff, H.J., 1995. Ultrastructure of the cell wall of *Schizosaccharomyces pombe* following treatment with various glucanases. J. Struct. Biol. 114, 140–152.

Kulkarni, P., Loira, I., Morata, A., Tesfaye, W., González, M.C., Suárez-Lepe, J.A., 2015. Use of non-*Saccharomyces* yeast strains coupled with ultrasound treatment as a novel technique to accelerate ageing on lees of red wines and its repercussion in sensorial parameters. LWT Food Sci. Technol. 64, 1255–1262.

Liu, L., Loira, I., Morata, A., Suárez-Lepe, J.A., González, M.C., Rauhut, D., 2016. Shortening the ageing on lees process in wines by using ultrasound and microwave treatments both combined with stirring and abrasion techniques. Eur. Food Res. Technol. 242, 559–569.

Loira, I., Vejarano, R., Morata, A., Ricardo-da-Silva, J.M., Laureano, O., González, M.C., et al., 2013. Effect of Saccharomyces strains on the quality of red wines aged on lees. Food Chem. 139, 1044–1051.

Magnelli, P., Cipollo, J.F., Abeijon, C., 2002. A refined method for the determination of *Saccharomyces cerevisiae* cell wall composition and β-1,6-glucan fine structure. Anal. Biochem. 301, 136–150.

Manners, D.J., Meyer, M.T., 1977. The molecular structures of some glucans from the cell walls of *Schizosaccharomyces pombe*. Carbohydr. Res. 57, 189–203.

Martínez, J.M., Cebrián, G., Álvarez, I., Raso, J., 2016. Release of mannoproteins during *Saccharomyces cerevisiae* autolysis induced by Pulsed Electric Field. Front. Microbiol. 7 (1435), 1–8.

Masino, F., Montevecchi, G., Arfelli, G., Antonelli, A., 2008. Evaluation of the combined effects of enzymatic treatment and aging on lees on the aroma of wine from Bombino bianco grapes. J. Agric. Food Chem. 56 (20), 9495–9501.

Morata, A., Suárez Lepe, J.A., 2016. New biotechnologies for wine fermentation and ageing. In: Ravishankar Rai, P.V. (Ed.), Advances in Food Biotechnology. John Wiley & Sons, Ltd, West Sussex, United Kingdom, pp. 293–295.

Morata, A., Gómez-Cordovés, M.C., Suberviola, J., Bartolomé, B., Colomo, B., Suarez, J.A., 2003. Adsorption of anthocyanins by yeast cell walls during the fermentation of red wines. J. Agric. Food Chem. 51, 4084–4088.

Nida, R.H., 2010. Application of High Pressure Technologies in Over Lees Aging of Red Wines and Repercussions in Sensorial Parameters. MS thesis. Vinifera-European Master of Viticulture and Enology.

Palomero, F., Bertani, P., Fernández de Simón, B., Cadahía, E., Benito, S., Morata, A., et al., 2015. Wood impregnation of yeast lees for winemaking. Food Chem. 171, 212–223.

Palomero, F., Morata, A., Benito, S., González, M.C., Suárez-Lepe, J.A., 2007. Conventional and enzyme-assisted autolysis during ageing over lees in red wines: Influence on the release of polysaccharides from yeast cell walls and on wine monomeric anthocyanin content. Food Chem. 105, 838–846.

Palomero, F., Morata, A., Benito, S., Calderón, F., Suárez-Lepe, J.A., 2009. New genera of yeasts for over-lees aging of red wine. Food Chem. 112, 432–441.

Palomero, F., Ntanos, K., Morata, A., Benito, S., Suárez-Lepe, J.A., 2011. Reduction of wine 4-ethylpohenol concentration using lyophilised yeast as a bioadsorbent: influence on anthocyanin content and chromatic variables. Eur. Food Res. Technol. 232, 971–977.

Popper, L., Knorr, D., 1990. Application of high pressure homogenization for food reservation. Food Technol. 44, 84–89.

Rodrigues, A., Ricardo-Da-Silva, J.M., Lucas, C., Laureano, O., 2012. Effect of commercial mannoproteins on wine colour and tannins stability. Food Chem. 131, 907–914.

Rodríguez-Bencomo, J.J., Ortega-Heras, M., Pérez-Magariño, S., 2010. Effect of alternative techniques to ageing on lees and use of non-toasted oak chips in alcoholic fermentation on the aromatic composition of red wine. Eur. Food Res. Technol. 230 (3), 485–496.

Rodríguez-Bencomo, J.J., Andújar-Ortiz, I., Moreno-Arribas, M.V., Simó, C., González, J., Chana, A., et al., 2014. Impact of glutathione-enriched inactive dry yeast preparations on the stability of terpenes during model wine aging. J. Agric. Food Chem. 62, 1373–1383.

Salmon, J.-M., Fornairon-Bonnefond, C., Mazauric, J.-P., Moutounet, M., 2000. Oxygen consumption by wine lees: impact on lees integrity during wine ageing. Food Chem. 71, 519–528.

Sanz, M., Fernández de Simón, B., Cadahía, E., Esteruelas, E., Muñoz, A.M., Hernández, M.T., et al., 2012. Polyphenolic profile as a useful tool to identify the wood used in wine ageing. Anal. Chim. Acta 732, 33–45.

Spillman, P.J., 1999. Wine quality biases inherent in comparisons of oak chip and barrel systems. Aust. N. Z. Wine Ind. J. 14, 25–33.

Strauss, M.L.A., Jolly, N.P., Lambrechts, M.G., Van Rensburg, P., 2001. Screening for the production of extracellular hydrolytic enzymes by non-*Saccharomyces* wine yeasts. J. Appl. Microbiol. 91 (1), 182–190.

Suárez-Lepe, J.A., Morata, A., 2009. Nuevo método de crianza sobre lías en vinos tintos. Oficina española de patentes y marcas, ES2311372 B2. October 07, 2009, Spanish Patent, p. 10905.

Suárez-Lepe, J.A., Morata, A., 2012. New trends in yeast selection for winemaking. Trends Food Sci. Technol. 23, 39–50.

Yang, N., Huang, K., Lyu, C., Wang, J., 2016. Pulsed electric field technology in the manufacturing processes of wine, beer, and rice wine: a review. Food Control 61, 28–38.

Further Reading

Escott, C., Vaquero, C., del Fresno, J.M., Bañuelos, M.A., Loira, I., Han, S.-Y., et al., 2017. Pulsed light effect in red grape quality and fermentation. Food Bioprocess Technol. 10 (8), 1540–1547.

Morata, A., Loira, I., Vejarano, R., González, C., Callejo, M.J., Suárez-Lepe, J.A., 2017. Emerging preservation technologies in grapes for winemaking. Trends Food Sci. Technol. 67, 36–43.

Palomero, F., Benito, S., Morata, S., Calderón, F., Suárez-Lepe, J.A., 2008. New yeast genera for over lees ageing in red wines. In: XXX World Congress of the International Organization of Grape and Wine (OIV). 16–19 June, Verona, Italy.

Suarez-Lepe, J.A., Morata, A., 2015. Levaduras para vinificación en tinto. AMV Ediciones. Madrid, Spain (Chapter 8) 277–291. in Spanish.

CHAPTER

12

Evolution of Proanthocyanidins During Grape Maturation, Winemaking, and Aging Process of Red Wines

António M. Jordão[1,2] and Jorge M. Ricardo-da-Silva[3]

[1]Department of Food Industries, Polytechnic Institute of Viseu (CI&DETS), Viseu, Portugal [2]CQ-VR, Chemistry Research Centre, Vila Real, Portugal [3]LEAF, Linking Landscape, Environment, Agriculture and Food, Higher Institute of Agronomy - ISA, University of Lisbon, Lisbon, Portugal

12.1 PROANTHOCYANIDINS: COMPOSITION, CONTENT, AND EVOLUTION DURING GRAPE MATURATION

12.1.1 General Composition and Content of Proanthocyanidins in Grapes

Grape proanthocyanidins (condensed tannins or flavanols) are oligomers and polymers of flavan-3-ols. They release the corresponding anthocyanidin on treatment with acid and alcohol at high temperature, by breaking the interflavan C—C bonds. Quantity, structure, and degree of polymerization of grape proanthocyanidins depend on their localization in grape tissues (Ricardo-da-Silva et al., 1991a; Souquet et al., 1996; De Freitas et al., 2000; Jordão et al., 2001a). These phenolic compounds are found in all grape cluster fractions, in particular in seeds, but also in the stems and skins. Grape pulp presents lower amounts of these compounds (Bourzeix et al., 1986; Sun et al., 2001; Monagas et al., 2003), with the pulps of the *teinturier* grapes being the richest in proanthocyanidins (Ricardo-da-Silva et al., 1992a, 1992b; Sun et al., 2001).

In grape seeds, these phenolic compounds are oligomers and polymers composed of the monomeric flavan-3-ols, (+)-catechin, (−)-epicatechin, and (−)-epicatechin gallate linked by C_4-C_8 and/or C_4-C_6 bonds (B type), while in skins, stems, and pulps it is also possible to detect (−)-epigallocatechin and trace amounts of (+)-gallocatechin and (−)-epigallocatechin gallate (Genebra et al., 2014). Several authors also found gallocatechins and prodelphinidins in grapes (Escribano-Bailón et al., 1995; Souquet et al., 1996).

Fig. 12.1 shows several examples of chemical structures of (+)-catechin and (−)-epicatechin monomers and B types of proanthocyanidins found in grapes. The A-type of proanthocyanidins present one supplementary linkage C_2-O-C_7 or C_2-O-C_5, in addition to linkage $C_4 \rightarrow C_8$ or $C_4 \rightarrow C_6$. A-type PAs, like A2 have also been detected in grapes (Glories et al., 1996; Passos et al., 2007).

The degree of polymerization is an important characteristic of the proanthocyanidins structure. According to their increasing degree of polymerization, proanthocyanidins are termed as follows: dimers, trimers, oligomers, and polymers (Glories, 1978; Fine, 2000). Grape and wine proanthocyanidins are presented essentially in polymeric forms (60%−80%), followed by oligomeric forms (15%−30%), while monomer flavan-3-ols [(+)-catechin and (−)-epicatechin] represent less than 10% of the total proanthocyanidins (Sun et al., 2001). According to Boido et al. (2011), 40% of the total flavan-3-ols quantified in Tannat grape seeds were galloylated compounds, whereas the flavan-3-ol profile in skins was characterized by the absence of galloylated forms. In addition, prodelphinidins in skins ranged between 30% and 35% with very low values for (−)-epigallocatechin.

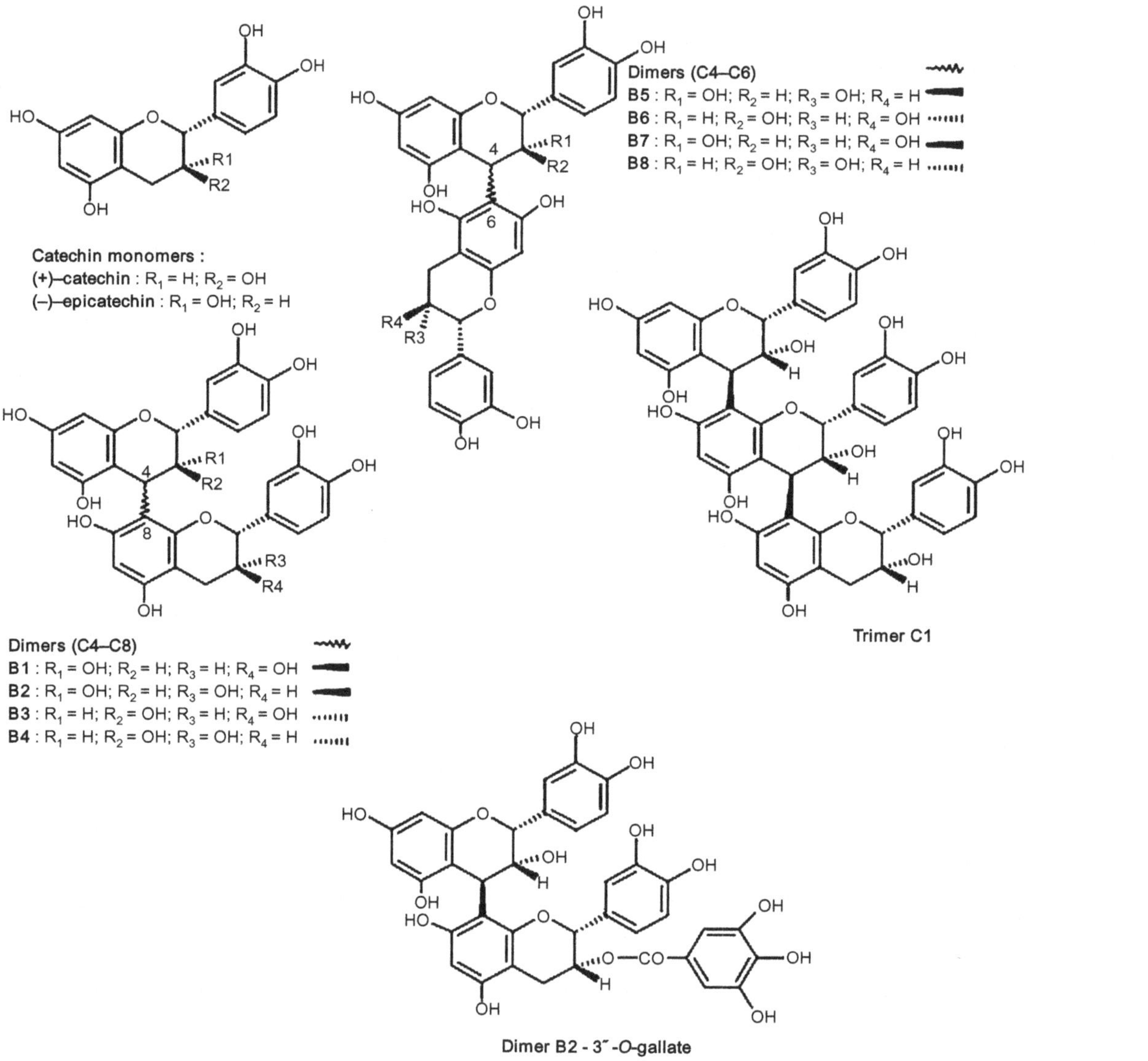

FIGURE 12.1 Different chemical structures of (+)-catechin and (−)-epicatechin monomers and B-type of proanthocyanidins detected in grapes. *Adapted from Jordão, A.M., Simões, S., Correia, A.C., Gonçalves, F.J., 2012a. Antioxidant activity evolution during Portuguese red wine vinification and their relation with the proanthocyanidin and anthocyanin composition. J. Food Process. Preserv. 36, 298–309.*

Several authors concluded that there is a high range of degree of polymerization of proanthocyanidins depending on grape fraction and also from the grape variety studied. Thus, Monagas et al. (2003) detected in grapes from Graciano, Tempranillo, and Cabernet Sauvignon, that the polymeric fraction represented 75%–81% of total flavan-3-ols in seeds and 94%–98% in skins and showed the mean degree of polymerization values of 6.4–7.3 in seeds and 33.8–85.7 in skins. A similar tendency was also described by Cosme et al. (2009), where different red grape varieties cultivated in Portugal were studied (Touriga Nacional, Trincadeira, Castelão, Syrah, and Cabernet Sauvignon). Thus, the polymeric fractions represented 77%–85% in seeds and 91%–99% in skins, while the distribution of the mean degree of polymerization of proanthocyanidins ranged from 2.8 to 12.8 for seeds and from 3.8 to 81.0 for skins. Previously, Prieur et al. (1994), reported in red grape seeds, a mean degree of polymerization values ranging from 2.3 to 16.7, while in red grape skins, Souquet et al. (1996) evaluated that mean degree of polymerization ranged from 3.4 to 83.3. Other authors reported similar trends for mean degree of polymerization of proanthocyanidins in grape seed (ranging from 2.4 to 31.5) and skin (ranging from 3.4 to 83.3) extracts (Prieur et al., 1994; Souquet et al., 1996; Sun et al., 1998; Chira et al., 2011).

It is important to note that also during grape maturation the mean degree of polymerization has significant changes. Thus, Bordiga et al. (2011) showed that mean the degree of polymerization of the polymeric fraction of seed proanthocyanidins generally decreased during ripening, starting with a maximum of 16.0 to a minimum of 7.8. Previously, Obreque-Slier et al. (2010) also observed a decrease of mean degree of polymerization from 10 to 3.8 in Carménère skin, whereas in Cabernet Sauvignon the mean degree of polymerization was almost constant varying from 6.4 to 7.1. For grape stems, other authors (Jordão et al., 2001a) also reported for different Portuguese grape varieties, two red (Castelão Francês and Touriga Francesa) and one white (Viosinho), a decrease throughout grape development for monomers and oligomers. For grape skins there have been reports of an increase in the mean polymerization degree of the proanthocyanidins during grape berry ripening (Hanlin and Downey, 2009; Kennedy et al., 2010, 2001).

As a result of different factors, such as grape variety, climatic conditions, and viticultural practices, there is a high range of the levels of proanthocyanidins quantified in the different fractions of grape bunch (seeds, skins, pulps, and stems). Thus, there are a great number of works that reported in a quantitative point of view the grape proanthocyanidin content. For example, Jordão et al. (2001b) and Sun et al. (2001), reported on average and on the basis of fresh weight, the following concentrations of proanthocyanidins: total monomers [(+)-catechin and (−)-epicatechin], 2–12 mg/g in seeds, 0.1–0.7 mg/g in skins and 0.2–11,6 mg/kg in pulps; total oligomers, 19–43 mg/g in seeds, 0.8–3.5 mg/g in skins, and 0.3–9.6 mg/kg in pulps and total polymers, 45–78 mg/g in seeds, 2–21 mg/g in skins, 0.9–90.8 mg/kg in pulps, and 28.0–35.8 mg/g in stems. In addition, the number and content of phenolic compounds in grape pulp, in particular flavan-3-ols, are significantly lower compared to skins and seeds. Pantelić et al. (2016) quantified for several grape varieties values that ranged from 0.37 to 0.46 mg/kg for (−)-epigallocatechin (for Cabernet Sauvignon and Petra grape varieties, respectively) and 1.68 to 1.95 mg/kg for (−)-epigallocatechin (for Welschriesling and Cabernet Franc grape varieties, respectively). Also Di Lecce et al. (2014) reported for Albariño white grape variety a lower concentration of flavan-3-ols (0.23 and 0.55 mg/100 g of fresh matter for (−)-epicatechin and (+)-catechin, respectively) and proanthocyanidins (0.53 mg/100 g of fresh matter for procyanidin B3) in grape pulp.

For the levels of individual proanthocyanidins quantified in several grape varieties from the different grape fractions a high range of values have been quantified in diverse published works. Thus, Table 12.1 shows the different flavan-3-ols and some individual proanthocyanidins quantified in several red grape varieties.

Considering the data shown in Table 12.1, it is easy to conclude that proanthocyanidin content and composition in different grape bunch fractions is strongly dependent of grape variety. Besides the authors mentioned in Table 12.1, many other authors report that grape varieties are decisive for the grape proanthocyanidin content (Ricardo-da-Silva et al., 1991a, 1991b, 1992a, 1992b; Sun et al., 1998; Jordão et al., 1998, 2001b; Fuleki and Ricardo-da-Silva, 1997, 2003; Monagas et al., 2003; Cosme et al., 2009; Bautista-Ortín et al., 2013).

According to Obreque-Slier et al. (2010), Carménère grape seeds and skins, for example, presented a higher mDP, a higher percentage of galloylation, compared to Cabernet Sauvignon seeds and skins. Other authors (Monagas et al., 2003), revealed that the monomeric and oligomeric content in Tempranillo seeds was the lowest one when compared to Graciano and Cabernet Sauvignon seeds. Nevertheless, Tempranillo skins showed higher content of monomeric, oligomeric, and polymeric flavanols than both Graciano and Cabernet Sauvignon skins. Boido et al. (2011) also reported that the content of flavan-3-ols in Tannat seeds was higher than that reported for a large number of other grape varieties. Recently Rice et al. (2017) confirmed this tendency, describing for several *Vitis vinifera* grape varieties (Marquette, Frontenac, and St. Croix) higher seed tannin concentration (ranging from 0.19 to 0.54 mg/berry in catechin equivalents) in comparison with skin tannins (ranging from 0.03 to 0.24 mg/berry in catechin equivalents). In addition, by the analysis of Table 12.1, in general it is also clear that procyanidin dimer B1 is the most abundant dimeric form in stems and skins, while procyanidins dimer B2 and B4 are found in highest concentration in seeds. According to De Freitas et al. (2000), procyanidin dimer B4 may be used as a chemical marker in musts and wines to quantify the contribution of the seeds. Similar tendencies for the different grape bunch fractions were also detected by other authors (Fuleki and Ricardo-da-Silva, 1997; Jordão et al., 1998, 2001b; De Freitas et al., 2000; Monagas et al., 2003). Furthermore, procyanidin dimer B7 was only detected in low concentration in the seeds of some grapes (Mateus et al., 2001). In general, grape skins have relatively low content of procyanidin dimers relative to that in seeds, the latter containing relatively high concentrations of procyanidin dimer B4, while in skins it is not possible to detect this dimer. However, before Lorrain et al. (2011) identified and quantified several oligomers (B1, B2, and B3) including also dimer B4 in seeds and skins at harvest in Merlot and Cabernet Sauvignon grapes from Bordeaux region.

12.1.2 Evolution of Proanthocyanidins During Grape Maturation

In general, there is a tendency for higher concentrations of flavanols, in particular proanthocyanidins in the early stages of grape berry development. The high values of these compounds may be related with their

TABLE 12.1 Flavan-3-Ols and Some Individual Procyanidins Quantified in Several Red Grape Varieties and From Different Grape Bunch Fractions

Red grape variety	(+)-Cat.	(−)-Epic.	Procy B3	Procy B1	Procy B4	Procy B2	References
			Grape fraction				
STEMS							
Merlot	60[a]						Souquet et al. (2000)
Saint Laurent	867[e]	81.4[e]					Balík et al. (2008)
Blauer Portugieser	1369[e]	156[e]					
Touriga Francesa	0.11[g]	1.33[g]	0.89[g]	6.59[g]	0.13[g]	0.27[g]	Jordão et al. (2001a)
Castelão Francês	0.07[g]	1.13[g]	0.04[g]	1.93[g]	0.14[g]	0.10[g]	
SKINS							
Merlot	25.0[b]	13.0[b]	35.0[b]	21.0[b]		2.2[b]	Montealegre et al. (2006)
Cencibel	22.0[b]	8.4[b]	39.0[b]	22.0[b]		1.5[b]	
Carménère	1.3[b]		0.60[b]				Obreque-Slier et al. (2010)
Cabernet Sauvignon	0.5[b]		0.70[b]				Montealegre et al. (2006)
Shiraz	8.5[b]	6.9[b]	16.0[b]	8.4[b]		0.75[b]	
Cabernet Sauvignon	1.8–6.2[d]						Nuñez et al. (2004)
Touriga Nacional	0.012–0.021[f]		0.013[f]	0.18–0.26[f]		0.020[f]	Mateus et al. (2001)
Touriga Francesa	0.012[f]	0.010[f]		0.09–0.13[f]		0.011–0.015[f]	
Autumn Royal	10.47[h]						Lutz et al. (2011)
Crimson Seedless	2.26[h]						
Red Globe	2.47[h]						
Ribier	2.84[h]						
Merlot	0.047[f]	0.030[f]	0.010[f]	0.021[f]			
Aglianico	0.151[g]	0.111[g]	0.026[g]	0.035[g]		0.036[g]	Rinaldi et al. (2014)
Merlot	0.167[g]	0.128[g]		0.034[g]		0.045[g]	
Syrah	0.07–0.34[i]		0.16–0.87[i]			0.04–0.19[i]	Kyraleou et al. (2016)
SEEDS							
Merlot	240.0[c]	210.0[c]	64.0[c]	170.0[c]	80.0[c]	37.0[c]	Montealegre et al. (2006)
Cencibel	82.0[c]	60.0[c]	43.0[c]	74.0[c]	39.0[c]	21.0[c]	
Cabernet Sauvignon	270.0[c]	130.0[c]	50.0[c]	150.0[c]	57.0[c]	41.0[c]	
Shiraz	120.0[c]	130.0[c]	55.0[c]	100.0[c]	33.0[c]	23.0[c]	
Touriga Francesa	3.30[g]	2.30[g]	0.33[g]	0.54[g]	0.59[g]	1.35[g]	Jordão et al. (2001a)
Castelão Francês	2.11[g]	4.55[g]	0.52[g]	0.08[g]	2.64[g]	2.02[g]	
Merlot	1.68[f]	2.18[f]	0.233[f]	0.100[f]	0.295[f]	0.530[f]	Lorrain et al. (2011)
Cabernet Sauvignon	1.73[f]	1.35[f]	0.172[f]	0.114[f]	0.655[f]	0.621[f]	
Aglianico	1.043[g]	0.870[g]	0.046[g]	0.065[g]		0.078[g]	Rinaldi et al. (2014)
Merlot	1.084[g]	1.558[g]	0.049[g]	0.059[g]		0.085[g]	
Syrah	6.15–8.93[i]	11.7–16.1[i]		0.30–0.38[i]		2.02–2.66[i]	Kyraleou et al. (2016)

(Continued)

TABLE 12.1 (Continued)

Red grape variety	(+)-Cat.	(−)-Epic.	Procy B3	Procy B1	Procy B4	Procy B2	References
			Grape fraction				
PULP							
Alicante Bouschet			0.001[j]	0.008[j]	0.004[j]	0.004[j]	Ricardo-da-Silva et al. (1992a)
Castelão	2.7–5.2[k]						Sun et al. (2001)
Ghara Shani	514[l]	234[l]					Farhadi et al. (2016)
Ghara Ghandome	354[l]	135[l]					
Not mentioned (cultivated in Thailand)	0.07[f]	0.05[f]					Wongnarat and Srihanam (2017)

[a] *mg/kg of stem.*
[b] *mg/kg of fresh grape skin.*
[c] *mg/kg of fresh grape seed.*
[d] *g (+)-catechin/kg of dried skin.*
[e] *mg/kg of fresh weight.*
[f] *mg/g of dry weight.*
[g] *mg/g.*
[h] *ppm.*
[i] *mg/g fresh weight.*
[j] *g/kg pulp.*
[k] *mg/kg fresh pulp.*
[l] *μg/g.*

metabolization throughout the ripening process. However, during grape maturation, in particular after *véraison*, several works describe a continual decrease of the proanthocyanidin content until grape harvest (Jordão et al., 1998; Kennedy et al., 2001; Downey et al., 2003; Ó-Marques et al., 2005; Hanlin and Downey, 2009; Bordiga et al., 2011; Jordão and Correia, 2012; Asproudi et al., 2015; Allegro et al., 2016; Rice et al., 2017). Fig. 12.2, shows an example of the evolution of the different seed and skin proanthocyanidins fractions during the grape ripening.

The decrease after *véraison* of proanthocyanidins in seeds and skins could be explained by oxidation reactions (Kennedy et al., 2000; Cadot et al., 2006) and also by a reduction of the extractability resulting from the conjugation of proanthocyanidins with other cellular components (Cheynier et al., 1997). However, for other authors the proanthocyanidin decrease is a consequence of different combined factors. Thus, for Valero et al. (1989) the proanthocyanidin concentration decrease during grape maturation is only a consequence of the increasing weight of the berries or seeds, while, for Bogs et al. (2005) the proanthocyanidin content during grape maturation is a result of a balance between the accumulation of proanthocyanidin through synthesis and decreased extractability. Previously, Baranac et al. (1997), reported that the proanthocyanidin decrease during grape maturation may be due to the deviation of the intermediate metabolites (cyanidin and delphinidin) toward the synthesis of anthocyanins as they share the same biosynthetic pathway or to little-known phenomena involving proanthocyanidin transformation and oxidation. Other research works have also shown that the proanthocyanidin content decreases in grape pulps and stems whatever the polymerization degree they presented (Ricardo-da-Silva et al., 1992a, 1992b; Jordão et al., 2001a,b; Ó-Marques et al., 2005).

Several research works establish that vintage (Tounsi et al., 2009; Lorrain et al., 2011; Kyraleou et al., 2016), environmental factors, geographical conditions (Obreque-Slier et al., 2010; Hernández et al., 2017), and also viticultural practices (Jordão et al., 1998; Genebra et al., 2014; Bogicevic et al., 2015) are decisive for the grape proanthocyanidin content and also for the evolution of these compounds during grape maturation. According to the results obtained by Fuleki and Ricardo-da-Silva (2003), the content of (+)-catechin, (−)-epicatechin, and nine procyanidins quantified in grape juice were influenced by different factors, namely by decreasing order of importance, by cultivar and vintage.

Environmental factors, geographical conditions, and the specific conditions that occur during each vintage could determine the flavonoid pathway and consequently have an impact on proanthocyanidin concentration and distribution. Gény et al. (2003) detected an increase of mean degree of polymerization from *véraison* to maturity in Cabernet Sauvignon seeds, whereas % gal stayed almost constant, while other authors for the same grape variety reported a progressive decrease in mean degree of polymerization of seed tannins during ripening and

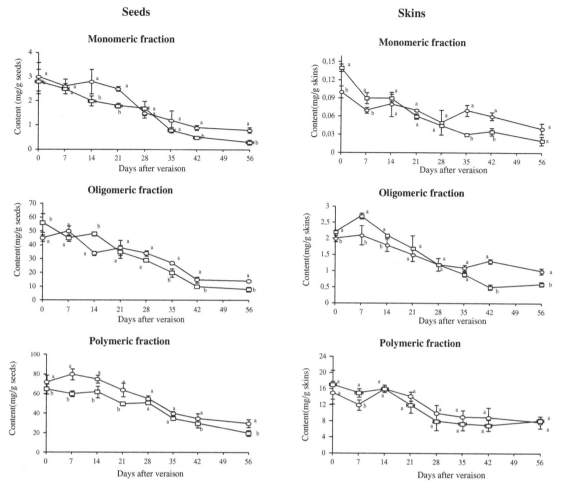

FIGURE 12.2 Evolution of the different seed and skin proanthocyanidin fractions during the ripening of two Portuguese red grape varieties (-o-Tinta Roriz, -□- Touriga Nacional). *From Jordão, A.M., Correia, A.C., 2012. Relationship between antioxidant capacity, proanthocyanidin and anthocyanin content during grape maturation of Touriga Nacional and Tinta Roriz grape varieties. S. Afr. J. Enol. Vitic. 33, 214–224.*

minimal levels at technological maturation in grapes from California and Chile (Kennedy et al., 2000; Obreque-Slier et al., 2010). Another example that exemplifies the impact of vintage on proanthocyanidin content is reported by Lorrain et al. (2011). Thus, these authors reported low values of mean degree of polymerization and concentrations of proanthocyanidins in seeds and skins in 2009 vintage in Bordeaux region, where the climatic conditions were characterized by high rainfall before flowering followed by a strong hydric stress during berry development until ripening. These conditions probably induced a lower activation of flavonoid pathway. Previously other authors (Hanlin and Downey, 2009) reported a relevant differences in the proportion of epigallocatechin between seasons in two red grape varieties (Shiraz and Cabernet Sauvignon), indicating that environmental conditions that occur in each vintage have an influence on the synthesis of (−)-epicatechin and (−)-epigallocatechin.

Among environmental factors, according to the several previous works mentioned, polyphenol composition depends largely on climate and the potential adverse environmental conditions. It is well known that abscisic acid is a key regulator of berry ripening and is strongly involved in the control of proanthocyanidin biosynthesis from the flowering stage. In addition, abscisic acid plays an important role in grape varieties' adaptation to adverse climatic conditions, playing an important role in grapes' phenols biosynthesis. Thus, according to Koyama et al. (2010) the presence of high levels of abscisic acid has an important impact on proanthocyanidins accumulation during grape ripening. More recently, Hernández et al. (2017) analyzed the influence of genetics and vintage in flavan-3-ol composition of grape seeds of a segregation *V. vinifera* population and concluded that (−)-epicatechin and (+)-catechin followed by proanthocyanidins A2 and B2 were the most abundant compounds and high correlations between vintage and each compound prove that the vintage has an important effect in proanthocyanidins content besides genetic factors.

Another viticultural practice that could be a factor that has a potential impact in grape proanthocyanidin content is the cluster thinning. Ó-Marques et al. (2005) reported low concentration of polymeric condensed tannin fractions in seeds, skins, and pulp in samples from vines without any pruning modality, when compared with two distinct cluster thinning modalities, in Cabernet Sauvignon and Tinta Roriz (Tempranillo) *V. vinifera* L. varieties. More recently, other authors (Bogicevic et al., 2015) reported an increase of proanthocyanidins from Vranac and Cabernet Sauvignon grapes varieties as a result of early leaf removal and cluster thinning treatments. Previously, Cortell et al. (2005) analyzed the influence of vine vigor in Pinot Noir grape variety on proanthocyanidins content and detected significant increases in skin proanthocyanidins, and average molecular mass of proanthocyanidins in fruit from zones with a reduction in vine vigor. Finally, Gil et al. (2013) for Syrah grape variety detected that the wine obtained from berry thinning grapes had a higher proanthocyanidin concentration than the control wine.

12.2 EVOLUTION OF PROANTHOCYANIDINS DURING FERMENTATIVE MACERATION OF RED WINES

After the grape harvest, the red winemaking process begins by including a high number of technical operations according to the grape characteristics and obviously according to the final red wine characteristics desired. During the winemaking process, the extraction of grape polyphenols into the grape juice is determined by biochemical and chemical phenomena and is also influenced by several technical operations. One of the red winemaking steps related to the evolution and extraction of proanthocyanidins from grape solid parts (skin, seeds, to a lower extent from pulps, and also occasionally from stems) during the fermentative process is related with the maceration process. Thus, maceration is one of the most important processes giving to the red wine its characteristics of taste and flavor profile, differentiating it from white wine.

In general, proanthocyanidins (in particular those from seeds and stems) are diffused more slowly and require longer maceration times than anthocyanins (Ribéreau-Gayon, 1964). According to several authors (González-Manzano et al., 2004; Canals et al., 2005; Del Llaudy et al., 2008; Hernández-Jiménez et al., 2012), proanthocyanidin extraction is favored by the presence of ethanol, because seed proanthocyanidins extraction requires a disorganization of the outer lipidic cuticle surrounding the seeds which is facilitated by the presence of high levels of ethanol in the final stage of the alcoholic fermentation. However, the skins' proanthocyanidins are more easily extracted due to their localization, and they are also solubilized, together with anthocyanins (Liu et al., 2016). Thus, the higher contribution to the proanthocyanidins comes mainly from the skins, and only for an extended maceration time is it possible to detect a large content of flavan-3-ols from seeds (González-Manzano et al., 2004; Gómez-Míguez et al., 2007; Lomolino et al., 2010). Sun et al. (1999) have shown that nearly half of the monomers and oligomeric proanthocyanidins in grape seeds were transferred into wine, but grape seeds did not contribute polymeric proanthocyanidins to the red wine, in an example of 10 days maceration time at 25°C, using a classical red winemaking technique where that cap was punched down three times daily until it remained submerged. In addition, the majority of the extractable monomeric, oligomeric, and polymeric proanthocyanidins were released into the same red wine. According to other authors (Monagas et al., 2003), red wine proanthocyanidins had a more similar profile to skin proanthocyanidins than to those of seeds.

Thus, it is obvious that maceration time is one of the most important factors that determine the proanthocyanidin extraction during the winemaking process. Table 12.2 shows an example of experimental data from the evolution of different proanthocyanidins fractions during the maceration process obtained by Jordão et al. (2012a).

Taking in account the data shown in Table 12.2, it is clear for both red grape varieties studied, that in general, the proanthocyanidin fractions increased progressively during the maceration process followed by a slight decrease of the values in the final days (in particular for oligomeric and polymeric fractions). In addition, the monomeric fraction appears to be extracted steadily into the wine during the maceration process.

Several authors studied the effect of maceration duration on proanthocyanidin extraction and consequently on red wine phenolic composition. In a paper published in 2008 by Cerpa-Calderón and Kennedy, a maximum proportion of proanthocyanidins extracted from skins was obtained after the 9th day of maceration followed by a stabilization of the concentration until day 18, after which no further extraction occurred (Cerpa-Calderón and Kennedy, 2008). For these authors, the proanthocyanidin content stabilization may reflect the continued adsorption of extracted proanthocyanidins to suspended cell wall material and, possibly, yeast cells. Previously, Gambuti et al. (2004) reported a maximum level of (+)-catechin and (−)-epicatechin after 12 days of maceration

TABLE 12.2 Extraction of the Different Proanthocyanidin Fractions During the Maceration Process of Two Portuguese Red Grape Varieties

	Proanthocyanidin fraction (mg/L)					
	Monomeric		Oligomeric		Polymeric	
Maceration time (days)	TR	TN	TR	TN	TR	TN
0	10.21^a ± 1.22	5.60^b ± 0.40	19.51^c ± 2.33	10.34^d ± 3.32	89.0^e ± 9.13	59.08^f ± 4.66
1	19.45^a ± 0.43	12.43^b ± 1.27	45.66^c ± 3.92	35.11^d ± 0.95	110.44^e ± 6.57	90.45^f ± 7.54
2	27.39^a ± 0.95	23.91^b ± 0.92	78.90^c ± 5.61	58.90^d ± 3.12	230.12^e ± 12.30	240.97^e ± 9.34
3	34.51^a ± 2.35	34.36^a ± 2.39	120.45^c ± 6.81	90.46^d ± 5.32	267.85^e ± 9.83	227.88^f ± 5.23
4	40.13^a ± 4.27	43.93^a ± 2.80	210.33^c ± 7.66	206.30^c ± 7.63	456.19^e ± 12.33	326.84^f ± 10.58
5	37.82^a ± 3.78	45.63^b ± 2.04	398.13^c ± 4.76	291.15^d ± 6.72^d	678.57^e ± 15.62	587.58^f ± 10.14
6	50.73^a ± 5.68	47.63^a ± 4.19	560.22^c ± 11.27	463.29^d ± 9.23	980.45^e ± 12.31	750.72^f ± 14.80
7	67.38^a ± 2.83	50.25^b ± 3.42	578.63^c ± 9.88	548.63^d ± 2.89	1,360.44^e ± 13.60	963.46^f ± 10.12
8	70.41^a ± 1.24	55.61^b ± 1.54	521.42^c ± 8.94	501.42^d ± 5.84	1,320.19^e ± 11.44	1,251.94^f ± 9.48
9	61.29^a ± 2.85	51.67^b ± 2.42	514.90^c ± 10.23	484.90^d ± 7.34	1,299.57^e ± 9.83	1,193.10^f ± 6.87

TR, Tinta Roriz; *TN*, Touriga Nacional; means of two replicates; ± standard deviation; mean values followed by different letter in row for each proanthocyanidin fraction analyzed between the two grape varieties studied indicate statistically differences according to Ducan's test ($\alpha = 0.05$).
From Jordão, A.M., Simões, S., Correia, A.C., Gonçalves, F.J., 2012a. Antioxidant activity evolution during Portuguese red wine vinification and their relation with the proanthocyanidin and anthocyanin composition. J. Food Process. Preserv. 36, 298–309.

for all red wines produced, although a decline of (−)-epicatechin occurred after maximum extraction. More recently, other authors (Kocabey et al., 2016) analyzed the effects of different maceration times (5, 10, and 15 days) on phenolic composition of red wines made from the *V. vinifera* L. Karaoglan. For these authors, extended maceration duration (in particular 15 maceration days) resulted in red wines with an increase in the concentration of phenolic compounds, namely, procyanidin B2, (+)-catechin, (−)-epicatechin.

The use of an extended maceration time, i.e., when the cap contact with the grape must is longer than the usual short period during which alcoholic fermentation occurs and instead extends up to several weeks, is another possibility studied by several authors for proanthocyanidin extraction. Thus, according to several works (Sacchi et al., 2005; Gil et al., 2012; Casassa et al., 2013; Casassa and Harbertson, 2014), this option is one of the principal methods by which an increase in extraction of proanthocyanidins can be achieved, in particular from seeds. After the alcoholic fermentation, the presence of ethanol in the medium facilitates proanthocyanidin extraction in particular from the seeds (González-Manzano et al., 2004; Canals et al., 2005).

Also other factors are determinants for proanthocyanidin extraction during the maceration process, such as cap management and the different technological solutions used to increase the contact between grape solid parts and grape must. Thus, Sacchi et al. (2005) reported that the use of rotary fermenters extracted more tannins than the punch-down method, which in turn showed greater extraction than the pumping over technique. Diverse conclusions were also obtained by other authors. For example, Ichikawa et al. (2012) reported less proanthocyanidin extraction by the use of punching down than submerged cap, while Bosso et al. (2012) described higher concentration of tannins in wines made by the submerged-cap method, but lower extraction occurred when the punch-down technique was used.

The fermentation temperature varies significantly depending on the wine style to be produce, as well as the final chemical composition and especially the sensory profile. However, to promote a higher phenolic compounds extraction, winemakers have different options, for example, by the use of application of higher temperatures. Thus, for example, flash *détente* and thermo *détente*, are in general two winemaking technologies that involve a preferrmentative heat treatment using a short contact time. In general, these technologies induce fast initial proanthocyanidin concentrations, as well as an increased tannins-to-anthocyanins ratio (González-Manzano et al., 2004; Morel-Salmi et al., 2006).

Finally, the use of maceration enzymes available in the market are a suitable option to increase the proanthocyanidin extraction from the grape cell wall. According to Smith et al. (2015), the degradation of the cell walls in the presence of macerating enzymes induced tannin extractability, but at same time, contributed to a reduction of tannins adsorption by the cell walls and wine and subsequent sedimentation. Thus, there is a potential increase of the proanthocyanidin extraction during the winemaking process by the use of enzymes, according to several results published (González-Neves et al., 2013; Ruiz-Garcia et al., 2014; Santos-Lima et al., 2015).

12.3 CHANGES ON PROANTHOCYANIDINS DURING RED WINE AGING IN CONTACT WITH WOOD

12.3.1 Natural Evolution of the Proanthocyanidins During Aging

Phenolic compounds contribute directly or indirectly to color, astringency, bitterness, and aroma of the wines, including those that aged in contact with wood (barrels or other alternative wood products, like chips, granulates, and staves).

The phenolic compounds of red wines extracted during fermentation/maceration are of undoubted importance during maturation and aging of the wines. Progressive changes, in particular on proanthocyanidins and anthocyanins, are inevitable because of the reactivity of these compounds, but the rate of phenolic interactions and degradations may be subject to many influences.

During the aging of red wines a decrease in the concentration of anthocyanins, responsible for the initial purple tints in the color of a young red wine, are displaced progressively, and irreversibly, by more stable colored polymeric pigments that increase in red wines (Somers, 1971). It was proposed by Jurd (1965, 1967) that the anthocyanins were condensed with other phenolic compounds, such as (+)-catechin or procyanidins, directly or mediated by aldehydes. Correspondingly, their structural modifications result in the cited characteristic variation of color, from purple-red color in young red wines to brick-red hue of the aged.

One of the oldest reference studies about this topic was performed by Haslam (1980) and his research team, where the role of the oligomeric procyanidins in the aging of red wines was shown, and he observed that at wine pH the familiar acid-catalyzed equilibration of procyanidins was occurring. Reactions between anthocyanins, proanthocyanidins, and additional compounds (hydrolytic degradation products of anthocyanins) were responsible for the precipitation of the proanthocyanidins from the liquid medium. The presence of acetaldehyde was of major importance in accelerating the chemical transformation of phenolic compounds, inducing the formation of new products (Timberlake and Bridle, 1976; Haslam, 1980; Dallas et al., 1996a, 1996b). As an example, (+)-catechin and (−)-epicatechin react more slowly than dimeric procyanidins B1, B2, and B3, while the trimeric procyanidin C1: (−)-epicatechin $C_4 \rightarrow C_8$ (−)-epicatechin $C_4 \rightarrow C_8$ (−)-epicatechin has presented the highest degradative reaction rate (Dallas et al., 1996a).

In particular for proanthocyanidins, during wine storage and aging, there are changes in the composition and structural characteristics of these phenolic compounds. Thus, these changes are mainly related to a decrease in the proanthocyanidins concentration associated to their precipitation and a decrease in mean degree of polymerization of proanthocyanidins linked to the subunit cleavage (Cheynier et al., 2006; Chira et al., 2012). These proanthocyanidin changes, including also the reactions with anthocyanins, induce in red wine relevant sensorial modifications, in particular for astringency intensity. Thus, according to the results obtained by Chira et al. (2012) in red wines from Cabernet Sauvignon, a slight astringency intensity was associated to mean degree of polymerization of native proanthocyanidins between 2 and 4 (for old wines), while the highest astringency was perceived in young red wines with mean degree of polymerization of proanthocyanidins of 7.6.

Because of the extreme complexity of chemical compounds involved, many investigations have been made using model solutions of known composition rather than wine. He et al. (2012) published an important overview of the new compounds formed and the principal reactions that may occur in wines, involving proanthocyanidins, anthocyanins, and derived compounds.

12.3.2 Effects of the Medium Factors on the Proanthocyanidin Evolution

Distinct studies have been done concerning the influence of environmental factors on the proanthocyanidin transformation that may happens in red wines, especially during the maturation.

Some procyanidins have been followed in a trial with real red wines, over time at different temperatures and SO_2 contents, and the degradative reaction of the procyanidins has been carried out. It was shown that all the procyanidins tested have produced regression lines with good linearity (Dallas et al., 1995). Thus, the degradation of the oligomers procyanidins at each treatment temperature considered (12, 22, 32, and 42°C) appears to be by first order kinetic reaction. In parallel in the same study, the effect of SO_2 on the procyanidins rate constants has also been tested and at each of the four previous temperatures, the results obtained show that at both concentrations (50 and 100 mg/L of SO_2) the procyanidins degradative rates exhibited for the studied wines were lower than those calculated for the control wine, without SO_2 addition. Comparing the results of the degradative rate in order to determine the influence of the storage temperature, it was observed that procyanidins decreased 5 or 10 times faster at the higher temperatures (32 and 42°C) (Dallas et al., 1995).

In conclusion, temperature exerts a marked influence on the progressive degradation of procyanidins, while the presence of SO_2 slows down the degradation (Dallas et al., 1995). Comparing their activation energies, the dimer procyanidins B1, B2, and B3 appear to be more stable to degradation, while trimer T2, which is: $(-)$-epicatechin $C_4 \rightarrow C_8$ $(-)$-epicatechin $C_4 \rightarrow C_8$ $(+)$-catechin, and procyanidin B2-3-O-gallate are the most reactive studied compound (Dallas et al., 1995, 2003).

Concerning the presence of acetaldehyde in the medium, a report has also demonstrated that the degradation of some studied procyanidins with low molecular weight followed apparent first-order kinetics, and the rate constants increased and the half-life times decreased with acetaldehyde content (0.0, 3.58, and 35.8 mM; normal concentrations found in red wines), showing an increasing procyanidin reactivity with increasing acetaldehyde concentration (Dallas et al., 2003).

Regarding the pH effect, a study in model wine solutions with separate pH of 2.0, 3.2, 3.6, 4.0, and 6.0, and using various dimeric and trimeric procyanidins, has shown that the disappearance of the procyanidins with pH was also found to be a first-order apparent reaction and that the degradative constant rates were higher and half-life times were lower at the lowest pH value (2.0) for all procyanidins studied, which seemed to be more stable at pH 3.2 than at higher or lower pH values (Dallas et al., 2003).

The effect of the wine macromolecules on the proanthocyanidin composition and content of red wines has also been studied by some research teams. Concerning the presence of polysaccharides in wines, the interaction between color compounds and tannins with mannoproteins may be of extreme importance as it can influence color stability and improve sensory qualities. Guadalupe et al. (2010) showed nevertheless that there was no positive interaction between mannoproteins and color compounds and that the interaction between mannoproteins and condensed tannins resulted in a decrease of wine tannin content, suggesting the precipitation of tannin and mannoprotein aggregates and a decrease in astringency, with an increase of the wine sweetness and roundness. In another study, it was observed that enriched mannoprotein commercial preparations did not have an effect on color stabilization of the studied red wines (Rodrigues et al., 2012). Actually, the evolution of color parameters with time was similar for all the experiments done, showing no efficient effect of the three commercial products studied on the color stability parameters of the wines, and the global tannin profile evolution did not show significant differences. However, it is possible that the commercial-based mannoprotein products may have some influence on the tannin aggregation evolution, contributing to the delay of tannin polymerization in red wines (Rodrigues et al., 2012). The same kind of effect was observed between proanthocyanidins from a red wine made from Touriga Nacional grape variety and external yeast lees (supposed to be rich in mannoproteins) arising from a white wine vinification performed by Rodrigues et al. (2013). During the time, after the initial decay, it appeared that there was a retarding effect on the proanthocyanidin polymerization reactions by the addition of external yeast lees. According to these authors, this effect could occur due to two factors: the rapid initial removal of the more polar proanthocyanidins by adsorption to yeast lees and the proanthocyanidin polymerization retarding effect promoted by the low and medium molecular weight mannoproteins.

The interactions between proteins (wine proteins, salivary glycoproteins, and proteins used in fining treatments) and polyphenols, particularly the condensed tannins, involving or not polysaccharides, are widely studied. Several studies have demonstrated the ability of some neutral and anionic polysaccharides to disrupt the binding of polyphenols to proteins (Ozawa et al., 1987; De Freitas et al., 2003). This mechanisms have an huge importance on the wine sensory astringency perception, on interactions and transformations/alterations during winemaking, and aging of wines, in particularly in the presence of wood, and during a current wine fining treatment with an enological commercial protein or, recently, with commercial polysaccharides.

Finally, different types of commercial enological tannins, mainly hydrolyzable (gallotannins and/or ellagitannins), condensed, and blends of the two, may also be used during winemaking and aging of the wines. Of course, when condensed tannins are present in those commercial preparations, an addition and correspondent increase in the proanthocyanidin content in wine is observed. Altogether, commercial enological tannins are highly diverse products that can differentially impact the physicochemical and sensorial characteristics of wines to which they are added (Vazallo-Valleumbrocio et al., 2017).

12.3.3 Wood Influence on Wine Proanthocyanidin Evolution

Many constituents can be extracted from staves during aging in barrels or in the presence of other wood alternatives products: ellagitannins (Peng et al., 1991; Viriot et al., 1994; Jordão et al., 2007) and other tannins, gallic, ferulic, vanillic and ellagic acids, vanillin, coumarins, and a great number of volatile compounds (Moutounet et al., 1989; Chatonnet et al., 1997; Perez-Prieto et al., 2002; Escalona et al., 2002; Jordão et al., 2005a, 2006a).

The environment evolving inside oak barrels during the maturation of wines provides conditions for further reactions, such as oxidation, hydrolysis and polymerization, involving several wood compounds and red wine phenolics, like proanthocyanidins and anthocyanins.

There are important reports on the formation of several oligomeric pigments resulting from reactions between malvidin-3-glucoside and (+)-catechin mediated by oak-derived furfural, methyl-furfural, and vanillin (Pissarra et al., 2005; Sousa et al., 2007). Based on experiments with model wine solutions at pH 3.5, Nonier et al. (2006) reported a complete structural study of the formation of condensed dimers from (+)-catechin and oak wood furfuraldehyde. In addition, under acidic conditions, new (+)-catechin derived pigments, called oaklins, are produced from the reaction of flavan-3-ols with sinapaldehyde or coniferaldehyde (De Freitas et al., 2004; Pissarra et al., 2005; Sousa et al., 2005). Oaklins have been detected in commercial red wines aged in oak barrels, and it has been reported that they contribute to color changes and astringency of red wines during the aging process (Sousa et al., 2005). In addition, C-glycosidic ellagitannins, such as vescalagin, are involved in stereoselective condensation with wine nucleophiles, in particular anthocyanins (Quideau et al., 2003). Chassaing et al. (2010) reported the UV–Vis and structural NMR properties of anthocyano-ellagitannin hybrid pigments and concluded that this molecular association contributes to color stability of red wines aging in oak barrels.

With regard to other nonvolatile wood constituents, in addition to the reported condensation reaction between C-glycosidic ellagitannins and malvidin-3-glucoside, other investigators reported a condensation reaction between the ellagitannin vescalagin, (+)-catechin and (−)-epicatechin (Quideau et al., 2003). As a result of these reactions, new compounds called acutissimins were also formed and identified. Quideau et al. (2005) showed an important aspect of the chemistry of oak-derived nonahydroxyterphenoyl (NHTP)-bearing C-glycosidic ellagitannins. These natural products are extracted by the wine solution during aging in barrels and have the capability to combine covalently by means of substitution reactions with a variety of grape-derived nucleophilic species, such as ethanol, flavanols, anthocyanins, and thiols. After this observation, Saucier et al. (2006) estimated the levels of four flavanoellagitannins and another identified wine polyphenol, β-1-O-ethylvescalagin, in Bordeaux red wine aged for 18 months in oak barrels. These five ellagitannin derivatives originate from the nucleophilic substitution reaction of vescalagin with grape flavan-3-ols (+)-catechin and (−)-epicatechin or ethanol.

Some studies performed using model wine solutions have also shown the influence of the presence of wood in anthocyanin and proanthocyanidin kinetics evolution. An experiment using oak wood chips has shown that the anthocyanin evolution was similar for all the wines assayed, except for malvidin-3-glucoside, in which the decrease was more evident for the wine aged in contact with oak wood chips (Jordão et al., 2006b) and seemed independent of the oxygen concentration (Jordão et al., 2006b). Contrarily, in an opposite direction, ellagic acid and/or oak wood extracts, including ellagitannins, slowed down the decline in the levels of (+)-catechin and procyanidin B1 (Jordão et al., 2008). Probably, other red wine proanthocyanidins would present similar behavior under wine aging conditions in the presence of wood. In addition, oak wood-derived volatile compounds such as furfural, vanillin, eugenol, and guaiacol that are transferred to wine matured in barrels were examined for their influence on the (+)-catechin content in a separate study (Jordão et al., 2008). A general decrease in (+)-catechin was observed in all the model wine solutions studied. This was more evident in the first 16 days of the study. The decrease, however, was less substantial in solutions containing (+)-catechin and furfural and more pronounced with (+)-catechin alone. In addition, Barrera-García et al. (2007) also reported similar changes in (+)-catechin and (−)-epicatechin in model wine solutions containing oak wood. Thus, according to these authors, a partial decrease of these compounds is a consequence of a sorption mechanism by oak wood.

In fact, as it was shown, during the maturation of red wines in oak barrels or using oak wood chips, and mainly in toasted barrels or chips, several changes in red wine phenolic compounds occur. An increase in the content of phenols extracted from the wood was already reported. The amount of phenolic compounds extracted into wine depends on aging time, oak type, size, and the possibility of the barrel having been used (Rous and Alderson, 1983; Jindra and Gallender, 1987; Laszlavik et al., 1995; Feuillat et al., 1997; Chatonnet et al., 1997; González-SanJosé and Revilla, 2001; Ho et al., 2001; Gambuti et al., 2010; Jordão et al., 2012b; Nunes et al., 2017). This last aspect has also been the subject of several studies in recent years, since the possibility of using oak barrels that have had multiyear use is a way to reduce costs in the wine industry, and obviously has a potential impact in ecological conservation of forests where the replacement of trees is not guaranteed. Thus, several works were also published about the potential influence of the use of used oak barrels on wine phenolic composition, including flavan-3-ols and proanthocyanidin content Gambuti et al. (2010) studied the effect of aging in new oak and one-year-used oak on phenolic evolution of three different monovarietal red wines (Piedirosso, Cabernet Sauvignon and Merlot) and didn't detect any variation of the proanthocyanidin content in Merlot and Piedirosso wines. However, for Cabernet Sauvignon wines aged in new oak barrels, lower condensed tannins content was detected. Probably, the use of new oak wood induces a greater quantity of oxygen penetrating through the porous wood and consequently determines a higher extent of polymerization and consequent precipitation of tannins (in that case quantified by vanillin reaction flavanols assay).

It seems that in red wines aged on American oak (*Quercus alba*), compared to the French oak (*Quercus petraea*), the proanthocyanidins react more slowly with anthocyanins, probably due to the lower content of ellagitannins in the American oak, as these ellagitannins are known to accelerate the reactions involving anthocyanins and proanthocyanidins (Vivas and Glories, 1996; Chatonnet et al., 1997). Another explanation may be the fact that the wood ultrastructure of the American oak is not favorable to the impregnation by the wine (Chatonnet et al., 1997) as well as the oxygen penetration occurring to a lesser extent. However, a report about wood alternatives (De Coninck et al., 2006) have shown that the oak wood chips species (*Q. petraea* L. and *Quercus pyrenaica* Willd.) and the chips concentration used (4 g/L) did not affect the proanthocyanidin contents in the wines during the time considered (13 weeks). In this approach, sensory results showed that the red wines aged in contact with Portuguese and French oak wood chips and the mixture of these two oak wood species (50/50) differed significantly from the control wine in several sensorial characteristics. Actually, the wines aged in contact with wood chips showed higher point values for intensity, toasted, wood and vanillin aroma, taste intensity and global appreciation. This positive effect was more evident for wines aged with *Q. pyrenaica* Willd oak wood chips (De Coninck et al., 2006).

Concerning the ellagitannins and ellagic acid when extracted from the wood to the wine during aging, an increase during the first weeks of extraction was observed, followed by a decrease (Jordão et al., 2005b). Under the extraction conditions examined in the study of Jordão et al. (2005b), the temperature was the main factor influencing ellagitannins and ellagic acid evolution. The results suggest that a decrease/degradation of these compounds is less noticeable at low temperatures (12°C). Moreover, after 104 extraction days the ellagitannins content in a model wine solution at 12°C was higher than the content of ellagitannins in model wine solutions at 20°C. On the other hand, the effect of alcoholic content and pH of the model wine solutions, on the extraction and evolution of the analyzed compounds (except for castalagin and vescalagin) seemed to be less important than the influence of temperature. In addition, it seems that proanthocyanidins and anthocyanins did not significantly affect the rate of the degradation of ellagitannins during the storage of a red wine in contact with the wood (Jordão et al., 2008).

Recently, some works reported the use of alternative wood species instead of the classical oak and chestnut, like acacia and cherry in wine aging on barrels/chips or other wood alternatives (De Rosso et al., 2009; Kozlovic et al., 2010; Sanz et al., 2012; Fernandez de Simon et al., 2014; Chinnici et al., 2011, 2015; Delia et al., 2017; Tavares et al., 2017). In a recent study from Tavares et al. (2017), wood chips from cherry (*Prunus avium*), acacia (*Robinia pseudoacacia*), and two oak species (*Q. petraea* and *Q. pyrenaica*) were added separately to a Portuguese red wine. Various global phenolic parameters and a sensory analysis of red wines were studied during the aging process (90 storage days). Only a few differences were detected between the wines. However, after 90 aging days, in general the wines aged in contact with cherry wood tended to have the lowest values for several phenolic parameters. For sensory parameters, the wine aged in contact with French oak chips showed significantly higher scores for several aroma descriptors, while for visual and taste descriptors no statistically significant differences were found between the wines. Previously, Chinnici et al. (2011) analyzed the changes in phenolic composition of red wines aged in cherry wood barrels and only detected significant changes in several phenolic compounds after 4 months of conservation. However, it was possible to detect between 30 and 45 aging days a slight increase of flavonoid compounds in wine aged in contact with chips from acacia wood compared to the other wines.

Finally, in addition the aging on lees involves maturation of the wine in contact with yeast cells after fermentation but, if combined with the presence of oak wood chips, it can soften the wood flavor and increase the aromatic complexity of a specific wine. A study has shown that distinct *Saccharomyces cerevisiae* strains have affected differently the tannic fractions (monomers, oligomers, and polymers) of the red wines giving different astringency sensations and bitterness taste (Loira et al., 2013). So, with this purpose, the choice of an adequate yeast strain is of the utmost importance in order to get the desired sensory profile.

12.4 FINAL REMARKS

A great number of diverse proanthocyanidins with different chemical structures and properties have been identified in grapes and wines. In the last decades, their identification and characterization has been studied in detail with the help of modern instrumental analysis. In addition, great improvements have been obtained in the investigation of the change of grape proanthocyanidins during the grape maturation and the relationship between their content in different grape bunch fractions (skins, seeds, pulps, and stems) and the diverse factors that determine the proanthocyanidin accumulation during the entire period of grape maturation.

Proanthocyanidins play an important role in red wine sensorial properties, contributing to the wine color and mouthfeel properties. Thus, substantial research has also been focused on the development of new winemaking operations, in particular during the fermentative maceration, to improve the proanthocyanidin extraction, and consequently to contribute to a development of a better red winemaking process. In addition, despite the importance of a great number of studies performed until now, the potential influence of red wine aging in contact with wood (including the use of other non-oak wood species) on the evolution of individual proanthocyanidins during red wine maturation has not been examined in complete detail, especially in regard to the formation of new compounds, even if some chemical structures have been detected, elucidated, and formally identified. Thus, the knowledge of all of these factors has an important impact on red wine composition and consequently in global quality.

References

Allegro, G., Pastore, C., Valentini, G., Muzzi, E., Filippetti, I., 2016. Influence of berry ripeness on accumulation, composition and extractability of skin and seed flavonoids in cv. Sangiovese (*Vitis vinifera* L.). J. Sci. Food Agric. 96, 4553–4559.

Asproudi, A., Piano, F., Anselmi, G., Di Stefano, R., Bertolone, E., Borsa, D., 2015. Proanthocyanidin composition and evolution during grape ripening as affected by variety: Nebbiolo and Barbera cv. J. Int. Sci. Vigne Vin. 49, 59–69.

Balík, J., Kyseláková, M., Vrchotová, N., Tříska, J., Kumšta, M., Veverka, J., et al., 2008. Relations between polyphenols content and antioxidant activity in vine grapes and leaves. Czech. J. Food Sci. 26, S25–S32.

Baranac, J.M., Petranovic, N.A., Dimitric-Markovic, J.M., 1997. Spectrophotometric study of anthocyan copigmentation reactions. 2. Malvin and the nonglycosidized flavone quercetin. J. Agric. Food Chem. 45, 1694–1697.

Barrera-García, V.D., Gougeon, R.D., Majo, D.D., Aguirre, C., Voilley, A., Chassagne, D., 2007. Different sorption behaviors for wine polyphenols in contact with oak wood. J. Agric. Food Chem. 55, 7021–7027.

Bautista-Ortín, A.B., Jiménez-Pascual, E., Busse-Valverde, N., López-Roca, J.M., Ros-García, J.M., Gómez-Plaza, E., 2013. Effect of wine maceration enzymes on the extraction of grape seed proanthocyanidins. Food Bioprocess Technol. 6, 2207–2212.

Bogicevic, M., Maras, V., Mugoša, M., Kodžulović, V., Raičević, J., Šućur, S., et al., 2015. The effects of early leaf removal and cluster thinning treatments on berry growth and grape composition in cultivars Vranac and Cabernet Sauvignon. Chem. Biol. Technol. Agric. 2, 13.

Bogs, J., Downey, M.O., Harvey, J.S., Ashton, A.R., Tanner, G.J., Robinson, S.P., 2005. Proanthocyanidin synthesis and expression of genes encoding leucoanthocyanidin reductase and anthocyanidin reductase in developing grape berries and grapevine leaves. Plant Physiol. 139, 652–663.

Boido, E., García-Marino, M., DellaCassa, E., Carrau, F., Rivas-Gonzalo, J.C., Escribano-Bailón, M.T., 2011. Characterization and evolution of grape polyphenol profiles of *Vitis vinifera* L. cv. Tannat during ripening and vinification. Aust. J. Grape Wine Res. 17, 383–393.

Bordiga, M., Travaglia, F., Locatelli, M., Coïsson, J.D., Arlorio, M., 2011. Characterisation of polymeric skin and seed proanthocyanidins during ripening in six *Vitis vinifera* L. cv. Food Chem. 127, 180–187.

Bosso, A., Panero, L., Petrozziello, M., Follis, R., Motta, S., Guaita, M., 2012. Influence of submerged-cap vinification on polyphenolic composition and volatile compounds of Barbera wines. Am. J. Enol. Vitic. 62, 503–511.

Bourzeix, M., Weyland, D., Heredia, N., 1986. Étude des catéchines et des procyanidols de la grappe de raisin, du vin et d'autres dérivés de la vigne. Bull. O.I.V 59, 1171–1254.

Cadot, Y., Minana-Castello, M.T., Chevalier, M., 2006. Anatomical, histological, and histochemical changes in grape seeds from *Vitis vinifera* L. cv Cabernet franc during fruit development. J. Agric. Food Chem. 54, 9206–9215.

Canals, R., Llaudy, M.C., Valls, J., Canals, J.M., Zamora, F., 2005. Influence of ethanol concentration on the extraction of color and phenolic compounds from the skin and seeds of Tempranillo grapes at different stages of ripening. J. Agric. Food Chem. 53, 4019–4025.

Casassa, L.F., Harbertson, J.F., 2014. Extraction, evolution, and sensory impact of phenolic compounds during red wine maceration. Annu. Rev. Food Sci. Technol. 5, 83–109.

Casassa, L.F., Larsen, R.C., Beaver, C.W., Mireles, M.S., Keller, M., Riley, W.R., et al., 2013. Impact of extended maceration and regulated deficit irrigation (RDI) in Cabernet Sauvignon wines: characterization of proanthocyanidin distribution, anthocyanin extraction, and chromatic properties. J. Agric. Food Chem. 61, 6446–6457.

Cerpa-Calderón, F.K., Kennedy, J.A., 2008. Berry integrity and extraction of skin and seed proanthocyanidins during red wine fermentation. J. Agric. Food Chem. 56, 9006–9014.

Chassaing, S., Lefeuvre, D., Jacquet, R., Jourdes, M., Ducasse, L., Galland, S., et al., 2010. Physicochemical studies of new anthocyano-ellagitannin hybrid pigments: about the origin of the influence of oak C-glycosidic ellagitannins on wine color. Eur. J. Org. Chem. 2010, 55–63.

Chatonnet, P., Ricardo-da-Silva, J.M., Dubourdieu, D., 1997. Influence de l'utilisation de barriques en chêne sessile européen (*Quercus petraea*) ou en chêne blanc américain (*Quercus alba*) sur la composition et la qualité des vins rouges. Rev. Fr. Oenol. 165, 44–48.

Cheynier, V., Prieur, C., Guyot, S., Rigaud, J., Moutounet, M., 1997. The structures of tannins in grapes and wines and their interaction with proteins. In: Watkins, T.R. (Ed.), Wine: Nutritional and Therapeutic Benefits. Am. Chem. Soc., Washington, DC, pp. 81–93.

Cheynier, V., Duenas-Paton, M., Salas, E., Maury, C., Souquet, J.M., Sarni-Manchado, P., et al., 2006. Structure and properties of wine pigments and tannins. Am. J. Enol. Vitic. 57, 298–305.

Chinnici, F., Natali, N., Sonni, F., Bellachioma, A., Riponi, C., 2011. Comparative changes in color features and pigment composition of red wines aged in oak and cherry wood casks. J. Agric. Food Chem. 59, 6575–6682.

Chinnici, F., Natali, N., Bellachioma, A., Versari, A., Riponi, C., 2015. Changes in phenolic composition of red wines aged in cherry wood. LWT Food Sci. Technol. 60, 977–984.

Chira, K., Lorrain, B., Ky, I., Teissedre, P.L., 2011. Tannin composition of Cabernet-Sauvignon and Merlot grapes from the Bordeaux area for different vintages (2006 to 2009) and comparison to tannin profile of five 2009 vintage Mediterranean grapes varieties. Molecules 16, 1519–1532.

Chira, K., Jourdes, M., Teissedre, P.L., 2012. Cabernet Sauvignon red wine astringency quality control by tannin characterization and polymerization during storage. Eur. Food Res. Technol. 234, 253–261.

Cortell, J.M., Halbleib, M., Gallagher, A.V., Righetti, T.L., Kennedy, J.A., 2005. Influence of vine vigor on grape (*Vitis vinifera* L. Cv. Pinot Noir) and wine proanthocyanidins. J. Agric. Food Chem. 53, 5798–5808.

Cosme, F., Ricardo-da-Silva, J.M., Laureano, O., 2009. Tannin profiles of *Vitis vinifera* L. cv. red grapes growing in Lisbon and from their monovarietal wines. Food Chem. 112, 197–204.

Dallas, C., Ricardo-da-Silva, J.M., Laureano, O., 1995. Degradation of oligomeric procyanidins and anthocyanins in a Tinta Roriz red wine during maturation. Vitis 34, 51–56.

Dallas, C., Ricardo-da-Silva, J.M., Laureano, O., 1996a. Interactions of oligomeric procyanidins in model wine solutions containing malvidin-3-glucoside and acetaldehyde. J. Sci. Food Agric. 70, 493–500.

Dallas, C., Ricardo-da-Silva, J.M., Laureano, O., 1996b. Products formed in model wine solutions involving anthocyanins, procyanidin B2 and acetaldehyde. J. Agric. Food Chem. 44, 2402–2407.

Dallas, C., Hipólito-Reis, P., Ricardo-da-Silva, J.M., Laureano, O., 2003. Influence of acetaldehyde, pH, and temperature on transformation of procyanidins in model wine solutions. Am. J. Enol. Vitic 54, 119–124.

De Coninck, G., Jordão, A.M., Ricardo-da-Silva, J.M., Laureano, O., 2006. Evolution of phenolic composition properties in red wine aged in contact with Portuguese and French oak wood chips. J. Int. Sci. Vigne Vin. 40, 25–34.

De Freitas, V., Sousa, C., González-Paramás, A.M., Santos-Buelga, C., Mateus, N., 2004. Synthesis of a new catechin-pyrylium derived pigment. Tetrahedron Lett. 45, 9349–9352.

De Freitas, V.A.P., Glories, Y., Monique, A., 2000. Developmental changes of procyanidins in grapes of red *Vitis vinifera* varieties and their composition in respective wines. Am. J. Enol. Vitic. 51, 397–403.

De Rosso, M., Panighel, A., Dalla Vedova, A., Stella, L., Flamini, R., 2009. Changes in chemical composition of a red wine aged in acacia, cherry, chestnut, mulberry, and oak wood barrels. J. Agric. Food Chem. 57, 1915–1920.

Del Llaudy, M.C., Canals, R., Canals, J.M., Zamora, F., 2008. Influence of ripening stage and maceration length on the contribution of grape skins, seeds and stems to phenolic composition and astringency in wine-simulated macerations. Eur. Food Res. Technol. 226, 337–344.

Delia, L., Jordão, A.M., Ricardo-da-Silva, J.M., 2017. Influence of different wood chips species (oak, acacia and cherry) used in a short period of aging on the quality of "Encruzado" white wines. Mitt. Klosterneuburg 67, 84–96.

Di Lecce, G., Arranz, S., Jáuregui, O., Tresserra-Rimbau, A., Quifer-Rada, P., Lamuela-Raventós, R.M., 2014. Phenolic profiling of the skin, pulp and seeds of Albariño grapes using hybrid quadrupole time-of-flight and triple-quadrupole mass spectrometry. Food Chem. 145, 874–882.

Downey, M.O., Harvey, J.S., Robinson, S.P., 2003. Analysis of tannins in seeds and skins of Shiraz grapes throughout berry development. Aust. J. Grape Wine Res. 9, 15–27.

Escalona, H., Birkmyre, L., Piggott, J.R., Paterson, A., 2002. Effect of maturation in small oak casks on the volatility of red aroma compounds. Anal. Chim. Acta 458, 45–54.

Escribano-Bailón, M.T., Guerra, M.T., Rivas-Gonzalo, J.C., Santos-Buelga, C., 1995. Proanthocyanidins in skins from different grape varieties. Z. Lebensm. Unters. Forsch 200, 221–224.

Farhadi, K., Esmaeilzadeh, F., Hatami, M., Forough, M., Molaie, R., 2016. Determination of phenolic compounds content and antioxidant activity in skin, pulp, seed cane and leaf of five native grape cultivars in West Azerbaijan province, Iran. Food Chem. 199, 847–855.

Fernandez de Simon, B., Sanz, M., Cadahia, E., Martinez, J., Esteruelas, E., Munoz, A.M., 2014. Polyphenolic compounds as chemical markers of wine ageing in contact with cherry, chestnut, false acacia, ash and oak wood. Food Chem. 143, 66–76.

Feuillat, F., Moio, L., Guichard, E., Marinov, M., Fournier, N., Puech, J.L., 1997. Variation in the concentration of ellagitannins and *cis*- and *trans*-β-methyl-γ-octalactone extracted from oak wood (*Quercus robur* L., *Quercus petraea* Liebl.) under model wine cask conditions. Am. J. Enol. Vitic. 48, 509–515.

Fine, A.M., 2000. Oligomeric proanthocyanidin complexes: history, structure, and phytopharmaceutical applications. Altern. Med. Rev. 5, 144–151.

De Freitas, V., Carvalho, E., Mateus, N., 2003. Study of carbohydrate influence on protein-tannin aggregation by nephelometry. Food Chem. 81, 503–509.

Fuleki, T., Ricardo-da-Silva, J.M., 1997. Catechin and procyanidin composition of seeds from grape cultivars grown in Ontario. J. Agric. Food Chem. 45, 1156–1160.

Fuleki, T., Ricardo-da-Silva, J.M., 2003. Effects of cultivar and processing method on the contents of catechins and procyanidins in grape juice. J. Agric. Food Chem. 51, 640–646.

Gambuti, A., Strollo, D., Ugliano, M., Lecce, L., Moio, L., 2004. *Trans*-resveratrol, quercetin, (+)-catechin, and (−)-epicatechin content in south Italian monovarietal wines: relationship with maceration time and marc pressing during winemaking. J. Agric. Food Chem. 52, 5747–5751.

Gambuti, A., Capuano, R., Lisanti, M.T., Strollo, D., Moio, L., 2010. Effect of aging in new oak, one-year-used oak, chestnut barrels and bottle on color, phenolics and gustative profile of three monovarietal red wines. Eur. Food Res. Technol. 231, 455–465.

Genebra, T., Santos, R.R., Francisco, R., Pinto-Marijuan, M., Brossa, R., Serra, A.T., et al., 2014. Proanthocyanidin accumulation and biosynthesis are modulated by the irrigation regime in Tempranillo seed. Int. J. Mol. Sci. 15, 11862–11877.

Gény, L., Saucier, C., Bracco, S., Daviaud, F., Glories, Y., 2003. Composition and cellular localization of tannins in grape seeds during maturation. J. Agric. Food Chem. 51, 8051–8054.

Gil, M., Kontoudakis, N., González, E., Esteruelas, M., Fort, F., Canals, J.M., et al., 2012. Influence of grape maturity and maceration length on color, polyphenolic composition, and polysaccharide content of Cabernet Sauvignon and Tempranillo wines. J. Agric. Food Chem. 60, 7988–8001.

Gil, M., Esteruelas, M., González, E., Kontoudatakis, N., Jiménez, J., Fort, F., et al., 2013. Effect of two different treatments for reducing grape yield in *Vitis vinifera* cv Syrah on wine composition and quality: berry thinning versus cluster thinning. J. Agric. Food Chem. 61, 4968–4978.

Glories, Y., 1978. Recherches sur la matière colorante des vins rouges. Ph.D. Thesis of State. Université de Bordeaux II.

Glories, Y., Saint-Cricq, N., Vivas, N., Augustin, M., 1996. Identification et dosage de la procyanidine A2 dans les raisins et les vins de *Vitis Vinifera* L. Cabernet Sauvignon et Merlot. In: Proceedings of the 18th International Conference of Polyphenols Group. Bordeaux, vol. 1, pp. 153.

Gómez-Míguez, M., Gonzáles-Miret, M.L., Hernanz, D., Fernández, M.A., Vicario, I.M., Heredia, F.J., 2007. Effects of prefermentative skin contact conditions on colour and phenolic content of white wines. J. Food Eng. 78, 238–245.

González-Manzano, S., Rivas-Gonzalo, J.C., Santos-Buelga, C., 2004. Extraction of flavan-3-ols from grape seed and skim into wine using simulated maceration. Anal. Chim. Acta 513, 283–289.

González-Neves, G., Gil, G., Favre, G., Baldi, C., Hernández, N., Traverso, S., 2013. Influence of winemaking procedure and grape variety on the colour and composition of young red wines. S. Afr. J. Enol. Vitic. 34, 138–146.

González-SanJosé, M.L., Revilla, I., 2001. Effect of different oak woods on aged wine color and anthocyanin composition. Eur. Food Res. Technol. 213, 281–285.

Guadalupe, Z., Martínez, L., Ayestarán, B., 2010. Yeast mannoproteins in red winemaking. Effect on polysaccharide, polyphenolic and colour composition. Am. J. Enol. Vitic. 61, 191–200.

Hanlin, R.L., Downey, M.O., 2009. Condensed tannin accumulation and composition in skin and Cabernet Sauvignon grapes during berry development. Am. J. Enol. Vitic. 60, 13–23.

Haslam, E., 1980. In vino veritas: oligomeric procyanidins and the aging of red wines. Phytochemistry 19, 1577–1582.

He, F., Liang, N.N., Mu, L., Pan, Q.H., Wang, J., Reeves, M.J., et al., 2012. Anthocyanins and their variation in red wines. II. Anthocyanin derived pigments and their color evolution. Molecules 17, 1483–1519.

Hernández, M.M., Song, S., Menéndez, C.M., 2017. Influence of genetic and vintage factors in flavan-3-ol composition of grape seeds of a segregating *Vitis vinifera* population. J. Sci. Food Agric. 97, 236–243.

Hernández-Jiménez, A., Kennedy, J.A., Bautista-Ortín, A.B., Gómez-Plaza, E., 2012. Effect of ethanol on grape seed proanthocyanidin extraction. Am. J. Enol. Vitic. 63, 57–61.

Ho, P., Silva, M.C.M., Hogg, T.A., 2001. Changes in colour and phenolic composition during the early stages of maturation of Porto in wood, stainless steel and glass. J. Sci. Food Agric. 81, 1269–1280.

Ichikawa, M., Hisamoto, M., Matsudo, T., Okuda, T., 2012. Effect of cap management technique on the concentration of proanthocyanidins in Muscat Bailey A wine. Food Sci. Technol. Res. 18, 201–207.

Jindra, J.A., Gallender, J.F., 1987. Effect of American and French oak barrels on the phenolic composition and sensory quality of Seyval blanc wine. Am. J. Enol. Vitic. 38, 133–138.

Jordão, A.M., Correia, A.C., 2012. Relationship between antioxidant capacity, proanthocyanidin and anthocyanin content during grape maturation of Touriga Nacional and Tinta Roriz grape varieties. S. Afr. J. Enol. Vitic. 33, 214–224.

Jordão, A.M., Ricardo-da-Silva, J.M., Laureano, O., 1998. Influência da rega na composição fenólica das uvas tintas da casta Touriga Francesa (*Vitis vinifera* L.). Cienc. Tecnol. Aliment. 2, 60–73.

Jordão, A.M., Ricardo-da-Silva, J.M., Laureano, O., 2001a. Evolution of proanthocyanidins in bunch stems during berry development (*Vitis vinifera* L.). Vitis 40, 17–22.

Jordão, A.M., Simões, S., Correia, A.C., Gonçalves, F.J., 2012a. Antioxidant activity evolution during Portuguese red wine vinification and their relation with the proanthocyanidin and anthocyanin composition. J. Food Process. Preserv. 36, 298–309.

Jordão, A.M., Correia, A.C., DelCampo, R., González-SanJosé, M.L., 2012b. Antioxidant capacity, scavenger activity and ellagitannins content from commercial oak pieces used in winemaking. Eur. Food Res. Technol. 235, 817–825.

Jordão, A.M., Ricardo-da-Silva, J.M., Laureano, O., 2001b. Evolution of catechins and oligomeric procyanidins during grape maturation of Castelão Francês and Touriga Francesa. Am. J. Enol. Vitic. 52, 230–234.

Jordão, A.M., Ricardo-da-Silva, J.M., Laureano, O., 2005a. Comparison of volatile composition of cooperage oak wood of different origins (*Quercus pyrenaica* vs *Quercus alba* and *Quercus petraea*). Mitt. Klosterneuburg. 52, 22–31.

Jordão, A.M., Ricardo-da-Silva, J.M., Laureano, O., 2005b. Extraction of some ellagic tannins and ellagic acid of oak chips (*Quercus pyrenaica* Willd) in model solutions: effect of time, pH, temperature and alcoholic content. S. Afr. J. Enol. Vitic. 26, 25–31.

Jordão, A.M., Ricardo-da-Silva, J.M., Laureano, O., Adams, A., Demyttenaere, J., Verhé, R., et al., 2006a. Volatile composition analysis by solid-phase microextraction applied to oak wood used in cooperage (*Quercus pyrenaica* and *Quercus petraea*): effect of botanical species and toasting process. J. Wood Sci. 56, 514–521.

Jordão, A.M., Ricardo-da-Silva, J.M., Laureano, O., 2006b. Effect of oak constituents and oxygen on the evolution of malvidin-3-glucoside and (+)-catechin in model wine. Am. J. Enol. Vitic. 57, 377–381.

Jordão, A.M., Ricardo-da-Silva, J.M., Laureano, O., 2007. Ellagitannins from Portuguese oak wood (*Quercus pyrenaica* Willd.) used in cooperage: influence of geographical origin, coarseness of the grain and toasting level. Holzforschung 61, 155–160.

Jordão, A.M., Ricardo-da-Silva, J.M., Laureano, O., Mullen, W., Crozier, A., 2008. Effect of ellagitannins, ellagic acid and volatile compounds from oak wood on (+)-catechin, procyanidin B1 and malvidin-3-glucoside content in model wines. Aust. J. Grape Wine Res. 14, 260–2270.

Jurd, L., 1965. Anthocyanins and related compounds.VIII. Condensation reaction of flavylium salts with 5,5-dimethyl-1,3-cyclohexanedionein acid solutions. Tetrahedron 21, 3707–3716.

Jurd, L., 1967. Anthocyanins and related compounds. XI Catechin-flavylium salts condensation reaction. Tetrahedron 23, 1057–1064.

Kennedy, J.A., Mattheus, M.A., Waterhouse, A.L., 2000. Changes in grape seed polyphenols during fruit ripening. Phytochemistry 55, 77–85.

Kennedy, J.A., Hayasaka, Y., Vidal, S., Waters, E.J., Jones, G., 2001. Composition of grape skin proanthocyanidins at different stages of berry development. J. Agric. Food Chem. 49, 5348–5355.

Kocabey, N., Yilmaztekin, M., Hayaloglu, A.A., 2016. Effect of maceration duration on physicochemical characteristics, organic acid, phenolic compounds and antioxidant activity of red wine from *Vitis vinifera* L. Karaoglan. J. Food Sci. Technol. 57, 3557–3565.

Koyama, K., Sadamatsu, K., Goto-Yamamoto, N., 2010. Abscisic acid stimulated ripening and gene expression in berry skins of the Cabernet-Sauvignon grape. Funct. Integr. Genomics 10, 367–381.

Kozlovic, G., Jeromel, A., Maslov, L., Pollnitz, A., Orlic, S., 2010. Use of acacia barrique barrels – Influence on the quality of Malvazija from Istria wines. Food Chem. 120, 698–702.

Kyraleou, M., Kotseridis, Y., Koundouras, S., Chira, K., Teissedre, P.L., Kallithraka, S., 2016. Effect of irrigation regime on perceived astringency and proanthocyanidin composition of skins and seeds of *Vitis vinifera* L. cv. Syrah grapes under semiarid conditions. Food Chem. 203, 292–300.

Laszlavik, M., Gal, L., Misik, S., Erdei, L., 1995. Phenolic compounds in two Hungarian red wines matured in *Quercus robur* and *Quercus petrea* barrels: HPLC analysis and diode array detection. Am. J. Enol. Vitic. 46, 67–74.

Liu, Y., Zhang, B., He, F., Duan, C.-Q., Shi, Y., 2016. The influence of pre fermentative addition of gallic acid on the phenolic composition and chromatic characteristics of Cabernet Sauvignon wines. J. Food Sci. 81, C1669–C1678.

Loira, I., Vejarano, A., Morata, A., Ricardo-da-Silva, J.M., Laureano, O., González, M.C., et al., 2013. Effect of *Saccharomyces* strains on the quality of red wines aged on lees. Food Chem. 139, 1044–1051.

Lomolino, G., Zocca, F., Spettoli, P., Zanin, G., Lante, A., 2010. A preliminary study on changes in phenolic content during Bianchetta Trevigiana winemaking. J. Food Comp. Anal. 23, 575–579.

Lorrain, B., Kleopatra, C., Teissedre, P.L., 2011. Phenolic composition of Merlot and Cabernet-Sauvignon grapes from Bordeaux vineyard for the 2009-vintage: comparison to 2006, 2007 and 2008 vintages. Food Chem. 126, 1991–1999.

Lutz, M., Jorquera, K., Cancino, B., Ruby, R., Henriquez, C., 2011. Phenolics and antioxidant capacity of table grape (*Vitis vinifera* L.) cultivars grown in Chile. J. Food Sci. 76, 1088–1093.

Mateus, N., Marques, S., Gonçalves, A.C., Machado, J.M., De Freitas, V., 2001. Proanthocyanidin composition of red *Vitis vinifera* varieties from the Douro Valley during ripening: influence of cultivation altitude. Am. J. Enol. Vitic. 52, 115–121.

Monagas, M., Gómez-Cordovés, C., Bartolomé, B., Laureano, O., Ricardo-da-Silva, J.M., 2003. Monomeric, oligomeric, and polymeric flavan-3-ol composition of wines and grapes from *Vitis vinifera* L. Cv. Graciano, Tempranillo, and Cabernet Sauvignon. J. Agric. Food Chem. 51, 6475–6481.

Montealegre, R.R., Peces, R.R., Vozmediano, J.L.C., Gascueña, J.M., Romero, E.G., 2006. Phenolic compounds in skins and seeds of ten grape *Vitis vinifera* varieties grown in a warm climate. J. Food Compos. Anal. 19, 687–693.

Morel-Salmi, C., Souquet, J.-M., Bes, M., Cheynier, V., 2006. Effect of flash release treatment on phenolic extraction and wine composition. J. Agric. Food Chem. 54, 4270–4276.

Moutounet, M., Rabier, P.H., Puech, J.-L., Verette, E., Barillere, J.-M., 1989. Analysis by HPLC of extractable substances in oak wood. Application to a Chardonnay wine. Sci. Aliments 9, 35–51.

Nonier, M.F., Pianet, I., Laguerre, M., Vivas, N., De Gaulejac, N.V., 2006. Condensation products derived from flavan-3-ol oak wood aldehydes reaction-1. Structural investigation. Anal. Chim. Acta 563, 76–83.

Nunes, P., Muxagata, S., Correia, A.C., Nunes, F., Cosme, F., Jordão, A.M., 2017. Effect of oak wood barrel capacity and utilization time on phenolic and sensorial profile evolution of an Encruzado white wine. J. Sci. Food Agric. 97, 4847–4856.

Nuñez, V., Monagas, M., Gomez-Cordovés, M.C., Bartolomé, B., 2004. *Vitis vinifera* L. cv. Graciano grapes characterized by its anthocyanin profile. Postharvest Biol. Technol. 31, 69–79.

Obreque-Slier, E., Pena-Neira, A., Lopez-Solis, R., Zamora-Martin, F., Ricardo-da-Silva, J.M., Laureano, O., 2010. Comparative study of the phenolic composition of seeds and skins from Carménère and Cabernet-Sauvignon grape varieties (*Vitis vinifera* L.) during ripening. J. Agric. Food Chem. 58, 3591–3599.

Ó-Marques, J., Reguinga, R., Laureano, O., Ricardo-da-Silva, J.M., 2005. Changes in grape seed, skin and pulp condensed tannins during berry ripening: effect of fruit pruning. Ciência Téc. Vitiv. 20, 35–52.

Ozawa, T., Lilley, T.H., Haslam, E., 1987. Polyphenol interaction: astringency and the loss of astringency in ripening fruit. Phytochemistry. 26, 2937–2942.

Pantelić, M.M., Zagorac, D.Č.D., Davidović, S.M., Todić, S.R., Bešlić, Z.S., Gašić, U.M., et al., 2016. Identification and quantification of phenolic compounds in berry skin, pulp, and seeds in 13 grapevine varieties grown in Serbia. Food Chem. 211, 243–252.

Passos, C.P., Cardoso, S.M., Domingues, M.R.M., Domingues, P., Silva, C.M., Coimbra, M.A., 2007. Evidence for galloylated type-A procyanidins in grape seeds. Food Chem. 105, 1457–1467.

Peng, S., Scalbert, A., Monties, B., 1991. Insoluble ellagitannins in *Castanea sativa* and *Quercus petraea* woods. Phytochemistry 30, 775–778.

Perez-Prieto, L.J., Lopez-Roca, J.M., Martinez-Cutillas, A., Pardo, M.F., Gomez-Plaza, E., 2002. Maturing wines in oak barrels. Effects of oak origin, volume and age of the barrel on the wine volatile composition. J. Agric. Food Chem. 50, 3272–3676.

Pissarra, J., Lourenço, S., González-Paramás, A.M., Mateus, N., Santos-Buelga, C., Silva, A.M.S., et al., 2005. Isolation and structural characterization of new anthocyanin-alkylcatechin pigments. Food Chem. 90, 81–87.

Prieur, C., Rigaud, J., Cheynier, V., Moutounet, M., 1994. Oligomeric and polymeric procyanidins from grape seeds. Phytochemistry 3, 781–784.

Quideau, S., Jourdes, M., Saucier, C., Glories, Y., Pardon, P., Baudry, C., 2003. DNA topoisomerase inhibitor acutissimin A and other flavano-ellagitannins in red wine. Angew. Chem. Int. Ed. 42, 6012–6014.

Quideau, S., Jourdes, M., Lefeuvre, D., Montaudon, D., Saucier, C., Glories, Y., et al., 2005. The chemistry of wine polyphenolic C-glycosidic ellagitannins targeting human topoisomerase II. Chem. Eur. J. 11, 6503–6513.

Ribéreau-Gayon, P., 1964. The phenolic composition of grapes and wine. Ann. Physiol. Vég. 6, 119–147.

Ricardo-da-Silva, J.M., Rigaud, J., Cheynier, V., Cheminat, A., Moutounet, M., 1991a. Procyanidin dimers and trimers from grape seeds. Phytochemistry 30, 1259–1264.

Ricardo-da-Silva, J.M., Bourzeix, M., Cheynier, V., Moutounet, M., 1991b. Procyanidin composition of Chardonnay, Mauzac and Grenache blanc grapes. Vitis 30, 245–252.

Ricardo-da-Silva, J.M., Belchior, A.P., Spranger, M.I., Bourzeix, M., 1992a. Oligomeric procyanidins of three grapevine varieties and wines from Portugal. Sci. Aliments 12, 223–237.

Ricardo-da-Silva, J.M., Rosec, J.P., Bourzeix, M., Mourgues, J., Moutounet, M., 1992b. Dimer and trimer procyanidins in Carignan and Mourvèdre grapes and red wines. Vitis 31, 55–63.

Rice, S., Koziel, J.A., Dharmadhikari, M., Fennell, A., 2017. Evaluation of tannins and anthocyanins in Marquette, Frontenac, and St. Croix Cold-Hardy grape cultivars. Fermentation 3, 47.

Rinaldi, A., Jourdes, M., Teissedre, P.L., Moio, L., 2014. A preliminary characterization of Aglianico (*Vitis vinifera* L. cv.) grape proanthocyanidins and evaluation of their reactivity towards salivary proteins. Food Chem. 164, 142–149.

Rodrigues, A., Ricardo-da-Silva, J.M., Lucas, C., Laureano, O., 2012. Effect of commercial mannoproteins on wine color and tannins stability. Food Chem. 131, 907–914.

Rodrigues, A., Ricardo-da-Silva, J.M., Lucas, C., Laureano, O., 2013. Effect of winery yeast lees on Touriga Nacional red wine color and tannin evolution. Am. J. Enol. Vitic. 64, 98–109.

Rous, C., Alderson, B., 1983. Phenolic extraction curves for white wine aged in French and American oak barrels. Am. J. Enol. Vitic. 34, 211–215.

Ruiz-Garcia, Y., Smith, P.A., Bindon, K.A., 2014. Selective extraction of polysaccharide affects the adsorption of proanthocyanidin by grape cell walls. Carbohydr. Polym. 114, 102–114.

Sacchi, K.L., Bisson, L.F., Adams, D.O., 2005. A review of the effect of winemaking techniques on phenolic extraction in red wines. Am. J. Enol. Vitic. 56, 197–206.

Santos-Lima, M., Dutra, M.C.P., Toaldo, I.M., Corrêa, L.C., Pereira, G.E., Dd, O., et al., 2015. Phenolic compounds, organic acids and antioxidant activity of grape juices produced in industrial scale by different processes of maceration. Food Chem. 188, 384–392.

Sanz, M., de Simon, F., Esteruelas, B., Munoz, E., Cadahia, A.M., Hernandez, E., et al., 2012. Polyphenols in red wine aged in acacia (*Robinia pseudoacacia*) and oak (*Quercus petraea*) wood barrels. Anal. Chim. Acta 732, 83–90.

Saucier, C., Jourdes, M., Glories, Y., Quideau, S., 2006. Extraction, detection, and quantification of flavano-ellagitannins and ethylvescalagin in a Bordeaux red wine aged in oak barrels. J. Agric. Food Chem. 54, 7349–7354.

Smith, P.A., Mcrae, J.M., Bindon, K.A., 2015. Impact of winemaking practices on the concentration and composition of tannins in red wine. Aust. J. Grape Wine Res. 21, 601–614.

Somers, T.C., 1971. The polymeric nature of wine pigments. Phytochemistry 10, 2175–2184.

Souquet, J.M., Cheynier, V., Brossaud, F., Moutounet, M., 1996. Polymeric proanthocyanidins from grape skins. Phytochemistry 43, 509–512.

Souquet, J.M., Labarbe, B., Le Guerneve, C., Cheynier, V., Moutounet, M., 2000. Phenolic composition of grape stems. J. Agric. Food Chem. 48, 1076–1080.

Sousa, C., Mateus, N., Perez-Alonso, J., Santos-Buelga, C., De Freitas, V., 2005. Preliminary study of oaklins, a new class of brick-red catechin-pyrylium pigments from the reaction between catechin and wood aldehydes. J. Agric. Food Chem. 53, 9249–9256.

Sousa, C., Mateus, N., Silva, A.M.S., González-Paramás, A.M., Santos-Buelga, C., De Freitas, V., 2007. Structural and chromatic characterization of a new malvidin 3-glucoside-vanillylcatechin pigment. Food Chem. 102, 1344–1351.

Sun, B.S., Leandro, C., Ricardo-da-Silva, J.M., Spranger, I., 1998. Separation of grape and wine proanthocyanidins according to their degree of polymerization. J. Agric. Food Chem. 46, 1390–1396.

Sun, B.S., Pinto, T., Leandro, M.C., Ricardo-da-Silva, J.M., Spranger, M.I., 1999. Transfer of catechins and proanthocyanidins from solid parts of grape cluster into wines. Am. J. Enol. Vitic. 50, 179–184.

Sun, B.S., Ricardo-da-Silva, J.M., Spranger, M.I., 2001. Quantification of catechins and proanthocyanidins in several Portuguese grapevine varieties and red wines. Ciência Téc. Vitiv 16, 23–234.

Tavares, M., Jordão, A.M., Ricardo-da-Silva, J.M., 2017. Impact of cherry, acacia and oak chips on red wine phenolic parameters and sensory profile. OENO-ONE 51, 329–342.

Timberlake, C.F., Bridle, P., 1976. Interaction between anthocyanins, phenolic compounds and acetaldehyde and their significance in red wines. Am. J. Enol. Vitic. 27, 97–105.

Tounsi, M.S., Ouerghemmi, I., Wannes, W.A., Ksouri, R., Zemni, H., Marzouk, B., et al., 2009. Valorization of three varieties of grape. Ind. Crops Prod. 30, 292–296.

Valero, E., Sánchez-Ferrer, A., Varón, R., García-Carmona, F., 1989. Evolution of grape polyphenol oxidase activity and phenolic content during maturation and vinification. Vitis 28, 85–95.

Vazallo-Valleumbrocio, G., Medel-Maraboli, M., Peña-Neira, A., López-Solís, R., Obreque-Slier, E., 2017. Commercial enological tannins: characterization and their relative impacto n the phenolic and sensory composition of Carmenère wine during bottle aging. LWT Food Sci. Technol. 83, 172–183.

Viriot, C., Scalbert, A., Hervé du Penhoat, C.L.M., Moutounet, M., 1994. Ellagitannins in woods of sessile oak and sweet chestnut dimerization and hydrolysis during wood ageing. Phytochemistry 36, 1253–1260.

Vivas, N., Glories, Y., 1996. Role of oak wood ellagitannins in the oxidation process of red wines during aging. Am. J. Enol. Vitic. 47, 103–107.

Wongnarat, C., Srihanam, P., 2017. Phytochemical and antioxidant activity in seeds and pulp of grape cultivated in Thailand. Orient. J. Chem. 33, 113–121.

CHAPTER

13

Wine Color Evolution and Stability

María Teresa Escribano-Bailón, Julián C. Rivas-Gonzalo and
Ignacio García-Estévez

Analytical Chemistry, Nutrition and Food Science, Faculty of Pharmacy, University of Salamanca, Salamanca, Spain

13.1 INTRODUCTION

The color of red wine depends, to a great extent, on its phenolic composition. During maceration different types of phenolics are extracted from the solid parts of the grape, especially hydroxycinnamic esters and flavonoids, the most significant of these being the anthocyanins, primarily responsible for color in young red wines, and the flavanols (catechins and proanthocyanidins), which are mainly responsible for astringency and influence the evolution of the wine color.

Anthocyanins are polyhydroxylated and/or methoxylated heterosides derived from flavylium ion (phenyl-2-benzopyrilium). The flavylium anthocyanin structure is highly unstable due to its deficiency on electrons and, therefore, the free forms of anthocyanins, namely anthocyanidins, are usually found as glycosylated forms in nature, as these are more stable. Moreover, the sugar moiety can be esterified or not by different organic acids.

The flavylium ion is an oxonium ion. However, in contrast to this kind of ion, that is not very stable due to the tetravalence of oxygen, flavylium ion is stabilized since it is a resonance hybrid in which a positive charge is delocalized over the whole structure (Fig. 13.1). Anthocyanin aglycons are hydroxylated and/or methoxylated pigments derived from flavylium ion. Five different structures of anthocyanidins can be found in grapes and wines: cyanidin, peonidin, delphinidin, petunidin, and malvidin (Fig. 13.1).

The 3-O-glucosides derived from the aglycons shown in Fig. 13.2 are major pigments found in red wines made from *Vitis vinifera* L. grapes, although the galactoside derivatives can also be detected in lower amounts. Moreover, acylated monoglucosides are also found, with acetic, coumaric, and caffeic acids being the main acids acylated in C-6 of the glucose moiety. Anthocyanin 3,5-diglucosides can also be detected but in *Vitis* American varieties of grapes.

13.2 ANTHOCYANIN STABILITY

Anthocyanins are unstable compounds both in plant tissues and once extracted (Markakis, 1982). Among the factors affecting anthocyanin stability, light and the presence of metal ions are less important with regards wine, since their influence during winemaking and aging is limited. The main factors determining anthocyanin stability in wines are:

13.2.1 Chemical Structure of Anthocyanins

The low stability of the flavylium nucleus of anthocyanin, due to its electronic deficiency, has an important consequence: the high reactivity of anthocyanins. Due to this high reactivity, anthocyanins can be involved in many reactions during winemaking and aging leading to their degradation or transformation. It has been reported that the stability of anthocyanins increases with the number of methoxyl groups linked to the B-ring,

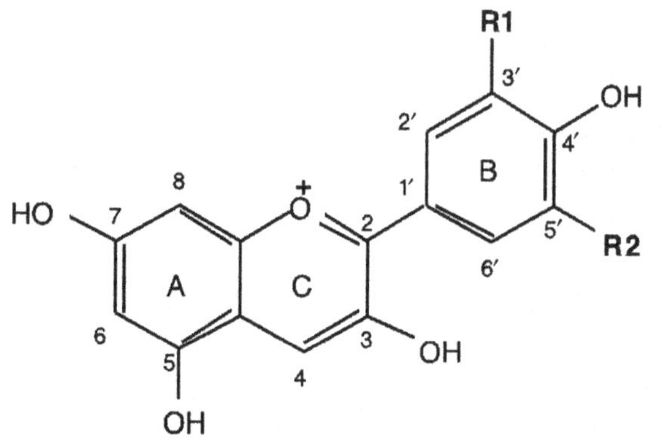

R1	R2	
OH	OH	**Delphinidin**
OH	H	**Cyanidin**
OH	OCH_3	**Petunidin**
OCH_3	H	**Peonidin**
OCH_3	OCH_3	**Malvidin**

FIGURE 13.1 Structure of the main anthocyanins detected in grapes and wines.

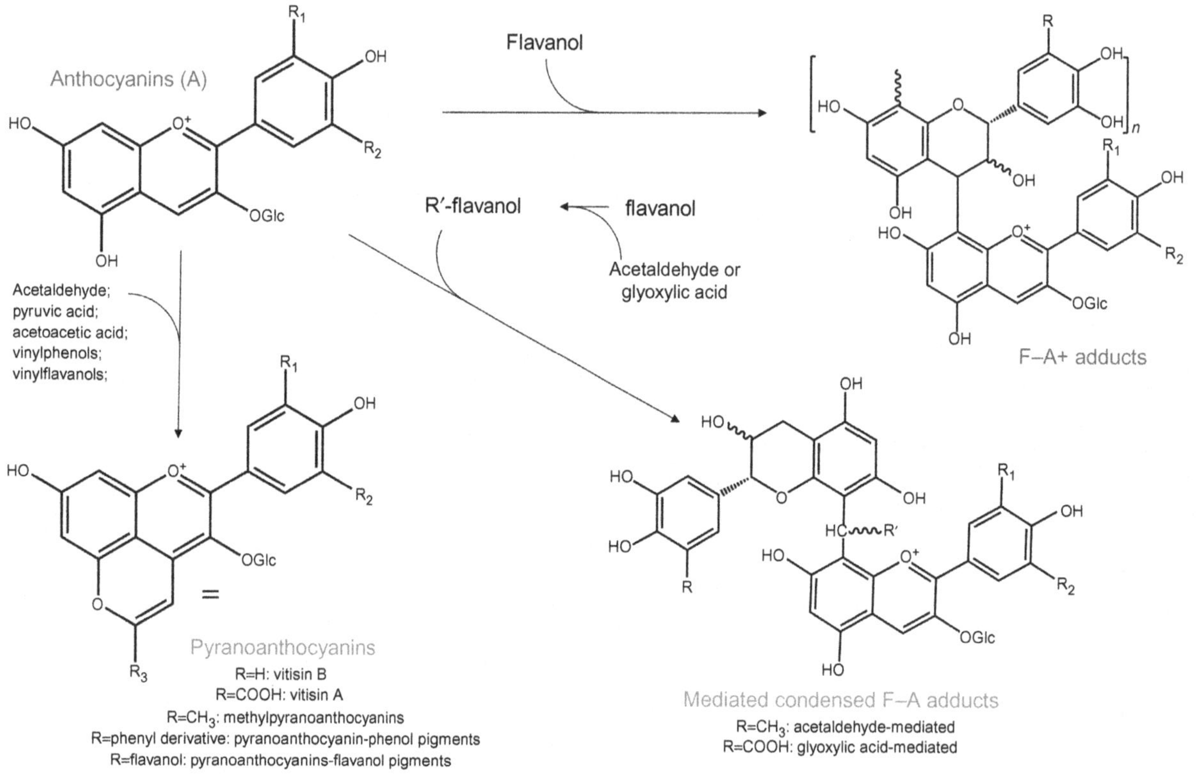

FIGURE 13.2 Main anthocyanin derivative pigments detected in red wines.

and decreases with the number of hydroxyl groups linked to this ring. Thus, the most stable grape anthocyanidin is malvidin, whereas delphinidin has been shown to be the least stable. Moreover, the presence of glycosylation in the anthocyanin structure increases its stability, so the higher the number of sugars residues in the structure, the higher the stability. As a consequence, anthocyanin diglucosides are more stable than the corresponding monoglucoside derivatives, although the former are more susceptible to being involved in browning reactions (Hrazdina, 1970).

The color of each anthocyanin is also determined by its structure. In the CIELAB color space, anthocyanins with two substituents in the B ring are placed in the orange hues, while those with three substituents are placed in the red–purple zone. The hue value is also influenced by the existence of methoxyl groups in the B-ring: the higher the number of methoxyl groups, the higher the displacement toward purple hues (Heredia et al., 1998).

Furthermore, the presence of acylation in the anthocyanin structure also increases its stability. For instance, acylated anthocyanins are more resistant against SO_2 bleaching and against pH changes than the corresponding monoglucosides. In fact, at pH values higher than 4, acylated anthocyanins showed a more important band of absorption in the visible spectra than the nonacylated ones, which are almost colorless at those pH values (Delgado-Vargas et al., 2000). The low contents of acylated anthocyanins in Pinot noir grapes could explain the light color of wines made from Pinot noir grapes and the difficulties of their aging. On the contrary, Tempranillo grapes show important levels of acylated anthocyanins, so wines made from this grape variety show darker colors.

13.2.2 Effect of pH

It is well established that anthocyanin color is pH-dependent. These pigments show red hues at very low pH values but change to blue hues when the pH is raised to neutral or basic values. Moreover, at mild-acid pH, these pigments are colorless, thus pointing out the existence of a colorless structure of anthocyanins (Hrazdina, 1982).

Hence, anthocyanins and related compounds, in mildly acidic aqueous solutions such as wine, are involved in a series of pH-dependent chemical reactions leading four different structures: flavylium cation (AH+), quinoidal base (A), carbinol base (hemiketal, B), and chalcone (C). At low pH values, the red flavylium cation is the most important. In aqueous media, when the pH is raised, the flavylium cation is involved in a hydration reaction leading to the formation of the colorless carbinol base. This species leads by a ring-opening tautomeric process to the yellow chalcones. At pH values close to 7, in nonaqueous media, the most important anthocyanin structure is the purple quinoidal base (Brouillard, 1982; Chen and Hrazdina, 1982). Thus, pH is a key factor for wine color stability. Low pH values of wines assure the predominance of the flavylium cation, which is colored, and avoid its hydration. At pH values between 3 and 6, the flavylium cation of anthocyanin 3-O-glucosides suffers the nucleophilic addition of H_2O on C-2 of anthocyanin, leading to the colorless carbinol base (hemiketal). This reaction is fast and the hemiketal is thermodynamically stable. Although the nucleophilic attack can also occur on C-4; this reaction is neither kinetically nor thermodynamically favored.

Anthocyanin aglycons can also be involved in that hydration reaction but the hemiketal product is less stable than that formed in the case of anthocyanin 3-O-glucosides. Moreover, those anthocyanin structures in which the C-4 position is not free are more stable since it makes the hydration reactions more difficult (Amic et al., 1990).

13.2.3 Effect of the Temperature

The deleterious effect of medium-high temperatures on the stability of anthocyanins is well known, although there are some discrepancies over the magnitude of this effect. In 1953, Meschter observed that the degradation rate of anthocyanins follows a logarithmic increase with increasing temperature (Meschter, 1953), while Calvi and Francis (1978) and Maccarone et al. (1985) indicated that the degradation rate doubles for every 10°C rise. Furthermore, thermic degradation of anthocyanins is influenced by pH, being more pronounced at high pH values. In wines, temperature needs to be controlled, especially during must fermentation, avoiding rises of temperature and preventing high temperatures for a long time.

13.2.4 Effect of the Bisulfite

Bisulfite (HSO_3^-) reacts reversibly with anthocyanins leading to the formation of colorless adducts.

At low pH values (pH 1) the anthocyanin-bisulfite adduct can dissociate, releasing the anthocyanin in its colored flavylium form and SO_2. Timberlake and Bridle (1967) suggested that SO_2 links to C-4 of the anthocyanin since flavylium salts possessing substitution in this position do not bleach. The SO_2 can also attack position C-2 of the anthocyanin. Discoloration can be irreversible if bisulfite or sulfur dioxide are added in high concentration. When discoloration is reversible, even if the anthocyanin loses its color temporarily, it increases its stability, since the formed adduct is less sensitive to hydration and polymerization reactions because reactive positions are occupied (Sarni-Manchado et al., 1996; Romero and Bakker, 1999). The heterocyclic of anthocyanins hydroxylated in A-ring (5 or 7 position) is less acidic than the one from nonhydroxylated anthocyanins, making the nucleophilic attack of SO_2 difficult. Nevertheless, anthocyanin 3,5-diglucosides are more prone to link bisulfite than 3-monoglucoside ones due to the higher acidity of the heterocycle (Timberlake and Bridle, 1967). The addition of bisulfite to anthocyanin solutions provokes not only discoloration, but also a displacement of the visible wavelength (hypsochromic in the absorption maximum and bathochromic in the minimum one) and of the ultraviolet (UV) region (bathochromic effect in the maximum).

13.2.5 Effect of Oxygen

Anthocyanins can be oxidized in contact with air, that is why it is essential that there is rigorous oxygenation control during winemaking and storage. In general, diglycosylated anthocyanins are more susceptible to oxidative browning than the monoglycosides. Some studies postulate the use of ascorbic acid as an antioxidant; nevertheless, it has been observed in model solutions (Ortega Meder, 1995) and in wines (Guerra-Sánchez Simón, 1997) that this compound increases anthocyanins degradation and leads to products like those obtained after thermic degradation.

13.3 COPIGMENTATION

Considering the anthocyanin equilibria in aqueous solution and the wine pH (pH 3–4), anthocyanin should be mainly present in a colorless form. Nevertheless, there are some mechanisms, known under the generic name of copigmentation, that contribute to stabilize the colored forms. Copigmentation can be defined as a spontaneous and exothermic process consisting of the stacking of a molecule called copigment on the planar flavylium ion or quinoidal forms of anthocyanins (Baranac et al., 1996). The planar structure of these highly conjugated anthocyanic forms allow their noncovalent association with the copigments (frequently colorless molecules), which present, at least, a portion of the molecule planar and are easily polarizable. Because of the copigmentation process, the colored forms of the anthocyanin are stabilized. In fact, copigmentation is the main color-stabilizing mechanism in plants and also in food products of vegetable origin like red wines (Escribano-Bailón and Santos-Buelga, 2012). Copigmentation leads to characteristic changes in the absorbance spectrum of the anthocyanin chromophore, consisting of an increase in the absorbance at the visible range (hyperchromic effect) and a positive displacement of the visible wavelength of maximum absorption (bathochromic effect). The main driving force leading to copigmentation is $\pi-\pi$ interaction (or $OH-\pi$ interaction) between pigment and copigment (Kunsági-Máté et al., 2006). Also, change transfer from the copigment to the comparatively electron-poor flavylium ion has been proposed (Ferreira da Silva et al., 2005). The final consequence is a hydrophobic environment that prevents the hydration of the flavylium form, reducing the formation of the colorless hemiketal and chalcone forms (Mazza and Brouillard, 1987).

In wine, copigmentation could account for almost 50% of the color in young red wines (Boulton, 2001) but a decrease is observed during storage suggesting that copigmentation in wines competes with other chemical reactions involving anthocyanins (Trouillas et al., 2016). Copigmentation, has also been postulated as the initial approaching mechanism that would be followed by the formation of more stable covalent links leading to the formation of complex anthocyanin derived pigments (Escribano-Bailón et al., 1996).

Different types of copigmentation can be distinguished:

Intermolecular copigmentation. It occurs when pigment and copigment are different molecular units. The copigmentation complex formed consists of a vertical stacking of alternate molecules of pigment and copigment maintained by van der Waals forces and hydrophobic interactions. Sugar molecules of anthocyanins disperse externally and interact by hydrogen bonds, forming a solvation shell that contributes to reinforce the structure (Goto and Kondo, 1991). The association of the anthocyanin with the copigment prevents the nucleophilic attack of the water molecule to the C-2 of the flavylium diminishing the formation of colorless forms and displacing the structural equilibrium of the anthocyanin toward the colored forms.

A particular case of intermolecular copigmentation is the homomolecular copigmentation, also known as self-association. It implies the formation of a copigmentation complex between anthocyanin molecules themselves and requires relatively concentrated solutions. In wines, this type of copigmentation could be responsible for a noticeable percentage of the visible absorbance (González-Manzano et al., 2008), most of all in wines deprived of the classical copigments (Di Stefano et al., 2005).

Intramolecular copigmentation. Anthocyanins with aromatic acyl residues have more stable colors and higher molar absorption coefficients in weakly acidic solutions than anthocyanins devoid of them. In such solutions, these last anthocyanins are almost totally as unstable colorless forms; but anthocyanins whose sugar residues are acylated by phenolic acids display unusual color stability. This can be explained by intramolecular copigmentation, in which the copigment is part of the anthocyanin itself (Brouillard, 1981). In such type of copigmentation, pigment and copigment are covalently bound, making the process more effective since it can take place even in diluted solutions.

Intramolecular copigmentation requires that the aromatic acyl residue interacts with the π-system of the pyrylium, that way protecting the chromophore against the nucleophilic attack of the water molecule. To allow the stacking of the aromatic acyl residue over the flavylium, a suitable spacer is necessary between the anthocyanin and the acyl group. This role is performed by the sugar moiety.

In wines, the anthocyanin acyl derivatives suitable to undergo intramolecular copigmentation are the hydroxycinnamic derivatives, mainly the coumaroyl derivatives since they are usually in the highest concentration from the hydroxycinnamic ones. Studies performed with malvidin-3-*O*-(6-coumaroylglucoside) suggest that this molecule could experience dual aggregation phenomenon: self-association and also intramolecular stacking of the acyl group with the pyrylium ring of the flavylium cation (Fernandes et al., 2015). This could explain the stability of the coumaroyl anthocyanin derivatives that have been observed during wine maturation (Alcalde-Eon et al., 2006).

Furthermore, some anthocyanin derived pigments like oligomeric pyrano anthocyanin–flavanol (A–F) pigments could also form intramolecular copigmentation complexes, which could be responsible for their high stability and molar extinction coefficient (He et al., 2010).

13.3.1 Factors Affecting Copigmentation

Some factors can affect the effectiveness of copigmentation in wines, between them, the structure of the pigment and the copigment. In anthocyanins from grapes, the capacity to form copigmentation complexes increases with the number of B-ring substituents (malvidin has higher capacity than cyanidin) (Mazza and Brouillard, 1990); but methoxylation in B-ring reduces the self-association process (Hoshino et al., 1981). It is also important that in the glycosylation of the anthocyanin, copigmentation is more favored in anthocyanin 3,5-diglucosides than in the 3-monoglucoside ones. As for the copigment, the capacity to stack with anthocyanins is related to the existence of a planar structure in the molecule. That is why flavanols, which present several asymmetric carbons conferring nonplanar spatial conformation, are worse copigments than other flavonoids like flavonols and even than hydoxycinnamic acids (Gómez-Míguez et al., 2006). Nevertheless, due to the lower amounts of flavonols and phenolic acids than flavanols in wines, these latter are postulated as the main factor stabilizing anthocyanin color in wines. It is worth noting that copigmentation is affected by the concentration of copigment and pigment, being favored at high molar ratio especially when dealing with poor copigments (Brouillard and Dangles, 1994).

As stated before, the color of anthocyanins in aqueous solution is largely dependent on the pH. Copigmentation optimum pH is in the range 3–5. In very acidic solutions (pH < 1) the anthocyanins are mainly in colored flavylium cation form, thus the presence of copigments does not lead to perceptible changes in the color of the solution. Copigmentation effects are easily observed in slightly acidic solutions like wine, where in the absence of copigments, hydration reactions are favored.

Copigmentation is linked to the presence of water molecules. For copigmentation to occur, the existence of high degree of organization is necessary, like the network formed by the water molecules linked by hydrogen bonds. The presence of organic solvents partially destroys that network, thus diminishing the intensity of the copigmentation process. Nevertheless, it still occurs in matrix where water is the principal component, like in wines.

Another factor affecting copigmentation is temperature. Since copigmentation is an exothermic process, an increase of the temperature reduces the interaction between pigment and copigment and the complexes partially dissociate (Brouillard and Dangles, 1994). Nevertheless, the complex is restored when cooling the system (Petrova and Gandova, 2016). The increase of temperature provokes the partial disruption of water structure, thus reducing the number of hydrogen bonds in the solvation shell and diminishing the efficiency of the copigmentation (Brouillard and Dangles, 1994).

13.4 RED WINE COLOR EVOLUTION

As red wine color is mainly determined by anthocyanins and its derivative pigments, the evolution of these compounds in wine is the key factor for the evolution of wine color. Native anthocyanins are not particularly stable and their levels fall during wine aging, since they can be involved not only in degradation reactions but also in reaction with other wine components leading to anthocyanin derivative pigments.

13.4.1 Anthocyanin Oxidation

As for organoleptic properties of red wine, controlled oxidation could be beneficial since it enhances and stabilizes color and may reduce astringency. However, oxidation can also provoke an important loss of anthocyanin and even a browning effect, which is not desirable in wines. In fact, it has been described that the degradation of anthocyanin significantly increases in the presence of oxygen (Shenoy, 1993). Moreover, in weak acidic alcoholic solutions such as wine, in which the flavylium and quinoidal forms of anthocyanins are the most abundant, the pH-dependent equilibria of anthocyanins can be displaced since the hydrated forms can react with o-diquinones (a product of phenolic oxidation) leading to unstable and colorless compounds, which implies a loss of color in wines (He et al., 2012a). On the other hand, anthocyanins can also act as substrate for the polyphenoloxidase (PPO), which in some cases has been pointed out as the main reason for the initial important losses of anthocyanins in some food products (Wesche-Ebeling and Montgomery, 1990). Although the ability of PPO to oxidize anthocyanins depends on the pigment structure, in all cases it entails pigment browning.

13.4.2 Formation of Anthocyanin Derivative Pigments

Several reactions, including condensation and cycloaddition reactions, occur during winemaking and aging involving anthocyanins and other wine components (mainly other phenolic compounds but also simpler acids or aldehydes) that lead to the formation of new anthocyanin derivative pigments (Fig. 13.2). In those reactions, due to their structure, anthocyanin can act as electrophile or as nucleophile. Due to the charge distribution of the anthocyanin structure, hydroxyl groups and C-6 and C-8 are usually the positions in the structure that act as nucleophiles, whereas C-2 and C-4 are the most electrophile positions (de Freitas and Mateus, 2006). As a result of those reactions, during winemaking and aging, pyranoanthocyanins and A—F condensation products can be formed (de Freitas and Mateus, 2006). Those new pigments usually show important differences in their physicochemical and colorimetric properties compared to native anthocyanins, so their formation during winemaking and aging entails changes in wine color from a purple-red hue, characteristic of young red wines, to a brick-red color, more related to aged wines (He et al., 2012b).

13.4.2.1 Flavanol—Anthocyanin Condensation Products

Anthocyanins and flavanols can condense either directly or by mediation of acetaldehyde or other compounds. As for flavanol—anthocyanin (F—A^+) direct condensation products, two different mechanisms have been proposed to explain their formation in wines. In that reactions, anthocyanin can act as an electrophile or as a nucleophile to form either A—F or F—A^+ adducts (Salas et al., 2004), respectively. The latter show red hues similar to their anthocyanin precursors (λ_{max} ca. 515—526 nm) and are the most abundant F—A^+ condensation products found in wines (Cheynier et al., 2006). The A—F adducts are colorless but they can be oxidized leading to A^+—F colored adducts that, in addition, are resistant to SO_2 bleaching, since they have C-4 position of the flavylium anthocyanin occupied (He et al., 2012b).

As for acetaldehyde-mediated F—A^+ pigments, anthocyanin acts as a nucleophile attacking the product of the reaction between flavanols and acetaldehyde (Es-Safi et al., 1999a). These pigments show purple hues (λ_{max} ca. 528—540 nm), which can be due to a stabilization of the blue quinoidal base of the anthocyanin (Timberlake and Bridle, 1976) and to the existence of an intramolecular copigmentation effect between the flavanol and the anthocyanin (Escribano-Bailón et al., 1996). Moreover, acetaldehyde-mediated F—A^+ pigments are more stable to water attack and bleaching by sulfite dioxide, which can be probably explained by a protective effect of the flavanol moiety thus protecting the anthocyanin from the nucleophilic attack of those bleaching agents (Escribano-Bailón et al., 2001). However, these pigments are more susceptible to degradation in aqueous solution than the native anthocyanins (Escribano-Bailón et al., 2001).

Furthermore, A—F condensation can be mediated by other wine compounds, such as glyoxylic acid (Fulcrand et al., 1997) or more complex aldehydes, namely vanillin or benzaldehyde, leading to the formation of new pigments that show an important bathochromic displacement in their visible spectra (Pissarra et al., 2003; Sousa et al., 2007). Likewise, anthocyanins can condense with other wine compounds such as ellagitannins, leading to anthocyanin—ellagitannin derivative pigments showing purple-blue hues (García-Estévez et al., 2013; Quideau et al., 2005).

Overall, those pigments can be involved in further reactions leading to other derivative pigments, such as xanthylium ions (yellow pigments; Dueñas et al., 2006) or act as intermediaries in the formation of complex pyranoanthocyanins (Pissarra et al., 2004).

13.4.2.2 Pyranoanthocyanins

Those derivative pigments are characterized by the presence of an additional pyran ring in the pigment structure resulting from the cycloaddition of wine nucleophiles at C4 and at the OH attached to C5 of the anthocyanin flavylium nucleus (Fulcrand et al., 1996). The presence of this fourth ring renders these pigments different chemical properties compared to their anthocyanin precursors, since the C-ring is no longer available to nucleophilic attacks through its carbon 2 and carbon 4 positions. To be precise, pyranoanthocyanins are more stable, resistant to sulfur dioxide bleaching and to oxidative degradations, and they show a higher stability in a wider pH range compared to their anthocyanin precursors (de Freitas and Mateus, 2011).

Some of these pigments are thought to arise from the reaction between anthocyanins and products derived from the yeast metabolism such as pyruvic acid, acetaldehyde, or acetoacetic acid, yielding to vitisin A, to vitisin B, and to methylpyranoanthocyanins, respectively (Fulcrand et al., 1998; Bakker and Timberlake, 1997; Hayasaka and Asenstorfer, 2002). In that reaction, the nucleophilic attack of the enol form of the corresponding acid or aldehyde on the C-ring of the anthocyanin, followed by a dehydration and an oxidation step leads to the formation of the pyranoanthocyanin (He et al., 2006). Those new pigments show great differences on their colorimetric properties compared to the original anthocyanins, since they show an important hypsochromic shift in the band of maximum absorption (λ_{max} ca. 510, 490, and 480 nm for vitisin A, vitisin B, and methylpyranonthocyanins, respectively) and a shoulder around 352–370 nm in their visible spectrum (Cameira-dos-Santos et al., 1996; Alcalde-Eon et al., 2006). Although the formation of these pyranoanthocyanins occurs mainly during alcoholic fermentation, when their precursors reach their maximum levels (Morata et al., 2003; Rentzsch et al., 2007), they can be also formed during the first steps of aging and they can be detected even after bottling (Alcalde-Eon et al., 2006). Moreover, vitisin A can be involved in different reactions leading to more complex pyranoanthocyanins (such as *Portisins*) which show more bluish hues (Mateus et al., 2006).

Other pyranoanthocyanins are formed from the reaction between anthocyanins and other wine constituents, such as the vinyl derivatives of flavanols (pyrano A–F) or hydroxycinnamic acids (pyrano anthocyanin–phenol, which can also be formed by direct reaction between anthocyanins and the hydroxycinnamic acids) (Schwarz et al, 2003; Francia-Aricha et al., 1997). The pyrano anthocyanin–phenol pigments show in their structure moieties derived from coumaric, cafeic, ferulic, or sinapic acid. These pigments show an orange-red hue at low pH (λ_{max} ca. 500 nm) and higher extinction coefficients than the corresponding original anthocyanins (Quijada-Morín et al., 2010). The formation of pyrano A–F pigments was attributed to the reaction between anthocyanins and vinyl–flavanol derivatives (Cruz et al., 2008), which can be found in wines due to the reaction between flavanols and acetaldehyde or to the degradation of ethyl-mediated flavanol–anthocyanin pigments (Es-Safi et al., 1999b). These pigments also show a high stability against pH changes and a hypsochromic shift in their maximum of absorption in the visible spectrum (λ_{max} ca. 490–510 nm) (Francia-Aricha et al., 1997), which depends on the characteristics of the flavanols linked to the pyranoanthocyanin (Mateus et al., 2003).

The aforementioned higher stability of these pigments compared to native anthocyanins has as a consequence that their levels remain quite stable during wine aging, whereas those of anthocyanins decrease, thus contributing to explain the changes in wine color during aging.

13.5 WINEMAKING PRACTICES FOR STABILIZING RED WINE COLOR

13.5.1 Technological Tools to Enhance Copigmentation in Wines

The addition of copigments at different stages of the winemaking process has been proposed to enhance the copigmentation phenomenon and hence to improve wine color stability. The incorporation of extra copigments in the wine can be achieved by external addition of phenolic extracts or enological tannins, by covinification of different grape varieties, each contributing with additional cofactors, and also by the addition of winery by-products like seeds or skins from white grapes, in order to provide supplementary sources of copigments. Experiments performed with enological tannins showed noticeable improvement of chromatic characteristics and stability in red wines (Alcalde-Eon et al., 2014a), although the effectiveness of this practice is conditioned to grape characteristics (Bautista-Ortín et al., 2007; Alcalde-Eon et al., 2014b). This can explain the absence of significant impact or even deleterious consequences observed in other studies (Harbertson et al., 2012). It has also been postulated that the covinification of some grape varieties could lead to wines with more balanced phenolic composition (Moreno-Arribas and Polo, 2008). In this regard, covinification of Tempranillo and Graciano varieties results in higher copigmentation values than in the corresponding monovarietal wines (García-Marino et al., 2013).

Skins and seeds from white varieties are a good source of nonanthocyaninc phenolic compounds with enological interest for their use in winemaking, especially when red grapes do not present good balance copigment/pigment (Kovac et al., 2005; Gordillo et al., 2014). Also, the addition of oak chips during grape maceration has been evaluated as an alternative technological application to modulate the phenolic composition and color characteristics of red wines, especially in warm climate regions (Gordillo et al., 2016).

13.5.2 Effect of Polysaccharide−Anthocyanin Interaction

Polysaccharides are biopolymers that can be found in wine due to two main sources. On one hand, they can be extracted from grape skin cell walls. On the other hand, they can be excreted by yeast during winemaking or released from yeast cell walls after yeast autolysis.

Yeast polysaccharides are also known as mannoproteins since they consist mainly of mannose (and small amounts of glucose), associated with a small percentage of protein in the structure (10%−20%). The possibility of interactions between anthocyanins and wine polysaccharides has been demonstrated (Gonçalvez et al., 2018), so these compounds can exert an important influence on wine color and stability.

13.5.2.1 Grape Skin Polysaccharides

Anthocyanins have shown the ability to interact with cell wall polysaccharides through hydrogen bonding and hydrophobic interactions (Fernandes et al., 2014), leading to the depletion of free anthocyanins (Padayachee et al., 2012). Due to this interaction, polysaccharides can hinder their extraction from grape skin. In fact, it has been reported that the composition of grape skin cell walls determines the anthocyanin extraction. Higher contents of total insoluble cell wall material, xyloglucans, homogalacturonans, and rhamnogalacturonans-I seem to make difficult the anthocyanin extraction, mainly in the case of the acylated ones, whereas cellulose and rhamnogalacturonans-II contents are directly related to anthocyanin extraction (Hernández-Hierro et al., 2012, 2014). Moreover, once in wine, anthocyanin and grape skin polysaccharides can interact, and an important percentage of anthocyanin may be adsorbed by the nonsolubilized cell wall material and, then, can precipitate (Fernandes et al., 2017).

13.5.2.2 Yeast Mannoproteins

There are many commercial products based on mannoproteins that are described to be used for color stabilization. In fact, several studies in literature have reported the ability of yeast mannoproteins for interacting with pigments in wine. However, most of those results are contradictory. Most of the commercial mannoproteins have been described to prevent tartrate salts precipitation, which, in turn, may avoid pigment precipitation. However, some studies using commercial mannoproteins or yeast lees show that there is no such effect (Rodrigues et al., 2012a, 2012b). In some cases, the use of this kind of product has a negative effect on wine color due to the adsorption of pigments (Rodrigues et al., 2012b). Other studies show that the practice of aging wine in the presence of lysated lees does not affect the monomeric anthocyanin content but can improve wine color if enzymes (pectinases and β-1,3 glucanases), which may release higher amounts of mannoproteins to wine, are used (Fernández et al., 2011). Guadalupe et al. (2007 and 2008) have reported that the addition of mannoproteins or the use of yeast strain overexpressed for mannoproteins neither help to maintain the extracted pigments in colloidal dispersion nor to ensure color stability.

On the contrary, other authors have pointed out that aging red wines over lees allows the monomeric anthocyanin content to be stabilized, favoring the stability of the wine's color (Palomero et al., 2007). Color studies when different yeast strains were used for winemaking pointed out a color stability improvement in wines with higher contents of mannoproteins, which was attributed to the possibility that these biopolymers retained anthocyanins, thereby preventing their precipitation (Escot et al., 2001). In fact, the study of the detailed pigment composition of wines have shown that the addition of mannoproteins can stabilize A-type vitisins and other derivative pigments, since it avoids their precipitation, thus pointing out the ability of mannoproteins for the stabilization of coloring matter of wines (Alcalde-Eon et al, 2014a).

Overall, the results reported indicate the possibility that mannoproteins can stabilize wine pigments in wines, but further studies are necessary in order to unravel the possible stabilization mechanisms and the relationship between mannoproteins characteristics and their ability to stabilize wine color.

Acknowledgments

The authors thank the Spanish MINECO (Project ref. AGL2014-58486-C02-R-1 and AGL2017-84793-C2-1-R co-funded by FEDER). IGE thanks FEDER-Interreg España-Portugal Programme (Project ref. 0377_IBERPHENOL_6_E) for postdoctoral contract.

References

Alcalde-Eon, C., Escribano-Bailón, M.T., Santos-Buelga, C., Rivas-Gonzalo, J.C., 2006. Changes in the detailed pigment composition of red wine during maturity and ageing: a comprehensive study. Anal. Chim. Acta 563, 238–254.

Alcalde-Eon, C., García-Estévez, I., Puente, V., Rivas-Gonzalo, J.C., Escribano-Bailón, M.T., 2014a. Color stabilization of red wines. A chemical and colloidal approach. J. Agric. Food. Chem. 62, 6984–6994.

Alcalde-Eon, C., García-Estévez, I., Ferreras-Charro, R., Rivas-Gonzalo, J.C., Ferrer-Gallego, R., Escribano-Bailón, et al., 2014b. Adding oenological tannin vs. Overripe grapes: effect on the phenolic composition of red wines. J. Food Compos. Anal. 34, 99–113.

Amic, D., Baranac, J., Vukadinovic, V., 1990. Reactivity of some flavilium cations and corresponding anhydrobases. J. Agric. Food Chem. 38, 936–940.

Bakker, J., Timberlake, C.F., 1997. Isolation, identification, and characterization of new color-stable anthocyanins occurring in some red wines. J. Agric. Food. Chem. 45, 35–43.

Baranac, J.M., Petranovic, N.A., Dimitric-Marcovic, J.M., 1996. Spectrophotometric study of anthocyanin copigmentation reactions. J. Agric. Food. Chem. 44, 1333–1336.

Bautista-Ortín, A.B., Fernández-Fernández, J.I., López-Roca, J.M., Gómez-Plaza, E., 2007. The effects of enological practices in anthocyanins, phenolic compounds and wine colour and their dependence on grape characteristics. J. Food Compos. Anal. 20, 546–552.

Boulton, R., 2001. The copigmentation of anthocyanins and its role in the color of red wine: a critical review. Am. J. Enol. Vitic. 52 (2), 67–87.

Brouillard, R., 1981. The chemistry of anthocyanin pigments. 7. Origin of the exceptional color stability of the Zebrina anthocyanin. Phytochemistry 20, 143–145.

Brouillard, R., 1982. Chemical structure of anthocyanins. In: Markakis, P. (Ed.), Anthocyanins as Food Colors. Academic Press, Nueva York, pp. 1–40.

Brouillard, R., Dangles, O., 1994. Anthocyanins molecular interactions: the first step in the formation of new pigments during wine aging? Food Chem. 51, 365–371.

Calvi, J.P., Francis, F.J., 1978. Stability of concord grape (*V. labrusca*) anthocyanins in model systems. J. Food Sci. 43, 1448–1456.

Cameira-dos-Santos, P.J., Brillouet, J.M., Cheynier, V., Moutounet, M., 1996. Detection and partial characterisation of new anthocyanin-derived pigments in wine. J. Sci. Food Agric. 70, 204–208.

Chen, L.J., Hrazdina, G., 1982. Structural transformation reactions of anthocyanins. Experientia 38, 1030–1032.

Cheynier, V., Dueñas, M., Salas, E., Maury, C., Souquet, J.M., Sarni-Manchado, P., et al., 2006. Structure and properties of wine pigments and tannins. Am. J. Enol. Vitic. 57, 298–305.

Cruz, L., Teixeira, N., Silva, A.M.S., Mateus, N., Rodríguez-Borges, J., de Freitas, V., 2008. The role of vynilcatechin in the formation of pyrano-malvidin-3-glucoside-(+)-catechin. J. Agric. Food Chem. 56, 10980–10987.

Delgado-Vargas, F., Jiménez, A.R., Paredes-López, O., 2000. Natural pigments: Carotenoids, anthocyanins, and betalains—characteristics, biosynthesis, processing, and stability. Crit. Rev. Food Sci. Nutr. 40, 173–289.

Di Stefano, R., Gentilini, N., Panero, L., 2005. Osservazioni Sperimentali sul Fenomeno della Copigmentazione. Riv. Vitic. Enol. 58, 35–50.

Dueñas, M., Fulcrand, H., Cheynier, V., 2006. Formation of anthocyanin–flavanol adducts in model solutions. Anal. Chim. Acta 563, 15–25.

Es-Safi, N.E., Fulcrand, H., Cheynier, V., Moutounet, M., 1999a. Competition between (+)-catechin and (−)-epicatechin in acetaldehyde-induced polymerization of flavanols. J. Agric. Food. Chem. 47, 2088–2095.

Es-Safi, N.E., Fulcrand, H., Cheynier, V., Moutounet, M., 1999b. Studies on the acetaldehyde-induced condensation of (−)-epicatechin and malvidin 3-O-glucoside in a model solution system. J. Agric. Food. Chem. 47, 2096–2102.

Escot, S., Feuillat, M., Dulau, L., Charpentier, C., 2001. Release of polysaccharides by yeasts and the influence of released polysaccharides on colour stability and wine astringency. Aust. J. Grape Wine Res. 7, 153–159.

Escribano-Bailón, M.T., Santos-Buelga, C., 2012. Anthocyanin copigmentation—evaluation, mechanisms and implications for the colour of red wines. Curr. Org. Chem. 16, 715–723.

Escribano-Bailón, T., Dangles, O., Brouillard, R., 1996. Coupling reactions between flavylium ions and catechin. Phytochemistry 41, 1583–1592.

Escribano-Bailón, M.T., Álvarez-García, M., Rivas-Gonzalo, J.C., Heredia, F.J., Santos-Buelga, C., 2001. Color and stability of pigments derived from the acetaldehydemediated condensation between malvidin 3-O-glucoside and (+)-catechin. J. Agric. Food. Chem. 49, 1213–1217.

Fernandes, A., Brás, N.F., Mateus, N., de Freitas, V., 2014. Understanding the molecular mechanism of anthocyanin binding to pectin. Langmuir 30, 8516–8527.

Fernandes, A., Brás, N., Mateus, N., Freitas, V., 2015. A study of anthocyanin self-association by NMR spectroscopy. New J. Chem. 39, 2602–2611.

Fernandes, A., Oliveira, O., Teixeira, N., Mateus, N., de Freitas, V., 2017. A review of the current knowledge of red wine colour. Oeno One 51, 1–21.

Fernández, O., Martínez, O., Hernández, Z., Guadalupe, Z., Ayestarán, B., 2011. Effect of the presence of lysated lees on polysaccharides, color and main phenolic compounds of red wine during barrel ageing. Food Res. Int. 44, 84–91.

Ferreira da Silva, P., Lima, J.C., Freitas, A.A., Shimizu, K., Maçanita, A.L., Quina, F.H., 2005. Charge-transfer complexation as a general phenomenon in the copigmentation of anthocyanins. J. Phys. Chem. A 109, 7329–7338.

Francia-Aricha, E.M., Guerra, M.T., Rivas-Gonzalo, J.C., Santos-Buelga, C., 1997. New anthocyanin pigments formed after condensation with flavanols. J. Agric. Food. Chem. 45, 2262–2266.

de Freitas, V., Mateus, N., 2006. Chemical transformations of anthocyanins yielding a variety of colours (review). Environ. Chem. Lett. 4, 175–183.

de Freitas, V., Mateus, N., 2011. Formation of pyranoanthocyanins in red wines: a new and diverse class of anthocyanin derivatives. Anal. Bioanal. Chem. 401, 1463–1473.

Fulcrand, H., Doco, T., Es-Safi, N.E., Cheynier, V., Moutounet, M., 1996. Study of the acetaldehyde induced polymerisation of flavan-3-ols by liquid chromatography-ion spray mass spectrometry. J. Chromatogr. A 752, 85–91.

Fulcrand, H., Cheynier, V., Oszmianski, J., Moutounet, M., 1997. An oxidized tartaric acid residue as a new bridge potentially competing with acetaldehyde in flavan-3-ol condensation. Phytochemistry 46, 223–227.

Fulcrand, H., Benabdeljalil, C., Rigaud, J., Cheynier, V., Moutounet, M., 1998. A new class of wine pigments generated by reaction between pyruvic acid and grape anthocyanins. Phytochemistry 47, 1401–1407.
García-Marino, M., Escudero-Gilete, M.L., Heredia, F.J., Escribano-Bailón, M.T., Rivas-Gonzalo, J.C., 2013. Color-copigmentation study by tristimulus colorimetry (CIELAB) in red wines obtained from Tempranillo and Graciano varieties. Food Res. Int. 51, 123–131.
García-Estévez, I., Jacquet, R., Alcalde-Eon, C., Rivas-Gonzalo, J.C., Escribano-Bailón, M.T., Quideau, S., 2013. Hemisynthesis and structural and chromatic charaterization of delphinidin 3-O-glucoside-vescalagin hybrid pigments. J. Agric. Food. Chem. 61, 11560–11568.
Gómez-Míguez, M., González-Manzano, S., Escribano-Bailón, M.T., Heredia, F.J., Santos-Buelga, C., 2006. Influence of different phenolic copigments on the color of malvidin 3-glucoside. J. Agric. Food. Chem. 54, 5422–5429.
Gonçalvez, F.J., Fernandes, P.A.R., Wessel, D.F., Cardoso, S.M., Rocha, S.M., Coimbra, M.A., 2018. Interaction of wine mannoproteins and arabinogalactans with anthocyanins. Food Chem. 243, 1–10.
González-Manzano, S., Santos-Buelga, C., Dueñas, M., Rivas-Gonzalo, J.C., Escribano-Bailón, M.T., 2008. Colour implications of self-association processes of wine anthocyanins. Eur. Food Res. Technol. 226, 483–490.
Gordillo, B., Cejudo-Bastante, M.J., Rodríguez-Pulido, F.J., Jara-Palacios, M.J., Ramírez-Pérez, P., González-Miret, M.L., et al., 2014. Impact of adding White pomace to red grapes on the phenolic composition and color stability of Syrah wines from a warm climate. J. Agric. Food. Chem. 62, 2663–2671.
Gordillo, B., Baca-Bocanegra, B., Rodríguez-Pulido, F.J., González-Miret, M.L., García-Estévez, I., Quijada-Morín, N., et al., 2016. Optimisation of an oak chips-grape mix maceration process. Influence of chip dose and maceration time. Food Chem. 206, 249–259.
Goto, T., Kondo, T., 1991. Structure and molecular stacking of antbocyanins flower colour variation. Angew. Chem. Int. Ed. 30, 17–33.
Guadalupe, Z., Ayestarán, B., 2008. Effect of commercial mannoprotein addition on polysaccharide, polyphenolic, and color composition in red wines. J. Agric. Food. Chem. 56, 9022–9029.
Guadalupe, Z., Palacios, A., Ayestarán, B., 2007. Maceration enzymes and mannoproteins: a possible strategy to increase colloidal stability and color extraction in red wines. J. Agric. Food. Chem. 55, 4854–4862.
Guerra-Sánchez Simón M.T., 1997. Estudio del ácido ascórbico en vinificación. Efecto sobre el color de los vinos tintos. Doctoral Thesis. Faculty of Pharmacy, University of Salamanca, Salamanca, Spain.
Harbertson, J.F., Parpinello, G.P., Heymann, H., Downey, M.O., 2012. Impact of exogenous tannin additions on wine chemistry and wine sensory character. Food Chem. 131, 999–1008.
Hayasaka, Y., Asenstorfer, R.E., 2002. Screening for potential pigments derived from anthocyanins in red wine using nanoelectrospray tandem mass spectrometry. J. Agric. Food. Chem. 50, 756–761.
He, F., Liang, N.N., Mu, L., Pan, Q.H., Wang, J., Reeves, M.J., et al., 2012a. Anthocyanins and their variation in red wines I. Monomeric anthocyanins and their color expression. Molecules 17, 1571–1601.
He, F., Liang, N.N., Mu, L., Pan, Q.H., Wang, J., Reeves, M.J., et al., 2012b. Anthocyanins and their variation in red wines II. Anthocyanin derived pigments and their color evolution. Molecules 17, 1483–1519.
He, J., Santos-Buelga, C., Silva, A.M.S., Mateus, N., de Freitas, V., 2006. Isolation and structural characterization of new anthocyanin-derived yellow pigments in aged red wines. J. Agric. Food. Chem. 54, 9598–9603.
He, J., Carvalho, A.R.F., Mateus, N., de Freitas, V., 2010. Spectral features and stability of oligomeric pyranoanthocyanin-flavanol pigments isolated from red wines. J. Agric. Food. Chem. 58, 9249–9258.
Heredia, F.J., Francia-Aricha, E.M., Rivas-Gonzalo, J.C., Vicario, I.M., Santos-Buelga, C., 1998. Chromatic characterization of anthocyanins from red grapes-I. pH effect. Food Chem. 63, 491–498.
Hernández-Hierro, J.M., Quijada-Morín, N., Rivas-Gonzalo, J.C., Escribano-Bailón, M.T., 2012. Influence of the physiological stage and the content of soluble solids on the anthocyanin extractability of Vitis vinifera L. cv. Tempranillo grapes. Anal. Chim. Acta 732, 26–32.
Hernández-Hierro, J.M., Quijada-Morín, N., Martínez-Lapuente, L., Guadalupe, Z., Ayestarán, B., Rivas-Gonzalo, J.C., et al., 2014. Relationship between skin cell wall composition and anthocyanin extractability of Vitis vinifera L. cv. Tempranillo at different grape ripeness degree. Food Chem. 146, 41–47.
Hoshino, T., Matsutmoto, U., Goto, T., 1981. Self-association of some anthocyanins in neutral aqueous solution. Phytochemistry 20, 1971–1976.
Hrazdina, G., 1970. Column chromatographic isolation of the anthocyanidin-3,5-diglucosides from grapes'. J. Agric. Food. Chem. 18 (2), 243–245.
Hrazdina, G., 1982. Anthocyanins. In: Harborne, J.B., Marbry, T.J. (Eds.), The Flavonoids. Advances in Research. Chapman and Hall, London, pp. 135–188.
Kovac, V., Alonso, E., Revilla, E., 2005. The effect of adding supplementary quantities of seeds during fermentation on the phenolic composition of wines. Am. J. Enol. Vitic. 46, 363–367.
Kunsági-Máté, S., Szabo, K., Nikfardjam, M.P., Kollar, L., 2006. Determination of the thermodynamic parameters of the complex formation between malvidin-3-O-glucoside and polyphenols. Copigmentation effect in red wines. J. Biochem. Bioph. Methods 69, 113–119.
Maccarone, E., Maccarone, A., Rapisarda, P., 1985. Stabilization of anthocyanins of blood orange fruit juice. J. Food Sci. 50, 901–904.
Markakis, P., 1982. Stability of anthocyanins in foods. In: Markakis, P. (Ed.), Anthocyanins as Food Colors. Academic Press, Nueva York, pp. 163–180.
Mateus, N., Carvalho, E., Carvalho, A.R.F., Melo, A., González-Paramás, A.M., Santos-Buelga, C., et al., 2003. Isolation and structural characterization of new acylated anthocyanin–vinyl–flavanol pigments occurring in aging red wines. J. Agric. Food. Chem. 51, 277–282.
Mateus, N., Oliveira, J., Pissarra, J., González-Paramás, A.M., Rivas-Gonzalo, J.C., Santos-Buelga, C., et al., 2006. A new vinylpyranoanthocyanin pigment occurring in aged red wine. Food Chem. 97, 689–695.
Mazza, G., Brouillard, R., 1987. Recent developments in the stabilization of anthocyanins in food products. Food Chem. 25, 207–225.
Mazza, G., Brouillard, R., 1990. The mechanism of co-pigmentation of anthocyanins in aqueous solutions. Phytochemistry 29, 1097–1102.
Meschter, E.E., 1953. Effects of carbohydrates and other factors on strawberry products. J. Agric. Food. Chem. 1, 574–579.
Morata, A., Gómez-Cordovés, M.C., Colomo, B., Suárez, J.A., 2003. Pyruvic acid and acetaldehyde productions by different strains of Saccharomyces cerevisiae: relationship with vitisin A and B formation in red wines. J. Agric. Food. Chem. 51, 7402–7409.
Moreno-Arribas, M.V., Polo, C., 2008. Wine Chemistry and Biochemistry. Springer, New York.

Ortega Meder, M.D., 1995. Efectos del ácido ascórbico sobre la estabilidad del 3-monoglucósido de malvidina en disoluciones modelo. Doctoral Thesis. Faculty of Pharmacy, University of Salamanca, Salamanca, Spain.

Padayachee, A., Netzel, G., Netzel, M., Day, L., Zabaras, D., Mikkelsen, D., et al., 2012. Binding of polyphenols to plant cell wall analogues—Part 1: Anthocyanins. Food Chem. 134, 155—161.

Palomero, F., Morata, A., Benito, S., González, M.C., Suárez-Lepe, J.A., 2007. Conventional and enzyme-assisted autolysis during ageing over lees in red wines: Influence on the release of polysaccharides from yeast cell walls and on wine monomeric anthocyanin content. Food Chem. 105, 838—846.

Petrova, I., Gandova, V., 2016. Thermodynamic and kinetic investigation in copigmentation reaction between strawberry anthocyanins and chlorogenic acid. Int. J. Innov. Sci. Eng. Technol. 3, 122—127.

Pissarra, J., Mateus, N., Rivas-Gonzalo, J., Santos Buelga, C., de Freitas, V., 2003. Reaction between malvidin 3-glucoside and (+)-catechin in model solutions containing different aldehydes. J. Food Sci. 68, 476—481.

Pissarra, J., Lourenço, S., González-Paramás, A.M., Mateus, N., Santos-Buelga, C., de Freitas, V., 2004. Formation of new anthocyanin-alkyl/aryl-flavanol pigments in model solutions. Anal. Chim. Acta 513, 215—221.

Quideau, S., Jourdes, M., Lefeuvre, D., Montaudon, D., Saucier, C., Glories, Y., et al., 2005. The chemistry of wine polyphenolic C-glycosidic ellagitannins targeting human topoisomerase II. Chemistry 11, 6503—6513.

Quijada-Morín, N., Dangles, O., Rivas-Gonzalo, J.C., Escribano-Bailón, M.T., 2010. Physico-chemical and chromatic characterization of malvidin 3-glucoside-vinylcatechol and malvidin 3-glucoside-vinylguaiacol wine pigments. J. Agric. Food. Chem. 58, 9744—9752.

Rentzsch, M., Schwarz, M., Winterhalter, P., Hermosín-Gutiérrez, I., 2007. Formation of hydroxyphenyl-pyranoanthocyanins in Grenache wines: precursor levels and evolution during aging. J. Agric. Food. Chem. 55, 4883—4888.

Rodrigues, A., Ricardo-Da-Silva, J.M., Lucas, C., Laureano, O., 2012a. Effect of commercial mannoproteins on wine colour and tannins stability. Food Chem. 131, 907—914.

Rodrigues, A., Ricardo-Da-Silva, J.M., Lucas, C., Laureano, O., 2012b. Effect of winery yeast lees on Touriga Nacional red wine color and tannin evolution. Am. J. Enol. Vitic. 64, 98—109.

Romero, C., Bakker, J., 1999. Interactions between grape anthocyanins and pyruvic acid, with effect of pH and acid concentration on anthocyanin composition and color in model solutions. J. Agric. Food Chem. 47, 3130—3139.

Salas, E., Atanasova, V., Poncet-Legrand, C., Meudec, E., Mazauric, J.P., Cheynier, V., 2004. Demonstration of the occurrence of flavanol—anthocyanin adducts in wine and in model solutions. Anal. Chim. Acta 513, 325—332.

Sarni-Manchado, P., Fulcrand, H., Souquet, J.M., Cheynier, V., Moutounet, M., 1996. Stability and color of unreported wine anthocyanin-derived pigments. J. Food Sci. 61, 938—941.

Schwarz, M., Wabnitz, T.C., Winterhalter, P., 2003. Pathway leading to the formation of anthocyanin—vinylphenol adducts and related pigments in red wines. J. Agric. Food Chem. 51, 3682—3687.

Shenoy, V.R., 1993. Anthocyanins-prospective food colours. Curr. Sci. 64, 575—579.

Sousa, C., Mateus, N., Silva, A.M.S., González-Paramás, A.M., Santos-Buelga, C., de Freitas, V., 2007. Structural and chromatic characterization of a new malvidin 3-glucosidevanillyl—catechin pigment. Food Chem. 102, 1344—1351.

Timberlake, C.F., Bridle, P., 1967. Flavylium salts, anthocyanidins and anthocyanins II: Reactions with sulphur dioxide. J. Sci. Food Agric. 18, 479—485.

Timberlake, C.F., Bridle, P., 1976. Interactions between anthocyanins, phenolic compounds, and acetaldehyde and their significance in red wines. Am. J. Enol. Vitic. 27, 97—105.

Trouillas, P., Sancho-García, J.C., de Freitas, V., Gierschner, J., Otyepka, M., Dangles, O., 2016. Stabilizing and modulating color by copigmentation: insights from theory and experiment. Chem. Rev. 116, 4037—4982.

Wesche-Ebeling, P., Montgomery, M.W., 1990. Strawberry polyphenoloxidase: its role in anthocyanin degradation. J. Food Sci. 55, 731—734.

Further Reading

Bridle, P., Timberlake, C.F., 1997. Anthocyanins as natural food colours—selected aspects. Food Chem. 58, 103—109.

CHAPTER 14

Polymeric Pigments in Red Wines

Joana Oliveira, Victor de Freitas and Nuno Mateus

REQUIMTE - LAQV - Department of Chemistry and Biochemistry, Faculty of Science, Oporto University, Oporto, Portugal

ABBREVIATIONS

HPLC–DAD–MS/MS	High performance liquid chromatography–diode array detection–mass spectrometry/mass spectrometry
LC–MS	Liquid chromatography–mass spectrometry
NMR	Nuclear magnetic resonance
LC/DAD–MS	Liquid chromatography–diode array detection–mass spectrometry
MALDI-TOF	Matrix-assisted laser desorption/ionization-time of flight
A	Anthocyanins
F	Flavanols
A-F	Anthocyanin-flavanol adducts
F-A	Flavanol-anthocyanin adducts
HRMS	High resolution mass spectrometry

14.1 INTRODUCTION

Anthocyanins are polyphenolic compounds derived from plant secondary metabolism that are present in vegetables, flowers, fruits, and some beverages such as red wines, being responsible for their red, violet, and blue colors. These pigments in aqueous solutions occur in different forms in equilibria that are dependent on the pH (Brouillard and Lang, 1990; Santos et al., 1993; Brouillard and Delaporte, 1977) (Fig. 14.1). At low pH values, anthocyanins are present in their red flavylium cation form. When the pH increases for values between 3 and 6, the flavylium cation form is hydrated yielding to the colorless hemiketal form that is in equilibrium with the pale yellow *cis*-chalcone form through tautomerization. Simultaneously, the flavylium cation is deprotonated to the respective violet neutral quinoidal base that at higher pH values can be deprotonated yielding the blue anionic quinoidal base (Brouillard and Delaporte, 1977). Considering all this, at wine pH (3–4) these pigments would be expected to be present mainly in their noncolored hemiketal form. However, flavylium cation is the main form present in red wines. This is the result of its stabilization by different copigmentation mechanisms such as self-association and interaction with other wine components (Liao et al., 1992; Trouillas et al., 2016; Brouillard and Dangles, 1994; Gonzalez-Manzano et al., 2009). In addition, copigmentation has been described as the first mechanism involved in the formation of polymeric anthocyanin-derived pigments in red wines during aging (Brouillard and Dangles, 1994).

Furthermore, the reactivity of anthocyanins is strongly affected by different parameters, namely their concentration, temperature, presence of oxygen, and pH, with this latter being the most relevant one. This is correlated with the fact that each equilibrium form of anthocyanins presents different activated positions, thereby affecting their reactivity. In fact, the flavylium cation form presents electrophilic characteristics at carbons C-2 and C-4 from the ring C that can undergo nucleophilic attack by water (Brouillard and Delaporte, 1977; Santos et al., 1993) or sulfur dioxide (Berké et al., 2000) (Fig. 14.2). At the same time, the hydroxyl group at carbon C-5 presents

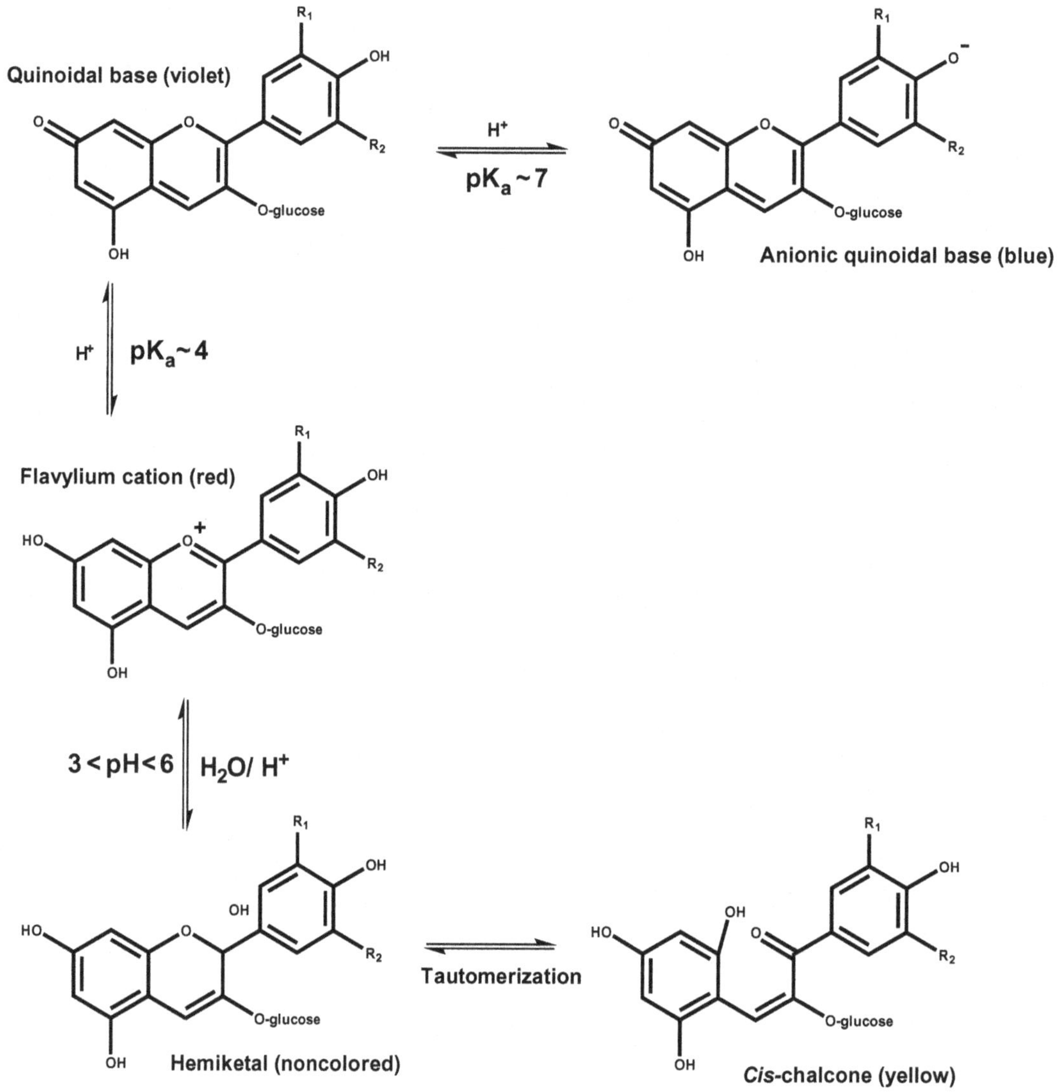

FIGURE 14.1 Equilibrium forms of anthocyanins in aqueous solutions at different pH values (R1 and R2 = H, OH or OMe). *Adapted from Brouillard, R., Lang, J., 1990. The hemiacetal-cis-chalcone equilibrium of malvin, a natural anthocyanin. Can. J. Chem. 68, 755–761.*

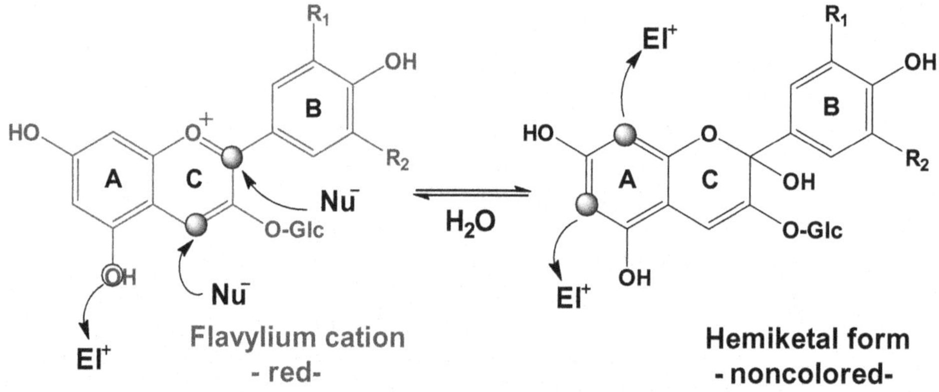

FIGURE 14.2 General scheme of the main reactive positions in anthocyanins according to the equilibrium form present (flavylium cation or hemiketal form) (R1 and R2 = H, OH or OMe).

nucleophilic features in the presence of electrophilic species (Oliveira et al., 2009a; He et al., 2006; Fulcrand et al., 1998). On the other hand, carbons C-6 and C-8 from the ring A can act as nucleophiles in the presence of electrophilic species when anthocyanins are present in their hemiketal form (Sousa et al., 2010). For instance, during red wine aging the color changes from an intense red color to a more brick-red hue due to the reactivity of anthocyanins with other phenolic compounds present in wines giving rise to the formation of more stable polymeric pigments displaying colors ranging from yellow to turquoise blue (He et al., 2010, 2006; Bakker and Timberlake, 1997; Fulcrand et al., 1996; Gomez-Alonso et al., 2012; Schwarz et al., 2003a, 2003b; Mateus et al., 2003a, 2003b; Oliveira et al., 2007, 2010, 2014b; Timberlake and Bridle, 1976; Dallas et al., 1996; Rivas-Gonzalo et al., 1995; Salas et al., 2003; Atanasova et al., 2002).

The formation of these pigments during wine aging will be discussed in this chapter alongside their stability in solution and their influence in wine color.

14.2 POLYMERIC PIGMENTS IN RED WINES

The intense red color displayed by young red wines is due to the presence of high concentrations of anthocyanins extracted from grapes during vinification. During wine aging and maturation, the concentration of anthocyanins starts to decrease leading to the formation of polymeric anthocyanin derivatives. Furthermore, polymeric pigments are described to play an important role in the long-term color stability of aged red wines (Boulton, 2001), although their full identity and origin in red wines is not completely determined.

14.2.1 Anthocyanin-Derived Pigments Found in Red Grapes and Wines

The presence of dimeric and trimeric anthocyanins in grape skins was first evidenced by Vidal and coworkers using mass spectrometry (MS) techniques (Vidal et al., 2004). Then, Pati et al. (2009) detected 22 anthocyanin dimers in grape skins from Cabernet Sauvignon, Montepulciano, and Malvazia varieties using high performance liquid chromatography—diode array detection (HPLC—DAD)—MS/MS. Dimeric structures were also reported by Salas et al. (2005) and Alcalde-Eon et al. (2007) in red wines using the same technique. A few years later, Oliveira et al. (2013a) established the presence of an A-type oenin trimer (Fig. 14.3) in a young Port wine using liquid chromatography (LC)—MS and nuclear magnetic resonance (NMR) spectroscopy. Moreover, the origin of these oligomeric pigments in red wine was postulated to result from their extraction from grape skins during the winemaking process, as they were detected in grape skins (Oliveira et al., 2013a; Vidal et al., 2004; Pati et al., 2009) and grape pomace (Oliveira et al., 2015) using LC/DAD—MS and/or Matrix-assisted laser desorption/ionization-time of flight spectrometry.

14.2.2 Anthocyanin-Derived Pigments Formed in Red Wines During Aging

Acetaldehyde is the main aldehyde (90%) present in wines as a result of yeast metabolism during the first stages of alcoholic fermentation, being also produced throughout the wine aging process from ethanol oxidation (Liu and Pilone, 2000). In fortified wines like Port wines, this compound and other aldehydes (propionaldehyde, 2-methylbutyraldehyde, isovaleraldehyde, methylglyoxal, benzaldehyde) are present in higher amounts due to the addition of wine spirit (40—260 mg/L of acetaldehyde) to stop the fermentation (Pissarra et al., 2005). Sherry wines also present high levels of acetaldehyde (90—500 mg/L) due to the fact that this wine style is produced under oxidative conditions (Zea et al., 2015). Moreover, acetaldehyde is extremely reactive and can react with different phenolic compounds present in grapes and wines (e.g., anthocyanins and flavanols) increasing the number of different chemical pathways that can occur in red wines starting from anthocyanins and/or flavanols to yield polymeric pigments (Pissarra et al., 2003; Jurd, 1969; Atanasova et al., 2002; Timberlake and Bridle, 1976; Garcia-Viguera et al., 1994; Rivas-Gonzalo et al., 1995). The acetaldehyde-mediated polymerization between either only flavanols or with anthocyanins is the most well documented reaction in the literature (Rivas-Gonzalo et al., 1995; Francia-Aricha et al., 1997; Es-Safi et al., 1999a, 1999b; Santos-Buelga et al., 1999; Escribano-Bailon et al., 2001; Remy-Tanneau et al., 2003; Salas et al., 2003; Pissarra et al., 2003, 2004b; Saucier et al., 1997). The condensation of anthocyanins (A) with flavanols (F) occurs directly (Remy-Tanneau et al., 2003) or is mediated by aldehydes (Pissarra et al., 2003, 2004b) (Fig. 14.4).

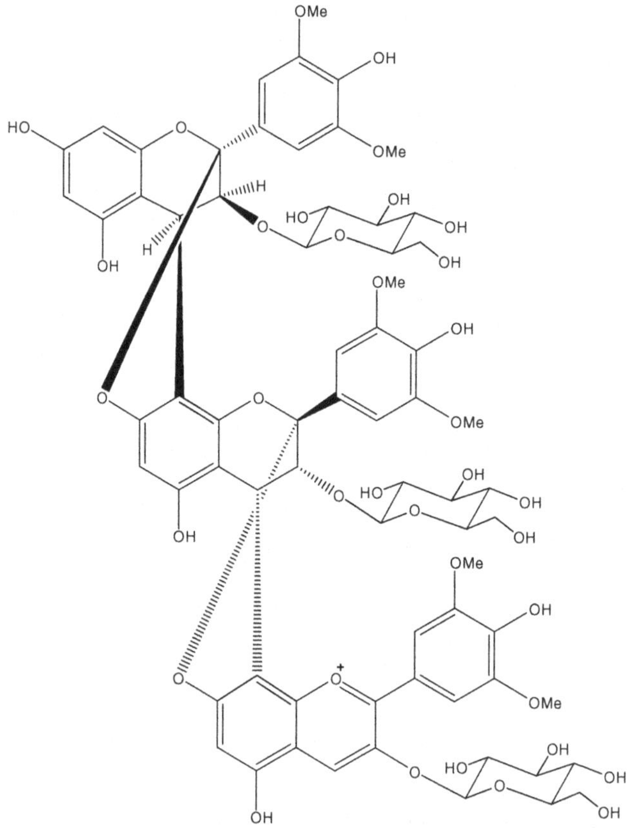

FIGURE 14.3 Structure of the malvidin-3-glucoside trimer isomer detected in a young red wine and in red grape skin. *Adapted from Oliveira, J., da Silva, M.A., Jorge Parola, A., Mateus, N., Brás, N.F., Ramos, M.J., et al. 2013a. Structural characterization of a A-type linked trimeric anthocyanin derived pigment occurring in a young Port wine. Food Chem. 141, 1987−1996.*

Two mechanisms were described in red wines for the direct condensation between anthocyanins and flavanols yielding flavanol-(4,8)-anthocyanin (F-A$^+$) and anthocyanin-(4,8)-flavanol (A$^+$-F) adducts. The first consists of the nucleophilic attack of the carbon C-6/C-8 of the anthocyanin (hemiketal form) to the electrophilic carbocation present at carbon C-4 of a flavanol unit produced during the cleavage of a proanthocyanidin in the slightly acidic conditions (Somers, 1971; Jurd, 1969; Salas et al., 2003). The occurrence of F-A$^+$ pigments in red wines was reported in the literature by Salas et al. (2005) using LC/DAD/electrospray ionization-mass spectrometry (ESI-MS) spectrometry (Salas et al., 2004a). F-A$^+$ acetaldehyde-derived pigments, namely F-methylmethine-F-A$^+$ and methylmethine-(F-A$^+$) were also found to occur in a 2-year-old Port wine using HPLC-ESI-MS (Nave et al., 2010b). In addition, the formation of A-F dimers in red wines is described to result from the nucleophilic attack of the carbon C-6/C-8 of a flavanol unit to the electrophilic carbon C-4 of an anthocyanin (flavylium cation form) giving rise to a flavene structure product (colorless) (Remy et al., 2000) that can evolve to the colorless bicyclic form [A-type, A-(O)-F] (Jurd, 1969; Remy et al., 2000; Remy-Tanneau et al., 2003) or undergo oxidation to give the red pigment A$^+$-F (Dueñas et al., 2005; Liao et al., 1992), which could dehydrate to the orange-yellow xanthylium salt. This latter was only detected by UV−vis spectroscopy in wine-like model solutions (Santos-Buelga et al., 1995, 1999).

On the other hand, the formation of aldehyde-mediated polymeric pigments starts with the addition of the protonated aldehyde (acidic conditions) to the nucleophilic carbon C-6/C-8 of a flavanol unit. Then, the formed adduct undergoes the nucleophilic attack of the carbon C-6/C-8 of an anthocyanin (hemiketal form). The last step includes a dehydration reaction to yield the anthocyanin-alkyl/aryl-flavanol dimer (Pissarra et al., 2003; Timberlake and Bridle, 1976; Sousa et al., 2007). Besides, these pigments were also reported to be at the origin of the pyranoanthocyanin-flavanol compounds found in red wines during aging (Francia-Aricha et al., 1997; Mateus et al., 2002, 2003a).

Moreover, a dimeric oenin acetaldehyde-mediated condensation product [malvidin-3-glucoside-(8,8)-malvidin-3-glucoside] was also found to occur in wine-like model solutions and in red wines (Atanasova et al., 2002).

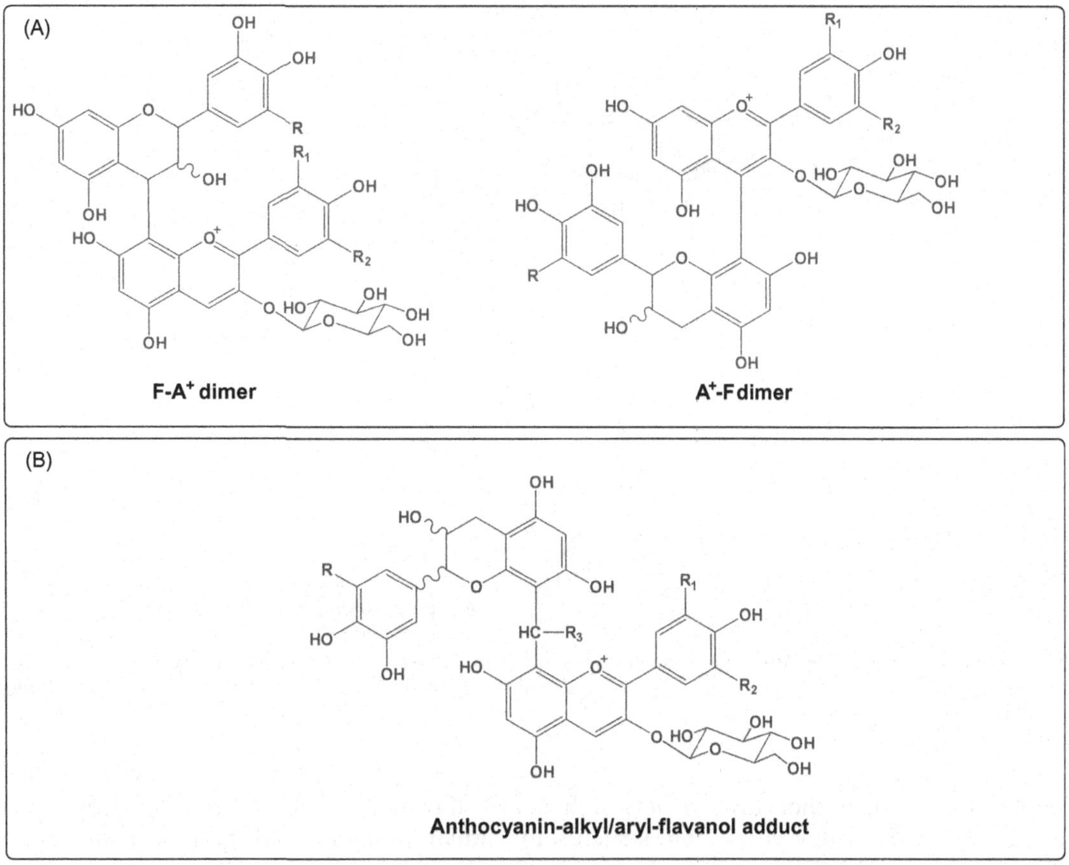

FIGURE 14.4 (A) General structure of the anthocyanin-flavanol (A^+-F) and flavanol–anthocyanin (F-A^+) adducts formed by the direct condensation of anthocyanins with flavanols (R_1 and R_2 = H, OH or OMe). (B) Anthocyanin-alkyl/aryl-flavanols adducts R_1 and R_2 = H, OH or OMe; R_3 = H; CH_3; $CH(CH_3)_2$; $CHCH_3CH_2CH_3$; $CH_2CH(CH_3)_2$; CH_2CH_3; Ph. *Adapted from Pissarra, J., Mateus, N., Rivas-Gonzalo, J., Buelga, C.S., de Freitas, V., 2003. Reaction between malvidin 3-glucoside and (+)-catechin in model solutions containing different aldehydes. J. Food Sci. 68, 476–481.*

14.2.3 A-Type Vitisin-Derived Pigments Formed in Red Wines During Aging

Over the years, different anthocyanin-derived compounds (pyranoanthocyanins) displaying orange colors have been described in the literature from the reaction of anthocyanins with small molecules produced by yeasts during fermentation, namely acetaldehyde (Bakker and Timberlake, 1997; Oliveira et al., 2009a), pyruvic acid (Fulcrand et al., 1996), oxaloacetic acid (Araujo et al., 2017), acetoacetic acid (He et al., 2006), diacetyl (Gomez-Alonso et al., 2012), vinyl-phenols (Fulcrand et al., 1996), and hydroxycinnamic acids (Schwarz et al., 2003b; Schwarz and Winterhalter, 2003). These compounds are likely to contribute to the red/orange hues observed in red wines during aging.

Carboxypyranoanthocyanins (A-type vitisins) are described in the literature as the main pyranoanthocyanins formed in red wines during aging (Mateus and de Freitas, 2001) from the reaction between anthocyanins and pyruvic (Fulcrand et al., 1996) and/or oxaloacetic (Araujo et al., 2017) acids. In red Port wines, the levels of A-type vitisins increase after wine fortification with wine spirit and start to decrease after around 100 days. The formation of these anthocyanin-pyruvic acid adducts occurs concomitantly with the decrease of anthocyanins (Mateus and de Freitas, 2001). Although these orange pigments are not polymeric by themselves, they may contribute to the polymeric fraction of red wine pigments, as they were reported in the literature to react with other wine components (Mateus et al., 2003b, 2004; Oliveira et al., 2007, 2010). Indeed, some years ago two anthocyanin-derived pigments, namely a vinylpyranomalvidin-3-glucoside-procyanidin dimer and the respective coumaroylated derivative (A-type portisins), presenting an unusual bluish color in acidic pH conditions were identified in a young Port red wine (Mateus et al., 2003b) (Fig. 14.5). Studies performed in wine model solutions suggested that these bluish pigments can be formed in red wines from the reaction between A-type vitisins and flavanols (in the presence of acetaldehyde) or vinyl-flavanols (Mateus et al., 2003b, 2004). The structure of A-type

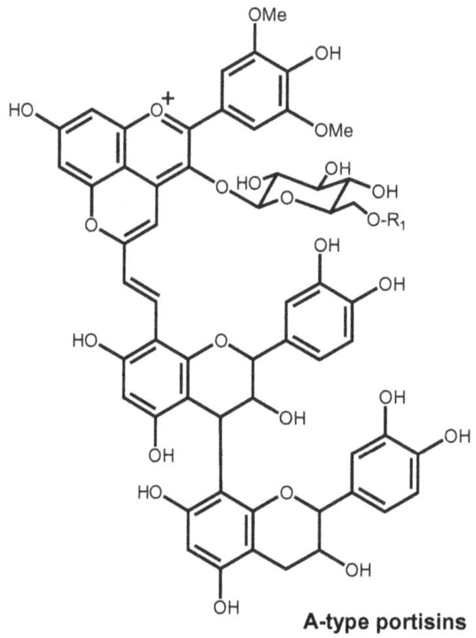

FIGURE 14.5 General structure of the A-type portisins found to occur in a 2-year-old Port wine (R_1 = H or, coumaroyl group). *Adapted from Mateus, N., Silva, A.M.S., Rivas-Gonzalo, J.C., Santos-Buelga, C., de Freitas, V., 2003b. A new class of blue anthocyanin-derived pigments isolated from red wines. J. Agric. Food Chem. 51, 1919–1923.*

portisins comprises a pyranoanthocyanin moiety linked to a flavanol (monomer or dimer) by a vinyl linkage (Mateus et al., 2003b, 2004). A few years later, structurally similar compounds (B-type portisins) were identified in Port red wines with a phenolic moiety replacing the flavanol one. B-type portisins were described to result from the reaction of A-type vitisins with vinyl-phenolics (Mateus et al., 2006) or hydroxycinnamic acids, such as *p*-coumaric, caffeic, ferulic, and sinapic acids (Oliveira et al., 2007).

The occurrence of A and B-type portisins in aged red wines points to a second-generation of anthocyanin derivatives in which the main precursors are no longer genuine anthocyanins but rather carboxypyranoanthocyanins (A-type vitisins) involved in the formation of polymeric pigments in the later stages of red wine aging.

Furthermore, a few years ago, Oliveira et al. (2010) identified in 9-year-old Port wine and in the respective lees, a family of pyranoanthocyanin dimers presenting an unusual turquoise blue color at acidic pH (Fig. 14.6). The origin of these pigments in red wines was demonstrated in wine-like model solutions to involve the reaction between an A-type vitisin and a methylpyranoanthocyanin (Oliveira et al., 2010). Even though the mechanism of formation is still undetermined, it has been proposed that the origin of pyranoanthocyanin dimers should start with the formation of a charge-transfer complex between the A-type vitisin and the methylpyranoanthocyanin similarly to what was reported by Chassaing et al. (2008) for related compounds.

All these anthocyanin-derived pigments may have an important contribution to the color hues and stability in red wine and the knowledge of their chemical pathways are crucial to better understand the color evolution of red wines during aging.

14.3 ANALYSIS OF POLYMERIC PIGMENTS

Red wine is unquestionably a very complex matrix which makes the identification of all its molecules and their chemical pathways more difficult to achieve. The main difficulties are associated with the limitation of the techniques used, their detection limits, and the increased complexity of the polymeric structures aimed to be identified.

The analytical methods commonly used to study grape and wine polyphenols involve reverse-phase LC coupled with spectrophotometry or MS (Di Stefano and Flamini, 2008; Flamini and de Rosso, 2008; Arapitsas et al., 2014, 2016; Ehrhardt et al., 2014). However, this technique combined with multiple mass spectrometry

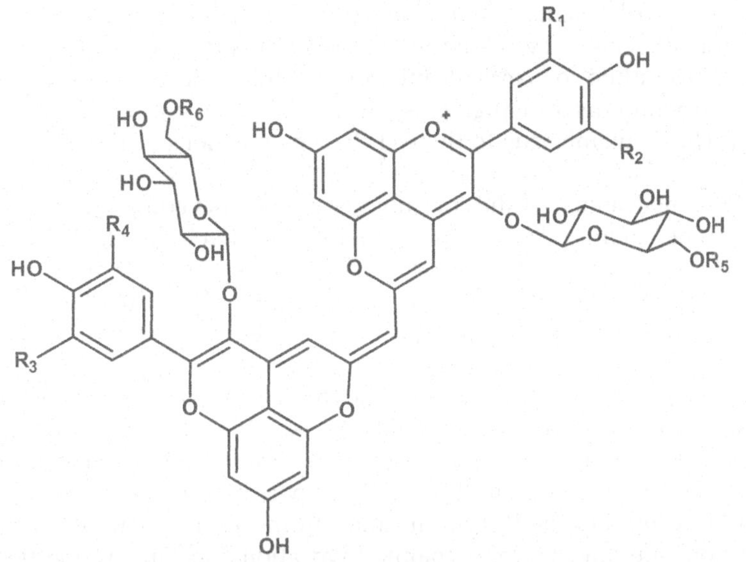

FIGURE 14.6 General structure of the pyranoanthocyanin dimers found to occur in aged Port wines and in respective lees (R_1, R_2, R_3 and R_4 = H, OH or OMe; R_5 and R_6 = H, acetyl, coumaroyl, independently of each other). *Adapted from Oliveira, J., Azevedo, J., Silva, A.M.S., Teixeira, N., Cruz, L., Mateus, N., et al., 2010. Pyranoanthocyanin dimers: a new family of turquoise blue anthocyanin-derived pigments found in Port wine. J. Agric. Food Chem. 58, 5154–5159.*

(MS/MS and MS^n) analysis is only effective for the structural characterization of low molecular weight compounds (Flamini et al., 2015; Arapitsas et al., 2012a).

In the last few years, the number of studies in the field of untargeted grape and wine polyphenolic LC−MS analysis has increased in the literature providing interesting information about various enological practices such as grape variety, geographical origin of wines, and their chemical age (Arapitsas et al., 2012a, 2014; Fulcrand et al., 2008; Cuadros-Inostroza et al., 2010). Untargeted methods have been found to have good resolution, high sensitivity, and high-throughput capacity being able to detect/identify a great number of possible compounds in a single run (Arapitsas et al., 2012b). Conversely, targeted analysis is used for the quantitative determination of specific molecules but with limited information on the overall sample metabolome (Cuadros-Inostroza et al., 2010; Vaclavik et al., 2011).

On the other hand, high-resolution mass spectrometry (HRMS) has been described as a powerful technique for the analysis of complex samples in many fields. With HRMS techniques it is possible to obtain the exact mass of molecules and to determine their elemental composition, which is crucial for the proper identification of the compounds (Vallverdu-Queralt et al., 2017a,b). Delcambre and Saucier (2012) demonstrated the application of Q-TOF-HRMS analysis of grapes and wines to be used as a footprint for the determination of the grape variety and the geographical origin of the wine.

The combination of untargeted analysis with Orbitrap-HRMS or Q-TOF-HRMS could be promising techniques for the identification of polymeric pigments in complex samples such as red wines. In fact, very recently Vallverdú-Queralt et al. (2017a,b), have demonstrated the ability of these kinds of methodologies to identify numerous polymeric pigments in wine-like model solutions containing oenin and (−)-epicatechin and acetaldehyde.

14.4 STABILITY IN SOLUTION AND INFLUENCE IN RED WINE COLOR

Color is one of the most important quality indicators of a red wine. During maturation and aging, the color of red wines changes from an intense red color to a more red-orange hue due to the chemical transformation of genuine anthocyanins extracted from the grape skins during fermentation forming polymeric pigments by the mechanisms discussed previously.

Sulfur dioxide (SO_2) is commonly used during the winemaking process as an antioxidant and antiseptic inhibiting the growth of undesirable microbial. However, in the case of red wines, SO_2 can increase the extractability of anthocyanins if added prior to fermentation and improve their stability when added at bottling (Burroughs, 1974). This is correlated to the reversible bleaching of anthocyanins that occurs in red wine due to the formation of colorless anthocyanin-2-bisulphite or anthocyanin-4-bisulphite adducts (Berké et al., 2000; Jurd, 1964) (Fig. 14.7). In general, anthocyanin-derived pigments are much more stable than the anthocyanin counterpart towards bleaching by SO_2. This is mainly due to the fact that the positions in the pigment structure at which SO_2 is likely to react are blocked.

Moreover, conversely to anthocyanins, studies of polymeric pigments equilibria in aqueous solutions and their contribution to the overall wine color are limited in the literature (Oliveira et al., 2006b, 2014a, 2014b; Cruz et al., 2010; Salas et al., 2004b; Asenstorfer et al., 2006; Pissarra et al., 2004a; Nave et al., 2010a; Sousa et al., 2007). The study of their chromatic characteristic at different pH values and the determination of the corresponding ionization constants is valuable information about their expected occurrence at red wine pH. In addition, the physicochemical features studied over the years for some polymeric pigments using UV–visible spectroscopy and NMR revealed a higher stability of these compounds towards the hydration reactions when compared to their anthocyanin precursors, which can contribute to the color stability of red wines during the aging process.

However, some exceptions are observed like the case of F-A$^+$ adducts. The study of these pigments performed in model solutions at different pH values using UV–visible spectroscopy showed that the flavanolic unit (catechin) shifts the absorption maximum of the flavylium cation from 518 to 535 nm but has no significant modification on the kinetic (hydration rate) and thermodynamic (hydration constant) properties compared to the oenin (Nave et al., 2010a).

Conversely, anthocyanin-alkyl/aryl-flavanol pigments that are described to contribute to the red/violet hues observed in young red wines during the first stages of wine maturation display a purple color with characteristic UV–visible spectra that present a λ_{max} in the visible region at 540 nm and a shoulder at 450 nm (Pissarra et al., 2004a, Sousa et al., 2007). Moreover, studies performed in aqueous solutions using UV–visible spectroscopy showed that when the pH increases from 2.2 to 5.5, oenin-methylmethine–catechin pigment solutions become gradually more violet, while similar solutions of the anthocyanin are almost colorless at pH 4.0 (Escribano-Bailon et al., 2001). This indicates a higher protection against water attack of the oenin moiety of the pigment when compared to the oenin alone (Escribano-Bailon et al., 2001). Similar results were obtained for other anthocyanin-alkyl/aryl-flavanol pigments (Sousa et al., 2007). However, these polymeric pigments are more prone to degradation in aqueous solution comparatively to anthocyanins with the cleavage of the methylmethine bridge yielding oenin as a major product (Escribano-Bailon et al., 2001).

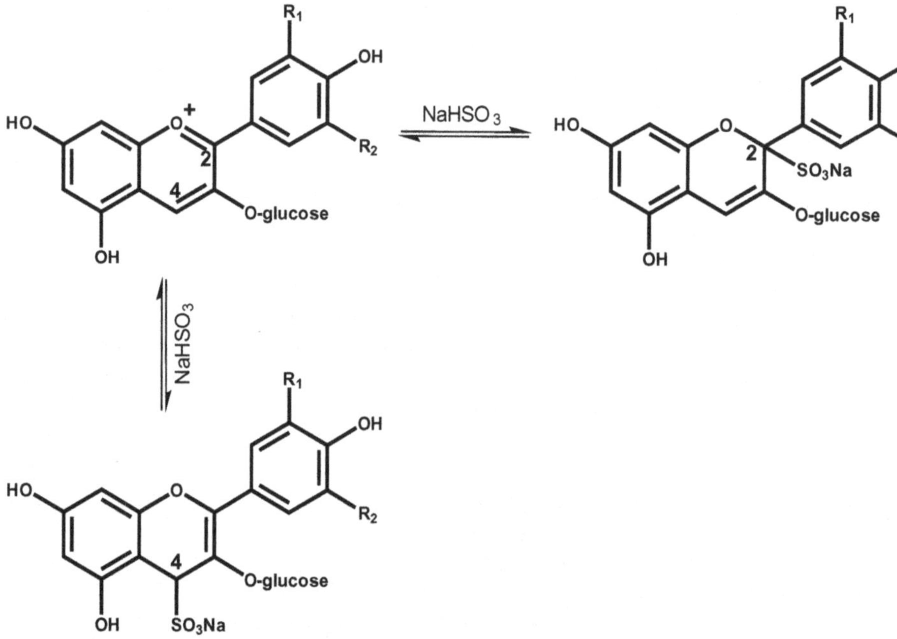

FIGURE 14.7 Bisulfite addition to anthocyanins (R_1 and R_2 = H, OH or OMe). *Adapted from Berké, B., Chèze, C., Deffieux, G., Vercauteren, J., Sulfur Dioxide Decolorization or Resistance of Anthocyanins: NMR Structural Elucidation of Bisulfite-Adducts. In: G.G. Gross, R.W. Hemingway, T. Yoshida, S.J. Branham (Eds.), Plant Polyphenols 2: Chemistry, Biology, Pharmacology, Ecology. Springer US: Boston, MA, 1999; pp 779–790.*

In addition, the titration of the oenin trimer showed that the multistate equilibria of this compound is strongly dominated by acid-base chemistry, with the reaction sequence hydration−tautomerization−isomerization accounting for less than 10% of the overall reactivity (Oliveira et al., 2014a). So, the lack of reactivity and the higher chromatic stability presented by these oligomeric compounds when compared to anthocyanin monomers can have a direct impact on the overall red wine color during aging.

Over the years, the equilibrium forms of pyranoanthocyanin pigments have been studied using NMR and UV−visible spectroscopy (Asenstorfer and Jones, 2007; Cruz et al., 2010; Oliveira et al., 2009b, 2011). In aqueous solutions these polymeric pigments have been shown to coexist under different equilibrium forms that are pH-dependent. Using NMR spectroscopy it was possible to postulate the absence of hydration reactions and that only proton transfer reactions occur when the pH change in pyranoanthocyanins. In fact, a number of studies based on UV−visible spectroscopy have already established that pyranoanthocyanin pigments are protected from the attack by water when compared with their anthocyanin precursors (Oliveira et al., 2006a, 2006b, 2009b, 2011, 2013b, 2014b; Asenstorfer and Jones, 2007; Cruz et al., 2010; Vallverdu-Queralt et al., 2016). An exception was described by Gomez-Alonso et al. (2012) for acetyl-pyranoanthocyanins that at wine pH are present in their noncolored hemiketal form (at carbon C-10).

All these polymeric pigments may have a direct or an indirect impact on the color of red wines during aging.

14.5 CONCLUSION

Red wine is a complex matrix, which makes the identification of all its molecules and their chemical formation pathways a rather difficult task. Over the years, different families of polymeric pigments have been described from the reaction of anthocyanins present in grapes with other wine components. After the first reactions involving genuine anthocyanins yielding newly-formed anthocyanin derivatives, another stage of pigment formation arises from the reaction of carboxypyranoanthocyanins (A-type vitisins) with other wine components. This has led to the identification of several anthocyanin polymeric pigments. However, this is only the tip of the iceberg on this matter and a lot remains to be done to fully determine the fraction of polymeric pigments present in red wines.

References

Alcalde-Eon, C., Escribano-Bailon, M.T., Santos-Buelga, C., Rivas-Gonzalo, J.C., 2007. Identification of dimeric anthocyanins and new oligomeric pigments in red wine by means of HPLC-DAD-ESI/MSn. J. Mass Spectrom. 42, 735−748.

Arapitsas, P., Perenzoni, D., Nicolini, G., Mattivi, F., 2012a. Study of Sangiovese wines pigment profile by UHPLC-MS/MS. J. Agric. Food Chem. 60, 10461−10471.

Arapitsas, P., Scholz, M., Vrhovsek, U., Di Blasi, S., Biondi Bartolini, A., Masuero, D., et al., 2012b. A metabolomic approach to the study of wine micro-oxygenation. PLOS ONE 7, e37783.

Arapitsas, P., Speri, G., Angeli, A., Perenzoni, D., Mattivi, F., 2014. The influence of storage on the "chemical age" of red wines. Metabolomics 10, 816−832.

Arapitsas, P., Corte, A.D., Gika, H., Narduzzi, L., Mattivi, F., Theodoridis, G., 2016. Studying the effect of storage conditions on the metabolite content of red wine using HILIC LC−MS based metabolomics. Food Chem. 197, 1331−1340.

Araujo, P., Fernandes, A., de Freitas, V., Oliveira, J., 2017. A new chemical pathway yielding A-type vitisins in red wines. Int. J. Mol. Sci. 18, 762.

Asenstorfer, R.E., Jones, G.P., 2007. Charge equilibria and pK values of 5-carboxypyranomalvidin-3-glucoside (vitisin A) by electrophoresis and absorption spectroscopy. Tetrahedron 63, 4788−4792.

Asenstorfer, R.E., Lee, D.F., Jones, G.P., 2006. Influence of structure on the ionisation constants of anthocyanin and anthocyanin-like wine pigments. Anal. Chim. Acta 563, 10−14.

Atanasova, V., Fulcrand, H., Le guerneve, C., Cheynier, V., Moutounet, M., 2002. Structure of a new dimeric acetaldehyde malvidin 3-glucoside condensation product. Tetrahedron Lett. 43, 6151−6153.

Bakker, J., Timberlake, C.F., 1997. Isolation, identification, and characterization of new color-stable anthocyanins occurring in some red wines. J. Agric. Food Chem. 45, 35−43.

Berké, B., Chèze, C., Deffieux, G., Vercauteren, J., 2000. Sulfur Dioxide decolorization or resistance of anthocyanins: NMR structural elucidation of bisulfite-adducts. In: Gross, G., Hemingway, R., Yoshida, T., Branham, S. (Eds.), Plant Polyphenols, 2. Springer, US.

Boulton, R., 2001. The copigmentation of anthocyanins and its role in the color of red wine: a critical review. Am. J. Enol. Vitic. 52, 67−87.

Brouillard, R., Dangles, O., 1994. Anthocyanin molecular interactions: the first step in the formation of new pigments during wine aging. Food Chem. 51, 365−371.

Brouillard, R., Delaporte, B., 1977. Chemistry of anthocyanin pigments. 2. Kinetic and thermodynamic study of proton transfer, hydration, and tautomeric reactions of malvidin 3-glucoside. J. Am. Chem. Soc. 99, 8461−8468.

Brouillard, R., Lang, J., 1990. The hemiacetal−cis-chalcone equilibrium of malvin, a natural anthocyanin. Can. J. Chem. 68, 755−761.

Burroughs, L.F., Sparks, A.H., 1964. The identification of sulphur dioxide-binding compounds in apple juices and ciders. J. Sci. Food Agric. 15, 176–185.

Chassaing, S., Isorez, G., Kueny-Stotz, M., Brouillard, R., 2008. En route to color-stable pyranoflavylium pigments—a systematic study of the reaction between 5-hydroxy-4-methylflavylium salts and aldehydes. Tetrahedron Lett. 49, 6999–7004.

Cruz, L., Petrov, V., Teixeira, N., Mateus, N., Pina, F., de Freitas, V., 2010. Establishment of the chemical equilibria of different types of pyranoanthocyanins in aqueous solutions: evidence for the formation of aggregation in pyranomalvidin-3-O-coumaroylglucoside-(+)-catechin. J. Phys. Chem. B 114, 13232–13240.

Cuadros-Inostroza, A., Giavalisco, P., Hummel, J., Eckardt, A., Willmitzer, L., Peña-Cortés, H., 2010. Discrimination of wine attributes by metabolome analysis. Anal. Chem. 82, 3573–3580.

Dallas, C., Ricardo-da-Silva, J.M., Laureano, O., 1996. Products formed in model wine solutions involving anthocyanins, procyanidin B2, and acetaldehyde. J. Agric. Food Chem. 44, 2402–2407.

Delcambre, A., Saucier, C., 2012. Identification of new flavan-3-ol monoglycosides by UHPLC-ESI-Q-TOF in grapes and wine. J. Mass Spectrom. 47, 727–736.

Di Stefano, R., Flamini, R., 2008. High performance liquid chromatography analysis of grape and wine polyphenols. Hyphenated Techniques in Grape and Wine Chemistry. John Wiley & Sons, Ltd, New Jersey, USA.

Dueñas, M., Salas, E., Cheynier, V., Dangles, O., Fulcrand, H., 2005. UV–visible spectroscopic investigation of the 8,8-methylmethine catechin-malvidin 3-glucoside pigments in aqueous solution: structural transformations and molecular complexation with chlorogenic acid. J. Agric. Food Chem. 54, 189–196.

Ehrhardt, C., Arapitsas, P., Stefanini, M., Flick, G., Mattivi, F., 2014. Analysis of the phenolic composition of fungus-resistant grape varieties cultivated in Italy and Germany using UHPLC-MS/MS. J. Mass Spectrom. 49, 860–869.

Es-Safi, N.E., Fulcrand, H., Cheynier, V., Moutounet, M., 1999a. Competition between (+)-catechin and (−)-epicatechin in acetaldehyde-induced polymerization of flavanols. J. Agric. Food Chem. 47, 2088–2095.

Es-Safi, N.E., Fulcrand, H., Cheynier, V., Moutounet, M., 1999b. Studies on the acetaldehyde-induced condensation of (−)-epicatechin and malvidin 3-O-glucoside in a model solution system. J. Agric. Food Chem. 47, 2096–2102.

Escribano-Bailon, T., Alvarez-Garcia, M., Rivas-Gonzalo, J.C., Heredia, F.J., Santos-Buelga, C., 2001. Color and stability of pigments derived from the acetaldehyde-mediated condensation between malvidin 3-O-glucoside and (+)-catechin. J. Agric. Food Chem. 49, 1213–1217.

Flamini, R., de Rosso, M., 2008. Polyphenols analysis by liquid–mass spectrometry. Hyphenated Techniques in Grape and Wine Chemistry. John Wiley & Sons, Ltd, New Jersey, USA.

Flamini, R., de Rosso, M., Bavaresco, L., 2015. Study of grape polyphenols by liquid chromatography-high-resolution mass spectrometry (UHPLC/QTOF) and suspect screening analysis. J. Anal. Methods Chem. 2015, 350259.

Francia-Aricha, E.M., Guerra, M.T., Rivas-Gonzalo, J.C., Santos-Buelga, C., 1997. New anthocyanin pigments formed after condensation with flavanols. J. Agric. Food Chem. 45, 2262–2266.

Fulcrand, H., dos Santos, P.-J.C., Sarni-Manchado, P., Cheynier, V., Favre-Bonvin, J., 1996. Structure of new anthocyanin-derived wine pigments. J. Chem. Soc., Perkin Trans. 1, 735–739.

Fulcrand, H., Benabdeljalil, C., Rigaud, J., Cheynier, V., Moutounet, M., 1998. A new class of wine pigments generated by reaction between pyruvic acid and grape anthocyanins. Phytochemistry 47, 1401–1407.

Fulcrand, H., Mané, C., Preys, S., Mazerolles, G., Bouchut, C., Mazauric, J.-P., et al., 2008. Direct mass spectrometry approaches to characterize polyphenol composition of complex samples. Phytochemistry 69, 3131–3138.

Garcia-Viguera, C., Bridle, P., Bakker, J., 1994. The effect of pH on the formation of coloured compounds in model solutions containing anthocyanins, catechin and acetaldehyde. Vitis 33, 37–40.

Gomez-Alonso, S., Blanco-Vega, D., Victoria Gomez, M., Hermosin-Gutierrez, I., 2012. Synthesis, isolation, structure elucidation, and color properties of 10-acetyl-pyranoanthocyanins. J. Agric. Food Chem. 60, 12210–12223.

Gonzalez-Manzano, S., Duenas, M., Rivas-Gonzalo, J.C., Escribano-Bailon, M.T., Santos-Buelga, C., 2009. Studies on the copigmentation between anthocyanins and flavan-3-ols and their influence in the colour expression of red wine. Food Chem. 114, 649–656.

He, J., Santos-Buelga, C., Silva, A.M.S., Mateus, N., de Freitas, V., 2006. Isolation and structural characterization of new anthocyanin-derived yellow pigments in aged red wines. J. Agric. Food Chem. 54, 9598–9603.

He, J., Oliveira, J., Silva, A.M.S., Mateus, N., de Freitas, V., 2010. Oxovitisins: a new class of neutral pyranone-anthocyanin derivatives in red wines. J. Agric. Food Chem. 58, 8814–8819.

Jurd, L., 1964. Reactions involved in sulfite bleaching of anthocyanins. J. Food Sci. 29, 16–19.

Jurd, L., 1969. Review of polyphenol condensation reactions and their possible occurrence in the aging of wines. Am. J. Enol. Vitic. 20, 191–195.

Liao, H., Cai, Y., Haslam, E., 1992. Polyphenol interactions. Anthocyanins: co-pigmentation and colour changes in red wines. J. Sci. Food Agric. 59, 299–305.

Liu, S.-Q., Pilone, G.J., 2000. An overview of formation and roles of acetaldehyde in winemaking with emphasis on microbiological implications. Int. J. Food Sci. Technol. 35, 49–61.

Mateus, N., de Freitas, V., 2001. Evolution and stability of anthocyanin-derived pigments during port wine aging. J. Agric. Food Chem. 49, 5217–5222.

Mateus, N., Silva, A.M.S., Santos-Buelga, C., Rivas-Gonzalo, J.C., de Freitas, V., 2002. Identification of anthocyanin-flavanol pigments in red wines by NMR and mass spectrometry. J. Agric. Food Chem. 50, 2110–2116.

Mateus, N., Carvalho, E., Carvalho, A.R.F., Melo, A., Gonzalez-Paramas, A.M., Santos-Buelga, C., et al., 2003a. Isolation and structural characterization of new acylated anthocyanin-vinyl-flavanol pigments occurring in aging red wines. J. Agric. Food Chem. 51, 277–282.

Mateus, N., Silva, A.M.S., Rivas-Gonzalo, J.C., Santos-Buelga, C., de Freitas, V., 2003b. A new class of blue anthocyanin-derived pigments isolated from red wines. J. Agric. Food Chem. 51, 1919–1923.

Mateus, N., Oliveira, J., Santos-Buelga, C., Silva, A.M.S., de Freitas, V., 2004. NMR structure characterization of a new vinylpyranoanthocyanin-catechin pigment (a portisin). Tetrahedron Lett. 45, 3455–3457.

Mateus, N., Oliveira, J., Pissarra, J., Gonzalez-Paramas, A.M., Rivas-Gonzalo, J.C., Santos-Buelga, C., et al., 2006. A new vinylpyranoanthocyanin pigment occurring in aged red wine. Food Chem. 97, 689–695.

Nave, F., Petrov, V., Pina, F., Teixeira, N., Mateus, N., de Freitas, V., 2010a. Thermodynamic and kinetic properties of a red wine pigment: catechin-(4,8)-malvidin-3-O-glucoside. J. Phys. Chem. B 114, 13487–13496.

Nave, F., Teixeira, N., Mateus, N., de Freitas, V., 2010b. The fate of flavanol-anthocyanin adducts in wines: study of their putative reaction patterns in the presence of acetaldehyde. Food Chem. 121, 1129–1138.

Oliveira, J., Fernandes, V., Miranda, C., Santos-Buelga, C., Silva, A., de Freitas, V., et al., 2006a. Color properties of four cyanidin-pyruvic acid adducts. J. Agric. Food Chem. 54, 6894–6903.

Oliveira, J., Santos-Buelga, C., Silva, A.M.S., de Freitas, V., Mateus, N., 2006b. Chromatic and structural features of blue anthocyanin-derived pigments present in Port wine. Anal. Chim. Acta 563, 2–9.

Oliveira, J., de Freitas, V., Silva, A.M.S., Mateus, N., 2007. Reaction between hydroxycinnamic acids and anthocyanin-pyruvic acid adducts yielding new portisins. J. Agric. Food Chem. 55, 6349–6356.

Oliveira, J., de Freitas, V., Mateus, N., 2009a. A novel synthetic pathway to vitisin B compounds. Tetrahedron Lett. 50, 3933–3935.

Oliveira, J., Mateus, N., Silva, A.M.S., de Freitas, V., 2009b. Equilibrium forms of vitisin B pigments in an aqueous system studied by NMR and visible spectroscopy. J. Phys. Chem. B 113, 11352–11358.

Oliveira, J., Azevedo, J., Silva, A.M.S., Teixeira, N., Cruz, L., Mateus, N., et al., 2010. Pyranoanthocyanin dimers: a new family of turquoise blue anthocyanin-derived pigments found in Port wine. J. Agric. Food Chem. 58, 5154–5159.

Oliveira, J., Petrov, V., Parola, A.J., Pina, F., Azevedo, J., Teixeira, N., et al., 2011. Chemical behavior of methylpyranomalvidin-3-O-glucoside in aqueous solution studied by NMR and UV–Visible spectroscopy. J. Phys. Chem. B 115, 1538–1545.

Oliveira, J., da Silva, M.A., Jorge Parola, A., Mateus, N., Brás, N.F., Ramos, M.J., et al., 2013a. Structural characterization of a A-type linked trimeric anthocyanin derived pigment occurring in a young Port wine. Food Chem. 141, 1987–1996.

Oliveira, J., Mateus, N., de Freitas, V., 2013b. Network of carboxypyranomalvidin-3-O-glucoside (vitisin A) equilibrium forms in aqueous solution. Tetrahedron Lett. 54, 5106–5110.

Oliveira, J., Bras, N.F., da Silva, M.A., Mateus, N., Parola, A.J., de Freitas, V., 2014a. Grape anthocyanin oligomerization: a putative mechanism for red color stabilization? Phytochemistry 105, 178–185.

Oliveira, J., Mateus, N., de Freitas, V., 2014b. Previous and recent advances in pyranoanthocyanins equilibria in aqueous solution. Dyes Pigm. 100, 190–200.

Oliveira, J., da Silva, M.A., Teixeira, N., de Freitas, V., Salas, E., 2015. Screening of anthocyanins and anthocyanin-derived pigments in red wine grape pomace using LC–DAD/MS and MALDI-TOF techniques. J. Agric. Food Chem. 63, 7636–7644.

Pati, S., Liberatore, M.T., Gambacorta, G., Antonacci, D., La Notte, E., 2009. Rapid screening for anthocyanins and anthocyanin dimers in crude grape extracts by high performance liquid chromatography coupled with diode array detection and tandem mass spectrometry. J. Chromatogr. A 1216, 3864–3868.

Pissarra, J., Mateus, N., Rivas-Gonzalo, J., Buelga, C.S., de Freitas, V., 2003. Reaction between malvidin 3-glucoside and (+)-catechin in model solutions containing different aldehydes. J. Food Sci. 68, 476–481.

Pissarra, J., Lourenco, S., Gonzalez-Paramas, A.M., Mateus, N., Buelga, C.S., Silva, A.M.S., et al., 2004a. Structural characterization of new malvidin 3-glucoside-catechin aryl/alkyl-linked pigments. J. Agric. Food Chem. 52, 5519–5526.

Pissarra, J., Lourenco, S., Gonzalez-Paramas, A.M., Mateus, N., Santos-Buelga, C., de Freitas, V., 2004b. Formation of new anthocyanin-alkyl/aryl-flavanol pigments in model solutions. Anal. Chim. Acta 513, 215–221.

Pissarra, J.I., Lourenço, S., Machado, J.M., Mateus, N., Guimaraens, D., de Freitas, V., 2005. Contribution and importance of wine spirit to the port wine final quality – initial approach. J. Sci. Food Agric. 85, 1091–1097.

Remy, S., Fulcrand, H., Labarbe, B., Cheynier, V., Moutounet, M., 2000. First confirmation in red wine of products resulting from direct anthocyanin-tannin reactions. J. Sci. Food Agric. 80, 745–751.

Remy-Tanneau, S., Le Guerneve, C., Meudec, E., Cheynier, V., 2003. Characterization of a colorless anthocyanin-flavan-3-ol dimer containing both carbon-carbon and ether interflavanoid linkages by NMR and mass spectrometry. J. Agric. Food Chem. 51, 3592–3597.

Rivas-Gonzalo, J.C., Bravo-Haro, S., Santosbuelga, C., 1995. Detection of compounds formed through the reaction of malvidin-3-monoglucoside and catechin in the presence of acetaldehyde. J. Agric. Food Chem. 43, 1444–1449.

Salas, E., Fulcrand, H., Meudec, E., Cheynier, V., 2003. Reactions of anthocyanins and tannins in model solutions. J. Agric. Food Chem. 51, 7951–7961.

Salas, E., Atanasova, V., Poncet-Legrand, C., Meudec, E., Mazauric, J.P., Cheynier, V., 2004a. Demonstration of the occurrence of flavanol-anthocyanin adducts in wine and in model solutions. Anal. Chim. Acta 513, 325–332.

Salas, E., Le Guerneve, C., Fulcrand, H., Poncet-Legrand, C., Cheynier, W., 2004b. Structure determination and colour properties of a new directly linked flavanol-anthocyanin dimer. Tetrahedron Lett. 45, 8725–8729.

Salas, E., Duenas, M., Schwarz, M., Winterhalter, P., Cheynier, V., Fulcrand, H., 2005. Characterization of pigments from different high speed countercurrent chromatography wine fractions. J. Agric. Food Chem. 53, 4536–4546.

Santos, H., Turner, D.L., Lima, J.C., Figueiredo, P., Pina, F.S., Macanita, A.L., 1993. Elucidation of the multiple equilibria of malvin in aqueous-solution by one-dimensional and 2-dimensional NMR. Phytochemistry 33, 1227–1232.

Santos-Buelga, C., Bravo-Haro, S., Rivas-Gonzalo, J.C., 1995. Interactions between catechin and malvidin-3-monoglucoside in model solutions. Z. Lebensm. Unters. Forsch. 201, 269–274.

Santos-Buelga, C., Francia-Aricha, E.M., de Pascual-Teresa, S., Rivas-Gonzalo, J.C., 1999. Contribution to the identification of the pigments responsible for the browning of anthocyanin-flavanol solutions. Eur. Food Res. Technol. 209, 411–415.

Saucier, C., Little, D., Glories, Y., 1997. First evidence of acetaldehyde-flavanol condensation products in red wine. Am. J. Enol. Vitic. 48, 370–373.

Schwarz, M., Winterhalter, P., 2003. A novel synthetic route to substituted pyranoanthocyanins with unique colour properties. Tetrahedron Lett. 44, 7583–7587.

Schwarz, M., Jerz, G., Winterhalter, P., 2003a. Isolation and structure of pinotin A, a new anthocyanin derivative from Pinotage wine. Vitis 42, 105–106.

Schwarz, M., Wabnitz, T.C., Winterhalter, P., 2003b. Pathway leading to the formation of anthocyanin − vinylphenol adducts and related pigments in red wines. J. Agric. Food Chem. 51, 3682−3687.

Somers, T.C., 1971. The polymeric nature of wine pigments. Phytochemistry 10, 2175−2186.

Sousa, A., Mateus, N., Soares Silva, A.M., Vivas, N., Nonier, M.-F., Pianet, I., et al., 2010. Isolation and structural characterization of anthocyanin-furfuryl pigments. J. Agric. Food Chem. 58, 5664−5669.

Sousa, C., Mateus, N., Silva, A.M.S., Gonzalez-Paramas, A.M., Santos-Buelga, C., de Freitas, V., 2007. Structural and chromatic characterization of a new malvidin 3-glucoside-vanillyl-catechin pigment. Food Chem. 102, 1344−1351.

Timberlake, C.F., Bridle, P., 1976. Interactions between anthocyanins, phenolic compounds and acetaldehyde and their significance in red wines. The effect of processing and other factors on the colour characteristics of some red wines. Am. J. Enol. Vitic. 27, 97−105.

Trouillas, P., Sancho-García, J.C., de Freitas, V., Gierschner, J., Otyepka, M., Dangles, O., 2016. Stabilizing and modulating color by copigmentation: insights from theory and experiment. Chem. Rev. 116, 4937−4982.

Vaclavik, L., Lacina, O., Hajslova, J., Zweigenbaum, J., 2011. The use of high performance liquid chromatography−quadrupole time-of-flight mass spectrometry coupled to advanced data mining and chemometric tools for discrimination and classification of red wines according to their variety. Anal. Chim. Acta 685, 45−51.

Vallverdu-Queralt, A., Biler, M., Meudec, E., Guerneve, C.L., Vernhet, A., Mazauric, J.P., et al., 2016. p-Hydroxyphenyl-pyranoanthocyanins: an experimental and theoretical investigation of their acid-base properties and molecular interactions. Int. J. Mol. Sci. 17, 1842.

Vallverdú-Queralt, A., Meudec, E., Eder, M., Lamuela-Raventos, R.M., Sommerer, N., Cheynier, V., 2017a. The hidden face of wine polyphenol polymerization highlighted by high-resolution mass spectrometry. ChemistryOpen 6, 336−339.

Vallverdú-Queralt, A., Meudec, E., Eder, M., Lamuela-Raventos, R.M., Sommerer, N., Cheynier, V., 2017b. Targeted filtering reduces the complexity of UHPLC-Orbitrap-HRMS data to decipher polyphenol polymerization. Food Chem. 227, 255−263.

Vidal, S., Meudec, E., Cheynier, V., Skouroumounis, G., Hayasaka, Y., 2004. Mass spectrometric evidence for the existence of oligomeric anthocyanins in grape skins. J. Agric. Food Chem. 52, 7144−7151.

Zea, L., Serratosa, M.P., Mérida, J., Moyano, L., 2015. Acetaldehyde as key compound for the authenticity of sherry wines: a study covering 5 decades. Compr. Rev. Food Sci. Food Saf. 14, 681−693.

CHAPTER

15

Spoilage Yeasts in Red Wines

Manuel Malfeito-Ferreira

Linking Lanscape, Environment, Agriculture and Food Research Centre (LEAF), University of Lisbon, Lisbon, Portugal

15.1 INTRODUCTION

Wine has been produced for millennia and it is common knowledge that it is enough to crush grapes to turn the grape juice into wine. With the work of Louis Pasteur on alcoholic fermentation, published between 1857 and 1860, it was possible to understand the intriguing fermentation process and recognize the essential role played by yeasts. The advances in fermentation technology made possible to recognize the advantage of inoculating fresh juice with well-performing *Saccharomyces cerevisiae* yeasts by the 1970s. A flourishing business of active dried yeast for the wine industry has never ceased to grow up to today. Presently, besides the almost mandatory use of active starters, a vast array of yeasts, not restricted to the classical species *S. cerevisiae* or *Saccharomyces bayanus*, has been directed to "answer" to the winemaker's demands. Ethanol tolerant strains for stuck fermentations, "aromatic" yeasts to produce perfumed white wines, "killer" yeasts to inhibit wild yeasts, "varietal" or "regional" yeasts to specific grape varieties or wine regions, body enhancing and after taste prolonging yeasts, yeasts with special enzymatic activities, or nonproducers of sulfur reduced compounds, illustrate the commercial creativity of yeast starter companies that cope with the search for "different" wine styles by winemakers. Without questioning the advantage of its utilization particularly for young wines, wines from damaged grapes, or for large-scale fermentations, we believe that the excessive discussion on its benefits has driven to a second place the dark side of yeast influence on wine quality. In fact, according to our experience in the industry, the central figure of wine yeasts in enology is much higher for its negative effects and economical losses than by the claimed quality imparted by specific yeast starters. In other words, once appropriate conditions are given to fermenting yeasts they fulfill their role without meriting special attention. On the contrary, spoilage yeasts are a permanent worry due to their possible detrimental effects. In fact, presently, wine spoilage yeasts represent one of the most significant concerns in modern enology, where the production of volatile phenols by *Dekkera/Brettanomyces bruxellensis* in red wines plays a major role (Schumaker et al., 2017).

15.1.1 Concept of Spoilage Yeasts

The manual of yeast taxonomy includes 761 yeast species (Boekhout et al., 2002) but the recent advances in molecular biology in species delineation have increased this number and rearranged yeast taxonomy (Hittinger et al., 2015). About a quarter of these species are food and beverage contaminants but only about a dozen are really detrimental to food quality. Among contamination yeasts, those surviving in foods but without the ability to grow are called *adventitious, innocuous,* or *innocent* yeasts. Those responsible for unwanted modifications of the processed product—visual, textural, or organoleptical (producing off-flavors or off-tastes)—are called *spoilage* yeasts. However, for technologists the concept of spoilage yeast is narrower and only fits to species able to adversely modify foods processed according to the standards of good manufacturing practices (GMPs) (Pitt and Hocking, 1985). These are the *sensu stricto* spoilage yeasts and represent the most resistant species to the stresses provoked by food or beverage processing (Loureiro and Querol, 1999; Malfeito-Ferreira, 2011).

In the wine industry, the definition of spoilage is not always obvious, because the microbial metabolites contribute to wine's, flavor and aroma and their pleasantness is driven by many subjective factors (e.g., habits, fashions, opinion makers' choices) that persuade the consumer taste. Therefore the definition of spoilage as "the alteration of food recognized by the consumer" (Stratford, 2006) is not fully satisfactory for wines. This situation is clearly demonstrated by the presence of volatile phenols in red wines, produced by the species *D. bruxellensis*. While some consumers and opinion makers prefer wines tainted by volatile phenols, particularly before knowing its origin, others do consider that, even in low concentration, these compounds depreciate wine quality due to diminished flavor complexity.

15.1.2 Significance and Occurrence of Wine-Related Yeast Species

Technological advances in winemaking and the improvement in winery GMPs has led to the decreasing importance of bacterial diseases like "amertume," "tourne" and lactic peak (for a description of these diseases see Ribéreau-Gayon et al., 2006). The acetic spoilage, due to acetic acid and ethyl acetate produced by acetic acid bacteria, is also much more infrequent today. Presently, winemakers' concerns have also been directed to food safety issues but without a direct effect on wine organoleptical quality. Ethyl carbamate, biogenic amines, and ochratoxyn A are new challenges motivated by their effect on consumer's health (Fugelsang and Edwards, 2007). In these cases yeasts play a minor role, if any at all. In wine, the only situation where yeasts may pose an indirect health threat is related to the explosion of bottles due to refermentation of sweet wines. This rare type of incident leading to eye injury has been described mainly in carbonated beverages (Kuhn et al., 2004). Therefore, the detrimental effects caused by yeasts are the most frequent problems of microbial origin related to wine quality.

The most common recognized symptoms of yeast spoilage are film formation in bulk wines, cloudiness, sediment formation, and gas production in bottled wines, and off-flavor production during all processing and storing stages (Loureiro and Malfeito-Ferreira, 2003a). The technological significance of wine or grape juice contamination yeasts has been thoroughly discussed in previous reviews (Thomas, 1993; Kunkee and Bisson, 1993; Loureiro and Malfeito-Ferreira, 2003a; Fugelsang and Edwards, 2007) which constitute the background for the following description. In Table 15.1 are listed the yeast species and their effects most relevant to wine spoilage according to authors' experience while Fig. 15.1 shows microphotographs of several common yeast species.

15.1.2.1 Grapes and Grape Juices

Common yeast contaminants of grapes and grape juices before fermentation include the genera *Candida*, *Cryptococcus*, *Debaryomyces*, *Hansenula*, *Kloeckera/Hanseniaspora*, *Metschnikowia*, *Pichia*, and *Rhodotorula* (Fleet et al., 2002). The yeast-like fungus *Aureobasidium pullulans* is also frequent in grapes. These grape yeasts do not cause problems in wines bottled according to GMPs but may be a cause of concern in the early stages of wine production. In grapes before harvesting, sour rot is a infection where a wide number of yeast species may contribute to the high concentration of acetic acid and unwanted modifications of juice composition before winemaking (Guerzoni and Marchetti, 1987; Barata et al., 2012a). The rotting process is not yet clear, being also influenced by acetic acid and lactic acid bacteria, and the main measures to adopt in winemaking to overcome problems involve grape selection, increased sulfur dioxide usage, and prompt inoculation of active starters (Barata et al., 2012a).

In juices, given the short period before fermentation, spoilage events are rare. In long red prefermentative maceration, film-forming yeasts (e.g., *Pichia anomala*) or apiculate yeasts may grow very fast. These species are easily controlled by adequate winemaking measures (low temperature, sulfur dioxide, hygiene). In principle these species are inhibited during fermentation but even for a short period, due to their fast growth, they may produce unwanted amounts of metabolites like ethylacetate (vinegar smell) or acetaldehyde (oxidized taint) (Romano, 2005), which can irremediably spoil the wine.

15.1.2.2 Wine Fermentation

Fermentation problems are related to the activity of fermenting yeasts (*S. cerevisiae* or *S. bayanus*). The production of off-flavors (sulfur reduced compounds) (Bell and Henschke, 2005) and acetic acid is due to nutritional imbalances or deficient fermenting conditions (e.g., high temperature) that may lead to stuck fermentations (Bisson and Butzke, 2000). These events are a result of the environment conditions and not of any particular spoilage characteristic and so the correct management of fermenting conditions overcomes the problem. The activity of other yeast species is limited or unknown, with stuck wines being highly susceptible to bacterial spoilage.

TABLE 15.1 Contamination Wine Yeast Species Associated With Confirmed Spoilage Events According to Authors' Experience

Group	Empirical denomination	Species	Occurrence	Spoilage events
Adventitious, innocent, or innocuous yeasts	Apiculate yeasts	Kloeckera/Hanseniaspora spp.	Grape juices	Spoilage by ethylacetate production
	Film-forming yeasts	Candida spp. Pichia spp.	Bulk wines	Film formation with production of off-flavors (e.g., ethylacetate)
		P. membranifaciens	Bottled wines	Spoilage by sediment formation
Sensu stricto spoilage yeasts	Fermenting yeasts	S. cerevisiae	Bottled dry wines	Spoilage by sediment or cloudiness formation
			Bottled or bag in box sweet wines	Spoilage by refermentation
	Contaminant yeasts	S. ludwigii	Bottled sweet wines	Spoilage by sediment or cloudiness formation
			Bulk sweet wines	Contamination without detected spoilage
			Corks	Contamination without detected spoilage
		Z. bailii	Bottled wines	Spoilage by sediment or cloudiness formation
			Desulfited grape juice, storage tanks and winery equipment	Contamination without detected spoilage
		Z. rouxii	Grape juice concentrate	Contamination without detected spoilage
		T. delbrueckii	Desulfited grape juice and storage tanks	Contamination without detected spoilage
		D. bruxellensis	Bulk, barrel matured, and bottled wines	Spoilage by 4-ethylphenol production
			Sparkling wine	Spoilage by cloudiness formation

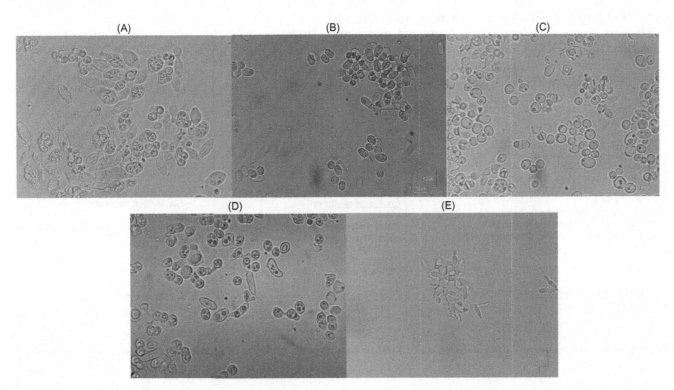

FIGURE 15.1 Typical cell morphologies of *Saccharomycodes ludwigii* (A), *Schizosaccharomyces pombe* (B), *Pichia anomala* (C), *Zygosaccharomyces bailii* (D), and *Brettanomyces bruxellensis* (E).

15.1.2.3 Bulk and Bottled Wine

During bulk wine storage, film-forming yeasts (e.g., *Candida* spp., *Pichia* spp.) may form pellicles on the wine surface and spoil wine by the production of odor active compounds. The absence of oxygen and proper sulfur dioxide usage together with appropriate hygienic measures prevent the emergence of films. These species are frequently encountered in bottled wine but, when in low number, tend to die, or to remain dormant, due to low resistance to the stress imposed by the bottled product.

The classical spoilage events due to *sensu stricto* spoilage yeasts occur in bottled wine. Their specific spoiling abilities are related to bottled sweet wine refermentation or growth in bottled dry wines. The typical members of this group are those capable of growing in bottled wines—*Z. bailii*, *S. cerevisiae*, *Schizosaccharomyces pombe*, and *Saccharomycodes ludwigii*. Albeit not frequent grape or winery contaminants, the stress resistances of these yeasts enable their survival and proliferation under conditions that are not tolerated by the other species. According to our experience, refermentation problems have increased in the last years in red wines because of the addition of concentrated grape juices to make softer wines in accordance to modern market demands. In bag-in-box wines this problem is easily recognized by swollen packages. In bottled wines, gas production and wine turbidity are also easily visible after bottle opening and pouring wine in the glass. In red wines the effect of sediment formation or haziness is less manifest, given the dark color.

The recognition of the role played by the species *D. bruxellensis* in red wine spoilage due to the production of odor active compounds in bulk or bottled wines has revealed a new challenge to winemakers since the last decade (Loureiro and Malfeito-Ferreira, 2006; Malfeito-Ferreira, 2011). Moreover, its effects are particularly notorious in premium red wines aged in costly oak barrels, which considerably increase the economical losses provoked by spoilage yeasts in the wine industry. Presently, this species is regarded as the main threat posed by yeasts to wine quality. The effect is not only direct, due to the production of volatile phenols, but also indirect due to the technological measures needed to control its activity and that may reduce wine attributes.

15.1.3 Factors Promoting the Dissemination of Spoilage Yeasts

The wine spoilage yeasts associated with wineries are disseminated by all surfaces with residues of nutrients where they can proliferate, namely wine residues in wooden barrels, valves, improperly cleaned tanks, hoses, fillers, corkers, filters, pumps, walls, floors, air, etc. They easily contaminate and grow in wine when environmental conditions are favorable. It is common knowledge that their incidence increases with low levels of hygiene, so that the best way to prevent their contamination is to avoid their propagation by sound hygiene procedures. However, the hygienic level is only a part of the problem. Fig. 15.2 shows the main factors affecting the dissemination and proliferation of yeasts in winery environments.

The first group of factors, poorly supported by scientific studies, is related to the primary source of spoilage yeasts and the way they enter inside the winery. These routes of contamination, for which it is not easy to

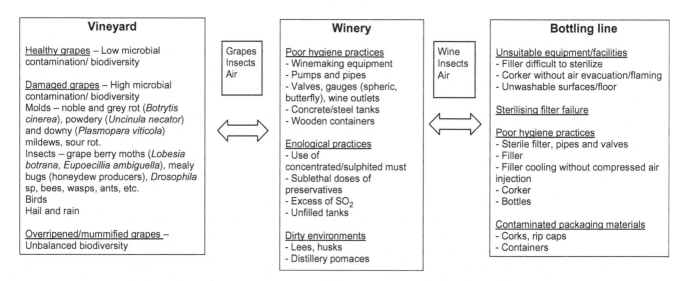

FIGURE 15.2 Factors promoting the contamination and colonization of spoilage yeasts in winemaking environments and bottled wine.

establish the degree of importance, include grapes (mainly those that are damaged), insects, air (mainly when the environment in the vicinity of the winery is dirty with winemaking residues), the additives used in vinification and/or in the preparation of sweet wines from dry wines, particularly those processed, such as concentrated juices, sulfited juice with sublethal preservative doses and sucrose (Loureiro and Malfeito-Ferreira, 2003a; Valero et al., 2005). The grapes are, probably, the main source for wine spoilage yeasts justifying the understanding of how yeasts survive along the year and reach the grapes in the vineyard. This issue is not well studied, given that most surveys performed did not include the study of vectors and rarely account for the minority and slow-growing populations. On the other hand, it is generally accepted that damaged grapes bear a wider diversity of species and in numbers of 5–6 orders of magnitude higher (Fleet et al., 2002; Barata et al., 2012b). Therefore, it seems legitimate to assume that rotten grapes are one of the most important sources of food spoilage yeasts. Recent studies showed that sour rot grapes bear a high concentration of *Z. bailii*, that survives through the fermentation and reaches high counts in the final wine (Barata et al., 2012b). Guerzoni and Marchetti (1987) also showed that *Brettanomyces* spp. were present in higher numbers in sour rot grapes and Renouf and Lonvaud-Funel (2007) recovered *D. bruxellensis* from sound grapes using an enrichment medium. These results show that it is essential to deepen the studies of grape microbial ecology, considering differences between sound and damaged material. Damage includes sour rot, gray rot, downy mildew, insects (*Lobesia botrana, Eupoecilia ambiguella*, honeydew producers, *Drosophila* spp., bees, wasps, ants, etc.), birds, hail, rain, and overripened/mummified grapes. Only then it will be possible to answerer to these essential questions: (1) Which are the vectors that carry spoilage yeasts to grapes? (2) Are there ecological niches (natural or artificial) visited by the vectors (insects or others) which are the origin of grape contamination? (3) In their absence how can we explain the origin of spoilage yeasts? (4) Can it be from wineries (or vinification residues in the vicinity) that insects spread and contaminate grapes in the vineyard? The answers to these questions will help to adopt measures to reduce the prevalence of spoilage yeasts in wineries and, consequently, in the wines.

Assuming that spoilage yeasts enter the winery and cannot be fully eradicated, we need to learn how to deal with them. Which winery practices favor their multiplication and dissemination? In first place stands, obviously the hygiene, conditioned not only by cleaning and disinfection programs but also by materials and status of the surfaces contacting wines, equipment design, many of them difficult to sanitize, and the technical skill of the staff. Many of these aspects have been poorly studied, particularly the microbiology of wooden barrels and the efficiency of cleaning and "disinfection" treatments, namely those with hot water and steam (Barata et al., 2013). Bearing in mind the increased risk of *D. bruxellensis* contamination with the age of oak barrels (Chatonnet et al., 1993), it is surprising the scarcity of studies on efficient measures to overcome the associated problems. In fact, wineries rarely monitor the contamination level of wine aged in oak barrels and, more rarely, the microbiological quality of wine put in the barrels.

Among the enological practices that disseminate yeasts are, naturally, all contaminated wine transfers, mainly when blends with wines aged in used barrels are required. Similarly, concentrated grape juices and sulfited juices with sublethal doses of sulfur dioxide are vectors and promote the dissemination of resistant strains of *Z. bailii* in the winery. These ingredients should be handled in restricted areas within the winery to decrease the incidence of infections. In addition, all enological practices that promote yeast growth like those leading to high levels of dissolved oxygen (e.g., micro-oxygenation, untopped tanks, racking with aeration), low sulfur dioxide or sorbate levels may enhance the proliferation of spoilage yeasts. Finally, improper management of wine residues (e.g., husks, lees, distillery pomaces) that lead to dirty winery vicinities may significantly contribute to yeast dissemination, mainly through insects (*Drosophila*) and air (Connel et al., 2002). Our experience tells us that wineries with dirty environments and without an efficient residue treatment system have, as a rule, higher incidence of spoilage yeasts. In some cases we observed that water in underground layers have heavy yeast populations, suggesting their contamination by winery effluents.

The bottling line is the last circuit run by the wine in the winery. The filling operation is the last opportunity for the wine to be contaminated by spoilage yeasts. Usually, the contamination of wine in this phase results from lack of hygiene and, particularly, from deficient disinfection of all the circuit, including surfaces contacting with the cork that closes the bottle. However, the lack of hygiene results most frequently from inadequate conditions to apply cleaning and disinfection programs—filler and corker design, microbiological quality of ambient air, equipment maintenance program and operator's training—than from the efficiency of sanitation programs. Thus, the specificities of each bottling line are determinant in the levels and kind of spoilage yeast contaminating it. From our experience, the main contamination sources are in the filler and in the corker and are more dangerous as equipment age increases and less maintenance is run (Loureiro and Malfeito-Ferreira, 2003b). When the fillers are poorly designed (or the sanitation procedures advised by the suppliers are not followed), their cleaning and disinfection may not reach all surfaces where there are wine residues, originating dangerous contamination sources, such as self-leveling systems, surfaces protected by o-rings, and valves of isobarometric fillers. The usage of steam disinfection is an alternative to the chemical disinfection given that is possible to kill microorganisms without directly contacting the cells. However, filler steaming may bring a problem frequently overlooked—the formation of a negative pressure in the filler interior during cooling—leading to air suction (with spoilage yeasts in suspension) and, as a rule, the immediate filler contamination. In these cases, the solution is to introduce sterile compressed air during cooling. The corker may also be an important contamination source when the surfaces contacting with the side part of the corks are heavily contaminated, such as the transport tube, the corker jaws, and the feeding bin, where the formation of water droplets condensation promotes the contamination and colonization by yeasts suspended in the air. In modern machines heating the jaws up to lethal temperatures prevents this type of contamination. Another frequent contamination source is the bottle rinser or the rinsing water. When bottles arrive from the factory in good packaging conditions, the contamination appears after rinsing (Donnelly, 1977b; Neradt, 1982). As a final remark, in many companies, either the fillers or corkers are located in a closed room aiming to reduce contaminations from the winery ambient air. When they are correctly designed—with overpressure, wet air evacuation, and correct hygiene—they are advantageous to proper bottling. However, in most cases they are poorly designed and become dangerous contamination sources, working as microbial incubators (Donnelly, 1977a). In fact, bottle breakage occurs during bottling, leading to wine dispersion through room surfaces and, consequently, to microbial growth, stimulated by the moist environmental conditions.

The few scientific studies on the ecology of bottling lines do not allow conclusions to be made on the most frequent type of contaminating yeasts. Film-forming species should prevail, but it is conceivable that dangerous contaminants are favored by developing on wine residues and by resisting sublethal doses of chemical or physical agents used in equipment cleaning and disinfection. From our experience, when concentrate grape juice is used for producing sweet wines, it is very frequent to detect yeasts of the genus *Zygosaccharomyces*, particularly *Z. bailii*, in the fillers.

15.2 DESCRIPTION OF THE MAIN YEAST GENERA/SPECIES INVOLVED IN WINE SPOILAGE

The species involved in wine spoilage are also known for affecting other food commodities and their taxonomical, physiological, or technological properties have been described in excellent textbooks (Deak and Beuchat, 1996; Boekhout and Robert, 2003; Blackburn, 2006; Querol and Fleet, 2006). The description of the species presented below is mainly concerned with their wine relevant properties.

15.2.1 Film-Forming Species

The denomination "film-forming yeasts" includes a group of species able to grow on the surface of wine developing pellicles. The species of the genus *Candida* and *Pichia* are regarded as the typical film-forming yeasts although *S. cerevisiae*, *D. bruxellensis*, or *Z. bailii* may also be recovered from wines pellicles (Ibeas et al., 1996; Farris et al., 2002). In the case of *S. cerevisiae* it is even a desirable feature for the race *beticus*, which is one of the agents of sherry-type wine production (Suárez-Lepe and Iñigo-Leal, 2004). The ability to form films by *Pichia* and *Candida* is probably explained by their aerobic nature and fast growth and so the other species are usually minor constituents of film microflora. In bulk wines they quickly cover the wine surface when air has not been removed from the top of storage vessels. Although strains of *Candida* spp. or *Pichia* spp. are tolerant to preservatives (Table 15.2), their control in wines is mainly due to their weak tolerance to low oxygen tensions, which enhances the inhibitory effect of ethanol or preservatives. In bottled wines they may cause cloudiness if the initial contamination load is high and so these species are regarded as indicators of poor GMPs. They may also produce, at the bottleneck, a film or a ring of cells adherent to the glass, if the closure does not prevent the diffusion of oxygen, the level of free sulfur dioxide is too low and the initial contamination is high.

15.2.2 *Zygosaccharomyces bailii* and Related Species

The genus *Zygosaccharomyces* comprises some of the most feared species in the industries of high sugar and high acidic food products. *Zygosaccharomyces rouxii* is mainly known for being highly osmotolerant while *Z. bailii*

TABLE 15.2 Yeast Tolerance to Antimicrobial Agents Used in Winemaking

Tolerance measure	Agent	D. bruxellensis	Pichia spp.	S. cerevisiae	S. pombe	S. ludwigii	Z. bailli	Reference
Maximum concentration allowing growth	Ethanol (% v/v)	15	15[a]	17	–	–	–	Dias et al. (2003a,b), Barata et al. (2006, 2007)
	PMB[b] (mg/L)	90	140[a]	200	–	–	–	Dias et al. (2003a,b), Barata et al. (2006, 2007)
	Total SO_2 (mg/L)	75	350[c]	50	250	250	500	Loureiro (1997)
	Sorbic acid (mg/L)	950	650[c]	300	–	–	600	Loureiro (1997)
Minimum concentration inducing complete death	DMDC[d] (mg/L) (500 CFU/mL)[e]	100	100[a]	100	100	25	25	Costa et al. (2008)
	DMDC (mg/L) (10^6 CFU/mL)[e]	300	300[a]	300	>300	200	150	Costa et al. (2008)
Minimum concentration inhibiting fermentation	DMDC (mg/L)	250	–	250	–	–	400	Delfini et al. (2002)
Maximum concentration allowing growth	CO_2 (dissolved volumes)	4.45	–	2.23	–	–	3.34	Ison and Gutteridge (1987)

[a]*P. guilliermondii.*
[b]*Potassium metabisulfite.*
[c]*P. membranifaciens.*
[d]*Dimethyl dicarbonate.*
[e]*Initial inoculum size.*

is notorious for its resistance to low pH, high concentration of organic acids, including preservatives, and high osmotolerance. *Zygosaccharomyces bisporus* has been described as having intermediate features between the former two species (Pitt and Hocking, 1985).

The most problematic species in wines is *Z. bailii*. Its activity includes visible sediment formation, cloudiness, or haziness in dry wines, and refermentation in sweet wines. Given the visual nature of the spoiling effect it is a greater concern in white wines. It may also produce undesirable odor active metabolites (Fugelsang and Edwards, 2007) but their relevance is lower that the former visual faults. One of the most relevant sources of *Z. bailii* are the grapes, particularly grapes damaged by sour rot (Barata et al., 2012b). These authors demonstrated that *Z. bailii* survived during wine fermentation being present in 10^5 CFU/mL in the fermented wine. Once established in the winery, the main risk is the contamination of wine after sterile filtration, due to improper sanitation, before bottling (Malfeito-Ferreira et al., 1997).

The species *Z. bisporus* has been isolated from grapes affected by honeydew and sour rotten grapes (Barata et al., 2008) but it was not detected at the end of fermentation (Barata et al., 2012a), which may explain its lower incidence in wines than *Z. bailii*. In spite of the ability of *Z. bisporus*, isolated from sherry film, to resist to sorbic acid and sulfur dioxide (Splitstoesser et al., 1978) and to produce odorous acyloins in sherry wines (Neuser et al., 2000), we are not aware of spoilage events by this species (Loureiro and Malfeito-Ferreira, 2003a).

The species *Z. rouxii* has been recovered from concentrated grapes juices. However, the importance of this yeast as a bottled wine spoiler is much lower than that of *Z. bailii*, probably due to its weaker resistance to low pH and chemical preservatives.

Recently, a new species—*Zygosaccharomyces lentus*—was recognized based on isolates from several food industries including one red wine. It is characterized by having similar stress tolerances to *Z. bailii* and *S. cerevisiae*, being distinguished by growing slowly at low temperature (4°C) (Steels et al., 1999). It remains to be seen if this species may be regarded as a wine spoiler.

Another species taxonomically closely related with *Z. bailii* is *Torulaspora delbrueckii* (Kurtzman and Fell, 1998). It contaminates concentrated grapes juices and its spoiling effects are related with growth in bottled wine, as described by Minarik (1983). We have detected this species, particularly in concentrated or sulfited grape juices, but in spoiled wines its incidence in bottled wines is much lower than that of *Z. bailii*, probably given its weaker resistance to preservatives.

The latest taxonomic rearrangements have created the genus *Lachancea* that includes *L. thermotolerans* (ex *Kluyveromyces thermotolerans*), *L. waltii* (ex *Kluyveromyces waltii*), *L. cidri* (ex *Zygosaccharomyces cidri*), and *L. fermentati* (ex *Zygosaccharomyces fermentati*) (Kurtzman and James, 2006). The latter species has been implicated in the production of odor active compounds in sherry-like medium (Freeman et al., 1977) but has not been implicated in wine spoilage.

15.2.3 *Saccharomyces cerevisiae* and Related Species

S. cerevisiae and *S. bayanus* are the desired agents of wine fermentation. However, they may also be responsible for wine spoilage. During fermentation, the occasional nutritional imbalance of grape juice may lead to off-flavor production imparted by sulfur-reduced compounds (Bell and Henschke, 2005). Modern winemaking systems with juice pumps over under anaerobic conditions tend to increase the problem, compared to old systems where juice aeration was present. If not treated in due time, these taints may persist during storage and in bottled wines. In finished wines the most frequent detrimental effects of this species are refermentation of sweet wines and sediment, cloudiness, or haziness formation. These effects are similar to those provoked by *Z. bailii* and some *S. bayanus* strains may be more dangerous given their higher tolerance to ethanol (Malfeito-Ferreira et al., 1990a,b).

15.2.4 *Saccharomycodes ludwigii* and *Schizosaccharomyces pombe*

S. ludwigii and *S. pombe* are notorious agents of wine spoilage thanks to their high resistance to stress conditions. Despite this feature their overall incidence is much lower than that of *Z. bailii* and *S. cerevisiae*, for which there is no obvious explanation. Both species have been isolated from grapes in vineyards but with reduced incidence (Florenzano et al., 1977; Combina et al., 2005) and in grape juices (Pardo et al., 1989), being also rare contaminants of winery environments. We hypothesize that their natural contamination sources are more restricted and/or their ability to survive in winery environments is lower than those of *Z. bailii* or *S. cerevisiae*. Their common effects result from cell growth in bottled wine leading to sediment or turbidity formation and refermentation. We currently isolate *S. ludwigii* in bulk white Vinho Verde wine, particularly when an excess of sulfur dioxide is used, and in sparkling wine plants using the Charmat system where their growth may clog stainless steel pipes. *S. ludwigii* is particularly resistant to sulfur dioxide probably due to its ability to produce high amounts of acetaldehyde. *S. pombe* has been exploited for the reduction of malic acid (Delfini and Formica, 2001) which, if adopted, must be followed by preventive measures to reduce the risk of post-treatment proliferation.

15.2.5 *Dekkera/Brettanomyces bruxellensis*

D. bruxellensis is presently the most notorious wine spoilage yeast due to the production of ethylphenols in red wines (Loureiro and Malfeito-Ferreira, 2006; Suárez et al., 2007). This species is long known as an undesirable contaminant but not as a recognized producer of these metabolites. The present widespread use of oak barrels to age premium red wine, where the ability to produce ethylphenols overwhelms the presence of other contaminants, contributed significantly to its notoriety. In addition, the controversy about its influence on wine quality involving winemakers, journalists, and consumers make this species the most prominent microbial wine spoilage subject.

D. bruxellensis is rather elusive yeast, being difficult to isolate from sources contaminated by other yeasts due to its low growth rates. Thus the use of selective media and long incubation periods are essential to its recovery. It has been rarely isolated from grapes (Guerzoni and Marchetti, 1987; Renouf and Lonvaud-Funel, 2007) and winery environments (Connel et al., 2002), being dominant in bottled red wines, as ethylphenols producers, or in sparkling wines, inducing cloudiness, when there is no occurrence of the other yeasts (Loureiro and Malfeito-Ferreira, 2006). In relative terms, it is not so tolerant to ethanol or preservatives as *S. cerevisiae* or *Z. bailii* (Table 15.2) but it has the ability to remain viable for long periods and proliferate when conditions become less severe (Renouf et al., 2007). The occasional detection in sparkling wines may be related with its resistance to carbon dioxide (Table 15.2) as observed also for *Dekkera naardenensis* in carbonated soft drinks (Esch, 1992). However, it is seldom isolated from still white wines for which there is yet no satisfactory explanation (Barata et al., 2007).

15.3 YEAST MONITORING

15.3.1 Microbiological Control

The conservative attitude of the wine industry and, namely, the absence of microbiological safety hazards, determine that the implementation of hazard analysis and critical control point (HACCP) and self-control plans, mandatory in most food industries, is not dealt with the desirable strictness. In fact, the microbial stability of most dry table wines—white, rosé, or red—attained when good winery practices are followed, leads to the absence of microbiological control by most producers. Exceptionally, commercial contracts with modern distributors (supermarket chains and others) or demanding clients may force the implementation of routine microbiological analysis. For this reason, microbiological control in wine industry is, as a rule, synonymous with microbiological assessment (particularly yeasts) of bottled sweet wines processing where the risk of refermentation is high. However, the present microbiological hazards of the wine industry should justify much more attention.

15.3.1.1 Grape and Grape Juice Monitoring

During wine fermentation it is neither easy nor justifiable to implement microbiological control plans to detect spoilage yeasts. Their influence in wine quality, as a rule, is irrelevant and possible corrective measures are practically absent. One of the few measures is to establish chemical indicators related to grape microbiological quality, already implemented in numerous wineries (particularly cooperatives or large companies to establish the price of grapes as a function of its health status), like laccase activity (indicator of grapes affected by gray rot) or volatile acidity and gluconic acid (indicator of grapes affected by sour rot). The utilization of costly Fourier-transform infrared spectroscopy (FTIR) instruments makes these determinations readily available, thus providing the possibility of separate processing according to raw material quality. In smaller dimension wineries, grape selection enables the removal or separate processing of poor quality grapes.

15.3.1.2 Bulk Wine Monitoring

After wine fermentation, most wineries measure qualitative or quantitative chemical indicators to control the activity of lactic acid bacteria (malic acid assessment) and acetic acid bacteria (volatile acidity). It is not current practice to monitor the presence of spoilage yeasts. However, it would be useful to screen spoilage yeasts or their secondary metabolites, such as 4-ethylphenol and ethylacetate, in wines produced from poor sanitary quality grapes. In this case, the prevalence of such yeasts seems to be high and the wine resistance to microbial colonization is reduced, creating conditions for product alteration. During this stage it is also important to monitor the presence of film-forming yeasts growing on the wine surface, mainly in large volume vessels or untopped tanks where it is not easy to avoid the presence of oxygen required by these yeasts. Microbiological analysis is not a requirement but visual inspection of tank tops every two weeks is a simple and effective practice. In white wines, particularly those with residual sugar, the specific detection of *Z. bailii* or *S. cerevisiae* should be considered, because they may cause refermentation problems during storage.

15.3.1.3 The Peculiar Case of D. bruxellensis

The relatively low demanding microbiological control during bulk wine storage is no longer advisable concerning red wines, particularly those deserving appropriate aging. Presently, the detection of *D. bruxellensis* is a prerequisite for wineries during all processing stages of premium red wines. In fact, it frequently appears in high levels just after the malolactic fermentation (Rodrigues et al., 2001) leading to premature "horse sweat" taint. During barrel aging, irrespective of grape quality, it is essential to monitor *D. bruxellensis* periodically, mainly in used barrels, which are a well known ecological niche of these yeasts. We have established, for many Portuguese wineries, microbiological criteria that have been giving adequate results so far, and are given here only as guidelines. In the first case, for bulk-stored wines, it is satisfactory to detect *D. bruxellensis* monthly, bimonthly, or even every 3 months (according to the type of wine and of container). The sample volumes are 1, 0.1, 0.01, and 0.001 mL, from a blend composed of wine from the interface air/liquid and from different depths of the container. When the result is positive for 1 mL, or less, and the level of 4-ethylphenol is higher than 150 µg/L, an immediate fine filtration is recommended, accompanied by sulfite addition. In following analysis, after filtration, it is sufficient to monitor the level of 4-ethylphenol, as a rule. For wines before bottling, the criteria are more stringent, and detection should be made on 100, 10, and 1 mL of wine, sampled as described above. When the result is positive in 1 or 10 mL, a very fine or sterilizing filtration is recommended. If positive detection is only obtained for 100 mL, it is acceptable to control viable cells only by the addition of preservatives (e.g., 1 mg/L of molecular sulfite). In this case, bottling must be technically correct and dissolved oxygen should be lowered to practically zero. Otherwise, a sterile filtration is recommended or, as an alternative, a soft thermal treatment of the wine to destroy viable cells.

15.3.1.4 Wine Bottling

Wine bottling is the main stage of conventional microbiological control, if adopted by wineries. Common procedures including analysis of bottles, rinsing water, closures (corks, rip caps), bottling and corking machines, and atmosphere. When properly applied this control enables the detection of contamination sources determining corrective measures. Most frequently, the contamination sources are located in the filling and corking machines (Loureiro and Malfeito-Ferreira, 2003b). The final analysis concerns the evaluation of bottled wine contamination (Loureiro and Malfeito-Ferreira, 2003a). Common microbial contaminants do not survive for long after bottling and if microbial counts are higher than the specifications, the product is retained until clearance is given (see below).

15.3.2 Tools Used in Microbiological Control in the Wineries

As a rule, yeast detection and enumeration methodologies are based on growth on plates containing a general-purpose culture medium, after membrane filtration of wine samples or rinsing solutions (Loureiro et al., 2004). The use of Most Probable Number (MPN) technique is not common, but according to our practical experience would be useful in some situations, particularly when an estimation of yeast contamination in bulk wine is desired, or in wines with a high percentage of suspended solids.

The utilization of selective and/or differential culture media has only somewhat increased in the last years, mainly owing to the problems with *D. bruxellensis*. This situation shows a clear distinction from the typical bacterial control of other food industries, where bacterial indicators, based on the use of a wide variety of selective/differential media, play a central role. Likewise, we presented the concept of zymological (zymo = yeast) indicators to be applied to the wineries, in order to increase the utility of the routine microbiological control (Loureiro and Querol, 1999).

The objective of using zymological indicators is to measure the hygienic quality of surfaces that have contact with wine and the spoilage risks involved, given that the microbiological safety is not an issue. Taking into account the formerly defined wine yeast groups, the hygienic quality of the wine processing may be assessed by the detection of film-forming yeasts by the MPN technique, using a general-purpose culture medium. To detect *sensu stricto* spoilage yeasts, selective/differential media have been developed directed to the most dangerous species—*Z. bailii* (Schuller et al., 2000) and *D. bruxellensis* (Rodrigues et al., 2001). *S. cerevisiae* could be indirectly estimated by the difference between counting on general purpose media in the absence or presence of cycloheximide or of lysine (Heard and Fleet, 1986). The presumptive results obtained with culture media could be further confirmed, if necessary, using biomarkers (biomolecule indicators) such as long-chain fatty acids (Malfeito-Ferreira et al., 1997) or molecular biological identification. In the case of long-chain fatty acid composition, a correlation was found between the proportions of C_{18} unsaturated fatty acids and the spoiling significance of yeast species. Currently, yeast identification is done by molecular methods, which are usually used by external laboratories and not by wineries, due to the degree of expertise and equipment required. In rare and special situations, particularly commercial conflicts, fine molecular typing techniques, adequate to source tracking (Giudici and Pulvirenti, 2002), may be used for forensic studies of wine contamination. Additionally, chemical indicators can also be used to monitor yeast activity in an easy and fast way. Among the molecules produced by yeasts, 4-ethylphenol is currently the most common indicator of *D. bruxellensis* activity and should be used together with microbiological detection. Given the slow growth of these yeasts, culture media only give results after more than 4–5 days and so early detection depends on the use of direct techniques. Presently, there are several real-time polymerase chain reaction (Real-Time PCR) protocols that provide results in about 4–6 hours and have high sensitivity (<10 cells/mL). They have two main drawbacks. One is the cost, very high for routine analysis (>60€/sample). The second is related with false positive responses given by DNA of dead cells. In this case, protocols must be adapted to remove this DNA from the samples (Vendrame et al., 2014).

15.3.3 Acceptable Levels of Yeasts

The establishment of acceptable levels of microorganisms in the final product is a common concern to many food industries. In foods harboring pathogenic microorganisms, the law regulates these levels and the technologists efforts are addressed to their compliance. Regarding yeast spoilage legislation is practically absent and the aim of the technologist is to establish levels that are attainable under industrial conditions and to ensure product stability during its shelf life (Loureiro and Malfeito-Ferreira, 2003a).

In the case of *Z. bailii*, one viable cell per bottle may cause spoilage (Davenport, 1986; Deak and Reichart, 1986; King et al., 1986; Thomas, 1993) but such a strict limit is difficult to attain in winery practice and is not at all appropriate when yeasts counts belong to innocent contaminants. Occasionally specifications are established as a function of the sugar content, assuming that sweet wines are more vulnerable than dry ones. However, Deak and Reichart (1986) did not find differences in the microbial stability of white, red, semidry, and semisweet wines and demonstrated that stability depends on the initial yeast population.

In the absence of sound scientific background to establish appropriate specifications, the industry empirically establishes its limits, which may be used for commercial purposes. A level of yeast counts as low as <1/500 mL or <1/mL are currently regarded as the maximum acceptable level (Andrews, 1992; Loureiro and Malfeito-Ferreira, 2003a) reflecting the caution on the prevention of spoilage events. As a rule, the estimation of contaminating flora is obtained after growth on general media and so results reflect the total flora and not the spoiling one. If this may be wise for highly virulent species, in most of the cases these values are too strict for innocuous contamination ones. Most wineries, when levels are higher than acceptable, hold the product for enough time to meet specifications or to rebottle it. This procedure gives an indication of the spoilage risk because the increase in yeast counts is a signal of contamination by spoiling yeasts. However, this may lead to long holding periods in the case of high initial yeast loads, which take time to be reduced to acceptable levels, even in the case of innocent contaminants. Their specific detection would give clearance to the final product sooner with obvious economical advantages. We believe that microbial guidelines based on sampling plans by attributes, similar to those applied in most food industries (Adams and Moss, 2000), would be an important improvement on wine microbiological control. Although these guidelines were devised for food pathogens we think they would also be appropriate to deal with spoilage situations. In an attribute sampling scheme analytical results are assigned into two or three classes. In the two-class scheme, a sample is defective if it contains more than a specified number of the targeted microorganism. In the three-class plans samples may be acceptable, marginally acceptable, and unacceptable. The two-class plans are more stringent and would be applied to the most dangerous and frequent species, *Z. bailli* and *D. bruxellensis*. The other dangerous species, the frequent *S. cerevisiae* and the rare *S. pombe* and *S. ludwigii*, may have specifications similar to *Z. bailli* once appropriate culture media are developed. When products are only contaminated by innocent yeasts the numbers tolerated are higher and a three-class plan is advised because they do not have the ability to grow in bottled wines. The analysis should also be performed 24 hours after bottling to give time for added preservatives to inactivate the microbial population. In Table 15.3 are presented the guidelines proposed by us to be used

TABLE 15.3 Proposal of Zymological Guidelines (CFU/100 mL) for Expedition of Bottled Wine to Market. Values are Based on Our Practical Experience and on Winery Inquiries Presented in Loureiro and Malfeito-Ferreira (2003a,b)

Product	Microorganism	Culture media[a]	Plan class	n[b]	m[c]	M[d]	c[e]
SWEET WINES (SUGAR >2 G/L)							
White, rosé, and red	Total yeasts	WLN	3	5	10^0	10^1	2
	Z. bailii	ZDM	2	5	0	–	0
DRY WINES (SUGAR <2 G/L)							
White and light rosé	Total yeasts	WLN	3	5	10^1	10^2	2
	Z. bailii	ZDM	2	5	0	–	0
Red and dark rosé	Total yeasts	WLN	3	5	10^2	10^3	2
Red (oak aged)	Total yeasts	ZDM	3	5	10^2	10^3	2
	D. bruxellensis	DBDM	2	10	0	–	0

[a]WLN, Wallerstein Laboratory Nutrient; ZDM, Zygosaccharomyces *differential medium*; DBDM, Dekkera/Brettanomyces *differential medium*.
[b]*Number of analyzed samples (bottles) to be taken from a lot.*
[c]*A count which separates good quality from marginal quality.*
[d]*A count which if exceeded by any of the tested bottles would lead to holding the product.*
[e]*Maximum number of bottles which may fall into the marginally acceptable category before the lot is hold.*

in industry to finished bottled product before expedition, reflecting our practical experience and the data provided by wineries (Loureiro and Malfeito-Ferreira, 2003a). The lot may be defined as the number of samples (bottles) per bottling day of a single wine. If a wine lot does not meet the specifications it should be hold and analyzed on a weekly basis until counts drop to acceptable values. If counts in any of the indicators increase the bottled wine should be reprocessed to prevent product recall from the market.

15.4 CONTROL OF YEAST POPULATIONS IN WINES

The enologist may use a wide range of measures to prevent yeast spoiling activities. They comprise inhibitory or lethal agents applied to wine, like chemical preservatives and thermal treatments. Other physical operations, although not directed to kill microorganisms, have a removal effect like the operations of clarification, fining, or filtration. All control operations must be accompanied by adequate hygienic procedures to prevent wine contamination by yeasts colonizing winery surfaces. In addition, it should be kept in mind that wine by itself is a stressful environment. The intrinsic properties of each wine (wine susceptibility or, the opposite concept, wine robustness) determine the efficiency of each control measure. For instance, the lack of nutrients makes wine less vulnerable to microbial growth and ethanol content increases wine robustness but ethanol concentration is not modulated having in mind the control of yeast activity. Likewise, carbonation, used in some types of wines is inhibitory to most yeasts but is not directed to yeast control. With the opposite effect, oxygen may be added to improve wine aging but it also stimulates yeast contaminants. Moreover, in the wine industry the concept of hurdle preservation is highlighted by the need to decrease the utilization of sulfur dioxide associated with human allergies and subjected to stricter legal limits. In conclusion, proper management of microbial contaminants is dependent on an integrated approach of all factors affecting yeast growth.

15.4.1 Hygiene

The control of microbial contaminations can not be achieved without proper hygienic procedures from grape picking to the bottling process. Principles of sanitation, regarded as the sum of cleaning and disinfecting operations, and the description respective sanitizing agents can be found elsewhere (Lelieveld et al., 2005; Fugelsang and Edwards, 2007). Here we only make some remarks necessary to minimize the proliferation of spoilage yeasts.

All yeast species are sensitive to common winery disinfectants (hydrogen peroxide, peracetic acid, iodophors, chlorine). Their tolerance may be increased when proliferation occurs in biofilms, which protect microorganisms from their killing effect. Therefore, adequate cleaning and disinfection processes ensure the prevention of contaminations in the surfaces of equipment, vessels, pipes, and hoses. The formation of biofilms in wineries is not deeply studied (Joseph et al., 2007) contrarily to other food industries. Hygiene stringency is particularly important after sterile filtration or flash pasteurization to avoid contamination, or cross-contaminations when concentrated or sulfited grape juices are used.

Nevertheless, winemakers are aware that, in practical terms, it is frequent to have situations where proper hygiene is not possible. The sanitizing efficiency decreases from materials like stainless steel, concrete, plastic, to rubber due to higher surface roughness. The most difficult, or practically impossible, surface to sanitize properly is the wood used for wine maturation. In modern winemaking, oak barrels are widely used particularly for high quality red wines and pose the main difficulty to prevent wine contamination by *D. bruxellensis*. Sanitizing agents with chlorine must be avoided to prevent the formation of trichloroanisoles responsible for "cork taint." Most common treatments use hot water, sulfite solutions, steam, and ozone as cleaning and disinfecting agents. However, their efficiency is very limited due to the porous nature of the wood. The contamination of the outer layers of the wood may be significantly reduced but the inner layers, soaked by wine, still harbor yeast population able to recontaminate wine after sanitation (Barata et al., 2013). The critical points are the grooves and the surfaces between staves, where the sanitizing agents do not reach the microbial cells embedded in the wood. Therefore, recovery of infected barrels must include dismantling and removal of all parts soaked by wine.

15.4.2 Clarification, Fining, and Filtration

During storage wines are subjected to several operations directed to the improvement of storage and aging conditions (Renouf and Lonvaud-Funel, 2004). Clarification by settling or centrifugation leads to the reduction of suspended material including microorganisms. High-speed centrifugation may achieve practically sterile wines just after fermentation. Fining agents are usually directed to improve wine organoleptic characteristics removing microorganisms during wine settling as well. Filtration by diatomaceous earths is currently done during wine aging and the tightest earths drastically reduce microbial numbers. The correct management of clarification, fining, and filtration operations favors the minimization of chemical or thermal treatments during storage. When wine is ready to bottle, the prebottling filtration is the most common procedure to achieve wine "sterilization." Several types of filtering media may be used, depending on the winemakers' decision, but the ultimate goal is to prevent microbial growth in bottled wines. If coarser pore sizes are used higher levels of wine stabilizing treatments should be used. Some wines, particularly stylish premium red wines, may not be filtered or pasteurized due to claimed quality constraints, and so require a higher dose of preservatives to avoid microbial development in the bottled product. However, according to our experience, these unfiltered or coarsely filtered wines are the most frequently affected by "horse sweat taint" and by refermentation in the bottle.

15.4.3 Oxygen and Storage Temperature

Spoilage yeasts are facultative aerobes, which are stimulated by small amounts of oxygen. The effect of air in contact with wine is well known by the winemaker. If vessels are not topped a microbial film develops at the wine surface together with the development of an oxidized taint due to acetaldehyde formation. However, low amounts of oxygen are required for wine maturation, especially for red wines, which have led to the development of the so-called "micro-oxygenation" processes. Therefore, adequate management of all operations introducing oxygen in wine is required to minimize spoilage yeast growth. Inadequate bottling machines may introduce oxygen in bottled wine, which exponentially stimulates the growth of *Z. bailli* (Malfeito-Ferreira et al., 1990a,b). In oak barriques, oxygen continuously diffuses up to 30 mg/L per year through the wood (Ribéreau-Gayon et al., 2006) and stimulates growth of *D. bruxellensis* (Malfeito-Ferreira et al., 2001; Du Toit et al., 2006). Unadjusted cork jaws may affect corks providing channels of air into the bottled wine, reducing free sulfur dioxide and stimulating yeast growth.

Storage at low temperatures acts by delaying microbial growth but should not be regarded as a lethal agent, because most microorganisms grow when temperature increases.

15.4.4 Chemical Preservatives

In winery practice the control of microbial populations depends most effectively on the maintenance of adequate levels of molecular sulfur dioxide. In wines it is present either in the free or combined form. The active form is the molecular sulfite, which is calculated by multiplying the free amount by the proportion of molecular sulfite as a function of pH. However, sulfur dioxide can not be continuously added because it is subjected to maximum legal limits. In the EU the limits are a function of wine type (Loureiro and Malfeito-Ferreira, 2003a) and in the United States the maximum level of total sulfite is 350 mg/L (Fugelsang and Edwards, 2007). After addition a part of sulfite is combined and loses its antimicrobial activity. All conditions leading to sulfite combination must be minimized. The use of high doses of sulfur dioxide before fermentation can increase the production of acetaldehyde by fermenting yeasts. The combination rate may be 50% or more of the added amount. Therefore additions should be controlled by sulfite measurement after treatment. To prevent microbial growth common advised levels are 0.5–0.8 mg/L of molecular sulfur dioxide (Fugelsang and Edwards, 2007) but yeasts vary on the resistance to this preservative (Table 15.2). In addition, growing populations are more resistant, sulfur dioxide being required at 1 mg/L to prevent the proliferation of *D. bruxellensis* (Barata et al., 2007; Chandra et al., 2015).

Sorbic acid is a weak acid, whose free form is present in higher proportions at lower pH values. The maximum legal limits are 200 mg/L in the EU and 300 mg/L in the USA (Fugelsang and Edwards, 2007). Due to higher solubility, potassium sorbate is used as the vehicle of sorbic acid. Its usage is advised, together with sulfur dioxide, at bottling of sweet wines to inhibit fermentation yeasts. It is metabolized by lactic acid bacteria originating the "geranium taint." At the maximum legal doses it is not effective against *D. bruxellensis* and *Z. bailii* (Table 15.2).

Given the low range of wine pH variations and the tolerance of most yeast species to acidic pH values, wine acidification with tartaric acid contributes to yeast control mainly by increasing the undissociated forms of acidic preservatives.

Dimethyl dicarbonate (DMDC) has been recently approved in the EU for use at the maximum amount of 200 mg/L at bottling of wines with more than 5 g/L of residual sugar. In the United States it may be used during

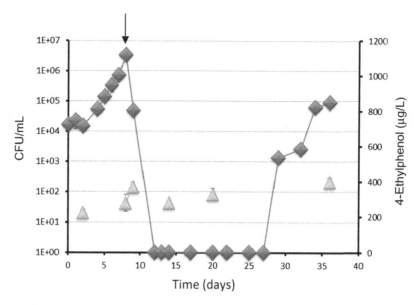

FIGURE 15.3 Effect of 0.1 g/L chitosan on the viability of growing cells *D. bruxellensis* (◇) and on 4-ethylphenol (△) production. Arrow: moment of chitosan addition. Absence of viable cells (CFU <1/mL) is indicated as 1 CFU/mL due to the logarithmic scale of the Y-axis (unpublished results).

the storage of wine in regular amounts up to the maximum level of 200 mg/L (Fugelsang and Edwards, 2007). Its efficiency depends on the initial microbial contamination, with it being advised to be used at a maximum of 500 viable cells/mL of wine. Yeasts vary in their susceptibility to DMDC (Daudt and Ough, 1980 and Table 15.2). *S. pombe* and *S. cerevisiae* were found to be more resistant than *D. bruxellensis* or *Z. bailii* (Costa et al., 2008). Bacteria are more resistant than yeasts and so this preservative should not be regarded as a sterilant when used alone (Costa et al., 2008). Therefore, in the winery routine DMDC should be used together with sulfite during wine storage or at bottling. Its activity depends on adequate homogenization achieved by a costly dosing apparatus. Another precaution is with regards to its human toxicity (Fugelsang and Edwards, 2007).

Chitosan has also been recently approved by the EU to be used in wines up to a maximum amount of 0.1 g/L. It has been claimed to be an efficient preservative against *D. bruxellensis* (Taillandier et al., 2015) but its efficiency depends on the initial contamination load and on the wine susceptibility to microbial growth. As yeasts may grow after an initial death phase (Fig. 15.3) the addition of sulfur dioxide as a complement is advised. The absence of culturable cells may indicate the presence of a "Viable but nonculturable" state but it may also only reflect viable populations below the detection threshold, as shown in Fig. 15.3.

15.4.5 Thermal Treatments

Several thermal treatments may be applied in wine processing with or without deliberate effects on microorganisms. Wine refrigeration may reduce microbial loads but the aim is to stabilize bitartrates and microbes are removed in the process. Thermovinification consists of heating crushed red grapes and separating the heavily colored juice to be fermented without maceration (Ribéreau-Gayon et al., 2006). The purpose is to extract coloring matter not to destroy contaminant microflora. However, concerning spoilage yeasts, this technique is specially appropriate for processing rotten grapes because it kills all contaminating microorganisms and enables fermentations dominated by *S. cerevisiae*.

Surprisingly, heat treatments are rarely used in the wine industry in comparison to other beverage industries (beer, juices, and soft drinks). In particular, heat treatments should be current options mainly for wines with residual sugar. In flash pasteurization the wine is heated and cooled in plate exchangers and may be sterile filtered before bottling to avoid recontaminations. In hot bottling, or thermolization, wine is heated and bottled at the desired temperature, cooling after bottling (Humbert, 1980). Wine microorganisms are heat sensitive and so relatively mild temperatures are enough to ensure product sterilization (Devéze and Ribéreau-Gayon, 1977). However, winemakers are generally reluctant to use them due to the claimed deleterious effects on wine organoleptical quality and longevity (shelf life).

15.5 FUTURE TRENDS

The present tendency to increase the adoption of quality management programs by the food industry will also be reflected in the wine industry despite the practical absence of food safety hazards. Then, these programs should be mainly directed to the prevention of wine spoilage by microorganisms where yeasts play the most important role. The wine industry has several flaws that make difficult the implementation of sound HACCP or other quality assurance programs which should be overcome in the near future by:

1. Defining microbiological indicators, which are the basis of microbiological criteria.
2. Standardizing microbiological criteria, namely sampling by attributes, using standard analytical methods to enable the definition of appropriate microbiological specifications.
3. Defining spoilage risks according to wine robustness that determines the choice of appropriate microbiological specifications.
4. Using rapid detection methods to enable prompt corrective measures.

From a scientific point of view, the primary future developments should include:

1. Deeper knowledge of the microbial ecology of grapes, particularly damaged grapes, of wineries, and of vectors is essential to have a full understanding of the origin and dissemination of spoilage yeasts in wine environments;
2. Establishment of different categories of wine robustness, based on the presence of substrates easily attacked by microorganisms, inhibitors, nutrient availability, or environmental stresses, to enable the development of predictive models of wine spoilage.

References

Adams, M.R., Moss, M.D., 2000. Food Microbiology. The Royal Society of Chemistry, Cambridge, UK.

Andrews, S., 1992. Specifications for yeasts in Australian beer, wine and fruit juice products. In: Samson, R.A., Hocking, A.D., Pitt, J.I., King, A.D. (Eds.), Modern Methods in Food Mycology. Elsevier, Amsterdam, The Netherlands, pp. 111–118.

Barata, A., Correia, P., Nobre, A., Malfeito-Ferreira, M., Loureiro, V., 2006. Growth and 4-ethylphenol production by the yeast *Pichia guilliermondii* in grape juices. Am. J. Enol. Vitic. 57, 133–138.

Barata, A., Caldeira, J., Botelheiro, R., Pagliara, D., Malfeito-Ferreira, M., Loureiro, V., 2007. Survival patterns of *Dekkera bruxellensis* in wines and inhibitory effect of sulphur dioxide. Int. J. Food Microbiol. 121, 201–207.

Barata, A., González, S., Malfeito-Ferreira, M., Querol, A., Loureiro, V., 2008. Sour rot-damaged grapes are sources of wine spoilage yeasts. FEMS Yeast Res. 8, 1008–1017.

Barata, A., Malfeito-Ferreira, M., Loureiro, V., 2012a. Changes in sour rotten grape berry microbiota during ripening and wine fermentation. Int. J. Food Microbiol. 154, 152–161.

Barata, A., Malfeito-Ferreira, M., Loureiro, V., 2012b. The microbial ecology of wine grape berries. Int. J. Food Microbiol. 153 (3), 243–259.

Barata, A., Laureano, P., D'Antuono, I., Martorell, P., Stender, H., Malfeito-Ferreira, M., et al., 2013. Enumeration and identification of 4-ethylphenol producing yeasts recovered from the wood of wine ageing barriques after different sanitation treatments. J. Food Res. 2, 140–149.

Bell, S.-J., Henschke, P.A., 2005. Implications of nitrogen nutrition for grapes, fermentation and wine. Aust. J. Grape Wine Res. 11, 242–295.

Bisson, L.F., Butzke, C.E., 2000. Diagnosis and rectification of stuck and sluggish fermentations. Am. J. Enol. Vitic. 51, 168–177.

Blackburn, C. (Ed.), 2006. Food Spoilage Microorganisms. Woodhead Pub. Ltd, Cambridge, England.

Boekhout, T., Robert, V. (Eds.), 2003. Yeasts in Food: Beneficial and Detrimental Aspects. B. Behr's Verlag, Hamburg.

Boekhout, T., Robert, V., Smith, M., Stalpers, J., Yarrow, D., Boer, P., et al., 2002. Yeasts of the World — Morphology, Physiology, Sequences and Identification. CD-ROM from ETI Information Services, Ltd, Wokingham, UK.

Chandra, M., Oro, I., Ferreira-Dias, S., Malfeito-Ferreira, M., 2015. Effect of ethanol, sulfur dioxide and glucose on the growth of wine spoilage yeasts using response surface methodology. PLoS ONE. Available from: https://doi.org/10.1371/journal.pone.0128702.

Chatonnet, P., Boidron, J., Dubourdieu, D., 1993. Influence des conditions d'élevage et de sulfitage des vins rouges en barriques sur le teneur en ácide acétique et en ethyl-phenols. J. Int. Sci. Vigne Vin. 27, 277–298.

Combina, M., Mercado, L., Borgo, P., Elia, A., Jofre, V., Ganga, A., et al., 2005. Yeasts associated to Malbec grape berries from Mendoza, Argentina. J. Appl. Microbiol. 98, 1055–1061.

Connel, L., Stender, H., Edwards, C., 2002. Rapid detection and identification of *Brettanomyces* from winery air samples based on peptide nucleic acid analysis. Am. J. Enol. Vitic. 53, 322–324.

Costa, A., Barata, A., Malfeito-Ferreira, M., Loureiro, V., 2008. Evaluation of the inhibitory effect of dimethyl dicarbonate (DMDC) against microorganisms associated with wine. Food Microbiol. 25, 422–427.

Daudt, C.E., Ough, C.S., 1980. Action of dimethyldicarbonate on various yeasts. Am. J. Enol. Vitic. 31, 21–23.

Davenport, R., 1986. Unacceptable levels for yeasts. In: King, A., Pitt, J., Beuchat, L., Corry, J. (Eds.), Methods for the Mycological Examination of Food. Plenum Press, New York, NY, pp. 214–215.

Deak, T., Beuchat, L., 1996. Handbook of Food Spoilage Yeasts. CRC Press, Boca Raton, FL, USA.

Deak, T., Reichart, O., 1986. Unacceptable levels of yeasts in bottled wine. In: King, A., Pitt, J., Beuchat, L., Corry, J. (Eds.), Methods for the Mycological Examination of Food. Plenum Press, New York, NY, pp. 215–218.

Delfini, C., Formica, J., 2001. Wine Microbiology: Science and Technology. Marcel Dekker, New York, NY, USA.

Delfini, C., Gaia, P., Schellino, R., Strano, M., Pagliara, A., Ambrò, S., 2002. Fermentability of grape must after inhibition with dimethyl dicarbonate (DMDC). J. Agric. Food Chem. 50, 5605–5611.

Devéze, M., Ribéreau-Gayon, P., 1977. Thermoresistance dês levures dans le vin application à la stabilisation biologique dês vins par la chaleur. Connaiss. Vigne Vin 11, 131–163.

Dias, L., Dias, S., Sancho, T., Stender, H., Querol, A., Malfeito-Ferreira, M., et al., 2003a. Identification of yeasts isolated from wine related environments and capable of producing 4-ethylphenol. Food Microbiol. 20, 567–574.

Dias, L., Pereira-da-Silva, S., Tavares, M., Malfeito-Ferreira, M., Loureiro, V., 2003b. Factors affecting the production of 4-ethylphenol by the yeast *Dekkera bruxellensis* in enological conditions. Food Microbiol. 20, 377–384.

Donnelly, D., 1977a. Airborne microbial contamination in a winery bottling room. Am. J. Enol. Vitic. 28, 176–181.

Donnely, D., 1977b. Elimination from table wines of yeast contamination by filling machines. Am. J. Enol. Vitic. 28, 182–184.

Du Toit, W.J., Lisjak, K., Marais, J., du Toit, M., 2006. The effect of micro-oxygenation on the phenolic composition, quality and aerobic wine-spoilage microorganisms of different South African red wines. S. Afr. J. Enol. Vitic. 27, 57–67.

Esch, F., 1992. Yeast in soft drinks and concentrated fruit juices. Brygmesteren 4, 9–20.

Farris, G., Zara, S., Pinna, G., Budroni, M., 2002. Genetic aspects of flor yeasts *Sardinian strains*, a case of study. In: Ciani, M. (Ed.), Biodiversity and Biotechnology of Wine Yeasts. Research Signpost, Kerala, India, pp. 71–83.

Fleet, G., Prakitchaiwattana, C., Beh, A., Heard, G., 2002. The yeast ecology of wine grapes. In: Ciani, M. (Ed.), Biodiversity and Biotechnology of Wine Yeasts. Research Signpost, Kerala, India, pp. 1–17.

Florenzano, G., Balloni, W., Materassi, R., 1977. Contributo alla ecologia dei lieviti *Schizosaccharomyces* sulle uve. Vitis 16, 38–44.

Freeman, B., Muller, C., Kepner, R., Webb, A., 1977. Some products of the metabolism of ethyl 4-oxobutanoate by *Saccharomyces fermentati* in the film form on the surface of simulated flor sherry. Am. J. Enol. Vitic. 28, 119–122.

Fugelsang, K., Edwards, C., 2007. Wine Microbiology. Springer Verlag, Berlin, Germany.

Giudici, P., Pulvirenti, A., 2002. Molecular methods for identification of wine yeasts. In: Ciani, M. (Ed.), Biodiversity and Biotechnology of Wine Yeasts. Research Signpost, Kerala, India, pp. 35–52.

Guerzoni, E., Marchetti, R., 1987. Analysis of yeast flora associated with grape sour rot and of the chemical disease markers. Appl. Environ. Microbiol. 53, 571–576.

Heard, G., Fleet, G., 1986. Evaluation of selective media for enumeration of yeasts during wine fermentation. J. Appl. Bacteriol. 60, 477–481.

Hittinger, C., Rokas, A., Bai, F.-Y., Boekhout, T., Gonçalves, P., Jeffries, T., et al., 2015. Genomics and the making of yeast biodiversity. Curr. Opin. Genet. Dev. 35, 100–109.

Humbert, C., 1980. Thermolisation: its effects on wine. Rev. Fr. Oenol. 16, 51–53.

Ibeas, J.I., Lozano, I., Perdigones, F., Jiménez, J., 1996. Detection of *Dekkera-Brettanomyces* strains in sherry by a nested PCR method. Appl. Environ. Microbiol. 62, 998–1003.

Ison, R., Gutteridge, C., 1987. Determination of the carbonation tolerance of yeasts. Lett. Appl. Microbiol. 5, 11–13.

Joseph, L., Kumar, G., Su, E., Bisson, L., 2007. Adhesion of biofilm production by wine isolates of *Brettanomyces bruxellensis*. Am. J. Enol. Vitic. 58, 373–378.

King, A., Pitt, J., Beuchat, L., Corry, J. (Eds.), 1986. Methods for the Mycological Examination of Food. Plenum Press, New York, NY.

Kuhn, F., Mester, V., Morris, R., Dalma, J., 2004. Serious eye injuries caused by bottles containing carbonated drinks. Br. J. Ophthalmol. 88, 69–71.

Kunkee, R., Bisson, L., 1993. Wine-making yeasts. In: Rose, A., Harrison, J. (Eds.), The Yeasts, vol. 5, second ed. Academic Press, London, pp. 69–127.

Kurtzman, C., Fell, J. (Eds.), 1998. The Yeasts, A Taxonomic Study, fourth ed. Elsevier, Amsterdam, The Netherlands.

Kurtzman, C., James, S., 2006. *Zygosaccharomyces* and related genera. In: Blackburn, C. (Ed.), Food Spoilage Microorganisms. Woodhead Publishers, Cambridge, UK, pp. 289–305.

Lelieveld, H., Mostert, M., Holah, J. (Eds.), 2005. Handbook of Hygiene Control in the Food Industry. Woodhead Publishing Limited, Cambridge, UK.

Loureiro, V., 1997. Spoilage yeasts in foods and beverages. Final Scientific Report of the European AIR Project CT93/830. DGXII-E-2, Bruxelles.

Loureiro, V., Malfeito-Ferreira, M., 2003a. Spoilage yeasts in the wine industry. Int. J. Food Microbiol. 86, 23–50.

Loureiro, V., Malfeito-Ferreira, M., 2003b. Yeasts in Spoilage. In: Caballero, B., Trugo, L., Finglas, P. (Eds.), Encyclopedia of Food Sciences and Nutrition, second ed Academic Press, London, pp. 5530–5536.

Loureiro, V., Malfeito-Ferreira, M., 2006. Spoilage activities of *Dekkera/Brettanomyces* spp. In: Blackburn, C. (Ed.), Food Spoilage Microorganisms. Woodhead Publishers, Cambridge, pp. 354–398.

Loureiro, V., Querol, A., 1999. The prevalence and control of spoilage yeasts in foods and beverages. Trends Food Sci. Technol. 10/11, 356–365.

Loureiro, V., Malfeito-Ferreira, M., Carreira, A., 2004. Detecting spoilage yeasts. In: Steele, R. (Ed.), Understanding and Measuring the Shelf-Life of Food. Woodhead Publishers, Cambridge, pp. 233–288.

Malfeito-Ferreira, M., 2011. Yeasts and wine off-flavours: a technological perspective. Ann. Microbiol. 61, 95–102.

Malfeito-Ferreira, M., Lopes, J., Loureiro, V., 1990a. Characterization of spoilage yeasts in Portuguese bottled dry white wines. In: Ribereau-Gayon, P., Lonvaud, A. (Eds.), Actualités Oenologiques 89, Comptes rendus du 4eme Symposium International d'Oenologie. Dunod, Bordeaux, Paris, pp. 293–296.

Malfeito-Ferreira, M., Wium St., H., Aubyn, A., Loureiro, V., 1990b. Rapid characterization of yeasts contaminants associated with sparkling wine production. Ind. Bevande 19, 504–506.

Malfeito-Ferreira, M., Tareco, M., Loureiro, V., 1997. Fatty acid profiling: a feasible typing system to trace yeast contaminations in wine bottling plants. Int. J. Food Microbiol. 38, 143–155.

Malfeito-Ferreira, M., Rodrigues, N., Loureiro, V., 2001. The influence of oxygen on the "horse sweat taint" in red wines. Ital. Food Beverage Technol. 24, 34–38.

Minarik, E., 1983. Levures de contamination des vins embouteillés. Bull. O.I.V. 56 (628), 414–419.

Neradt, F., 1982. Sources of reinfections during cold-sterile bottling of wine. Am. J. Enol. Vitic. 33, 140–144.

Neuser, F., Zorn, H., Berger, R., 2000. Generation of odorous acyloins by yeast pyruvate decarboxylases and their occurrence in sherry and soy sauce. J. Agric. Food Chem. 48, 6191–6195.

Pardo, I., García, M.J., Zúniga, M., Uruburu, F., 1989. Dynamics of microbial populations during fermentation of wines from the Utiel-Requena Region of Spain. Appl. Environ Microbiol. 55, 539–541.

Pitt, J., Hocking, A., 1985. Fungi and Food Spoilage. Academic Press, Sydney, Australia.

Querol, A., Fleet, G. (Eds.), 2006. Yeasts in Food and Beverages. Springer-Verlag, Berlin, Germany.

Renouf, V., Lonvaud-Funel, A., 2004. Racking are key stages for the microbial stabilization of wines. J. Int. Sci. Vigne Vin 38, 219–224.

Renouf, V., Lonvaud-Funel, A., 2007. Development of an enrichment medium to detect *Dekkera/Brettanomyces bruxellensis*, a spoilage wine yeast, on the surface of grape berries. Microbiol. Res. 162, 154–157.

Renouf, V., Perello, M.-C., Revel, G., Lonvaud-Funel, A., 2007. Survival of wine microorganisms in the bottle during storage. Am. J. Enol. Vitic. 58, 379–386.

Ribéreau-Gayon, P., Dubourdieu, D., Donèche, B., Lonvaud, A., 2006. Handbook of Enology, The Microbiology of Wine and Vinification, vols. 1 and 2. John Wiley and Sons, Ltd, Chichester, UK.

Rodrigues, N., Gonçalves, G., Pereira-da-Silva, S., Malfeito-Ferreira, M., Loureiro, V., 2001. Development and use of a new medium to detect yeasts of the genera *Dekkera/Brettanomyces* spp. J. Appl. Microbiol. 90, 588–599.

Romano, P., 2005. Proprietà technologiche e di qualità delle specie di lieviti vinari. In: Vicenzini, M., Romano, P., Farris, G. (Eds.), Microbiologia del vino. Casa Editirice Ambrosiana, Milano, Italy, pp. 101–131.

Schuller, D., Côrte-Real, M., Leão, C., 2000. A differential medium for the enumeration of the spoilage yeast *Zygosaccharomyces bailii* in wine. J. Food Prot. 63, 1570–1575.

Schumaker, M., Chandra, M., Malfeito-Ferreira, M., Ross, C., 2017. Influence of *Brettanomyces* ethylphenols on red wine aroma evaluated by consumers in the United States and Portugal. Food Res. Int. 100 (1), 161–167.

Splittstoesser, D.F., Queale, D.T., Mattick, L.R., 1978. Growth of *Saccharomyces bisporus* var. *bisporus*, a yeast resistant to sorbic acid. Am. J. Enol. Vitic. 29, 272–276.

Steels, H., James, S., Roberts, I., Stratford, M., 1999. *Zygosaccharomyces lentus*: a significant new osmophilic, preservative-resistant spoilage yeast, capable of growth at low temperature. J. Appl. Microbiol. 87, 520–527.

Stratford, M., 2006. Food and beverage spoilage yeats. In: Querol, A., Fleet, G. (Eds.), Yeasts in Food and Beverages. Springer-Verlag, Berlin, Germany, pp. 335–379.

Suárez, R., Suárez-Lepe, J., Morata, A., Calderón, F., 2007. The production of ethylphenols in wine by yeasts of the genera *Brettanomyces* and *Dekkera*: a review. Food Chem. 102 (1), 10–21.

Suárez-Lepe, J., Iñigo-Leal, B., 2004. Microbiología Enológica: Fundamentos de Vinificación, third ed Ediciones Mundi-Prensa, Madrid, Spain.

Taillandier, P., Joannis-Cassan, C., Jentzer, J.-B., Gautier, S., Sieczkowski, N., Granes, D., et al., 2015. Effect of a fungal chitosan preparation on *Brettanomyces bruxellensis*, a wine contaminant. J. Appl. Microbiol. 118, 123–131.

Thomas, D.S., 1993. Yeasts as spoilage organisms in beverages. In: Rose, A., Harrison, J. (Eds.), The Yeasts, vol. 5. Academic Press, London, pp. 517–561.

Valero, E., Schuller, D., Cambon, B., Casal, M., Dequin, S., 2005. Dissemination and survival of commercial wine yeast in the vineyard: a large-scale, three-years study. FEMS Yeast Res. 5, 959–969.

Vendrame, M., Manzano, M., Comi, G., Bertrand, J., Lacumin, L., 2014. Use of propidium monoazide for the enumeration of viable *Brettanomyces bruxellensis* in wine and beer by quantitative PCR. Food Microbiol. 42, 196–204.

CHAPTER

16

Red Wine Clarification and Stabilization

Aude Vernhet

Montpellier SupAgro, Institute for Higher Education in Vine and Wine Sciences, Joint Research Unit Sciences for Enology, Montpellier, France

Wines after alcoholic and malolactic fermentation are turbid and unstable media that need to be clarified and stabilized to preserve their quality until their consumption. Beside solutes and macromolecules, they include a wide diversity of particles, responsible for hazes and deposits. These particles are mainly microorganisms (yeast and bacteria), tartrate crystals, grape skin and pulp cell debris, and aggregates of molecules/macromolecules that have formed during fermentation and maceration (Fig. 16.1). Within these particles, it is important to distinguish between colloidal particles, that form colloidal dispersions, and larger ones, that form suspensions and deposits.

Clarification operations allow to achieve limpidity and brightness, which are of importance as the first visual impression strongly impacts the whole perception of wine quality. Hazes and sediments, even when they do not affect taste, have a detrimental impact: most of the consumers will reject the product. Clarification also contributes to decrease microorganism populations. Stabilization treatments are necessary to preserve the characteristics and the quality of wines from their bottling to their consumption, regardless of transport and storage conditions. They have different objectives: avoid the formation of hazes or deposits in bottled wines, and thus preserve limpidity, and prevent qualitative alterations of taste, flavor, or color related to spoilage by microorganisms or to negative chemical changes.

This chapter will focus on the origin of the main physicochemical instabilities in red wines and on the tests used to assess the risks, on the different clarification and stabilization treatments and on their positioning. Colloids, colloidal interactions, and colloidal equilibria, which play a determinant part in physicochemical stability as well as in the efficiency and/or selectivity of stabilization and clarification treatments, will be briefly presented first. Specific risks related to microbial spoilage, as well as the role of SO_2 or alternative additives on microbial and chemical stabilization, addressed in other chapters of this book, won't be detailed here. The treatments needed and their extent usually strongly differ depending on the aging conditions and lengths, owing to spontaneous clarification and stabilization of wines. Final clarification and stabilization, performed during the weeks before bottling, must then be reasoned depending on the wine type and style as well as on the market specifications.

16.1 COLLOIDS AND COLLOIDAL INSTABILITIES IN RED WINES

16.1.1 Colloids and Colloidal Interactions

The term colloids refers to macromolecules and finely subdivided particles that have some linear dimensions between 10^{-9} and 10^{-6} m (1 nm to 1 μm). Whereas macromolecules, as dissolved components, form colloidal solutions (one phase), colloidal particles form biphasic systems called dispersions. Colloidal macromolecules in wines are polysaccharides, oligomeric and polymeric polyphenols and proteins. Colloidal particles can be small aggregates arising from the precipitation of molecules/macromolecules, grape cell debris, bacteria, etc. Contrarily to that observed with largest particles, Brownian motion has a more important effect on the behavior of colloidal particles than sedimentation: they do not settle spontaneously and develop high specific surface areas that make

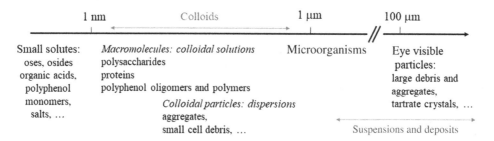

FIGURE 16.1 Size range of wine constituents and particles.

them, as macromolecules, very sensitive to physicochemical interactions (Hiemenz, 1986), in solutions or at interfaces.

The stability or instability of colloidal systems is governed by the long-range physicochemical interactions (between a few to a few tens of nm) that develop between macromolecules and/or particles. These interactions can be described first for a very simple system, i.e., spherical colloidal particles in an aqueous solvent and separated by a distance d (in nm). According to the extended Derjaguin, Landau, Verwey, and Overbeek theory, the total interaction potential between these colloids is the sum of the potentials related to Lifshitz–van der Waals, electrostatic, and polar hydrophobic/hydrophilic interactions (Hiemenz, 1986; Israelachvili, 1992; van Oss, 1994).

Lifshitz–van der Waals forces between colloids are usually attractive and small in aqueous solvents: they lead to an attractive interaction potential that is maximum at contact and decreases with the separation distance d. Electrostatic interactions take place when colloids carry surface charges, related to the presence of ionized groups or to ion adsorption. These charges induce an accumulation of counterions in the liquid that surrounds the colloidal particles, and the existence of an electrical double layer. Electrostatic interactions are related to the overlapping of these electrical double layers when the colloidal particles get closer. They are repulsive if they carry the same charge and attractive otherwise. For a given surface charge, electrostatic interactions and especially their range strongly depend on the ionic strength of the medium, and thus on its ionic composition. Their impact decreases as the ionic strength increases, owing to the effect of this parameter on the double layer thickness. Within the wine ionic strength range (0.01 to 0.1 M), different ionic compositions may lead to different impacts of electrostatic interactions on colloidal equilibria. In aqueous solvents, polar interactions are related to H-bonds. Hydrophilic colloids have a strong affinity for water, leading to the existence of a water layer bound to the surface. When two hydrated surfaces get closer, this bound water creates an "hydrophilic repulsion" that varies with the distance according to an exponential law. By contrast, water molecules orient themselves around hydrophobic colloids in a way which is unfavorable from a thermodynamic point of view, i.e., more ordered than in the bulk. Water molecules thus tend to spontaneously exclude themselves from such surfaces, which leads to a long-range "hydrophobic" attraction. When dealing with wines, which are hydroalcoholic solutions, the presence of ethanol strongly decreases the cohesion of the solvent, and modulates the impact of "hydrophilic/hydrophobic" polar interaction forces (Poncet et al., 2003).

The respective impact of van der Waals, electrostatic, and polar "hydrophilic/hydrophobic" interactions depends on the physicochemical properties of the interacting species (charge, polarity), on their shape and size (radius for a spherical particle), on the suspending medium (pH, ionic strength, ethanol content), and on the temperature. Depending on this impact, the total interaction potential U may evolve as a function of the separation distance (d) in very different ways (Fig. 16.2). The result can be: (1) a strong attraction ($U(d) < 0$), leading to fast aggregation; (2) a strong repulsion ($U(d) > 0$), leading to stability; (3) a more complex interaction potential with the existence of a secondary minimum at a finite distance, of a more or less high energy barrier at smaller distances and of a strong primary minimum at very short distances. In this latter case the stability of the system, and for unstable systems aggregation kinetics, will be dependent on the energy barrier height and of the ability of the colloidal particles to overcome it. Aggregation kinetics also depend on the concentration of colloids and of their probability to collide. In diluted systems such as wine, they may develop according to very slow kinetics.

The same primary long-range interactions are responsible for the colloidal behavior of macromolecules (aggregation, adsorption at interfaces). The latter however can hardly be described as spherical particles with given surface properties. Their physicochemical interactions are related to the presence of interaction sites or areas (hydrophilic/hydrophobic, charged or not) and are strongly dependent on their structural characteristics. Among these characteristics are: (1) their dimensions (molecular weight), structure (linear versus branched polymers)

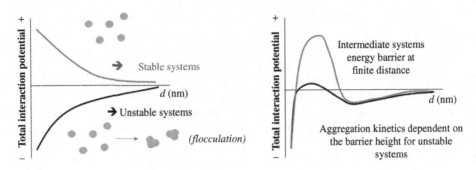

FIGURE 16.2 Evolution of the total interaction potential between two colloidal particles as a function of the distance and of the respective impact of long-range Lifshitz–van der Waals, electrostatic, and polar interactions. A positive interaction potential results in repulsion between particles whereas a negative interaction potential results in attraction.

and conformation in solution (more or less extended and flexible); and (2) the nature and distribution of their constitutive units (hydrophilic/hydrophobic units and their distribution, charged groups, ...).

Attractive interactions between macromolecules may lead to the formation of soluble "macromolecular complexes" or to that of colloidal aggregates, which can be assimilated to particles (Mekoue Nguela et al., 2016; Poncet et al., 2003; Riou et al., 2002; Zanchi et al., 2007). Aggregates may stop growing at finite submicronic sizes due to the presence of repulsive interactions at a given separation distance. Whether or not an eye-visible haze appears will depend on the average size and concentration of these particles. Contrarily, particle size may keep on growing due to the flocculation of the previously formed aggregates, and then start settling to form deposits. For systems such as those encountered in wine-making, in which both macromolecules and colloidal particles are present, interactions between macromolecules and particles can participate in the stabilization of particles, the macromolecules acting as "protective colloids" or contrarily trigger their flocculation.

16.1.2 Colloidal Instabilities in Red Wines and Their Prevention

Colloidal instabilities in red wines are often attributed to the aggregation of colloidal coloring matter, as hazes and deposits present a red color. Such aggregations may develop during winemaking and aging, or latter in the bottled wine. To assess their exact origin and the structural characteristic of the involved components is however a challenging task and there is so far only limited information. This is related to the difficulties encountered to fully solubilize and analyze these hazes or deposits, especially for polyphenols, to the structural diversity of red wine macromolecules, and to the variability of the wine matrix.

Colloidal macromolecules in red wines are mostly polyphenol oligomers and polymers and polysaccharides. It has long been considered that must-soluble grape proteins were fully precipitated by tannins during the winemaking process and removed with lees, and then that red wines were devoid of proteins. However, there is some evidence that residual proteins may be found in red wines, even if only in very low amounts (Mainente et al., 2014; Wigand et al., 2009). Both polysaccharides and polyphenols constitute complex mixtures of macromolecules with different structures (Cheynier et al., 2006; Vidal et al., 2003), the composition of which is dependent on the grape variety and maturity, as well as on the winemaking process. Maceration lengths, use of enzymes, choice of the strains used to perform the alcoholic and malolactic fermentations, oxygen management or aging on lees are for example operations that strongly impact wine composition in polysaccharides and polyphenols and that contribute to increase its diversity and the complexity of colloidal equilibria. When dealing with polyphenols, the polydispersity of the oligomeric and polymeric compounds is increased during winemaking and aging as these compounds undergo several biochemical and chemical changes, leading to the formation of so-called derived pigments and tannins (Fulcrand et al., 2006). These changes are of importance as they induce strong modifications of the chemical structure of the polyphenol units, of the molecular weight distribution of tannins and pigments and of their conformation (de Freitas and Mateus, 2011; Mouls and Fulcrand, 2012; Poncet-Legrand et al., 2010) that are likely to affect their solubility as well as their interactions with other wine colloidal components.

Development of colloidal instabilities related to colloidal coloring matter can then be simply related to the formation of less soluble species that tend to coaggregate progressively during wine aging, according to kinetics dependent on their concentration and on physicochemical conditions. Colloidal instabilities may also involve interactions with other wine components, such as polysaccharides and proteins. Thus, Prakash et al. (2016) have shown that precipitates related to colloidal unstable compounds in barrel-aged red wines include polyphenols and pectic polysaccharides, along with solutes such as organic acids and free amino acids that must be associated with the aggregates. These findings differ from previous ones, where thin lacquer-like colloidal deposits formed on the inner surface of bottled red wines were shown to involve grape proteins (around 10%) associated with pigmented polyphenols, but no polysaccharides (Waters et al., 1994).

The most widely used way to assess the risks of colloidal instability development in bottled red wines are cold tests. Low temperatures accelerate the aggregation and precipitation of sparingly soluble species. A preclarified wine sample is placed several days (between 2 and 6 days) at low temperatures (around 4°C) and the turbidity measured after the test correlates to colloidal coloring matter instability. Other time-temperature conditions can be found in the literature. Heat-tests have also been proposed to check instabilities related to proteins and polyphenols (Peng et al., 1996a): wines are heat-treated at 84°C for 16 hours, cooled at 23°C, and the turbidity measured after 4 hours, as in cold tests. Unstable wines can be stabilized by fining, cold treatments or the use of an additive.

Fining with proteins such as gelatin and egg albumin may be sufficient to remove unstable pigments and stabilize the wines. Protein fining when dealing with red wines involves the development of attractive physico-chemical interactions between proteins and polyphenols, especially oligomeric and polymeric tannins and pigments, leading to their aggregation and precipitation (Haslam, 1998; Maury et al., 2001, 2003). Interactions in this case are mainly driven by hydrophobic attraction, and strengthened by H-bond formations between the H-donor hydroxyl groups of polyphenols and protein H-acceptor groups when at short distances. Protein fining contributes then to decrease the wine content of tannins and polymerized pigments. However, for wines with high content of colloidal coloring matter, the use of bentonite is recommended. Bentonites are negatively charged clay particles and their interactions with wine constituents involve electrostatic attraction with positively charged compounds. Their effect on anthocyanins and pigments is usually higher than the one observed with proteins (Ghanem et al., 2017; Gonzales-Neves et al., 2014). Besides fining, cold treatments, usually performed to prevent the crystallization of tartaric salts, are also effectives to prevent precipitations of colloidal coloring matter in bottled wines. Their effectiveness strongly depends on the wine, the applied temperature and the treatment length. Thus, between 1 and 5 days at −4°C or 15 days at 2°C have been proposed to prevent the formation of lacquer-like deposits (Peng et al., 1996b). Another possibility is the use of an additive, Arabic gum, added at bottling at doses between 10 and 20 g/hL. The impact of this protective colloid is attributed to its ability to coat colloidal particles, preventing their growth by flocculation and thus the formation of visible hazes or of deposits.

16.2 WINE CLARIFICATION

The objective of wine clarification is to remove particles responsible for hazes and deposits, while preserving the overall quality of the wine, related to small solutes and to macromolecules. Indeed, colloidal macromolecules and colloidal equilibria in wines, if they can be involved in the formation of hazes and precipitates and affect the efficiency of clarification and stabilization treatments (Gautier, 2015; Gerbaud et al., 1997; Maury et al., 2016; Vernhet and Moutounet, 2002), also play a determinant part in the wine quality owing to their impact on both mouthfeel and aroma perception (Diako et al., 2016; Scollary et al., 2012; Villamor and Ross, 2013). The level of particles in wines, and their limpidity or brightness, is evaluated in practice by turbidity measurements (nephelometer). Young red wines after fermentation present turbidities of several hundred nephelometric turbidity units (NTU), related to high wet suspended solids contents (0.5%−2% weight/weight). Target values to obtain bright red wines are in the order of or below two NTU.

16.2.1 Clarification by Settling, With or Without Fining Aids

According to Stokes' law, the settling velocity v_p of a spherical particle in a diluted suspension (no interferences between particles) and a fluid at rest (lack of convection motions) is given by:

$$v_p (\text{m/s}) = \frac{D^2}{18\mu} \cdot \left(\rho_p - \rho_f\right) \cdot g \tag{16.1}$$

where D is the particle diameter (m), ρ_p and ρ_f the particle and fluid mass densities (kg/m^3), μ the fluid dynamic viscosity (Pa.s), and g the acceleration related to gravity (m/s^2). Though only indicative owing to the hypotheses made, Stokes' law underlines the critical parameters that affect the effectiveness of settling for wine clarification. Of particular importance is the particle diameter: v_p is decreased by a factor of 100 for a 1-μm particle by comparison to a 10-μm one, and by a factor 10,000 for a 0.1-μm particle. Thus, as stated before, submicronic colloidal particles do not settle spontaneously unless they flocculate to form large aggregates. Other important parameters are the lack of convection motions such as CO_2 release, which interfere with settling, and of course the time and the container height: to be separated by racking, particles must reach the bottom of the container. The clarification levels obtained by natural settling and racking are usually higher for wines aged in barrels than for wines stored in vats, owing to their smaller size. In the best conditions, natural settling allows the removal of tartrate crystals, of large aggregates, and of the majority of yeasts (90%) but not that of colloidal size range debris/aggregates and bacteria. To this end fining aids may be used.

Fining consists of introducing in a turbid wine a substance that flocculates and settles, dragging down other suspended particles. This treatment involves physicochemical interactions between the fining aid and wine components and/or between the fining aid and residual particles. These interactions induce an aggregation, followed by an increase of particle size (flocculation) that favors their settling. The clarified wine is then separated from fining lees by racking. The exact mechanisms involved in clarification by fining are not clearly established and are likely diverse. Colloidal and suspended particles may be trapped in the network formed by the interacting species and/or included in the aggregates owing to direct interactions with the fining aid. They may also settle by simple entrainment (nondiluted suspensions). Beside their clarifying effect, fining aids are also used to remove substances from wine for the purposes of enhancing color, taste, and/or stability. This implies specific physicochemical interactions between fining aids and some of the wine components, and the removal of these components. Thus in red wines, fining is used to clarify but also to soften tannic aggressiveness/intensity and improve mouthfeel, or to stabilize with regards to the precipitation of colloidal coloring matter.

When dealing with red wines, fining aids are mainly proteins and the success of the treatment strongly relies on their physicochemical interactions with polyphenols, especially tannins, and polymerized pigments. Traditionally gelatines, egg white, or egg albumin were used. However, increasing interest for other nonanimal proteins for wine fining has been triggered over the last 20 years by the bovine spongiform encephalopathy crisis and by the potential allergenicity of egg proteins (Gambuti et al., 2012; Marchal et al., 2002; Maury et al., 2003; Tolin et al., 2012). New protein fining agents have been studied and are now available, including plant proteins but also yeast protein extracts. All these proteins present a wide diversity in terms of molecular mass distribution, amino acid composition, and conformation in solution (extended and linear versus globular conformation), which affect both their physicochemical properties and polyphenol accessibility to interaction sites. These parameters modulate the extent and specificity of their interactions with wine polyphenols and eventually particles, and their clarifying effect. Interactions and clarification are also strongly impacted by the wine phenolic composition, by its pH and ionic strength and by the presence of cosolutes such as for example polysaccharides (Maury et al., 2016; Soares et al., 2012). Thus, fining treatments remain quite empirical in winemaking and fining trials are required to define the best fining aid and dose according to the objectives. Settling and fining treatments may be hampered in wines by the presence of polysaccharides such as pectins or β-glucans. In such cases, enzymatic treatments often allow to solve the problem (Canal-Lalaubères, 1993; Villettaz et al., 1984).

16.2.2 Centrifugation and Wine Clarification

Clarification rates, with or without the use of fining aids, can be strongly accelerated by the use of continuous disk-stack centrifuges (Ribéreau-Gayon et al., 2006). In centrifuges, the acceleration related to gravity [Eq.(16.1)] is replaced by a centrifugal force, linked to the rotation of the centrifuge bowl in which the wine circulates around a fixed axis (feeding and recovery of the clarified wine). This centrifugal force increases g by a factor of Z, which can range from 7000 to $15,000 \times g$ (high-performance clarifiers) for continuous disk-stack centrifuges used in enology. This acceleration factor depends on the bowl rotation speed and on the distance to the axis of rotation. It is maximum at the periphery of the bowl for a given rotation speed. Centrifuges are then continuously and centrally fed by the turbid wine, which is distributed at the periphery of the bowl where the centrifugal acceleration is maximum. The wine is then distributed among the disks and goes up toward the center of the bowl, where it is evacuated. The compartmentalization of the bowl by the disk-stack makes it possible to: (1) considerably reduce the height of fall necessary to obtain a separation of the particles (distance of the order of 2 mm); and (2) create laminar flow conditions, without turbulences likely to interfere with particle settling, even at high feed rates. The largest particles accumulate at the periphery of the bowl, in the sludge chamber, whereas

16.2.3 Filtration

Beside other clarification and stabilization techniques, filtration is an essential separation process to get the targeted limpidity and microbial stabilization of wines. In filtration, suspended solids are separated from the liquid by interposing a porous medium to fluid flow through which the liquid can pass, but the solids (or at least part of the solids) are retained. The driving force to achieve this separation is a pressure gradient (Fig. 16.3). Important characteristics of filtration media are their porosity, which is the percentage of empty space in a porous structure in relation with the total volume, and their permeability, which represents the ability of a medium to transmit fluids through at more or less high rates. The permeability is often expressed in Darcy (1 Da = 9.8×10^{-13} m^2). The flux rate J of fluids through a porous medium is given by Darcy's law.

In practice, wines are mainly filtered during their preparation for bottling. This corresponds to very different situations, depending on the wine characteristics and turbidity (young wine vs. aged wine), the winemaker choices, and the market specifications. Different filtration modes can be used: dead-end filtration or cross-flow microfiltration (CF-MF).

16.2.3.1 Dead-End Filtration

Wine particles are characterized by a wide polydispersity in terms of size and shape and are compressible under the application of a pressure (Fig. 16.4A). Their accumulation at the surface of a porous medium in dead-end filtration leads to the formation of compact and nonporous deposits (Fig. 16.4B). Beside a sharp decrease of fluxes, this clogging increases the risks of excessive retentions in macromolecules of interest for wine quality. Dead-end filtration of turbid wines is then achieved either by the use of filtration aids (earth filtration) or by the use of depth filtration media (Gautier, 2015; Ribéreau-Gayon et al., 2006).

Earth filtration is based on the addition to the wine of diatomaceous earths, also called kieselguhrs, which are porous and rigid siliceous skeletons of microscopic water algae. A protective layer of filter aid, the precoat, is first built up on the filtering support by recirculation of an earth suspension. Its role is to prevent the clogging of the support by wine particles and to obtain immediate clarity of the filtered wine. Small amounts of filter aids are then regularly added to the wine to be filtered. Along with wine suspended solids, their retention at the surface of the precoat as the filtration proceeds forms a deposit (filtration cake), the thickness of which progressively increases. Earth particles form porous and rigid deposits that entrap suspended wine particles, allowing the liquid to pass through (Fig. 16.5A). In depth filtration, wine particles are entrapped within the porous medium (Fig. 16.5B). As the fluid flows through the medium, they are slowed down and eventually retained by the tortuous path within the filter medium, or by means of interactions with the porous surface. The filtering medium consists of a three-dimensional network of cellulose fibers, and also includes different proportions of kieselguhrs and perlite, depending on the required filtration grade, and synthetic polymers to strengthen the structure of the

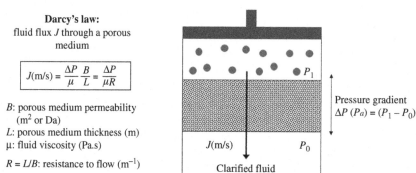

FIGURE 16.3 Clarification by filtration and Darcy's law.

16.2 WINE CLARIFICATION

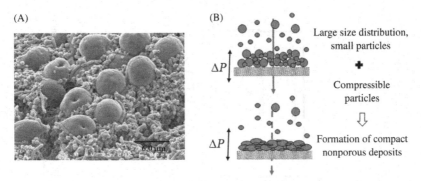

FIGURE 16.4 (A) Scanning electron microscopy observation of wine particles in a red wine (initial turbidity 80 NTU) showing the polydispersity of suspended particles (yeast cells, lactic bacteria, and colloidal amorphous aggregates). (B) Clogging related to their accumulation at the surface of a porous medium.

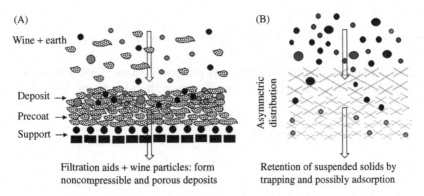

FIGURE 16.5 (A) Dead-end surface filtration of wine using filtration aids (diatomaceous earths). (B) Retention of wine particles in depth filtration.

porous material. The filter pores are distributed asymmetrically, with the largest on the input side. Depth-filtration media can be found under the form of sheets for press and frame type filters, or of lenticular modules.

In both cases, the particle size distribution and dosage of kieselguhrs or the sheet grade must be adapted to the wine content in suspended solids to avoid clogging. There are thus kieselguhrs/sheets adapted to the coarse filtration of young and highly turbid wines, for which filtration media with high permeability are needed, and kieselguhrs/sheets adapted to the clarification of wines after several months of aging and/or after fining or centrifugation (fine or polishing filtration), with much smaller permeability. Filtrations on sterilizing sheets will be performed to reduce the residual number of viable germs: when properly conducted, wines with less than one viable germ per 100 mL can be obtained. Depending on the preclarification operations and of their effectiveness, several successive filtration steps will then be needed to obtain the turbidity value and reduction level in microorganisms expected for bottling.

The assessment of wine filterability and the development of filterability tests are of importance in winemaking when choosing the adequate filtration medium or operating conditions (Alarcon-Mendez and Boulton, 2001; Romat, 2007). Indeed, turbidity is not always a good indicator of wine filterability as it provides no indication on the size distribution and on the nature of the involved particles. In addition, the presence of pectins or β-glucans can strongly affect the performances of the process (Villettaz et al., 1984; Wucherpfennig and Dietrich, 1989). Filtration criteria, based on turbidity measurements combined with filterability tests have been established (Romat and Reynou, 2007). These filtration criteria can be used (1) to select the right filtration medium and grade, (2) to assess the feasibility of a filtration on a given medium, or (3) to evaluate the ability of a pretreatment (such as the use of enzymes or fining) to improve filterability.

16.2.3.2 Cross-Flow Microfiltration

In CF-MF, the liquid to be filtered circulates tangentially to the filtering medium, a microporous membrane with an average pore size of the order of 0.2 μm in enology (Fig. 16.6). The liquid is kept circulating in a loop:

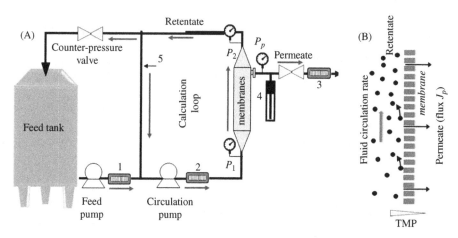

FIGURE 16.6 (A) Cross flow-microfiltration implementation: (1), (2), and (3) flowmeters; P_1, P_2, and P_P: pressure sensors for the control of the TMP applied across the membrane; (4) back-pulse; (5) heat-exchanger to control the temperature in the circulation loop. (B) The fluid circulation along the membrane (controlled by the circulation pump) creates a shear stress that prevents particle deposition upon the effect of the fluid flux through the membrane, generated by the TMP. Particles are progressively concentrated in the feed. TMP, transmembrane pressure.

part of it (permeate) crosses over the membrane through the application of a pressure gradient. The transmembrane pressure (TMP), which governs the permeate flux J_p across the membrane [Eq. (16.2)] and the transport of particles at the membrane surface, is given by:

$$\text{TMP (Pa)} = \frac{P_1 - P_2}{2} - P_p \tag{16.2}$$

where P_1 and P_2 are the pressures at the entrance and the exit, respectively, of the filtration module and P_p the pressure on the permeate side. At the same time, the liquid circulation creates a shear stress τ (Pa) at the membrane surface that prevents the deposition of the retained particles and cake formation. The suspended solids thus progressively concentrate in the feed. The shear stress τ is dependent on the pressure drop along the membrane, $P_1 - P_2$, and thus on the fluid velocity and of the geometry of the filtration modules.

CF-MF allows then the one-step clarification and microbial stabilization of even very turbid fluids. The choice of appropriate microfiltration membranes and the implementation of suitable hydrodynamic conditions are of primary importance for the process efficiency (Belleville et al., 1992; Boissier et al., 2008; Cameira dos Santos, 1995; Li et al., 1998, 2003; Vernhet et al., 2003; Vernhet and Moutounet, 2002). Membrane materials used for wine clarification have been selected to limit physicochemical interactions between wine components and the membrane surface, and their average pore size is chosen to prevent internal pore blockage by the smallest particles. Physicochemical interactions and internal fouling not only impact fluxes; due to their irreversible character, they also affect the effectiveness of cleaning procedures. Hydrodynamic conditions are of critical importance to control external fouling. As stated before, particle deposition at the membrane surface is governed by the ratio between the permeate flux J_p (controlled by the TMP) and the shear stress τ. To prevent cake formation, J_p/τ must be lower than a critical value which is dependent on the particle size and shape (Li et al., 1998, 2003; Ripperger and Altmann, 2002). This critical value decreases with the particle size, so that the finest suspended solids are preferentially deposited in CF-MF, inducing an additional resistance to flux related to the formation of a cake at the membrane surface. When dealing with wines, it has been demonstrated that some deposition of colloidal particles and bacteria occurs from very low J_p/τ ratios (Boissier et al., 2008). Though of finite thickness, this deposit decreases the process performances. For this reason, back-pulsing, which consists of very short and periodic flux inversions by reversing the TMP, has been associated with the process (Cadot, 2001; Cameira dos Santos, 1995). During the back-pulse, the filtered wine is forced back through the membrane to the feed side. Deposited particles are dislodged and carried out of the membrane by the tangential flow rate. Progresses realized in the adaptation of CF-MF membranes and implementation to the specificity of wine clarification and microbiological stabilization have allowed its development and its use has experienced a continuous progression over several years. Both organic hollow fiber and ceramic multichannel membranes, associated with different cross-flow filter designs, are proposed now to winemakers.

The main interest of CF-MF is that bright and sterile wines are obtained in one unit operation from even very turbid products, whereas different successive operations are required with traditional processes (fining/settling and dead-end filtrations). Beside the great simplification of the wine clarification and stabilization process, this strongly limits the wine losses and SO_2 adjustments associated with treatments. The process presents other advantages, including the elimination of earth use and the automation of filtration and cleaning. CF-MF units are proposed with different levels of automation: from the minimal level, corresponding to the control of the operating parameters (TMP, fluid velocity, back-pulses, temperatures, etc.), to much more advanced levels, with a complete management of filtration cycles and of cleaning procedures. This allows substantial time and labor-saving and increases plant productivity. All these advantages make up for the higher investment costs and the lower flow rates by comparison to dead-end filtration modes.

16.2.3.3 Filtration and Microbial Stabilization of Wines

In cases where sterile (less than one viable germ per bottle) or low germ (less than one viable germ per 100 mL) bottling is required, a final filtration on membranes is performed just ahead the bottling line. Contrarily to sheets, microfiltration membranes act as sieves: all particles with diameter above the pore diameter will be retained. Membranes with pore diameters of 0.45 or 0.65 μm are used for the retention of yeasts and bacteria. In the dead-end mode, membrane filtration can only be applied on perfectly bright and prefiltered wines to avoid an excessive surface fouling and ensure a flow rate compatible with the constant flow rate of the bottle filler. Filterability essays using 0.65-μm membranes of 25-mm diameter and well-defined pressure conditions are thus always performed when a final membrane filtration is required. These trials allow the calculation of a fouling index (*FI*) or of a maximum filterable volume (V_{max}), the values of which are used to assess the feasibility of the final membrane filtration.

16.3 STABILIZATION WITH REGARDS TO THE CRYSTALLIZATION OF TARTARIC SALTS

16.3.1 Mechanisms and Stability Assessment

The crystallization of potassium hydrogen tartrate (KHT) or calcium tartrate (CaT) salts may occur in wines in supersaturation conditions, i.e., when their concentration product (CP) exceeds their solubility product (SP) (Berg and Keefer, 1958a,b). The CPs are given by:

$$CP_{KHT} = \gamma.[HT^-][K^+]; \quad CP_{CaT} = \gamma'.[T^{--}][Ca^{++}] \tag{16.3}$$

They depend on the pH, which determines the respective concentrations of H_2T, HT^-, and T^{--}, and on the wine composition. In Eq. (16.3), γ and γ' represent activity coefficients, i.e., the ion concentrations in wine really available for crystallization (Balakian and Berg, 1968; Gerbaud et al., 1997). The SP corresponds to the maximum amounts of tartaric salts that can be dissolved at equilibrium. SP strongly depends on the ethanol concentration: it decreases when the ethanol content increases, so that spontaneous crystallizations are usually observed during fermentation. For a given ethanol content, SP also decreases with the temperature, and further crystallization occurs during wine storage at low temperatures.

In supersaturated conditions, the crystallization of tartrate salts develops in two steps (Dunsford and Boulton, 1981). The first one, nucleation, corresponds to the formation of stable crystal germs, called nuclei. Nucleation is immediately followed by crystal growth (Fig. 16.7A). Crystal growth rate is controlled by the transport of ions by diffusion and/or convection from the solution to the crystal surface and by their integration into crystals. Crystal growth continues until ion concentrations reach values such that their CP equals their SP in the considered conditions (matrix, temperature, and ethanol). From a kinetics point of view, nucleation is the limiting step in the crystallization process. The induction time of crystallization (t_{ind}) is the period experimentally observed between the setting up of supersaturation conditions and the first changes in physicochemical parameters related to nuclei formation and crystal growth. It is usually followed by the measurement of the wine conductivity. Crystal growth rate is dependent on the supersaturation level, on the surface area developed by crystals and on the stirring conditions. As musts are naturally rich in potassium and as HT^- is the major form of tartaric acid at wine pH, KHT crystallization is the most common. However, the solubility of CaT is much lower than that of KHT and its crystallization, in supersaturated conditions, develops over much longer periods (Abguéguen and Boulton, 1993; Gerbaud et al., 2010; McKinnon et al., 1994).

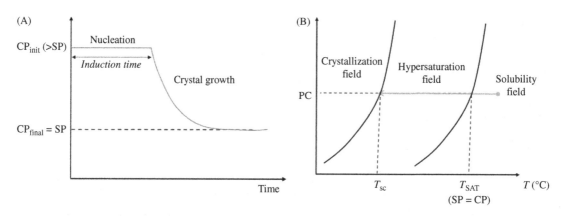

FIGURE 16.7 (A) The two steps of crystallization in supersaturation conditions. *CP*, concentration product; *SP*, solubility product. (B) Schematic representation of the solubility, hypersolubility, and crystallization fields of tartaric salts in a wine. T_{SAT}, saturation temperature; T_{SC}, spontaneous crystallization temperature.

In practice, the risks of tartaric salt crystallization is dependent on the wine composition and cannot be predicted on the basis of simple analytical data and of the knowledge of SP values established in model ethanolic solutions (Berg and Keefer, 1958a,b). The relationship between the KHT CP, the temperature, and KHT crystallization or solubility is illustrated in Fig. 16.7B (Maujean et al., 1985). For a given wine matrix, it is possible to determine a solubility, a hypersaturation, and an instability field. The solubility field corresponds to conditions where concentrations and temperature are such than CP is lower than SP: the solution is undersaturated and a crystal added will dissolve. Considering a given CP and starting from the solubility field, a decrease in temperature induces a decrease of SP. Below a given temperature, called the saturation temperature T_{SAT}, the solution is supersaturated: it is not possible to dissolve an exogenous crystal and the latter will grow. However, spontaneous crystallization, which involves nucleation, does not occur. Spontaneous crystallization (crystallization field) is observed at a temperature T_{SC} well below T_{SAT}. The width of the hypersaturation field is dependent on the wine composition: some components enhance the solubility of tartaric salts by comparison to that observed in model hydroalcoholic solutions (Balakian and Berg, 1968; Gerbaud et al., 1997; McKinnon et al., 1995, 1996; Pilone and Berg, 1965; Rodriguez-Clemente and Correa-Gorospe, 1988). These components can form complexes with the involved ions, thus decreasing the supersaturation level. They can also slow down or prevent crystal growth. This effect is attributed to their adsorption on the crystal faces, blocking the integration of new units in the crystal lattice. Inhibitory effects on KHT crystals nucleation and growth have been evidenced for some wine polysaccharides and for proanthocyanidins (Balakian and Berg, 1968; Gerbaud et al., 1997; Maujean et al., 1985; Tanahashi et al., 1987). Due to the strong impact of proanthocyanidins, the inhibition of KHT crystallization is usually much more pronounced in red wines than in white ones.

Specific tests have been developed to assess wine stability or instability and to monitor stabilization technologies. In the cold test, the wine is simply kept at negative temperatures during a defined period (2 weeks at $-2°C$ or 6 days at $-4°C$) and crystallization assessed visually. Though representative of crystallization risks as they account for both the nucleation and growth steps, such tests are time-consuming. In addition, much longer periods (for example 1 month at $-2°C$) are needed if the risk is related to CaT (Vallée et al., 1995). Other tests have then been designed, by reference to cold tests, to decrease the time required and provide a better evaluation of the wine degree of instability. In the minicontact test, and considering KHT crystallization, the wine is cooled and seeded with 4 g/L KHT crystals. The nucleation step is then avoided. The wine is maintained under constant stirring and its stability/instability is assessed by the analysis of crystal growth over 4 hours. Crystal growth is followed by weighing (Muller-Späth, 1977) or more simply by conductimetry (Dunsford and Boulton, 1981). The drop in conductivity is positively correlated with wine instability toward the crystallization of tartaric salts. In the instability degree test (DIT), crystal growth at $-4°C$ is analyzed during 4 hours and the conductivity at equilibrium (final conductivity, reached for concentrations such as CP = SP) is determined by extrapolation of the curve at "infinite" time (Moutounet et al., 2010). The degree of instability is calculated from the initial and final conductivity values and represents the drop in conductivity expressed in % of the initial conductivity:

$$\mathrm{DIT}(\%) = 100 * \frac{C_{\text{initial}}^{-4°C} - C_{\text{final}}^{-4°C}}{C_{\text{initial}}^{-4°C}} \tag{16.4}$$

By reference to a cold test at −4°C and over 6 days, wines are considered as stable if their DIT values are lower than 5%. Values between 5% and 10% indicate moderate risks of crystallization, whereas values higher than 10% indicate unstable wines. The critical stability index (ISTC 50) differs from the previous tests as it is based on the analysis of the induction time. 0.5 g/L KHT crystals are dissolved in the wine previously heated at 36°C and kept under stirring. The temperature is then decreased at −4°C and the conductivity followed. Red wines are considered as stable if the conductivity drop is less than 5 μS/cm in 240 minutes. Other stability tests are based on the determination of the saturation temperature T_{SAT} (Maujean et al., 1985; Vallée et al., 1995). As stated before, T_{SAT} differs from T_{SC} due to the presence in wines of compounds that inhibit nucleation and crystal growth. Rules between T_{SAT} and wine stability have been established by reference to a cold test (−2°C, 15 days). As polyphenols strongly enhance the width of the hypersaturation field, these rules account for the measurement of the total polyphenol index (TPI) for red wines. Thus, it is usually admitted that wines with TPI below 50 are stable if their T_{SAT} is under 22°C, whereas wines with TPI above 50 are stable if their T_{SAT} is under 24°C.

16.3.2 Stabilization Technologies

Basically, technologies available to stabilize wines can be divided into three categories: cold stabilization processes, based on the induction of crystallization by means of cooling the wine; processes based on the removal of ions in excess, such as electrodialysis; and the use of additives that inhibit crystallization.

16.3.2.1 Cold Stabilization

The aim of cold stabilization processes is to provoke the crystallization of tartaric salts in excess (Blouin, 1982). This can be achieved without or with addition of tartrate crystals. In the first case, the wine is quickly cooled at a temperature close to its freezing point to reduce the solubility of tartaric salts and stored in an insulated tank for 1 week and under stirring to allow nucleation and crystal growth. A quick decrease in temperature, provided by adequate heat exchangers, and a continuous stirring of the wine are essential for the effectiveness of the treatment. The thermal shock favors the nucleation and the formation of numerous nuclei, which increases the integration surface. Along with stirring, this enhances crystal growth rates. After the cold treatment, the wine is filtered (insulated filter) to remove crystals and brought back to positive temperatures. The treatment length is dependent on the supersaturation level and stabilization is assessed by stability tests. Seeding the wine with tartrate cream, as in stability tests, suppresses the nucleation step. Tartrate cream consists of finely divided KHT crystals (50−100 μm), added to the wine at a concentration of 4 g/L. This contact method (Muller-Späth, 1977) provides results similar to the conventional cold treatment for temperatures in the order of 0−1°C and strongly decreases the treatment length (several hours). KHT crystals can be reused twice for red wines but beyond, the effectiveness of the treatment is reduced due to the adsorption of wine colloidal components on the crystal surfaces. Cold treatment results are dependent on the wine content in colloids and they are usually performed on preclarified wines. The advantages of cold stabilization treatments are that they are well experienced techniques in wineries and that they may also contribute to wine stabilization toward the precipitation of colloidal coloring matter. However, this last point is strongly dependent on the treatment length and colloidal stabilization is less effective when the contact method is chosen. Their disadvantages are their energetic costs and the need of additional filtrations.

16.3.2.2 Electrodialysis

Electrodialysis is a membrane separation process which allows the selective transport of cations and anions through ion permeable membranes under the application of a potential gradient. Electrodialysis membranes are sheets of cross-linked synthetic polymers (100−200 μm width) carrying negatively (cation exchange membranes, CEM) or positively (anion exchange membranes, AEM) charged groups. The fixed charges on CEM and AEM membranes allow the selective transport of counterions between two solutions whereas co-ions are rejected by Donnan exclusion. Electrodialysis in the dilution-concentration mode is used to achieve wine stabilization with regards to the crystallization of tartaric salts (Moutounet et al., 1999): alternating cation and anion-exchange membranes delimit concentration and dilution compartments (Fig. 16.8). The wine circulates in the dilution compartments. Under the application of a potential gradient, cations (for example K^+ and Ca^{++}) migrate toward the cathode and leave the dilution compartment by crossing the CEM. As they cannot cross the AEM, they are then confined in the concentration compartments where they accumulate. Simultaneously, anions (for example TH^- and T^{--}) migrate toward the anode and also accumulate in the next concentration compartment. A basic

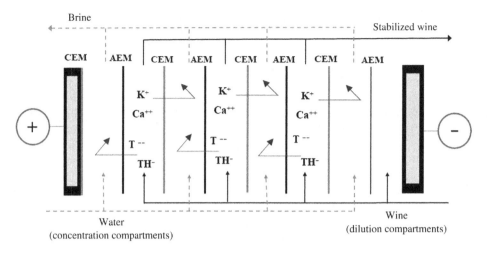

FIGURE 16.8 Principle of electrodialysis for the tartrate stabilization of wines. *CEM*, cation exchange membrane; *AEM*, anion exchange membrane.

electrodialysis cell consists of a CEM and an AEM separated by a frame spacer. An electrodialysis stack consists of hundreds of basic cells, each compartment being fed independently. Electrodes are placed at the ends in independent compartments. The process is monitored in-line by a conductivity probe, the target conductivity required to achieve wine stabilization being calculated from the DIT. Electrodialysis is automated and can be used in the continuous mode, the rate of ion removal being controlled by the potential gradient applied.

Electrodialysis membranes have been selected to allow a large extraction of potassium and of tartrate ions, along with some extraction of calcium, without significant changes of the physicochemical equilibrium and of the organoleptic characteristics of the wine. As the transfer rate of potassium exceeds that of the tartrate ions, a decrease in pH can be observed. It depends on the considered wine and of its degree of instability. Decreases in the order of 0.2 units are observed only for the most unstable wines (DIT of 20%–25%).

16.3.2.3 Additives

The additives used for the tartrate stabilization of red wines are metatartaric acid and yeast mannoproteins. Metatartaric acid is a mixture of mono and diesters of tartaric acids, also including few polyesters, obtained through the fusion at 170°C of a tartaric acid powder. Metatartaric acid inhibits crystal growth. Its inhibitory power is dependent on its esterification index (IE), and an IE value higher than 32% is required by the Enological Codex. It is added at bottling at a maximum concentration of 10 g/hL. Though very effective in the short term, the disadvantage of metatartaric acid is its instability: its progressive hydrolysis in wines, favored by elevated temperatures, limits its use to wines destined for a fast consumption. Specific and selected yeast mannoprotein fractions are also proposed as inhibitors of the crystallization of KHT (Moine-Ledoux et al., 1997). Even at high doses, these additives are however insufficient for the stabilization of very unstable wines.

16.4 MICROBIOLOGICAL STABILIZATION

The microbiological stabilization of red wines needs to be reasoned depending on their physicochemical characteristics, i.e., pH, ethanol, and residual sugars, which strongly affect the risk of microorganism developments. Microbiological stabilization of dried red wines before bottling can usually be achieved by low-germ filtration and the adequate use of SO_2. However, special attention must be paid to red wines with high pH as the proportion of molecular SO_2, which is the form that is active against microorganisms, strongly decreases with the pH in the usual wine range. In addition, microorganisms vary considerably in their sensitivity to SO_2 (Du Toit and Pretorius, 2000). Sterile filtration on membranes or flash-pasteurization may be required for wines with residual sugars or that present specific risks. Sterile membrane filtration in the dead-end mode can only be performed on perfectly prefiltered wines and its application is then restricted to bottling. By contrast, if a microbiological deviation appears during aging and if an immediate stabilization is required, this can be achieved by CF-MF regardless the initial turbidity of the wine. An alternative to sterile filtration is flash-pasteurization (Couto et al., 2005;

Lonvaud-Funel, 1999). This treatment allows the heat-destruction of vegetative forms of microorganisms, which are the only ones present at the pH of wines. The temperature–time treatments applied to ensure the microbial stabilization depend on the wine characteristics but also on the heat-resistance of the concerned microorganisms. Usually, heat-treatments in the order of 20 seconds and at temperature around 70°C are sufficient to stabilize dry wines.

16.5 CONCLUSION

Winemakers have a set of different tools at their disposal to clarify and stabilize their wines. The choice of one or other of the techniques depends on the organization of the cellar, the volumes to be treated, and in some cases the specificities of the market. If the winemaker will carry out final clarification and stabilization operations before bottling, the clarification and stabilization of the wines actually begin at the end of fermentation and continue throughout post-fermentation operations: successive racking after settling to preclarify the wines, prevention of microbial risks by keeping them protected from oxygen associated with use of SO_2, and possibly early filtration, aggregation, and spontaneous flocculation over time, crystallization in case of storage at low temperature, etc. The treatments to be carried out before bottling will therefore be influenced not only by the initial composition of the wine but also by the aging methods adopted. They will generally be more important for young wines. Moreover, the order in which they are positioned depends on the technologies chosen. Generally speaking, the preclarification or stabilization operations by fining and decantation, whether or not combined with centrifugation, are carried out first in order to limit the number of filtration operations needed to obtain clear wines. Cold stabilization operations are performed on clarified wines and will lead to additional filtration for the elimination of crystals. Filtration on sterilizing sheets or with membranes will be necessary to obtain microbiological stabilization. CF-MF, if chosen, allows the clarification and microbiological stabilization of wines in a single operation, regardless of their initial turbidity. Combined with electrodialysis, this technique allows a significant simplification of the wine processing chain. CF-MF and electrodialysis can be performed continuously. Additives, if they are chosen to obtain stabilization, are added just before bottling.

References

Abguéguen, O., Boulton, R.B., 1993. The crystallization kinetics of calcium tartrate from model solutions and wines. Am. J. Enol. Vitic. 44, 65–75.
Alarcon-Mendez, A., Boulton, B., 2001. Automated measurement and interpretation of wine filterability. Am. J. Enol. Vitic. 52, 191–197.
Balakian, S., Berg, H.W., 1968. The role of polyphenols in the behaviour of potassium bitartrate in red wines. Am. J. Enol. Vitic. 19, 91–100.
Belleville, M., Brillouet, J., Tarodo de la Fuente, B., Moutounet, M., 1992. Fouling colloids during microporous alumina membrane filtration of wine. J. Food Sci. 57, 396–400.
Berg, H.W., Keefer, R.M., 1958a. Analytical determination of tartrate stability in wine. I Potassium bitartrate. Am. J. Enol. Vitic. 9, 180–199.
Berg, H.W., Keefer, R.M., 1958b. Analytical determination of tartrate stability in wine. II Calcium tartrate. Am. J. Enol. Vitic. 10, 105–109.
Blouin, J., 1982. Les techniques de stabilisation tartrique des vins par le froid. Connaiss. Vigne Vin 16, 63–77.
Boissier, B., Lutin, F., Moutounet, M., Vernhet, A., 2008. Particle deposition during the cross-flow microfiltration of red wines—incidence of the hydrodynamic conditions and of the yeast to fines ratio. Chem. Eng. Process. 47, 276–286.
Cadot, Y., 2001. New developments in crossflow filtration. Aust. Grapegrower Winemaker 2001, 101–105.
Cameira dos Santos, P., 1995. Colmatage en microfiltration tangentielle : mise en évidence d'interactions entre les polysaccharides et les polyphénols d'un vin et des membranes polymériques. Université de Montpellier II, Montpellier.
Canal-Lalaubères, R.-M., 1993. Enzymes in winemaking. In: GH, F. (Ed.), Wine Microbiology and Biotechnology. Harwood Academic Publisher, Switzerland.
Cheynier, V., Dueñas-Paton, M., Salas, E., Maury, C., Souquet, J.-M., Sarni-Manchado, P., et al., 2006. Structure and properties of wine pigments and tannins. Am. J. Enol. Vitic. 57, 298–305.
Couto, J.A., Neves, F., Campos, F., Hogg, T., 2005. Thermal inactivation of the wine spoilage yeasts Dekkera/Brettanomyces. Int. J. Food Microbiol. 104, 337–344.
de Freitas, V., Mateus, N., 2011. Formation of pyranoanthocyanins in red wines: a new and diverse class of anthocyanin derivatives. Anal. Bioanal. Chem. 401, 1463–1473.
Diako, C., McMahon, K., Mattinson, S., Evans, M., Ross, C., 2016. Alcohol, tannins, and mannoprotein and their interactions influence the sensory properties of selected commercial merlot wines: a preliminary study. J. Food Sci. 81, S2039–S2048.
Dunsford, P., Boulton, R.B., 1981. The kinetics of potassium bitartrate crystallization from table wines. II Effect of particle size, particle surface area and agitation. Am. J. Enol. Vitic. 32, 100–105.
Du Toit, M., Pretorius, I.S., 2000. Microbial spoilage and preservation of wine: using weapons from nature's own arsenal—a review. S. Afr. J. Enol. Vitic. 21, 76–96.
Fulcrand, H., Dueñas, M., Salas, E., Cheynier, V., 2006. Phenolic reactions during winemaking and aging. Am. J. Enol. Vitic. 57, 289–297.

Gambuti, A., Rinaldi, A., Moio, L., 2012. Use of patatin, a protein extracted from potato, as alternative to animal proteins in fining of red wine. Eur. Food Res. Technol. 235, 753–765.

Gautier, B., 2015. Practical aspects of wine filtration, Œnoplurimedia, Avenir Œnologie.

Gerbaud, V., Gabas, N., Blouin, J., Pellerin, P., Moutounet, M., 1997. Influence of wine polysaccharides and polyphenols on the crystallization of potassium hydrogen tartrate. J. Int. Sci. Vigne Vin 31, 65–83.

Gerbaud, V., Gabas, N., Blouin, J., Crachereau, N., 2010. Study of wine tartaric salts stabilization by addition of carboxymethylcellulose (CMC): comparison with the "protective colloid effect". J. Int. Sci. Vigne Vin 44, 231–242.

Ghanem, C., Taillandeir, P., Rizk, M., Rizk, Z., Nehme, N., Souchard, J., et al., 2017. Analysis of the impact of fining agents types, enological tannins and mannoproteins and their concentrations on the phenolic composition of red wine. LWT Food Sci. Technol. 83, 101–109.

Gonzales-Neves, G., Favre, G., Gil, G., 2014. Effect of fining on the colour and pigment composition of young red wines. Food Chem. 157, 385–392.

Haslam, E., 1998. Molecular recognition - phenols and polyphenols. Practical Polyphenolics: From Structure to Molecular Recognition and Physiological Action. Cambridge University Press, Cambridge, pp. 138-177.

Hiemenz, P.C., 1986. Principles of Colloids and Surface Chemistry, second ed. Marcel Dekker, New York.

Israelachvili, J., 1992. Intermolecular and Surface Forces. Academic Press, London.

Li, H., Fane, A., Coster, H., Vigneswaran, S., 1998. Direct observation of particle deposition on the membrane surface during cross-flow microfiltration. J. Membr. Sci. 83–97.

Li, H., Fane, A.G., Coster, H.G.L., Vigneswaran, S., 2003. Observation of deposition and removal behaviour of submicron bacteria on the membrane surface during crossflow microfiltration. J. Membr. Sci. 217, 29–41.

Lonvaud-Funel, A., 1999. Principes de la stabimlisation microbiologique des vins apr la chaleur. Traitements physicques des moûts et des vins, n° hors série. J. Int. Sci. Vigne Vin 147–153.

Mainente, F., Zoccatelli, G., Lorenzini, M., Cecconi, D., Vincenzi, S., Rizzi, C., et al., 2014. Red wine proteins: two dimensional (2-D) electrophoresis and mass spectrometry analysis. Food Chem. 164, 413–417.

Marchal, R., Marchal-Delahaut, L., Lallement, A., Jeandet, P., 2002. Wheat gluten used as a clarifying agent of red wines. J. Agric. Food. Chem. 50, 177–184.

Maujean, A., Sausy, L., Vallée, D., 1985. Détermination de la sursaturation en bitartratye sz potassium d'un vin; quantification des effets colloïdes protecteurs. Revue Française d'Oenologie 104, 34–41.

Maury, C., Sarni-Manchado, P., Lefebvre, S., Cheynier, V., Moutounet, M., 2001. Influence of fining with different molecular weight gelatins on proanthocyhanidin composition and perception of wines. Am. J. Enol. Vitic. 52, 140–145.

Maury, C., Sarni-Manchado, P., Lefebvre, S., Cheynier, V., Moutounet, M., 2003. Influence of fining with plant proteins on proantjhocyanidin composition of red wines. Am. J. Enol. Vitic. 56, 140–145.

Maury, C., Sarni-Manchado, P., Poinsaut, P., Cheynier, V., Moutounet, M., 2016. Influence of polysaccharides and glycerol on proanthocyanidin precipitation by protein fining agents. Food Hydrocolloids 60, 598–605.

McKinnon, A.J., Scollary, G.R., Solomon, D.H., Williams, P.J., 1994. The mechanism of precipitation of calcium L(+)-tartrate in a model wine solution. Colloids Surf., A: Physicochem. Eng. Aspects 82, 225–235.

McKinnon, A.J., Scollary, G.R., Solomon, D.H., Williams, P.J., 1995. The influence of wine components on the spontaneous precipitation of calciul L(+)-tartrate in a model wine solution. Am. J. Enol. Vitic. 46, 509–517.

McKinnon, A.J., Williams, P.J., Scollary, G.R., 1996. Influence of uronic acids on the spontaneus precipitation of calcium (L +)-tartrate in a model wine solution. J. Agric. Food. Chem. 44, 1382–1386.

Mekoue Nguela, J., Poncet-Legrand, C., Sieczkowski, N., Vernhet, A., 2016. Interactions of grape tannins and wine polyphenols with a yeast protein extract, mannoproteins and Œ≤-glucan. Food Chem. 210, 671–682.

Moine-Ledoux, V., Perrin, A., Paladin, I., Dubourdieu, D., 1997. First results on tartaric stabilization by adding mannoproteins. J. Int. Sci. Vigne Vin 31, 23–33.

Mouls, L., Fulcrand, H., 2012. UPLC–ESI–MS study of the oxidation markers released from tannin depolymerization: toward a better characterization of the tannin evolution over food and beverage processing. J. Mass Spectrom. 47, 1450–1457.

Moutounet, M., Battle, J.L., Saint-Pierre, B., Escudier, J.-L., 1999. Stabilisation tartrique. Détermination du degré d'instabilité des vins. Mesure de l'efficacité d'inhibiteurs de la cristallisation. In: Lonvaud-Funel, A. (Ed.), Oenologie 99. Lavoisier Tec&Doc, Bordeaux.

Moutounet, M., Bouisson, D., Escudier, J.L., 2010. Determination of the degree of tartaric instability: principles and applications. Revue Française d'Oenologie 242, 24–28.

Muller-Späth, H., 1977. Un nouveau procédé de prévention des précipitations de bitartrate de potassium: Le procédé contact. Ann. Technol. Agric. 27, 333–337.

van Oss, C.J., 1994. Interfacial Forces in Aqueous Media. Marcel Dekker Inc, New-York.

Peng, Z., Waters, E.J., Pocock, K.F., Williams, P.J., 1996a. Red wine bottle deposit, I: a predictive assay and an assesment of some factors affecting deposit formation. Aust. J. Grape Wine Res. 2, 25–29.

Peng, Z., Waters, E.J., Pocock, K.F., Williams, P.J., 1996b. Red wine bottle deposit, II: cold stabilization is an effective procedure to prevent deposit formation. Aust. J. Grape Wine Res. 2, 30–34.

Pilone, B.F., Berg, H.W., 1965. Some factors affecting tartrate stability in wine. Am. J. Enol. Vitic. 16, 195–212.

Poncet, C., Cartalade, D., Putaux, J.L., Cheynier, V., Vernhet, A., 2003. Flavan-3-ol aggregation in model ethanolic solutions: incidence of polyphenol structure, concentration, ethanol content and ionic strength. Langmuir 19, 10563–10572.

Poncet-Legrand, C., Cabane, B., Bautista-Ortin, A.B., Carrillo, S., Fulcrand, H., Perez, J., et al., 2010. Tannin oxidation: infra-versus intermolecular reactions. Biomacromolecules 11, 2376–2386.

Prakash, S., Irtumendi, N., Grelard, A., Moine, V., Dufourc, E., 2016. Quantitative analysis of Bordeaux red wine precipitates by solid-state NMR: role of tartrate and polyphenols. Food Chem. 229–237.

Ribéreau-Gayon, P., Glories, Y., Maujeau, A., Dubourdieu, D., 2006. Clarifying wines by centrifugation and filtration, Handbook of Enology, second ed. Jon Wiley & Sons, Ltd, Chichester.

Riou, V., Vernhet, A., Doco, T., Moutounet, M., 2002. Aggregation of grape seed tannins in model wine—effect of wine polysaccharides. Food Hydrocolloids 16, 17–23.

Ripperger, S., Altmann, J., 2002. Cross-flow microfiltration—state of the art. Sep. Purif. Technol. 26, 19–31.

Rodriguez-Clemente, R., Correa-Gorospe, I., 1988. Structural, morphological and kinetic aspects of potassium hydrogen tartrate precipitation from wines and ethanolic solutions. Am. J. Enol. Vitic. 169–179.

Romat, H., 2007. Coefficient de colmatage, une nouvelle approche de la filtrabilité des vins—Part 1. Revue des œnologues no. 123, 31–33.

Romat, H., Reynou, G., 2007. Proposition de critères de filtration en application du coefficient de colmatage. Revue des œnologues no. 124, 36–38.

Scollary, G.R., Pasti, G., Kallay, M., Blackman, J., Clark, A.C., 2012. Astringency response of red wines: potential role of molecular assembly. Trends Food Sci. Technol. 27, 25–36.

Soares, S., Mateus, N., de Freitas, V., 2012. Carbohydrates inhibit salivary proteins precipitation by condensed tannins. J. Agric. Food. Chem. 60, 3966–3972.

Tanahashi, H., Nishino, H., Tohjo, K., Nakai, T., 1987. Nucleation and crystal growth of potassium bitartrate in model solutions, white wines and red wines. In: Nyvlt, J., Zacek, S. (Eds.), Industrial Crystallization. 10th Symposium. Elsevier Science Publishers, Amsterdam, Bechne.

Tolin, S., Pasini, G., Curioni, A., Arrigoni, G., Masi, A., Mainente, F., et al., 2012. Mass spectrometry detection of egg proteins in red wines treated with egg white. Food Control 23, 87–94.

Vallée, D., Bagard, A., Salva, G., Raoulx-Pantacacci, N., Bourde, L., Lavergne, C., et al., 1995. Determination of the temperature of saturation, of the field of supersaturation and of the "rules" of stability towards the calcium tartrate in wines. J. Int. Sci. Vigne Vin 29, 143–158.

Vernhet, A., Moutounet, M., 2002. Fouling of organic microfiltration membranes by wine constituents: importance, relative impact of wine polysaccharides and polyphenols and incidence of membrane properties. J. Membr. Sci. 103–122.

Vernhet, A., Cartalade, D., Moutounet, M., 2003. Contribution to the understanding of fouling build-up during microfiltration of wines. J. Membr. Sci. 211, 357–370.

Vidal, S., Williams, P., Doco, T., Moutounet, M., Pellerin, P., 2003. The polysaccharides of red wine: total fractionation and characterization. Carbohydr. Polym. 439–447.

Villamor R.R., Ross C.F., Wine matrix compounds affect perception of wine aromas. Annu. Rev. Food Sci. Technol. 4(1), 1–20.

Villettaz, J.-C., Steiner, D., Trogus, H., 1984. The use of a beta glucanase as an enzyme in wine clarification and filtration. Am. J. Enol. Vitic. 35, 253–256.

Waters, E.J., Peng, Z., Pocock, K.F., Jones, G.P., Clarke, P., Williams, P.J., 1994. Solid-state 13C NMR investigation into insoluble deposits adhering to the inner glass surface of bottled red wine. J. Agric. Food. Chem. 1761–1766.

Wigand, P., Tenzer, S., Schild, H., Decker, H., 2009. Analysis of protein composition of red wine in comparison with rosé and white wines by electrophoresis and high-pressure liquid chromatography—mass spectrometry (HPLC–MS). J. Agric. Food. Chem. 57, 4328–4333.

Wucherpfennig, K., Dietrich, H., 1989. The importance of colloids for clarification of musts and wines. Vitic. Enol. Sci. 44, 1–12.

Zanchi, D., Vernhet, A., Poncet-Legrand, C., Cartalade, D., Tribet, C., Schweins, R., et al., 2007. Colloidal dispersions of tannins in water-ethanol solutions. Langmuir 23, 9949–9959.

CHAPTER 17

Sensory Analysis of Red Wines for Winemaking Purposes

Pablo Ossorio[1] and Pedro Ballesteros Torres[2]

[1]Winemaker and Oenology Consultant, Bodegas Hispano-Suizas, Ctra. Nacional 322, Valencia, Spain
[2]Agronomical Engineer, Master of Wine, Av Bourgmestre Herinckx 16, Brussels, Belgium

17.1 TASTING OF GRAPES

The main aim of tasting grapes is to determine the best moment to harvest. Since grapes are solid and differentiated, they cannot blend, it is of particular importance to apply a sound sampling procedure.

17.1.1 Field Sampling

By their nature, grapes tend to have relevant degrees of heterogeneity among them. Since they get together in the vats during winemaking, it is crucial for the decision maker to have as precise as possible a picture of the average quality, state of health, and ripeness of the plot of land.

As a previous step, sampling will be more precise and useful if there is good knowledge of soils, exposures, vines diversity, and in general all key vineyard descriptors. It will also benefit installed capacities to harvest by uniform plots. A compromise between an ideal situation in which all grapes come from an uniform plot, the possibilities in the winery to separate lots, and the expected commercial value of the wine needs to be reached in each occasion.

Once the plots have been classified, up to whichever level of detail is possible and convenient, a sampling protocol can be applied. It is crucial that the sampling protocol is consistent from year to year, in reference to operations as well as the sampling points. By keeping consistency, the vine grower will build the vineyard's track record and gain the necessary experience to understand the vineyard.

A minimum of 100 grapes are collected for each tasting exercise and each plot. Those grapes shall come from all parts in the bunch—shoulders, middle, and lower tip—in a predetermined and fixed proportion (to be decided upon cluster shape and tightness, principally). In general, grapes tend to ripen earlier when they are closer to the vine trunk. Grapes from the tip of the bunch take some more time to ripen.

17.1.2 Grape Tasting

The objective of the sensory analysis of the grapes in the field is to assess their composition and balance, as well as their ripeness. Tasting is undertaken in four consecutive phases: visual (with the eyes), tactile (with hands and palate), flavors (palate), and smell (nose and palate).

The visual phase's main objective is to assess the sanitary quality of the grape—mostly that it is not infected by fungus or harmed by pests, and that it does not show any viral or structural malformation. It is also useful to ascertain the color and its intensity.

The tactile phase is exclusive to grape tasting, because of the solid nature of grapes, and provides crucial information about the grape's maturity. A number of indicators must be observed and registered.

First, when collecting the grapes, the operators can have an idea on ripening degree by assessing the resistance offered to its separation from the pedicel (more resistance less maturity, obviously). The state of lignification of the stalks and the pedicel should also be registered (colors tending toward red and brown indicate ripeness, lignified stalks are also indicative of ripeness). It is very important to keep in mind that those indications are relative. Some grape varieties' stalks tend to ripen quite easily, while others never reach full lignification, although the fruit is in full maturation.

Crushing the grape with the finger is the first really tactile operation. When ripe, grapes tend to be softer and less resilient to pressure, breaking easily. Again, this is a relative measure, which makes sense only if compared to similar exercises with the same variety in a similar plot or, even better, if compared with previous vintages in the same plot. Vintage variability makes things more difficult. In warm and drought-affected seasons, for instance, grapes tend to be more flaccid, while other conditions do not change.

Tasting in palate is the most relevant phase in terms of the information to be acquired. The operation should provide data on the grape's quality potential in terms of fruit and aroma precursors, the acidity, the sweetness, the bitterness, and the tannicity.

In order to taste adequately, it is first of all necessary to separate pulp, skins, and seeds since each one of those grape components contributes to the wine in a different way. The physical process of separating skin, sedes, and pulp provides useful information: the easier the separation, the riper the grape (upon ceteris paribus conditions: all other conditions equal).

The pulp should be tasted first, with a carefully applied protocol, aimed to ensure that all grapes are tasted the same. Normally, it is chewed in a homogeneous way, four or five times. The taster should then note her evaluation concerning sugar concentration in the pulp, any mineral impression and the acid tasting, with discernment between malic and tartaric acid. Malic acid is a key indicator of ripeness. Depending upon varieties, wine types, and climates, malic acid should be well below 2 g/L when harvesting, so that its presence in grapes should be barely detectable for grape tasters.

The pulp's sweetness is determined by the concentration of glucose and fructose, but it is partially compensated when tasting by the acidity (mostly tartaric) in the grape. It is of crucial importance not to underestimate the sweetness of grapes because of this cross-effect, since an error in this appreciation can result in jammy overripe high-pH wines. This is of particular importance in the present times, when the still not well known effects of climate change result in gaps between commercial, phenolic, and acid maturity. The pulp tends to be riper and sweeter in the central part, while it is slightly greener closer to the skins and closer to the seeds.

Skin tasting is paramount to determine the grape's state in what concerns polyphenolic maturation. As for the pulp, the taster should chew the skins in a homogeneous way, three to four times, following exactly the same operation for each grape in the tasting lot.

The most relevant characteristics to note and register are the skins tasting profile, flavor intensity, and astringency. When tannins are not ripe, grapes' skins tend not to be intensely flavored, give a strong feeling of dryness, and show herbaceous notes. After chewing, it is most convenient to note the color of the skins, and how quickly it gets out, which tends to give a good indication of its extractability. Besides, the riper the grape, the more intense the color found in the skins once chewed, and the quicker the color extraction when chewing.

The grape seeds provide a lot of information about its state of phenological maturity. Tasting the seeds is essential to define the aging profile for the wine. It is also necessary to consider, when tasting seeds, that the proportion of seeds is variable. Such a proportion is to be noted each year, since it can complicate winemaking if excessive.

The seeds, which are kept apart from pulp and skins, should be bitten with the teeth. Ripe seeds tend to be crunchy, not to be very astringent, and to release nutty flavors when bitten. Green seeds are hard to crack and very drying, releasing herbaceous and green aromas when open.

The grape tasting always has to be accompanied by the chemical analysis of the different components of the grape, and with all the parameters the winemaker will be the one who decides at what time the harvest should start depending on the type of wine she wants to make.

17.2 TASTING IN THE PRODUCTION OF RED WINE

Once grape are harvested and brought to the cellars, for fermentation and maceration, daily tasting of each vat becomes a very important activity enabling the winemakers to assess and keep under control the transformation of the must in the wine that is intended.

As for the grapes, a clear tasting protocol, to be respected scrupulously, is of vital importance. A most important operation is to make each sample representative of its vat, by homogenizing the liquid, by pumping over or any other available system. The first tasting, before fermentation starts, should confirm the results of grape tasting in terms of quality and potential. Fresh fruit aromas, and any other notes typical of the variety or the place, as well as the initial color indications, should be consistent with the results in the vineyard.

Once yeasts are added, or when natural fermentation starts, tasting the must becomes a less useful evaluation, since the organoleptic characteristics are hidden by the yeast activity. When the fermentation is well under way, the must in transformation recovers and gets new aromas as a consequence of the activity of the yeasts.

The cold maceration of the first days of fermentation results in color intensification and the appearance of newly formed aromatic components, but at that moment it is quite difficult to get a precise idea on developments through tasting, because sugar is very high. Besides, tannins seem to be much softer than in the final wine, because of the very low concentration of alcohol in the first stages of fermentation (tannins extracted in water taste softer than in hydroalcoholic solution).

Tasting during the beginning of the fermentation should be perceived mainly in the mouth, through a slight sensation of carbon in the tip of the tongue. The tasting of the must-wine at the beginning of the fermentation is usually very appealing, since the fermentation frees primary aromas and synthesizes secondary aromas that are more volatile with the presence of carbon dioxide in the liquid, and so more intense to the nose.

It is easily perceived, on a day to day basis, how sweetness reduces and acids and tannins appear more prominent. If there is insufficient ripeness in the grapes, it shows up already in these early stages, with herbaceous and/or vegetal notes that become more nuanced later on in the winemaking process.

Once the tumultuous fermentation begins, tasting becomes again complicated, because there is a lot of background "noise" that can interfere in the perception of aromatic profile and structure of the wine. In particular, sulfur compounds that are formed by lack of oxygen and easily assimilable nitrogen in wines during fermentation are quite prominent.

It is crucial to pay attention to flaws that may appear during fermentation. Some of them are transient (can be corrected during fermentation), others are structural, and may result detrimental to wine quality.

Flaws produced by sulfides and sulfur compounds are easily identifiable because of their rotten eggs, gas, dirty water notes. These compounds are produced by a high dose of sulfur at the beginning of fermentation, a high pH in the medium, or because of high fermentation temperatures that result in a lot of biomass, or lack of assimilable nitrogen and organic residues that absorb oxygen. The application of organic and inorganic nutrients together with oxygen serves to eliminate those compounds during fermentation, so that the off-odors disappear and the aromatic character of the base wine appears again.

If not properly managed, those sulfur flaws can result in mercaptanoethanol, which gives rag, gunpowder, farmyard odors. Those odors are permanent; in order to eliminate them it is necessary to treat with copper sulfate or silver chloride.

At tasting during fermentation, it is also possible to detect amyl notes, often described as pharmacy or medicinal, mainly due to excess of inorganic ammonium, at high pH and high fermentation temperatures. This aroma is very persistent and will remain in the wine.

Pyrazines, particularly typical in wines made with cabernet and merlot varieties, can be avoided by allowing over ripeness in the grapes.

Vegetal aromas of the type of freshly cut grass, green beans, crushed leaves, etc., are also present during the fermentation of grapes from plots with lack of phenological maturity or excessive yields. The treatment with protein of animal, vegetable, or fish origin decreases the effect significantly.

A more serious problem is the detection through tasting and analysis of acetic acid and ethyl acetate. They can be produced at moderate quantities during the fermentation, with neutral or positive effects, but when there are fermentation arrests those problems can become grave, and difficult to eliminate from the wine. Sour smells, solvent, nail polish note, etc. are indicative of those problems.

Deciding when to rack off the fermentation vats is crucial in defining the type of wine to be produced. Depending upon oxygenation strategies during fermentation, wines to be sold young are racked after a shorter period of time than wines that will be aged in oak. The key tasting notes to consider for racking decisions are based upon the tactile profile, indicating the tannin extraction in the wine and the color intensity. Wines racked at high densities, above 1020, are normally made as young wines, built upon soft harmonious round tannins, rather than more abundant firmer tannins typical of wines for aging.

Just after the finalization of the alcoholic fermentation, the red wine is very complicated to taste, because there is far too much organic matter with strong taste, together with malic acid that highlights the tannin's astringency.

However, tasting becomes crucial immediately after, if it is decided to apply microoxygenation before malolactic fermentation. The most accurate way of deciding ethanal contents before the second fermentation is sensory analysis.

17.3 TASTING DURING MALOLACTIC FERMENTATION

The malolactic fermentation is the second fermentation that takes place in a red wine to ensure its stability and harmony in the mouth. Lactic bacteria carry out this fermentation, which should be tightly controlled in order to avoid possible organoleptic alterations in the wines. Red wines with malolactic effect are much more stable aromatically and microbiologically over time.

If the red wine is high pH, low alcohol, and low SO_2, malolactic fermentation is highly likely to trigger spontaneously in uncontrolled conditions. Upon those conditions, it is probable that the wine acquires aromas of putrefaction, which are associated with molecules such as Putrescine, or sweat aromas, stagnant water, which are indicative of Cadaverine. Both Putrecine and Cadaverine are organic molecules derived from the enzymatic decarboxylation of some amino acids.

Lactic bacteria will use up all acetaldehyde produced during microoxygenation for those red wines that have been produced under this process. The wines will then develop from initial aromas of freshly cut apple and a certain bitterness in the mouth to a marked ripe fruit character, with softer tannins and less green impression.

The end of the malolactic fermentation is indicated by the presence of diacetyl, which in low proportions reduces the fruit character of the wine and in higher quantity releases notes of vanilla, milk, and caramel.

Some bacteria strains can transform cinnamic acids present in the grapes into vinyl phenols, which are the substrate for the synthesis of ethyl phenols, which release odors of barnyard, leather, horse sweat, in the case of ethyl phenol, and aromas of burnt gum and spices, in the case of ethyl guaiacol.

17.4 CONCLUSIONS

The use of tasting at all the steps of the winemaking process and in the control of grape quality is an indispensable tool to modulate wine flavor and to correct organoleptic deviations. Moreover, it is critical to take the suitable enological decisions to reach the wine type one is intending. It needs the sensory ability but also winemaking experience to predict how the evolution of the wine will proceed in order to reach the sensory quality that consumers are expecting.

CHAPTER 18

Management of Astringency in Red Wines

Alvaro Peña-Neira

Department of Agro-Industry and Enology, Faculty of Agronomical Sciences, University of Chile, Santiago, Región Metropolitana, Chile

18.1 INTRODUCTION

Wine contains many phenolic substances, most of which originate in the grape berry. These phenolics influence the nutrition, color, and other sensory properties of wine. One of the most important sensory characteristics of red wine is astringency. Among the phenolic compounds in wine, condensed tannins produce sensations of astringency in food and drink and form the "structure" or "body" of red wine.

Tannins, a name that originates from their use in the tanning of leather, have been utilized by humans over decades for their capability to bind with proteins (Gawel, 1998). Condensed tannins (also called proanthocyanidins or PAs) are often incorporated by flavan-3-ol monomers which are biosynthesized via a complicated pathway beginning with phenylalanine (less often tyrosine) and ending in the polymerization of PAs (Ali et al., 2010).

The oral sensation referred to as "astringency," most commonly described as "drying," "roughing," and "puckering," is a primary mouthfeel attribute in red table wine (Breslin et al., 1993). From a perceptual point of view, there is little or no consensus about whether astringency is a single sensation or a general category made up of multiple subquality sensations. According to the definition of American Society for Testing and Materials, astringency refers to "the complex of sensations due to shrinking, drawing or puckering of the epithelium as a result of exposure to substances such as alums or tannins" (ASTM, 2004).

The sensation of astringency is thought to be primarily a result of aggregation of protein—tannin complexes and subsequent increases in friction of wine during ingestion (Dinnella et al., 2009; Baxter et al., 1997a,b), however these interactions are only part of the complex sensation (Gawel et al., 2000; Hufnagel and Hofmann, 2008; Gawel, 1998). Astringency has been shown to be a multiple perceptual phenomenon. Using focus groups, Lee and Lawless (1991) generated the orthogonal terms "drying," "puckering," "sour," "astringent," "bitter," and "rough" to describe the sensations arising from the astringent compounds alum, gallic acid, and tartaric acid. Later, Lawless et al. (1994) developed a vocabulary to describe the mouthfeel characteristics of the astringents alum, gallic acid, catechin, citric acid, and their mixtures. The sensory categories were defined as: (1) "drying," the lack of lubrication or moistness resulting in friction between oral surfaces; (2) "roughing," the unsmooth texture in the oral cavity marked by inequalities, ridges, and/or projections felt when oral surfaces come in contact with one another; (3) "puckery," the drawing or tightening sensation felt in the mouth, lips and/or cheeks; and (4) "astringency," a complex sensation involving attributes of the three preceding categories. The spectrum of subtle differences in astringency sensations was compiled as a "red wine mouthfeel wheel" by Gawel et al. (2000), which includes such descriptors as "powder," "adhesive," and "aggressive."

It is widely acknowledged that high quality red wines have a balanced level of astringency. When in excess, the astringency detracts from other components, while with too little, the wines may taste "flat," insipid and uninteresting. The perception of astringency is a highly dynamic process that changes continuously during ingestion, especially following expectoration or swallowing (Noble, 1995). Since astringency influences the quality of red wine (Boselli et al., 2004; Landon et al., 2008), knowledge of the structures of astringent compounds in a wine matrix and the impact of these structures on the sensory properties are important aspects of winemaking.

18.2 ASTRINGENCY IN WINES

18.2.1 Mechanisms of Astringency

Due to its importance, astringency has been comprehensively studied in wine over the last decades and some important reviews have been published (Gawel, 1998; Bajec and Pickering, 2008; de Freitas and Mateus, 2012; Ma et al., 2014; Upadhyay et al., 2016; Garcia-Estevez et al., 2017). Although the exact mechanisms of astringency are not well understood, astringency is generally considered to be a tactile sensation caused by a loss of lubricity in oral saliva. Salivary proteins covalently bind to oral mucosal cells and form a layer surrounding the soft structure of the mouth. When tannins pass by the mucosal layer, they bond to salivary proteins to form insoluble tannin--protein precipitates, thus increasing friction and reducing lubrication in the oral cavity (Baxter et al., 1997a,b; Dinnella et al., 2009; Luck et al., 1994). This process causes a drying and grainy sensation in the mouth. This sensation has been shown to differ based on the size and concentration, as well as the hardness or softness, of the precipitate (de Wijk and Prinz, 2005; Zanchi et al., 2008). Any remaining unbound tannins may interact with other oral surfaces. Experiments by Payne et al. (2009) demonstrated that tannins also interact with oral epithelial cells or salivary proteins adhered to buccal mucosal cells (Nayak and Carpenter, 2008).

18.2.1.1 Salivary Proteins

Saliva is approximately 99% water and 1% protein and salts. Saliva primarily functions as lubrication to protect the oral mucosa, however it also aids in the digestion of food and cleansing of the oral cavity, reduces the damaging effects of acid on oral tissue by buffering action, maintains supersaturated calcium phosphate concentrations, promotes antimicrobial action, and helps to facilitate speech. Normal daily salivary flow ranges from 0.5 to 1.5 L. Müller et al. (2010) observed a panel of 13 participants and found that this group has between 230 and 1310 μL ($\chi = 696 \pm 312$) of saliva in their mouth at a given moment. Salivary flow rate, viscosity, and protein composition vary between people and the latter two variables have been shown to have a significant effect on perceived astringency (Demiglio and Pickering, 2008; Dinnella et al., 2009; Condelli et al., 2006). Higher concentrations of particular salivary proteins and a higher flow rate of saliva have been shown to generally reduce the sensation of astringency (Kallithraka et al., 2001; Condelli et al., 2006). Even though the sensation of astringency is due to a decrease in salivary viscosity, baseline salivary viscosity levels do not directly influence astringency perception as is true for salivary protein composition and flow rate (de Wijk and Prinz, 2005).

Several studies have been conducted on the binding of tannins to salivary proteins because of the potential importance of these interactions as a biological defense against the deleterious effects of tannins on digestive system enzymes. These studies have indicated that saliva contains a complex mixture of protein families and isoforms. Protein families found in saliva include proline-rich proteins (PRPs), statherin, cystatins, P-B peptide, histatins are histidine-rich-peptides (HRPs), and mucins. There appears to be a significant amount of functional redundancy among salivary proteins. The reason for this functional redundancy is not clear, however it may be a mechanism to ensure that a given function is always present. Specifically, functional redundancy may help to ensure protein expression under a broad range of physiological conditions which would not be possible with a single protein. This functional redundancy is well-represented by two families of salivary proteins, PRPs and histatins, which, among other activities, share an ability to readily precipitate tannins (Bennick, 2002; Messana et al., 2008).

PRPs seem to be the most important family of salivary proteins concerning astringency (Gawel, 1998; Bennick, 2002). Most proteins in human parotid saliva and many proteins in submandibular/sublingual saliva (about 70% of the proteins) belong to a unique multigene family of PRPs (Azen and Maeda, 1988). More than 20 PRPs are present in human saliva, and they are divided into acidic PRPs (aPRPs) (Mr 12–16 kDa) (Bennick and Connel, 1971; Oppenheim et al., 1971; Friedman and Merritt, 1975), basic PRPs (bPRPs) (Mr 6–9 kDa), and glycosylated PRPs (Mr 36–78 kDa) (Levine et al., 1969; Armstrong, 1971; Friedman et al., 1971; Levine and Keller, 1977; Arneberg, 1974; Degand et al., 1975; Manconi et al., 2016). Characteristically, proline, glycine, and glutamine together account for 70%–88% of all amino acids in PRP proteins, with a high content of proline residues (25%–42%) (Bennick, 2002; Manconi et al., 2016).

The bPRP family is the most studied family among the PRPs and is reported as having the highest affinity for tannins (Bennick, 2002; Lu and Bennick, 1998). However, recent in vitro and in vivo studies have reported a stronger interaction between polyphenols and aPRPs or P-B peptide than with bPRPs (Quijada-Morín et al., 2016; Brandao et al., 2014).

HRPs comprise a group of structurally related, small histidine-rich proteins found only in saliva. Histidine accounts for about 25% of all residues and, together with basic amino acids arginine and lysine, makes up

30%–75% of total amino acids. Twelve HRPs, named HRP 1–12, have been isolated from human saliva and their primary structures determined (Troxler et al., 1990). The most prominent members are HRPs 1, 3, and 5, and they account for 85%–90% of proteins of this family in saliva. All three HRPs have been found to bind and precipitate hydrolyzable tannins. HRPs only constitute 2.6% of total salivary proteins and tend to have lower molecular weight than PRPs (Yan and Bennick, 1995; Naurato et al., 1999).

18.2.1.2 Phenolic Compounds

Wines that are astringent are in common parlance termed "tannic." The term "tannin" was first used by Seguin in 1796 to describe the compound present in oak galls responsible for the formation of leather when hide was treated with an aqueous infusions of the galls (Gawel, 1998). Swain and Bate-Smith (1962) provided the first useful phytochemical definition of tannins as "water-soluble phenolic compounds, having molecular weights lying between 500 and 3000, which have the ability to precipitate alkaloids, gelatin and other proteins."

Hydrolyzable tannins (gallo- and ellagitannins) and condensed tannins are the most important polyphenolic compounds present in wines that are able to interact with salivary proteins. As such, these compounds present in wine are the most related to astringency perception. However, other wine phenolic compounds, such as flavonols, phenolic acids, and anthocyanins, can also play an important role in astringency perception (Thorngate, 1993; Thorngate and Noble, 1995; Gawel, 1998).

18.2.1.2.1 Proanthocyanidins

Monomeric flavan-3-ols are widely considered to combine to produce condensed tannins. Flavan-3-ol monomers (also known as catechins) are responsible for bitterness in wine and may also be associated with astringency. The major flavan-3-ol monomers found in grapes and wine include (+)-catechin, (−)-epicatechin, and (−)-epicatechin-3-O-gallate (Su and Singleton, 1969). (+)-Catechin and (−)-epicatechin differ around the two stereo centers of the flavan-3-ols, (+)-catechin having the 2,3-*trans* configuration and epicatechin being 2,3-*cis*. These features of the molecules become important when considering their biosynthesis. Early work found these compounds to be particularly high in seed material and that flavan-3-ol monomers are produced before the onset of ripening (*veraison*) and change during the fruit ripening process (Romeyer et al., 1986). Grape seeds have higher amounts of flavan-3-ol than skins; only procyanidins [(+)-catechin and (−)-epicatechin] have been found in the former, while in the latter prodelphinidins (gallocatechin and epigallocatechin) are also present (Escribano-Bailon et al., 1992, 1995). Grape skin tannins consist of long polymeric chains composed of procyanidins and prodelphinidins (Brossaud et al., 2001; Vidal et al., 2004a,b; Cheynier et al., 1997; Kennedy et al., 2001; Cosme et al., 2009; Souquet et al., 1996). The trihydroxylated prodelphinidin subunits consist mainly of epigallocatechin, but with trace amounts of gallocatechin and epigallocatechin 3-O-gallate (Souquet et al., 1996). Qualitative and quantitative differences in PA content have been observed between skins and seeds from the same variety.

PAs are a class of compounds that have been variously described as anthocyanogens, leucoanthocyanidins, flavan-3,4-diols, condensed tannins, and tannins. In current grape and wine literature, these compounds are generally called tannins or PAs. PAs impart astringency to red wines and are extracted from the skin, seeds, and stems of the grape berry. These compounds were the last of the major phenolic compounds (based upon abundance) to be determined structurally (Creasy and Swain, 1965), but their sensory properties have been studied much longer (McRae and Kennedy, 2011). The chemical structure of the flavan-3-ols, (+)-catechin and (−)-epicatechin, compared to the known structures of tannin polymers, suggests a precursor product relationship; that is, the monomeric flavan-3-ols are widely considered to combine to produce tannins. However, the way in which the 4–6 and 4–8 interflavan bonds are formed remains an important unknown in plant biology.

Indeed, there are at least two closely related problems of tannin biosynthesis from the corresponding flavan-3-ols. The first is how interflavan bonds are formed and in the second is in which compartment of the plant cell the reaction takes place. If the bond is formed in the vacuoles where the tannins are ultimately located, then the issue of how they are transported across the tonoplast does not arise. However, if the interflavan bonds are formed in the cytoplasm or in organelles such as plastids, then the question of how tannins are transported across membranes becomes important (Zhao et al., 2010). It has been shown that mutants of *Arabidopsis* with a defective plasma membrane H^+-ATPase are unable to accumulate tannins in the vacuoles of the seed coat endothelial cells (Baxter et al., 2005), suggesting that tannin accumulation requires proper endomembrane trafficking and that at least one of the requisite H^+-ATPases plays a specific role in tannin biogenesis and vacuole formation. This research suggests that understanding how tannins are transported across membranes is essential to knowing how they accumulate in grape skin hypodermal cells and the soft parenchyma of the outer seed coat.

Since PAs are polymers of flavan-3-ol subunits, a wide range in molecular weight is possible. Progress in fully understanding PA structure has been hindered because of limited subunit variation. The seed and skin PAs have medium degree of polymerization (mDP) of around 10 and 30 subunits, respectively. Epicatechin, and to a lesser extent epigallocatechin, is the extension unit and catechin is the terminal unit in grapevine PAs (Terrier et al., 2009).

Kennedy et al. (2001) showed that the mDP of grape skin tannins increased from 7.3 subunits in green berries of Shiraz 3 weeks after fruit set to 11.3 subunits in red berries at *veraison*. By commercial harvest, the mDP of skin PAs in Shiraz fruit reached 27 subunits. Seed tannins have a lower average degree of polymerization than skin tannins and are composed mainly of catechin and epicatechin subunits. Seed tannins have a greater proportion of galloylated units (13%−29%) compared with skin tannins (3%−6%) (Cheynier, 2006; Riou et al., 2002). The pattern of polymerization of seed tannins is less clear than that in skins. mDP determinations by thiolysis shows a decline of seed tannins from 8.29 subunits at *veraison* to 5.63 during ripening, whereas estimates made by HPLC predicted an mDP between 11 and 15 subunits as the fruit ripened (Kennedy et al., 2000a).

Differences in extractable PAs according to the degree of polymerization also exist between grape skins and seeds. In the cultivar cv. Cabernet Sauvignon, seeds have higher concentrations of monomers, oligomers, and polymers of flavan-3-ol than the skins. In both skins and seeds, polymers are present in the highest percentages ($\approx$95% and 78% for skins and seeds, respectively) followed by the oligomers ($\approx$3% and 18%, respectively) and the monomers ($\approx$2% and 4%, respectively) (Cáceres-Mella et al., 2012).

Other authors have reported that the content of monomers [(+)-catechin and (−)-epicatechin], procyanidin dimers (B3, B1, B4, and B2), trimers (T2 and C1), and dimer gallates (B2-3-O-gallate, B2-3′-O-gallate, and B1-3-O-gallate) ranged from 2.30 to 8.21 mg/g in grape seeds, and from 0.14 to 0.38 mg/g in grape skins. In grapes, the polymeric fraction represented 75%−81% of total flavan-3-ols in seeds and 94%−98% of total flavan-3-ols in skins. In these studies, seeds and skins had mDP values of 6.4−7.3 and 33.8−85.7, respectively (Monagas et al., 2003).

There are key structural characteristics that govern the ability of PAs to interact with proteins and affect astringency, namely the molecular weight, the interflavanic bond, mDP, protein conformation, the presence of galloyl groups (galloylation), the substitution pattern of the B ring, and the specific linkage of tannins subunits and their stereochemistry.

With regard to mDP, the mechanism of PA polymerization is unknown, but it is thought to occur in the vacuole under the acidic conditions induced by tonoplast proton pumps (H^+-ATPases) (Zhao et al., 2010; Davies, 1997). Following transport of glycosylated flavan-3-ol precursors from the cytoplasm through the tonoplast (Zhao and Dixon, 2009), glucose is cleaved and the subsequent condensation of a flavan-3-ol terminal subunit to a quinone methide or protonated carbocation form would occur under acidic pH conditions, yielding a PA molecule (He et al., 2008; Hemingway and Foo, 1983). In relation to the molecular weight and mDP, it seems that the more polymerized, the more astringent the PAs (Baxter et al, 1997a,b; Hufnagel and Hofmann, 2008). Soares et al. (2007) observed that the protein binding affinity of polyphenol compounds increases with their molecular weight. In grape tannins extract, highly polymerized tannins predominantly precipitate together with salivary proteins, while the condensed tannins remaining in solution are low molecular weight polymers (Sarni-Manchado et al., 1999).

Higher degrees of tannin polymerization attract more hydrophobic groups to the tannins, which in turn increases the binding affinity between tannins and proteins (Chira et al., 2009; Scollary et al., 2012). However, when the polymerization degree of condensed tannins is too high, self-precipitation is induced (Scollary et al., 2012).

Galloylation is the esterification of PAs with the addition of gallic acid. Galloylation plays a significant role in the protein binding capability resulting in a rise of astringency intensity (Scharbert et al., 2004; Ferrer-Gallego et al., 2010; Kawamoto et al., 1996; Soares et al., 2007). The rise in astringency can be attributed to the ability of galloyl groups to establish hydrogen bonds with the proteins, thus strengthening and stabilizing the interaction (Charlton et al., 2002a,b; Li and Hagerman, 2014). Furthermore, the rise of astringency intensity induced by galloylation is probably due to more aromatic and hydroxyl groups, which enhances the hydrophobic interaction and formation of hydrogen bonds (Charlton et al., 2002b).

PAs can be divided in two groups according to their B-ring substitution pattern: the dihydroxylated group, composed of catechins which polymerize to create procyanidins; and the trihydroxylated group, composed of gallocatechins which are present in prodelphinidin structure. Hagerman (1992) showed that the pattern of hydroxylation on either the A- or B-rings of condensed tannins influences protein interactions. For example, prodelphinidins, which contain three o-hydroxy groups on the B-ring, have a higher affinity for proteins than procyanidins, which have only two such groups (Schwarz and Hofmann, 2008). A number of simple phenols that possess only a single dihydroxyphenyl residue, including (+)-catechin, (−)-epicatechin, 5-O-caffeoylquinic acid, and salicylic acid, elicit astringency at high concentrations (Clifford, 1997).

Both prodelphinidins and procyanidins are present in grape skins, whereas only procyanidins are present in seeds. The differences in reactivity and astringency between these two groups of condensed tannins affect the levels of these compounds in the final wine product, therefore, present an important consideration in the winemaking and extraction processes (Llaudy et al., 2008).

The hydroxylation on the A-ring strongly influences stereochemistry at C4. Procyanidins and prodelphinidins with 2,3-cis stereochemistry frequently adopt 3,4-trans stereochemistry during interflavanoid bond formation with C4−C8 or C4−C6 interflavanoid linkages, occurring in relative proportion of 3:1. Dimers with these types of linkages are termed, B′-type and include the following: procyanidin B-1 [(−)-epicatechin-(4β→8)-(+)-catechin], B-2 [(−)-epicatechin-(4β→8)-(−)-epicatechin], B-3 [(+)-catechin-(4α→8)-(+)-catechin], B-4 [(+)-catechin (4α→8)-(−)-epicatechin] and B-5, a regioisomer of the B-2 dimer that possesses a C4→C6 interflavan linkage. Typically 2,3-trans stereochemistry dominates the chain extender units, although a minority exist in the 2,3-cis formation. PA oligomers most commonly terminate with (+)-catechin subunits (Ricardo da Silva et al., 1992; Peyrot des Gachons and Kennedy, 2003; Herderich and Smith, 2005).

The subunits of B-type PAs are mainly linked by two kinds of interflavanoid bonds (C4−C6 or C4−C8), which have different protein binding ability according to conformational changes. Dimers linked through C4−C8 have consistently greater tannin specific activity (TSA) for PRPs than their counterparts with a C4−C6 linkage (Ricardo da Silva et al., 1992; de Freitas and Mateus, 2001).

18.2.2 Methods for Astringency Analysis

A number of methodologies have been used to study astringency perception. The most effective method appears to be sensory evaluation (Gambuti et al., 2006). Perceived astringency intensity, time intensity (recording multiple parameters such as start time, end time, maximum intensity, time to maximum intensity, total duration, area under the curve, etc.), and descriptive analysis (description of sensation details such as mouthfeel and astringent subqualities) have all been employed for astringency sensory evaluation (Gawel et al., 2000; Ma et al., 2014).

Given the complexity of evaluating astringency perception, in vitro methods have been proposed to study astringency. Hagerman and Butler (1981) proposed a method based on the amount of radioiodinated bovine serum albumin precipitated by tannins, resulting in a measure referred to as the TSA. Subsequently, a saliva precipitation index and particle size distribution measurement were also introduced to evaluate the astringency intensity of tannins by the means of SDS−PAGE electrophoresis and dynamic light scattering methods, respectively (Pascal et al., 2007; Rinaldi et al., 2010). Cáceres-Mella et al. (2013) compared four methodologies for measuring PAs (two precipitation methods, a methylcellulose precipitation assay and a protein precipitation assay, two colorimetric methods, Bate-Smith and vanillin assays) and their relationship with perceived astringency. The authors observed that precipitation methods had a higher correlation with perceived astringency than colorimetric methods. Although precipitation assays have showed high correlations with perceived astringency, studies on tannin−protein interactions conducted on cellulose membranes demonstrated that a tannin−protein *interaction* is more closely associated with astringency than tannin−protein *precipitation* (López-Cisternas et al., 2007; Obreque-Slier et al., 2010a,b).

18.3 INFLUENCE OF WINEMAKING TECHNOLOGY IN WINE ASTRINGENCY

Winemaking technology is used to enhance wine quality attributes through the modulation of wine composition and its physicochemical characteristics. Several mouthfeel sensations such as color, bitterness, and astringency change depending on enological practices. Grape composition at harvest and winemaking technologies such as extraction techniques during alcoholic fermentation, stabilization, and fining treatment, among others, influence astringency.

18.3.1 Grape Ripening

In viticulture, ripeness is the completion of the maturation process of wine grapes on the vine which signals the beginning of harvest. A harvesting decision for "optimal ripeness" requires adequate knowledge of grape compositional factors relevant to achieve a targeted wine style, taking into consideration the grape cultivar, climate, topography, seasonal weather conditions, and vineyard management practices (Wu-Dai et al, 2011).

The physical and chemical components of the grape which will influence a wine's quality are essentially set upon harvest; berry composition does not change once harvested from the plant. However, the polyphenolic compositions of grape skin, seed and pulp change during the ripening process and consequently influence red wine astringency. Therefore, selection of the optimal moment of grape ripeness may be considered the most crucial decision in winemaking (Bindon et al., 2013).

There are several factors that contribute to the ripeness of the grape. As the grapes go through *veraison*, sugars in the grapes will continue to rise as acid levels fall. The balance between sugar (as well as the potential alcohol level) and acids is considered one of the most critical aspects of producing quality wine. In addition to total acidity measures, must soluble solids and the pH of the grapes are evaluated to determine ripeness.

Towards the end of the 20th century, winemakers began focusing on the concept of achieving "phenolic ripeness" in the grapes (Jackson and Lombard, 1992; Perez-Magarino and Gonzalez-San Jose, 2006). Phenolic ripeness is described as a higher evolution of tannins and other polyphenolic compounds like anthocyanins (colored compounds).

In general terms, tannin accumulation in grape seed and skin occurs early in berry development. Tannin accumulation ceases when the ripening process starts and tannin content remains unchanged or decreases after *veraison* (Quijada-Morín et al., 2016; Bogs et al., 2007).

In general, research has shown that gene expression for the biosynthesis of flavan-3-ol precursors to PAs is active up until the *veraison*, and is downregulated by the hormone abscisic acid which signals the onset of *veraison* (Bogs et al., 2005; Bogs et al., 2007; Lacampagne et al., 2010).

PA levels also decrease following *veraison* in both skin and seeds, and expression of genes encoding the enzymes that regulate the synthesis of the monomers of flavan-3-ols, leucoanthocyanidin reductase (LAR) and anthocyanidin reductase (ANR) are no longer detected. These results indicate that PA accumulation occurs early in grape development and is completed upon the onset of ripening. Both ANR and LAR contribute to PA synthesis in fruit, and the tissue and temporal-specific regulation of the genes encoding ANR and LAR determine PA accumulation and composition during grape berry development (Bogs et al., 2005).

PA localization in grapes occurs within the cytoplasmic compartment (vacuole) and is associated with cell-wall material (Gagne et al., 2006; Geny et al., 2003; Conn et al., 2010). Grape skin PAs in the vacuole are thought to be less polymerized than those associated with the cell wall. Grape skin PAs experience a shift in subcellular organization from the vacuole to the cell wall during the later stages of berry ripening (Gagne et al., 2006). Studies on other plant species indicate that this shift could be mediated by vesicle trafficking, in which PAs are transported via vesicles from the vacuole to the cell wall, later fusing with it (Zhao and Dixon, 2009; Zhao et al., 2010; Davies, 1997).

Grape skin tannins change little from *veraison* to harvest, thus, tannins in the hypodermal tissue appear to be made early in berry development. However, the concentration of skin tannin (mg/g fresh weight or mg/L) declines during ripening (phase III) in proportion to berry growth. Quantitative changes in tannin content have been observed, but these measures do not consider any qualitative changes (such as mDP) in tannin composition that might occur from *veraison* to harvest (Adams, 2006). The tannins extracted from the skin of commercially ripe grapes consist of a portion of anthocyanins covalently bound to the oligomeric condensed tannins (Bindon et al., 2010).

According to Kennedy et al. (2000a,b), in grape seeds of the cv. Shiraz, procyanidins are highest in the 3 weeks prior to *veraison* and increase little during the same period. On the other hand, the amount of flavan-3-ol monomers increases fivefold during the 3 weeks prior to *veraison*, which indicates that the procyanidins and the flavan-3-ol monomers accumulate at different stages. Grape seed polyphenols decrease dramatically during ripening; flavan-3-ol monomers decrease by 90% and procyanidins decrease by 60%. The decrease in flavan-3-ol monomers follows the order epicatechin-3-*O*-gallate > (+)-catechin > (−)-epicatechin (Kennedy et al., 2000a). At the onset of berry development, seeds are green, plump, and have pliable seed coats. However, beginning at *veraison*, the seeds turn brown in color, become desiccated, and the seed coats hardens. Polyphenol changes after *veraison* could be explained by oxidation and this oxidation could explain the changes in the seed color. Kennedy et al. (2000b) used electron paramagnetic resonance spectroscopy to follow the potential development of radical species in the developing seeds. They observed that the concentration of radicals remained low until *veraison*, increased beginning at that point, reached a maximum 3 weeks later, and then slowly declined thereafter. Changes in radical intensity, together with other documented changes in the seed are consistent with the hypothesis of an oxidative event occurring during fruit ripening.

To date, a clear relationship between the tannin concentration of grapes, and the resulting tannin concentration in wines is inconsistent, in particular with respect to grape ripeness (Fournand et al., 2006). Work by Ristic et al. (2010) demonstrated a strong relationship between grape skin tannin concentration and wine tannin concentration, while seed tannins showed a poor relationship. The relationship between skin tannin concentrations and commercial wine allocation grading highlights the need for a greater understanding of tannin partitioning from grape solids to wine as it relates to grape ripening (Kassara and Kennedy, 2011).

Canals et al. (2005) studied how grape ripeness and ethanol concentration affect the extraction of color and phenolic compounds from skins and seeds during the maceration/fermentation process. They observed that at all the stages of ripening (10 days after complete *veraison*, until complete ripeness), there were marked increase in ethanol concentrations (0%–13.0%) and in the total phenolic compound extracts in seeds and skins. In addition, these authors observed that higher ethanol concentrations in the maceration media was correlated with a higher the astringency of PAs. Therefore, the evolution of the ethanol concentration in the must and final wine product may be one of the main factors affecting the quality and quantity of phenolic extraction. Furthermore, these data suggest that the final extraction phenol content is dependent on the stage of grape ripening in regard to soluble solids concentration.

Choosing the optimal ripening date to control wine astringency affects not only phenolic ripeness but also other compounds, namely carbohydrates as part of the cell-wall material. Recent research suggests that astringency perception could be also related to these compounds (Watrelot et al., 2017; Boulet et al., 2016; Gil Cortiella and Peña-Neira, 2017).

Research suggests that polysaccharides influence astringency through either a direct mechanism, in which cell-wall polysaccharides bind to tannins reducing their extractability, or through indirect mechanisms via a competition process at a sensory level. PAs from pre-*veraison* skin or seed cell-wall material has greater affinity for higher molecular mass polymers, irrespective of their origin within the grape berry. However, PAs from ripe grape skins of higher molecular mass show reduced affinity for skin cell-wall material (Bindon et al., 2012). This observation suggests that the changes in molecular mass and binding properties with cell-wall material that take place following *veraison* may have implications for their extractability during vinification (Garcia-Estevez et al., 2017).

Several structural and chemical changes occur during ripening, leading to the softening of the cellular structure of many fruits. The most notable changes occur in hemicelluloses, pectins, and cellulose microfibrils of the cell wall (Fasoli et al., 2016). Pectin modification is mainly responsible for the progressive loss of firmness during ripening. Fruit softening allows the release of small pectin-soluble fragments of the cellular structure during ripening. Furthermore, the xyloglucan content of hemicelluloses decreases significantly during berry growth and ripening, whereas cellulose levels are only marginally affected (Bindon et al., 2012; Fasoli et al., 2016).

Post-*veraison* grape skin cell wall material has reduced affinity for higher molecular mass PAs than grapes at pre-*veraison*. Furthermore, the affinity of the cell wall for PAs generally decreases with ripeness Greater levels of PA polymerization during of extraction from riper grapes may therefore be expected due to a poor affinity of greater molecular mass PAs to skin cell wall material. It is of significance to note, however, that despite showing small decreases in the affinity for skin PAs from riper grapes, the adsorption of PAs by flesh cell wall material remains high. This means that the presence of flesh material in must during vinification may represent a significant sink for extracted PAs, preferentially adsorbing those of higher molecular mass. The interaction between skin and flesh cell wall material in the selective adsorption of PAs is therefore complex and wine PA properties such as astringency will be dependent upon both aspects (Bindon et al., 2012; Fasoli et al., 2016; Garcia-Estevez et al., 2017).

Considering the difficulty of assessing the impacts of ripeness on astringency, several winemaking techniques have been used to correct for final wine composition deficiencies derived from grapes harvested at suboptimal stages of ripeness.

18.3.2 Maceration and Fermentation

Although phenolics are present in grape juice, the amount is much greater in wine because of their greater solubility in ethanol than in purely aqueous solutions. The difference in solubility is especially true for tannins. This phenomenon leads to increased astringency.

Because of the large influence of phenolic compounds on red wine quality, many winemaking techniques have been developed to influence the extraction of these compounds during vinification. Most winemaking techniques have been directed at enhancing the extraction of phenolics. Winemaking variables and techniques such as pectolytic enzyme treatments, prefermentation juice runoff (also known as *saignée*), and maceration and fermentation temperature have been reported to increase phenolic concentration (Sacchi et al., 2005).

Prior to discussing these processes, it is worth noting that different factors limit the extraction of different classes of phenolic compounds. For example, extraction of tannins appears to be hindered by limited solubility in water at the start of the fermentation process, as indicated by saturation behavior over time (Watson et al., 1995; Gao et al., 1997). Extraction of tannins increases with increasing alcohol content, sulfur dioxide exposure, temperature, and skin contact time, which in turn increase the amount of tannin soluble in wine (Berg and Akiyoshi, 1958; Singleton and Draper, 1964; Ribéreau-Gayon, 1974; Ozmianski et al., 1986). Furthermore, there appears to be a synergistic effect with combinations of alcohol, sulfur dioxide, and temperature (Ozmianski et al., 1986). Thus far, evidence suggests that tannins are extracted from both skins and seeds.

Pectolytic enzymes have been used to increase wine color due to their ability to break down skin cell walls, thereby releasing pigments located inside the vacuoles (Ducruet et al., 1997; Castro-López et al., 2016; Bautista-Ortín et al., 2016a). The use of this maceration enzyme also results in an increase in the PA content of must and wine by favoring the extraction of these compounds from skin cell vacuoles and by promoting a lower adsorption of PAs on cell walls (Castro-López et al., 2016; Bautista-Ortín et al., 2016a). Due to their ability to increase the extraction of PAs, the use of pectolytic enzymes during maceration could promote higher astringency wines.

Prefermentation juice runoff, which is also referred to as *saignée*, is a process by which juice is removed before fermentation, thus increasing the skin to juice ratio. Since the anthocyanins and tannins extracted during fermentation are found in the skins and seeds and not in the juice, this practice should theoretically increase anthocyanin and tannin concentration in the finished wine. However, *saignée* could also have the opposite effect. If extraction is strictly solubility limited, the effect of *saignée* would be to decrease anthocyanin and tannin content since there would be less liquid in which the phenolics could be dissolved. Experiments over three vintages of Malbec showed that wines produced with *saignée* were characterized by more and larger tannins and by more polymeric pigment in all vintage (Zamora et al., 1994). This research suggests that *saingée* leads to increased extraction of these compounds, at least in Malbec wines.

Tannin extractability can vary considerably during winemaking, even under identical fermentation conditions. Winemaking techniques, such as punch-downs and punch-overs, have been developed to influence the extraction of phenolic compounds.

During fermentation of grape skins, a cap of solids develops over the liquid wine surface as carbon dioxide evolution causes the grape solids to rise to the top of the fermentation vessel. Separation of the liquids and solids in this manner has two potentially negative consequences. First, heat is trapped in the cap thereby affecting fermentation temperature, and second, contact between the bulk juice and the skins and seeds is reduced thereby reducing extraction efficiency. To overcome these problems, the skins and juice are usually mixed several times a day, either by pushing the skins below the juice surface (punch-down) or by pumping juice out and spraying it over the top of the skins (pump-over). Fermentation temperature and skin particle size and their ability to pass through the pump change throughout the fermentation process, therefore, the effect of a pump-over on phenolic content is dependent upon the timing of its occurrence during fermentation. The temperature the cap is allowed to attain prior to a pump-over is also an important variable, but its influence has not been adequately investigated. Since some type of juice mixing is always included in red wine fermentations, studies have focused on comparing the effects of different methods. Generally, studies indicate that the effect of juice mixing depends on the variety (Sacchi et al., 2005).

Early work showed a small increase in the color and tannin concentration with pump-overs *versus* punch-downs in Cabernet Sauvignon must fermentation (Ough and Amerine, 1961). A study with Pinotage comparing manual punch-down, pump-over, and rotary mixing found that extraction rates of total flavonoids, total tannins, and anthocyanins were slower for pump-overs and that lower concentrations were achieved as compared with the other methods (Marais, 2003a,b). Analysis of the total phenols content of the resulting wines showed that the highest concentrations were obtained with rotary mixing, followed by punch-down, then pump-over.

Pump-over has also been compared to *delestage* for the Italian variety Montepulciano d'Abruzzo (Bosso et al., 2001). *Delestage*, also referred to as rack and return, is a technique where all of the juice is drained from the tank and then poured back over the skins. As compared with *delestage*, pump-overs yielded higher anthocyanins, polymeric pigments, and tannins with the largest differences observed for tannins. Of course these results are likely dependent on the type of pump-over operation performed and the length of time it is performed.

At the end of fermentation process, the practice of extended maceration prolongs skin contact after the must has fermented to dryness. Based on what is known about anthocyanin and tannin extraction profiles, extended maceration would be expected to increase tannin, but not anthocyanin content. A study compared extended maceration of Cabernet Sauvignon for 7, 13, and 21 days and found that total phenols, gallic acid, and flavanols all increased with skin contact time (Auw et al., 1996). Several experiments with Pinot Noir indicate that extraction of flavanols and polymeric phenols, as well as the levels of total phenols, increases with extended skin contact (Watson et al., 1994, 1995). A study with Syrah also showed an increase in total pigment with a 3-day extended maceration, which likely reflects an increase in polymeric pigment because of greater tannin extraction (Reynolds et al., 2001).

Anthocyanins have also been described to affect tannin extractability (Bautista-Ortín et al., 2016b). In this way, anthocyanins could have an indirect influence on wine astringency. Recent studies suggest that the incorporation of anthocyanins in polymeric procyanidins increases the amount of tannins retained in wine. This phenomenon may help explain the larger quantities of tannins found in red wines versus white wines fermented with skins and seeds, and their higher astringency extractability (Bautista-Ortín et al., 2016b). Two mechanisms may be responsible for the higher tannin content: (1) the higher solubility of polymers formed from anthocyanins and condensed tannins or (2) the competition between anthocyanins and tannins by cell-wall binding sites (Garcia-Estevez et al., 2017).

Yeast selection may also affect the phenolic profile of red wines due to differences in production of ethanol by the yeast, anthocyanin adsorption by yeast lees and the differences in astringency modulation by polysaccharide liberation during alcoholic fermentation (Vasserot et al., 1997; Morata et al., 2003; del Barrio-Galán et al., 2015).

Saccharomyces and non-*Saccharomyces* yeasts influence the ethanol content in wine. Non-*Saccharomyces* yeasts have shown potential for producing wines with lower ethanol content than *Saccharomyces* yeasts. The ethanol formed during fermentation by yeast strains not only modifies the solubility of must and wine compounds (mainly tannins), but is also directly related to astringency. As the ethanol level increases in model solution, a decrease in perceived astringency and in some astringent subqualities is observed (Vidal et al., 2004a,b). According to Lee and Lawless (1991) alcohol may increase the solubility of the hydrophobic phenol–protein complexes, thereby decreasing astringent sensations in the wine matrix; overall astringency usually decreases with increasing ethanol concentrations from 0% to 15% ethanol. Meillon et al. (2009) also showed the astringency sensation is reduced by increases in ethanol content in red wine. However, in contrast to previous findings, Obreque-Slier et al. (2010a,b) observed in model solution that enological concentrations of 13% ethanol exacerbate astringency and salivary protein–tannin interactions, compared with lower ethanol concentrations.

In relation to polysaccharide liberation during alcoholic fermentation, several scientific studies have focused on testing the role of yeast cell wall polysaccharides from *Saccharomyces cerevisiae* in the technological and sensory characteristics of red wines. Certain studies assert that these compounds can reduce wine astringency and bitterness (Mekoue Nguela et al., 2016).

Mannoprotein polysaccharides are extracted from yeast cell walls either during alcoholic fermentation or later by enzymatic action during wine contact with yeast lees (Pérez-Través et al., 2016). The presence of mannoproteins in the winemaking process can help to decrease astringency in the final wine product. For this reason, several vinification yeast strains of *S. cerevisiae* have been commercialized in the last few years with the objective of releasing greater amounts of mannoproteins into wines during alcoholic fermentation (del Barrio-Galán et al., 2015).

A variety of commercial products rich in yeast cell wall polysaccharides from *S. cerevisiae* have been developed and supplied by manufacturers to provide the same benefits as the aging of wines on lees, but in a shorter period of time (Del Barrio-Galán et al., 2011, 2012). In red wines, dry yeast preparations mainly reduce green tannins which increases the softness of the wine on the palate. Therefore, yeast products could be useful in order to improve the roundness and softness of wines in the mouth, especially in young wines that are more astringent (Del Barrio-Galán et al., 2012). Supplementation of grape juice with commercial inactive yeasts during winemaking and oak aging can significantly decrease the PA content of red wines and coincide with a decrease in high molecular weight mannoproteins (Del Barrio-Galán et al., 2016). This suggests that the co-aggregate mannoprotein–tannin precipitates during treatment (González-Royo et al., 2017). Furthermore, the use of these commercial dry yeast preparations improve some sensory characteristics of white and red wines probably due to the increase of neutral polysaccharides (Del Barrio-Galán et al., 2012). Interaction of an inactive dry yeast product with either salivary proteins or a PA-rich extract (binary mixtures) or with both of them (ternary mixtures) was verified (López-Solís et al., 2017).

In winemaking, fining is the process where a substance (known as a fining agent) is added to the wine to create an adsorbent, enzymatic, or ionic bond with the suspended solid particulates, producing larger molecules and particles that will precipitate out of the wine more readily and rapidly. Unlike filtration which can only remove larger, insoluble particulates such as dead yeast cells and grape fragments, fining can remove soluble substances such as polymerized tannins, coloring phenols and proteins. The reduction of tannins through fining can reduce astringency in red wines intended for early drinking.

Many substances have historically been used as fining agents, but today there are two general types of fining agents, organic compounds and solid or mineral materials (Ribéreau-Gayon et al., 2006).

The proteins from animal origin most commonly used as fining agents are gelatin, β-lactoglobulin, ovalbumin, and casein. Different types of gelatin remove different amounts of tannins (from 9% to 16%) depending on the wine and gelatin composition (Smith et al., 2015). General observations show that highly polymerized tannins as well as highly galloylated tannins are preferentially removed during fining (Maury et al., 2001). The interactions between PAs and fining agents depend on the molecular weight and amino acid composition of the proteins used (Maury et al., 2001).

Ricardo-da-Silva et al. (1991) studied the interactions of grape seed procyanidin dimers and trimers, galloylated or not, with various proteins. Proteins poly-L-proline, gelatin, casein, dried blood, and grape arabinogalactan-protein were studied in wine-like model solutions. Except for casein, protein–procyanidin complexes were produced within the first 8 hours of protein contact. All poly-L-prolines presented very high affinity towards grape seed procyanidins. The extent of pro-cyanidin–protein interaction increased with the degree of procyanidin polymerization, the rate of galloylation, and with protein concentration. When the same authors applied various protein fining treatments to a young red wine, dimeric and trimeric procyanidin levels were not affected, although total phenolic levels were reduced.

Proteins from vegetable origin have also been evaluated for fining wines. Maury et al. (2003) evaluated five plant protein preparations (two whole wheat glutens, two hydrolyzed wheat glutens, and one preparation from white lupine, *Lupinus albus*) for their capacity to precipitate condensed tannins as wine fining agents. Fining of two wines and a model wine with these proteins showed that precipitation always occurred, although the level of precipitation was low.

Several authors have also observed that the addition of polysaccharides influences the clarification process (Maury et al., 2001; Vidal et al., 2004a,b). Polysaccharides influence the fining process by affecting the aggregation of tannins and proteins. The aggregations that can potentially occur may involve PA–PA, PA–protein, PA–polysaccharide, and PA–protein–polysaccharide interactions (Scollary et al., 2012).

Filtering prior to bottling contributes to removing residual fining agents as well as provides wine with esthetic and microbial stability. Different types of filtration processes and membrane types can be used. Although the purpose of filtering is to maintain clarity and stability, filtering can result in several sensory changes and membrane fouling (Smith et al., 2015). The concentration of tannins has been found to decrease significantly after fining with cross-flow membranes, which suggests that adherence of tannins to the membrane may be involved in membrane fouling. Higher concentrations of tannins in wine also significantly decrease the flow rate of wines passing through a ceramic cross-flow membrane, which implies that high concentration of tannins in wine (greater than 1.25 g/L) could potentially foul membranes (Smith et al., 2015). Cross-flow filtration leads to a decrease in tannin mDP by up to 25% (Ulbricht et al., 2009; Oberholster et al., 2013).

Polarity is usually measured in ultrafiltration membranes in terms of relative hydrophobicity; polarity of the membrane influences the way the wine is cleaned. Membranes with greater polarity have stronger interactions with phenolic substances because of greater hydrogen bonding capacity (Schroën et al., 2010, 2016). Ulbricht et al. (2009) observed that model wines filtered with polyethersulfone membranes had lower fluxes and a greater proportion of adsorbed tannins and polysaccharides than those filtered with polypropylene membranes. Schroën et al. (2010) demonstrated that tannins bind in lesser frequencies to polyethersulfone membranes than to those made with polyvinylpyrrolidone.

The impact of filtration on tannin and polysaccharide concentration and composition has been verified (Arriagada-Carrazana et al., 2005), however, this process only resulted in negligible difference in mouthfeel perception (Buffon et al., 2014). Further research in the area of filtration is required to better understand its role in wine colloid structure and mouthfeel (Garcia-Estevez et al., 2017).

18.4 FUTURE OUTLOOK

Astringency is a fundamental sensory attribute of red wines that contributes to their character and quality. This chapter reinforces the contribution of condensed tannin structure to astringency perception in wine and describes the chemical binding mechanisms between polyphenols and some salivary proteins and the factors influencing these interactions. Although much information is known about the biochemical sensation of astringency, details on the physical changes to saliva that are required to elicit astringency, and the relationships between the chemical processes and their physical manifestations, are unknown. Astringency perception in wine takes on many subtle forms, having broad subqualities of "roughing," "drying," and "puckering," and in the minds of many tasters, these attributes can be split into even finer subqualities. Whether these subtle differences are the result of the different phenolic profiles exhibited by wines or a result of perceptual modification of phenol-derived astringency by other wine components is yet unknown. Polysaccharides, ethanol, pH, and ionic concentration can influence salivary lubrication, both by influencing phenol–salivary protein binding and by directly reducing salivary viscosity. Although not explicitly discussed in this chapter, it is important to recognize that a full understanding of astringency cannot be achieved without considering the subjectivity of astringency as the appreciation of wine is ultimately a perceptual phenomenon.

References

Adams, D., 2006. Phenolics and ripening in grape berries. Am. J. Enol. Vitic. 57, 249–256.

Ali, K., Maltese, F., Choi, Y.H., Verpoorte, R., 2010. Metabolic constituents of grapevine and grape-derived products. Phytochem. Rev. 9, 357–378.

Armstrong, W.G., 1971. Characterization studies on the specific human salivary proteins adsorbed in vitro to hydroxyapatite. Caries Res. 5, 215–227.

Arneberg, P., 1974. Partial characterization of five glycoprotein fractions secreted by the human parotid glands. Arch. Oral. Biol. 19, 921–928.

Arriagada-Carrazana, J.P., Sáez-Navarrete, C., Bordeu, E., 2005. Membrane filtration effects on aromatic and phenolic quality of Cabernet-Sauvignon wines. J. Food Eng. 68, 363–368.

ASTM, 2004. Standard definitions of terms relating to sensory evaluation of materials and products. Annual Book of ASTM Standards. American Society for Testing and Materials, Philadelphia, PA.

Auw, J.M., Blanco, V., O'Keefe, S.F., Sims, C.A., 1996. Effect of processing on the phenolics and color of Cabernet Sauvignon, Chambourcin, and Noble wines and juices. Am. J. Enol. Vitic. 47, 279–286.

Azen, E.A., Maeda, N., 1988. Molecular genetics of human salivary proteins and their polymorphisms. Adv. Hum. Genet. 17, 141–199.

Bajec, M.R., Pickering, G., 2008. Astringency: mechanisms and perception. Crit. Rev. Food Sci. Nutr. 48, 858–875.

Bautista-Ortín, A., Ben Abdallah, R., Castro-López, L., Jiménez-Martínez, M., Gómez-Plaza, E., 2016a. Technological implications of modifying the extent of cell wall-proanthocyanidin interactions using enzymes. Int. J. Mol. Sci. 17, 123.

Bautista-Ortín, A.B., Martínez-Hernández, A., Ruiz-García, Y., Gil-Muñoz, R., Gómez-Plaza, E., 2016b. Anthocyanins influence tannin–cell wall interactions. Food Chem. 206, 239–248.

Baxter, I.R., Young, J.C., Armstrong, G., Foster, N., Bogenschutz, N., Cordova, T., et al., 2005. A plasma membrane H^+-ATPase is required for the formation of proanthocyanidins in the seed coat endothelium of Arabidopsis thaliana. Proc. Natl. Acad. Sci. U.S.A. 102 (15), 5635.

Baxter, N.F., Lilley, T.H., Haslam, E., Williamson, M.P., 1997a. Multiple interactions between polyphenols and a salivary proline-rich protein repeat results in complexation and precipitation. Biochemistry 36, 5566–5577.

Baxter, N.J., Lilley, T.H., Haslam, E., Williamson, M.P., 1997b. Multiple interactions between polyphenols and a salivary proline-rich protein repeat result in complexation and precipitation. Biochemistry 36, 5566–5577.

Bennick, A., 2002. Interaction of plant polyphenols with salivary proteins. Crit. Rev. Oral. Biol. Med. 13, 184–196.

Bennick, A., Connel, G.E., 1971. Purification and partial characterization of four proteins from human parotid saliva. Biochem. J. 123 (3), 455–464.

Berg, H.W., Akiyoshi, M., 1958. Further studies of the factors affecting the extraction of color and tannin from red grapes. Food Res. 23, 511–517.

Bindon, K., Smith, P., Kennedy, J.A., 2010. Interaction between grape-derived proanthocyanidins and cell wall material. 1. Effect on proanthocyanidin composition and molecular mass. J. Agric. Food Chem. 58, 2520–2528.

Bindon, K., Varela, C., Kennedy, J., Holt, H., Herderich, M., 2013. Relationships between harvest time and wine composition in Vitis vinifera L. cv. Cabernet Sauvignon 1. Grape and wine chemistry. Food Chem. 138, 1696–1705.

Bindon, K.A., Bacic, A., Kennedy, J.A., 2012. Tissue-specific and developmental modification of grape cell walls influences the adsorption of proanthocyanidins. J. Agric. Food Chem. 60, 9249–9260.

Bogs, J., Downey, M.O., Harvey, J.S., Ashton, A.R., Tanner, G.J., Robinson, S.P., 2005. Proanthocyanidin synthesis and expression of genes encoding leucoanthocyanidin reductase and anthocyanidin reductase in developing grape berries and grapevine leaves. Plant Physiol. 139, 652–663.

Bogs, J., Jaffé, F.W., Takos, A.M., Walker, A.R., Robinson, S.P., 2007. The grapevine transcription factor VvMYBPA1 regulates proanthocyanidin synthesis during fruit development. Plant Physiol. 143, 1347–1361.

Boselli, E., Boulton, R.B., Thorngate, J.H., Frega, N.G., 2004. Chemical and sensory characterization of doc red wines from Marche (Italy) related to vintage and grape cultivars. J. Agric. Food Chem. 52, 3843–3854.

Bosso, A., Panero, L., Guaita, M., Marulli, C., 2001. La tecnica del delestage nella vinificazione del Montepulciano d'Abruzzo. Enologo 37, 87–96.

Boulet, J.C., Trarieux, C., Souquet, J.M., Ducasse, M.A., Caillé, S., Samson, A., et al., 2016. Models based on ultraviolet spectroscopy, polyphenols, oligosaccharides and polysaccharides for prediction of wine astringency. Food Chem. 190, 357–363.

Brandao, E., Soares, S., Mateus, N., de Freitas, V., 2014. In vivo interactions between procyanidins and human saliva proteins: effect of repeated exposures to procyanidins solution. J. Agric. Food Chem. 62, 9562–9568.

Breslin, P.A.S., Gilmore, M.M., Beauchamp, G.K., Green, B.G., 1993. Psychophysical evidence that oral astringency is a tactile sensation. Chem. Senses 18, 405–417.

Brossaud, F., Cheynier, V., Noble, A.C., 2001. Bitterness and astringency of grape and wine polyphenols. Aust. J. Grape Wine Res. 7, 33–39.

Buffon, P., Heymann, H., Block, D.E., 2014. Sensory and chemical effects of cross-flow filtration on white and red wines. Am. J. Enol. Vitic. 65, 305–314.

Cáceres-Mella, A., Peña-Neira, A., Galvez, A., Obreque-Slier, E., López-Solís, R., Canals, J.M., 2012. Phenolic compositions of grapes and wines from cultivar Cabernet Sauvignon produced in Chile and their relationship to commercial value. J. Agric. Food Chem. 60 (35), 8694–8702.

Cáceres-Mella, A., Peña-Neira, A., Narváez-Bastias, J., Jara-Campos, C., López-Solís, R., Canals, J.M., 2013. Comparison of analytical methods for measuring proanthocyanidins in wines and their relationship with perceived astringency. Int. J. Food Sci. Technol. 48 (12), 2588–2594.

Canals, R., Llaudy, M.C., Valls, J., Canals, J.M., Zamora, F., 2005. Influence of ethanol concentration on the extraction of color and phenolic compounds from the skin and seeds of Tempranillo grapes at different stages of ripening. J. Agric. Food Chem. 53, 4019–4025.

Castro-López, Ld.R., Gómez-Plaza, E., Ortega-Regules, A., Lozada, D., Bautista-Ortín, A.B., 2016. Role of cell wall deconstructing enzymes in the proanthocyanidin–cell wall adsorption–desorption phenomena. Food Chem. 196, 526–532.

Charlton, A.J., Baxter, N.J., Khan, M.L., Moir, A.J.G., Haslam, E., Davies, A.P., et al., 2002a. Polyphenol/peptide binding and precipitation. Aust. J. Grape Wine Res. 50, 1593–1601.

Charlton, A.J., Haslam, E., Williamson, M.P., 2002b. Multiple conformations of the proline-rich protein/epigallocatechin gallate complex determined by time-averaged nuclear Overhauser effects. J. Am. Chem. Soc. 124, 9899–9905.

Cheynier, V., 2006. Flavonoids in wine. In: Andersen, O.M., Markham, K.R. (Eds.), Flavonoids—Chemistry, Biochemistry and Applications. Taylor & Francis, Boca Raton, FL, USA.

Cheynier, V., Prieur, C., Guyot, S., Rigaud, J., Moutounet, M., 1997. The structures of tannins in grapes and wines and their interactions with proteins. In: Watkins, T.R. (Ed.), Wine—Nutritional and Therapeutic Benefits. ACS, Washington, DC, USA, pp. 81–93.

Chira, K., Schmauch, G., Saucier, C., Fabre, S., Teissedre, P.L., 2009. Grape variety effect on proanthocyanidin composition and sensory perception of skin and seed tannin extracts from Bordeaux wine grapes (Cabernet Sauvignon and Merlot) for two consecutive vintages (2006 and 2007). Aust. J. Grape Wine Res. 57, 545–553.

Clifford, M.N., 1997. Astringency. In: Tomas-Barberon, F.A., Robins, R.J. (Eds.), Phytochemistry of Fruit and Vegetables. Clarendon Press, Oxford, pp. 87–107.

Condelli, N., Dinnella, C., Cerone, A., Monteleone, E., Bertuccioli, M., 2006. Prediction of perceived astringency induced by phenolic compounds II: criteria for panel selection and preliminary application on wine samples. Food Qual. Preference 17, 96–107.

Conn, S., Franco, C., Zhang, W., 2010. Characterization of anthocyanic vacuolar inclusions in *Vitis vinifera* L. cell suspension cultures. Planta. 231, 1343–1360.

Cosme, F., Ricardo-Da-Silva, J.M., Laureano, O., 2009. Tannin profiles of *Vitis vinifera* L. cv. red grapes growing in Lisbon and from their monovarietal wines. Food Chem. 112, 197–204.

Creasy, L.L., Swain, T., 1965. Structure of condensed tannins. Nature 208, 151.

Davies, J.M., 1997. The bioenergetics of vacuolar H^+ pumps. Adv. Bot. Res. 25, 339–363.

de Freitas, V., Mateus, N., 2001. Structural features of procyanidin interactions with salivary proteins. Aust. J. Grape Wine Res. 49, 940–945.

de Freitas, V., Mateus, N., 2012. Protein/polyphenol interactions: past and present contributions. Mechanisms of astringency perception. Curr. Org. Chem. 16, 724–746.

de Wijk, R.A., Prinz, J.F., 2005. The role of friction in perceived oral texture. Food Qual. Preference 16, 121–129.

Degand, P., Boersman, A., Roussell, P., Richet, C., Biserte, G., 1975. The peptide moiety of the region carrying the carbohydrate chains of the human parotid glycoprotein. FEBS Lett. 54, 189–192.

del Barrio-Galán, R., Cáceres-Mella, A., Medel-Marabolí, M., Peña-Neira, A., 2015. Effect of selected *Saccharomyces cerevisiae* yeast strains and different aging techniques on the polysaccharide and polyphenolic composition and sensorial characteristics of Cabernet Sauvignon red wines. J. Sci. Food Agric. 95, 2132–2144.

Del Barrio-Galán, R., Pérez-Magariño, S., Ortega-Heras, M., 2011. Techniques for improving or replacing ageing on lees of oak aged red wines: the effects on polysaccharides and the phenolic composition. Food Chem. 127, 528–540.

Del Barrio-Galán, R., Pérez-Magariño, S., Ortega-Heras, M., Guadalupe, Z., Ayestarán, B., 2012. Polysaccharide characterization of commercial dry yeast preparations and their effect on white and red wine composition. LWT Food Sci. Technol. 48, 215–223.

Del Barrio-Galán, R., Medel-Marabolí, M., Peña-Neira, Á., 2016. Effect of different ageing techniques on the polysaccharide and phenolic composition and sensorial characteristics of Chardonnay white wines fermented with different selected *Saccharomyces cerevisiae* yeast strains. Eur. Food Res. Technol. 242, 1069–1085.

Demiglio, P., Pickering, G.J., 2008. The influence of ethanol and pH on the taste and mouthfeel sensations elicited by red wine. J. Food Agric. Environ. 6, 143–150.

Dinnella, C., Recchia, A., Fia, G., Bertuccioli, M., Monteleone, E., 2009. Saliva characteristics and individual sensitivity to phenolic astringent stimuli. Chem. Senses. 34, 295–304.

Ducruet, J., Dong, A., Canal-Llauberes, R.M., Glories, Y., 1997. Influence des enzymes pectolytiques sélectionées pour l'oenologie sur la qualité et la composition des vins rouges. Rev. Fr. Oenol. 155, 16–19.

Escribano-Bailon, M.T., Gutierrez-Fernandez, Y., Rivas-Gonzalo, J.C., Santos-Buelga, C., 1992. Characterisation of procyanidins of *Vitis vinifera* variety Tinta del Pais grape seeds. J. Agric. Food Chem. 40, 1794–1799.

Escribano-Bailón, M.T., Guerra, M.T., Rivas-Gonzalo, J.C., Santos-Buelga, C., 1995. Proanthocyanidins in skins from different grape varieties. Z. Lebensm.-Unters. Forsch. 200 (3), 221–224.

Fasoli, M., Dell'Anna, R., Dal Santo, S., Balestrini, R., Sanson, A., Pezzott, I.M., et al., 2016. Hemicelluloses and celluloses show specific dynamics in the internal and external surfaces of grape berry skin during ripening. Plant Cell Physiol. 57, 1332–1349.

Ferrer-Gallego, R., Garcia-Marino, M., Hernandez-Hierro, J.M., Rivas-Gonzalo, J.C., Escribano-Bailon, M.T., 2010. Statistical correlation between flavanolic composition, colour and sensorial parameters in grape seed during ripening. Anal. Chim. Acta. 660, 22–28.

Fournand, D., Vicens, A., Sidhoum, L., Souquet, J.-M., Moutounet, M., Cheynier, V., 2006. Accumulation and extractability of grape skin tannins and anthocyanins at different advanced physiological stages. J. Agric. Food Chem. 54, 7331–7338.

Friedman, R., Merritt, A.D., 1975. Partial purification and characterization of a polymorphic protein (Pa) in human parotid saliva. Am. J. Hum. Genet. 27, 304–314.

Friedman, R., Merritt, A.D., Bixler, D., 1971. Immunological and chemical comparison of heterogenous basic glycoproteins in human parotid saliva. Biochim. Biophys. Acta. 230, 599–602.

Gagne, S., Saucier, C., Geny, L., 2006. Composition and cellular localization of tannins in Cabernet Sauvignon skins during growth. J. Agric. Food Chem. 54, 9465–9471.

Gambuti, A., Rinaldi, A., Pessina, R., Moio, L., 2006. Evaluation of aglianico rape skin and seed polyphenol astringency by SDS–PAGE electrophoresis of salivary proteins after the binding reaction. Food Chem. 97, 614–620.

Gao, L., Girard, B., Mazza, G., Reynolds, A.G., 1997. Changes in anthocyanins and color characteristics of Pinot noir wines during different vinification processes. J. Agric. Food Chem. 45, 2003–2008.

Garcia-Estevez, I., Perez-Gregorio, R., Soares, S., Mateus, N., de Freitas, V., 2017. Oenological perspective of red wine astringency. OENO One 51 (3), 237–249.

Gawel, R., 1998. Red wine astringency: a review. Aust. J. Grape Wine Res. 4, 74–95.

Gawel, R., Oberholster, A., Francis, I.L., 2000. A 'mouth-feel wheel': terminology for communicating the mouth-feel characteristics of red wine. Aust. J. Grape Wine Res. 6, 203–207.

Geny, L., Saucier, C., Bracco, S., Daviaud, F., Glories, Y., 2003. Composition and cellular localization of tannins in grape seeds during maturation. J. Agric. Food Chem. 51, 8051–8054.

Gil Cortiella, M., Peña-Neira, Á., 2017. Extraction of soluble polysaccharides from grape skins. Cienc. Invest. Agrar. 44 (1), 83–93.

González-Royo, E., Esteruelas, M., Kontoudakis, N., Fort, F., Canals, J.M., Zamora, F., 2017. The effect of supplementation with three commercial inactive dry yeasts on the colour, phenolic compounds, polysaccharides and astringency of a model wine solution and red wine. J. Sci. Food Agric. 97, 172–181.

Hagerman, A.E., 1992. Tannin–protein interactions. In: Ho, C.T., Lee, C.Y., Huang, M.T. (Eds.), Phenolic Compounds in Food and their Effects on Health I. Analysis, Occurrence and Chemistry. ACS, Washington, pp. 236–247.

Hagerman, A.E., Butler, L.G., 1981. The specificity of proanthocyanidin–protein interactions. J. Biol. Chem. 256, 4494–4497.

He, F., Pan, Q.H., Shi, Y., Duan, C.Q., 2008. Chemical synthesis of proanthocyanidins in vitro and their reactions in aging wines. Molecules 13, 3007–3032.

Hemingway, R.W., Foo, L.Y., 1983. Condensed tannins: quinone methide intermediates in procyanidin synthesis. J. Chem. Soc. Chem. Commun. 18, 1035–1036.

Herderich, M.J., Smith, P.A., 2005. Analysis of grape and wine tannins: methods, applications and challenges. Aust. J. Grape Wine. Res. 11, 1–10.

Hufnagel, J.C., Hofmann, T., 2008. Orosensory-directed identification of astringent mouthfeel and bitter-tasting compounds in red wine. J. Agric. Food Chem. 56, 1376–1386.

Jackson, D.I., Lombard, P.B., 1992. Environmental and management practices affecting grape composition and wine quality—a review. Am. J. Enol. Vitic. 44, 409–430.

Kallithraka, S., Bakker, J., Clifford, M.N., Vallis, L., 2001. Correlations between saliva protein composition and some T-I parameters of astringency. Food Qual. Preference 12, 145–152.

Kassara, S., Kennedy, J.A., 2011. Relationship between red wine grade and phenolics. 2. Tannin composition and size. J. Agric. Food Chem. 59, 8409–8412.

Kawamoto, H., Nakatsubo, F., Murakami, K., 1996. Stoichiometric studies of tannin–protein co-precipitation. Phytochemistry 41, 1427–1431.

Kennedy, J.A., Troup, G.J., Pilbrow, J.R., Hutton, D.R., Hewitt, D., Hunter, C.R., et al., 2000a. Development of seed polyphenols in berries from Vitis vinifera L. cv. Shiraz. Aust. J. Grape Wine Res. 6 (3), 244–254.

Kennedy, J.A., Matthews, M.A., Waterhouse, A.L., 2000b. Changes in grape seed polyphenols during fruit ripening. Phytochemistry 55, 77–85.

Kennedy, J.A., Hayasaka, Y., Vidal, S., Waters, E.J., Jones, G.P., 2001. Composition of grape skin proanthocyanidins at different stages of berry development. J. Agric. Food Chem. 49, 5348–5355.

Lacampagne, S., Gagne, S., Geny, L., 2010. Involvement of abscisic acid in controlling the proanthocyanidin biosynthesis pathway in grape skin: new elements regarding the regulation of tannin composition and leucoanthocyanidin reductase (LAR) and anthocyanidin reductase (ANR) activities and expression. J. Plant. Growth Regul. 29, 81–90.

Landon, J.L., Weller, K., Harbertson, J.F., Ross, C.F., 2008. Chemical and sensory evaluation of astringency in Washington state red wines. Am. J. Enol. Vitic. 59, 153–158.

Lawless, H.T., Corrigan, C.J., Lee, C.B., 1994. Interactions of astringent substances. Chem. Senses 19, 141–154.

Lee, C.B., Lawless, H.T., 1991. Time-course of astringent sensations. Chem. Senses 16, 225–238.

Levine, M., Keller, P.J., 1977. The isolation of some basic proline-rich proteins from human parotid saliva. Arch. Oral. Biol. 22, 37–41.

Levine, M.J., Weill, J.C., Ellison, S.A., 1969. The isolation and analysis of a glycoprotein from parotid saliva. Biochim. Biophys. Acta. 188, 165–167.

Li, M., Hagerman, A.E., 2014. Role of the flavan-3-ol and galloyl moieties in the interaction of (−)-epigallocatechin gallate with serum albumin. J. Agric. Food Chem. 62, 3768–3775.

Llaudy, M.C., Canals, R., Canals, J.M., Zamora, F., 2008. Influence of ripening stage and maceration length on the contribution of grape skins, seeds and stems to phenolic composition and astringency in wine-simulated macerations. Eur. Food Res. Technol. 226, 337–344.

López-Cisternas, J., Castillo-Díaz, J., Traipe-Castro, L., López-Solís, R., 2007. A protein dye-binding assay on cellulose membranes for tear protein quantification: use of conventional Schirmer strips. Cornea. 26 (8), 970–976.

López-Solís, R., Duarte-Venegas, C., Meza-Candia, M., del Barrio-Galán, R., Peña-Neira, Á., Medel-Marabolí, M., et al., 2017. Great diversity among commercial inactive dry-yeast based products. Food Chem. 219, 282–289.

Lu, Y., Bennick, A., 1998. Interaction of tannin with human salivary proline-rich proteins. Arch. Oral. Biol. 43, 717–728.

Luck, G., Liao, H., Murray, N.J., Grimmer, H.R., Warminski, E.E., Williamson, M.P., et al., 1994. Polyphenols, astringency and proline-rich proteins. Phytochemistry 37, 357–371.

Ma, W., Guo, A., Zhang, Y., Wang, H., Liu, Y., Li, H., 2014. A review on astringency and bitterness perception of tannins in wine. Trends Food Sci. Technol. 40, 6–19.

Manconi, B., Castagnola, M., Cabrasa, T., Olianasa, A., Vitali, A., Desiderio, C., et al., 2016. The intriguing heterogeneity of human salivary proline-rich proteins: short title: salivary proline-rich protein species. J. Proteomics. 134 (16), 47–56.

Marais, J., 2003a. Effect of different wine-making techniques on the composition and quality of Pinotage wine. I. Low-temperature skin contact prior to fermentation. S. Afr. J. Enol. Vitic. 24, 70–75.

Marais, J., 2003b. Effect of different wine-making techniques on the composition and quality of Pinotage wine. II. Juice/skin mixing practices. S. Afr. J. Enol. Vitic. 24, 76–79.

Maury, C., Sarni-Manchado, P., Lefebvre, S., Cheynier, V., Moutounet, M., 2001. Influence of fining with different molecular weight gelatins on proanthocyanidin composition and perception of wines. Am. J. Enol. Vitic. 52, 140–145.

Maury, C., Sarni-Manchado, P., Lefebvre, S., Cheynier, V., Moutounet, M., 2003. Influence of fining with plant proteins on proanthocyanidin composition of red wines. Am. J. Enol. Vitic. 54, 105–111.

McRae, J.M., Kennedy, J.A., 2011. Wine and grape tannin interactions with salivary proteins and their impact on astringency: a review of current research. Molecules 16, 2348–2364.

Meillon, S., Urbano, C., Schlich, P., 2009. Contribution of the temporal dominance of sensations (TDS) method to the sensory description of subtle differences in partially dealcoholized red wines. Food Qual. Preference 20, 490–499.

Mekoue Nguela, J., Poncet-Legrand, C., Sieczkowski, N., Vernhet, A., 2016. Interactions of grape tannins and wine polyphenols with a yeast protein extract, mannoproteins and β-glucan. Food Chem. 210, 671–682.

Messana, I., Inzitari, R., Fanali, C., Cabras, T., Castagnola, M., 2008. Facts and artifacts in proteomics of body fluids. What proteomics of saliva is telling us? J. Sep. Sci. 31, 1948–1963.

Monagas, M., Gómez-Cordovés, C., Bartolomé, B., Laureano, O., Ricardo da Silva, J.M., 2003. Monomeric, oligomeric, and polymeric flavan-3-ol composition of wines and grapes from *Vitis vinifera* L. cv. Graciano, Tempranillo, and Cabernet Sauvignon. J. Agric. Food Chem. 51, 6475–6481.

Morata, A., Gómez-Cordovés, C., Subervolia, J., Bartolomé, B., Colomo, B., Suarez, J.A., 2003. Adsorption of anthocyanins by yeast cell walls during fermentation of red wines. J. Agric. Food Chem. 51, 4084–4088.

Müller, K., Figueroa, C., Martínez, C., Medel, M., Obreque, E., Peña-Neira, A., et al., 2010. Measurement of saliva volume in the mouth of members of a trained sensory panel using a beetroot (*Beta vulgaris*) extract. Food Qual. Preference 21 (5), 569–574.

Naurato, N., Wong, P., Lu, Y., Wroblewski, K., Bennick, A., 1999. Interaction of tannin with human salivary histatins. J. Agric. Food Chem. 47, 2229–2234.

Nayak, A., Carpenter, G.H., 2008. A physiological model of tea-induced astringency. Physiol. Behav. 95, 290–294.

Noble, A.C., 1995. Application of time-intensity procedures for the evaluation of taste and mouthfeel. Am. J. Enol. Vitic. 46, 128–133.

Oberholster, A., Carstens, L.M., Du Toit, W.J., 2013. Investigation of the effect of gelatine, egg albumin and cross-flow microfiltration on the phenolic composition of Pinotage wine. Food Chem. 138, 1275–1281.

Obreque-Slier, E., López-Solís, R., Peña-Neira, A., Zamora-Marın, F., 2010a. Tannin–protein interaction is more closely associated with astringency than tannin–protein precipitation: experience with two oenological tannins and a gelatin. Int. J. Food Sci. Technol. 45, 2629–2636.

Obreque-Slier, E., Pena-Neira, A., Lopez-Solis, R., 2010b. Enhancement of both salivary protein–enological tannin interactions and astringency perception by ethanol. J. Agric. Food Chem. 58, 3729–3735.

Oppenheim, F.G., Hay, D.I., Franzblau, C., 1971. Proline-rich proteins from human parotid saliva. I. Isolation and partial characterization. Biochemistry 10, 4233–4238.

Ough, C.S., Amerine, M.A., 1961. Studies on controlled fermentation. V. Effects on color, composition, and quality of red wines. Am. J. Enol. Vitic. 12, 9–19.

Ozmianski, J., Romeyer, F.M., Sapis, J.C., Macheix, J.J., 1986. Grape seed phenolics: extraction as affected by some conditions occurring during wine processing. Am. J. Enol. Vitic. 37, 7–12.

Pascal, C., Poncet-Legrand, C., Imberty, A., Gautier, C., Sarni-Manchado, P., Cheynier, V., et al., 2007. Interactions between a non glycosylated human proline-rich protein and flavan-3-ols are affected by protein concentration and polyphenol/protein ratio. Aust. J. Grape Wine Res. 55, 4895–4901.

Payne, C., Bowyer, P.K., Herderich, M., Bastian, S.E.P., 2009. Interaction of astringent grape seed procyanidins with oral epithelial cells. Food Chem. 115, 551–557.

Perez-Magarino, S., Gonzalez-San Jose, M.L., 2006. Polyphenols and colour variability of red wines made from grapes harvested at different ripeness grade. Food Chem. 96, 197–208.

Pérez-Través, L., Querol, A., Pérez-Torrado, R., 2016. Increased mannoprotein content in wines produced by *Saccharomyces kudriavzevii* × *Saccharomyces cerevisiae* hybrids. Int. J. Food Microbiol. 237, 35–38.

Peyrot des Gachons, C., Kennedy, J.A., 2003. Direct method for determining seed and skinproanthocyanidin extraction into red wine. J. Agric. Food Chem. 51, 5877–5881.

Quijada-Morín, N., García-Estévez, I., Nogales-Bueno, J., Rodríguez-Pulido, F.J., Heredia, F.J., Rivas-Gonzalo, J.C., et al., 2016. Trying to set up the flavanolic phases during grape seed ripening: a spectral and chemical approach. Talanta 160, 556–561.

Reynolds, A., Cliff, M., Girard, B., Kopp, T.G., 2001. Influence of fermentation temperature on composition and sensory properties of Semillon and Shiraz wines. Am. J. Enol. Vitic. 52, 235–240.

Ribéreau-Gayon, P., 1974. The chemistry of red wine color. In: Webb, A.D. (Ed.), Chemistry of Winemaking. American Chemical Society, Washington, DC, p. 81.

Ribéreau-Gayon, P., Dubourdieu, D., Donèche, B., Lonvaud, A., 2006. Handbook of Enology, The Chemistry of Wine—Stabilization and Treatments, vol. 2, second ed., John Wiley & Sons, UK, p. 521.

Ricardo da Silva, J.M., Rosec, J.P., Bourzeix, M., Mourgues, J., Moutounet, M., 1992. Dimer and trimer procyanidins in Carignan and Mourvèdre grapes and red wines. Vitis 31, 55–63.

Ricardo-da-Silva, J.M., Cheynier, V., Souquet, J.M., Moutounet, M., Cabanis, J., Bourzeix, M., 1991. Interaction of grape seed procyanidins with various proteins in relation to wine fining. J. Sci. Food Agric. 57 (1), 111–125.

Rinaldi, A., Gambuti, A., Moine-Ledoux, V., Moio, L., 2010. Evaluation of the astringency of commercial tannins by means of the SDSePAGE-based method. Food Chem. 122, 951–956.

Riou, V., Vernhet, A., Doco, T., Moutounet, M., 2002. Aggregation of grape seed tannins in model wine-effect of wine polysaccharides. Food Hydrocolloids 16, 17–23.

Ristic, R., Bindon, K., Francis, I.L., Herderich, M.J., Iland, P.G., 2010. Flavonoids and C13-norisoprenoids in *Vitis vinifera* L. cv. Shiraz: relationships between grape and wine composition, wine colour and wine sensory properties. Aust. J. Grape Wine Res. 16, 369–388.

Romeyer, F.M., Macheix, J.J., Sapis, J.C., 1986. Changes and importance of oligomeric procyanidins during maturation of grape seeds. Phytochemistry 25, 219.

Sacchi, K.L., Bisson, L.F., Adams, D.O., 2005. A review of the effect of winemaking techniques on phenolic extraction in red wines. Am. J. Enol. Vitic. 56 (3), 197–206.

Sarni-Manchado, P., Cheynier, V., Moutounet, M., 1999. Interactions of grape seed tannins with salivary proteins. J. Agric. Food Chem. 47 (1), 42–47.

Scharbert, S., Jezussek, M., Hofmann, T., 2004. Evaluation of the taste contribution of theaflavins in black tea infusions using the taste activity concept. Eur. Food Res. Technol. 218, 442–447.

Schroën, C.G.P.H., Roosjen, A., Tang, K., Norde, W., Boom, R.M., 2010. In situ quantification of membrane foulant accumulation by reflectometry. J. Membr. Sci. 362, 453–459.

Schroën, K., Ferrando, M., de Lamo-Castellví, S., Sahin, S., Güell, C., 2016. Linking findings in microfluidics to membrane emulsification process design: the importance of wettability and component interactions with interfaces. Membranes 6, 26.

Schwarz, B., Hofmann, T., 2008. Is there a direct relationship between oral astringency and human salivary protein binding? Eur. Food Res. Technol. 227, 1693–1698.

Scollary, G.R., Pásti, G., Kállay, M., Blackman, J., Clark, A.C., 2012. Astringency response of red wines: potential role of molecular assembly. Trends Food Sci. Technol. 27, 25–36.

Singleton, V.L., Draper, D.E., 1964. The transfer of polyphenolic compounds from grape seeds into wines. Am. J. Enol. Vitic. 15, 34–40.

Smith, P.A., McRae, J.M., Bindon, K.A., 2015. Impact of winemaking practices on the concentration and composition of tannins in red wine. Aust. J. Grape Wine Res. 21, 601–614.

Soares, S., Mateus, N., de Freitas, V., 2007. Interaction of different polyphenols with bovine serum albumin (BSA) and human salivary α-amylase (HSA) by fluorescence quenching. J. Agric. Food Chem. 55, 6726–6735.

Souquet, J.-M., Cheynier, V., Brossaud, F., Moutounet, M., 1996. Polymeric proanthocyanidins from grape skins. Phytochemistry 43, 509–512.

Su, C.T., Singleton, V.L., 1969. Identification of three flavan-3-ols from grapes. Phytochemistry 8 (8), 1553–1558.

Swain, T., Bate-Smith, E.C., 1962. Flavonoid compounds. In: Florkin, M., Mason, H.S. (Eds.), Comparative Biochemistry. Academic Press, London, pp. 755–809.

Terrier, N., Torregrosa, L., Ageorges, A., Vialet, S., Verries, C., Cheynier, V., et al., 2009. Ectopic expression of VvMybPA2 promotes proanthocyanidin biosynthesis in grapevine and suggests additional targets in the pathway. Plant Physiol. 149, 1028–1041.

Thorngate, J.H., 1993. Flavan-3-ols and their polymers: analytical techniques and sensory considerations. In: Gump, B.H. (Ed.), Beer and Wine Production: Analysis, Characterization and Technological Advances. ACS, San Francisco, CA, pp. 98–109.

Thorngate, J.H., Noble, A.C., 1995. Sensory evaluation of bitterness and astringency of 3R(−)-epicatechin and 3S(+)-catechin. J. Sci. Food Agric. 67, 531–535.

Troxler, R.F., Offner, G.D., Xu, T., Vanderspeck, J.C., Oppenheim, F.G., 1990. Structural relationship between human salivary histatins. J. Dent. Res. 69, 2–6.

Ulbricht, M., Ansorge, W., Danielzik, I., König, M., Schuster, O., 2009. Fouling in microfiltration of wine: the influence of the membrane polymer on adsorption of polyphenols and polysaccharides. Sep. Purif. Technol. 68, 335–342.

Upadhyay, R., Brossard, N., Chen, J., 2016. Mechanisms underlying astringency: introduction to an oral tribology approach. J. Phys. D: Appl. Phys. 49, 104003.

Vasserot, Y., Caillet, S., Maujean, A., 1997. Study of anthocyanin adsorption by yeast lees. Effect of some physicochemical parameters. Am. J. Enol. Vitic. 48, 433–437.

Vidal, S., Francis, L., Williams, P., Kwiatkowski, M., Gawel, R., Cheynier, V., et al., 2004a. The mouth-feel properties of polysaccharides and anthocyanins in a wine like medium. Food Chem. 85, 519–525.

Vidal, S., Courcoux, P., Francis, L., Kwiatkowski, M., Gawel, R., Williams, P., et al., 2004b. Use of an experimental design approach for evaluation of key wine components on mouth-feel perception. Food Qual. Preference 15, 209–217.

Watrelot, A.A., Schulz, D.L., Kennedy, J.A., 2017. Wine polysaccharides influence tannin–protein interactions. Food Hydrocolloids 63, 571–579.

Watson, B.T., Price, S.F., Chen, H.P., Valladao, M., 1994. Pinot noir processing effects on wine color and phenolics. Abstr. Am. J. Enol. Vitic. 45, 471–472.

Watson, B.T., Price, S.F., Valladao, M., 1995. Effect of fermentation practices on anthocyanin and phenolic composition of Pinot noir wines. Abstr. Am. J. Enol. Vitic. 46, 404.

Wu-Dai, Z., Ollat, N., Gomès, E., Decroocq, S., Tandonnet, J.P., Bordenave, L., et al., 2011. Ecophysiological, genetic, and molecular causes of variation in grape berry weight and composition: a review. Am. J. Enol. Vitic. 62, 413–425.

Yan, Q., Bennick, A., 1995. Identification of histatins as tannin-binding proteins. Biochem. J. 311, 341–347.

Zamora, F., Luengo, G., Margalef, P., Magriña, M., Arola, L., 1994. Nota. Efecto del sangrado sobre el color y la composición en compuestos fenólicos del vino tinto. Rev. Esp. Cienc. Tecnol. Aliment. 34, 663–671.

Zanchi, D., Poulain, C., Konarev, P., Tribet, C., Svergun, D.I., 2008. Colloidal stability of tannins: astringency, wine tasting and beyond. J. Phys.: Condens. Matter 20, 494224.

Zhao, J., Dixon, R.A., 2009. MATE transporters facilitate vacuolar uptake of epicatechin-3-*O*-glucoside for proanthocyanidin biosynthesis in *Medicago truncatula* and *Arabidopsis*. Plant Cell 21, 2323–2340.

Zhao, J., Pang, Y., Dixon, R.A., 2010. The mysteries of proanthocyanidin transport and polymerization. Plant Physiol. 153, 437–443.

Further Reading

Quijada-Morín, N., Regueiro, J., Simal-Gandara, J., Tomas, E., Rivas-Gonzalo, J.C., Teresa Escribano-Bailon, M., 2012. Relationship between the sensory determined astringency and the flavanolic composition of red wines. J. Agric. Food Chem. 60, 12355–12361.

CHAPTER

19

Aromatic Compounds in Red Varieties

Doris Rauhut and Florian Kiene

Department of Microbiology and Biochemistry, Hochschule Geisenheim University, Geisenheim, Germany

19.1 INTRODUCTION

Hundreds of volatile substances have been identified in grapes, musts, and wines, which is evidence for the high complexity. The qualitative and quantitative composition of aromatic compounds with various chemical structures and properties and their interaction within different wine matrices is leading to wines with individual flavor and sensory impressions. The specific composition of aromatic compounds in each wine is influenced by the grape variety, the region of origin, viticultural and enological techniques and treatments, filling techniques, and closure types, as well as the storage condition and time after bottling. This chapter reviews the current knowledge of the most important aromatic compounds in wines of red varieties, in particular varietal thiols, methoxypyrazines, C_{13}-norisoprenoides, and a collection of high aroma-active compounds that are responsible for specific or typical aromas in red wines. Information about the impact on aromatic compounds at special steps during the whole red wine production process, especially the impact of microorganisms and fermentation processes is given in the other related chapters.

19.2 SELECTION OF AROMATIC COMPOUNDS WITH DISTINCT IMPACT

19.2.1 Sulfur Compounds

19.2.1.1 Dimethyl Sulfide

Dimethyl sulfide (DMS) concentrations up to 474 µg/L were found by Loubser and Du Plessis (1976) in wine. The flavor resulting from DMS is described as reminiscent of asparagus, corn, and molasses (Goniak and Noble, 1987) and it has a perception threshold in red wine of 27 µg/L (Anocibar Beloqui et al., 1996) and an odor threshold of 60 µg/L (de Mora et al., 1987) (Table 19.1). It was demonstrated that an addition of DMS can contribute to truffle or hay notes or in high concentrations to a green olive aroma, which was described as more unpleasant (Anocibar Beloqui et al., 1996). In addition, Ségurel et al. (2004) described black olive and truffle notes in Grenache and Syrah wines.

Several research studies showed that DMS can be formed by yeast metabolism during fermentation from cysteine or glutathione (de Mora et al., 1986; Schreier et al., 1974) or dimethyl sulfoxide (DMSO) by yeast reduction (Anness and Bamforth, 1982; Anocibar Beloqui, 1998; Bamforth and Anness, 1981). Only small amounts of DMS are formed during fermentation. Furthermore it is assumed that most of the yeast-derived DMS is lost due to its high volatility with the CO_2-release (Baumes, 2009). Young wines have only traces or very small amounts of DMS. Their amounts increase during storage time and with increasing temperature (Marais, 1979; Simpson, 1979). Goto and Takamuro (1987) analyzed in red wines, aged for 1 year, 3.2–6.0 µg/L DMS. A maximum amount of DMS was observed in wines after 5–10 years of aging and then it seemed to decrease. Ségurel et al. (2004, 2005) investigated the potential DMS (PDMS) in wine by chemical degradation of DMS precursors. They drew the conclusion on the basis of their results that DMS formation during wine storage cannot be explained by DMSO (Baumes, 2009). Finally, S-methylmethionine (SMM) was identified as main DMS precursor in grapes and

TABLE 19.1 Structure, odor, concentration in wine, and perception threshold of important volatile sulfur compounds in wine

Name	Structure	Odor	Concentration in wine/(ng/L)[a]	Perception threshold/(ng/L)	Literature
Dimethyl sulfide (DMS)	(structure)	Truffle[a]	23,000–146,000[b]	1740[c],[1] 27,000[d],[2]	[a] San-Juan et al. (2011) [b] Picard et al. (2015) [c] Lytra et al. (2014) [d] Anocibar Beloqui et al. (1996)
2-Methyl-3-sulfanylbutan-1-ol (2M3SB)	(structure)	Raw onion[a]	10–70[b]	10[c],[3]	[a] Floch et al. (2016) [b] Darriet and Pons (2017) [c] Acuña et al. (2003)
3-Methyl-3-sulfanylbutan-1-ol (3MSB)	(structure)	Onion, leek[a]	20–150[b]	1300[c],[3] 1500[c],[1]	[a] Sarrazin et al. (2007) [b] Ribéreau-Gayon et al. (2006) [c] Tominaga et al. (1998)
3-Sulfanylpentan-1-ol (3SP)	(structure)	Grapefruit[a]	90–380[a]	620[a],[3] 950[a],[1]	[a] Sarrazin et al. (2007)
4-Methyl-4-sulfanylpentan-2-ol (4MSPOH)	(structure)	Citrus zest[a]	15–150[b]	20[a],[3] 55[a],[1]	[a] Tominaga et al. (1998) [b] Ribéreau-Gayon et al. (2006)
3-Sulfanylhexanol (3SH)	(structure)	Grapefruit (R isomer), passion fruit (S isomer)[a]	10–5000[b]	17[c],[3] 60[c],[1]	[a] Styger et al. (2011) [b] Bouchilloux et al. (1998b) [c] Tominaga et al. (1998)
3-Sulfanylheptan-1-ol (3SHp)	(structure)	Grapefruit[a]	0–80[b]	10[a],[3] 35[a],[1]	[a] Sarrazin et al. (2007) [b] Darriet and Pons (2017)
4-Methyl-4-sulfanylpentan-2-one (4MSP)	(structure)	Blackcurrant[a] Box tree[b]	0–120[c]	0.07[d],[3] 0.8[d],[4]	[a] Rigou et al. (2014) [b] Sarrazin et al. (2007) [c] Ribéreau-Gayon et al. (2006) [d] Darriet et al. (1995)
Ethyl-2-sulfanylpropionate (E2SP)	(structure)	Fruity[a]	50–800[b]	500[c],[5]	[a] Ribéreau-Gayon et al. (2006) [b] Tominaga et al. (2003b) [c] Blanchard (2000)
Ethyl-3-sulfanylpropionate (E3SP)	(structure)	Meaty[a]	40–12,000[b]	200[c],[3]	[a] Roland et al. (2011) [b] Tominaga et al. (2003b) [c] Kolor (1983)

(Continued)

TABLE 19.1 (Continued)

Name	Structure	Odor	Concentration in wine/(ng/L)[a]	Perception threshold/(ng/L)	Literature
3-Sulfanyl-hexylacetate (3SHA)		Grapefruit, passion fruit[a]	1–200[b]	2.3[c],[3]	[a] Floch et al. (2016)
				4.2[c],[6]	[b] Bouchilloux et al. (1998b)
					[c] Tominaga et al. (1996)
2-Methyl-3-furanthiol		Fried[a]	50–145[c]	5[d],[7]	[a] Ferreira et al. (2002)
					[b] Sarrazin et al. (2007)
		Meaty[b]			[c] Tominaga and Dubourdieu (2006)
					[d] Ferreira et al. (2002)
2-Furfurylthiol		Roast coffee[a]	2–5500[b]	0.4[c],[5]	[a] Tominaga and Dubourdieu (2006)
					[b] Tominaga et al. (2003b)
					[c] Picard et al. (2015)
Benzylthiol		Smoky[a]	10–400[b]	0.3[a],[1]	[a] Tominaga et al. (2003a)
					[b] Tominaga et al. (2003b)
Vanillylthiol		Clove, smoke[a]	50–8300[b]	3800[a],[1]	[a] Floch et al. (2016)

[a] General concentration range in wines; no distinction is drawn between white, red, or rosé wines.
Threshold determined in: [1] water/ethanol 12% (v/v) + 5 g/L tartaric acid, pH = 3.5; [2] red wine; [3] water; [4] water/ethanol 10% (v/v) + 10% sucrose; [5] water/ethanol 12% (v/v); [6] hydro alcoholic solution; [7] water/ethanol 10% (v/v) + 7 g/L glycerol + 1 g/L tartaric acid, pH = 3.2.

the corresponding musts and wines (Loscos et al., 2008). DMS is mainly released by chemical degradation during wine storage. A biochemical transformation of SMM to DMS by yeasts is also discussed, because PDMS is decreasing during alcoholic fermentation and only a small amount of it is remaining in the young wines (Loscos et al., 2008). Most of the DMS produced during fermentation is probably lost with the fermentation gas (Baumes, 2009). The results on PDMS in various grape musts showed high discrepancies, e.g., concentrations below the perception threshold in Grenache and distinct higher values in Petit Manseng (Ségurel et al., 2005; Dagan, 2006). Apart from the variation of the PDMS levels in different varieties, also other factors influence the concentrations, for example, a considerable increase was noticed with overripening (Baumes, 2009). The PDMS amounts in musts were always higher than the quantities in the corresponding wines, but a well-defined correlation couldn't be demonstrated. Also de Mora et al. (1987) found no relationship between storage time and DMS concentration of Australian wines, particularly in Cabernet wines from the Coonawara region in South Australia. In those wines extremely high concentrations up to 910 µg/L were determined. Vintage, environmental and microclimate impact, viticultural and enological techniques are discussed as influencing factors. Spedding and Raut (1982) reported positive effects with low amounts of DMS on the flavor of wines. Several authors assumed that the occurrence of DMS at increased concentrations contributes to bottle age bouquet and late harvest wines (Du Plessis and Loubser, 1974; Simpson, 1979).

In 1987 de Mora et al. noticed that the intensification of fruitiness in Cabernet Sauvignon red wines correlates with increasing levels of DMS. This was later verified in other red wine samples from Ségurel et al. (2004) and in premium red wines from Escudero et al. (2007). It was also found that DMS is participating in the modulation of blackberry fruit notes and intensification of blackcurrant aroma (Lytra et al., 2014).

San-Juan et al. (2011) related the vegetal character in a collection of high quality Spanish red wines to the combination of DMS, 1-hexanol, and methanethiol. PDMS increased in berries from vines exposed to moderate water deficit and also in the corresponding musts. This could probably contribute to higher fruity notes in the wines during storage (Royer Dupré et al., 2014). Picard et al. (2017) demonstrated that a moderate-to-severe water deficit has an impact on the bouquet of aged red Bordeaux wines, in particular on mint and truffle notes.

19.2.1.2 Thiols

Volatile thiols are extremely aroma active compounds. The three most prominent thiols which are involved in the varietal aromas of wine are 3-sulfanylhexanol (3SH), 3-sulfanylhexyl acetate (3SHA), and 4-methyl-4-sulfanylpentan-2-one (4MSP). Their formation, occurrence, concentration in wines from several grape varieties (in particular Sauvignon blanc), their fate during the whole wine production process and during storage has been investigated over almost three decades which is reflected in several reviews with different focuses (Belda et al., 2017; Dubourdieu and Tominaga, 2009; Parker et al., 2018; Roland et al., 2011). Apart from exotic fruits like grapefruit and passion fruit (3SH), also descriptions such as box tree and broom (3SHA and 4MSP) or blackcurrant, citrus zest, roast coffee, smoky, onion etc. are used to characterize the odor and flavor notes of the different thiols. More details about the descriptions, their concentrations in wine, and perception thresholds are given in Table 19.1. Several of these thiols have been detected and investigated in red wine from different varieties, particularly in Cabernet Sauvignon, Merlot, Cabernet franc, or Grenache wines (Blanchard, 2000; Bouchilloux et al., 1998a, 1998b; Ferreira et al., 2002). Investigations showed that 3SH and furfurylthiol can participate in the flavor of red Bordeaux wines (Blanchard, 2000). Several research groups demonstrated the contribution of 3SH to the fruity rosé character in wines from Cabernet Sauvignon, Merlot, and Grenache (Culleré et al., 2007; Ferreira et al., 2002; Murat et al., 2001). 3SH amounts at the end of alcoholic fermentation of red wines decreased during malolactic fermentation and barrel aging. Blanchard et al. (2004) showed in a model solution that phenolic fractions influence the evolution of 3SH in the presence of oxygen and that higher levels of free SO_2 lower the decrease of 3SH, which preserves the fruity notes in red wines supported by this thiol.

Rigou et al. (2014) demonstrated a distinct contribution of 4MSP in red wines from the Languedoc Roussillon region with typical blackcurrant aromas. Furthermore they found out that 3SH and 3SHA enhance the blackcurrant impression if they occur in high quantities in connection with 4MSP. A concentration of more than 16 ng/L of 4MSP seems to be necessary to achieve the typical blackcurrant character in red wines. The authors pointed out that the thiol amounts varied by the factor of 10 for 4MSP, 20 for 3SH, and 30 for 3MHA in the investigated red wines.

In addition, it can be assumed that several other powerful thiols which have been first identified in Sauvignon blanc wines and then also in wines from other white and red grape varieties contribute to the varietal flavor of different red wines. In particular, 4-methyl-4-sulfanylpentan-2-ol (citrus zest flavor) and 2-methyl-3-sulfanylbutan-1-ol and 3-methyl-3-sulfanylbutan-1-ol with odor impressions reminiscent of vegetables like leek or onions have to be mentioned (Blanchard et al., 2004; Bouchilloux et al., 1998b; Tominaga et al., 1998).

2-Methyl-3-furanthiol offers cooked meat aromas and was detected far above the perception threshold. Its contribution to cooked meat aromas in red wines is highly supposed (Tominaga and Dubourdieu, 2006).

Benzylthiol has an extremely low perception threshold and an aroma reminiscent of smoke or a gunflint odor. Tominaga et al. (2003b) measured the lowest amounts of benzylthiol in red wines. Nevertheless it was assumed that benzylthiol probably contributes to the empyreumatic character of red wines due to its very low perception threshold.

2-Furfurylthiol (Tominaga et al., 2000; Tominaga and Dubourdieu, 2006) has a roast coffee aroma and was also detected in red wines far above its low perception threshold. Therefore this high aroma-active thiol can participate in the roast-coffee notes of barrel aged red wines. Furfural from toasted oak wood in addition with sulfur metabolism of yeast and lactic acid bacteria are discussed for its formation (Darriet and Pons, 2017).

Picard et al. (2015) predicted that DMS, 2-furfurylthiol and 3SH concentrations correlated with the typicality of aged red Bordeaux wines and have an impact on flavor attributes like undergrowth, truffle, and empyreumatic.

Vanillylthiol was first identified by Floch et al. (2016) in young red and dry white wines. This thiol contributes to clove- and smoke-like notes. The highest concentrations of vanillylthiol were detected in red wines aged 12 months in new oak barrels. Therefore the authors supposed that vanillylthiol is probably involved in the clove-like and spicy flavor of red wines aged in oak barrels.

19.2.1.3 Methoxypyrazines

Methoxypyrazines are very powerful nitrogen heterocyclic compounds with low perception thresholds at low nanogram per liter range, in particular 2-methoxy-3-isobutylpyrazine (IBMP), 2-methoxy-3-isopropylpyrazine, and 2-methoxy-3-sec-butylpyrazine have been identified in a wide range of wines from different varieties (Table 19.2). They contribute also to wines from several red varieties like Cabernet Sauvignon, Cabernet franc,

TABLE 19.2 Structure, odor, concentration in wine, and perception threshold of IBMP and (−)-rotundone

Name	Structure	Odor	Concentration in wine/(ng/L)	Perception threshold/(ng/L)	Literature
3-Isobutyl-2-methoxypyrazine (IBMP)		Green peas[a]	1–25[c]	2[d],[1]	[a] Murray et al. (1970)
		Green bell pepper[b]		2[e],[2]	[b] Roujou de Boubée et al. (2000)
					[c] Alberts et al. (2009)
					[d] Buttery et al. (1969)
					[e] Zhao et al. (2017)
(−)-Rotundone		Peppery[a]	2–560[b]	8[a],[1]	[a] Wood et al. (2008)
				16[a],[3]	[b] Herderich et al. (2013), Mattivi et al. (2011)

Threshold determined in: [1] water; [2] water/ethanol 10% (w/w); [3] red wine.

Merlot, Carmenère, Pinot noir, Syrah, Pinotage, Grenache, Tempranillo, and others to green pepper and green peas aromas and vegetable-like notes (Allen et al., 1994; Bayonove et al., 1975; Culleré et al., 2009; Escudero et al., 2007; Ferreira et al., 2002; Roujou de Boubée et al., 2000). IBMP is the most abundant in grapes and wine. In red wines of the variety Carmenère concentrations could be measured up to 160 ng/L (Belancic and Agosin, 2007). The fate of IBMP during the whole wine production process was studied very intensively. It could be demonstrated that the methoxypyrazines that are mainly occurring in the berry skin are easily extracted during pressing and maceration. The achieved amounts of IBMP in the musts are reduced to a maximum of 30% during alcoholic fermentation and also the aging of wines (Maggu et al., 2007; Roujou de Boubée et al., 2002). Lower IBMP amounts were noticed in grapes with access to higher temperature during ripening (Allen and Lacey, 1993). IBMP levels vary in wines received from grapes obtained from different altitudes (Falcão et al., 2007). Viticultural and physiological parameters as ultraviolet (UV) light intensity, yields, availability of water and nitrogen, as well as climate-related changes affect IBMP and other methoxypyrazine development and concentration in grapes (Blanchard et al., 2004; Darriet and Pons, 2017; Roujou de Boubée et al., 2000; Ryona et al., 2008; Sala et al., 2005).

19.2.1.4 Sesquiterpene—(−)-Rotundone

(−)-Rotundone is a bicyclic sesquiterpene which is formed from its precursor α-guaiene by autoxidation (Huang et al., 2014, 2015). It has been identified in Australian Shiraz grapes for the first time. Rotundone was highlighted to contribute to spicy notes and also to black and white pepper aromas (Wood et al., 2008) (Table 19.2). Its concentrations in the analyzed Shiraz grapes ranged from 10 to 620 ng/kg. The highest detected amount of rotundone in the corresponding Shiraz wines was 145 ng/L. Low concentrations up to 20 ng/L were detected in Cabernet Sauvignon wines and in a blend of Cabernet Sauvignon, Merlot, and Cabernet franc. A very good correlation of the measured rotundone concentrations and the rating of black pepper aroma intensity was determined, although about 20% of the tasters were insensitive to this compound. Apart from Shiraz rotundone was also found in the French red varieties Duras and Pineau d'Anuis, it was also present in Graciano and Gamay (Herderich et al., 2013) as well as in Schioppettino and Vespolina (Mattivi et al., 2011).

The analysis of white and red wines with strong peppery characteristics led to concentrations up to 561 ng/L (Mattivi et al., 2011). Caputi et al. (2011) measured in Vespolina grapes up to 5.44 μg/kg of rotundone, which is considerably higher than levels reported for Shiraz grapes. It could be shown that rotundone is accumulating mainly in the berry skins (exocarp), but only 10% of its concentration in the grapes was measured after alcoholic fermentation and a further decrease to only 6% was detected in the bottled wines.

Herderich et al. (2013) pointed out that rotundone concentrations in grapes increase close to harvest time. They observed significantly higher amounts of rotundone in Shiraz grapes in certain vintages, vineyards, and from grapes grown in cooler climates. Furthermore, clonal effects may have an impact on the final concentrations of rotundone in the wines as well as the interaction of the grapevine genome with its environment. The impact of clonal selection, viticultural and enological techniques have to be studied in the future to achieve the understanding how the concentrations of rotundone in grapes and the corresponding wines could be influenced. This will assist to create certain wine styles with different pepper aromas and to meet consumer preferences (Herderich et al., 2013).

TABLE 19.3 Structure, odor, concentration in wine, and perception threshold of C_{13}-norisoprenoides in wine

Name	Structure	Odor	Concentration in wine/(ng/L)	Perception threshold/(ng/L)	Literature
β-Damascenone		Baked apple[a]	Grenache: 1000–4000[b]	50[d],[1]	[a] Escudero et al. (2007)
			Merlot noir: 250–1300[c]		[b] Sabon et al. (2002)
					[c] Kotseridis et al. (1998)
					[d] Peng et al. (2013)
β-Ionone		Violet, balsamic, roses[a]	Grenache: 100–300[b]	90[d],[1]	[a] Peng et al. (2013)
			Merlot noir: 100–200[c]		[b] Sabon et al. (2002)
					[c] Kotseridis et al. (1998)
					[d] Peng et al. (2013)

Threshold determined in: [1] water/ethanol 12% (v/v) + 5 g/L tartaric acid, pH = 3.2.

19.2.1.5 C_{13}-Norisoprenoides

Norisoprenoids with 13 carbon atoms (C_{13}-norisoprenoids) were formed through the oxidative degradation of carotenoids. β-Damascenone is one of the received derivatives (Table 19.3). It has a very low perception threshold in water (low nanogram level) and offers different odor impressions (baked apple or tropical fruit) depending on its concentration and the matrix. The perception threshold of β-damascenone in red wine is considerably higher (low microgram level). In addition, the published thresholds vary in a wide range (Darriet and Pons, 2017; Pineau et al., 2007). β-Damascenone occurs in wines from nearly all varieties. Its amount in red wines is about 1–4 μg/L (Pineau et al., 2007; Sabon et al., 2002). In this concentration β-damascenone acts more in a synergistic effect or as an aroma enhancer (Culleré et al., 2009; Escudero et al., 2007; Ferreira, 2010; Pineau et al., 2007). Pineau et al. (2007) demonstrated in a hydroalcoholic model solution that β-damascenone enhanced fruity aromas of ethyl cinnamate and caproate and masked the herbaceous note of IBMP.

β-Ionone, another C_{13}-norisoprenoid, has an odor of violet. Its perception threshold in water is about 90 ng/L (Peng et al., 2013). β-Ionone was analyzed in various wines from different varieties (concentration range in red wines: 100–300 ng/L) and its contribution to the flavor of wines has been shown and discussed (Kotseridis et al., 1998; Mendes-Pinto, 2009; Peng et al., 2013; Sabon et al., 2002).

19.2.1.6 Esters

Several research groups draw attention to specific esters and their contribution as enhancers for certain aroma nuances in wines. Van Wyk et al. (1979) and Ferreira et al. (2000) already pointed out that isoamyl acetate is a characteristic aroma compound in wines produced with Pinotage and Tempranillo. Also ethyl esters of fatty acids and higher alcohol acetates can have a high impact on red wine fruity aroma, even when they occur in concentrations below their odor detection thresholds, due to synergistic effects (Lytra et al., 2012, 2013; Pineau et al., 2009).

Falcão et al. (2012) identified ethyl 2-hydroxy-4-methylpentanoate for the first time in red and white table wines and highlighted that it is involved in fresh blackberry aroma notes. Its perception threshold in red wine is around 0.9 μg/L (dearomatized red wine) and the average amount in various analyzed wines was about 400 μg/L. Additional studies of Lytra et al. (2012) underlined that ethyl 2-hydroxy-4-methylpentanoate enhances blackberry and fresh-fruit aromas in red wine extracts. There is also evidence that other red wine esters with similar chemical structures are involved in these fruity flavor notes (Lytra et al., 2013).

19.2.1.7 Miscellaneous Aroma Compounds

A few interesting flavor compounds which contribute to specific aromas in red wine and cannot be grouped in the abovementioned chemical categories will be shortly introduced.

1,8-Cineole is responsible for minty and eucalyptus aromas in certain red wines (Capone et al., 2011; Sun et al., 2011; Wedler et al., 2015). This monoterpenoid with a perception threshold of 1.1 μg/L (Table 19.4) is also known as eucalyptol and was first reported in wine by Herve et al. (2003).

TABLE 19.4 Structure, odor, concentration in wine, and perception threshold of 1,8-cineole and piperitone

Name	Structure	Odor	Concentration in wine/(ng/L)	Perception threshold/(ng/L)	Literature
1,8-Cineole		Minty[a]	White wine: <800[b]	1100[c],[1]	[a] Sun et al. (2011)
		Eucalyptus[b]	Red wine: <19,600[b]		[b] Capone et al. (2011)
					[c] Herve et al. (2003)
Piperitone		Minty[a]	120–1100[a]	900[b],[2]	[a] Picard et al. (2016a)
				70,000[b],[3]	[b] Pons et al. (2016)

Threshold determined in: [1] water; [2] water/ethanol 12% (v/v) + 5 g/L tartaric acid, pH = 3.2; [3] red wine.

It was proposed that vineyards producing wines with intensive eucalyptus aromas are surrounded by *Eucalyptus* sp. trees, which may transfer their essence to the grapes (Herve et al., 2003). Fariña et al. (2005) demonstrated that 1,8-cineole can also be built by chemical transformation of limonene and α-terpineol. They proposed that this reaction is probably an explanation for the eucalyptus-like aromas in Tannat wines from vines not growing near eucalyptus trees.

Poitou et al. (2017) quantified 1,8-cineole in French Cabernet Sauvignon and Merlot grapes during berry development and noticed high amounts in unripe berries and a decrease during grape maturation. In addition, sensory tests confirmed that 1,8-cineole has an impact on the menthol nuance and overall green perception in wines, particularly in connection with IBMP due to an additive effect.

Piperitone, a monoterpene ketone with minty aroma notes, was detected by Picard et al. (2016a) for the first time in aged red Bordeaux wines with pleasant mint notes (Table 19.4). The concentration of piperitone ranged from 120 to 1100 ng/L in 51 red wines with different mint aroma intensities. Its participation in minty aromas was proved on the basis and connection of sensory and chemical data. In addition, an impact on the piperitone concentrations in wines was detected due to the geographical origin within the Bordeaux region. One of the two enantiomeric forms [(+)-(6S) enantiomer] was measured in significantly higher amounts in wines with a strong aging bouquet (Picard et al., 2016b). Recently, various other limonene-derived mint aroma compounds were analyzed by Picard et al. (2018).

3-Methyl-2,4-nonanedione was identified in aged red wine for the first time and suggested to be also responsible for the characteristic prune aroma of prematurely aged red wines by Pons et al. (2008). Its perception threshold in an aqueous ethanolic model solution is 16 ng/L. Furthermore the authors detected that γ-nonalactone and β-damascenone can additionally contribute to an intensive prune aroma.

19.3 CONCLUSION

Comprehensive flavor research and highly sophisticated analytical techniques to analyze high aroma-active compounds at extremely low concentrations have offered the opportunity to study certain compounds or substance groups in a very detailed form within the last decades, e.g. varietal thiols. More and more important is the understanding of interactions between the wine matrix and volatile compounds and the impact on the overall flavor as well as on typical or specific flavors in wines. In addition, additive, synergistic, and masking effects have to be studied in greater detail (Darriet and Pons, 2017). The gained knowledge about the behavior of certain aromatic compounds has then to be applied during the different steps of the wine production process to create, develop, and to optimize specific wine styles and to meet consumer preferences.

References

Acuña, G., Gautschi, M., Kumli, F., Schmid, J., Zsindely, J., 2003. Mercapto-alkanol Flavor Compounds, US6610346B1.

Alberts, P., Stander, M.A., Paul, S.O., Villiers, A. de, 2009. Survey of 3-alkyl-2-methoxypyrazine content of South African Sauvignon blanc wines using a novel LC–APCI–MS/MS method. J. Agric. Food Chem. 57 (20), 9347–9355. Available from: https://doi.org/10.1021/jf9026475.

Allen, M.S., Lacey, M.J., 1993. Methoxypyrazine grape flavour: influence of climate, cultivar and viticulture. Wein-Wiss. 48, 211–213.

Allen, M.S., Lacey, M.J., Boyd, S., 1994. Determination of methoxypyrazines in red wines by stable isotope dilution gas chromatography–mass spectrometry. J. Agric. Food Chem. 42 (8), 1734–1738. Available from: https://doi.org/10.1021/jf00044a030.

Anness, B.J., Bamforth, C.W., 1982. Dimethyl sulphide—a review. J. Inst. Brew. 88 (4), 244–252.

Anocibar Beloqui, A., 1998. Les composés soufrés volatils des vins rouges. Thesis. Université de Bordeaux 2, Bordeaux.

Anocibar Beloqui, A., Kotseridis, Y., Bertrand, A., 1996. Détermination de la teneur en sulfure de diméthyle dans quelques vins rouges. J. Int. Sei. Vigne Vin 30 (03), 167–170.

Bamforth, C.W., Anness, B.J., 1981. The role of dimethyl sulphoxide reductase in the formation of dimethyl sulphide during fermentations. J. Inst. Brew. 87 (1), 30–34.

Baumes, R., 2009. Chapter 8A: wine aroma precursors. In: Moreno-Arribas, M.V., Polo, M.C. (Eds.), Wine Chemistry and Biochemistry. Springer New York, New York, NY, pp. 251–274.

Bayonove, C., Cordonnier, R., Dubois, P., 1975. Etude d'une fraction caractéristique de l'arôme du raisin de la variété Cabernet-Sauvignon: mise en évidence de la 2-méthoxy-3-isobutylpyrazine. C.R. Acad. Sci 60, 1321–1329.

Belancic, A., Agosin, E., 2007. Methoxypyrazines in grapes and wines of Vitis vinifera cv. Carmenere. Am. J. Enol. Vitic. 58 (4), 462–469.

Belda, I., Ruiz, J., Esteban-Fernández, A., Navascués, E., Marquina, D., Santos, A., et al., 2017. Microbial contribution to wine aroma and its intended use for wine quality improvement. Molecules 22 (2). Available from: https://doi.org/10.3390/molecules22020189.

Blanchard, L., 2000. Recherche sur la contribution de certains thiols volatils à l'arôme des vins rouges: étude de leur genèse et de leur stabilité. Thesis. Université Bordeaux 2, Bordeaux.

Blanchard, L., Darriet, P., Dubourdieu, D., 2004. Reactivity of 3-mercaptohexanol in red wine: impact of oxygen, phenolic fractions, and sulfur dioxide. Am. J. Enol. Vitic. 55 (2), 115–120.

Bouchilloux, P., Darriet, P., Dubourdieu, D., 1998a. Identification du 2-methyl-3-furanthiol dans les vins. Vitis 37, 177–180.

Bouchilloux, P., Darriet, P., Henry, R., Lavigne-Cruège, V., Dubourdieu, D., 1998b. Identification of volatile and powerful odorous thiols in Bordeaux red wine varieties. J. Agric. Food Chem. 46 (8), 3095–3099. Available from: https://doi.org/10.1021/jf971027d.

Buttery, R.G., Seifert, R.M., Guadagni, D.G., Ling, L.C., 1969. Characterization of some volatile constituents of bell peppers. J. Agric. Food Chem. 17 (6), 1322–1327. Available from: https://doi.org/10.1021/jf60166a061.

Capone, D.L., van Leeuwen, K., Taylor, D.K., Jeffery, D.W., Pardon, K.H., Elsey, G.M., et al., 2011. Evolution and occurrence of 1,8-cineole (eucalyptol) in Australian wine. J. Agric. Food Chem. 59 (3), 953–959. Available from: https://doi.org/10.1021/jf1038212.

Caputi, L., Carlin, S., Ghiglieno, I., Stefanini, M., Valenti, L., Vrhovsek, U., et al., 2011. Relationship of changes in rotundone content during grape ripening and winemaking to manipulation of the 'peppery' character of wine. J. Agric. Food Chem. 59 (10), 5565–5571. Available from: https://doi.org/10.1021/jf200786u.

Culleré, L., Cacho, J., Ferreira, V., 2007. An assessment of the role played by some oxidation-related aldehydes in wine aroma. J. Agric. Food Chem. 55 (3), 876–881. Available from: https://doi.org/10.1021/jf062432k.

Culleré, L., Escudero, A., Campo, E., Cacho, J., Ferreira, V., 2009. Multidimensional gas chromatography-mass spectrometry determination of 3-alkyl-2-methoxypyrazines in wine and must. A comparison of solid-phase extraction and headspace solid-phase extraction methods. J. Chromatogr. A. 1216 (18), 4040–4045. Available from: https://doi.org/10.1016/j.chroma.2009.02.072.

Dagan, L., 2006. Potentiel aromatique des raisins de Vitis vinifera L. cv. Petit Manseng et Gros Manseng. Contribution à l'arôme des vins de pays Côtes de Gascogne. Doctoral Dissertation. University of Montpellier II, Montpellier.

Darriet, P., Pons, A., 2017. Chapter 8: Wine. In: Buettner, A. (Ed.), Springer Handbook of Odor. Springer, Cham, pp. 143–170.

Darriet, P., Tominaga, T., Lavigne, V., Boidron, J.-N., Dubourdieu, D., 1995. Identification of a powerful aromatic component of Vitis vinifera L. var. Sauvignon wines: 4-mercapto-4-methylpentan-2-one. Flavour Fragr. J 10 (6), 385–392.

Du Plessis, G.J., Loubser, G.J., 1974. The bouquet of "late harvest" wine. Agrochemophysica 6 (3), 49–52.

Dubourdieu, D., Tominaga, T., 2009. Chapter 8B: Polyfunctional thiol compounds. In: Moreno-Arribas, M.V., Polo, M.C. (Eds.), Wine Chemistry and Biochemistry. Springer New York, New York, NY, pp. 275–293.

Escudero, A., Campo, E., Fariña, L., Cacho, J., Ferreira, V., 2007. Analytical characterization of the aroma of five premium red wines. Insights into the role of odor families and the concept of fruitiness of wines. J. Agric. Food Chem. 55 (11), 4501–4545. Available from: https://doi.org/10.10.1021/jf0636418.

Falcão, L.D., Revel, G., de, Perello, M.C., Moutsiou, A., Zanus, M.C., Bordignon-Luiz, M.T., 2007. A survey of seasonal temperatures and vineyard altitude influences on 2-methoxy-3-isobutylpyrazine, C13-norisoprenoids, and the sensory profile of Brazilian Cabernet Sauvignon wines. J. Agric. Food Chem. 55 (9), 3605–3612.

Falcão, L.D., Lytra, G., Darriet, P., Barbe, J.-C., 2012. Identification of ethyl 2-hydroxy-4-methylpentanoate in red wines, a compound involved in blackberry aroma. Food Chem. 132 (1), 230–236. Available from: https://doi.org/10.1016/j.foodchem.2011.10.061.

Fariña, L., Boido, E., Carrau, F., Versini, G., Dellacassa, E., 2005. Terpene compounds as possible precursors of 1,8-cineole in red grapes and wines. J. Agric. Food Chem. 53 (5), 1633–1636. Available from: https://doi.org/10.1021/jf040332d.

Ferreira, V., Lopez, R., Cacho, J.F., 2000. Quantitative determination of the odorants of young red wines from different grape varieties. J. Sci. Food Agric. 80 (11), 1659–1667.

Ferreira, V., 2010. Volatile aroma compounds and wine sensory attributes. In: Reynolds, A.G. (Ed.), Viticulture and Wine Quality. Woodhead Pub. Ltd, Boca Raton, FL, Oxford.

Ferreira, V., Ortín, N., Escudero, A., López, R., Cacho, J., 2002. Chemical characterization of the aroma of Grenache rosé wines: aroma extract dilution analysis, quantitative determination, and sensory reconstitution studies. J. Agric. Food Chem. 50 (14), 4048–4054. Available from: https://doi.org/10.1021/jf0115645.

Floch, M., Shinkaruk, S., Darriet, P., Pons, A., 2016. Identification and organoleptic contribution of vanillylthiol in wines. J. Agric. Food Chem. 64 (6), 1318–1325. Available from: https://doi.org/10.1021/acs.jafc.5b05733.

Goniak, O.J., Noble, A.C., 1987. Sensory study of selected volatile sulfur compounds in white wine. Am. J. Enol. Vitic. 38 (3), 223–227.

Goto, S., Takamuro, R., 1987. Concentration of dimethyl sulphide in wine after different storage times. Hakkokogaku 65, 53–57.

Herderich, M.J., Siebert, T.E., Parker, M., Capone, D.L., Mayr, C., Zhang, P., et al., 2013. Synthesis of the ongoing works on rotundone, an aromatic compound responsible of the peppery notes in wines. Internet J. Enol. Vitic. 6 (1), 1–6.

REFERENCES

Herve, E., Price, S., Burns, G., 2003. Eucalyptol in wines showing a "eucalyptus" aroma. In: Lonvaud-Funel, A., Revel, G., de, Darriet, P. (Eds.), Oenologie. Lavoisier, Tec. & Doc, Paris, pp. 598–600.

Huang, A.-C., Burrett, S., Sefton, M.A., Taylor, D.K., 2014. Production of the pepper aroma compound, (−)-rotundone, by aerial oxidation of α-guaiene. J. Agric. Food Chem. 62 (44), 10809–10815. Available from: https://doi.org/10.1021/jf504693e.

Huang, A.-C., Sefton, M.A., Sumby, C.J., Tiekink, E.R.T., Taylor, D.K., 2015. Mechanistic studies on the autoxidation of α-guaiene: structural diversity of the sesquiterpenoid downstream products. J. Nat. Prod. 78 (1), 131–145. Available from: https://doi.org/10.1021/np500819f.

Kolor, M.G., 1983. Identification of an important new flavor compound in Concord grape: ethyl-3-mercaptopropionate. J. Agric. Food Chem. 31 (5), 1125–1127. Available from: https://doi.org/10.1021/jf00119a052.

Kotseridis, Y., Anocibar Beloqui, A., Bertrand, A., Doazan, J.P., 1998. An analytical method for studying the volatile compounds of merlot noir clone wines. Am. J. Enol. Vitic. 49 (1), 44–48.

Loscos, N., Ségurel, M., Dagan, L., Sommerer, N., Marlin, T., Baumes, R., 2008. Identification of S-methylmethionine in Petit Manseng grapes as dimethyl sulphide precursor in wine. Anal. Chim. Acta 621 (1), 24–29. Available from: https://doi.org/10.1016/j.aca.2007.11.033.

Loubser, G.J., Du Plessis, C.S., 1976. The quantitative determination and some values of dimethyl sulphide in white table wines. Vitis 15 (4), 248–252.

Lytra, G., Tempere, S., Revel, G., de, Barbe, J.-C., 2012. Impact of perceptive interactions on red wine fruity aroma. J. Agric. Food Chem. 60 (50), 12260–12269. Available from: https://doi.org/10.1021/jf302918q.

Lytra, G., Tempere, S., Le Floch, A., Revel, G., de, Barbe, J.-C., 2013. Study of sensory interactions among red wine fruity esters in a model solution. J. Agric. Food Chem. 61 (36), 8504–8513. Available from: https://doi.org/10.1021/jf4018405.

Lytra, G., Tempere, S., Zhang, S., Marchand, S., Revel, G., de, Barbe, J.-C., 2014. Olfactory impact of dimethyl sulfide on blackcurrant aroma expression. J. Int. Sei. Vigne Vin 48 (1), 75–85.

Maggu, M., Winz, R., Kilmartin, P.A., Trought, M.C.T., Nicolau, L., 2007. Effect of skin contact and pressure on the composition of Sauvignon Blanc must. J. Agric. Food Chem. 55 (25), 10281–10288.

Marais, J., 1979. Effect of storage time and temperature on the formation of dimethyl sulphide and on white wine quality. Vitis 18, 254–260.

Mattivi, F., Caputi, L., Carlin, S., Lanza, T., Minozzi, M., Nanni, D., et al., 2011. Effective analysis of rotundone at below-threshold levels in red and white wines using solid-phase microextraction gas chromatography/tandem mass spectrometry. Rapid Commun. Mass Spectrom. 25 (4), 483–488. Available from: https://doi.org/10.1002/rcm.4881.

Mendes-Pinto, M.M., 2009. Carotenoid breakdown products the—norisoprenoids—in wine aroma. Arch. Biochem. Biophys. 483 (2), 236–245. Available from: https://doi.org/10.1016/j.abb.2009.01.008.

de, Mora, S.J., Eschenbruch, R., Knowles, S.J., Spedding, D.J., 1986. The formation of dimethyl sulphide during fermentation using a wine yeast. Food Microbiol. 3 (1), 27–32.

de, Mora, S.J., Knowles, S.J., Eschenbruch, R., Torrey, W.J., 1987. Dimethyl sulphide in some Australian red wines. Vitis 26, 79–84.

Murat, M.-L., Tominaga, T., Dubourdieu, D., 2001. Assessing the aromatic potential of cabernet sauvignon and merlot musts used to produce rose wine by assaying the cysteinylated precursor of 3-mercaptohexan-1-ol. J. Agric. Food Chem. 49 (11), 5412–5417. Available from: https://doi.org/10.1021/jf0103119.

Murray, K.E., Shipton, J., Whitfield, F.B., 1970. 2-Methoxypyrazines and the flavour of green peas (*Pisum sativum*). Chem. Ind. 27, 897–898.

Parker, M., Capone, D.L., Francis, I.L., Herderich, M.J., 2018. Aroma precursors in grapes and wine: flavor release during wine production and consumption. J. Agric. Food Chem. 66 (10), 2281–2286. Available from: https://doi.org/10.1021/acs.jafc.6b05255.

Peng, C.-T., Wen, Y., Tao, Y.-S., Lan, Y.-Y., 2013. Modulating the formation of Meili wine aroma by prefermentative freezing process. J. Agric. Food Chem. 61 (7), 1542–1553. Available from: https://doi.org/10.1021/jf3043874.

Picard, M., Thibon, C., Redon, P., Darriet, P., Revel, G., de, Marchand, S., 2015. Involvement of dimethyl sulfide and several polyfunctional thiols in the aromatic expression of the aging bouquet of red bordeaux wines. J. Agric. Food Chem. 63 (40), 8879–8889. Available from: https://doi.org/10.1021/acs.jafc.5b03977.

Picard, M., Lytra, G., Tempere, S., Barbe, J.-C., Revel, G., de, Marchand, S., 2016a. Identification of piperitone as an aroma compound contributing to the positive mint nuances perceived in aged red bordeaux wines. J. Agric. Food Chem. 64 (2), 451–460. Available from: https://doi.org/10.1021/acs.jafc.5b04869.

Picard, M., Tempere, S., Revel, G., de, Marchand, S., 2016b. Piperitone profiling in fine red bordeaux wines: geographical influences in the bordeaux region and enantiomeric distribution. J. Agric. Food Chem. 64 (40), 7576–7584. Available from: https://doi.org/10.1021/acs.jafc.6b02835.

Picard, M., van Leeuwen, C., Guyon, F., Gaillard, L., Revel, G., de, Marchand, S., 2017. Vine water deficit impacts aging bouquet in fine red bordeaux wine. Front. Chem. 5, 56. Available from: https://doi.org/10.3389/fchem.2017.00056.

Picard, M., Franc, C., Revel, G., de, Marchand, S., 2018. Dual solid-phase and stir bar sorptive extraction combined with gas chromatography—mass spectrometry analysis provides a suitable tool for assaying limonene-derived mint aroma compounds in red wine. Anal. Chim. Acta 1001, 168–178. Available from: https://doi.org/10.1016/j.aca.2017.11.074.

Pineau, B., Barbe, J.-C., van Leeuwen, C., Dubourdieu, D., 2007. Which impact for beta-damascenone on red wines aroma?. J. Agric. Food Chem. 55 (10), 4103–4108. Available from: https://doi.org/10.1021/jf070120r.

Pineau, B., Barbe, J.-C., van Leeuwen, C., Dubourdieu, D., 2009. Examples of perceptive interactions involved in specific "red-" and "blackberry" aromas in red wines. J. Agric. Food Chem. 57 (9), 3702–3708. Available from: https://doi.org/10.1021/jf803325v.

Poitou, X., Thibon, C., Darriet, P., 2017. 1,8-Cineole in French red wines: evidence for a contribution related to its various origins. J. Agric. Food Chem. 65 (2), 383–393. Available from: https://doi.org/10.1021/acs.jafc.6b03042.

Pons, A., Lavigne, V., Eric, F., Darriet, P., Dubourdieu, D., 2008. Identification of volatile compounds responsible for prune aroma in prematurely aged red wines. J. Agric. Food Chem. 56 (13), 5285–5290. Available from: https://doi.org/10.1021/jf073513z.

Pons, A., Lavigne, V., Darriet, P., Dubourdieu, D., 2016. Identification and analysis of piperitone in red wines. Food Chem. 206, 191–196. Available from: https://doi.org/10.1016/j.foodchem.2016.03.064.

Ribéreau-Gayon, P., Dubourdieu, D., Donèche, B., Lonvaud, A., 2006. Handbook of Enology: The Microbiology of Wine and Vinifications, 2nd ed John Wiley, Chichester.

Rigou, P., Triay, A., Razungles, A., 2014. Influence of volatile thiols in the development of blackcurrant aroma in red wine. Food Chem. 142, 242−248. Available from: https://doi.org/10.1016/j.foodchem.2013.07.024.

Roland, A., Schneider, R., Razungles, A., Cavelier, F., 2011. Varietal thiols in wine: discovery, analysis and applications. Chem. Rev. 111 (11), 7355−7376. Available from: https://doi.org/10.1021/cr100205b.

Roujou de Boubée, D., van Leeuwen, C., Dubourdieu, D., 2000. Organoleptic impact of 2-methoxy-3-isobutylpyrazine on red Bordeaux and Loire wines. Effect of environmental conditions on concentrations in grapes during ripening. J. Agric. Food Chem. 48 (10), 4830−4834. Available from: https://doi.org/10.1021/jf000181o.

Roujou de Boubée, D., Cumsille, A.M., Pons, M., Dubourdieu, D., 2002. Location of 2-methoxy-3-isobutylpyrazine in Cabernet Sauvignon grape bunches and its extractability during vinification. Am. J. Enol. Vitic. 53 (1), 1−5.

Royer Dupré, N., de, Schneider, R., Payan, J.C., Salançon, E., Razungles, A., 2014. Effects of vine water status on dimethyl sulfur potential, ammonium, and amino acid contents in Grenache Noir grapes (Vitis vinifera). J. Agric. Food Chem. 62 (13), 2760−2766. Available from: https://doi.org/10.1021/jf404758g.

Ryona, I., Pan, B.S., Intrigliolo, D.S., Lakso, A.N., Sacks, G.L., 2008. Effects of cluster light exposure on 3-isobutyl-2-methoxypyrazine accumulation and degradation patterns in red wine grapes (Vitis vinifera L. Cv. Cabernet Franc). J. Agric. Food Chem. 56 (22), 10838−10846. Available from: https://doi.org/10.1021/jf801877y.

Sabon, I., Revel, G., de, Kotseridis, Y., Bertrand, A., 2002. Determination of volatile compounds in Grenache wines in relation with different terroirs in the Rhone Valley. J. Agric. Food Chem. 50 (22), 6341−6345. Available from: https://doi.org/10.1021/jf025611k.

Sala, C., Busto, O., Guasch, J., Zamora, F., 2005. Contents of 3-alkyl-2-methoxypyrazines in musts and wines from Vitis vinifera variety Cabernet Sauvignon: influence of irrigation and plantation density. J. Sci. Food Agric. 85 (7), 1131−1136.

San-Juan, F., Ferreira, V., Cacho, J., Escudero, A., 2011. Quality and aromatic sensory descriptors (mainly fresh and dry fruit character) of Spanish red wines can be predicted from their aroma-active chemical composition. J. Agric. Food Chem. 59 (14), 7916−7924. Available from: https://doi.org/10.1021/jf1048657.

Sarrazin, E., Shinkaruk, S., Tominaga, T., Bennetau, B., Frérot, E., Dubourdieu, D., 2007. Odorous impact of volatile thiols on the aroma of young botrytized sweet wines: identification and quantification of new sulfanyl alcohols. J. Agric. Food Chem. 55 (4), 1437−1444. Available from: https://doi.org/10.1021/jf062582v.

Schreier, P., Drawert, F., Junker, A., 1974. Gaschromatographisch-massenspektrometrische Untersuchung flüchtiger Inhaltsstoffe des Weines. Z. Lebensm. Unters. Forsch. 154 (5), 279−284. Available from: https://doi.org/10.1007/BF01083423.

Ségurel, M.A., Razungles, A.J., Riou, C., Salles, M., Baumes, R.L., 2004. Contribution of dimethyl sulfide to the aroma of Syrah and Grenache Noir wines and estimation of its potential in grapes of these varieties. J. Agric. Food Chem. 52 (23), 7084−7093. Available from: https://doi.org/10.1021/jf049160a.

Ségurel, M.A., Razungles, A.J., Riou, C., Trigueiro, M.G.L., Baumes, R.L., 2005. Ability of possible DMS precursors to release DMS during wine aging and in the conditions of heat-alkaline treatment. J. Agric. Food Chem. 53 (7), 2637−2645. Available from: https://doi.org/10.1021/jf048273r.

Simpson, R.F., 1979. Aroma composition of bottle aged white wine. Vitis 18, 148−154.

Spedding, D.J., Raut, P., 1982. The influence of dimethyl sulphide and carbon disulphide in the bouquet of wines. Vitis 21, 240−246.

Styger, G., Prior, B., Bauer, F.F., 2011. Wine flavor and aroma. J. Ind. Microbiol Biotechnol. 38 (9), 1145−1159. Available from: https://doi.org/10.1007/s10295-011-1018-4.

Sun, Q., Gates, M.J., Lavin, E.H., Acree, T.E., Sacks, G.L., 2011. Comparison of odor-active compounds in grapes and wines from Vitis vinifera and non-foxy American grape species. J. Agric. Food Chem. 59 (19), 10657−10664. Available from: https://doi.org/10.1021/jf2026204.

Tominaga, T., Dubourdieu, D., 2006. A novel method for quantification of 2-methyl-3-furanthiol and 2-furanmethanethiol in wines made from Vitis vinifera grape varieties. J. Agric. Food Chem. 54 (1), 29−33. Available from: https://doi.org/10.1021/jf050970b.

Tominaga, T., Darriet, P., Dubourdieu, D., 1996. Identification de l'acétate de 3-mercaptohexanol, composé à forte odeur de buis, intervenant dans l'arôme des vins de Sauvignon. Vitis 35 (4), 207−210.

Tominaga, T., Furrer, A., Henry, R., Dubourdieu, D., 1998. Identification of new volatile thiols in the aroma of Vitis vinifera L. var. Sauvignon blanc wines. Flavour Fragr. J. 13 (3), 159−162. Available from: https://doi.org/10.1002/(SICI)1099-1026(199805/06)13:3(159::AID-FFJ709)3.0.CO;2-7.

Tominaga, T., Blanchard, L., Darriet, P., Dubourdieu, D., 2000. A powerful aromatic volatile thiol, 2-furanmethanethiol, exhibiting roast coffee aroma in wines made from several Vitis vinifera grape varieties. J. Agric. Food Chem. 48, 1799−1802.

Tominaga, T., Guimbertau, G., Dubourdieu, D., 2003a. Contribution of benzenemethanethiol to smoky aroma of certain Vitis vinifera L. wines. J. Agric. Food Chem. 51 (5), 1373−1376. Available from: https://doi.org/10.1021/jf020756c.

Tominaga, T., Guimbertau, G., Dubourdieu, D., 2003b. Role of certain volatile thiols in the bouquet of aged champagne wines. J. Agric. Food Chem. 51 (4), 1016−1020. Available from: https://doi.org/10.1021/jf020755k.

Van Wyk, C.J., Augustyn, O.P.H., de Wet, P., Joubert, W.A., 1979. Isoamyl Acetate — a Key Fermentation Volatile of Wines of Vitis Vinifera CV Pinotage. Am. J. Enol. Vitic. 30 (3), 167−173.

Wedler, H.B., Pemberton, R.P., Tantillo, D.J., 2015. Carbocations and the complex flavor and bouquet of wine: mechanistic aspects of terpene biosynthesis in wine grapes. Molecules 20 (6), 10781−10792. Available from: https://doi.org/10.3390/molecules200610781.

Wood, C., Siebert, T.E., Parker, M., Capone, D.L., Elsey, G.M., Pollnitz, A.P., et al., 2008. From wine to pepper: rotundone, an obscure sesquiterpene, is a potent spicy aroma compound. J. Agric. Food Chem. 56 (10), 3738−3744. Available from: https://doi.org/10.1021/jf800183k.

Zhao, P., Gao, J., Qian, M., Li, H., 2017. Characterization of the key aroma compounds in Chinese Syrah wine by gas chromatography-olfactometry−mass spectrometry and aroma reconstitution studies. Molecules 22 (7). Available from: https://doi.org/10.3390/molecules22071045.

CHAPTER

20

The Instrumental Analysis of Aroma-Active Compounds for Explaining the Flavor of Red Wines

Laura Culleré, Ricardo López and Vicente Ferreira

Laboratory for Flavor Analysis and Enology, Instituto Agroalimentario de Aragón (IA2), Department of Analytical Chemistry, Faculty of Sciences, Universidad Zaragoza, Zaragoza, Spain

20.1 INTRODUCTION

This chapter seeks to define what in analytical chemistry is known as the "analytical problem" of red wine aroma. That is, our goals are to define which ones are the "analytes" (the targets to quantify), at what concentrations levels should they be quantified, what chemical or physicochemical properties do they have that are useful from the analytical point of view, and finally and taking into account all the previous questions, what solutions have been given or could potentially be given to solve such analytical problem.

It is important to note that our interest in those molecules that have or may have some impact on the aroma perception, or better, on all the aroma perceptions, elicited during the consumption of red wine. This means that our targets are those compounds present in the wine at concentrations in the liquid phase enough to produce vapor phases, during any time of wine consumption, containing the compounds at concentrations detectable by our olfactory receptors.

Thus, we need to understand first that there is not a single wine aroma perception, but a continuum of perceptions during all the contact that we have with the wine during its enjoyment: from the first sniff of the vapors in the unstirred glass containing the recently poured wine, to the last molecules of aroma compounds remaining in the mouth after swallowing the last drops of wine and which are further released to the nasal cavity imparting aftertaste and retronasal aroma perceptions (Meillon et al., 2009). The period during which wine is consumed is relatively large. Usually, we can have the wine in the glass for more than 15 minutes, often subject to vigorous shaking. During this time, the aroma composition changes: some very volatile aroma compounds can be completely lost (Franco-Luesma et al., 2016; Lytra et al., 2016) and some chemical equilibria may shift releasing odor molecules that at the beginning were bonded in odorless adducts (Wen et al., 2018). Scientists deal with this complexity by averaging. Leaving aside the different sensory techniques to measure the evolution during consumption of the sensory perception (Labbe et al., 2009; Castura et al., 2016; Schlich, 2017), most sensory techniques measure the averaged sensory perceptions of a group of panelists. The score of a given sensory descriptor in a given product, in fact is a time and panel average.

Having this in mind, there are just two possibilities to make a preliminary assessment of the potential relevance of an odorant in the aroma perceptions associated to the consumption of wine: a comparison of its concentration with the odor threshold, and the gas-chromatography-olfactometry screening of an extract resembling the composition of the vapor phases emanated from the product (San-Juan et al., 2010; Escudero et al., 2014). As odor thresholds have inevitably a relatively large uncertainty, and sometimes even important biases, they should be used prudently and not too strictly. With these disclaimers, it can be said that the purpose of the chapter is to discuss how to analyze aroma compounds which can be present at concentrations higher than their thresholds in red wines.

20.2 ANALYTES AND AN ANALYTICAL CLASSIFICATION

Tables 20.1 and 20.2 summarize 75 aroma compounds which, according to present knowledge, can be found above threshold in all or in some specific type of red wine. Compounds in Table 20.1 are those considered as "easy to analyze," since most of them are present in nearly all wines at concentrations in general above the μg/L, they are not very reactive, and they do not form strong complexes with matrix components. Within the table, compounds are arranged by origin. All compounds in this table can be easily analyzed by GC with Flame Ionization Detector (FID) or Mass Spectrometry (MS) detectors, as will be later discussed.

TABLE 20.1 "Easy to analyze" wine aroma compounds. Compounds which only very exceptionally will be present above threshold are given between brackets

1. **Grape origin**
   1.1. Terpenols: linalool (1), geraniol (2), (α-terpineol, citronelol, nerol)
   1.2. Nor-isoprenoids: β-damsascenone (3), β-ionone (4), (α-ionone)
   1.3. Micellaneous: ethyl cinnamate (5), ethyl dihydrocinnamate (6), (hexanol), Z-3-hexenol (7), (E-2-hexenol)
2. **Fermentative origin**
   2.1. Ethanol (8)
   2.2. Fusel alcohols: isobutanol (9), 2 and 3-methylbutanol (10) "isoamylalcohol," β-phenylethanol (11), methionol (12)
   2.3. Fusel alcohol acetates: (isobutyl acetate), isoamyl acetate (13), (hexyl acetate), phenylethyl acetate (14)
   2.4. Fatty acids and their ethyl esters: acetic acid (15), butyric acid (16), hexanoic acid (17), octanoic acid (18), decanoic acid (19); ethyl acetate (20), ethyl butyrate (21), ethyl hexanoate (22), ethyl octanoate (23), ethyl decanoate (24)
   2.5. Branched acids and their ethyl esters: isobutyric acid (25), 2-methylbutyric acid (26), 3-methylbutyric acid (27) "isovaleric acid"; ethyl isobutyrate (28), ethyl 2-methylbutyrate (29), ethyl 3-methylbutyrate (30) "ethyl isovalerate"
   2.6. Carbonyls and lactones: diacetyl (31), 2,3-pentanedione (32), (acetoine, γ-butyrolactone)
   2.7. Ethyl esters of major acids: ethyl lactate (33), (diethyl succinate)
3. **Wood origin**
   3.1. Lactones: Z-whiskylactone (34), (E-whiskylactone)
   3.2. Volatile phenols: (o and m-cresol), guaiacol (35), (4-methylguaiacol), eugenol (36), E-isoeugenol (37), (2,6-dimethoxyphenol, 4-allyl-2,6-dimethoxyphenol)
   3.3. Vanillins: vanillin (38), acetovanillone (39), (propiovanillone, ethylvanillate, methylvanillate)
   3.4. (Furfural, 5-methylfurfural)
4. **Brett phenols:** 4-ethylphenol (40), 4-ethylguaiacol (41), 4-propylguaiacol (42)
5. **Miscellaneous or unclear origin**
   5.1. γ-lactones: γ-octalactone (43), γ-nonalactone (44), γ-decalactone (45), (γ-undecalactone, γ-dodecalactone)
   5.2. 4-vinylphenol (46), 4-vinylguaiacol(47)

TABLE 20.2 Families of red wine aroma compounds requiring specific analytical methods. Compounds which only very exceptionally will be present above threshold are given between brackets

1. **Key pivotal compounds:** acetaldehyde (48) and SO_2 (49)
2. **Volatile sulfur compounds (VSCs):** hydrogen sulfide "H_2S" (50), methanethiol "MeSH" (51), ethanethiol "EtSH" (52), dimethylsulfide "DMS" (53), (dimethyltrisulfide, ethyl thioacetate "EtSAc," methyl thioacetate "MeSAc")
3. **Strecker aldehydes:** isobutanal (54), (butanal), 2-methylbutanal (55), 3-methylbutanal (56), methional (57), phenylacetaldehyde (58)
4. **Highly polar compounds:**
   4.1. Enolones and furanones: 4-hydroxy-2,5-dimethyl-3(2H)-furanone "furaneol" (59), 2-ethyl-4-hydroxy-5-methylfuran-3-one "ethylfuraneol or homofuraneol"(60), 3-hydroxy-2-methyl-4H-pyran-4-one "maltol," 3-hydroxy-4,5-dimethylfuran-2-(5H)-one "sotolon" (61)
   4.2. Trace hydroxyesters: ethyl 2-hydroxy-4-methylpentanoate (62), (ethyl 2-hydroxy-3-methylpentanoate, ethyl 3-hydroxybutyrate, ethyl 3-hydroxy-3-methylbutyrate, ethyl 2-hydroxy-2-methylbutyrate)
5. **Polyfunctional mercaptans:**
   5.1. Varietal: 4-methyl-4-mercapto-pentan-2-one (63), 3-mercaptohexanol (64), 3-mercaptohexyl acetate (65), (3-mercapto-2-methylpropanol)
   5.2. Fermentative/Aging related: furfurylthiol (66), benzylmercaptan (67), 2-methyl-3-furanthiol (68), vanillylthiol (69)
6. **Alkylmethoxypyrazines:** 2-methoxy-3-isopropylpyrazine (70), 2-methoxy-3-isobutylpyrazine (71), (2-methoxy-3-secbutylypyrazine)
7. **Miscellaneous positive ultratrace odorants**
   7.1. Peppery: rotundone (72)
   7.2. Fruity esters: ethyl cyclohexanoate (73), ethyl 4-methylpentanoate (74), (ethyl 3-methylpentanoate, ethyl 2-methylpentanoate)
   7.3. Minty, fresh: p-menth-1-en-3-one "piperitone" (75), (3,6-dimethyl-5,6,7,7a-tetrahydro-4H-1-benzofuran-2-one "mintlactone")
8. **Miscellaneous negative ultratrace odorants**
   8.1. Corky-moldy: 2,4,6-trichloroanisol, (2,4,6-tribromoanisol, 2-methoxy-3, 5-dimethylpyrazine, 2-methylisoborneol, geosmin)
   8.2. Papery, cardboard: E-2-nonenal, (E,E-2,4-nonadienal, E,E-2,4-decadienal)
   8.3. Fungus type: 1-octen-3-one, 1-nonen-3-one
   8.4. Pruneau, green, metallic, geranium: 3-methyl-2,4-nonanadione, (Z)-1,5-octadien-3-one, 2-ethoxyhexa-3,5-diene
   8.5. Untypical aging: o-aminoacetophenone

By contrast, compounds in Table 20.2 require specific methods of analysis because of a number of reasons. The table further classifies compounds into eight different nonhomogeneous categories related well to similarity of origin, chemical structure, or analytical difficulty:

1. All compounds in the first, second, and third categories (acetaldehyde and SO_2, VSCs, and Strecker aldehydes) form complexes with different wine components and hence their analysis requires breaking the complexes and using speciation strategies. Additionally, VSCs require specific detection strategies since they cannot be properly detected using FID or MS detectors.
2. The highly polar compounds in the fourth category are compounds from diverse origins having in common a high polarity which hampers their isolation from wine.
3. Finally, compounds in the fifth to eighth categories (polyfunctional mercaptans, alkymethoxypyrazines, and miscellaneous compounds) are extremely powerful aroma compounds present at sub µg/L levels in many cases, and hence, requiring highly selective and sensitive analytical strategies.

20.3 THE ANALYSIS OF "EASY" AROMA COMPOUNDS

Compounds in Table 20.1 can be classified into two categories simply attending to their levels:

1. Major aroma compounds, present at levels above 0.2 mg/L (0.2–400 mg/L).
2. Minor aroma compounds, present in the range 0.1–200 µg/L.

The group of major aroma compounds includes the major secondary metabolites of alcoholic fermentation, such as higher alcohols and their acetates, volatile fatty acids and their ethyl esters, branched fatty acids and their ethyl esters, acetoin, diacetyl, and acetaldehyde. Most of these aroma compounds are naturally present at levels well above their olfactory thresholds. This complex pool of volatile compounds displays a characteristic odor defined as "vinous" or "fermented." Furthermore, different families of aroma compounds forming that mixture play different roles on wine aroma perception. San-Juan et al. (2011) demonstrated that fatty acids are essential for the perception of fresh fruit. Branched fatty acids have been found to mask the animal character of ethyl phenols (Romano et al., 2009) while ethyl esters and fusel alcohol acetates are essential for wine fruitiness (Ferreira et al., 1995, 2016; Lytra et al., 2013). More recently, the suppressing effects of aliphatic fusel alcohols on wine fruity and woody characters has been demonstrated, while the effects of β-phenylethanol and methionol were found to be negligible (de-la-Fuente-Blanco et al., 2016).

Major compounds present in wines and distillates can be directly determined by GC-FID using many different preconcentration techniques. Even direct injection has been proposed repeatedly by different authors (Villen et al., 1995; Lopez-Vazquez et al., 2010; Cacho et al., 2012; Kim et al., 2017), and can be applied even though ghost peaks are expected if the injector inlet is not cleaned or replaced frequently. Nearly all types of approaches have been used to quantify these easy aroma compounds and it will be out of place to give a complete coverage of all the proposals. In our lab we have been using a simple liquid–liquid microextraction with dichloromethane in which wine is diluted with brine and four different internal standards are used to correct for possible matrix effects (Ortega et al., 2001a). The method gives satisfactory analytical results for 21 main compounds in Table 20.1 (compounds 7, 9–27, 31, and 33), plus another eight secondary compounds (between brackets in the table). Other strategies using less internal standards, higher levels of solvents and slightly more complicated working setup provide easily data about 10–15 aroma compounds in Table 20.1 (Louw et al., 2009). Table 20.3 summarizes quantitative average data obtained in different sets of red wine samples.

Minor volatile compounds (C <0.2 mg/L) can also be determined by direct GC–MS analysis using a similar array of sample preparation techniques. Old approaches used liquid–liquid extractions typically with dichloromethane, which can extract a wide range of analytes and has a very low solubility in wine. For achieving quantitative results, the wine was typically extracted with large volumes of solvent (around 1:1 volume ratio) (Kotseridis and Baumes, 2000) or with three consecutive smaller volumes of solvent (around 1:10 or 1:20 volume ratios), which were then pooled together and further evaporated to produce a concentrated but very often "dirty" extract, containing relatively large amounts of nonvolatile material. These techniques are no longer used because of environmental, personal safety, and cost-effectiveness reasons and have been replaced by:

1. Liquid–liquid microextractions.
2. Solid-phase extraction techniques.
3. Solventless automated strategies.

TABLE 20.3 "Easy to analyze" wine aroma compounds: range of occurrence in red wines, odor threshold values, and aromatic descriptors

No.	Compound	Range (min–max) (μg/L)	Odor threshold[a] (μg/L)	Aromatic descriptor
1	Linalool	1.7–10.1[b]	25	Flowery, muscat
		<0.170–16.4[d]	25	
		2–18[f]		
2	Geraniol	0.91–44.4[b]	20	Flowery
		<0.010–9.10[d]		
3	β-Damascenone	0.29–4.7[b]	0.05	Baked apple
		<0.20–10.5[c]		
		<0.200–10.5[d]		
4	β-Ionone	0.032–0.24[b]	0.09	Violets
		<0.089–0.55[c]		
		<0.089–1.17[d]		
5	Ethyl cinnamate	0.11–8.89[b]	1.1	Flowery
		<0.032–1.86[d]		
6	Ethyl dihydrocinnamate	0.21–3.02[b]	1.6	Plum, flowery
		<0.210–2.35[d]		
7	Z-3-Hexenol	7.2–651[b]	400	Green
		<4.47–290[d]		
		22–69[e]		
8	Ethanol	$99 \times 10^6 – 115 \times 10^6$	24,900	Alcohol
9	Isobutanol	25,700–86,900[b]	40,000	Fusel, alcohol
		4472–13,595[g]		
10	Isoamyl alcohol	83,953–333,032[b]	30,000	Harsh, fusel
		111,000–353,379[d]		
11	β-Phenylethanol	40,093–153,269[b]	14,000	Rose
		18,700–83,100[d]		
		3010–40,099[g]		
12	Methionol	166–2398[b]	1000	Plastic, green
		737–2399[g]		
13	Isoamyl acetate	118–3371[b]	30	Banana
		110–370[c]		
		110–906[d]		
		8–148[g]		
14	Phenylethyl acetate	0.54–157[b]	250	Honey, flowery
		20.2–1008[d]		
		48–316[g]		
15	Acetic acid	69,110–313,310[b]	300,000	Vinegar
		512,500–950,000[c]		

(Continued)

TABLE 20.3 (Continued)

No.	Compound	Range (min–max) (μg/L)	Odor threshold[a] (μg/L)	Aromatic descriptor
		164,892–950,000[d]		
16	Butyric acid	434–4719[b]	173	Cheese, Rancid
		<54.8–1850[c]		
		<54.8–2310[d]		
		235–1175[g]		
17	Hexanoic acid	853–3782[b]	420	Fatty, cheese
		390–2120[c]		
		390–4396[d]		
		595–1390[g]		
18	Octanoic acid	562–4667[b]	500	Rancid, harsh
		180–1020[c]		
		180–2720[d]		
		406–2062[g]		
19	Decanoic acid	62.1–857[b]	1000	Fatty
		110–940[c]		
		62.2–940[d]		
		106–550[g]		
20	Ethyl acetate	45,134–127,801[d]	12,300	Solvent, fruity
21	Ethyl butyrate	69.2–371[b]	125	Fruity
		70.0–270[c]		
		70.0–320[d]		
		20–85[g]		
22	Ethyl hexanoate	153–622[b]	62	Green, apple
		70.0–210[c]		
		70.0–3 50[d]		
		161–275[g]		
23	Ethyl octanoate	138–783[b]	580	Sweet, fruit soap
		50.0–210[c]		
		26.4–230[d]		
		186–362[g]		
24	Ethyl decanoate	14.5–215[b]	200	Sweet, fruit
		<4.03–81.1[c]		
		<4.03–163[d]		
		34–82[g]		
25	Isobutyric acid	434–2345[b]	2300	Acid, cheese
		1240–2450[c]		
		550–1869[d]		
		503–1451[g]		

(Continued)

TABLE 20.3 (Continued)

No.	Compound	Range (min–max) (μg/L)	Odor threshold[a] (μg/L)	Aromatic descriptor
26	2-Methylbutyric acid	88.4–365[c]	33	Acid, cheese
		88.4–469[d]		
27	3-Methylbutyric acid (isovaleric acid)	305–1151[b]	33	Acid, cheese
		55.4–430[c]		
		20.2–1104[d]		
28	Ethyl isobutyrate	9.8–94[b]	15	Sweet fruit
29	Ethyl 2-methylbutyrate	1.1–29.9[b]	18	Sweet fruit
		6.50–82.6[c]		
		4.41–82.6[d]		
30	Ethyl 3-methylbutyrate (ethyl isovalerate)	2.2–36.1[b]	3	Fruity
		10.9–131[c]		
		10.9–131[d]		
31	Diacetyl	200–1840[b]	100	Cream, butter
32	2,3-Pentanedione	24–413[g]	900	Cream, butter
33	Ethyl lactate	11,317–49,550[g]	154,000	
34	Z-Whiskylactone	<0.130–668[c]	67	Coconut
		<0.130–668[d]		
		0–369[g]		
		396–1216[f]		
35	Guaiacol	1.1–10.9[b]	9.5	Smoky, spicy
		<0.026–38.1[d]		
		325–783[g]		
		25.3–76.1[f]		
36	Eugenol	0.88–15.6[b]	6	Clove
		<0.074–56.9[c]		
		<0.074–56.9[d]		
		22.7–96.2[f]		
37	E-Isoeugenol	<0.011–8.27[c]	6	Flowery
		<0.011–33.2[d]		
		26.8–120[f]		
38	Vanillin	<0.120–116[d]	995	Vanilla
		342–854[f]		
39	Acetovanillone	14.8–297[d]	1000	Vanilla, caramel
		102–309[f]		
40	4-Ethylphenol	<0.540–1214[c]	35	Animal, leather
		<0.54–1214[d]		
		0–872[g]		
		10.1–251[f]		

(Continued)

TABLE 20.3 (Continued)

No.	Compound	Range (min–max) (µg/L)	Odor threshold[a] (µg/L)	Aromatic descriptor
41	4-Ethylguaiacol	0–116[b]	33	Phenolic
		<0.035–167[c]		
		<0.035–167[d]		
		0–107[g]		
		4.98–105[f]		
42	4-Propylguaiacol	<0.048–18.0[d]	10	Clove, phenolic
		1.75–33.5[f]		
43	γ-Octalactone	0.067–8.37[d]	7	Coconut
44	γ-Nonalactone	3.3–40.8[b]	25	Peach
		0.430–87.0[d]		
45	γ-Decalactone	0.67–2.9[b]	0.7	Peach
		2.5–52.1[d]		
46	4-Vinylphenol	<1.00–272[d]	180	Medicinal
		1406–3028[f]		
47	4-Vinylguaiacol	0.2–15[b]	40	Phenolic, pleasant
		<0.83–44.9[d]		
		14–293[g]		
		68.3–409[f]		

[a]Odor threshold values were reported by Ferreira et al. (2000a,b), Peinado et al. (2004), San Juan et al. (2012).
[b]Ferreira et al. (2000a,b).
[c]San-Juan et al. (2011).
[d]San Juan et al. (2012).
[e]Farina et al. (2015).
[f]de Simon et al. (2008) (red wine aged 12 months).
[g]Bueno et al. (2016).

A good example of the first is a microextraction with Freon 113 after a preconcentration by demixing by salting out to avoid matrix effects, which provided highly quantitative albeit tedious methods (Ferreira et al., 1998, 2000a).

Solventless automated sample preparation techniques such as Sorptive Bar Solid Extraction (SBSE) or Solid Phase Microextraction (SPME) present some interesting advantages, since they are very easy to use and are relatively cost-effective. However, they require dedicated instruments, a significant investment in specific autosamplers and their analytical performance maybe not as straightforward as one could deduce from the nice chromatograms they provide, particularly in the case of SPME.

It has to be taken into account that in the most often used SPME procedures, the SPME fiber is exposed to the vapors present on a relatively large volume of wine 5–10 mL to which relatively large amounts of salt have been often added. The problem is that wine vapors in these conditions are extremely concentrated in ethanol, fusel alcohols, and fatty acid ethyl esters. The amounts of these compounds present in the headspace exceed by far the extraction capacity of the fibers. This implies that the fiber becomes saturated, which brings as a consequence that the amount of compound extracted becomes matrix-dependent. This problem can be particularly intense when adsorption fibers are used, but in any case, obtaining a good calibration providing unbiased results can become a really difficult task. There are two suitable solutions: strong dilution and using good enough internal standards. It has been demonstrated that after diluting wine 1:50 in a brine there is still good sensitivity so that 31 of the numbered compounds in Table 20.1 can be satisfactorily quantified by GC–MS (Ferreira et al., 2015). This same work gives a good idea of the quantification problem, since even after the strong dilution, single internal standards provided robust calibration for just 15 out of 47 compounds. Best results were obtained by using "multivariate internal standards" built by combining results from the 13 different internal standards. From this point of view, a radical solution to overcome matrix effects is using isotopically labeled internal standards. The landmark in this regards is the work presented by Siebert et al. (2005), which robustly quantified 31 fermentation-derived compounds using the 31 isotopically labeled analogous compounds as internal standards (22 numbered compounds in Table 20.1). However, the effort required for synthesizing the deuterated molecules and integrating 31 additional peaks questions the cost-effectiveness of this strategy in comparison with other alternatives. The difficulty in properly calibrating SPME is recognized by Haggerty et al. (2016), which after a careful optimization conclude that it is adequate as a general semiquantitative high-throughput strategy for metabolomics studies of not very different wines.

SBSE in comparison to SPME has as major advantages the much increased mass of sorbent (up to 100 times higher than some thin SPME fibers), and the possibility to directly extract liquid samples. These two characteristics make it much more resistant to matrix effects and hence much easier to calibrate. In contrast, the chromatographic process is much more complicated for several reasons. First, thermal desorption cannot be carried out in the standard chromatographic injector, but it requires an external thermal desorption unit. Second, desorption takes a longer time, and hence, either split injection is used or the system has to be equipped with a cryofocusing unit. In this last case, it should be noted that the mass of compounds extracted by the "twister" can be so high that the chromatographic column can be completely overloaded. Another drawback, is that all material is desorbed, which leaves the laboratory without the chance to reanalyze the sample. The technique has been used since its commercial introduction (Fang and Qian, 2006; Tredoux et al., 2008; Weldegergis and Crouch, 2008) to quantify 18–22 of the numbered compounds present in Table 20.1 (plus 10–20 others between brackets) and it has been used extensively to quantify a more limited range of aroma compounds, as it will be later detailed. A most recent work reports the quantitative determination of up to 38 numbered compounds in Tables 20.1 and 20.2, using two different twisters (the standard polydimethylsyloxane and a new more polar EG-Silicone/Polydimethylsiloxane (PDMS)) (Zhao et al., 2017).

An alternative, less easily automated, but providing cleaner and more purified extracts and not requiring additional instruments is Solid Phase Extraction (SPE). This technique was introduced in the analytical laboratory in the 1980s to replace the costly liquid–liquid extractions. The first commercial materials were mostly based on Silica-C18, which unfortunately, does not have very good extraction ability from hydroalcoholic solutions. Much better results were obtained using polystyrene-divinylbenzene macroreticular copolymers (Ferreira et al., 2000b; Ortega et al., 2001b), but the rigid-particle version suitable for quick SPE was only available at the end of the 1990s. The first comprehensive method using LiChrolut-EN resins was proposed in 2002 (Lopez et al., 2002), and it has been advantageously used in our laboratory since then (Cullere et al., 2004b; Escudero et al., 2007; San Juan et al., 2012; Geffroy et al., 2015; Herrero et al., 2016). In the present version, 15 mL of wine are extracted in a 50 mg-cartridge and most aroma compounds are quantitatively recovered in just 0.6 mL of solvent. The method makes it possible to quantify satisfactorily 23 numbered compounds in Table 20.1, complementing the information obtained with the L–L microextraction GC-FID approach.

20.4 THE SPECIFIC ANALYSIS OF VOLATILE PHENOLS

Two phenols, 4-ethylphenol and 4-ethylguaiacol are responsible for a well-known off-odor of red wines described as animal, leather, horse, smoky, ink, and medicinal (Chatonnet et al., 1992). Both phenols are produced by *Brettanmoyces/Dekkera spp.* contaminant yeasts (Chatonnet et al., 1992). These microbes are efficiently controlled by the correct dosage of sulfur dioxide. However, the actual tendency to reduce the use of this additive leads to wines being more susceptible to *Brettanmoyces* contamination. There is an overwhelming amount of research looking for methods to prevent the apparition of the contamination or to reduce the levels of the phenols once the wine has become tainted. The revision of those papers is out of the scope of the present paper, and the reader is directed to some reviews on the issue (Suarez et al., 2007; Wedral et al., 2010; Malfeito-Ferreira, 2011; Kheir et al., 2013).

Not surprisingly, the aromatic and industrial relevance of these compounds has stirred the development of many specific analytical methods, even if the quantitative determination of these compounds is quite easy and straightforward. Different authors have proposed the synthesis of deuterated analogues for their accurate quantitation (Pollnitz et al., 2000; Rayne and Eggers, 2007), while others very soon developed headspace-SPME methods (Martorell et al., 2002; Pollnitz et al., 2004). There is also a proposal in which the headspace SPME is preceded by a derivatization reaction in order to include 4-ethylcatechol which is poorly volatile (Carrillo and Tena, 2007).

Other authors have developed methods based on HPLC–MS–MS and HPLC–Diode Array Detector (DAD)-fluorescence (Caboni et al., 2007; Nicolini et al., 2007) or High Performance Liquid Chromatography (HPLC) with coulometric array detector (Larcher et al., 2007). Apart from this, there is an array of proposals using different "experimental" sample preparation strategies such as dispersive liquid–liquid microextraction (Farina et al., 2007), multiple solid phase microextraction (Pizarro et al., 2007), acetylation followed by extraction in a silicone sorbent (Carpinteiro et al., 2010), or ultrasound-assisted emulsification microextraction (Pizarro et al., 2011) for ulterior GC–MS determination, or more recently, a proposal uses a QuEChERS extraction procedure followed by HPLC–UV-fluorescence determination (Valente et al., 2013). In most cases, the determination of the two phenols is used more as proof of concept than to provide a satisfactory solution to an easy problem.

20.5 THE ANALYSIS OF "DIFFICULT" AROMA COMPOUNDS IN RED WINE

20.5.1 Acetaldehyde and Sulfur Dioxide

It may seem surprising that a chapter like this is paying specific attention to these two basic molecules, whose smell should not be detectable in red wine unless the wine has suffered a strong oxidation or has received a too-large dose of SO_2. However, these two molecules play outstanding roles in wine flavor chemistry. On the one hand, both molecules determine the kinetics and effects of oxygen consumption by red wine (Carrascon et al., 2015, 2018; Marrufo-Curtido et al., 2018); on the other hand, many odor-active carbonyls form odorless adducts with SO_2, and the evaporation of free SO_2 in the wine glass, or its losses by oxidation, determines the release of those adducts, the major one of which is formed with acetaldehyde. In addition, the analysis of these two molecules in wine is not as straightforward as the existence of "official" methods of analysis would suggest. This is because both molecules are involved in different chemical equilibria, some of which are slow enough to allow the existence of different species (Table 20.4). In the case of SO_2, acid–base equilibrium is so fast that it is not possible to obtain different signals for the three aqueous species (H_2SO_3, HSO_3^-, and SO_3^{2-}) whose relative concentration is just deduced from the corresponding acid dissociation constants in hydroethanolic solution (Waterhouse et al., 2016b). However, sulfite and acetaldehyde react to form 1-hydroxyethylsulfonate, in a reversible reaction which can be expressed as follows:

$$CH_3CHO + HSO_3^- \leftrightarrow CH_3CHOHSO_3^-$$

The reagents and products of this reaction are also in chemical equilibrium regulated by the corresponding dissociation constants, but the equilibrium is slow enough so that free and bonded forms of both acetaldehyde and SO_2 can be clearly separated.

In the case of acetaldehyde, the determination of total forms (free + bonded) requires the cleavage of the α-hydroxysulfonates. Three possibilities arise:

1. A thermal cleavage directly in the hot GC-injector (Cacho et al., 1995).
2. Alkaline hydrolysis, which has to be carried out avoiding oxidation (Elias et al., 2008b; Jackowetz and de Orduna, 2013).
3. Acid hydrolysis, also taking caution to avoid oxidation (Han et al., 2015).

In the first case, acetaldehyde can be directly determined by GC-FID. The method has been seldom used, although there is a recent work reporting acceptable recoveries in a quite limited analytical validation (Bueno et al., 2018). In the second and third strategies, acetaldehyde is further derivatized with 2,4-dinitrophenylhydrazine, and further determined by HPLC, together with 2-ketoglutaric acid, pyruvic acid, and acetoin. The proposal by Han (Han et al., 2015) is the simpler and quicker.

TABLE 20.4 Sulfur dioxide and acetaldehyde

No.	Compound	Range (min–max) (mg/L)
48	Total acetaldehyde	13.90–20.33[a]
49	Free sulfur dioxide	2.6–33.7[a]
49	Total sulfur dioxide	24–77[a]

[a]Carrascon et al. (2018).

Regarding SO_2, there are increasing evidences (Coelho et al., 2015; Waterhouse et al., 2016a; Carrascon et al., 2017) that the fraction of free SO_2 determined by the aeration-oxidation method, recommended by the Association of Official Analytical Chemists and by the International Organization of Vigne and Wine, contains in fact a subfraction of SO_2 which in fact is forming weak complexes with anthocyanins whose antioxidant activity is not really known. In this regard, there are two recent proposals providing analytical procedures, both based on the analysis of SO_2 present in the headspace in equilibrium with the undisturbed (not acidified, not agitated) wine. In that from Coelho (Coelho et al., 2015), SO_2 is determined by using a gas detection tube, while in that of Carrascon, the measurement is carried out by GC–MS (Carrascon et al., 2017).

20.5.2 Volatile Sulfur Compounds

Within the group of VSCs there are some highly relevant aroma compounds (Table 20.5). Hydrogen sulfide (H_2S), methanethiol (MeSH), and to a lesser extent ethanethiol (EtSH), and eventually some other mercaptans, such as ethyl and methyl thioacetates (MeSAc and EtSAc), are responsible for the so-called reductive off-odors (Siebert et al., 2010). As H_2S and MeSH are ubiquitous compounds and are present at very low levels even in nonreduced wines (and any naturally fermented product), it can be inferred that at those low levels, close or slightly above the threshold, they are part of the wine aroma signature. Another Volatile Sulfur Compounds (VSC) with most remarkable odor properties is dimethyl sulfide (DMS), whose origin and aroma properties are completely different. DMS comes from specific precursors from grapes (Segurel et al., 2004; Loscos et al., 2008), its levels increase during aging, and in many red wines play an outstanding role in supporting, enhancing, and modulating wine fruitiness (Escudero et al., 2007; Lytra et al., 2014) with no chemical or sensory relationship to reduction (Franco-Luesma and Ferreira, 2016).

The analysis of these compounds would not represent a major challenge if H_2S and mercaptans did not form strong complexes with metals (Franco-Luesma and Ferreira, 2014) and would not be also in equilibria with oxidized forms (Vela et al., 2018). However, these compounds form strong complexes with $Cu(II)$ and $Cu(I)$, and H_2S also forms complexes with $Fe(II)$ and $Zn(II)$. As the complexes are strong enough, and the reactions are not fast enough, this makes it possible to identify a free and a bound fraction of these compounds. Furthermore, oxidized forms of H_2S and mercaptans, which seem to be mixed disulfides, polysulfides, and hydropolysulfides with the wine major mercaptans, glutathione and cysteine (Jastrzembski et al., 2017; Kreitman et al., 2017), are in equilibria with free and bound forms (reduced forms). The distribution between oxidized and reduced forms, and within these last ones, between free and bound, is determined by the wine redox potential (Ferreira et al., 2017) as illustrated in Fig. 20.1.

Analytically, VSCs can be best isolated in headspace fractions, since they are very volatile. As the MS signals are very poor, because they are small molecules yielding small ions, they require specific detectors. All this means that the preferred technique for the quantitative analysis of these molecules is static headspace gas chromatography with sulfur selective detection, which was proposed 20 years ago (Rauhut et al., 1998). The key

TABLE 20.5 Volatile sulfur compounds (VSCs)

No.	Compound	Range (min–max) (µg/L)	OT[a] (µg/L)	Aromatic descriptor
50	Hydrogen sulfide	<2.00–141[b]	1.1	Rotten egg
		nd–8.7[c]		
51	Methanethiol	<0.200–18.0[b]	1.8	Rotten cabbage
		nd–5.0[c]		
52	Ethanethiol	<3.49[b]	1.1	Onion, rubbery
		nd–0.7[c]		
53	Dimethylsulfide	2.95–208[b]	25	Cooked cabbage, asparagus
		48.0–379.5[c]		

[a]Odor threshold values estimated in red wine samples. These values were found in Siebert et al. (2010).
[b]San Juan et al. (2012).
[c]Siebert et al. (2010).

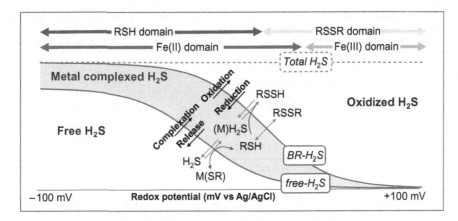

FIGURE 20.1

question here is how to ensure that the signals correspond exclusively to free forms of the analytes, or to free plus bonded forms. From this point of view, methods can be classified into two categories:

1. Methods for measurement of free forms. These methods should avoid any shift in the equilibrium. For that, ideally, the method should analyze exclusively the VSCs present in the equilibrated headspace on the undisturbed and undiluted wine. Any change in pH, ionic force or any preconcentration step, such as SPME should be avoided, since it would shift the equilibrium. To the best of our knowledge only two of the proposed procedures meet this requisite (Rauhut et al., 1998; Franco-Luesma and Ferreira, 2014).
2. Methods for the measurement of free + bonded forms (brine releasable forms or BR-forms). For this, the prerequisite is that all the complexes should be broken. This is not as straightforward as it could seem, since complexes are strong. The most satisfactory solution consists of the dilution of the wine sample with a concentrated brine (Franco-Luesma and Ferreira, 2014; Chen et al., 2017), and because of that this fraction is known as BR-forms. Chloride anions are required to complex Cu cations, while dilution seems to be required to cleave complexes with Fe(II) and Zn(II) (Ontanon et al., 2018). A dilution 1:10 with the concentrated brine seems to provide an optimal solution (Chen et al., 2017; Ontanon et al., 2018). Again only two published procedures fulfill these requisites: the one developed by Franco-Luesma and Ferreira preconcentrates the vapors in a SPME fiber and uses a GC with a pulsed flame photometric detector; the proposal by Chen et al. (2017) uses a gas detection tube providing information only about H_2S.

Any other procedure using salt but not dilution (Siebert et al., 2010), or using a SPME preconcentration (Herszage and Ebeler, 2011) will inevitably provide results including free forms plus an undetermined amount of bonded forms. It is not clear whether the use of a standard addition strategy (Herszage and Ebeler, 2011) can completely sort out the problem, since different complexes take different times to form. While copper complexes are immediately formed, Fe(II) and Zn(II) complexes seem to require a longer time (Franco-Luesma and Ferreira, 2014).

Finally, and taking into account the distribution of species shown in Fig. 20.1, it can be observed that any measurement of free and BR-forms at a not very negative redox potential, will provide information exclusively about the fractions of H_2S and mercaptans which are as reduced forms at such redox potential. The reduction of the redox potential can be carried out using an accelerated reductive aging procedure, in which the wine is stored at 50°C in complete anoxia for two weeks (Franco-Luesma and Ferreira, 2016), or could be attained using a chemical reductive agent, such as TCEP [*tris*(2-carboxyethyl)phosphine] (Chen et al., 2017; Kreitman et al., 2017), albeit these last procedures have not been properly validated.

20.5.3 Strecker Aldehydes and Other Odor-Active Aldehydes

Many volatile aldehydes have remarkable odor properties (van Gemert and Nettenbreijer, 1977) and in general, carbonyls in wine can be responsible for a large diversity of aroma descriptors, such as vanilla, butter, caramel, honey, potato, rancid oil, paper, cardboard, lemon, orange, violets, cider, or plum (Table 20.6). Some

TABLE 20.6 Levels of concentration found in free and bound forms of aldehydes in 16 Spanish red wines (Bueno et al., 2014)

No.	Compound	Free forms (min−max) (μg/L)	Total forms (min−max) (μg/L)	Odor threshold[a] (μg/L)	Aromatic descriptor
54	Isobutanal	6.87−47.5	9.60−55.1	6	Fruity, harsh
55	2-Methylbutanal	2.53−15.7	3.94−18.2	16	Fatty, green
56	3-Methylbutanal	5.75−27.3	33.8−111	4.6	Fatty, peach
57	Methional	4.80−22.4	26.3−93.7	0.5	Cooked vegetables
58	Phenylacetaldehyde	2.06−16.1	19.9−66.2	1.0	Honey

[a]Odor threshold values reported in Cullere et al. (2007) and Escudero et al. (2000a,b).

aldehydes have been identified as the most important contributors to oxidative off-flavors (Escudero et al., 2000a, b; Ferreira et al., 2002a, 2003a; Cullere et al., 2007). The most powerful oxidation-related odorants are methional (with a boiled potato odor) (Escudero et al., 2000a,b) and phenylacetaldehyde (honeylike) (Ferreira et al., 2002a). At low levels, these carbonyl compounds may participate in the complexity of a wine, but at higher levels, they are responsible for the development of specific oxidation off-flavors (Cullere et al., 2007) and for the loss of freshness (San-Juan et al., 2011).

Due to the poor mass spectrometric and chromatographic properties of aldehydes and to their reactivity, many analytical methods are based on the formation of chemical derivatives which are more stable and more easily extractable and detectable. For the subsequent GC−MS or GC-Electronic Capture Detector (ECD) analysis of the derivatives, the most used reagent has been O-(2,3,4,5,6-pentafluorobenzyl)hydroxylamine hydrochloride (Cullere et al., 2004a; Ferreira et al., 2004). The procedure developed by the authors consists of the preconcentration of the aldehydes in a SPE extraction cartridge, where the derivatization takes directly place. This procedure makes it possible to obtain very low detection limits for many different carbonyls, including 1-octen-3-one (see Table 20.2) (Cullere et al., 2006). Analytical aspects of the method have been thoroughly studied (Ferreira et al., 2006; Zapata et al., 2010), including the in-fiber SPME derivatization and extraction and Negative Chemical Ionization (NCI)-MS determination. More recently, Australian researchers have used the same reagent, although in this case the derivatization reaction is directly carried out in wine, the derivatives are further isolated by SPE and are finally analyzed by GC−MS/MS (Mayr et al., 2015). The procedure uses deuterated internal standards to improve accuracy and provide very low detection limits for aldehydes. Some polar compounds (see the next section) are also determined in the same run, but with poor sensitivities.

Some other derivatization agents have been also proposed for the analysis of aldehydes, although to the best of our knowledge, not for the analysis of oxidation-related aldehydes. Cysteamine was proposed as derivatization agent (Lau et al., 1999), but it was only applied to the analysis of aliphatic aldehydes. The other common derivation agent, 2,4-dinitrophenylhydrazine, widely used for subsequent HPLC analysis of the corresponding hydrazones (Nascimento et al., 1997) has only been applied to the analysis of the most concentrated wine carbonyls (Elias et al., 2008a).

A highly relevant question in the analysis of aldehydes is the known fact that these compounds can establish strong interactions with other species present in wine. Cullere et al. (2011) examined the existence of bonded or complexed forms of aliphatic aldehydes in wine (octanal-undecanal). Three different fractions were identified: (1) free extractable aldehydes; (2) aldehydes bound in hydrophilic complexes (extractable in the presence of acetaldehyde); and (3) aldehydes bound in hydrophobic complexes (extractable in the presence of heptanal). The hydrophilic fraction was predominant in some wines and is mainly composed of the α-hydroxyalkylsulfonates that most aldehydes and α-dicarbonyls form by reversible reaction with sulfur dioxide (Osborne et al., 2006; de Azevedo et al., 2007). The existence of this fraction means that wine in fact contains a pool of strong-smelling aldehydes and dicarbonyls under odorless nonvolatile forms. Since aldehydes and α-hydroxyalkylsulfonates are in chemical equilibrium:

$$R-CHO + HSO_3^- \leftrightarrow R-CHOH-SO_3^-$$

The exhaustion of SO_2 by oxidation will bring about the displacement of the equilibrium to the left, releasing the aldehyde which will then become odor-active. For studying this issue, Bueno et al. (2014) have developed a method for the specific measurement of free and bonded forms of aldehydes (these last are estimated). The method uses a set of surrogates which are different aldehydes not naturally present in wine, and similar in structure and properties to the analytes. For instance, for methional−methylthiopropanal, methylthiobutanal is used.

The method also has an internal standard whose concentration in the headspace was found to be independent on the wine (methyl 2-methylbutyrate). Surrogates and standards are added and left to equilibrate in complete anoxia for 12 hours in order to ensure equilibration. Analytes are then determined in the headspace of the unstirred sample using a PDMS/Divinylbenzene (DVB) fiber. The area of analyte (normalized to the IS) is used to determine its free levels, while the ratios of the areas of the surrogates (normalized to the IS) measured in the sample to those measured in the calibration solutions, are used to estimate the proportion under bound forms. With this strategy, it was demonstrated that important aroma-active carbonyls, such as methional, phenylacetaldehyde, isovaleraldehyde or decanal, are mostly (80%–95%) as odorless forms in normal bottled wines and that these bonded forms can effectively release free forms upon wine oxidation (Bueno et al., 2016). A second possibility is cleaving first adducts and then analyzing all the aldehydes released. In this case, the proposed strategy for breaking the adducts consists of a 6-hour incubation with glyoxal at 50°C in complete anoxia (Bueno et al., 2018).

20.5.4 Highly Polar Compounds

Two different families of highly polar compounds are found in wine, and both of them are poorly known and seldom analyzed (Table 20.7). The first family is known as the "burnt sugar family" and encloses a series of oxygen containing heterocyclic compounds with different chemical origins and relatively similar and outstanding odor properties: sotolon [4,5-dimethyl-3-hydroxy-2(5H)-furanone], maltol (3-hydroxy-2-methyl-4H-pyran-4-one), furaneol [2,5-dimethyl-4-hydroxy-3(2H)-furanone], and homofuraneol [2 (or 5)-ethyl-4-hydroxy-5(or 2)-methyl-3 (2H)-furanone]. Sotolon has a strong sweet odor remining of curry and fenugreek, and the others have sweet aromas described as "burnt sugar, caramel, and maple." Sotolon and furaneol have very low odor thresholds in water—ethanol −5 µg/L in both cases (Martin et al., 1992; Ferreira et al., 2002b). Homofuraneol is less powerful (threshold 125 µg/L), but it also can be found at higher levels (Ferreira et al., 2002b). Maltol has not been reported to be above threshold (5000 µg/L) (Cutzach et al., 1999).

Sotolon is an oxidation-related compound with multiple and not completely known origins. It is particularly rich in wines derived from botrytized grapes (Campo et al., 2008a) or in wines aged under flor yeast (Martin et al., 1992), where it is likely biochemically formed from threonine. And it is also found in sweet wines (Schneider et al., 1998; Cutzach et al., 1999; Campo et al., 2006b; Noguerol-Pato et al., 2012) and in prematurely aged white wines (Du Toit et al., 2006), where it is most likely formed from native α-ketobutyric acid or formed by degradation of ascorbic acid (Pons et al., 2010). Regarding furaneol, this compound was proposed some time ago as a potential marker for the detection of forbidden hybrids (*Vitis vinifera* × non-*vinifera*) for making wine (Depinho and Bertrand, 1995). In these varieties it can be present at levels above 1 mg/kg, while it rarely will reach 0.05 mg/kg in *vinifera* wines (Rapp and Engel, 1995). Its glucopyranoside and the gene encoding the Uridine diphosphate (UDP)-glycoside have been recently identified in Muscat Bailey A [*Vitis labrusca* (Bailey) × *V. vinifera* (Muscat Hamburg)] (Sasaki et al., 2015a,b), but the compound was detected by olfactometry in the aroma profile of hydrolysates extracted from glycosidic precursors of garnacha grapes (Lopez et al., 2004), which

TABLE 20.7 Highly polar compounds

No.	Compound	Range (min−max) (µg/L)	Odor threshold[a] (µg/L)	Aromatic descriptor
59	Furaneol	<10.0−62.6[b]	5	Cotton candy
		46−199[c]		
60	Homofuraneol	<180−725[b]	125	Cotton candy
		38−86[c]		
61	Sotolon	<50.0−<50.0	5	Curry, spicy
		5−12[c]		
62	Ethyl 2-hydroxy-4-methylpentanoate	112−672[d]	300	Blackberry

[a]*Odor threshold values reported in Martin et al. (1992), Ferreira et al. (2002a,b), Falcao et al. (2012).*
[b]*San-Juan et al. (2011).*
[c]*data corresponding to the analyses of red dry wines, not published.*
[d]*Falcao et al. (2012).*

suggests that it is a normal component of grapes. The presence of its precursor is consistent with the increase during aging of this compound (Jarauta et al., 2005). Furaneol is also a normal product of dehydration of sugars and it is also found in toasted wood (Cutzach et al., 1997), which can be a significant source of this compound in red wines (de Simon et al., 2014). From the sensory point of view, the contribution of sotolon to the aroma of dry red wines is expected to be low. Furaneol, however, may play a quite relevant role. Together with homofuraneol, it exerts a great impact in fruity and caramel notes of some rosé wines (Ferreira et al., 2002b). It also has been found to be positively related to red wine quality (Ferreira et al., 2009), and to have a neat influence on red wine fruitiness (Ferreira et al., 2016).

These compounds can be directly determined by GC–MS of extracts obtained from wine with polar solvents such as dichloromethane (Camara et al., 2004) or by direct SPE extraction (Mayr et al., 2015) but most often this strategy provides too-high detection limits and may even cause some artifacts due to the presence of nonvolatile material. For a more sensitive determination, dual GC was proposed time ago (Rapp and Engel, 1995). A convenient alternative is a selective SPE isolation procedure in which nonpolar wine aroma compounds are washed out of the cartridge by a nonpolar solvent (Ferreira et al., 2003b). This procedure was later redesigned (San Juan et al., 2012) and provides clean extracts and detection limits well below the odor thresholds of the compounds. In the particular case of sotolon, some recent proposals for white wines, make use of ultrahigh-pressure liquid chromatography coupled with UV detection and Ultra Performance Liquid Chromatography (UPLC)–MS/MS (Gabrielli et al., 2015a,b).

The second group of relevant polar aroma compounds is formed by some hydroxyesters. Several compounds belonging to this family have been reported in wine, although only a very limited number have shown relevant roles on red wine aroma. This is the case of ethyl 2-hydroxy-4-methylpentanoate which has been associated with "fresh blackberry aroma" in red wines (Falcao et al., 2012).

The analysis of this family of compounds is demanding due to their polarity and uncharacteristic mass spectra, frequently requiring the use of multidimensional chromatography (MDGC). Campo et al. (2006a) isolated 2-hydroxy-4-methylpentanoate and 2-hydroxy-4-methylpentanoate thanks to heart-cut MDGC combined with large volume injections. MDGC–MS was also used to obtain quantitative data of 2-hydroxy-4-methylpentanoate from 15 Bordeaux red wines (Falcao et al., 2012).

20.5.5 Polyfunctional Mercaptans

Polyfunctional mercaptans (Table 20.8), are amongst the strongest smelling molecules in nature and may play an important role in the aroma of many products (Tominaga et al., 1998). Several compounds belonging to this family have been reported in wine, and at least seven of them, shown in Table 20.2, are key aroma compounds in many different wines. Three of them, 4-Mercapto-4-methylpentan-2-one (4MMP), 3-mercaptohexan-1-ol (3MH), 3-mercaptohexyl acetate (3MHA), have a varietal origin while 2-methyl-3-furanthiol (2M3F) seems to be a yeast derivative, and benzylmercaptan (BM), furfurylthiol (FFT), and the recently identified vanillylthiol (Floch et al., 2016) are formed by reaction during aging. These compounds are often found at concentrations far above their olfactory perception thresholds, 60.4 ng/L for 3MH, 4 ng/L for 3MHA, 0.8 ng/L for 4MMP (Tominaga et al., 1998), 0.3 ng/L for BM (Tominaga et al., 2003), 0.4 ng/L for FFT (Tominaga et al., 2000), and 1 ng/L for 2M3F.

The varietal polyfunctional mercaptans are more relevant in white wines, in some of which they are really impactful compounds (Darriet et al., 1995; Guth, 1997; Petka et al., 2006; Sarrazin et al., 2007; Campo et al., 2008a, b), but even in neutral wines they seem to play an outstanding role in aroma perception (Escudero et al., 2004; Mateo-Vivaracho et al., 2010). In red varieties, the key role played by 3MH and eventually 3MHA in rosé wines has been demonstrated (Ferreira et al., 2002c; Murat, 2005). There are few reports about levels of these compounds in red wines, but some of them were in fact first identified in red wines from Bourdeaux (Bouchilloux et al., 1998). Levels reported in Spanish red wines (San Juan et al., 2012) for the three compounds, while lower than those founds in whites, were well above thresholds and maximum levels (23, 47, and 671 ng/L for 4MMP, 3MHA, and 3MH, respectively) are close to the impact area. More recently, Capone et al. (2015) found also maxima levels of 4MMP close to 20 ng/L in Pinot Noir wines. Regarding the thiols of fermentative or aging-related origin, the aromatic role of 2M3F is not clear, while FFT is the main active responsible for coffee and toffee nuances often observed in red wines. Benzemethanthiol has also an unclear impact on the toasted and empyreumatic aroma character of red wine.

The analysis of these compounds is one of the major challenges for analytical chemists. The difficulties lie not only in the low levels which should be attained, but in the poor spectrometric properties and the high reactivity

TABLE 20.8 Polyfunctional mercaptans

No.	Compound	Range (min–max) (ng/L)	OT[a] (ng/L)	Aromatic descriptor
63	4-Methyl-4-mercapto-pentan-2-one	<0.6–10.7[b]	0.8	Box tree, passion fruit
		<0.6–23[c]		
64	3-Mercaptohexanol	115–1343[b]	60	Grapefruit, passion fruit
		45–671[c]		
		422–775[d]		
65	3-Mercaptohexyl acetate	<2.0–64.9[b]	4	Grapefruit, passion fruit
		5–47[c]		
		8.4–22.3[d]		
66	Furfurylthiol	1.5–206[b]	0.4	Roast coffee, toasty
		7–112[c]		
		nd–38.7[d]		
67	Benzylmercaptan	2.68–12.6[b]	0.3	Smoke, burnt wood
		3–45[c]		
68	2-Methyl-3-furanthiol	124–613[b]	1.0	Meaty
		67–687[c]		
69	Vanillylthiol	<50–8300[e]	3.8	Clove, smoke

[a]*Odor threshold values were reported by Tominaga et al. (1998), Tominaga et al. (2000), Tominaga et al. (2003), Floch et al. (2016).*
[b]*Mateo-Vivaracho (2009).*
[c]*San Juan et al. (2012).*
[d]*Pavez et al. (2016).*
[e]*Floch et al. (2016).*

of these molecules, which can be lost by oxidation during the sample setup, or by irreversible interaction with any adsorptive spot in the chromatographic systems. Broadly speaking, there are three different alternatives for their quantitative determination:

1. Highly selective separations and preconcentration schemes for the analysis of the underivatized mercaptans.
2. Use of derivatives with enhanced sensitivity for GC–MS analysis.
3. Use of derivatives with enhanced sensitivity for HPLC–MS analysis.

First proposals made use of different highly selective preconcentration techniques, most often using mercury salts. The most common reagent was *p*-hydroxymercurybenzoate, first introduced by Tominaga (Tominaga et al., 1998; Tominaga and Dubourdieu, 2006). The methods were extremely tedious and difficult to carry out. Some more simplified procedures used a combination of the mercury salt and SPE sorbents (Ferreira et al., 2007) or what was called as covalent chromatography, consisting of a hydrophilic sorbent containing Hg-salts, able to concentrate thiols present in an organic extract (Schneider et al., 2003). Still a more sophisticated and miniaturized method using the *p*-hydroxymercurybenzoate previously retained in a micro-SPE cartridge (20 mg) was later developed as general procedure for the selective isolation of underivatized mercaptans (Mateo-Vivaracho et al., 2009).

Secondly, and because of the poor "detectability" of underivatized mercaptans, different strategies based on the formation of derivatives with enhanced detectability were further developed. For GC–MS analysis, the most used reagent is 2,3,4,5,6-pentafluorobenzyl bromide (PFBBr) (Mateo-Vivaracho et al., 2006, 2007, 2008; Rodriguez-Bencomo et al., 2009), whose derivatives give very intense signals in GC-NCI-MS mode. In the optimized procedure (Mateo-Vivaracho et al., 2008, 2010), only 20 mL of wine are required. 4MMP is first oximated with *O*-methylhydroxylamine, the mercaptans are further retained in a BondElut-ENV cartridge where the derivatization reaction is carried out, followed by a number of cleaning operations. The strategy provides detection limits below odor thresholds and it is robust enough to support the routine analysis of several hundreds of samples per year. In a variant of the procedure using deuterated standards, the final extract is evaporated in a SPME vial, and the derivatized mercaptans are finally concentrated in a Divinylbenzene Carboxen (DVBCAR)/PDMS SPME fiber and directly injected in the GC–MS system (Rodriguez-Bencomo et al., 2009). More recently, ethyl propiolate has been also proposed as derivatization agent (Herbst-Johnstone et al., 2013) for GC—MS analysis. The derivatization with this reagent is easier than that of PFBBr, but so far, the proposed method does not reach the required limits of detection. Finally, there is a recent report in which 4MMP is quantified by Headspace (HS) HS—SPME—GC—MS/MS just after oximation. Inconveniently, the method is not suitable for the other mercaptans.

In recent years, the huge developments in LC—MS/MS instrumentation have made possible achieving amazing sensitivities with these instruments. Because of a major efficiency in ionization, a nearly null ion fragmentation in the ion source, and because of the narrow peaks provided by the 1.5—1.7 μm-based columns, the absolute sensitivities are clearly above those attained by GC—MS and close to those obtained by GC-NCI-MS. In addition, working procedures in LC—MS/MS are in general easier and more robust than those of GC, in which adsorptivity in the inlets and columns due to nonvolatile or reactive material constitute more than a nuisance. Still, the mercaptans have to be derivatized to achieve enough sensitivity, and this can only be obtained in the most powerful and advanced instruments, but the derivatization reaction no longer needs nonpolar reagents and getting rid of the excess reagent is much less problematic. Different reagents have been tried, such as 4,4'-dithiodipyridine (DTDP) (Capone et al., 2015), o-phtalaldehyde (Piano et al., 2015), ebselen (Vichi et al., 2015), or N-phenylmaleimide (Roland et al., 2016), all of which provide a quite satisfactory sensitivity and relatively easy setups. For instance, in the method proposed by Capone (Capone et al., 2015), the derivatization takes place directly without prior pH adjustment in 30 minutes. Derivatives are then isolated by SPE on a C18 cartridge, the extract is dried, redissolved, and directly injected. The limits of detection obtained are worse than those obtained by GC-NCI-MS, particularly for 4MMP, 3MHA, and FFT, but with deuterated internal standards, a quite robust operation is obtained. Detection limits reported by Roland et al. (Roland et al., 2016) for the two single compounds determined, MH and 3MHA, are however much better, suggesting that with a powerful instrument HPLC—MS/MS can provide robust quantitative methods for these challenging compounds.

20.5.6 Alkylmethoxypyrazines

3-alkyl-2-methoxypyrazines (Table 20.9) are a family of compounds whose presence in wine has been the focus of much research for their contribution to herbaceous and vegetative aromas in wine. The first component of the family reported in grape was 3-isobutyl-2-methoxypyrazine (IBMP) (Bayonove et al., 1975), followed, a few years later, by the identification of 3-isopropyl-2-methoxypyrazine (IPMP) in Sauvignon blanc wine (Augustyn et al., 1982). Other alkylmethoxypyrazines have been reported in wine during the last decades, among them 3-sec-butyl-2-methoxypyrazine (SBMP) (Lacey et al., 1991) and 3,5-dimethyl-2-methoxypyrazine (MDMP) and its isomers (Simpson et al., 2004; Ferreira et al., 2009; Slabizki et al., 2015) are likely the more relevant.

The origin of IBMP, SBMP, and IPMP is mainly varietal, associated to specific grape varieties like Cabernet Sauvignon, Sauvignon blanc, Cabernet franc, Merlot, Carmenere, or Fer Servadou (Allen et al., 1991; Augustyn et al., 1982; Belancic and Agosin, 2007; Geffroy et al., 2015; Roujou de Boubée et al., 2000). IBMP and IPMP appear in grapes at early mature stages, being their levels highly correlated with vine vigor and shade conditions (Gregan and Jordan, 2016; Ryona et al., 2008). However, the biochemical pathways that lead to their synthesis and degradation in grapes are still not resolved. Amino acids have been proposed as the initial source of alkylmethoxypirazynes (leucine for IBMP and valine for IPMP), with a final step involving O-methyltransferase in the synthesis mechanism (Dunlevy et al., 2013). The family of O-methyltransferase genes (VvOMT) are expressed in Cabernet Sauvignon, but not in Pinot, suggesting that VvOMT plays a major role in alkylmethoxypirazine synthesis (Dunlevy et al., 2013). There is also a known exogenous origin of IPMP from several species of coleopteran insects that can infest grapes (Botezatu et al., 2014). The presence of 3,5-dimethyl-2-methoxypyrazine in wine has been associated with tainted corks (Simpson et al., 2004; Slabizki et al., 2015).

TABLE 20.9 Alkylmethoxypyrazines

No.	Compound	Range (min—max) (ng/L)	OT[a] (ng/L)	Aromatic descriptor
70	3-Isopropyl-2-methoxypyrazine	<0.5—38[b]	1	Bell pepper
71	3-Isobutyl-2-methoxypyrazine	<0.5—56[c]	15	Bell pepper

[a]Odor threshold values were reported by Roujou de Boubée et al. (2000), Pickering et al. (2007).
[b]Sala et al. (2002).
[c]Allen et al. (1994).

3-alkyl-2-methoxypyrazines vegetative aroma is usually considered detrimental for the quality of wine (Allen et al., 1991; Belancic and Agosin, 2007; Ferreira et al., 2009), although in some instances such odors could be appreciated as a marker of typicality for certain regions (Parr et al., 2007). Due to their extremely low sensory detection thresholds, 3-alkyl-2-methoxypyrazines in concentrations of few ng/L can impact the aroma of wine. This is especially frequent in wines elaborated with the abovementioned varieties in cooler climates, where IBMP can be found in quantities of more than 50 ng/L (Allen et al., 1994). The threshold in red wine for IBMP was reported to be 10 ng/L (Kotseridis et al., 1998), IPMP usually appears in lower concentrations in wines, but it has a lower threshold (2.3 ng/L in red wine (Pickering et al., 2007)). Furthermore, these two compounds can act additively (Campo et al., 2005) or can interact with other compounds to enhance vegetative aroma (Escudero et al., 2007). Finally, MDMP has a threshold in red wine of 31 ng/L and a characteristic odor that combines vegetative with musty descriptors (Botezatu and Pickering, 2012). It seems that MDMP can be involved in some natural cork taint problems (Chatonnet et al., 2010).

The low levels at which these compounds can impact aroma perception (few ng/L), poses an analytical challenge. Early determination strategies involved laborious methods of sample preparation based on distillation and selective isolation in cation-exchange resins (Lacey et al., 1991). Methods developed in the last decade are simpler and mainly based on HS–SPME (Callejon et al., 2016), SPE (Cullere et al., 2009; Lopez et al., 2011), or SBSE (Gamero et al., 2013; Hjelmeland et al., 2016; Zhao et al., 2017).

However, it has been acutely pointed out (Schmarr et al., 2010) that HS–SPME, SPE, and SPE have limited extraction selectivity in a complex matrix such as wine. Therefore, one-dimensional GC–MS involves a high risk of coelution in critical cases (Schmarr et al., 2010). For this reason, various authors have developed different strategies to tackle this problem. Before the general availability of this MDGC, the determination of 3-alkyl-2-methoxypyrazines in wine made use of two specific characteristics of these compounds: their slightly basic properties and the presence of the two nitrogen atoms. The latter approach led to the use of the specific nitrogen detector in one-dimensional (Sala et al., 2002) or MDGC (Ryan et al., 2005), while the former looked for selective isolation combining distillation and ion-exchange resins (Allen et al., 1994) or SPE with mixed-mode phase (Lopez et al., 2011). However, there is no doubt that hyphenated chromatographic techniques are preferable due to their reduced sample preparation requirements. In 2005, Ryan et al. described the application of GC × GC coupled with time of flight MS or Nitrogen-phosphorus detector (NPD) for the determination of IBMP (Ryan et al., 2005). Schmarr also applied single quadrupole GC × GC–MS to the analysis of alkylmethoxypyrazines in wine (Schmarr et al., 2010). In spite of its very high separation efficiency, GC × GC is probably not the simplest approach for a limited number of target analytes as is the case with alkylmethoxypyrazines. Due to its simpler experimental setup and easiness for data processing, heart-cut MDGC has also been applied. Cullere et al. (2009) used MDGC combined with SPE to achieve very low detection limits for alkylmethoxypyrazines in wine. While alkylmethoxypyrazines extraction with HS–SPME–MDGC was faster, it provided slightly higher detection limits (Botezatu et al., 2014; Koegel et al., 2015). Even more selectivity could be obtained by combining on-line liquid chromatography with the MDGC–MS technique (Schmarr et al., 2010), or through tandem mass spectrometry, MDGC–MS/MS (Legrum et al., 2014). Ochiai et al. proposed a MDGC–MS combined with olfactometry and with preparative fraction collection for the determination of IBMP among other off-flavors (Ochiai and Sasamoto, 2011). Finally, a recent comparison of different extraction techniques combined with GC–MS/MS has also proved the viability of tandem mass spectrometry without MDGC (Hjelmeland et al., 2016).

20.5.7 Other Compounds

Rotundone (72) is a bicyclic sesquiterpene responsible for a potent aromatic note of black pepper. Its odor threshold has been estimated in red wine and in water, being values of 16 and 8 ng/L, respectively (Wood et al., 2008) (Table 20.10). The main difficulty for analyzing this compound is related to the lack of a commercial standard of rotundone, which is required to be synthesized (Siebert et al., 2008; Wood et al., 2008; Mattivi et al., 2011). The methodology proposed in the literature is based on a SPE followed by SPME–GC–MS strategy (Siebert et al., 2008; Caputi et al., 2011; Mattivi et al., 2011; Geffroy et al., 2014) and uses stable isotope dilution analysis with d5-rotundone as internal standard. Cullere et al. (2016) proposed a simpler strategy based on the isolation of this molecule using a SPE strategy, followed by GC–MS quadrupole analysis, which allows the quantification of this aromatic compound at levels of a few ng/L.

Some minor ethyl esters, such as the ethyl esters of 2-methyl and 3-methylbutyric acids (Table 20.10) may also play some role, due to their extremely low odor thresholds, ranging from 0.001 to 0.01 µg/L (in water), well

TABLE 20.10 Miscellaneous positive ultratrace

No.	Compound	Range (min–max) (ng/L)	Odor threshold[a] (ng/L)	Aromatic descriptor
72	Rotundone	<0.6–162.5[b]	16	Black pepper
73	Ethyl cyclohexanoate	<0.8–50[c] nd–5.5[d]	1	Strawberry
74	Ethyl 4-methylpentanoate	<0.5–934[c] 175–262[d]	10	Strawberry
75	Piperitone	170–1091[e]	n.f.	Minty

[a]Odor threshold values were reported by Campo et al. (2007), Wood et al. (2008).
[b]Cullere et al. (2016).
[c]San Juan et al. (2012).
[d]Campo et al. (2007).
[e]Picard et al. (2016).
nd, not determined; n.f., data not found.

below the odor threshold of their major linear cousin, ethyl hexanoate (1–5 μg/L). Main compounds within this group are ethyl cyclohexanoate (73) and ethyl 4-methylpentanoate (74). 2- and 3-methylpentanoates are also present. This group of molecules exhibits pleasant strawberry-liquorice-like odors. Campo et al. (2007) developed a method for the quantification of these four powerful aromatic ethyl esters based on a selective SPE and further multidimensional gas-chromatographic analysis in wines and other beverages. According to the analyses carried out by Campo et al. (2007) in 31 wine samples set, the levels of ethyl 2- and 3-methyl pentanoate range from less than 2 ng/L to more than 1 μg/L. Ethyl 4-methylpentanoate was found in a range from 14 to 2724 ng/L and ethyl cyclohexanoate was found at lower levels than the rest, whose maximum was only 85 ng/L in one brandy.

Finally, Picard et al. (2016) identified the molecule of piperitone (Table 20.10), a monoterpene ketone, as a contributor to the positive mint aroma of aged red Bordeaux wines. Quantification was proposed based on a HS–SPME–GC–MS according to the method published previously (Antalick et al., 2015) with some modifications. The range of concentration of this molecule in a set of 12 red wines from Bordeaux was 121–1091 ng/L. According to these authors, the addition of piperitone at levels found in wines produced an increase in the perceived intensity of the minty character.

20.6 FINAL CONSIDERATIONS

One of the main conclusions of this chapter is that the analytical problem of red wine flavor is nowadays well defined, meaning that nearly all molecules potentially relevant in red wine flavor are known and can be quantified. This should make it possible to develop specific programs to optimize some key aspects of wine flavor chemistry and wine flavor development, and we will see significant improvements in many aspects related to specific wine aroma vectors. It should be also acknowledged, however, that we are yet far from an optimal solution, since getting all the required analytical information to provide a comprehensive view of wine flavor is yet a titanic task. This could change in a close future, however, since the development of new analytical instrumentation could provide instruments sensitive and selective enough to get in just a couple of runs analytical data from most relevant compounds. New developments in ionization sources for volatile compounds, together with the amazing sensitivities of high resolution mass spectrometers (Time of Flights and Orbitraps) and maybe with the extra-selectivity provided by comprehensive gas chromatography, suggest that commercial solutions could be really close. In the meantime, there is much work pending to do. We have just begun to understand the dynamic aspects of wine perception. Relevant questions in this area are how interactions with mucosa, polyphenols and saliva influence the profile of odorant release during consumption; which is the role in such profile of release of in mouth enzymatic activities or of the different chemical processes able to induce complex cleavage during the time that wine is kept in the mouth. And of course, there is also a lot or work to do trying to understand how the different aroma vectors interact to form the different features comprising wine aroma perception. Hopefully, we will learn a lot about all these issues in the next years.

References

Allen, M.S., Lacey, M.J., Harris, R.L.N., Brown, W.V., 1991. Contribution of methoxypyrazines to Sauvignon blanc wine aroma. Am. J. Enol. Vitic 42 (2), 109–112.

Allen, M.S., Lacey, M.J., Boyd, S., 1994. Determination of methoxypyrazines in red wines by stable-isotope dilution gas-chromatography mass-spectrometry. J. Agric. Food Chem. 42 (8), 1734–1738.

Antalick, G., Tempere, S., Suklje, K., Blackman, J.W., Deloire, A., de Revel, G., et al., 2015. Investigation and sensory characterization of 1,4-cineole: a potential aromatic marker of Australian cabernet sauvignon wine. J. Agric. Food Chem. 63 (41), 9103–9111.

Augustyn, O.P.H., Rapp, A., van Wyk, C.J., 1982. Some volatile aroma components of Vitis vinifera cv. Sauvignon blanc. A. Afr. J. Enol. Vitic. 3 (2), 53–60.

Bayonove C., Cordonnier R., Dubois P., Etude d"une fraction caractéristique de l"arôme du raisin de la variété Cabernet-Sauvignon: mise en évidence de la 2-méthoxy-3-isobutylpyrazine, C. R. Acad., Sci. Paris Ser. D. 281, 1975, 75–78.

Belancic, A., Agosin, E., 2007. Methoxypyrazines in grapes and wines of Vitis vinifera cv. Carmenere. Am. J. Enol. Vitic. 58 (4), 462–469.

Botezatu, A., Pickering, G.J., Kotseridis, Y., 2014. Development of a rapid method for the quantitative analysis of four methoxypyrazines in white and red wine using multi-dimensional gas chromatography–mass spectrometry. Food Chem. 160, 141–147.

Botezatu, A., Pickering, G.J., 2012. Determination of ortho- and retronasal detection thresholds and odor impact of 2,5-dimethyl-3-methoxypyrazine in wine. J. Food Sci. 77 (11), S394–S398.

Bouchilloux, P., Darriet, P., Henry, R., Lavigne-Cruege, V., Dubourdieu, D., 1998. Identification of volatile and powerful odorous thiols in Bordeaux red wine varieties. J. Agric. Food Chem. 46 (8), 3095–3099.

Bueno, M., Zapata, J., Ferreira, V., 2014. Simultaneous determination of free and bonded forms of odor-active carbonyls in wine using a headspace solid phase microextraction strategy. J. Chromatogr. A 1369, 33–42.

Bueno, M., Carrascon, V., Ferreira, V., 2016. Release and formation of oxidation-related aldehydes during wine oxidation. J. Agric. Food Chem. 64 (3), 608–617.

Bueno, M., Marrufo-Curtido, A., Carrascon, V., Fernandez-Zurbano, P., Escudero, A., Ferreira, V., 2018. Formation and accumulation of acetaldehyde and Strecker aldehydes during red wine oxidation. Front Chem. 6, 1–19.

Caboni, P., Sarais, G., Cabras, M., Angioni, A., 2007. Determination of 4-ethylphenol and 4-ethylguaiacol in wines by LC–MS–MS and HPLC–DAD-fluorescence. J. Agric. Food Chem. 55 (18), 7288–7293.

Cacho, J., Castells, J.E., Esteban, A., Laguna, B., Sagrista, N., 1995. Iron, copper, and manganese influence on wine oxidation. Am. J. Enol. Vitic. 46 (3), 380–384.

Cacho, J., Cullere, L., Moncayo, L., Palma, J.C., Ferreira, V., 2012. Characterization of the aromatic profile of the Quebranta variety of Peruvian pisco by gas chromatography-olfactometry and chemical analysis. Flavour Fragrance J. 27 (4), 322–333.

Callejon, R.M., Ubeda, C., Ríos-Reina, R., Morales, M.L., Troncoso, A.M., 2016. Recent developments in the analysis of musty odour compounds in water and wine: a review. J. Chromatogr. A 1428, 72–85.

Camara, J.S., Marques, J.C., Alves, M.A., Ferreira, A.C.S., 2004. 3-hydroxy-4,5-dimethyl-2(5H)-furanone levels in fortified Madeira wines: relationship to sugar content. J. Agric. Food Chem. 52 (22), 6765–6769.

Campo, E., Ferreira, V., Escudero, A., Cacho, J., 2005. Prediction of the wine sensory properties related to grape variety from dynamic-headspace gas chromatography – olfactometry data. J. Agric. Food Chem. 53 (14), 5682–5690.

Campo, E., Cacho, J., Ferreira, V., 2006a. Multidimensional chromatographic approach applied to the identification of novel aroma compounds in wine – identification of ethyl cyclohexanoate, ethyl 2-hydroxy-3-methylbutyrate and ethyl 2-hydroxy-4-methylpentanoate. J. Chromatogr. A 1137 (2), 223–230.

Campo, E., Ferreira, V., Escudero, A., Marques, J.C., Cacho, J., 2006b. Quantitative gas chromatography-olfactometry and chemical quantitative study of the aroma of four Madeira wines. Anal. Chim. Acta 563 (1–2), 180–187.

Campo, E., Cacho, J., Ferreira, V., 2007. Solid phase extraction, multidimensional gas chromatography mass spectrometry determination of four novel aroma powerful ethyl esters – assessment of their occurrence and importance in wine and other alcoholic beverages. J. Chromatogr. A 1140 (1–2), 180–188.

Campo, E., Cacho, J., Ferreira, V., 2008a. The chemical characterization of the aroma of dessert and sparkling white wines (Pedro Ximenez, Fino, Sauternes, and Cava) by gas chromatography-olfactometry and chemical quantitative analysis. J. Agric. Food Chem. 56 (7), 2477–2484.

Campo, E., Do, B.V., Ferreira, V., Valentin, D., 2008b. Aroma properties of young Spanish monovarietal white wines: a study using sorting task, list of terms and frequency of citation. Aust. J. Grape Wine Res. 14 (2), 104–115.

Capone, D.L., Ristic, R., Pardon, K.H., Jeffery, D.W., 2015. Simple quantitative determination of potent thiols at ultratrace levels in wine by derivatization and high-performance liquid chromatography-tandem mass spectrometry (HPLC–MS/MS) analysis. Anal. Chem. 87 (2), 1226–1231.

Caputi, L., Carlin, S., Ghiglieno, I., Stefanini, M., Valenti, L., Vrhovsek, U., et al., 2011. Relationship of changes in rotundone content during grape ripening and winemaking to manipulation of the 'peppery' character of wine. J. Agric. Food Chem. 59 (10), 5565–5571.

Carpinteiro, I., Abuin, B., Rodriguez, I., Ramil, M., Cela, R., 2010. Sorptive extraction with in-sample acetylation for gas chromatography–mass spectrometry determination of ethylphenol species in wine samples. J. Chromatogr. A 1217 (46), 7208–7214.

Carrascon, V., Fernandez-Zurbano, P., Bueno, M., Ferreira, V., 2015. Oxygen consumption by red wines. Part II: differential effects on color and chemical composition caused by oxygen taken in different sulfur dioxide-related oxidation contexts. J. Agric. Food Chem. 63 (51), 10938–10947.

Carrascon, V., Ontanon, I., Bueno, M., Ferreira, V., 2017. Gas chromatography–mass spectrometry strategies for the accurate and sensitive speciation of sulfur dioxide in wine. J. Chromatogr. A 1504, 27–34.

Carrascon, V., Vallverdú-Queralt, A., Meudec, E., Sommerer, N., Fernandez-Zurbano, P., Ferreira, V., 2018. The kinetics of oxygen and SO_2 consumption by red wines. What do they tell about oxidation mechanisms and about changes in wine composition. Food Chem. 241, 206–214.

Carrillo, J.D., Tena, M.T., 2007. Determination of ethylphenols in wine by in situ derivatisation and headspace solid-phase microextraction-gas chromatography–mass spectrometry. Anal. Bioanal. Chem. 387 (7), 2547–2558.

Castura, J.C., Antunez, L., Gimenez, A., Ares, G., 2016. Temporal Check-All-That-Apply (TCATA): a novel dynamic method for characterizing products. Food Qual. Preference 47, 79–90.

Chatonnet, P., Dubourdieu, D., Boidron, J.N., Pons, M., 1992. The origin of ethylphenols in wines. J. Sci. Food Agric. 60 (2), 165–178.

Chatonnet, P., Fleury, A., Boutou, S., 2010. Origin and incidence of 2-methoxy-3,5-dimethylpyrazine, a compound with a "fungal" and "corky" aroma found in cork stoppers and oak chips in contact with wines. J. Agric. Food Chem. 58 (23), 12481–12490.

Chen, Y., Jastrzembski, J.A., Sacks, G.L., 2017. Copper-Complexed Hydrogen Sulfide in Wine: Measurement by Gas Detection Tubes and Comparison of Release Approaches. Am. J. Enol. Vitic. 68 (1), 91–99.

Coelho, J.M., Howe, P.A., Sacks, G.L., 2015. A headspace gas detection tube method to measure SO inf 2 /inf in wine without disrupting SO inf 2 /inf equilibria. Am. J. Enol. Vitic. 66 (3), 257–265.

Cullere, L., Cacho, J., Ferreira, V., 2004a. Analysis for wine C5–C8 aldehydes through the determination of their O-(2,3,4,5,6-pentafluorobenzyl)oximes formed directly in the solid phase extraction cartridge. Anal. Chim. Acta 524 (1–2), 201–206.

Cullere, L., Escudero, A., Cacho, J., Ferreira, V., 2004b. Gas chromatography-olfactometry and chemical quantitative study of the aroma of six premium quality Spanish aged red wines. J. Agric. Food Chem. 52 (6), 1653–1660.

Cullere, L., Cacho, J., Ferreira, V., 2006. Validation of an analytical method for the solid phase extraction, in cartridge derivatization and subsequent gas chromatographic-ion trap tandem mass spectrometric determination of 1-octen-3-one in wines at ng L(−1) level. Anal. Chim. Acta 563 (1–2), 51–57.

Cullere, L., Cacho, J., Ferreira, V., 2007. An assessment of the role played by some oxidation-related aldehydes in wine aroma. J. Agric. Food Chem. 55 (3), 876–881.

Cullere, L., Escudero, A., Campo, E., Cacho, J., Ferreira, V., 2009. Multidimensional gas chromatography–mass spectrometry determination of 3-alkyl-2-methoxypyrazines in wine and must. A comparison of solid-phase extraction and headspace solid-phase extraction methods. J. Chromatogr. A 1216 (18), 4040–4045.

Cullere, L., Ferreira, V., Cacho, J., 2011. Analysis, occurrence and potential sensory significance of aliphatic aldehydes in white wines. Food Chem. 127 (3), 1397–1403.

Cullere, L., Ontanon, I., Escudero, A., Ferreira, V., 2016. Straightforward strategy for quantifying rotundone in wine at ng L^{-1} level using solid-phase extraction and gas chromatography-quadrupole mass spectrometry. Occurrence in different varieties of spicy wines. Food Chem. 206, 267–273.

Cutzach, I., Chatonnet, P., Henry, R., Dubourdieu, D., 1997. Identification of volatile compounds with a "toasty" aroma in heated oak used in barrelmaking. J. Agric. Food Chem. 45 (6), 2217–2224.

Cutzach, I., Chatonnet, P., Dubourdieu, D., 1999. Study of the formation mechanisms of some volatile compounds during the aging of sweet fortified wines. J. Agric. Food Chem. 47 (7), 2837–2846.

Darriet, P., Tominaga, T., Lavigne, V., Boidron, J.N., Dubourdieu, D., 1995. Identification of a powerful aromatic component of *Vitis vinifera* L. var sauvignon wines: 4-Mercapto-4-methylpentan-2-one. Flavour Fragance J. 10 (6), 385–392.

de Azevedo, L.C., Reis, M.M., Motta, L.F., da Rocha, G.O., Silva, L.A., de Andrade, J.B., 2007. Evaluation of the formation and stability of hydroxyalkylsulfonic acids in wines. J. Agric. Food Chem. 55 (21), 8670–8680.

de Simón, B., Cadahía, E., Sanz, M., Poveda, P., Perez-Magariño, S., Ortega-Heras, M., Gonzalez-Huerta, C., et al., 2008. Volatile compounds and sensorial characterization of wines from four Spanish denominations of origin, aged in Spanish Rebollo (Quercus pyrenaica Willd.) oak wood barrels. J. Agric. Food Chem. 56 (19), 9046–9055.

de Simon, B.F., Martinez, J., Sanz, M., Cadahia, E., Esteruelas, E., Munoz, A.M., 2014. Volatile compounds and sensorial characterisation of red wine aged in cherry, chestnut, false acacia, ash and oak wood barrels. Food Chem. 147, 346–356.

de-la-Fuente-Blanco, A., Saenz-Navajas, M.P., Ferreira, V., 2016. On the effects of higher alcohols on red wine aroma. Food Chem. 210, 107–114.

Depinho, P.G., Bertrand, A., 1995. Analytical determination of furaneol (2,5-dimethyl-4-hydroxy-3(2H)-furanone) - application to differentiation of white wines from hybrid and various vitis-vinifera cultivars. Am. J. Enol. Vitic. 46 (2), 181–186.

Du Toit, W.J., Marais, J., Pretorius, I.S., Du Toit, M., 2006. Oxygen in must and wine: a review. S. Afr. J. Enol. Vitic. 27, 76–94.

Dunlevy, J.D., Dennis, E.G., Soole, K.L., Perkins, M.V., Davies, C., Boss, P.K., 2013. A methyltransferase essential for the methoxypyrazine-derived flavour of wine. Plant J. 75 (4), 606–617.

Elias, R.J., Laurie, V.F., Ebeler, S.E., Wong, J.W., Waterhouse, A.L., 2008a. Analysis of selected carbonyl oxidation products in wine by liquid chromatography with diode array detection. Anal. Chim. Acta 626 (1), 104–110.

Elias, R.J., Laurie, V.F., Wong, J.W., Ebeler, S.E., Waterhouse, A.L., 2008b. A method for the accurate measurement of free and sulfite-bound wine carbonyls by HPLC. Am. J. Enol. Vitic. 59 (3), 332A.

Escudero, A., Cacho, J., Ferreira, V., 2000a. Isolation and identification of odorants generated in wine during its oxidation: a gas chromatography-olfactometric study. Eur. Food Res. Technol. 211 (2), 105–110.

Escudero, A., Hernandez-Orte, P., Cacho, J., Ferreira, V., 2000b. Clues about the role of methional as character impact odorant of some oxidized wines. J. Agric. Food Chem. 48 (9), 4268–4272.

Escudero, A., Gogorza, B., Melus, M.A., Ortin, N., Cacho, J., Ferreira, V., 2004. Characterization of the aroma of a wine from Maccabeo. Key role played by compounds with low odor activity values. J. Agric. Food Chem. 52 (11), 3516–3524.

Escudero, A., Campo, E., Farina, L., Cacho, J., Ferreira, V., 2007. Analytical characterization of the aroma of five premium red wines. Insights into the role of odor families and the concept of fruitiness of wines. J. Agric. Food Chem. 55 (11), 4501–4510.

Escudero, A., San-Juan, F., Franco-Luesma, E., Cacho, J., Ferreira, V., 2014. Is orthonasal olfaction an equilibrium driven process? Design and validation of a dynamic purge and trap system for the study of orthonasal wine aroma. Flavour Fragance J. 29 (5), 296–304.

Falcao, L.D., Lytra, G., Darriet, P., Barbe, J.C., 2012. Identification of ethyl 2-hydroxy-4-methylpentanoate in red wines, a compound involved in blackberry aroma. Food Chem. 132 (1), 230–236.

Fang, Y., Qian, M.C., 2006. Quantification of selected aroma-active compounds in pinot noir wines from different grape maturities. J. Agric. Food Chem. 54 (22), 8567–8573.

Farina, L., Boido, E., Carrau, F., Dellacassa, E., 2007. Determination of volatile phenols in red wines by dispersive liquid-liquid microextraction and gas chromatography–mass spectrometry detection. J. Chromatogr. A 1157 (1–2), 46–50.

Farina, L., Villar, V., Ares, G., Carrau, F., Dellacassa, E., Boido, E., 2015. Volatile composition and aroma profile of Uruguayan Tannat wines. Food Res. Int. 69, 244–255.

Ferreira, A.C.S., de Pinho, P.G., Rodrigues, P., Hogg, T., 2002a. Kinetics of oxidative degradation of white wines and how they are affected by selected technological parameters. J. Agric. Food Chem. 50 (21), 5919–5924.

Ferreira, A.C.S., Hogg, T., de Pinho, P.G., 2003a. Identification of key odorants related to the typical aroma of oxidation-spoiled white wines. J. Agric. Food Chem. 51 (5), 1377–1381.

Ferreira, V., Fernandez, P., Pena, C., Escudero, A., Cacho, J.F., 1995. Investigation on the role played by fermentation esters in the aroma of young Spanish wines by multivariate analysis. J. Sci. Food Agric. 67 (3), 381–392.

Ferreira, V., Lopez, R., Escudero, A., Cacho, J.F., 1998. Quantitative determination of trace and ultratrace flavour active compounds in red wines through gas chromatographic ion trap mass spectrometric analysis of microextracts. J. Chromatogr. A 806 (2), 349–354.

Ferreira, V., Lopez, R., Cacho, J.F., 2000a. Quantitative determination of the odorants of young red wines from different grape varieties. J. Sci. Food Agric. 80 (11), 1659–1667.

Ferreira, V., Ortega, L., Escudero, A., Cacho, J.F., 2000b. A comparative study of the ability of different solvents and adsorbents to extract aroma compounds from alcoholic beverages. J. Chromatogr. Sci. 38 (11), 469–476.

Ferreira, V., Ortin, N., Escudero, A., Lopez, R., Cacho, J., 2002b. Chemical characterization of the aroma of Grenache rose wines: aroma extract dilution analysis, quantitative determination, and sensory reconstitution studies. J. Agric. Food Chem. 50 (14), 4048–4054.

Ferreira, V., Pet'ka, J., Aznar, M., 2002c. Aroma extract dilution analysis. Precision and optimal experimental design. J. Agric. Food Chem. 50 (6), 1508–1514.

Ferreira, V., Jarauta, I., Lopez, R., Cacho, J., 2003b. Quantitative determination of sotolon, maltol and free furaneol in wine by solid-phase extraction and gas chromatography-ion-trap mass spectrometry. J. Chromatogr. A 1010 (1), 95–103.

Ferreira, V., Cullere, L., Lopez, R., Cacho, J., 2004. Determination of important odor-active aldehydes of wine through gas chromatography–mass spectrometry of their O-(2,3,4,5,6-pentafluorobenzyl)oximes formed directly in the solid phase extraction cartridge used for selective isolation. J. Chromatogr. A 1028 (2), 339–345.

Ferreira, V., Cullere, L., Loscos, N., Cacho, J., 2006. Critical aspects of the determination of pentafluorobenzyl derivatives of aldehydes by gas chromatography with electron-capture or mass spectrometric detection – validation of an optimized strategy for the determination of oxygen-related odor-active aldehydes in wine. J. Chromatogr. A 1122 (1–2), 255–265.

Ferreira, V., Ortin, N., Cacho, J.F., 2007. Optimization of a procedure for the selective isolation of some powerful aroma thiols – development and validation of a quantitative method for their determination in wine. J. Chromatogr. A 1143 (1–2), 190–198.

Ferreira, V., San Juan, F., Escudero, A., Cullere, L., Fernandez-Zurbano, P., Saenz-Navajas, M.P., et al., 2009. Modeling quality of premium Spanish red wines from gas chromatography-olfactometry data. J. Agric. Food Chem. 57 (16), 7490–7498.

Ferreira, V., Herrero, P., Zapata, J., Escudero, A., 2015. Coping with matrix effects in headspace solid phase microextraction gas chromatography using multivariate calibration strategies. J. Chromatogr. A 1407, 30–41.

Ferreira, V., Saenz-Navajas, M.P., Campo, E., Herrero, P., de la Fuente, A., Fernandez-Zurbano, P., 2016. Sensory interactions between six common aroma vectors explain four main red wine aroma nuances. Food Chem. 199, 447–456.

Ferreira, V., Franco-Luesma, E., Vela, E., Lopez, R., Hernandez-Orte, P., 2018. Elusive Chemistry of Hydrogen Sulfide and Mercaptans in Wine. J. Agric. Food Chem. 66 (10), 2237–2246. acs.jafc.7b02427-02431.

Floch, M., Shinkaruk, S., Darriet, P., Pons, A., 2016. Identification and organoleptic contribution of vanillylthiol in wines. J. Agric. Food Chem. 64 (6), 1318–1325.

Franco-Luesma, E., Ferreira, V., 2014. Quantitative analysis of free and bonded forms of volatile sulfur compounds in wine. Basic methodologies and evidences showing the existence of reversible cation-complexed forms. J. Chromatogr. A 1359, 8–15.

Franco-Luesma, E., Ferreira, V., 2016. Reductive off-odors in wines: formation and release of H_2S and methanethiol during the accelerated anoxic storage of wines. Food Chem. 199, 42–50.

Franco-Luesma, E., Saenz-Navajas, M.P., Valentin, D., Ballester, J., Rodrigues, H., Ferreira, V., 2016. Study of the effect of H_2S, MeSH and DMS on the sensory profile of wine model solutions by Rate-All-That-Apply (RATA). Food Res. Int. 87, 152–160.

Gabrielli, M., Buica, A., Fracassetti, D., Stander, M., Tirelli, A., du Toit, W.J., 2015a. Determination of sotolon content in South African white wines by two novel HPLC–UV and UPLC–MS methods. Food Chem. 169, 180–186.

Gabrielli, M., Fracassetti, D., Tirelli, A., 2015b. UHPLC quantification of sotolon in white wine. J. Agric. Food Chem. 62 (21), 4878–4883.

Gamero, A., Wesselink, W., de Jong, C., 2013. Comparison of the sensitivity of different aroma extraction techniques in combination with gas chromatography–mass spectrometry to detect minor aroma compounds in wine. J. Chromatogr. A 1272, 1–7.

Geffroy, O., Dufourcq, T., Carcenac, D., Siebert, T., Herderich, M., Serrano, E., 2014. Effect of ripeness and viticultural techniques on the rotundone concentration in red wine made from Vitis vinifera L. cv. Duras. Aust. J. Grape Wine Res. 20 (3), 401–408.

Geffroy, O., Lopez, R., Serrano, E., Dufourcq, T., Gracia-Moreno, E., Cacho, J., et al., 2015. Changes in analytical and volatile compositions of red wines induced by pre-fermentation heat treatment of grapes. Food Chem. 187, 243–253.

Gregan, S.M., Jordan, B., 2016. Methoxypyrazine accumulation and o-methyltransferase gene expression in Sauvignon blanc grapes: the role of leaf removal, light exposure, and berry development. J. Agric. Food Chem. 64 (11), 2200–2208.

Guth, H., 1997. Quantitation and sensory studies of character impact odorants of different white wine varieties. J. Agric. Food Chem. 45 (8), 3027–3032.

Haggerty, J., Bowyer, P.K., Jiranek, V., Taylor, D.K., 2016. Optimisation and validation of a high-throughput semi-quantitative solid-phase microextraction method for analysis of fermentation aroma compounds in metabolomic screening studies of wines. Aust. J. Grape Wine Res. 22 (1), 3–10.

Han, G.M., Wang, H., Webb, M.R., Waterhouse, A.L., 2015. A rapid, one step preparation for measuring selected free plus SO_2-bound wine carbonyls by HPLC–DAD/MS. Talanta 134, 596–602.

Herbst-Johnstone, M., Piano, F., Duhamel, N., Barker, D., Fedrizzi, B., 2013. Ethyl propiolate derivatisation for the analysis of varietal thiols in wine. J. Chromatogr. A 1312, 104–110.

Herrero, P., Saenz-Navajas, P., Cullere, L., Ferreira, V., Chatin, A., Chaperon, V., et al., 2016. Chemosensory characterization of Chardonnay and Pinot Noir base wines of Champagne. Two very different varieties for a common product. Food Chem. 207, 239–250.

Herszage, J., Ebeler, S.E., 2011. Analysis of volatile organic sulfur compounds in wine using headspace solid-phase microextraction gas chromatography with sulfur chemiluminescence detection. Am. J. Enol. Vitic. 62 (1), 1–8.

Hjelmeland, A.K., Wylie, P.L., Ebeler, S.E., 2016. A comparison of sorptive extraction techniques coupled to a new quantitative, sensitive, high throughput GC−MS/MS method for methoxypyrazine analysis in wine. Talanta 148, 336−345.

Jackowetz, J.N., de Orduna, R.M., 2013. Survey of SO_2 binding carbonyls in 237 red and white table wines. Food Control 32 (2), 687−692.

Jarauta, I., Cacho, J., Ferreira, V., 2005. Concurrent phenomena contributing to the formation of the aroma of wine during aging in oak wood: an analytical study. J. Agric. Food Chem. 53 (10), 4166−4177.

Jastrzembski, J.A., Allison, R.B., Friedberg, E., Sacks, G.L., 2017. Role of elemental sulfur in forming latent precursors of H_2S in wine. J. Agric. Food Chem. 65, 10542−10549.

Kheir, J., Salameh, D., Strehaiano, P., Brandam, C., Lteif, R., 2013. Impact of volatile phenols and their precursors on wine quality and control measures of *Brettanomyces/Dekkera* yeasts. Eur. Food Res. Technol 237 (5), 655−671.

Kim, H.M., Yang, G., Kim, J.Y., Yoon, S.J., Shin, B.K., Lee, J., et al., 2017. Simultaneous determination of volatile organic compounds in commercial alcoholic beverages by gas chromatography with flame ionization detection. J. AOAC Int. 100 (5), 1492−1499.

Koegel, S., Botezatu, A., Hoffmann, C., Pickering, G., 2015. Methoxypyrazine composition of Coccinellidae-tainted Riesling and Pinot noir wine from Germany. J. Sci. Food Agric. 95 (3), 509−514.

Kotseridis, Y., Beloqui, A.A., Bertrand, A., Doazan, J.P., 1998. An analytical method for studying the volatile compounds of Merlot noir clone wines. Am. J. Enol. Vitic. 49 (1), 44−48.

Kotseridis, Y., Baumes, R., 2000. Identification of impact odorants in Bordeaux red grape juice, in the commercial yeast used for its fermentation, and in the produced wine. J. Agric. Food Chem. 48 (2), 400−406.

Kreitman, G.Y., Danilewicz, J.C., Jeffery, D.W., Elias, R.J., 2017. Copper(II)-mediated hydrogen sulfide and thiol oxidation to disulfides and organic polysulfanes and their reductive cleavage in wine: mechanistic elucidation and potential applications. J. Agric. Food Chem. 65 (12), 2564−2571.

Labbe, D., Schlich, P., Pineau, N., Gilbert, F., Martin, N., 2009. Temporal dominance of sensations and sensory profiling: a comparative study. Food Qual. Preference 20 (3), 216−221.

Lacey, M.J., Allen, M.S., Harris, R.L.N., Brown, W.V., 1991. Methoxypyrazines in Sauvignon blanc grapes and wines. Am. J. Enol. Vitic. 42 (2), 103−108.

Larcher, R., Nicolini, G., Puecher, C., Bertoldi, D., Moser, S., Favaro, G., 2007. Determination of volatile phenols in wine using high-performance liquid chromatography with a coulometric array detector. Anal. Chim. Acta 582 (1), 55−60.

Lau, M.N., Ebeler, J.D., Ebeler, S.E., 1999. Gas chromatographic analysis of aldehydes in alcoholic beverages using a cysteamine derivatization procedure. Am. J. Enol. Vitic. 50 (3), 324−333.

Legrum, C., Gracia-Moreno, E., Lopez, R., Potouridis, T., Langen, J., Slabizki, P., et al., 2014. Quantitative analysis of 3-alkyl-2-methoxypyrazines in German Sauvignon blanc wines by MDGC−MS or MDGC−MS/MS for viticultural and enological studies. Eur. Food Res. Technol. 239 (4), 549−558.

Lopez, R., Aznar, M., Cacho, J., Ferreira, V., 2002. Determination of minor and trace volatile compounds in wine by solid-phase extraction and gas chromatography with mass spectrometric detection. J. Chromatogr. A 966 (1−2), 167−177.

Lopez, R., Ezpeleta, E., Sanchez, I., Cacho, J., Ferreira, V., 2004. Analysis of the aroma intensities of volatile compounds released from mild acid hydrolysates of odourless precursors extracted from Tempranillo and Grenache grapes using gas chromatography-olfactometry. Food Chem. 88 (1), 95−103.

Lopez, R., Gracia-Moreno, E., Cacho, J., Ferreira, V., 2011. Development of a mixed-mode solid phase extraction method and further gas chromatography mass spectrometry for the analysis of 3-alkyl-2-methoxypyrazines in wine. J. Chromatogr. A 1218 (6), 842−848.

Lopez-Vazquez, C., Bollain, M.H., Berstsch, K., Orriols, I., 2010. Fast determination of principal volatile compounds in distilled spirits. Food Control 21 (11), 1436−1441.

Loscos, N., Segurel, M., Dagan, L., Sommerer, N., Marlin, T., Baumes, R., 2008. Identification of S-methylmethionine in Petit Manseng grapes as dimethyl sulphide precursor in wine. Anal. Chim. Acta 621 (1), 24−29.

Louw, L., Roux, K., Tredoux, A., Tomic, O., Naes, T., Nieuwoudt, H.H., et al., 2009. Characterization of selected South African young cultivar wines using FTMIR spectroscopy, gas chromatography, and multivariate data analysis. J. Agric. Food Chem. 57 (7), 2623−2632.

Lytra, G., Tempere, S., Le Floch, A., de Revel, G., Barbe, J.C., 2013. Study of sensory interactions among red wine fruity esters in a model solution. J. Agric. Food Chem. 61 (36), 8504−8513.

Lytra, G., Tempere, S., Zhang, S., Marchand, S., de Revel, G., Barbe, J.-C., 2014. Olfactory impact of dimethyl sulfide on red wine fruity esters aroma expression in model solution. J. Int. Sci. Vigne Vin 48 (1), 75−85.

Lytra, G., Tempere, S., Marchand, S., de Revel, G., Barbe, J.C., 2016. How do esters and dimethyl sulphide concentrations affect fruity aroma perception of red wine? Demonstration by dynamic sensory profile evaluation. Food Chem. 194, 196−200.

Malfeito-Ferreira, M., 2011. Yeasts and wine off-flavours: a technological perspective. Ann. Microbiol. 61 (1), 95−102.

Marrufo-Curtido, A., Carrascon, V., Bueno, M., Ferreira, V., Escudero, A., 2018. A procedure for the measurement of oxygen consumption rates and some observations about the influence of wine initial chemical composition. Food Chem. 248, 37−45. Available from: https://doi.org/10.1016/j.foodchem.2017.1012.1028.

Martin, B., Etievant, P.X., Lequere, J.L., Schlich, P., 1992. More clues about sensory impact of sotolon in some flor sherry wines. J. Agric. Food Chem. 40 (3), 475−478.

Martorell, N., Marti, M.P., Mestres, M., Busto, O., Guasch, J., 2002. Determination of 4-ethylguaiacol and 4-ethylphenol in red wines using headspace-solid-phase microextraction-gas chromatography. J. Chromatogr. A 975 (2), 349−354.

Mateo-Vivaracho, L., Ferreira, V., Cacho, J., 2006. Automated analysis of 2-methyl-3-furanthiol and 3-mercaptohexyl acetate at ng L^{-1} level by headspace solid-phase microextracion with on-fibre derivatisation and gas chromatography-negative chemical ionization mass spectrometric determination. J. Chromatogr. A 1121 (1), 1−9.

Mateo-Vivaracho, L., Cacho, J., Ferreira, V., 2007. Quantitative determination of wine polyfunctional mercaptans at nanogram per liter level by gas chromatography-negative ion mass spectrometric analysis of their pentafluorobenzyl derivatives. J. Chromatogr. A 1146 (2), 242−250.

Mateo-Vivaracho, L., Cacho, J., Ferreira, V., 2008. Improved solid-phase extraction procedure for the isolation and in-sorbent pentafluorobenzyl alkylation of polyfunctional mercaptans optimized procedure and analytical applications. J. Chromatogr. A 1185 (1), 9−18.

Mateo-Vivaracho, L., Cacho, J., Ferreira, V., 2009. Selective preconcentration of volatile mercaptans in small SPE cartridges: quantitative determination of trace odor-active polyfunctional mercaptans in wine. J. Sep. Sci. 32 (21), 3845–3853.

Mateo-Vivaracho, L., Zapata, J., Cacho, J., Ferreira, V., 2010. Analysis, occurrence, and potential sensory significance of five polyfunctional mercaptans in white wines. J. Agric. Food Chem. 58 (18), 10184–10194.

Mattivi, F., Caputi, L., Carlin, S., Lanza, T., Minozzi, M., Nanni, D., et al., 2011. Effective analysis of rotundone at below-threshold levels in red and white wines using solid-phase microextraction gas chromatography/tandem mass spectrometry. Rapid Commun. Mass Spectrom. 25 (4), 483–488.

Mayr, C.M., Capone, D.L., Pardon, K.H., Black, C.A., Pomeroy, D., Francis, I.L., 2015. Quantitative analysis by GC–MS/MS of 18 aroma compounds related to oxidative off-flavor in wines. J. Agric. Food Chem. 63 (13), 3394–3401.

Meillon, S., Urbano, C., Schlich, P., 2009. Contribution of the temporal dominance of sensations (TDS) method to the sensory description of subtle differences in partially dealcoholized red wines. Food Qual. Preference 20 (7), 490–499.

Murat, M.L., 2005. Recent Findings on Rosé Wine Aromas. Part 1: Identifying Aromas Studying the Aromatic Potential for Grape Juice. The Australian and New Zealand Grapegrower and Winemaker 497a (Annual Technical Issue), pp. 64–65, 69–71, 73–74, 76.

Nascimento, R.F., Marques, J.C., Neto, B.S.L., DeKeukeleire, D., Franco, D.W., 1997. Qualitative and quantitative high-performance liquid chromatographic analysis of aldehydes in Brazilian sugar cane spirits and other distilled alcoholic beverages. J. Chromatogr. A 782 (1), 13–23.

Nicolini, G., Larcher, R., Bertoldi, D., Puecher, C., Magno, F., 2007. Rapid quantification of 4-ethylphenol in wine using high-performance liquid chromatography with a fluorimetric detector. Vitis 46 (4), 202–206.

Noguerol-Pato, R., Gonzalez-Alvarez, M., Gonzalez-Barreiro, C., Cancho-Grande, B., Simal-Gandara, J., 2012. Aroma profile of Garnacha Tintorera-based sweet wines by chromatographic and sensorial analyses. Food Chem. 134 (4), 2313–2325.

Ochiai, N., Sasamoto, K., 2011. Selectable one-dimensional or two-dimensional gas chromatography-olfactometry/mass spectrometry with preparative fraction collection for analysis of ultra-trace amounts of odor compounds. J. Chromatogr. A 1218 (21), 3180–3185.

Ontanon, I., Vela, E., Hernandez-Orte, P. and Ferreira, V., 2018. Advances in the analytical determination of free and bonded VSCs and of truly free SO_2. In preparation.

Ortega, C., Lopez, R., Cacho, J., Ferreira, V., 2001a. Fast analysis of important wine volatile compounds Development and validation of a new method based on gas chromatographic-flame ionisation detection analysis of dichloromethane microextracts. J. Chromatogr. A 923 (1–2), 205–214.

Ortega, L., Lopez, R., Cacho, J., Ferreira, V., 2001b. Use of solid-liquid distribution coefficients to determine retention properties of Porapak-Q resins – determination of optimal conditions to isolate alkyl-methoxypyrazines and beta-damascenone from wine. J. Chromatogr. A 931 (1–2), 31–39.

Osborne, J.P., Dube Morneau, A., Mira de Orduna, R., 2006. Degradation of free and sulfur-dioxide-bound acetaldehyde by malolactic lactic acid bacteria in white wine. J. Appl. Microbiol 101 (2), 474–479.

Parr, W., Green, J., White, K., Sherlock, R., 2007. The distinctive flavour of New Zealand Sauvignon blanc: sensory characterisation by wine professionals. Food Qual. Preference 18 (6), 849–861.

Pavez, C., Agosin, E., Steinhaus, M., 2016. Odorant screening and quantitation of thiols in Carmenere red wine by gas chromatography-olfactometry and stable isotope dilution assays. J. Agric. Food Chem. 64 (17), 3417–3421.

Peinado, R.A., Moreno, J., Medina, M., Mauricio, J.C., 2004. Changes in volatile compounds and aromatic series in sherry wine with high gluconic acid levels subjected to aging by submerged flor yeast cultures. Biotechnol. Lett. 26 (9), 757–762.

Petka, J., Ferreira, V., Gonzalez-Vinas, M.A., Cacho, J., 2006. Sensory and chemical characterization of the aroma of a white wine made with Devin grapes. J. Agric. Food Chem. 54 (3), 909–915.

Piano, F., Fracassetti, D., Buica, A., Stander, M., du Toit, W.J., Borsa, D., et al., 2015. Development of a novel liquid/liquid extraction and ultra-performance liquid chromatography tandem mass spectrometry method for the assessment of thiols in South African Sauvignon Blanc wines. Aust. J. Grape Wine Res. 21 (1), 40–48.

Picard, M., Lytra, G., Tempere, S., Barbe, J.C., de Revel, G., Marchand, S., 2016. Identification of piperitone as an aroma compound contributing to the positive mint nuances perceived in aged red Bordeaux wines. J. Agric. Food Chem. 64 (2), 451–460.

Pickering, G.J., Karthik, A., Inglis, D., Sears, M., Ker, K., 2007. Determination of ortho- and retronasal detection thresholds for 2-isopropyl-3-methoxypyrazine in wine. J. Food Sci. 72 (7), S468–S472.

Pizarro, C., Perez-del-Notario, N., Gonzalez-Saiz, J.M., 2007. Determination of Brett character responsible compounds in wines by using multiple headspace solid-phase microextraction. J. Chromatogr. A 1143 (1–2), 176–181.

Pizarro, C., Saenz-Gonzalez, C., Perez-del-Notario, N., Gonzalez-Saiz, J.M., 2011. Ultrasound-assisted emulsification microextraction for the sensitive determination of Brett character responsible compounds in wines. J. Chromatogr. A 1218 (50), 8975–8981.

Pollnitz, A.P., Pardon, K.H., Sefton, M.A., 2000. Quantitative analysis of 4-ethylphenol and 4-ethylguaiacol in red wine. J. Chromatogr. A 874 (1), 101–109.

Pollnitz, A.P., Pardon, K.H., Sykes, M., Sefton, M.A., 2004. The effects of sample preparation and gas chromatograph injection techniques on the accuracy of measuring guaiacol, 4-methylguaiacol and other volatile oak compounds in oak extracts by stable isotope dilution analyses. J. Agric. Food Chem. 52 (11), 3244–3252.

Pons, A., Lavigne, V., Landais, Y., Darriet, P., Dubourdieu, D., 2010. Identification of a sotolon pathway in dry white wines. J. Agric. Food Chem. 58 (12), 7273–7279.

Rapp, A., Engel, L., 1995. Determination and detection of furaneol (2,5-dimethyl-4-hydroxy-3-furanon) in wines from vitis-vinifera varieties. Vitis 34 (1), 71–72.

Rauhut, D., Kurbel, H., Macnamara, K., Grossmann, M., 1998. Headspace GC-SCd monitoring of low volatile sulfur-compounds during fermentation and in wine. Analusis 26 (3), 142–145.

Rayne, S., Eggers, N.J., 2007. Quantitative determination of 4-ethylphenol and 4-ethyl-2-methoxyphenol in wines by a stable isotope dilution assay. J. Chromatogr. A 1167 (2), 195–201.

Rodriguez-Bencomo, J.J., Schneider, R., Lepoutre, J.P., Rigou, P., 2009. Improved method to quantitatively determine powerful odorant volatile thiols in wine by headspace solid-phase microextraction after derivatization. J. Chromatogr. A 1216 (30), 5640–5646.

Roland, A., Delpech, S., Dagan, L., Ducasse, M.A., Cavelier, F., Schneider, R., 2016. Innovative analysis of 3-mercaptohexan-1-ol,3-mercaptohexylacetate and their corresponding disulfides in wine by stable isotope dilution assay and nano-liquid chromatography tandem mass spectrometry. J. Chromatogr. A 1468, 154–163.

Romano, A., Perello, M.C., Lonvaud-Funel, A., Sicard, G., de Revel, G., 2009. Sensory and analytical re-evaluation of "Brett character.". Food Chem. 114 (1), 15–19.

Roujou de Boubée, D., van Leeuwen, C., Dubourdieu, D., 2000. Organoleptic impact of 2-methoxy-3-isobutylpyrazine on red bordeaux and loire wines. Effect of environmental conditions on concentrations in grapes during ripening. J. Agric. Food Chem. 48 (10), 4830–4834.

Ryan, D., Watkins, P., Smith, J., Allen, M., Marriott, P., 2005. Analysis of methoxypyrazines in wine using headspace solid phase microextraction with isotope dilution and comprehensive two-dimensional gas chromatography. J. High Resolut. Chromatogr. 28 (9–10), 1075–1082.

Ryona, I., Pan, B.S., Intrigliolo, D.S., Lakso, A.N., Sacks, G.L., 2008. Effects of cluster light exposure on 3-isobutyl-2-methoxypyrazine accumulation and degradation patterns in red wine grapes (Vitis vinifera L. Cv. Cabernet Franc). J. Agric. Food Chem. 56 (22), 10838–10846.

Sala, C., Mestres, M., Marti, M.P., Busto, O., Guasch, J., 2002. Headspace solid-phase microextraction analysis of 3-alkyl-2-methoxypyrazines in wines. J. Chromatogr. A 953 (1–2), 1–6.

San-Juan, F., Pet'ka, J., Cacho, J., Ferreira, V., Escudero, A., 2010. Producing headspace extracts for the gas chromatography-olfactometric evaluation of wine aroma. Food Chem. 123 (1), 188–195.

San-Juan, F., Ferreira, V., Cacho, J., Escudero, A., 2011. Quality and aromatic sensory descriptors (Mainly fresh and dry fruit character) of Spanish red wines can be predicted from their aroma-active chemical composition. J. Agric. Food Chem. 59 (14), 7916–7924.

San Juan, F., Cacho, J., Ferreira, V., Escudero, A., 2012. Aroma chemical composition of red wines from different price categories and its relationship to quality. J. Agric. Food Chem. 60 (20), 5045–5056.

Sarrazin, E., Shinkaruk, S., Tominaga, T., Bennetau, B., Frerot, E., Dubourdieu, D., 2007. Odorous impact of volatile thiols on the aroma of young botrytized sweet wines: identification and quantification of new sulfanyl alcohols. J. Agric. Food Chem. 55 (4), 1437–1444.

Sasaki, K., Takase, H., Kobayashi, H., Matsuo, H., Takata, R., 2015a. Molecular cloning and characterization of UDP-glucose: furaneol glucosyltransferase gene from grapevine cultivar Muscat Bailey A (*Vitis labrusca* × *V. vinifera*). J. Exp. Bot. 66 (20), 6167–6174.

Sasaki, K., Takase, H., Tanzawa, F., Kobayashi, H., Saito, H., Matsuo, H., et al., 2015b. Identification of furaneol glucopyranoside, the precursor of strawberry-like aroma, furaneol, in Muscat Bailey A. Am. J. Enol. Vitic. 66 (1), 91–94.

Schlich, P., 2017. Temporal dominance of sensations (TDS): a new deal for temporal sensory analysis. Curr. Opin. Food Sci. 15, 38–42.

Schmarr, H.-G., Ganß, S., Koschinski, S., Fischer, U., Riehle, C., Kinnart, J., et al., 2010. Pitfalls encountered during quantitative determination of 3-alkyl-2-methoxypyrazines in grape must and wine using gas chromatography–mass spectrometry with stable isotope dilution analysis. Comprehensive two-dimensional gas chromatography–mass spectrometry and on-line liquid chromatography-multidimensional gas chromatography–mass spectrometry as potential loopholes. J. Chromatogr. A 1217 (43), 6769–6777.

Schneider, R., Baumes, R., Bayonove, C., Razungles, A., 1998. Volatile compounds involved in the aroma of sweet fortified wines (Vins Doux Naturels) from Grenache noir. J. Agric. Food Chem. 46 (8), 3230–3237.

Schneider, R., Kotseridis, Y., Ray, J.L., Augier, C., Baumes, R., 2003. Quantitative determination of sulfur-containing wine odorants at sub parts per billion levels. 2. Development and application of a stable isotope dilution assay. J. Agric. Food Chem. 51 (11), 3243–3248.

Segurel, M.A., Razungles, A.J., Riou, C., Salles, M., Baumes, R.L., 2004. Contribution of dimethyl sulfide to the aroma of Syrah and Grenache Noir wines and estimation of its potential in grapes of these varieties. J. Agric. Food Chem. 52 (23), 7084–7093.

Siebert, T.E., Smyth, H.E., Capone, D.L., Neuwohner, C., Pardon, K.H., Skouroumounis, G.K., et al., 2005. Stable isotope dilution analysis of wine fermentation products by HS–SPME–GC–MS. Anal. Bioanal. Chem. 381 (4), 937–947.

Siebert, T.E., Wood, C., Elsey, G.M., Pollnitz, A.P., 2008. Determination of rotundone, the pepper aroma impact compound, in grapes and wine. J. Agric. Food Chem. 56 (10), 3745–3748.

Siebert, T.E., Solomon, M.R., Pollnitz, A.P., Jeffery, D.W., 2010. Selective determination of volatile sulfur compounds in wine by gas chromatography with sulfur chemiluminescence detection. J. Agric. Food Chem. 58 (17), 9454–9462.

Simpson, R.F., Capone, D.L., Sefton, M.A., 2004. Isolation and identification of 2-methoxy-3,5-dimethylpyrazine, a potent musty compound from wine corks. J. Agric. Food Chem. 52 (17), 5425–5430.

Slabizki, P., Fischer, C., Legrum, C., Schmarr, H.-G., 2015. Characterization of atypical off-flavor compounds in natural cork stoppers by multidimensional gas chromatographic techniques. J. Agric. Food Chem. 63 (35), 7840–7848.

Suarez, R., Suarez-Lepe, J.A., Morata, A., Calderon, F., 2007. The production of ethylphenols in wine by yeasts of the genera *Brettanomyces* and *Dekkera*: a review. Food Chem. 102 (1), 10–21.

Tominaga, T., Dubourdieu, D., 2006. A novel method for quantification of 2-methyl-3-furanthiol and 2-furanmethanethiol in wines made from *Vitis vinifera* grape varieties. J. Agric. Food Chem. 54 (1), 29–33.

Tominaga, T., Murat, M.L., Dubourdieu, D., 1998. Development of a method for analyzing the volatile thiols involved in the characteristic aroma of wines made from *Vitis vinifera* L. cv. Sauvignon Blanc. J. Agric. Food Chem. 46 (3), 1044–1048.

Tominaga, T., Blanchard, L., Darriet, P., Dubourdieu, D., 2000. A powerful aromatic volatile thiol, 2-furanmethanethiol, exhibiting roast coffee aroma in wines made from several *Vitis vinifera* grape varieties. J. Agric. Food Chem. 48 (5), 1799–1802.

Tominaga, T., Guimbertau, G., Dubourdieu, D., 2003. Contribution of benzenemethanethiol to smoky aroma of certain *Vitis vinifera* L. wines. J. Agric. Food Chem. 51 (5), 1373–1376.

Tredoux, A., de Villiers, A., Majek, P., Lynen, F., Crouch, A., Sandra, P., 2008. Stir bar sorptive extraction combined with GC–MS analysis and chemometric methods for the classification of South African wines according to the volatile composition. J. Agric. Food Chem. 56 (12), 4286–4296.

Valente, I.M., Santos, C.M., Moreira, M.M., Rodrigues, J.A., 2013. New application of the QuEChERS methodology for the determination of volatile phenols in beverages by liquid chromatography. J. Chromatogr. A 1271, 27–32.

van Gemert, L.J., Nettenbreijer, A.H., 1977. Compilation of Odour Threshold Values in Air and Water. Voorburg, The Netherlands.

Vela, E., Hernandez-Orte, P., Franco-Luesma, E., Ferreira, V., 2018. Micro-oxygenation does not eliminate hydrogen sulfide and mercaptans from wine; it simply shifts redox and complex-related equilibria to reversible oxidized species. Food Chem. 243, 222–230.

Vichi, S., Cortes-Francisco, N., Caixach, J., 2015. Analysis of volatile thiols in alcoholic beverages by simultaneous derivatization/extraction and liquid chromatography-high resolution mass spectrometry. Food Chem. 175, 401–408.

Villen, J., Senorans, F.J., Reglero, G., Herraiz, M., 1995. Analysis of wine aroma by direct-injection in gas-chromatography without previous extraction. J. Agric. Food Chem. 43 (3), 717–722.

Waterhouse, A.L., Frost, S., Ugliano, M., Cantu, A.R., Currie, B.L., Anderson, M., et al., 2016a. Sulfur dioxide-oxygen consumption ratio reveals differences in bottled wine oxidation. Am. J. Enol. Vitic. 67 (4), 449–459.

Waterhouse, A.L., Sacks, G.L., Jeffery, D.W., 2016b. Understanding Wine Chemistry. Wiley, Chichester, UK.

Wedral, D., Shewfelt, R., Frank, J., 2010. The challenge of *Brettanomyces* in wine. LWT Food Sci. Technol. 43 (10), 1474–1479.

Weldegergis, B.T., Crouch, A.M., 2008. Analysis of volatiles in Pinotage wines by stir bar sorptive extraction and chemometric profiling. J. Agric. Food Chem. 56 (21), 10225–10236.

Wen, Y., Lopez, R., Ferreira, V., 2018. An automated gas chromatographic–mass spectrometric method for the quantitative analysis of the odor-active molecules present in the vapors emanated from wine. J. Chromatogr. A 1534, 130–138.

Wood, C., Siebert, T.E., Parker, M., Capone, D.L., Elsey, G.M., Pollnitz, A.P., et al., 2008. From wine to pepper: rotundone, an obscure sesquiterpene, is a potent spicy aroma compound. J. Agric. Food Chem. 56 (10), 3738–3744.

Zapata, J., Mateo-Vivaracho, L., Cacho, J., Ferreira, V., 2010. Comparison of extraction techniques and mass spectrometric ionization modes in the analysis of wine volatile carbonyls. Anal. Chim. Acta 660 (1–2), 197–205.

Zhao, P.T., Gao, J.X., Qian, M., Li, H., 2017. Characterization of the Key aroma compounds in Chinese Syrah wine by gas chromatography-olfactometry–mass spectrometry and aroma reconstitution studies. Molecules 22 (7), 1045–1058.

CHAPTER

21

SO$_2$ in Wines: Rational Use and Possible Alternatives

Simone Giacosa, Susana Río Segade, Enzo Cagnasso, Alberto Caudana, Luca Rolle and Vincenzo Gerbi

Department of Agricultural, Forestry and Food Sciences, University of Turin, Grugliasco (TO), Italy

21.1 SULFUR DIOXIDE: USE IN THE WINEMAKING PROCESS AND LEGAL LIMITS

Sulfur dioxide (SO$_2$, M_r: 64.058, Fig. 21.1) is a colorless gas, with a pungent smell, that is very soluble in water (up to 39 volumes per volume of water), and whose antiseptic and preservative power has been known for millennia. Even Homer in the Odyssey, book XXII, talks about sulfur and fire for the fumigation of houses. Sulfur dioxide has been systematically used in enology since the 19th century, but already in previous centuries the sulfurization of barrels was practiced to improve the conservation of wine during transport. Pasteur (1866), suggests to burn the sulfur in the barrels to prevent deterioration. In winemaking the expressions sulfur dioxide or SO$_2$ and sulfites are generally used as synonymous.

For its many functions it is still today a virtually irreplaceable additive in the winemaking process. The functions of sulfur dioxide can be summarized as follows:

1. Antiseptic: the two main antiseptic activities are the selective action of the microflora of musts and the antimicrobial action in the conservation of wines during their aging.
2. Antioxidant: it inhibits the action of dissolved oxygen; this slow process allows to protect wines from chemical oxidation, such as the oxidation of some polyphenols and some odorous substances.
3. Antioxidase: it inhibits the effect, and sometimes causes the denaturation of oxidasic enzymes (polyphenoloxidase, PPO) in the must. The result is a protection for musts from prefermentation oxidations.
4. Solubilizing: in contact with the skins, at high doses, it favors the diffusion of the coloring substances contained in the vacuoles, by means of small holes caused on the cell walls, thus favoring the release of the anthocyanins.
5. Binding agent: improves the olfactory and gustatory qualities of wines, as it combines with some substances of pungent odor or taste, such as acetaldehyde and pyruvic acid, making them no longer perceptible to taste.
6. Fining agent: it has a light clarifying action, as it promotes the coagulation of colloidal substances, increasing the phenomenon of spontaneous precipitation of the lees.

Nowadays, wine consumers move on the market driven by different motivations: some are looking for the best price performance, but the number of those interested in wines increases due to their cultural, historical, anthropological contents, willing to spend to satisfy their need for knowledge, as well as for the pleasure related to the consumption of the product. Moderate consumption of wine is an integral part of the lifestyle of these consumers. All consumers, however, are attentive to the pursuit of quality and food safety. The disappearance from the market of defective, or even altered wines, is explained by the progress of the winemaking technique, by the improvement of the hygienic conditions and by the generalized use of an antiseptic, sulfur dioxide, which has remained the undisputed "additive" for over a century for wine. In recent years, producers have felt the need to reduce or eliminate the use of sulfites, to increase the image of naturalness of wines, but also to prevent health damages in consumers sensitive to their presence. For the European Community legislation, producers have the obligation to indicate on the label the sentence "contains sulfites" if the content is higher than 10 mg/L. In fact, allergic reactions in sensitive individuals are the main concerns, since the effects on other subjects, due to an eventual enzyme deficiency in the liver or to the alterations of the metabolism of vitamin B1, can cause acute symptoms, such as headache, but not serious and permanent risks. In any case, the World Health Organization has set the safety limit for daily intake of sulfites in 0.7 mg/kg of body weight.

Taking into account this value, in an individual weighting 60 – 80 kg, the admissible daily dose is between 42 and 56 mg per day, values that would be easily exceeded by those who drank even only half a liter of wine containing sulfites at the maximum admitted level. In Europe this limit is set by law: 150 mg/L of total SO_2 for red wines, 200 mg/L for white wines, but in other countries it can be different (e.g., 350 ppm in the USA, 250 mg/L in Australia; Table 21.1). There are then exceptions and higher limits for sweet wines, with limited differences regulated at national level.

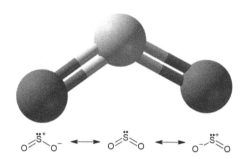

FIGURE 21.1 Molecular structure and resonance system of sulfur dioxide.

TABLE 21.1 Total SO_2 limits for red wine in the main winemaking countries

Country	Sugar content	Limit[a]	Regulatory—Notes
OIV acceptable limit	≤ 4 g/L	150 mg/L	Compendium of international methods of must and wine analysis 2017 edition
	>4 g	300 mg/L	
Australia	<35 g/L	250 mg/L	ANZFSC 4.5.1: Clause 5(5)(a)
	≥ 35 g/L	300 mg/L	
Canada	All	350 ppm	Can. Food & Drug Reg. B.02.100(b)(viii)(B)
		70 ppm[b]	
European Union	<5 g/L	150 mg/L	Reg. (EC) No 606/2009 (Annex I B)
	≥ 5 g/L	200 mg/L[c]	
	<2 g/L	100 mg/L	Reg. (EU) No 203/2012 (Annex VIII bis)—Organic Wines
	≥ 2 g/L	120 mg/L	
	<5 mg/L		
	≥ 5 g/L	170 mg/L	
New Zealand	<35 g/L	250 mg/kg	ANZFSC 4.5.1: Clause 5(5)(a)
	≥ 35 g/L	400 mg/kg	
South Africa	<5 g/L	150 mg/L	Liquor Products Act 60 of 1989 Regulations Regulation 32 (Table 8)
	≥ 5 g/L	200 mg/L	
United States of America	All	350 ppm	27 CFR 4.22(b)(1)
		100 ppm	Wine labeled as "Made with Organic Grapes"

[a]Maximum limit expressed as SO_2.
[b]Limit of free SO_2.
[c]Some exceptions exist for particular wines with limits as high as 300, 350, 400 mg/L.

The stability induced by sulfites in wines is so well known that even the recent European legislation concerning organic wines [Commission Implementing Regulation (EU) No 203/2012] admits their presence, but with a limit of 100 mg/L for red wines. The equivalence agreement concluded between the EU, United States, and Canada recognizes the possibility of labeling for European organic wines as National Organic Program or Canada Organic Regime, under conditions that respect some more restrictive limits (e.g., maximum limit 100 mg/L, added exclusively in gaseous form). Although the US FDA indicates that only 1% of the population has manifested sensitivity to sulfites, these have assumed a symbolic meaning of "unnatural preservative," so it has become interesting for all producers to eliminate, or at least reduce, the sulfite content of their wines.

From this derives the efforts made by producers to limit the presence of sulfites in wines, even if reaching and declaring the total absence of sulfites is very difficult, since yeasts in the fermentation process produce sulfites (use of sulfured amino acids for their nitrogen metabolism, reduction of sulfates), often in excess of 10 mg/L.

The ability to produce sulfites has been known and studied for long time (Eschenbruch, 1974), but this capacity in the past century was considered positively, due to the contribution to the stability of the wines that the most productive strains could give (Suzzi et al., 1985). The ability to produce sulfites during alcoholic fermentation is a genetic feature of the yeast strain, but the concentration of sulfites produced is also influenced by the composition of the must, which can influence the production of sulfites by a yeast strain by as much as threefold (Werner et al., 2009).

It is therefore plausible, to declare the absence of added sulfites, as required in the United States for wines labeled as "organic wine," but it is rare to be able to affirm the complete absence (less than 10 mg/L), a possibility achievable only with the use of yeasts selected for this specific genetic character.

The more or less justified distrust of consumers toward sulfites, together with the tendency to naturalness (organic wine, biodynamic culture), have convinced many producers to adopt, in the management of fermentation processes, a return to the "past enology" with the use of spontaneous fermentation, also considered capable of inducing a greater gustatory and olfactory complexity. Although justifiable, this belief is not supported by objective data on the real superiority of wines produced with spontaneous fermentations. In the face of a risk related to stuck or slow fermentations, which are more and more frequent due to the higher sugar content of the grapes, to agronomic treatments, and to the tendency to climate warming that characterizes the current historical period (Mira de Orduña, 2010), it seems reasonable to support the usefulness of the native yeast strains, but selected and prepared in an appropriate form to ensure a regular course of fermentation. The risk of fermentation stops, also in terms of food safety, should not be overlooked due to the onset of abnormal lactic fermentations (Guo et al., 2015; Beneduce et al., 2010).

Truly innovative compared to traditional ones, but really natural, are the technologies that can allow a reduction of the use of sulfites in the winemaking process, reserving them for an antiseptic and antioxidant role only in wine bottling, where they can guarantee, even at very low doses, an extension of the shelf life of the product.

Since the mid-20th century, producers have used sulfites to select microorganisms at the time of grape crushing, then during the storage of wine in barrel, to avoid bacterial alterations, and at bottling, to maintain the stability of wines and lengthen the shelf life of the product (Fig. 21.2).

21.2 DIFFERENT FORMS TO USE SO_2

The addition of sulfur dioxide is possible in different forms:

1. by combustion of sulfur;
2. in gaseous form;
3. in aqueous solution;
4. by salts that release SO_2.

The combustion of sulfur is the oldest and most known form of sulfurization to sanitize closed environments. In the enological field, it is still used nowadays in the sulfurization of the barrels, after their thorough cleaning and drying, to destroy unwanted microorganisms present on the inner surface of the oak or in the wood cavities affected by the penetration of the wine. The observation of the staves during the disassembly of the barrels allows us to observe that the wine penetrates into the depth of the wood for some millimeters. In this portion microorganisms can nest, a phenomena feared especially if they are altering agents such as lactic and acetic acid bacteria, or *Brettanomyces* spp. yeast, which can produce off-flavor compounds.

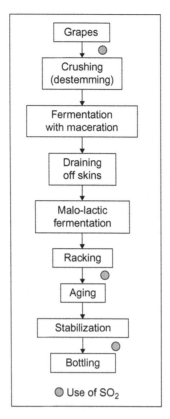

FIGURE 21.2 Flow-chart of red winemaking. The orange dot indicates the main phases where SO_2 is generally used.

The gaseous sulfur dioxide released by the combustion of sulfur is capable of reaching the porosity of the oak and can be used in empty containers even at high doses. The released SO_2 is twice the weight of the burnt sulfur (the relative atomic mass of sulfur is 32, equal to that of oxygen relative molecular mass), but in practice the yield is lower, due to impurities or partial combustion, therefore a gas production equal to the weight of burned sulfur can be considered. So, it is easy to calculate, given the volume of the barrel, how much sulfur is needed for sulfurization. This sulfur type is available in different shapes (wicks, candles, disks), but the most used are the perforated disks which, when lit, are suspended in the center of the barrel with a metal wire. This form of sulfurization is particularly effective for preventing and combating the presence of *Brettanomyces* (Chatonnet et al., 1993) and for this reason it is still widely used and recommended in the production process of red wines, in which the defects caused by the presence of 4-ethylphenols are particularly feared. The combustion of sulfur involves risks for the operators, who must be properly instructed and equipped with personal protections, because it can cause corrosion. Part of SO_2 present in the sulfured barrel can be dissolved in the wine at the time of filling, therefore it is always necessary to analyze the sulfite content of the wines in storage.

The use of SO_2 in gaseous form is possible by dispensing it from tanks that contain it in liquid form, under pressure (2 – 5 bar, depending on the temperature), from which it is possible to dispense in gas or liquid form (with the tank always upside down; Fig. 21.3). The measurement of the dispensed fluid is usually done by weighing the tank, hence it is a form of use suitable for large volumes or to producers that have sophisticated measurement and delivery systems. The liquefied product is pure, therefore the addition does not involve other substances or combustion impurities to the wine. In gaseous form it is used for the sanitization of the containers, instead of the combustion of the sulfur, but requires adequate precautions for the protection of the cellar personnel. In liquid form it is suitable for on-line additions along the wine handling pipes, allowing an optimal mixing, but, since it can be very aggressive, requires equipment made of resistant materials and adequate protections to prevent accidental leaks. For these reasons, in medium and small cellars producers prefer to use aqueous solutions or salts that are easier to manage and dispense. In the sulfitation of crushed grapes for the production of red wines, the most used form of addition is generally as liquid solution, because it allows an easy dosage and homogenization in the mass. This last aspect is important to avoid parts of the containers with excessive or deficient sulfite concentrations. SO_2 aqueous solutions in water, usually 5%–10%, are pure and easy to dose, but

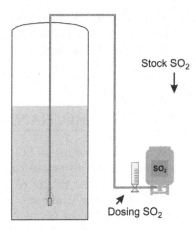

FIGURE 21.3 SO_2 dosage in wine from pressurized tanks.

have a pungent odor and are not very chemically stable, therefore it is more common to use potassium hydrogen sulfite ($KHSO_3$, potassium bisulfite) solutions, sometimes with the addition of SO_2, with a content ranging from 10% to 25%. They smell less aggressively than pure SO_2 solutions and are more chemically stable, but have higher densities and require more attention for optimal mixing. Lastly, they increase the concentration of the potassium ion. However, using dosages of 30–50 mg/L of SO_2, commonly used during the pressing, this is not a problem in the vinification of grapes that already have strong concentrations of potassium ions, sometimes more than 1 g/kg.

The ammonium hydrogen sulfite (NH_4HSO_3, ammonium bisulfite) solutions are also widely used in the prefermentation phase, with a very variable SO_2 content, generally 15%. They are also easily dosed, fairly stable, and involve the simultaneous addition of about 22% of nitrogen, which can constitute a useful supplement for the nitrogenous yeast nutrition. For this reason, ammonia salts are used only in the prefermentative phase, since it is not advisable to add ammonia salts in wines for reasons of biological stability and food safety. The use of solid salts that release SO_2 is practically compatible with all the phases of winemaking, and it is certainly preferable in wine sulfitation, since it does not involve the addition of water. Further, the dosage is precise, given that the salts are easily weighable and more stable over time compared to aqueous solutions, when stored in closed containers and in dry environments. The most used salt is potassium disulfite ($K_2S_2O_5$, potassium metabisulfite) which has a theoretical 57% SO_2 content, slightly lowered in commercial enological products to about 50%.

In the sulfitation of wines in small barrels, the use of effervescent tablets is also used, because of their simplicity in dosage that does not require any weighing. They consist of a mixture of potassium metabisulfite (67%) and potassium bicarbonate (33%). The effervescence produced by potassium bicarbonate facilitates the mixing of SO_2. This technique is particularly indicated for modest additional sulfite additions during storage.

The choice of the SO_2 form to be used is influenced by the phase of the process in which it is added and by the possibilities of dispensing that are available in cellar. As illustrated above, in a winemaking aimed at the reduction of sulfites in wine, there is a tendency not to add SO_2 to the must before alcoholic fermentation. However, when the health of the grapes is compromised and it is necessary to add SO_2, the solutions of salts, particularly ammonia, are the most suitable. The usage dose must be commensurate with the degree of infection of the grapes, also taking into account the pH that conditions the antiseptic action. In the case of red wines, an important addition of SO_2 is done after malolactic fermentation, to avoid alterations and spoilage of wines, especially if aged in oak. For this aim the use of salts, or effervescent tablets for small containers, appears the most suitable. Finally, the undoubtedly more frequent addition, and still largely irreplaceable, is at bottling. In this case, to avoid precipitation phenomena due to potassium ions, the best choice appears to be that of the aqueous solution of sulfur dioxide, or, in large wineries, with liquid SO_2 pressurized tanks.

The first two additions of sulfites can be considered no longer indispensable, because the use of starter cultures, selected or indigenous, make the use of sulfites superfluous to win the competition with the unwanted microorganisms, if the grapes are not affected by alterations of the cluster. In the storage or aging phase, the use of inert gases can be noted, together with the evolution of the shape and the materials of the winemaking containers that, since they improve cleaning and sanitization, have greatly reduced the risk of microbial alterations. The storage/aging in bottle is the phase where it is more difficult to protect the wine without the use of sulfites, even if in the filling operation it is nowadays possible to create aseptic conditions and limit the dissolution of oxygen. These technological choices have allowed a strong reduction of sulfites in commercial wines, but it should be emphasized that a better knowledge of the mechanism of action of sulfur dioxide has allowed the rationalization of their use.

21.3 SO₂ ACTION MECHANISMS

The fundamental contributions to the rationalization of the use of SO_2 was possible thanks to researchers who, in the second half of the 20th century, clarified the role of substances capable of steadily combining SO_2 (Blouin, 1966; Burroughs and Sparks, 1973; Delfini et al., 1980) and the role of pH in regulating the presence of the molecular fraction of SO_2 (Sudraud and Chauvet, 1985; Usseglio-Tomasset, 1985, 1986; Ribèreau-Gayon et al., 2006). The sensory threshold in this form in wines is considered to be about 2 mg/L (Ribèreau-Gayon et al., 2006). The above mentioned studies have investigated the ionization equilibrium of sulfur dioxide in aqueous solution, showing the influence of alcoholic degree, quantifying the ability to combine with anthocyanins, and confirming the ability of acetaldehyde and other carbonyl compounds to permanently combine it, creating a compound of unnecessary addition both for antiseptic and antioxidant purposes.

To recall the fundamentals of the use of sulfur dioxide, it should be remembered that this gas, in aqueous solutions such as must and wine, behaves like a weak diprotic acid (the H_2SO_3 species has not been isolated in aqueous solution, due to the high instability; Fig. 21.4A) in equilibrium with hydrogensulfite ions (HSO_3^-, bisulfite) and sulfite (SO_3^{2-}). The ionization equilibria of SO_2 are pH-dependent, as shown in Fig. 21.4B.

The wine alcohol content and temperature increase produces an overall effect of reducing the acidic strength of sulfur dioxide compared to an aqueous solution with same ionic strength (Usseglio-Tomasset and Bosia, 1984). These latter authors calculated the values of pK_{a1} in wine conditions, obtaining values close to 2 (Table 21.2; Usseglio-Tomasset and Bosia, 1984; OIV, 2017a).

The second ionization, with the formation of SO_3^{2-} ion, is negligible at wine pH (3.00 − 4.00) because the relative constant K_{a2} has a too small value ($pK_{a2} \approx 7$) and the fraction of the sulfite ion is less than 1‰ of the

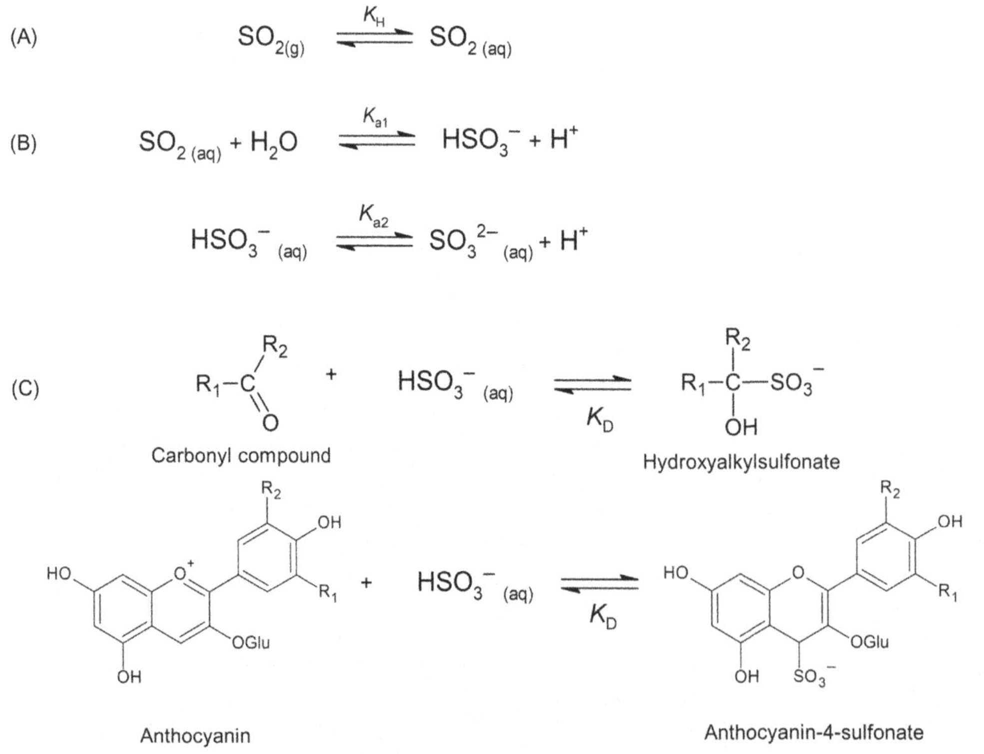

FIGURE 21.4 Sulfur dioxide equilibriums in aqueous solution: (A) solubilization (Henry's law); (B) acid-base ionization; (C) hydrogensulfite-adduct formation with carbonylic compounds and anthocyanins.

TABLE 21.2 Molecular SO$_2$ fraction in must and wine-like conditions at different alcohol strength ($T = 20$ °C, $I = 0.038$ M)

pK_{a1}:	1.72	1.84	1.95
		Alcohol strength/%	
pH	0	6	12
2.80	7.73	9.88	12.38
2.90	6.24	8.01	10.09
3.00	5.02	6.47	8.18
3.10	4.03	5.21	6.61
3.20	3.23	4.18	5.32
3.30	2.58	3.35	4.28
3.40	2.06	2.68	3.43
3.50	1.64	2.14	2.74
3.60	1.31	1.71	2.19
3.70	1.04	1.36	1.75
3.80	0.83	1.08	1.39
3.90	0.66	0.86	1.11
4.00	0.53	0.69	0.88

Data expressed as percentage. Since SO$_3^{2-}$ fraction is considered negligible, the HSO$_3^-$ fraction accounts approximately for the complementary percentage.
Data from Usseglio-Tomasset and Bosia, 1984; OIV, 2017a.

concentration of free SO$_2$. Therefore, considering only the equilibrium of the first ionization, it is possible to easily calculate, as a function of the pH, the fraction of bisulfite (HSO$_3^-$) and molecular (SO$_2$) forms. This latter form is responsible for the antiseptic action of sulfur dioxide, especially with regard to bacteria and yeasts, as already highlighted in the 1930s by Moreau and Vinet (1937a,b). Considering the sulfur dioxide at the wine pH as a monoprotic acid, it is possible to calculate, from the equation of Henderson–Hasselbach derived from the mass action law, the fraction of the molecular SO$_2$ depending on the pH:

$$\frac{[\text{Molecular SO}_2]}{[\text{Free SO}_2]} = \frac{1}{1 + 10^{(\text{pH}-\text{p}K_{a1})}}$$

By varying the pH of one unit (e.g., from 2.80 to 3.80), the molecular sulfur dioxide fraction decreases about 10 times (Table 21.2). Fig. 21.5 shows the different forms of sulfur dioxide in wines. It is evident that, in order to keep the total sulfur level low, it is necessary to limit the combined part and to have pH conditions which allow the presence of sulfur dioxide in a molecular form.

The hydrogen sulfite ion (HSO^{3-}) is a fair nucleophile, therefore it reacts with electrophilic compounds containing a carbonyl group and with anthocyanins (Blouin, 1966; Burroughs and Sparks, 1973; Brouillard and El Hage Chahine, 1980; Fig. 21.4C). The stability of the adduct compounds is expressed in terms of apparent dissociation constant of adduct (K_D) according to mass action law (Burroughs and Sparks, 1973), the values related to the main binders present in the wine are indicated in Table 21.3. Combination compounds with K_D less than 3×10^{-6} are considered particularly stable, since less than 1% of the binder is in free form in the presence of 20 mg/L of free SO$_2$ (Ribèreau-Gayon et al., 2006).

Recent studies on aged wines shown also the presence of sulfonated derivatives of phenolic compounds such as epicatechin and procyanidin B2, influenced by storage condition and temperature (Mattivi et al., 2015; Arapitsas et al., 2018).

Since the substance that shows the greatest combination capacity is acetaldehyde (Table 21.3), it is evident that it is possible to significantly limit the final content of sulfites in wines if the steadily combined SO$_2$ fraction is reduced (Jackowetz and Mira de Orduña, 2013). All authors agree that the maximum concentration of acetaldehyde occurs at the beginning of alcoholic fermentation. The SO$_2$ present in that phase will be completely

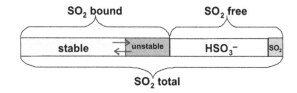

FIGURE 21.5 Representation of the fractions of sulfur dioxide in wine.

TABLE 21.3 Hydrogensulfite adducts in wines: apparent dissociation constants (K_D)

Electrophilic agent	K_D/M	pK_D	Note
Acetaldehyde	$1.5 \cdot 10^{-6}$	5.82	a
	$2.06 \cdot 10^{-6}$	5.686	b
Anthocyanin	$9.1 \cdot 10^{-6}$	5.04	c
2,5-Diketogluconic acid	$4.5 \cdot 10^{-4}$	3.35	a
Formaldehyde	$1.09 \cdot 10^{-7}$	6.963	b
Fructose	1.5	−0.18	d
Galacturonic acid	$1.6 \cdot 10^{-2}$	1.80	a
Glucose	0.9	0.046	d
Hexanal	$3.45 \cdot 10^{-6}$	5.462	b
2-Ketoglutaric acid	$4.9 \cdot 10^{-4}$	3.31	a
β-Ionone	$2.07 \cdot 10^{-4}$	3.684	b
Pyruvic acid	$1.4 \cdot 10^{-4}$	3.85	a

[a]*Burroughs and Sparks (1973) (pH = 3).*
[b]*De Azevedo et al. (2007).*
[c]*Brouillard and El Hage Chahine (1980) (cyanidin-3-O-glucoside).*
[d]*Blouin (1966).*

combined and therefore not able to protect the future wine. Operating with extreme care during the fermentation processes, scrupulously handling the topping of the barrels, with protection from oxygen, and hygiene in the cellar, it is also possible to produce wines without adding sulfur dioxide, reaching the end of the production phase with extremely low levels of total SO_2. If the storage is very long, as in aged wines, it is very difficult to avoid sulfites without causing a more rapid decay of the sensory characteristics of the wine.

The longest and uncontrolled part of storage takes place after bottling. Therefore, avoiding sulfur dioxide at bottling is much more difficult because of antiseptic action, which prevents re-fermentations and bacterial alterations. Moreover, their antioxidant action prolongs the shelf life of the wine for the time necessary for its commercialization. Some authors point out that to obtain a fair antiseptic action an amount of 0.6−0.8 mg/L of molecular SO_2 is needed, while 20−40 mg/L as free SO_2 are useful for their antioxidant effect (Waterhouse et al., 2016). The first action can be replaced by adopting preventive microbial stabilization techniques (microfiltration, pasteurization), followed by aseptic filling in perfectly sanitized and pollution-protected systems. These techniques are now widespread, even if not always adaptable to small productions or long-aging wines. Furthermore, some replacement additives may partially exert this action.

In the case of antioxidant action, the problem is more complex and, at present, fewer alternatives are known, especially in the case of younger red wines, which are characterized by fruity and fermentation aromas. This sensory aspect is difficult to defend and maintain in the absence of an antioxidant such as sulfur dioxide. There are also wines that owe their olfactory characteristics to oxidative phenomena (e.g., Porto wines), but in most red wines oxidation traits are not desired. Bottling with pre-evacuation of the air and nitrogen compensation to eliminate oxygen in the bottle and the elimination of oxygen in the bottle headspace, may minimize the oxygen supply in the bottle.

The adoption of alternative closures, such as screw caps, can also prevent oxygen intake in the bottle. However, the oxidation in the bottle is not exclusively linked to the presence of oxygen, but is also connected to the action of oxidizing-reducing pairs of the natural components of the wine, such as traces of metal ions and the polyphenolic substances that catalyze the effect of oxygen (Li et al., 2008; Oliveira et al., 2011). The presence of sulfur dioxide, especially if added during the bottling phase, allows to delay the maderization phenomena (Fig. 21.6A): the antioxidant action of sulfur dioxide is based on the reaction with the intermediate reduction compounds of oxygen and on the reduction of o-quinones to the corresponding o-diphenols (Danilewicz, 2007; Danilewicz and Wallbridge, 2010). The use of ascorbic acid, a natural antioxidant, allows a strong reduction of the dose of SO_2, but not a total replacement due to the formation of hydrogen peroxide, a strong oxidizing agent which forms during the reaction with oxygen (Fig. 21.6B).

A number of natural substances are under study, often obtainable from enological by-products or from the agro-food chain such as phenolic acids and tannins, which could be a possible alternative to antioxidant action of sulfites. Also the glutathione, a natural tripeptide present in the grape, has captured the attentions of the researchers for its remarkable antioxidant capacities. These SO_2 possible replacements will be discussed in Section 21.4.

In addition to SO_2 replacement products, processing techniques based on high hydrostatic pressures (HHP, or high pressure processing, HPP) were also proposed for the reduction of sulfite content in winemaking. These emerging techniques were previously tested during wine production for antimicrobial purposes or for the improvement of phenolic compound extraction (Morata et al., 2017). With regards to a possible SO_2 substitution, HPP techniques were useful for the inactivation of PPO enzyme activities even using a treatment time as short as 10 minutes, reducing by about 85% these enzyme activities when considering a white grapes must treated at high pressures (900 MPa; Castellari et al., 1997). HPP was also found useful in *Brettanomyces bruxellensis* inactivation on wine: 200 MPa pressure treatments for 3 minutes caused a 3 to about 6 log reduction of this yeast population, depending on the alcohol content of the tested wine (van Vyk and Silva, 2017). However, early HHP trials on sulfite-free wines demonstrated that these treatments may have an adverse effect on the wine volatiles composition, in particular concerning an acceleration of Maillard reaction causing a faster aging during storage (Santos et al., 2015).

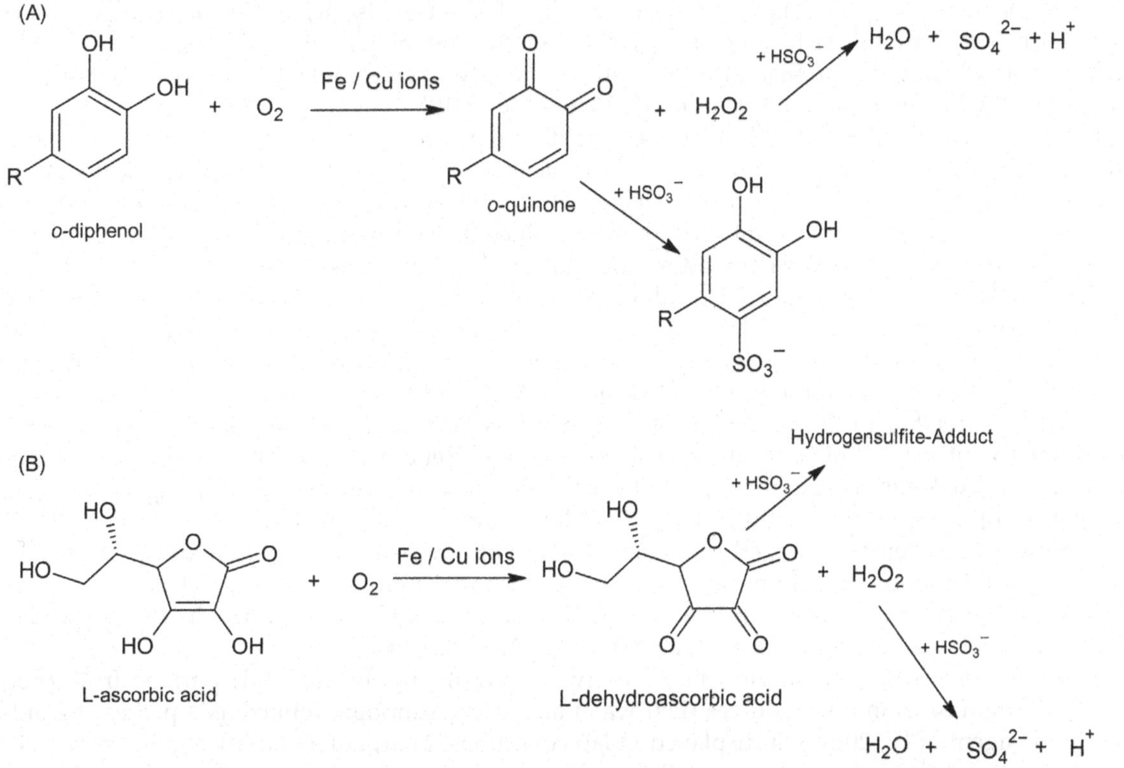

FIGURE 21.6 Antioxidant mechanisms of SO_2 in wine: (A) reactions with the intermediate reduction compounds of oxygen and with the reduction of o-quinones to the corresponding o-diphenols; (B) reaction with redox products derived from L-ascorbic acid.

21.4 SO$_2$ REPLACEMENT PRODUCTS FOR RED WINE PRODUCTION

In recent years, the ability to reduce sulfur dioxide contents from red wine was investigated by means of the use of substitute or replacement wine additives. Plenty of different molecules are allowed by the wine legislation, and some of them were tested as a replacement or in a synergistic way. In addition to those, new proposed products are continuously tested in order to try to compensate SO$_2$ actions. To understand SO$_2$ substitutes we need to refer to the actions that must be conducted by these replacement products, namely antimicrobial and antioxidant activities. Some other SO$_2$ activities previously described, such as solubilizing activity, are scarcely expressed in wine conditions. In particular, referring to this latter activity all the enological products that can aid the extraction of secondary metabolites from the grape pomace can be hardly describable as direct or indirect replacements for SO$_2$. To the best of our knowledge, no additive is available for wine production which may exert both antimicrobial and antioxidant actions as SO$_2$ does with limited doses.

21.4.1 Antimicrobial Activity Substitutes

The antimicrobial activity is carried out by SO$_2$ in a very broad way, and no other additives can completely replicate this trait. The main allowed substitutes for this activity are lysozyme and sorbic acid, two products active against Gram-positive bacteria or yeast proliferation, respectively. More recently, also dimethyl dicarbonate (DMDC) was allowed as an additive in the European Union [Commission Regulation (EU) No. 643/2006], while newer studies on experimental products are beginning to appear in scientific literature.

Lysozyme is an enzyme useful for the control of lactic acid bacteria in wine. When compared with SO$_2$, depending on the time of addition, the use of this product may slightly improve the color of red wines due to the absence of bleaching effects. However, in red wine there are many drawbacks related to the use of this additive, such as the extraction and presence of phenolic compounds, especially low molecular weight proanthocyanidins, which limited the action of lysozyme in red wine production (Guzzo et al., 2011). In addition, the presence of grape pomace reduced the efficacy of lysozyme during alcoholic fermentation to a complete absence of this trait at the end of the process, and lysozyme activity was also found to be different depending on the lactic acid bacteria strain considered (Azzolini et al., 2010). Regarding color, it can also have a negative impact on color density when used before bottling (Bartowsky et al., 2004). Due to possible allergens related to egg derivatives, the European Union legislation impose that all wines with residual lysozyme contents must state the presence of this compound in the label [Commission Regulation (EU) No. 1266/2010].

Sorbic acid is an antifungal agent which can be added as potassium sorbate. It was generally used in bottled wines presenting residual sugars and low to medium alcohol content (such as partly fermented products), to avoid the proliferation of yeast (mainly *Saccharomyces* spp.) without the need of high doses of SO$_2$. Sorbic acid cannot be used as the sole antiseptic additive in the wine, since the concentrations used of this compound are not sufficient to inhibit lactic acid bacteria spreading (Edinger and Splittstoesser, 1986). Furthermore, these bacteria can degrade sorbic acid itself leading to the production of 2-ethoxy-3,5-hexadiene, an off-flavor compound with a distinctive scent of geranium leaves (Maarse, 1991). To prevent these issues, sorbic acid was generally used together with SO$_2$ to assure the microbiological stability in sweet wines. Nowadays, the evolution of filtration techniques and of the production processes limited the use of this additive.

DMDC activity is related to the inhibition of enzymes involved in the glycolysis of yeast, while the action against bacteria is limited (Delfini et al., 2002; Costa et al., 2008). Renouf et al. (2008) also evidenced the action of DMDC against *B. bruxellensis* yeast in red wine, however it can be used only as a curative agent as a consequence of the mechanism of action of this additive. Indeed, DMDC action is very quick in wines, and within 5 hours (at 10 °C; depending also on temperature, ethanol, and presence of other substances) its hydrolysis leads to the production of several compounds including methanol (Stafford and Ough, 1976; Delfini et al., 2002), a risk factor subjected to legal limits. In the European Union, DMDC is allowed only for addition in wines presenting more than 5 g/L of residual sugars [Commission Regulation (EU) No. 643/2006].

Another product presenting an antimicrobial activity is chitosan, a polysaccharide derived from processing of chitin of fungal origin or from other sources (Petrova et al., 2016). Although defined as a processing aid and proposed as a fining agent (OIV, 2017b), it displayed in lab conditions a particular activity against some wine-related microorganisms, such as *Oenococcus oeni* and *B. bruxellensis*, while *Saccharomyces cerevisiae* evidenced a stronger resistance (Bağder Elmacı et al., 2015). Other experiments in red wine conditions evidenced a three-log reduction in culturability of *B. bruxellensis* in the samples treated with chitosan but not a complete eradication, either in bottle or in barrel (Petrova et al., 2016). Moreover, the fungistatic effect is limited in time and not definitive. The same authors evidenced also that the source of chitosan compound influenced the treatment effectiveness.

The experimental use of colloidal silver complex (CSC, CAgC) was tested by Izquierdo-Cañas et al. (2012) for the substitution of SO_2 in red winemaking. The authors noted that CSC exerted a similar action against microorganisms as SO_2, while lowering the ethanol content. Furthermore, a strong reduction in phenolic content, especially anthocyanins and flavonols, was found in CSC-treated samples with relation to SO_2-added wines. A further study of Garde-Cerdán et al. (2014) on Tempranillo wine production confirmed this trend on anthocyanins. To our knowledge, this additive is not allowed for winemaking production at this time.

The use of antimicrobial additives is sometimes tested in conjunction with sulfur dioxide: for instance, recent studies have proposed a joint use of SO_2 and lysozyme or DMDC to reduce SO_2 content in red wines while retaining its antioxidant action. The results evidenced a reduction of biogenic and volatile amines contents in joint trials of DMDC + SO_2 and lysozyme + SO_2 with respect to sulfur dioxide alone (Ancín-Azpilicueta et al., 2016), with also an increase in volatile compounds and overall sensory quality (Nieto-Rojo et al., 2015).

21.4.2 Antioxidant Activity Substitutes

The antioxidant activity is already present in red wine, depending on grape cultivar and terroir, production methods, extraction techniques, and oxygen exposure. The ability to increase the antioxidant capacity may be reached by using additives other than SO_2, with different characteristics and side effects. Phenolic compounds are, by quality and quantity, the most important antioxidant compounds related to red wine. Among them, tannins have a key role in determining astringency, bitterness, anthocyanin combination, and antioxidant protection. According to Waterhouse et al. (2016), the compounds of this latter class with an enological interest have generally high molecular mass (more than 1200 u) and may be ascribable to condensed or hydrolyzable tannins. Red grapes are naturally rich of condensed tannins, which are extracted by the solid parts (skins and seeds) during maceration, along with smaller, monomeric or oligomeric flavanols. Hydrolyzable tannins may derive from sources other than grape, including oak or commercial additives, and are generally constituted by gallic or ellagic acid.

Red wine is particularly rich in tannins and hence less prone to oxidation with respect to white or rosé wines. Furthermore, during the winemaking process it is possible to enrich wines with tannins derived either from grape or from oak, thus containing condensed or hydrolyzable tannins. This feature may be exerted by the oak contact during large oak cask or barrel aging, or by direct addition of tannin in wines, an enological practice allowed in several countries including the EU and United States (Versari et al., 2013).

Although red wines have characteristic antioxidants present, the unique antioxidant action of SO_2 may be difficult to replicate only with grape tannin or reinforcement by addition of them. In red wine, tannins have notable sensory effects, such as bitterness and astringency, and they also may mediate the color stabilization with anthocyanins, or influence the color directly toward orange notes due to an increase of yellow color (Versari et al., 2013). An incorrect addition of them may, therefore, modify extensively the wine sensory traits.

Pascual et al. (2017) conducted a trial in model wine solution evidencing some different trends in oxygen consumption rate by SO_2 and enological tannins extracted from different sources: SO_2 action was quick even at lower dosage, surpassed only by ascorbic acid. When considering the relative antioxidant capacity, ellagitannins value is comparable with respect to SO_2, therefore good to be considered as a possible SO_2 complement or alternative, depending on doses and wine characteristics. Among tannins, ellagitannins showed the highest oxygen consumption rates, with lower values found in decreasing order by quebracho tannins, skin tannins, seeds tannins, and gallotannins, with an influence of the storage temperature.

Leaves extracts and other natural products were also tested for their antioxidant activity, however their availability, price, and legal regulations, depending on the product, may limit their use as possible SO_2 replacements. One recent example of this new possibility is represented by grapevine shoot extracts. Raposo et al. (2016) tested a product containing 29% of stilbenes in comparison to a control added with SO_2. Cv. Syrah wines produced using grapevine shoot extract showed at the end of the wine production process an improved color intensity and better sensory judgements than the control (SO_2), however after 12 months in bottle an increase of the oxidation notes, coupled with a decrease in color intensity and higher color hue, was found with respect to SO_2-treated wines.

Ascorbic acid is another antioxidant used in wine production. Its capacity to scavenge molecular oxygen before the oxidation of phenolic compounds is a key trait of this additive in white wines, however it is jointly used with SO_2 to capture the hydrogen peroxide previously produced with the oxidation of ascorbic acid (Bradshaw et al., 2011; Fig. 21.6B). The use of this antioxidant is not recommended without the presence of adequate doses of free SO_2. In red wine production the abundance of tannins and the lower free SO_2 available are a serious obstacle to the use of ascorbic acid.

The use of glutathione was also proposed for the reduction of SO$_2$ contents due to its antioxidant activity, however the importance of this additive is related more to white wine production rather than red wine. Conversely, the presence of glutathione in model and real red wine experiments caused an increase of the degradation of malvidin-3-glucoside when oxidation occurred (Gambuti et al., 2017), a trait that may pose unexpected issues to the possible use of this additive in red wine production. Further studies are necessary to evaluate the feasibility of glutathione addition during the wine production process for red wine improvement.

21.4.3 Considerations on SO$_2$ Replacement Additives

Although the quantitative importance of SO$_2$ in wines is small, its actions and effectiveness toward mycobiota and oxidations remain unparalleled especially for long storage. However, some alternative techniques may be considered for very short storage times. With this aim, a comprehensive comparison between several additives used in the production of SO$_2$-free red Tempranillo wines was conducted by Ferrer-Gallego et al. (2017), without evidencing notable differences in sensory quality between control (SO$_2$) and sulfite-free trials when analyzed 3 months after bottling (storage temperature 12 °C; 7 months after the end of alcoholic and malolactic fermentation in total). This is an encouraging result for ready-to-drink wines requiring short storage times, but not a definitive solution for wines needing long shelf life periods.

It would therefore seem easier to keep red wines without SO$_2$, rather than white wines, since there have been reported differences in the perception of the fruity characters and the freshness descriptors when the level of sulfur dioxide decreases as a consequence of the behavior of different bottle closures (Godden et al., 2001).

Perhaps we are one step away from the resolution of the problem of sulfites in wines, but the process may be long: even if a definitive substitute is found, we will have to wait for the validation and the necessary authorizations. For now, it is realistic to propose to increase the naturalness of the wine without compromising the perceivable quality. Practicing a drastic reduction of the sulfite content could be easily obtainable by increasing the attention to the quality of the grapes and adopting a rational winemaking process based on a deep theoretical and practical knowledge of natural phenomena, and keeping an eye on the product shelf life.

References

Ancín-Azpilicueta, C., Jiménez-Moreno, N., Moler, J.A., Nieto-Rojo, R., Urmeneta, H., 2016. Effects of reduced levels of sulfite in wine production using mixtures with lysozyme and dimethyl dicarbonate on levels of volatile and biogenic amines. Food Addit. Contam., A 33, 1518–1526.

Arapitsas, P., Guella, G., Mattivi, F., 2018. The impact of SO$_2$ on wine flavanols and indoles in relation to wine style and age. Sci. Rep. 8 (858), 1–13.

Azzolini, M., Tosi, E., Veneri, G., Zapparoli, G., 2010. Evaluating the efficacy of lysozyme against lactic acid bacteria under different winemaking scenarios. S. Afr. J. Enol. Vitic. 31, 99–105.

Bartowsky, E.J., Costello, P.J., Villa, A., Henschke, P.A., 2004. The chemical and sensorial effects of lysozyme addition to red and white wines over six months' cellar storage. Aust. J. Grape Wine Res. 10, 143–150.

Bağder Elmacı, S., Gülgör, G., Tokatlı, M., Erten, H., İşci, A., Özçelik, F., 2015. Effectiveness of chitosan against wine-related microorganisms. Antonie Van Leeuwenhoek 107, 675–686.

Beneduce, L., Romano, A., Capozzi, V., Lucas, P., Barnavon, L., Bach, B., et al., 2010. Biogenic amine in wines. Ann. Microbiol. 60, 573–578.

Blouin, J., 1966. Contribution à l'étude des combinaisons de l'anydride sulfureux dans les moûts et les vins. Ann. Technol. Agricole 15 (223–287), 359–401.

Bradshaw, M.P., Barril, C., Clark, A.C., Prenzler, P.D., Scollary, G.R., 2011. Ascorbic acid: a review of its chemistry and reactivity in relation to a wine environment. Crit. Rev. Food Sci. Nutr. 51, 479–498.

Brouillard, R., El Hage Chahine, J.M., 1980. Chemistry of anthocyanin pigments. 6. Kinetic and thermodynamic study of hydrogen sulfite addition to cyanin. Formation of a highly stable meisenheimer-type adduct derived from a 2-Phenylbenzopyrylium salt. J. Am. Chem. Soc. 102, 5375–5378.

Burroughs, L.F., Sparks, A.H., 1973. Sulphite binding power of wines and ciders. I. Equilibrium constants for the dissociation of carbonyl-bisulfite compounds. J. Sci. Food Agric. 24, 187–198.

Castellari, M., Matricardi, L., Arfelli, G., Rovere, P., Amati, A., 1997. Effects of high pressure processing on polyphenoloxidase enzyme activity of grape musts. Food Chem 60, 647–649.

Chatonnet, P., Boidron, J.N., Dubordieu, D., 1993. Influence des conditions d'élevage et de sulfitage des vins rouges en barriques sur leur teneur en acide acétique et en éthyl-phenols. J. Int. Sci. Vigne Vin 27, 277–298.

Costa, A., Barata, A., Malfeito-Ferreira, M., Loureiro, V., 2008. Evaluation of the inhibitory effect of dimethyl dicarbonate (DMDC) against wine microorganisms. Food Microbiol. 25, 422–427.

Danilewicz, J.C., 2007. Interaction of sulfur dioxide, polyphenols, and oxygen in a wine-model system: central role of iron and copper. Am. J. Enol. Vitic. 58, 53–60.

Danilewicz, J.C., Wallbridge, P.J., 2010. Further studies on the mechanism of interaction of polyphenols, oxygen, and sulfite in wine. Am. J. Enol. Vitic. 61, 166–175.

REFERENCES

De Azevedo, L.C., Reis, M.M., Motta, L.F., Da Rocha, G.O., Silva, L.A., De Andrade, J.B., 2007. Evaluation of the formation and stability of hydroxyalkylsulfonic acids in wines. J. Agric. Food Chem. 55, 8670–8680.

Delfini, C., Castino, M., Ciolfi, G., 1980. L'aggiunta di tiamina ai mosti per ridurre i chetoacidi ed accrescere l'efficacia dell'SO_2 nei vini. Riv. Vitic. Enol. 33, 572–589.

Delfini, C., Gaia, P., Schellino, R., Strano, M., Pagliara, A., Ambrò, S., 2002. Fermentability of grape must after inhibition with dimethyl dicarbonate (DMDC). J. Agric. Food Chem. 50, 5605–5611.

Edinger, W.D., Splittstoesser, D.F., 1986. Sorbate tolerance by lactic acid bacteria associated with grapes and wine. J. Food. Sci. 51, 1077–1078.

Eschenbruch, R., 1974. Sulfide and sulfite formation during winemaking – a review. Am. J. Enol. Vitic. 25, 157–161.

Ferrer-Gallego, R., Puxeu, M., Nart, E., Martín, L., Andorrà, I., 2017. Evaluation of Tempranillo and Albariño SO_2-free wines produced by different chemical alternatives and winemaking procedures. Food Res. Int. 102, 647–657.

Gambuti, A., Picariello, L., Rolle, L., Moio, L., 2017. Evaluation of the use of sulfur dioxide and glutathione to prevent oxidative degradation of malvidin-3-monoglucoside by hydrogen peroxide in the model solution and real wine. Food Res. Int. 99, 454–460.

Garde-Cerdán, T., López, R., Garijo, P., González-Arenzana, L., Gutiérrez, A.R., López-Alfaro, I., et al., 2014. Application of colloidal silver versus sulfur dioxide during vinification and storage of Tempranillo red wines. Aust. J. Grape Wine Res. 20, 51–61.

Godden, P., Leigh, F., Field, J., Gishen, M., Coulter, A., Valente, P., et al., 2001. Wine bottle closures: physical characteristics and effect on composition and sensory properties of a Semillon wine 1. Performance up to 20 months post-bottling. Aust. J. Grape Wine Res. 7 (2), 64–105.

Guo, Y.Y., Yang, Y.P., Peng, Q., Han, Y., 2015. Biogenic amines in wine: a review. Int. J. Food Sci. Technol. 50, 1523–1532.

Guzzo, F., Cappello, M.S., Azzolini, M., Tosi, E., Zapparoli, G., 2011. The inhibitory effects of wine phenolics on lysozyme activity against lactic acid bacteria. Int. J. Food Microbiol. 148, 184–190.

Izquierdo-Cañas, P.M., García-Romero, E., Huertas-Nebreda, B., Gómez-Alonso, S., 2012. Colloidal silver complex as an alternative to sulphur dioxide in winemaking. Food Control 23, 73–81.

Jackowetz, J.N., Mira de Orduña, R., 2013. Survey of SO_2 bindings carbonyls in 237 red and white table wines. Food Control 32, 687–692.

Li, H., Guo, A., Wang, H., 2008. Mechanisms of oxidative browning of wine. Food Chem. 108, 1–13.

Maarse, H., 1991. Volatile Compounds in Foods and Beverages. Marcel Dekker Inc, New York, NY.

Mattivi, F., Arapitsas, P., Perenzoni, D., Guella, G., 2015. Influence of storage conditions on the composition of red wines. Advances in Wine Research. American Chemical Society, Washington, DC, USA.

Mira de Orduña, R., 2010. Climate change associated effects on grape and wine quality and production. Food Res. Int. 43, 1844–1855.

Morata, A., Loira, I., Vejarano, R., González, C., Callejo, M.J., Suárez-Lepe, J.A., 2017. Emerging preservation technologies in grapes for winemaking. Trends Food Sci. Technol. 67, 36–43.

Moreau, L., Vinet, E., 1937a. Sur la determination du pouvoir antiseptique reel de l' acide sulfureux dans les moths et les vins par la methude de l'index iode. C. R. Acad. Agric. Fr. 23, 570–576.

Moreau, L., Vinet, E., 1937b. Variations du pouvoir antiseptique reel de 1'acide sulfureux dans Ie mouts et les vins suivant l'acidite du milieu. C. R. Acad. Agric. Fr. 23, 599–607.

Nieto-Rojo, R., Luquin, A., Ancín-Azpilicueta, C., 2015. Improvement of wine aromatic quality using mixtures of lysozyme and dimethyl dicarbonate, with low SO_2 concentration. Food Addit. Contam., A 32, 1965–1975.

OIV, 2017a. Molecular sulfur dioxide. Compendium of International Methods of Must and Wine Analysis 2017 Edition. Organisation Internationale de la Vigne et du Vin, Paris, France.

OIV, 2017b. Fining using chitosan. International Code of Oenological Practices 2017 Issue. Organisation Internationale de la Vigne et du Vin, Paris, France.

Oliveira, M.C., Silva Ferreira, A.C., De Freitas, V., Silva, A.M.S., 2011. Oxidation mechanisms occurring in wines. Food Res. Int. 44, 1115–1126.

Pascual, O., Vignault, A., Gombau, J., Navarro, M., Gómez-Alonso, S., García-Romero, E., et al., 2017. Oxygen consumption rates by different oenological tannins in a model wine solution. Food Chem. 234, 26–32.

Pasteur, L., 1866. Études sur le vin. Imprimerie Impériale, Paris, France.

Petrova, B., Cartwright, Z.M., Edwards, C.G., 2016. Effectiveness of chitosan preparations against *Brettanomyces bruxellensis* grown in culture media and red wines. J. Int. Sci. Vigne Vin 50, 49–56.

Raposo, R., Ruiz-Moreno, R.J., Garde-Cerdán, T., Puertas, B., Moreno-Rojas, J.M., Gonzalo-Diago, A., et al., 2016. Grapevine-shoot stilbene extract as a preservative in red wine. Food Chem. 197, 1102–1111.

Renouf, V., Strehaiano, P., Lonvaud-Funel, A., 2008. Effectiveness of dimethlydicarbonate to prevent *Brettanomyces bruxellensis* growth in wine. Food Control 19, 208–216.

Ribèreau-Gayon, P., Glories, Y., Maujean, A., Dubourdieu, D., 2006. Handbook of enology, second ed. The Microbiology of Wine and Vinification, vol. I. John Wiley & Sons, Chichester, United Kingdom.

Santos, M.C., Nunes, C., Rocha, M.A.M., Rodrigues, A., Rocha, S.M., Saraiva, J.A., et al., 2015. High pressure treatments accelerate changes in volatile composition of sulphur dioxide-free wine during bottle storage. Food Chem. 188, 406–414.

Stafford, P.A., Ough, C.S., 1976. Formation of methanol and ethyl methyl carbonate by dimethyl dicarbonate in wine and model solutions. Am. J. Enol. Vitic. 27, 7–11.

Sudraud, P., Chauvet, S., 1985. Activité antilevure de l'anhydride sulfureux moléculaire. Connaissance Vigne Vin 19, 31–40.

Suzzi, G., Romano, P., Zambonelli, C., 1985. *Saccharomyces* strain selection in minimizing SO_2 requirement during vinification. Am. J. Enol. Vitic. 36, 199–202.

Usseglio-Tomasset, L., 1985. Chimica Enologica, II edizione. AEB, Brescia, Italia.

Usseglio-Tomasset, L., 1986. Le azioni dell'anidride solforosa sui vini. Vini d'Italia 1, 29–34.

Usseglio-Tomasset, L., Bosia, P.D., 1984. La prima costante di dissociazione dell'acido solforoso. Vini d'Italia 26 (5), 7–14.

Versari, A., du Toit, W., Parpinello, G., 2013. Oenological tannins: a review. Aust. J. Grape Wine Res. 19, 1–10.

van Vyk, S., Silva, F.V.M., 2017. High pressure processing inactivation of *Brettanomyces bruxellensis* in seven different table wines. Food Control 81, 1–8.

Waterhouse, A.L., Sacks, G.L., Jeffery, D.W., 2016. Understanding Wine Chemistry. John Wiley and Sons, Inc, Chichester, United Kingdom.

Werner, M., Rauhut, D., Cottereau, P., 2009. Yeasts and natural production of sulphites. Infowine Intern. J. Vitic. Enol. 12 (3), 1–5. Available from: www.infowine.com.

CHAPTER

22

Red Wine Bottling and Packaging

Mark Strobl

Hochschule Geisenheim University, Geisenheim, Germany

22.1 GLASS BOTTLES

22.1.1 History and Developments

Since the time that bottles could be produced by machines (late 19th century), bottles started to replace barrels more and more. The main advantages were:

- The wine drinker did not need to go to the wine barrel at the winery or to the pub anymore. A small portable amount could be transported to the table in restaurants, taverns, or to the homes.
- The selling units became affordable, in some regions even for normal workers.
- A wine bottle could be emptied. Barrels, when tapped, had an uptake of oxygen, as air replaced the tapped volume. This resulted in an oxidation of aroma components, the development of yeast and bacteria, such as *Brettanomyces*, flor yeasts, and lactic and acetic acid bacteria, so that the last wine to come out of the barrel was sour and mucilaginous (Rhein, 2012).
- The wine could be seen through the transparent glass of the bottles. Turbidity was quickly recognized, hinting at bad quality or a too young wine, that has had no time to mature. Thus, the filtration technique followed the industrial glass bottle production. Filtered wines became quality standard, even if clarity had not been achieved by maturation. Maturation in the bottle was necessary in some cases, to develop the biochemical processes to a stable status, which could be kept for years. In some cases the wine still improves in the bottle.
- Sparkling wine became possible and affordable for most people. There was no other possibility to keep high amounts of carbon dioxide in the fluids in wooden barrels or amphoras made of clay. For a long period, bottle fermentation was the only way to get wines and beers with more than 2 g CO_2/L.
- Bottles made brands possible. Emblems and names on the bottle's surface and, later printed on the paper labels were used to inform the customer about the origin of the bottle.
- Closures developed from cloth to wood to cork. Nowadays we have screw caps, crown corks, plastic caps, or glass closures with plastic sealing. Closures should assure that the contents do not drip out of the bottle and that the contents were not altered or falsified by air, microorganisms, or extortioners. Closures can seal the wine and can be additionally sealed by labels or capsules (Fig. 22.1).

22.2 TARGETS TODAY

Nowadays, bottles are transport packaging and sales packaging. Presented as a selling unit at the point of sale, a bottle should look attractive and appealing—especially in restaurants and at retailers—when compared to the bottles of the competitors. The bottle has to ensure the shelf life of the wines, which means that the look, smell, and taste of the wine will meet the consumer's expectations.

It is necessary to distinguish between old fashioned red wines with ripening, micro-oxygenation, development in the bottle on the one hand and bulk wines on the other, which are produced and sold in a defined

FIGURE 22.1 The bottling and packaging process—wine, bottle, closures, labels, carton, palette.

quality. The latter are bottled when the aroma quality is at its height and preserved as long as possible over the period of the shelf life.

22.2.1 Traditional Fashioned Red Wines Ripening in the Bottle

Particularly red wines produced using classical manufacturing processes change their aromas and taste while ripening (compare previous chapters). This process can be continued to a certain extent after bottling. If living yeast is still present at bottling, it reduces the oxygen content. Bacteria and yeasts may still ferment residual sugars to alcohols and acids. Alcohols and acids form esters, which may in the end be more pleasant than those in the wine before bottling. Polyphenols which cause bitterness, should have been oxidized during fermentation or maturation. Wooden barrels, which are not gas tight, allow the exchange of carbon dioxide out of the wine and the uptake of oxygen from the air outside a barrique. A gas exchange similar to the gas exchange through the barrel wood is possible if the bottle is sealed with natural cork. As these processes take time, some wines are kept bottled for 5 years and longer, before being sold.

The bottle should be kept horizontally, so that the cork will not dry out and embrittle. Nontight bottles with leakage can cause stain. This should preferably happen in the wine maker's cellar, rather than on the retailer's shelf or in the customer's home. The bottles at that stage should not be labeled or encapsulated.

There will be losses in water, alcohol, and volatile substances from the bottle through the cork. In the case of very long storage, the bottle should be opened at the latest after 30 years, topped up—if possible with the same wine, if not with a sulfurized similar wine or glass balls—and be closed with a new safe cork. This cork should ensure that the wine gets no cork taint after all those years. Suppliers offer corks, that are treated with carbon dioxide or microwaves to avoid anisol, the agent, causing cork taint beside halogenic atoms (compare Section 5.1).

After storage the bottles are dirty and have to be cleaned with water and brushes, before they are labeled and capsuled.

The advantage of this method is that it results in classical wines with individual characters. But each bottle can develop differently.

Key disadvantages are: a lot of time is required, fixed capital, a risk that the whole effort does not turn out in a positive way. The bottles will have a deposit of yeast, bacterials, proteins, tartaric acid, insoluble sediments, which again need to meet the consumer's expectation. The whole process is expensive and the target group must be willing to pay for all of this.

22.2.2 Modern Ready to Drink Wines and Shelf Life

Modern wines are produced in big batches. The aim is not individual bottles, but constant quality—in the case of brands, over years—without noticeable changes. Consumers' expectations have to be met and no variations are allowed. This means flavor, taste, and appearance always have to be the same. Malodor—in some cases even dominant flavors which might diminish the number of potential consumers—must be avoided.

22.3 BOTTLING LINES

Filling line processes (Fig. 22.2) developed during the last 100 years. The bottles may be set automatically on belt conveyers made of stainless steel and plastic or by hand. The bottles should be rinsed with water. Bottle

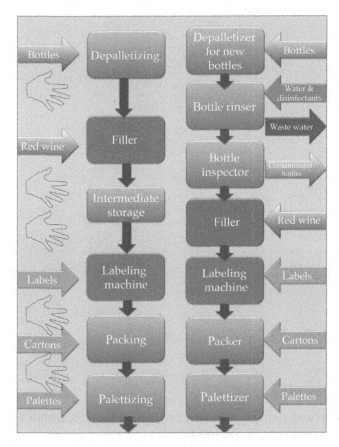

FIGURE 22.2 Bottling process in a small (left) and in a big winery (right)

rinsers wash the bottles by rinsing the inside with water or air and active ingredients. The bottle is turned bottom up. In bigger lines the bottles are fixed by a clamp arm to the radius of a big turning ring cylinder 1–5 m in diameter, so that enough time for the treatment and the rinsing off of the bottle is assured.

The bottle should be inspected empty, before passing into the filler—best by inspection machines with camera systems to detect false matter.

The filler (Fig. 22.3) itself usually is a ring bowl, filled with the wine. The wine runs by gravity into the bottles underneath. The bottles are positioned under a filling valve, pressed by lifting elements onto a gas tight sealing. The bottle can be pre-evacuated by vacuum, or prerinsed with nitrogen, before a valve opens the connection to the ring bowl with the wine. The wine runs by gravity into the bottle. It can be filled isobarometric with normal pressure or with overpressure (especially necessary for sparkling wines). There also can be a slight vacuum above the wine (-0.05 bar) in the ring bowl for filling level adjustment, where the overfilled wine is sucked up from the bottle back into the bowl.

The bottle is released, lowered, and passed to the *corker*, where the bottle is closed. After that the bottle might be stored intermediately until labeling and selling, or labeled and packed into cartons directly.

The output varies from 1000 to 50,000 btl/hour. –Two to eight employees are needed to supply and discharge the bottles, boxes, and palettes from the running line. The machines are expensive, the employees needed afford a good and equal degree of capacity utilization.

22.4 HAZARDS IN BOTTLING RED WINE

22.4.1 Microbiology

Hazard No 1 is a microbiological contamination. Microorganisms—even the mainly used *cerevisiae*—may metabolize ingredients into unwanted components and form haze by propagation. Microorganisms can be seen

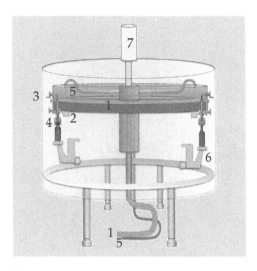

FIGURE 22.3 Main parts of a mechanical filler:
(1) *wine inlet and ring bowl*
(2) *vacuum canal for preevacuation*
(3) *crank arm for valve control*
(4) *centering and fixing of the bottle*
(5) *air or nitrogen inlet*
(6) *lifting unit*
(7) *electronics*

at a certain stage of propagation and some of them can be smelled by their metabolites. Customers will complain about such faults or just change the brand or the supplier.

To reduce the probability of microbiological growth lots of tests need to be carried out.

The wine should be produced at a *hygienic standard*, which makes the survival of potential yeast or bacteria less probable. Tanks, pipes, hoses, valves, everything in contact with the wine, should be rinsed to reduce nutrition for biofilms. This includes the periphery from the cellars to the bottling plant.

Dry wines have less sugar, so less potential spoilage is caused by bacteria and yeasts. Residual sugars are more dangerous.

Less oxygen diminishes the number of potential spoilage species that might cause problems. On the other hand, some strict anaerobic bacteria might cause sulfuric aromas, recognized as reductive notes (Back, 2008).

High alcohol amounts and low pH diminish the risk of potential growth of unwanted microorganisms. On the other hand, alcohol-free or alcohol-reduced wines are more vulnerable.

Filtration and membrane filtration decreases the amount of potential microorganisms in the wine before filling. Dead yeast cells could themselves be nutrition for wine spoilage bacteria and yeast, and should be removed by kieselghur or cross flow filtration. Cardboard filters or membrane filters, shortly before the wine is bottled, can ensure cell amounts near to zero. Membrane filtration with 0.65 μm pore size will remove the yeasts. Membranes with 0.45 μm should remove wine spoilage bacteria, if present.

Flash pasteurization uses heat for a short time to kill most wine spoilage bacteria at temperatures beneath 80°C within less than a minute, depending on ethanol concentration and pH of the wine. If oxygen is avoided (beneath 0.05 mg O_2/L) this treatment cannot be tasted. Another advantage is that all enzymes added or produced during wine growth or by microorganisms during fermentation or maturation will be inactivated. Thus, unwanted changes to the wine in the bottle can be avoided.

Plate heat exchangers are used, which recover nearly all the energy. The inflowing wine is heated up by the pasteurized wine. Flash pasteurizers are continuously working equipment and work best around the clock with big batches. Big bottlers use these techniques with success.

Very safe methods for wine export or the bottling of alcohol-reduced or alcohol-free wines are:

Hot filling: The wine is heated up to 80°C or more, and filled into the bottles. The wine with its hot alcohol also disinfects the bottle and the closure. There is definitely a change in the aroma towards a marmalade and jam odor. Oxygen—if present—will cause brown colors, Maillard products, and caramel flavors. Corks cannot be used, the vacuum after the cooling period would suck the cork into the bottleneck. Red wines are less complicated for pasteurization than white wine. White wines usually have some CO_2 for a fresh appearance, which is lost if hot filling is applied.

Pasteurization of the filled bottle: The filled and closed bottle is heated up to pasteurization temperature. To avoid thermic damage of the wine, it should be chilled again immediately after the heating process. In smaller wineries, batches of wine are put into small pasteurization cupboards or heated up in vats with hot water. Bigger producers have tunnel pasteurizers, where the bottles are heated up and chilled back again in a defined time by rinsing hot and cold water over the bottles. The wine and the bottle have to be heated up. Most of the heat energy is lost. This method is similar to hot filling. Cork cannot be used as it would be pressed out of the bottle during the heating process and would not return to a defined position after the cooling down of the pasteurized bottle.

Chemical stabilization can be done by SO_2, sorbic acid, or di-methyl-di-carbonate (DMDC, E242, Velcorin) (Dahmen, 2017). Chemical preservatives may have an influence on the aroma, for example, SO_2 binds the acetaldehyde. Sorbic acid can cause off flavors, like geranium notes. Preservatives should be dosed/added to the wine, at best, shortly before filling—in line with local laws—and should be distributed very evenly in the wine to get the optimum contact with the microorganisms to be killed. On the other hand, toxic concentrations must be avoided to protect employees and customers. The staff must be trained in how to handle and dose the chemicals safely.

Acceptable daily intake expositions/exposure must be respected, as DMDC, for example, is volatile and toxic when inhaled and harmful if swallowed (Unsupported source type (InternetSite) for source Lan17.). It must have a higher temperature while dosing it into the wine. Otherwise it will sink to the bottom and stay a poison in the bottle for the consumer, if not dosed properly at defined temperature above 20°C but beneath 30°C (Lanxess, 2015). So special dosing and tempering equipment is necessary.

Disinfection of the bottles can be done by rinsing machines. If new glass is delivered, it is probably not spoiled with wine spoilage microorganisms, depending on the package of the new glass bottles. If the palettes are opened, they should be used immediately, to avoid contamination by the air. If already opened palettes are kept for a while, the bottles must be rinsed to achieve microbiological and physical security. The bottle is conveyed into a filler like rondel. It is turned bottom up. A nozzle sprays fluids or air from underneath to the bottom of the bottle, so that the rinsing media is deviated to the walls and washes or blows the unwanted particles down to the bottleneck and out of the bottle.

Rinsing media can be: water of good quality, and preferable deionized to avoid water spots on the surface. The water should be microbiologically filtered with membranes <0.45 μm pore size, treated with ultraviolet (UV) light or chlorine-dioxide, to avoid infection of the bottles. As a *disinfectant* chlorine dioxide can be used, sulfurized water is popular in the wine industry, as residues would not be detectable beside the normal SO_2 contents in the wine. The disinfectants must be rinsed out of the bottle with water according to the ethics—or the local laws—but also to avoid contact and unwanted reactions with the product. *Steam* rinsers have advantages. Only water condensate is a residue. Steam kills microorganisms with heat, without rinsing it. Therefore steam rinsers should be combined with a water rinsing before steam treatment. Some rinsers just use pressurized air, to blow out the dust, which might have fallen into the bottles. This saves water and waste water costs. A disinfection with gas is possible with *ozone*, which is very toxic. It produces no wastewater. The waste air has to be evaporated from the hall, so as not to harm the employees working in that area.

Rinsers need time for rinsing, chemical, and thermal reactions. The bottle must be treated for at least a minute. At the end, most of the rinsing water should run off. This is achieved with huge rinsers with diameters depending on the quantity of bottles to be rinsed.

22.4.2 Chemical Contaminations

A hazard in modern bottling plants is *chemical contamination* from the cellars. Tanks, pipes, pumps, and hoses are cleaned with caustics, acids, disinfectants. Filling large quantities of product—one after another—may lead to accidentally connected tanks which contain chemicals. The craftsmen on the bottling lines usually should not drink alcohol, so the checking of the wine into the filler might be reduced to a quick look or is simply forgotten. In particular, caustic substances can, if they get into the wine bottle and not be recognized by the customer, cause alkali burn, acid burn (in the case of acid), intoxications in the case of disinfectants. Usually, red wine drinkers pour the wine into a glass, smell and review the wine before drinking it, but these potential accidents have to be examined if hazard analysis of critical control points (CCPs) concepts and *certified management systems* are practiced. There should be an exclusion of fluids other than wine by a conductivity measurement in the pipe shortly before the wine gets into the filler. Conductivity also helps to differentiate between the wines when the filling program changes from one wine to another. The conductivity also helps to optimize the cleaning programs, showing, when caustic, acid, or water is running. *Disinfectants* are a little bit more difficult, usually used in lower concentrations with less or no conductivity. Test strips are useful, like those for the detection of peracetic acid, to ensure, that all of the disinfectants are rinsed out of the pipes, before wine is filled. Certainly, a good germ-free water has to be used, to rinse out the disinfectants after their reaction time.

22.4.3 Physical Contamination

In certified enterprises that produce beverages containing alcohol, there is one point where the *consumer's security* and health really can be harmed physically. This is by foreign objects in the wine. This can just *dust* particles from transport, *insects*, *metal* from machinery or *glass splinters*.

When bottles break during transport, cleaning, or bottling and open bottles are around, it may happen that a piece of broken glass falls into one of the bottles before closing seals them.

Also, damages to the bottleneck during rinsing, filling, and corking can cause *sharp edges*, which might hurt the consumers lips, if not pouring the wine into a wine glass.

Checking the bottles just by *visual examination* before putting them on the conveyer is not feasible anymore due to the huge number of bottles filled by modern bottling lines. Machines do this much more accurately. The full UV and IR light spectrum can be used in favor of detection by the human eye. At a speed of more than 2000 btl/hours, human eyes and alertness are not good enough to find physical contaminations in the bottles.

To detect foreign matter, *inspection machines* with camera systems are offered. The problem is how to detect transparent glass splinters in a transparent glass bottle before bottling. With flash lights the shadows of splinters might be detected by the cameras and the bottle can be separated from the other ones.

Although the probability of glass contamination while filling the bottles in red wine bottles is less than with sparkling wines during pressurized filling, it might result in an accidental personal injury or just a legal dispute if broken glass leaves the winery in an original sealed bottle.

Bottle burst in the course of filling the bottle causes glass splinters whirling over open bottles. A splinter falling into an open bottle is probable. Glass splinters may land on top of the filler and move to the edge over the open bottles due to the centrifugal forces of the turning filler and then fall into the open bottles underneath, long after the bottle burst. The only method to overcome a risk of splinters in the bottles is to remove all the bottles that were in the area of the filler and had not already been closed. After the filler is emptied, it should be rinsed with water, so that no splinter remains on the fillers surface. Splinters sticking in gaskets, sticking on the filling tubes or centering units have to be considered. Some producers of fillers offer automatic *bottle burst programs* that run these steps automatically, if a bottle burst happened.

After the bottling only an X-ray check can detect foreign matter in the filled liquid (Heuft, 2016). Purchasing such a machine becomes more and more interesting, the higher the equipment performance becomes.

22.4.4 Avoidance of Oxygen With Filling Technology

Red wine might be improved by small amounts of oxygen. Oxidation of food compounds is normally considered detrimental to food quality and conservation. Considering wine, this evolution may be beneficial or detrimental. During bottle aging, the characteristics of wine, primarily based on fruit and fermentation-originated flavors, tend to evolve, leading to the appearance of the so-called developed characters as a result of the occurrence of numerous acid-catalyzed reactions. Nevertheless, wine may benefit from a certain exposure to oxygen as it contributes to color stabilization, astringency reduction, or aroma improvement. This is especially true for red wines as they contain a higher level of phenolic compounds which are the main reactants with oxygen (Silva et al., 2011). To achieve a stable product, the necessary oxidations of red wines should be finished before bottling. Once in the bottles, oxidation is more difficult to control, especially if the bottle is already on the way to the customers. At the runtime of shelf life, temperature and light can form unpleasant oxidative aromas, aldehydes, sherry-like notes and browning colors. Oxygen also enhances the probability of aerobic bacteria and yeasts.

To remove oxygen, filters, pipes, and tanks should be prerinsed with degassed water, or pressurized with nitrogen, before wine is pumped in, to avoid contact between the wine and the oxygen in the air.

In modern fillers it is possible to pre-evacuate the bottles with a vacuum, to rinse the bottles with nitrogen, and then fill the wines into the bottles. 0.05 mg O_2/L are possible. Sulfur helps, but can be reduced to a necessary minimum with consequent oxygen management.

Long tube fillers have a long filling tube, which releases the wine into the bottle about 2 cm above the bottom. When the outlet is covered by wine, only the horizontal surface of the wine has contact to the gas phase in the bottle. The wine displaces the gas out of the bottle. Lowering the bottle takes out the filling tube of the wine. The volume of the tube is partly replaced by the contents of the filling tube.

Bottles, filled with long tube fillers, have to be brought up a longer way to the sealing, as the bottle has to move up the whole length of the long tube up to the centering bell and the sealing. Also, it takes time to lower the bottle at the end of the process. If 20.000.000 bottles/year are filled, 0.02 seconds cost more than 100 hours per year. To decrease the lowering time, some wine fillers (e.g., the GmbH, Germany (KHS) Innofill Normaldruck, Rechnergesteuert Trinox (NRT)) lower the bottle while filling it. An electrical contact controls the filling level in the bottle. If the tube is drawn out of the wine throughout filling it, the tube's volume does not have to be replaced by the end of the filling process with the volume that is in the filling tube. Like in a pipette, the red wine stays in the tube and is filled into the next bottle. This saves time and avoids oxygen contact (Müller and Schmoll, 2010).

Long tube fillers are good in avoiding oxygen uptake, but undesirable if bottle shapes are often changed. Then the tubes have to be replaced with tubes that match the bottles that follow in the filling program. A further disinfection of the filler, after replacing the filling tubes by hand, will be necessary. These operations take time.

Short tube fillers just have a tube in the bottleneck, which does not fill the wine, but exhausts the suppressed gas, when the wine flows into the bottle, as it flows in a film along the wall of the bottle. This has the advantage, that the tube does not have to be replaced when bottle shapes are changed during the filling program, because they fit to most bottlenecks. The disadvantage is that the surface of the wine flowing into the bottle is greater, compared to filling with long tube fillers. So, the contact with the gas phase and oxygen in the bottle is more intensive. Short tube fillers can do preevacuation and an inert gas pressurization in order to achieve oxygen values similar to those of long tube fillers, if necessary.

SO_2 or ascorbic acid can solve the problem, but they have an impact on pH and taste, which might not be wanted. It is possible to produce wines without SO_2, but then acetaldehyde, if not masked by SO_2, will become a more dominant aroma.

Before corking *nitrogen droppers* can supersede the air still in the bottleneck above the wine (Fig. 22.4). A little drop of liquid nitrogen is dripped into the wine and evaporates immediately on the surface of the wine. The volume increases and the air in the bottleneck is replaced by the nitrogen, before the cork is pushed into the bottle or the screw cap is turned on the thread. Alternatively, a vacuum can remove most of the air, before the cork is pressed (and—in this case sucked) in.

Balloon-style bottling fillers have been introduced. A balloon is blown up in the bottle to displace the air. Then the wine displaces the gas in the balloon, so that the wine has contact with the balloon but not with the air. Long-term experiences are not available yet (Horneber, 2015).

FIGURE 22.4 Nitrogen dropper before closing the bottle.

22.4.5 Adjusting the Filling Level

Adjusting the filling level can be achieved by adjusting the filling height in the bottle. *Level controlled fillers* use the container's volume as the measuring vessel. This requires a constant bottle quality with a defined volume in a certain range.

Low vacuum fillers use a slight vacuum 0.03–0.05 bar below atmospheric pressure (Henning, 2004) in the ring bowl, that can suck the overfilled wine back into the ring bowl, until the exhaust tube is above the wine level.

Overpressure, that comes from a special canal in the filler (Trinox canal at fillers, produced by KHS) and pressurizes the wine's surface in the bottle can also press the overfilled wine up the exhaust tube into the ring bowl again, until the level in the bottle is under the level of the exhaust tube. The valves can then be closed, the pressure is released, and the bottle will be lowered and passed to the corker (Fig. 22.5).

Filling tubes, that cause an electrical circuit between an anode and cathode, with an insulation at the filling level are an electric solution. The ascending wine causes a circuit between the two electrodes on the tube. This electrical signal, caused by the ions of the wine closing a contact, gives a signal to the filling valve to close. The filling process is controlled by the filling height in the bottle and is not dependent on the pressure counterbalances or the position of the bottle in the filler.

Volumetric filling systems dose a defined volume of wine into the bottle. They divide the product into appropriately sized portions before filling it into the bottle. The wine rinses from a supply tank into a measuring cup of specified content, thus filling it (Blüml and Fischer, 2004). The metered quantity of wine flows into the bottle underneath.

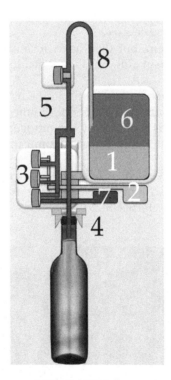

FIGURE 22.5 Filling of a wine bottle with a Trinox canal filler (by KHS):
(1) wine in the ring bowl
(2) vacuum canal for preevacuation of the bottle
(3) valves controlled by pressurized air
(4) centering unit, short filling tube in the middle. The filling tube adjusts the filling level
(5) replaced air or inert gas goes up in the ring bowl
(6) inert gas nitrogen$_2$ above the wine
(7) after reaching the filling tube, the Trinox canal puts pressure in the bottleneck, so that overfilled wine is pressed
(8) upwards into the ring bowl again

Volumetric fillers used to consist of a piston with the filling volume. This was a very accurate filling method, but not very flexible when changing the type of bottle and volumes. Moving parts and gaskets need maintenance and good cleaning.

Modern volumetric fillers use *electromagnetic flow meters*, which open a valve, measure the amount of wine flowing into the bottle and then close the valve, when the required amount has passed the measuring system.

Fillers with electric flow meters are able to measure the amount of wine by inductivity inline and close the filling valve electronically, when the targeted amount of wine is in the bottle. The needed volume can easily be altered by computers, if bottles are changed in the filling program. Furthermore, these fillers are good when red wine is filled into plastic bottles, because these are elastic and expandable and the volume has to be defined, before filling it into the bottle.

22.4.6 Filling Speed

The *filling speed* is determined by:

- the hydrostatic pressure;
- the pressure tube diameter;
- the geometry of the return gas path (Blüml and Fischer, 2004); and
- the pressure differences between ring bowl and bottle, if not isobarometric.

The amount of bottles produced per hour is determined by:

- the diameter of the filler;
- the rotation speed and the time for the filling process (limited by centrifugal forces); and
- the amount of filling valves that fit on the ring bowl (depending on the diameter of the bottles).

The amount of wine flowing into the bottle is limited by the number and the diameter of the exhaust pipes opened.

22.4.7 Automation Standards Mechanic Fillers—Electronic Fillers

Mechanical rinsers and fillers are being replaced more and more by electronic devices. While mechanical rinsers and fillers are controlled by the rotation of the filling organ and its crank arms, that open or close the valves at each turn at the same position. The time the process lasts depends on the speed of the filler and may alter. Problems occur when hydraulic shocks caused by opening valves, tanks with a high hydrostatic volume, or just by the stop and go of the filler have an influence on the filling level. This may cause overfilled or underfilled bottles.

Electronic rinsers and fillers have pneumatic controlled valves, that open or close to let or not let through the required gas or fluid, independent from the speed or the position of the filler.

The advantages of electronic controlled machines are:

- electronically controlled rinsing and filling of the bottles is independent of the speed and the position of the machine.
- each bottle can be treated for exactly the same time, with the same pressure, which leads to an optimized and consistent product quality.
- a change of bottles or product is done by an electronic signal.
- times and pressures are easily adjusted to the individual needs of the products, even while filling.

22.4.8 Packaging Materials

Packaging materials have to fulfill very different consumer expectations. Additional the ability to recycle has to be ensured in modern consumption societies. The availability of some raw materials is, in some cases, limited. Since the time bottles became the main alternative to barrels and amphoras 150 years ago, new packaging methods for wine have been introduced.

22.4.9 Glass Bottles

Glass bottles are beside the bulk wines the most used containers for red wines. This is because glass is an inert material which does not react with the wine. There is no gas migration at the glass bottle walls, only at the closure. Glass can be recycled to make new glass bottles again. Glass has a blank surface, so it can be cleaned easily. It can be sterilized and pasteurized. The wine can be seen, because glass is more or less transparent. This, however, is also a disadvantage, as light is energy and might alter some components of the wine to unpleasant aromas. Glass is heavier than other containers. Weight is expected to be valuable, so it is necessary for high price/quality wines. Weight is certainly a disadvantage during the carriage. The cracking of glass also can cause splinters that might appear inside a wine bottle. These last two factors have led to alternative packaging for red wines especially at the cheaper end of the spectrum.

22.4.10 PET Bottles

In line with distribution and recycling systems some red wines are offered in PET (polyethylene terephthalate) bottles. PET bottles can be filled similarly to glass bottles. Advantages are in mass production: the empty bottles are delivered as a small preform to the bottler. Stretch blow machines heat up the preforms and blow them to their final shape and volume using pressurized air. This saves transport volume. The empty bottle can be transported by neck handling and air conveyers to the filler. Volumetric fillers assure the filling volume. Closures are usually plastic screw caps.

The PET bottles are not intended for long storage of the wine. As the PET bottle and the plastic screw are not gas tight, water, ethanol, and aroma vaporize through the walls and the cap. This leaves a vacuum, the bottle shrinks in the course of storage. Additionally, oxygen may pass through the walls and cause unwanted changes in the aroma and solubility of some compounds which will result in clouding and sediment.

The PET bottles can and should be recycled. A certain percentage of new PET bottles can be produced out of recycled PET material. Due to the rising levels of PET bottle consumption worldwide, new materials based on sustainable raw materials are being tested, for example poly lactic acid, or Polyethylene Furanoate (Nünning, 2017).

22.4.11 Carton Packaging

Carton packaging is technically a good option for beverages like red wine. The packaging is sterile, oxygen-free, light protected, maintains a good transport volume empty and filled, and has some environmental advantages in that the cardboard is a renewable raw material (Tetrapak, 2017). Tetra Pak delivers coils of the printed packages, cardboard laminated with polyethylene and aluminum foil. The cardboard is sterilized by boiling H_2O_2. A continuous tube is formed. While the tube is welded together the wine is filled into it. The surrounding space that has to be kept sterile is small in comparison to bottle fillers. This is achieved by HEPA[1] filtered air. In the dairy industry this system fulfills the required *hygienic standard*s, so it will do the same for red wine with lower pH, alcohol, and SO_2 as preservatives. The polyethylene layer of the packages is welded by ultrasonically or electrically heated grippers and the tube is cut into brick- or prisma-shaped packages (Tetrapak, 2017). The edges are folded to the brick and fixed with the heated polyethylene of the outer layer.

The Tetra Pak system is strongly dependent on the packaging material delivered by Tetra or Combibloc GmbH (SIG), Italpack, or Elopak[2], who all offer similar packaging systems. The printing and laminating needs high accuracy. To adjust the desired colors printing can be done with rotogravure or offset printing. This demands big editions of packaging and big volumes of wine.

22.4.12 Bag in Box

The *Bota de Vino* used to be the traditional transport container especially for red wine in Hispanic countries. Animal skin is sewn together and used to transport wine in a flexible hose. The interior is coated with pine pitch for water-proofing (Sierra and Sierra, 2017). The wine can be drunk directly out of the bota by putting pressure on the bota.

A Bag in Box (BIB) contains a plastic bag instead of the animal skin, into which the wine is filled. The bags can be sterilized by gamma rays (Blüml and Fischer, 2004) or ethylene oxide. To stabilize the bag and to ensure safe transport, the filled bag is put in a carton, which may be used as a transport and selling unit. Additionally, a tap may be placed at the bottom of the box, so that the consumer only needs to open the sealed tap.

---

[1]HEPA-Filter = high efficiecy particulate air filter.
[2]SIG and Elopak have systems, where the folded container is unfolded, welded, and filled similar to bottles.

The big advantage is the filling of an empty bag, which is inflated by the wine. This has less oxygen uptake. Also the emptying needs no air to replace the tapped volume, the bags shrinks with the remaining wine (Shea and Vimont, 2017). Contamination or oxidation while tapping the wine, is unlikely. Aerobic bacteria and yeasts like acetic acid bacteria, flor yeasts, or *Brettanomyces* can be avoided for a long time.

Unlike the PET bottles, the bags may be coated with an aluminum layer, similar to the Tetra Paks. This prevents water, ethanol, and aroma losses from the wine by evaporation and migration and hinders the uptake of oxygen. The boxes can be packed directly on pallets and transported with a good volume to contents ratio. The wasted bags and boxes should be recycled. The cardboard and the plastic material can be recycled separately.

Filling machines for BIB are not very expensive for small wineries, and can be automized for bigger demands. The packaging material must be of a high quality, to avoid leakage and spoilage of the boxes.

The size of BIB system may go up to 25,000 L. Flexitainer bags use a transport container as the box. The Flexitainer system can be used to ship large amounts of wine with the advantages similar to the BIB system (Schöllhammer, 2010).

22.5 CLOSURES

22.5.1 Cork

Since bottles have been available in larger amounts, the traditional closure for red wine has been natural cork. As a result cork has had to be supplied in higher amounts since bottles are popular. Cork comes from the bark of oak trees and is therefore dependent on longer-term calculations to fulfill the consumer's demands—cork oaks need 12–15 years of growth before harvest can start.

Cork is dried and cut to fit into the bottlenecks. As a natural product it might have wrinkles and holes that might lead to leakage. It is also probable that it will not be sterile. In the last century, chlorine was used to disinfect corks and corking machines. This could lead to cork taint, as chlorine reacts spontaneously with anisol which might be present in the wine. To avoid cork taint (2,4,6-trichloranisol, TCA), it is good to avoid anisol in the cellars. Anisol is produced by mold and fungi that thrive on the volatile aromas and humidity which evaporates from wooden wine barrels. Hygiene helps. On the other hand, it is essential to avoid chlorine, and the other halogens, which all produce unpleasant aromas with anisol. Corks can be preselected from barks without mold and treated with microwaves or extracted with hypercritical carbon dioxide to reduce the possibility of TCA formation.

Natural cork is not gas tight, so there will be evaporation of water, alcohol, and volatile aroma components and the uptake of oxygen. This may be good for the development of, in particular, unfiltered wine in the bottle, especially for red wines produced in a traditional way. Micro-oxygenation continues and might develop the bitterness to a smoother sensation.

Technically, the cork is compressed in a compression box with clamping jaws, so it will fit into the bottleneck. The compressed cork is moved over the bottleneck. The headspace in the bottleneck may be evaporated by vacuum. This reduces the oxygen contents in the bottle and the pressure needed to push the cork with a plunger into the bottle. It is also possible, to add nitrogen on top of the wine in gas form, or with a drop of liquid nitrogen, which evaporates on the surface of the wine and supersedes the oxygen in the bottleneck, before the cork is pushed into the bottle by a plunger.

Extruded agglomerated corks are a cheaper cork option. The reason is that they are not very strong, so more likely to break when uncorking a bottle and giving a less reliable seal to the bottle. Molded corks (which are always micro-agglomerated) are a higher quality option and are becoming increasingly common, because they are much stronger and their reliability/consistency is much better than extruded corks (Kevin Crouvisier-Urion, 2018).

Multipiece natural stoppers are manufactured from two or more halves of natural cork glued together with an adhesive approved for use with food. These are stoppers made from thinner cork that would be insufficient to make natural stoppers from a single piece. These stoppers are of higher density (Jung, Rainer, and Zürn, 2012).

Technical stoppers were designed for bottling of wines to be consumed, in general, within 2 or 3 years. These consist of a very dense agglomerated cork body with discs of natural cork glued to its top or both ends.

Corks have to glide into the bottleneck and therefore have wax on the surface. This is also important for the removal of the cork. There are differences in the frictions and variable forces are needed to close and to open the bottles. Also, the gas migration varies and has to be specified by the cork supplier. The aim is to adjust the permeability to the amount of oxygen, so that it improves the wine by micro-oxygenation and does not harm it by oxidation. To find out the right time to enjoy the wine is not always successful.

22.5.2 Screw Caps

Screw caps are usually made of aluminum. The conveyed bottle passes a hold ready aluminum deep drawn cylinder cap with a compound mass, as a sealing in the cover. It is set on to the thread top of the bottleneck. For oxygen management the cap should be rinsed with nitrogen, before encasing the possible air in the cap into the bottle. The aluminum cap is pressed by rotating steel wheels onto the thread. The compound mass in the cap is the gasket. It is pressed on the sealing face with defined torque. The compound has little contact with the surrounding, so that gas diffusion is very slow, compared to plastic screw caps or cork. This saves the wine from impact from the surroundings, but also hinders substances from evaporating out of the bottle. All sulfuric compounds stay in the bottle and might create or keep reductive notes.

The compound mass has a valve function and releases pressure if fermentation of residual sugars takes place or the bottle is stored under warmer conditions, so that pressure might increase. As red wine has no carbon dioxide this is normally not a problem, but knocks or pressure due to the weight of pallets stacked on top of bottles might activate the valve function of the compound in the caps. The bottle is then open for air and microorganisms and does not close again. The thickness of the metal sheet that the caps are made of must be checked from time to time.

22.5.3 Vinolok

Vinolok is a glass or plastic closure, which fits in the bottleneck and is sealed by a silicone gasket. It is just put on the bottle and has to be fixed with a capsule, by pressing or shrinking the capsule onto the bottleneck. As the vinolok can be used to close the wine bottle again, the capsule is also the seal of the bottle. The gas exchange is limited by the surface of the gasket, so this is comparable to the screw caps.

22.6 PREPARING THE WINE FOR MARKET

22.6.1 Checking the Filling Level

To ensure the filling level, camera systems check the filling height in the bottle by visual control. If bottle volumes are too different they sort out underfilled bottles by cameras. Alternatively scales can be integrated in the conveyers. Gamma rays can be used to check the filling level, even if the containers are not transparent, for example in cans or kegs.

22.6.2 Treatment of the Closed Bottles

After corking and labeling some wineries put the bottles into lattice box pallets in a horizontal position. This is done if the wine should develop in the bottle for a certain time. It is also done, to detect leakages of the closures, before the wines are labeled or on the way to the customer. A further advantage for smaller wineries is that, if the wine is not sold immediately, wine mistakes or infections, which might become apparent over the period of shelf life, can still be detected in the winery before selling it and losing customers. The bottles in some cases collect dust, and can be cleaned before labeling with water and rotating brushes to improve the appearance.

In bigger wineries, labeling is done immediately after filling and closing of the bottle. The bottle is conveyed to a labeling machine.

It is necessary to ensure proper labeling, not only for marketing reasons. Particularly if alcohol-free or alcohol-reduced red wines are being produced, no other red wine should be filled into the bottles by accident. It is possible to check the labels with camera systems, so that the contents of the wine running to the filler has to match the label, glued on the bottles, to avoid unwanted problems with consumers. Also for tax reasons in some bottling plants, labeling is a CCP.

22.6.3 Labeling

The purpose of labeling is:

- to fulfill consumer's expectations, label, status, information of the origin, differentiation to other products;
- to comply with legal requirements, like the declaration of ingredients, warnings, contents, alcohol percentage;
- to advertise the brand or the vineyard;
- to make the wine retraceable, brand, vintage, vineyard, variety of grapes;
- to seal the closure to guarantee original filling before opening, i.e., avoid manipulation; and
- to advertise the wine at the point of sale.

22.6.4 Paper Labels

Paper labels are the usual way of labeling red wines. Labels used to be glued on to the bottle by hand. Nowadays, there are fast running machines that put the labels on the bottles with a speed up to 72,000 labels per hour (Krones, 2017).

The paper for high value impression and fast machines needs to have a printable surface, with a certain wet— and rupture strength. It has to be able to absorb glue, water and to be cut burr free without punching edges. The fiber direction has to be crosswise to the axis of the bottle, so that the label "grabs" the bottle. It has to be resistant to abrasion. Paper labels should be stored at 20–25°C with 60%–70% hygroscopic moisture.

The glue usually has a low viscosity at higher temperatures. The contact of the glue to the bottle surface chills the glue and has to increase the viscosity at once, so that the label adheres to the bottle. Depending on the temperature, dextrine glues are used at higher temperatures. At lower temperatures, casein glue with specific casein contents is used. The glues should have a short curing time, the amount of glue should not be too high, and they should not become fluid again, just by water condensate on the bottle.

To put the glue on the label, the casein glue is heated up to 28–32°C in a circulation pump that supplies a glue carrier on a labeling machine. The surface of the cylinder is moistened by the glue, a scraper ensures the thickness of the glue on the cylinder. The rest of the glue flows back into the glue bucket and is heated up and pumped to the cylinder again.

Labeling machines are longitudinal or rotating (for higher speeds). A glue carrier, coated with rubber material is rolled onto the glue cylinder's surface and is dampened with a thin film of glue. It moves to the label magazine, where it sticks a label with adhesive forces to the glue. Now the glue is on the back of the label, usually in stripes, so that the gaps between the stripes of glue provide space for the glue to spread into the gaps once the label is fixed to the bottle.

The label, glued to the carrier is removed by a gripper, which takes over the label from the carrier. The glue is now on the outer side, while the printed side is facing a soft sponge. The gripper cylinder rolls the label onto the bottle's surface. The sponge presses the label onto the surface of the bottle, the viscosity of the glue increases, because the temperature of the bottle's surface is lower than that of the glue, heated by the glue pump. The gripper releases the label, which is now fixed to the bottle. To fix the edges of the label to the bottle, the bottle is fixed with a centering cone and turned left and right by a carrier. Brushes mechanically spread the glue between the label and bottle surface. The glue increases viscosity and fixes the label. Now the humidity of the glue has time to evaporate through the paper and along the edges of the label.

22.6.5 Self-Adhesive Labels

Self-adhesive labels were already being used 100 years ago. They were paper labels with a rubber glue on the back, which was activated by the condensate of cold bottles coming from the cellars. The labels were fixed by hand. Nowadays, self-adhesive labels are applied by machines. Self-adhesive labels are placed on a carrier paper, which is passed over a sharp edge. As the carrier paper is drawn away from the stiffer label, the label with the glue on it is released from the carrier and juts out over the bottle conveyer. A bottle is rolled along the label, so that the label is fixed to the surface of the bottle as it leaves the machine. Just by rolling the bottle along brushes, the label is passed from the carrier feeding paper onto the bottle's surface and fixed.

With self-adhesive labels speeds of up to 60,000 labels/hours are possible (KHS, 2017). Self-adhesive labels can be produced with embossing or with transparent parts (no label look) if the label is made of plastic material.

22.6.6 Hot Glue Labeling

Hot glue labeling is more expensive, but very good for sealing the product, as the glue fixes the label to a higher extent. Nozzles spray 180°C hot resin glue onto the place where the label is immediately pressed. Especially for tax labels and seals which block the closure, this method ensures that the seal has to be broken before getting to the contents of the wine bottle (Gernep, 2017).

22.6.7 Sleeves

Sleeves are printed transparent polyethylene hoses from a coil. The hose is opened by grippers, sliced and drawn over the bottle. Here it may be fixed by water with adhesive forces. Finally it can be shrunk by hot air to the shape of the bottle. The whole bottle can be covered.

22.6.8 Alternative Labeling Systems

In the case of PET bottles, it might be advantageous to avoid glue aromas diffusing into the product. So shrinking sleeves on the bottle or wrapping the label 380° around the bottle and just glueing the overlapping parts together are possible solutions.

Direct printing without paper or plastic onto the surface of glass or PET bottles reduces paper and plastic waste and is possible at up to 36,000 btl/hours (NMP Systems, 2017).

22.7 PACKAGING

The filled and labeled containers can be the selling unit, but for transport they are packed as a bigger selling unit or into a transport unit that can be handled by forklift trucks or automatic loading systems. They must be transportable and, ever more frequently, storable in high rack warehouses with automatic pallet transport.

Transport package is:

- protection from light, dust, microorganisms, temperature, contamination;
- transport protection, logistic transport unit;
- sealing and protection against manipulation;
- presentation of the brand, advertisement; and
- selling unit.

Transport package has to assure:

- declaration of the content (machine readable) of batch and in some countries of best before and traceability, usually printed on the label by ink jet printers, or burned on the label or bottle by laser systems as a bar- or QR code;
- recycling ability.

22.7.1 Boxing and Wrapping Machines for Bottles

The wooden boxes, that were used for red wine bottles a 100 years ago, have been replaced by cartons. Cardboard has different qualities, and needs to be stable, even if a bottle leaks and the pallet gets wet.

The bottles are put into the cartons by hand, or by packers, that grab the bottles at the neck. A cone like grabber is positioned over the bottleneck, lowered and an inflatable rubber membrane inside the cone fixes the bottle by pressurized air. The bottle is lifted, placed into a carton, and the pressure of the membrane is released, leaving the bottle in the carton. Dividers can be placed between the bottles. They are made of cardboard that keeps the bottles from knocking against each other and scratching throughout the transport.

The carton lid is dotted with hot glue, the carton lid is bowed by rods over the conveyers, pressing the cardboard cover in such a way that the glue fixes the lid on the carton. The box is sealed by the glue. For continuous production the carton is wrapped around a pack of portioned wine bottles on the conveyers, kinked, and fixed by glue.

Boxes are stacked like brickwork on pallets. Overlapping of the cartons helps to ensure a cohesion of the packed unit. In some wineries the top layer is fixed to the second layer by spots of hot melted glue. The boxes should match the size of the pallets, otherwise gaps between the pallets may cause displacement of the boxes while being transported.

To give the cartons protection and stability, and to seal the unit, the whole pallet can be shrink-wrapped in a plastic sleeve or wrapped with a stretch foil. In both cases the pallet has to be included in the shrinking or wrapping process, to fix the cartons on the pallet for the transport, especially, when the lorry/truck or forklift truck has to brake or turn a corner sharply. The unit can be labeled with bar codes for high rack storage warehouses (Fig. 22.6).

22.8 ECONOMY

Bottling plants are expensive and can only be run economically if the amount of bottles to be filled is large and constant. For small wineries it might be more efficient to share a bottling plant with others as a cooperative, or to order service providers. This could be a small rinsing and bottling unit on a truck that comes to the winery, or transportation of the wine to a bottling center. Transport in tanks is certainly risky with regard to microbiological and oxygen uptake, but the risk can be limited by cleaning and disinfection of the tanks and by using nitrogen as a shielding gas, or topping up the tanks. Flexitainers, big plastic bags in containers, may also be a solution, if the material of the bags is safe against styrene tones.

Running a bottling plant means high capital costs, so it needs to be used to the maximum extent. Technical equipment runs best with well-trained employees. Essentially, a winemaker or a skilled person should always supervise the filling process and pay due attention to what is best for the red wine.

Bottling plants tend to be loud, because of empty bottles clashing together. Electronically controlled conveyers can transport bottles with a gap between them so that noise is avoided. A hygienic environment hinders the build-up of biofilms, and, if hygiene levels are maintained, the risk of contamination of the bottles, corks, and wines is minimized. Smooth/frictionless surfaces are needed for hygienic reasons but they reflect noise. So a good planning of the plants and their buildings should be carried out by specialists with experience in this genre. Baffle ceilings, noise encasing of the machines and conveyers may be necessary. The employees should—in some cases must—wear ear protection.

Logistics have to be considered. Empty bottles need nearly as much space as filled bottles (less for PET bottles or Tetra Paks). Thus, nearly the same amount of pallets has to be transported with empty bottles as with filled ones. Even if the machines run continuously, the logistics need a buffer. Empty bottle supply might be buffered outdoors, if well packed and shrinked, and the climate is mild. The filled and packed wines have to be stocked, commissioned, and distributed.

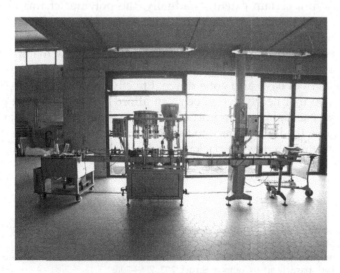

FIGURE 22.6 Filling equipment of a 12 ha winery. From the left: rinser, filler, and corker.

FIGURE 22.7 Parameters in red wine filling.

22.9 ECOLOGY

The type of packaging depends on the distribution and the size of and distance to target group. Packaging can have different impacts on the environment.

Return bottles are interesting if only one type of bottle is used in an area about 100 km maximum distance from the bottler to the customer.

One-way systems with glass recycling offer in most cases ecological advantages over return bottles. Return transport at full bottle volume, sorting and washing of the bottles is not always ecological (IFEU, 2017).

Labels can be recycled, if removed from the bottles. In a glass return systems, they burn in the glass melt and add some energy into the fluid glass. Glass recycling is real recycling, where glass bottles become glass bottles again. The recycled bottles help to save energy at glass production. PET bottles, Bag in Box, or cardboard packages can be recycled only to a certain extent. Gradually, the polymer chains of the cellulose or the plastic materials decrease, so that the recycled papers or plastics have different, less valuable characteristics.

Wine quality is certainly much better today, than 200 years ago. The success of the bottles carries on, not only in glass, but also in plastic. Red wine mostly comes in a glass bottle, but new materials are tested. Quality can be planned and maintained for a long time, so that consumer's expectations can be fulfilled for years, and bad surprises are very rare nowadays. Prices have also gone down. Effective bottling and distribution increased competition during the last 100 years, so that nearly everybody can afford and enjoy a nice good tasting red wine from all over the world. Bottling and packaging are important tools in becoming better than the competitors in quality and economy. In many cases, it is the last interface to the consumers that decides on the success of a winery and, finally the consumer decides, which red wine package and which contents best meet his or her expectations (Fig. 22.7).

References

Back, W., 2008. Mikrobiologie der Lebensmittel Bd.5: Getränke. Behr's Verlag, Hamburg.
Blüml, Susanne, Fischer, Sven, 2004. Manual of Filling Technology. Behr's Verlag, Hamburg.
Crouvisier-Urion, K., Bellat, J.-P., Gougeon, R.D., Karbowiak, T., 2018. Mechanical properties of agglomerated cork stoppers for sparkling wines: influence of adhesive and cork particle size. Compos. Struct. 203, 789–796.
Dahmen, M., 2017. Innovation und Investition im Einklang. Brauindustrie 09, 16–18.

Gernep, 2017. Hot Glue Labelling from the Magazine. ⟨http://www.gernep.de/en/labelling-systems/hot-glue-labelling.html⟩ (accessed 2017-12-10).
Henning, J., 2004. In: von Blüm, S., Fischer, S. (Eds.), Manual of Filling Technology. Behr's Verlag, Hamburg, pp. 275–285.
Heuft, 2016. "Heuft Examiner II." HEUFT SYSTEMTECHNIK GMBH. ⟨https://heuft.com/en/product/beverage/full-containers/foreign-object-inspection-heuft-examiner-ii-xac-bev⟩ (accessed 2017-11-20).
Horneber, Annette, 2015. "Fast and environmentally friendly bottling". Beer Brew. Int. 4, 26–28.
Jung, R., Zürn, F., 2012. Geisenheimer Testmethoden: Teil I Prüf-Vorschriften für die Qualitäts-Kontrolle von Weinkorken; Teil II Vorschriften für die Handhabung und Verarbeitung von Korken durch die Weinwirtschaft; Teil III Geisenheimer Prüfsiegel.
KHS, 2017. "KHS". KHS Innoket Neo SK Labeling Machine. KHS. ⟨https://www.khs.com/en/products/detail/khs-innoket-neo-sk-labeling-machine/⟩ (accessed 2017-12-10).
Krones, 2017. Krones Modular Labellers. ⟨https://www.krones.com/en/products/machines/modular-labellers.php⟩ (accessed 2017-12-10).
Lanxess Safety Data Sheet Velcorin 00673404 Version 4.01, 2015. Avialable from: < http://www.scottlab.com/uploads/documents/downloads/87/VELCORIN%20SDS%20SHEET%202014.pdf > (approved 26.07.17.).
Müller, L., Schmoll, W., 2010. New filling techniques innofill DRF merkmale. In: Intervitis IVIF-Kongress. Stuttgart.
NMP Systems, 2017. ⟨https://nmpsystems.khs.com/direct-print/⟩ (accessed 2017-12-10).
Nünning, J.ürgen, 2017. Die durchsichtige Flasche wird grün. Getränkeindustrie 11, 14–17. S.
Rhein, Stephan, 2012. Der Wein ist gesegnet. Gesellschaft für Geschichte des Weines e.V, Wiesbaden.
Schöllhammer, L., 2010. Bag in box. In: Scholle Europe GmbH (Ed.), Intervitis. Stuttgart.
Shea, P., Vimont, F., 2017. Technical specifications of wine BIB packaging. MOZ. Available from: < http://www.b-i-b.com/bib/web/downloads/PerfBIBspecTechJul07Eng.pdf > (accessed 10.12.17.).
Sierra, Lisa, Sierra, Tony, 2017. Spanish Botas – Wineskins. The Spruce -02-18 https://www.thespruce.com/spanish-botas-wineskins-3083094 (accessed 2017-12-10).
Silva, M.A., Julien, M., Jourdes, M., Teissedre, P.L., 2011. Impact of closures on wine post-bottling development: a review. Eur. Food Res. Technol. 19 (10), 905–914.
Tetrapak, 2017. Wine, A Global Love Story. ⟨https://www.tetrapak.com/us/findbyfood/wine-and-spirits/wine⟩ (accessed 2017-10-12).

Further Reading

Apcor, 2015. Apcor. ⟨https://www.apcor.pt/en/products/cork-stoppers/multi-piece-natural-cork-stoppers/⟩ (accessed 2017-12-01).
Chloe & Wine, 2014-07-14. ⟨http://www.chloeandwines.com/2014/07/defects-in-wine-really-en.html⟩ (accessed 2018-01-07).
CorkLink, 2015. CorkLink. CorkLing, 3885-482 Esmoriz, Portugal. ⟨http://www.corklink.com/index.php/agglomeratdor-natural-corks/⟩ (accessed 2017-12-01).
IFEU, 2010. Zusammenfassung der Handreichung zur Diskussion um Einweg- und Mehrweggetränkeverpackungen. ifeu - Institut für Energie- und Umweltforschung Heidelberg GmbH. 2017-08-24 ⟨http://www.umweltbundesamt.de/umwelttipps-fuer-den-alltag/essen-trinken/mehrwegflaschen#textpart-1⟩ (accessed 2017-11-19).
Lanxess, 2017. Lanxess. Velcorin. ⟨http://velcorin.com/hazards-identificatio-velcorin/⟩ (accessed 2017-12-15).

CHAPTER

23

Red Winemaking in Cool Climates

Belinda Kemp[1,2], Karine Pedneault[3,4], Gary Pickering[1,2,5,6], Kevin Usher[7] and James Willwerth[1,2]

[1]Cool Climate Oenology and Viticulture Institute (CCOVI), Brock University, St Catharines, ON, Canada
[2]Department of Biological Sciences, Brock University, St Catharines, ON, Canada [3]Département des Sciences, Université Sainte-Anne, NS, Canada [4]Institut de Recherche en Biologie Végétale, Jardin Botanique de Montréal, Montreal, QC, Canada [5]Charles Sturt University, Wagga Wagga, NSW, Australia [6]Sustainability Research Centre, University of the Sunshine Coast, Sippy Downs, QLD, Australia [7]Agriculture and Agri-food Canada, Summerland Research and Development Centre, Summerland, BC, Canada

23.1 INTRODUCTION

Cool climate regions are diverse in terms of climate and rainfall, growing season, soil types, the grape varieties grown, and wine styles. Cool climate regions generally have four seasons, diurnal temperature variations, cool autumn temperatures, and variable rainfall periods. Therefore, cool climates can limit fruit maturity in any given year due to growing season conditions and in the main, have some risk of freeze/frost injury. Regions can differ within the cool climate category ranging from cool maritime regions such as Champagne or the United Kingdom, to warmer maritime regions like Bordeaux or most of New Zealand. Continental regions such as the Niagara Peninsula and the Okanagan Valley typically have warm summers and cold winters (Shaw, 2005).

Viticulture decisions in cool climate regions that are paramount include appropriate site and cultivar selection. When planting vineyards in these regions, the viticulturist must select a site carefully, using slopes or southern aspects (Becker, 1977; Falcetti, 1994; Jackson and Lombard, 1993), with cold-resistant or early-ripening varietals, to coincide with climatic conditions and length of growing season. High wine quality is generally associated with early, even budburst, flowering and development as a result of warm springs; optimal fruit maturation with low diurnal fluctuations in temperature near harvest; and low frost damage as a result of mild winters (Gladstones, 1992; Jones and Davis, 2000). In terms of growing season requirements, the most suitable grape varieties for a cool climate region should match the length of the growing season so that fruit maturation will occur during the cool portion of the season but with enough warmth to continue accumulation of sugars and flavor development in the berries. Grapes should ideally mature at the end of the growing season in order to obtain terroir expression (Jackson and Lombard, 1993; Van Leeuwen and Seguin, 2006). Canopy management is critical in cool climate regions to improve canopy microclimate, fruit exposure, and disease control. Some common viticulture practices across cool climate regions include leaf removal (Kemp et al., 2011), shoot positioning, and hedging (Reynolds and Wardle, 1989) to meet these objectives. In recent years, more attention has been paid to improve plant material selection (Van Leeuwen et al., 2012) as well as advancements in freeze protection technology (i.e., wind machines) to mitigate damaging low temperatures and the impacts of climate change on cool climate regions (Willwerth et al., 2014).

23.1.1 Classifying Cool Climate Regions

The Winkler Index classifies regions based on the accumulation of heat summation units by adding up hours above 10°C during the growing season (Amerine and Winkler, 1944). Through this index, cool climate regions have growing degree-days (GDD) <1648 during the growing season. Huglin (1978) developed the Huglin Index (HI), a bioclimatic heat index for viticulture regions using heliothermic potential, which calculates the temperature sum above 10°C from April until September. The calculation takes into consideration daily maximum and average temperature and, slightly modifies the calculated total with the latitude of the location. Tonietto (1999) proposed a classification of viticultural climates based on the HI calculation and applied this to many viticulture areas worldwide. Very cool regions were classified as having an HI <1500, Cool regions, 1500 < HI < 1800, and Temperate regions 1800 < HI < 2100. The HI has also been utilized to match grape variety to potential ripening of berries with a content of 180–200 g/L sugar. Cool climate red cultivars ranges include Pinot noir with an HI of 1700, Cabernet franc 1800, whereas Cabernet sauvignon and Merlot have an HI of 1900 (Huglin, 1978). The Latitude Temperature Index (LTI) developed by Jackson and Cherry (1988) dealt with some shortcomings with degree-days using an index based on the latitude and mean temperature of the warmest month. It is a proxy indicator of the amount of solar energy that areas are likely to receive during the growing season. Typically, cool climate regions fall within <380 LTI. The biologically effective degree-day (BEDD) index is another variant on calculating heat summation (Gladstones, 1992). It incorporates both an adjustment for diurnal temperature range and a day length correction similar to HI. Generally, cool regions fit within the range of 1000–1600 based on BEDD calculations. The Growing Season Temperature (GST) index correlates broadly to the maturity potential for grape cultivars grown across many wine regions (Jones, 2006), and provides the basis for zoning viticultural areas in both hemispheres (Hall and Jones, 2009; Jones, 2004). A growing season average temperature index (GST) is calculated by taking the average of the growing season (Apr–Oct) and classifying the result into five groups based on maturity groups (Hall and Jones, 2009; Jones, 2004). In general, GSTs between 13 and 21°C are considered suitable for quality wine grape production where cool regions are considered to have GSTs between 13 and 15°C (Jones, 2006). This bioclimatic index has been used to study cool climate regions in Australia (Hall and Jones, 2009), the United States (Jones, 2004), and the United Kingdom (Nesbitt et al., 2016).

23.2 COOL CLIMATE GRAPE VARIETIES IN THE NORTHERN AND SOUTHERN HEMISPHERE

Cool climate short season wine regions produce red wine from red grape varieties that are primarily *Vitis vinifera* or French–American hybrids. Hybrids are generally more cold tolerant than *V. vinifera* so they are a clear choice for some colder regions with lethal or damaging winter temperatures or low GDD. Many *V. vinifera* varieties perform well in cool climates achieving optimal ripeness. Marginal production areas growing *V. vinifera* may also adjust wine style for achievable ripeness. A variety grown within its optimal climatic range will often produce higher quality table wine at the cooler end of the range and with shorter rather than longer growing seasons. For this reason, many *V. vinifera* varieties perform better in cooler coastal regions, such as of Australia, California, and South Africa, compared to the hotter inland regions, or at elevation in typically warm countries like Spain compared to the warm lower elevations (Johnson, 1985).

International recognition for *V. vinifera* wines is greater than hybrid wines so they are more easily marketed, and achieve higher price points. Grape growers in cool climates often push limits, accepting significant risk in producing *V. vinifera* varieties under marginal conditions where there is low risk of cold/freeze injury. Pinot noir is the most commonly planted red *V. vinifera* variety in cool climate wine production due to early ripening and acceptable levels of bud hardiness. Cabernet franc is a cold hardy red *V. vinifera* variety but has a longer ripening period so the number of frost-free days and GDD are important considerations (Jones, 2006). Other important cool climate red *V. vinifera* varieties are Merlot, Cabernet sauvignon, Syrah, and Gamay. Regions on the warmer side (>1500 GDD) of the cool climate spectrum can ripen other varieties but they might not be economical due to the required yield reduction for ripening fruit.

23.3 CHEMICAL COMPOSITION OF GRAPES IN COOL AND WARM CLIMATES

23.3.1 Sugar/Alcohol

The chemical composition of grapes grown in cool climates differs from warm climates with the most obvious being acidity levels followed by sugar accumulation, and thus potential alcohol. Sugar content or soluble solids (SS) is the initial indicator of ripeness and appropriate levels can be difficult to achieve with shorter growing

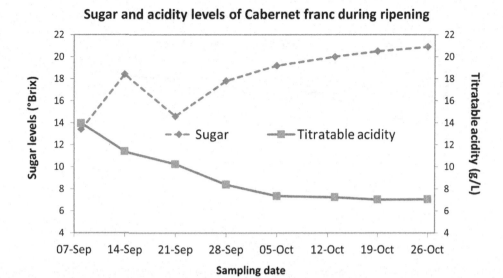

FIGURE 23.1 The increase in sugar level (°Brix) and decrease in acidity levels (titratable acidity, TA g/L) levels in Cabernet franc grape berries during ripening in 2015 in the Niagara Peninsula (Lincoln Lakeshore subappellation), Ontario, Canada.

seasons and limited heat accumulation. This is often the first determinant for cool climate production of high quality wines and achieving appropriate SS will depend on grape variety and wine style. Warm regions may have the opposite issue where SS reaches high levels before flavor, aroma, and tannin maturity is optimal. This may result in high alcohol wines where alcohol reduction may be used (Jackson and Lombard, 1993).

23.3.2 Acid

Tartaric and malic acids are the primary grape acids. Malic acid plays a central role in determining final acid concentration, pH, and sensory perception in wine. Total acid concentration typically ranges from 5 to 16 g/L and the acids accumulate from berry set to véraison (Fig. 23.1). The optimum temperature for malic acid biosynthesis is between 20 and 25°C and malate concentrations are at a maximum just prior to véraison (Conde et al., 2007). During pre-véraison, cool climate grapes produce more acid because temperatures are closer to optimum for biosynthesis. Retention of acidity post-véraison is due to cool temperatures that limit malic acid catabolism. The reduction of acidity is primarily from the consumption of malate through respiration, which uses malate. Malate is an energy compound, a central metabolite in respiration and is consumed to concentrations of approximately 2–3 g/L by maturity but malate levels can be significantly lower in warm climates (Jackson and Lombard, 1993). Retaining natural acidity in cool climate wines has advantages of reduced input costs due to acid additions, better sensory profiles without added tartrate, and color stability and retention.

Malolactic fermentation (MLF) is common practice in cool climate red wine production to reduce acidity, stabilize wine, and create body and mouthfeel (Conde et al., 2007). In cool climates, malic acid levels can be high enough that when converted to lactic acid through MLF there can be a negative sensory impact. A double salt deacidification may be required prior to the MLF to reduce malate to acceptable levels. Malic acid may also confer a green taste to fruit and wine. Co-inoculation of lactic acid bacteria with yeast for alcoholic fermentation of Cabernet franc on commercial scale wines in Switzerland reduced fermentation time (Versari et al., 2016). The authors reported important changes in the chemical parameters of wines, especially color intensity and volatile compounds. Additionally, there was a greater perception of red and ripe fruits in the co-inoculation wine, while notes of spice and herbs dominated the control wine.

23.3.3 Flavor and Aroma

The impact of temperature and light on grape flavor and aroma chemistry is not well understood. However, nonphenolic flavor and aroma compounds, such as terpenes, thiols, alkaloids, and alkane alcohols, are unquestionably dependent on temperature. In general, cooler temperatures promote biosynthesis and reduce

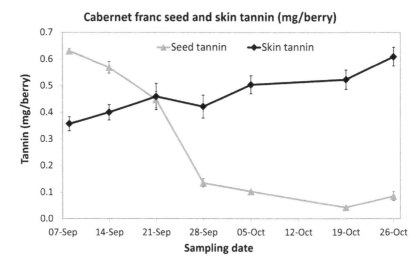

FIGURE 23.2 Skin and seed tannin concentrations during Cabernet franc grape berry ripening in 2015 (corresponding to Fig. 23.1), in a Niagara Peninsula Vineyard (Lincoln Lakeshore subappellation), Ontario, Canada. Grape skins and seeds were analyzed using the using the methylcellulose (MCP) from Mercurio et al. (2007) and expressed as epicatechin units mg berry.

degradation. An example are the methoxypyrazines (MPs) where lower temperatures promote pre-véraison accumulation and slow down post-véraison degradation (Roujou de Boubée et al., 2000; Lacey et al., 1991). The sesquiterpene pepper aroma rotundone, characteristic of Syrah and present in other red varieties, is influenced by temperature and light exposure. Light is important for the production of monoterpenes from véraison to maturity (Friedel et al., 2016). In cooler vineyards, total volatile terpenes increased at a slower rate but were at higher concentrations at maturity (Ewart, 1987). The aroma composition of grapes influenced by climate, soil texture/composition, management practices, and fermentation remains one of the most complex systems in the wine industry.

Important phenolic compounds for wine quality are tannins, anthocyanins, and flavonols that impart color, astringency, bitterness, and phenolic stability to wine. Color development is maximal between 20 and 30°C, and it is often not difficult to achieve good color development but attaining phenolic ripeness especially tannin, can be difficult in cool climates. This is due to short seasons or low heat accumulation often resulting in bitterness and astringency due to unripe tannins (Pirie, 1977; Kliewer and Torres, 1972; Spayd et al., 2002). Technological seed maturity, when grape seeds turn brown, is a critical factor in cool climates and it can often limit quality since seeds are a significant source of tannin variability in red wine production (Fredes et al., 2017). Seasonal temperature variation plays an important role in seed maturity and the complement of seed tannins.

In general, grape skin tannin concentrations increase during berry ripening, and seed tannin decreases (Fig. 23.2), although this might not be true for all sites or grape varieties due to many factors including climate and vineyard practices. Sugar levels increase and acidity levels decrease during ripening but their concentrations have not been found to correspond or be related to skin or seed tannin concentrations. Although Cadot et al. (2012) reported that advancing grape maturity of Cabernet franc was positively correlated with wine tannin and epigallocatechin, and suggested that extractability of tannins is influenced by ripening (Bindon et al., 2013).

23.3.4 "Greenness" in Red Wines

One of the challenges of red winemaking in cooler climates is the potential for high levels of green characters in the wines, mainly due to vinifying grapes that have not achieved optimal levels of physiological maturity prior to their harvest. Such green aromas and flavors are generally undesirable in red wine. At low concentrations and in a very limited number of varietal wines, such as Carménère, they may contribute to the expected sensory profile, but generally greenness is rejected by consumers. The main compounds implicated are MPs, hexanol, and some thiols, with the former the most important and the focus of this section.

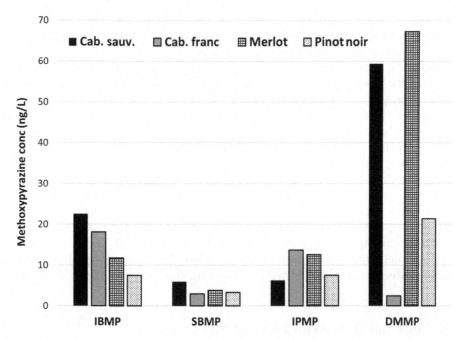

FIGURE 23.3 Methoxypyrazine composition of red wine varies with grape cultivar. Average values shown for Cabernet sauvignon (47), Cabernet franc (12), Merlot (23), and Pinot noir (31) wines. *IBMP*, *iso*butyl-methoxypyrazine; *SBMP*, *sec*butyl-methoxypyrazine; *IPMP*, *iso*propyl-methoxypyrazine; *DMMP*, dimethyl-methoxypyrazine. *Adapted from Botezatu, A., Kotseridis, Y., Inglis, D., Pickering, G.J., 2016a. A survey of methoxypyrazines in wine. J. Food Agric. Environ. 14(1), 24–29.*

23.3.5 Sources of Green Compounds

MPs are secondary metabolites that accumulate in grapes early during berry development, reaching maximum levels between pre-véraison and the end of the first stage of ripening, after which their concentration decreases, particularly during the earlier stages of ripening (Ryona et al., 2008; Sidhu et al., 2015). A second source of MPs is infestation of grapes prior to harvest from members of the *Coccinellidae* (ladybeetle) family. If either *Harmonia axyridis* Multicolored Asian Lady Beetle (MALB) or *Coccinella septempunctata* ("C7") are inadvertently incorporated at harvest along with the fruit into the processing stream, MPs from their hemolymph can produce greenness in the final wine, a fault known as "ladybug taint."

Four MPs that can be present at levels above their human detection threshold have been identified to date in red grapes and wines; *iso*butyl- (IBMP), *sec*butyl- (SBMP), *iso*propyl- (IPMP), and dimethyl- (DMMP) MP (Botezatu et al., 2016a). The composition of MPs varies with the source, with, for instance, IBMP dominant in wines made from underripe grapes, while IPMP is the most prevalent in wines affected by ladybug taint. Levels in red wine are also strongly influenced by the grape cultivar, with for instance, Cabernet sauvignon, Cabernet franc, Merlot, and Carménère containing relatively high concentrations. Fig. 23.3 gives an indication of this variability for four common red varietal wines.

While all MPs found in red wines elicit greenness or herbaceousness, their specific contribution can be nuanced, with, for instance, IBMP smelling like green peppers, whereas IPMP is often described as more peanut-like. Importantly, these MPs are particularly potent, with detection thresholds in red wine ranging from 1 ng/L for IPMP to 31 ng/L for DMMP (Pickering et al., 2007a; Botezatu and Pickering, 2012). These low values raise some analytical challenges with respect to measuring such trace compounds, but also provide a target for both viticultural and enological practices seeking to prevent or remove greenness from wine.

23.3.6 Preventing Greenness in the Vineyard

The available literature is surprisingly equivocal in regard to the impact of vineyard conditions and practices on the accumulation and degradation of MPs during berry development, with light exposure, trellising, pruning, canopy density, bud number, yield, vine vigor, temperature, humidity, and soil type all suggested to influence grape MP concentrations at harvest. Perhaps the only recommendation that can be made with some surety is that techniques applied pre-véraison that promote cluster exposure, such as leaf removal, may reduce the accumulation of some MPs. Less equivocal is the general finding that the riper the fruit, the lower the concentration of MPs. Sidhu et al. (2015) provides a comprehensive review of these studies and considerations.

With respect to MPs contributed by ladybeetles, some insecticidal spays can be effective at reducing beetle densities prior to harvest, including synthetic pyrethroids (e.g., Cypermethrin) and organophosphates (e.g., Malathion). A limitation of these sprays is the mandatory pre-harvest intervals after their use, which can allow for reinfestation to occur (Ker and Pickering, 2006). Application of potassium metabisulfite as a spray at a rate of 10 g/L has also been shown to be effective at reducing MALB density in vines by approximately a half (Glemser et al., 2012). After the grapes have been harvested, there is still an opportunity to separate the beetles from the clusters before vinification begins. Shaker tables have been particularly effective in this regard; grapes and other harvested material move along a vibrating metal mesh conveyor which "shakes" beetles but not grapes through the mesh to a collection area below, where they are disposed of. A significant limitation is that the process can only be used effectively with hand-harvested fruit, and the throughput is not high. Alternatively, some wineries have reported that grape clusters can be soaked in water, with the beetles floating to the surface where they can be removed. However, this approach may also not be suitable for machine-harvested fruit, and the resultant dilution of grape constituents, including sugars, is not ideal for quality wine. Whichever approach is used, interventions should aim to reduce beetle numbers to below approx. 1300–1900/t of red grapes; at densities greater than this ladybug taint can be evident in the finished wines (Pickering et al., 2007b; Galvan et al., 2007).

23.3.7 Remediating Must and Wine With Elevated MP Levels

Regardless of the origin of the MPs, there are several opportunities during vinifying and finishing the wine for the winemaker to manage MP levels. The distribution of MPs within the berry varies dramatically, and this information can be used to minimize their extraction. For instance, 53% of IBMP content in Cabernet sauvignon is located in the stems, 45% in the skins, 2% in the seeds, and less than 1% in the flesh (Roujou de Boubée et al., 2002). Thus, it is critical to destem prior to fermentation to minimize IBMP, SBMP, and IPMP concentrations in the final wine (Hashizume and Umeda, 1996; Roujou de Boubée et al., 2002). Given the high proportion of MPs localized in the skins of red grapes, it may be worth considering in the case of excessively underripe fruit pressing the whole grapes immediately, and making a white or blush-style wine. Thermovinification—the practice of heating red musts for a short time to 60–80°C—can result in wines with significantly lower MP concentrations, and can be considered where such technology is available.

Roujou de Boubee (2004) reported a decrease of up to 67% for IBMP in red wine from thermovinification. While not so dramatic, Kögel et al. (2015) showed reductions in all four key MPs after thermovinification of Pinot noir (Fig. 23.4). While the treatment does not remove all of the MPs, it may be sufficient, particularly if used in conjunction with other vinification options, to reduce greenness to below detection levels.

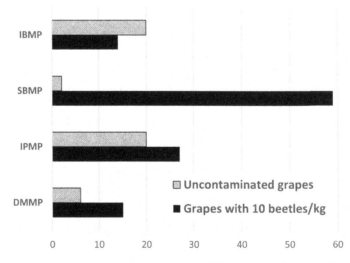

FIGURE 23.4 Reduction in methoxypyrazine content of Pinot noir wine (%) after must thermovinification of both healthy and *Harmonia axyridis* (MALB)-infested grapes. *Adapted from Kögel, S., Botezatu, A., Hoffmann, C., Pickering, G.J., 2015. Methoxypyrazine composition of Coccinellidae-tainted Riesling and Pinot noir wine from Germany. J. Sci. Food Agric. 95, 509–514.*

The choice of yeast strain for the alcoholic fermentation is one of the first and key decisions for the red wine vintner with respect to flavor and quality of the final product. Unfortunately, to date, there is no evidence that commercial yeast strains can degrade MPs during fermentation. Indeed, one study reported a significant *increase* in IPMP concentration in Cabernet sauvignon wines fermented with the Lalvin BM45 *Saccharomyces cerevisiae* yeast (Pickering et al., 2008), suggesting it should be avoided in musts with elevated MPs levels. In the same study, Lalvin D21 produced wines that were the least green in aroma and flavor, probably attributable to a masking effect from the greater fruitiness produced by this strain. Most MPs are extracted from the grapes during the first 24 hours of fermentation, so the duration of maceration may not play a large part in final MP levels (Sala et al., 2004), although in musts that contain ladybeetles this remains to be determined.

After the alcoholic fermentation, there may be further opportunities to influence greenness. Microoxygenation, the injection of small and controlled quantities of oxygen into wine, primarily to moderate tannin content, has anecdotally been reported to reduce greenness. Surprisingly, there is little evidence from the peer-reviewed literature supporting this, but the observation is widespread enough amongst winemakers to suggest it is real. Whether this is due to a reduction in MPs, thiols, and/or hexanol remains to be determined. Alternatively, the purported effect might be predominantly psychological; by reducing the "harsh" tannins associated with fruit that has not achieved optimal ripeness, the greenness that normally accompanies such wine is also perceived to be lessened. MLF does not appear to affect MP levels (Sala et al., 2004). Several fining agents have been examined for their capacity to reduce greenness in red wines, with limited success (Pickering et al., 2006). Treatment with activated charcoal resulted in a modest (11%) reduction in IPMP. This corresponded to lower intensity of asparagus flavor in the wine, but several other green attributes were not significantly changed by the charcoal treatment. Instead, the use of oak chips gave the most consistent reduction across the spectrum of green/herbaceous aromas and flavors in this wine. This is likely due to a masking effect from oak-derived volatile compounds, and the judicious use of well-seasoned oak is indicated in wines with elevated greenness.

Some less conventional materials hold promise, with silicone-treatment of both red musts (Ryona et al., 2012) and wine (Botezatu and Pickering, 2015) yielding large reductions in several MPs, and importantly, with minimal change to nontarget volatile compounds (Ryona et al., 2012; Botezatu et al., 2016b). Similar results have been observed in a Merlot wine with elevated levels of IBMP, SBMP, and IPMP when treated with a biodegradable polylactic acid-based polymer (Botezatu et al., 2016b). How these polymers can best be integrated into the winemaking process remains to be determined, but they do show considerable promise. There are final opportunities for winemakers to influence wine composition through decisions on packaging, closure types, and storage. Compared with using bottles, red wine finished in Tetra Pak cartons contained significantly lower MP concentrations across an 18-month storage trial (Blake et al., 2009), although the rapid loss of free sulfur dioxide observed with Tetra Pak should factor into packaging and finishing decisions. In the same study, closing Cabernet franc wine with molded synthetic corks resulted in a greater decrease in MPs than wine closed with natural cork or screw-cap, particularly later in the storage trial, presumably due to sorptive processes. Finally, simple aging of bottled red wine can reduce IBMP and IPMP by up to 30% (Pickering et al., 2005; Blake et al., 2009, 2010), possibly through binding with polyphenols (Sidhu et al., 2015), although a corresponding loss of greenness may not be observed (Pickering et al., 2005) due to concurrent changes in other volatile constituents during aging. Given these limitations in remediating green wines, prevention in the form of viticultural techniques and interventions that promote optimal grape ripeness and prevent ladybeetle infestation are key.

23.4 INNOVATIONS IN COOL CLIMATE WINEMAKING

Several winemaking practices have been shown to increase the concentration of phenolic compounds in finished wines including: higher fermentation temperatures, thermovinification protocols that involve short-term heating of skins to 60–70°C, must freezing and treatment with pectolytic enzymes, freezing with dry ice and steam blanching, and reducing grape skins to 6% of their original size without damaging seeds (Accentuated Cut Edges, ACE) (Rio Segade et al., 2014; Sacchi et al., 2005; Sparrow et al., 2016; Harrison, 2017). Developed in the cool climate region of Tasmania, the innovative ACE process involves grape skin cut into fragments to assist with the extraction of phenolic compounds from the subsequent broken edges (Sparrow et al., 2016) tested at the laboratory scale with Pinot noir grapes, is now available on a commercial scale. Judicial use on thick-skinned red grapes with higher skin phenolic content than Pinot noir may be required by alterations to its severity or timing.

Pinot noir winemaking can be challenging due to low skin anthocyanin and tannin content, which has led to not only ACE, but also the study of microwave maceration as a novel thermovinification method to improve phenolic composition in the wine. Microwave maceration with early pressing has been suggested as a promising alternative to Pinot noir fermentation on pomace for 7 days (Carew et al., 2014). Nevertheless, the method has yet to be adopted on a commercial scale likely due to equipment restrictions. Although used by winemakers to assist with phenolic extraction, cold soaking of Pinot noir must after crushing and destemming reduces effective winery throughput, and escalates energy use, thus production costs are increased (Carew et al., 2014). Lukić et al. (2017) reported that cold prefermentation maceration lowered seed tannin extraction from Teran grapes in Croatia, and post-fermentation heating of wine increased total phenols, flavonoids, monomeric, oligomeric, and polymeric flavanols, color intensity and hue. Saignée (removing juice before fermentation) and prefermentation heating of grapes/must with extended maceration lowered total phenols compared to cold prefermentation maceration (Harrison, 2017). Fischer et al. (2000) used manual punch-down, mechanical punch-down, pumpover, and fermentation in a spiral rotor tank with Pinot noir, Dornfelder, and Portugieser grapes in Germany. The impact the fermentation technology had on phenolic compounds was dependent upon the grape variety. For example, the total monomeric anthocyanin extraction was greater for mechanical punch-down than mechanical pumpover for Pinot noir and Dornfelder than for Portugieser grapes (Setford et al., 2017). Chittenden et al. (2015) found that the mechanical disruption of the fermenting cap of Merlot grapes extracted less phenolic material than no mechanical disruption of the cap. These studies demonstrate that grape varieties vary in their response to phenolic management techniques due to their grape phenolic compositional differences.

Optical sorting, thermovinification, flash détente, micro-oxygenation, prefermentation enzymes, specific yeasts for phenolic extraction and/or flavor complexity, stem or whole berry inclusion, mannoprotein additions, and tannin addition are also used in cool climate red winemaking (Smith et al., 2015). Although depending on grape variety, the rate and timing of addition and specific product content (i.e., oak, skin, or seed tannins), wines could result in higher astringency than is sensorially acceptable. Grape variety, timing, duration, and temperature of the maceration are the most important factors that modify the diffusion of varietal aromas and phenolic compounds into must and juice (Sacchi et al., 2005).

23.4.1 Appassimento-Style Red Wines

Post-harvest grape drying/dehydration is a traditional enological technique, referred to as "Appassimento" technique in Canada. It is used to complement red grape maturation, increase potential alcohol, and concentrate chemical composition/flavor (Inglis et al., 2016; Panceri et al., 2013). The severity of chemical changes during drying depend upon variety, maturity stage at harvest, cultural practices, and the management of the dehydration process (Rio Segade et al., 2016). Using a controlled method to dry grapes can substantially shorten the time for commercially ripe grapes to reach the desired sugar level (depending on environmental factors) (Inglis et al., 2016). Due partially to the decline of the Ontario tobacco industry, the use of kilns for red grape drying has increased in popularity. However, several drying methods have been investigated, including barns, greenhouses, natural sunlight, closed chambers with warm or cool air, and specifically designed drying chambers (Inglis et al., 2016; Chen et al., 2016; Panceri et al., 2013; Barbanti et al., 2008; Bellincontro et al., 2004; López de Lerma et al., 2013). The resulting prefermentation must chemical composition is dependent upon the grape variety (not suitable for thin-skinned varieties), but requires a high alcohol tolerant yeast for fermentation. Differences in grape dehydration kinetics and final chemical composition have been reported in studies from the different techniques due to the drying method, duration of drying, grape variety, temperature, and humidity levels. Post-fermentation, wines can either be blended with a percent of red wine produced from nondried grapes or be bottled as a standalone wine. The resultant red wines tend to be fuller-bodied with higher alcohol and phenolic compounds than their counterparts made from grapes picked at commercial ripeness without drying.

23.4.2 Red Icewine

Icewine is produced from red grapes that have been frozen on the vine and pressed while still frozen, resulting in a dessert wine highly concentrated in sugar and flavor compounds (Soleas and Pickering, 2007). Known as Eiswein in Germany and Austria, its production is most dominant in Canada, with smaller quantities produced in other wine regions including China (Li et al., 2016). Compared with dry red wines there is a low phenolic concentration in red Icewines due to not destemming or crushing the grapes, and less contact with berry skins and seeds during fermentation (Li et al., 2016). Without the pre- and/or post-fermentation maceration process, phenolic extraction from skins and seeds into the grape juice is limited although color compounds are released. The term "Icewine" is trademarked and refers to ice wine that has been produced in Canada according to strict standards of quality control, enforced by the Vintners Quality Alliance (VQA), which came into force in 2000. Quality parameters such as temperature at harvest (<−8°C), juice soluble solids (average >35 °Brix) and minimum wine residual sugar (125 g/L) are regulated under VQA, and these standards have contributed significantly to the success of the style (Pickering, 2006). Cabernet franc and Cabernet sauvignon are the two dominant varieties used for red Icewine, although Syrah is also used in British Columbia. Red Icewines have similar compositional values for juice and wine analytes as white varieties, except for the concentration of phenolic compounds (Pickering, 2006). Titratable acidity (TA g/L) is elevated during berry dehydration, with final levels in the wine of around 11 g/L. Nevertheless, acidulation of the juice or wine with up to 4 g/L of acid is permitted in Canada, as some varieties require small additions to improve the sweetness:sourness balance, and to a lesser extent to maintain an acceptable pH (Pickering, 2006, Pickering and Inglis, 2013).

23.5 MAKING WINE FROM RED INTERSPECIFIC HYBRID AND FUNGUS-RESISTANT VARIETIES

Recent focus on pesticide reduction in viticulture increased the interest for interspecific hybrid (IH) varieties especially those resistant to fungal diseases (Pedneault and Provost 2016). IH result from interspecific crossbreeding between the Mediterranean species *V. vinifera*, and North American and Asian *Vitis* spp. such as *Vitis riparia*, *Vitis amurensis*, and *Vitis rupestris* that carry high resistance to fungal diseases, including powdery and downy mildews, and gray rot (Pedneault and Provost, 2016; Rice et al., 2017). Issued from traditional breeding, the first fungus-resistant IH grape varieties carried a significant percentage of non-*V. vinifera* species in their genetic material (Sivčev et al., 2010).Recently, marker-assisted selection combined with multiple backcrossing with *V. vinifera* varieties allowed the development of fungus-resistant varieties (FRG) carrying both disease-resistance genes and a significant percentage (more than 85%) of *V. vinifera* in their pedigree; those are generally referred to as "PIWI" rather than IH (from German: Pilzwiderstandsfähige, "disease resistant") and are accepted as *V. vinifera* varieties in European catalogues (Sivčev et al., 2010). In some cases though, "PIWI" indistinctly refers to both interspecific hybrids and newer "fungal-resistant *V. vinifera*" varieties (Pedneault and Provost, 2016).

The particular biochemistry of IH results in specific challenges in wine production (Ehrhardt et al. 2014; Pedneault and Provost, 2016). Indeed, the introduction of fungus resistance in grape varieties involves the addition of certain biological features such as thicker berry skin and high production of pathogenesis-related proteins such as thaumatin-like proteins and chitinases that may affect the development of the sensory attributes of wine (Van Sluyters et al., 2009). IH cultivars can be organized in two categories: (1) winter hardy varieties from Midwestern, and Northeastern United States and parts of Canada (i.e., Quebec and Nova Scotia); and (2) hot climate varieties from Southern and Western United States. IHs developed for cold-climate wine production have specific characteristics like the ability to be winter hardy in temperatures of −20 to −30°C i.e., Marquette, Frontenac, and ripen early (Rice et al., 2017; Willwerth, 2016; Provost et al., 2012; Pedneault and Provost, 2016; Akkurt et al., 2006). Those additional genetic characters may also affect berry and wine composition.

23.5.1 Acidity, pH and Potassium

Juice acidity has both a cultivar- and a climate-specific component. Certain cultivars such as St. Croix usually have low acidity (TA 5–7 g/L, in tartaric ac. eq.), whereas others such as Frontenac generally have acidity values in the higher range (TA ≥10 g/L, in tartaric ac. eq.) (Pedneault and Provost, 2016). Acidity from malic acid may vary according to ripening conditions but different ratios of tartaric to malic acid as well as variable concentration levels of those acids are also found within IH grape cultivars (Haggerty, 2013). The challenge with acidity management in IH winemaking is that high acidity often couples with high pH value, which is partly attributable to the tendency of many IH to accumulate high levels of potassium (Dharmadhikari and Mansfield, 2012). Attempts to manage acidity and pH in IH berries using viticulture techniques have often been unreliable from 1 year to another (Pedneault and Provost, 2016). Thus, deacidification has become an essential tool in IH-based cool-climate wine production. Deacidification can be carried out chemically, to remove tartrates using carbonate salts (e.g., $KHCO_3$, K_2CO_3, $CaCO_3$) or biologically, to remove malic acid through microbial metabolism (Jackson, 2008a; 2008b). Biological deacidification can be achieved with deacidifying yeast strains i.e., *Schizosaccharomyces pombe*, but it is generally carried out with MLF (Dharmadhikari and Mansfield, 2012; Jackson, 2008b). Recently, the selection of non-*Saccharomyces* yeast strains (*Pichia kudriavzevii*) with deacidification capacity suggested promising new venues for IH wine (Del Mónaco et al., 2014). In Eastern Canada (i.e., Quebec, Nova Scotia), biological deacidification (i.e., malolactic fermentation) is widely used for red IH wine production. Chemical deacidification is also used but limited by the fact that this technique is complex and necessitates a high level of accuracy in order to lower acidity while maintaining pH. One of the simplest and most utilized ways to manage acidity among small wineries is to blend high-acid cultivars with low-acid ones.

23.6 YEAST ASSIMILABLE NITROGEN

YAN level (mg N/L) of IH varieties generally ranges between 150 and 300 (Pedneault and Provost, 2016), but studies by Stewart (2013), Mansfield (2015), and Slegers et al. (2015) showed that YAN may range from 89 to 938 mg N/L depending on environmental conditions, year, location, berry maturity, and cultivar. YAN values in the upper range favor fast fermentation that are generally assumed to result in simpler wine bouquet. High YAN fermentation may also lead to unwanted ethyl carbamate, a potential carcinogen compound, and biogenic amines such as histamine and phenylethylamine, that can contribute to cause headache and nausea (Stewart, 2013). For these reasons analyzing YAN concentration in must and avoiding systematic addition of yeast nutrient as often occurs in *V. vinifera* winemaking, is highly recommended in IH winemaking. Again, to avoid these issues, small wineries sometimes choose to blend high-YAN with low-YAN musts prior alcoholic fermentation.

23.7 TANNINS AND ANTHOCYANIN

Tannin and anthocyanin concentration and profile are among the main differences between wines made from IH and *V. vinifera* varieties. Tannin concentration is generally higher than anthocyanin concentration in red *V. vinifera* wine, whereas the opposite occurs in red IH varieties (Pedneault and Provost, 2016). Despite these compositional differences, typical red winemaking protocol can, in theory, lead to some tannin extraction in red IH wine. However, it is well known among winemakers that wine made from most IH varieties have very low astringency (Pickering and Pour Nikfardjam, 2007; Pour Nikfardjam and Pickering 2008). Indeed, it has been shown that most traditional winemaking processes aimed at tannin extraction lead to limited tannin but high anthocyanin concentration in red IH wine (Table 23.1) (Manns et al., 2013).

Recent development have revealed that the key to astringency in IH wine might be tannin retention rather than extraction from berries (Springer and Sacks, 2014). Springer and Sacks (2014) showed that proteins and pectins from cell walls highly contribute to tannin binding and precipitation in model wine, and that such interactions are critical to tannin retention in IH wine. The protein concentration in the juice of seven IH varieties (52 to 199 mg/L) was found to be significantly higher than that of *V. vinifera* juices (11 to 123 mg/L) (Springer et al., 2016). In the same study, winemaking treatments such as heating or flash freezing the juice, adding tannins (0.6 g/L), or adding bentonite successfully decreased the level of protein in Maréchal Foch juice compared to the control, but no improvement was reported regarding tannin retention (Table 23.1) (Springer et al., 2016).

Color intensity and stability is another well-known issue in IH varieties. The color instability of IH wine is related to their low concentration in polymeric tannins and their high concentration in anthocyanin diglucosides (Manns et al., 2013; Nicolle et al., 2018a). In Midwestern and Northeastern United States, issues with color stability have been reported for a number of IH cultivars, including Maréchal Foch, De Chaunac, and Marquette (Coquard Lenerz, 2012). In a study combining various proportions of white pomace (*Vitis sp.* "Vidal") cofermented with red pomace (RP) and juice (*Vitis sp.* "Frontenac"), Nicolle et al. (2018a) found that a proportion of 30% RP improved color stability in the resulting wine analyzed 1 year after bottling. Manns et al. (2013) suggested that phenolic compounds such as hydroxycinnamic acids and flavonols could play a significant role in the stabilization of diglucoside anthocyanin in IH wine. Nicolle et al. (2018) also observed a decrease in color intensity depending on the proportion of white pomace added, suggesting that IH wine color could be modulated using this technique.

23.7.1 Aroma

Since the very establishment of IH development, issues with atypical wine aroma (e.g., "foxy," "labrusca" flavors) were pointed out (Ribéreau-Gayon et al., 2006). Fuleki (1974) showed that carbonic maceration reduced

TABLE 23.1 Impact of different winemaking processes on chemistry (phenolic and volatile compounds) and sensory perception of red wines made from interspecific hybrid grape varieties

Treatments	Variety	Yeast strain	Impacts on wine — Phenolic and volatile compounds	Sensory perception	Ref.
Phenolic compound extraction using the following treatments: • Cold-soak • Enzyme addition at crush • Tannins addition at crush • Hot press • Skin fermentation (7 days)	Maréchal Foch	Lalvin ICV-GRE	Hot press increased the level of nonanthocyanin phenolic compounds and the level of anthocyanins; Tannin addition and hot press increased tannin concentration but had no impact on their mean degree of polymerization		Manns et al. (2013)
	Corot noir	Lalvin ICV-GRE	Tannin addition increased tannin concentration but hot press decreased it; Neither treatment impacted the mean degree of polymerization		Manns et al. (2013)
Phenolic compound extraction using the following treatments: • Hot press • Skin fermentation (7 days)	Marquette	Lalvin ICV-GRE	Hot press increased anthocyanin monoglucosides and tannin concentrations		Manns et al. (2013)
Phenolic compound extraction using the following treatments: • Immediate press for juice • Immediate press for wine • Hot press for juice • Hot press for wine • Skin fermentation (7, 13, and 21 days)	Chambourcin	Prise de mousse	Hot press increased color intensity (A520 nm); 13 and 21 days skin fermentation decreased color intensity compared to 7 days; Immediate press for both juice and wine increased browning; Hot press increased total phenol concentration		Auw et al. (1996)
Winemaking trials: • CM at 15 or 27°C, for 1 or 2 weeks • Skin fermentation • Hot press	Concord	No starter in CM; Unknown starter used in other trials	Carbonic maceration reduced the occurrence of bluish tones, the concentration in total phenols and decreased the concentration in methyl anthranilate	Carbonic maceration decreased the intensity of typical *V. labrusca* flavor	Fuleki (1974)
Two distinct winemaking experiments: 1. Standard fermentation with varying percentage of whole berries; 2. CM with whole clusters (2, 4, 8 days)	Maréchal Foch	Epernay II	100% whole berries and whole clusters decreased total phenols and color; Extended CM (8 days) increased total phenols		Miller and Howell (1989)
Addition of OptiRed (inactivated yeast derivative) at the beginning of fermentation (19 days, on-skin) compared to control (without)	Baco noir	EC1118	OptiRed increased the concentration of dimers procyanidins	OptiRed had no perceptible impact on wine mouthfeel	Pour Nikfardjam and Pickering (2008), Pickering and Pour Nikfardjam (2007)
	Maréchal Foch	EC1118	OptiRed increased the concentration of dimer and trimers procyanidins	OptiRed had no perceptible impact on wine mouthfeel	Pour Nikfardjam and Pickering (2008), Pickering and Pour Nikfardjam (2007)

(Continued)

TABLE 23.1 (Continued)

Treatments	Variety	Yeast strain	Impacts on wine		Ref.
			Phenolic and volatile compounds	Sensory perception	
Four treatments carried on juice, prior adding grape pomace back and carry alcoholic fermentation, compared to control: • Heating the juice to 95°C • Flash freezing the juice on dry ice • Tannin addition (0.6 g/L) • Bentonite addition (6%)	Maréchal Foch	Lalvin ICV-GRE	All treatments significantly reduce the protein content of juice and bottled wine; Tannin addition increased tannin content from "undetectable" ($\geq$100 mg/L) to 145 mg/L		Springer et al. (2016)
Co-fermentation of different ratio of WP and RP in red juice: • 50% w/w RP in juice (control) • 30% w/w RP and 6% w/w WP in red juice; • 30% w/w of RP and 12% w/w WP in red juice; • 30% w/w of RP and 18% w/w of WP in red juice; • 23% w/w WP in red juice	Frontenac (white pomace from Vidal)	Anchor NT50	WP addition increased the concentration of monomeric and oligomeric tannins in the wines Different WP/RP ratio modulated the anthocyanin profile of wine and successfully faded red IH wine A ratio of 30% RP to 6% WP improved color stability Different proportion of white to red pomace modulated wine aroma profile		Nicolle et al. (2018)
Kinetic of protein, tannin and anthocyanin extraction/retention in red wine following pre-fermentative maceration, using a factorial design: • Must treatment (untreated, bentonite-treated and heat-treated musts) • pomace (must fermented with or without pomace) • Tannin addition at 0, 3, 6, and 9 g/L	Frontenac	Lalvin BM 4 × 4	Bentonite addition significantly decreased the protein content in juice (day 4 of alcoholic fermentation) and wine Treatments fermented without pomace showed a better tannin retention than those fermented with pomace Must treated with heat had higher total pigment than other treatments, either with or without pomace		
MLF trial:Noninoculated control compared to inoculation with the Leuconostoc oenos strains ML-34, PSU-1, and LS-5A	Chancellor, DeChaunac, Maréchal Foch	Montrachet		Wines fermented with PSU-1 and LS-5A were preferred over ML-34 wines; All inoculated wines were preferred to the non inoculated control	Giannakopoulos et al. (1984)

CM, carbonic maceration; IH, interspecific hybrid; MLF, malolactic fermentation; RP, red pomace; WP, white pomace.
Adapted and updated from Pedneault, P., Provost, C., 2016. Fungus resistant grape varieties as a suitable alternative for organic wine production: benefits, limits, and challenges. Sci. Hortic. 208, 57–77.

foxy flavous in IH wine. In recently bred IH varieties, especially cold-hardy ones, foxy flavous have been largely removed by favouring *V. riparia* over *V. labrusca* to bring cold-hardiness characters to off springs (Sun et al., 2009). Yet the herbaceous character is frequently found in cold-hardy IH varieties. Slegers et al., (2015) showed that the concentration of herbaceous compounds derived from unsaturated fatty acids in berries is highly correlated with the concentration of herbaceous compounds in the resulting wine. Both ripening and growing conditions (e.g., temperature) have a significant impact on the concentration of herbaceous compounds in berries (Pedneault et al., 2013). But the volatile profile of IH varieties, including their potential in developing herbaceous and/or fruity notes in wine shows large variation from one cultivar to another (Slegers et al., 2015). Consequently, blending is an asset to improve the quality of IH wines. The improvement of wine quality is a constant target for the wine industry, especially for IH varieties that are mostly unknown to consumers. The particular biochemical composition of IH suggests that many traditional winemaking processes may be unsuited for the production of quality wines from those varieties. Thus, further research is needed to fill this gap and improve wine production in northern areas.

References

Akkurt, M., Welter, L., Maul, E., Töpfer, R., Zyprian, E., 2006. Development of SCAR markers linked to powdery mildew (*Uncinula necator*) resistance in grapevine (*Vitis vinifera* L. and *Vitis* sp.). Mol. Breed. 19 (2), 103–111.

Amerine, M., Winkler, A., 1944. Composition and quality of musts and wines of California grapes. Hilgardia 15 (6), 493–674.

Auw, J.M., Blanco, V., O'keefe, S.F., Sims, C.A., 1996. Effect of processing on the phenolics and color of Cabernet sauvignon, Chambourcin, and Noble wines and juices. Am. J. Enol. Vitic. 47 (3), 279–286.

Barbanti, D., Mora, B., Ferrarini, R., Tornielli, G.B., Cipriani, M., 2008. Effect of various thermo-hygrometric conditions on the withering kinetics of grapes used for the production of Amarone and Recioto wines. J. Food Eng. 85, 350–358.

Becker, N., 1977. Experimental research on the influence of microclimate on grape constituents and on the quality of the crop. The OIV Symposium on Quality of the Vintage, Oenological and Viticulture Research Institute, Cape Town, South Africa.

Bellincontro, A., De Santis, D., Botondi, R., Villa, I., Mencarelli, F., 2004. Different postharvest dehydration rates affect quality characteristics and volatile compounds of Malvasia, Trebbiano and Sangiovese grapes for wine production. J. Sci. Food Agric. 84, 1791–1800.

Bindon, K., Varela, C., Kennedy, J., Holt, H., Herderich, M., 2013. Relationships between harvest time and wine composition in *Vitis vinifera* L. cv. Cabernet sauvignon 1. Grape and wine chemistry. Aus. J. Grape Wine Res. 138, 1696–1705.

Blake, A., Kotseridis, Y., Brindle, I.D., Inglis, D., Sears, M., Pickering, G.J., 2009. Effect of closure and packaging type on 3-Alkyl-2-methoxypyrazines and other impact odorants of Riesling and Cabernet Franc wines. J. Agric. Food Chem. 57, 4680–4690.

Blake, A., Kotseridis, Y., Brindle, I.D., Inglis, D., Pickering, G.J., 2010. Effect of light and temperature on 3-alkyl-2-methoxypyrazine concentration and other impact odourants of Riesling and Cabernet Franc wine during bottle ageing. Food Chem. 119, 935–944.

Botezatu, A., Pickering, G.J., 2012. Determination of ortho- and retronasal detection thresholds and odor impact of 2,5-dimethyl-3-methoxypyrazine in wine. J. Food Sci. 77 (11), S394–S398.

Botezatu, A., Pickering, G.J., 2015. Application of plastic polymers in remediating wine with elevated alkyl-methoxypyrazine levels. Food Addit. Contam. 32 (7), 1199–1206.

Botezatu, A., Kotseridis, Y., Inglis, D., Pickering, G.J., 2016a. A survey of methoxypyrazines in wine. J. Food Agric. Environ. 14 (1), 24–29.

Botezatu, A., Kemp, B.S., Pickering, G.P., 2016b. Chemical and sensory evaluation of silicone and polylactic acid-based remedial treatments for elevated methoxypyrazine levels in wine. Molecules 21, 1238.

Cadot, Y., Caille, S., Samson, A., Barbeau, G., Cheynier, V., 2012. Sensory representation of typicality of Cabernet franc wines relate to phenolic composition: impact of ripening stage and maceration time. Anal. Chim. Acta 732, 91–99.

Carew, A., Gill, W., Close, D.C., Dambergs, R.G., 2014. Microwave maceration with early pressing improves phenolics and fermentation kinetics in Pinot noir. Am. J. Enol. Vitic. 65 (3), 401–405.

Chen, Y., Martynenko, A., Mainguy, M., 2016. Wine grape dehydration kinetics: effect of temperature and sample arrangement. CSBE/SCGAB 2016 Annual Conference, Halifax, Nova Scotia, Canada, July 3–6, 2016.

Conde, C., Silva, P., Fontes, N., Dias, A.C.P., Tavares, R.M., Sousa, M.J., et al., 2007. Food 1 (1), 1–22.

Coquard Lenerz, C., 2012. Phenolic Extraction from Red Hybrid Winegrapes. MSc Thesis, Cornell University, USA.

Del Mónaco, S.M., Barda, N.B., Rubio, N.C., Caballero, A.C., 2014. Selection and characterization of a Patagonian *Pichia kudriavzevii* for wine deacidification. J. Appl. Microbiol. 117 (2), 451–464.

Dharmadhikari, M., Mansfield, A.K., 2012. Managing Acidity in Cold Climate Winemaking Presented at the Northern Grapes Project Webinar Series. Retrieved from http://youtu.be/GE8FPuywBNE.

Ehrhardt, C., Arapitsas, P., Stefanini, M., Flick, G., Mattivi, F., 2014. Analysis of the phenolic composition of fungus-resistant grape varieties cultivated in Italy and Germany using UHPLC-MS/MS. J. Mass Spectrom. 49 (9), 860–869.

Ewart, A.J.W., 1987. Influence of vineyard site and grape maturity on juice and wine quality of *Vitis vinifera*, cv. Riesling. In: Lee TH (Ed.), Proceedings of Sixth Australian Wine Industry Conference, Adelaide, 71–74.

Falcetti, M., 1994. Le Terroir. Qu'est-ce qu'un Terroir? Pourquoi l'étudier? Pourquoi l'enseigner? Bull. l'OIV 67 (757-58), 246–275.

Fredes, C., Mora, M., Carrasco-Benavides, C., 2017. An analysis of seed colour during ripening of Cabernet sauvignon grapes. S. Afr. J. Enol. Vitic. 38 (1), 38–45.

Friedel, M., Frotscher, J., Nitsch, M., Hofmann, M., Bogs, J., Stoll, M., et al., 2016. Light promotes expression of monoterpene and flavonol metabolic genes and enhances flavour of winegrape berries (*Vitis vinifera* L. cv. Riesling). Aus. J. Grape Wine Res 22 (3), 409–421.

Fuleki, T., 1974. Application of carbonic maceration to change the bouquet and flavour characteristics of red table wines made from concord grapes. J. Inst. Can. Sci. Technol. Aliment. 7, 269–273.

Galvan, T., Burkness, E., Vickers, Z., Stenberg, P., Mansfield, A., Hutchison, W., 2007. Sensory-based action threshold for multicolored Asian lady beetle-related taint in wine grapes. Am. J. Enol. Vitic. 58, 518–522.

Giannakopoulos, P.I., Markakis, P., Howell, G.S., 1984. The influence of malolactic strain on the fermentation on wine quality of three Eastern red wine grape cultivars. Am. J. Enol. Vitic. 35 (1), 1–4.

Gladstones, J., 1992. Viticulture and Environment. Winetitles, Adelaide, Australia.

Glemser, E.J., Dowling, L., Hallett, R.H., Inglis, D., Pickering, G.J., McFadden-Smith, W.H., et al., 2012. A novel method for controlling *Harmonia axyridis* (Coleoptera: Coccinellidae) in vineyards. Environ. Entomol. 5, 1169–1176.

Haggerty, L.L., 2013. Ripening Profile of Grape Berry Acids and Sugars in University of Minnesota Wine Grape Cultivars, Select *Vitis vinifera*, and Other Hybrid Cultivars. Master thesis, University of Minnesota, 90 p.

Hall, A., Jones, G.V., 2009. Effect of potential atmospheric warming on temperature-based indices describing Australian winegrape growing conditions. Aus. J. Grape Wine Res 15 (2), 97–119.

Harrison, R., 2017. Practical interventions that influence the sensory attributes of red wines related to the phenolic composition of grapes: a review. Int. J. Food Sci. Technol. 1–16.

Hashizume, K., Umeda, N., 1996. Methoxypyrazine content of Japanese red wines. Biosci. Biotechnol. Biochem. 60, 802–805.

Huglin, P. 1978. Nouveau mode d'évaluation des possibilités héliothermiques d'un milieu viticole. Comptes rendus des seances.

Inglis, D.L., Kemp. B. Dowling. L., Kelly, J. Diprofio, F., Pickering, G. 2016. A new wine style for Canada's cool climate wine regions based on the appassimento technique. International Cool Climate Wine Symposium (ICCWS 2016), Brighton, United Kingdom. May 2016.

Jackson, D., Cherry, N., 1988. Prediction of a district's grape-ripening capacity using a latitude-temperature index (LTI). Am. J. Enol Vitic. 39 (1), 19–28.

Jackson, D.I., Lombard, P.B., 1993. Environmental and management practices affecting grape composition and wine quality—a review. Am. J. Enol Vitic. 44 (4), 409–430.

Jackson, R.S., 2008a. Fermentation. In Wine Science, Third edition Elsevier Inc, San Diego, pp. 332–417.

Jackson, R.S., 2008b. Post-fermentation Treatments and Related Topics., In Wine Science, Third edition Elsevier Inc, San Diego, pp. 418–519.

Johnson, H., 1985. The World Atlas of Wine. Mitchell Beazley, London, UK, p. 320.

Jones, G., Davis, R., 2000. Climate influences on grapevine phenology, grape composition, and wine production and quality for Bordeaux, France. Am. J. Enol. Vitic. 51 (3), 249–261.

Jones, G.V., 2004. Climate change in the western United States grape growing regions. VII International Symposium on Grapevine Physiology and Biotechnology, 689.

Jones, G.V., 2006. Climate and terroir: impacts of climate variability and change on wine. in fine wine and terroir—the geoscience perspective. In: Macqueen, R.W., Meinert, L.D. (Eds.), Geoscience Canada Reprint Series Number 9. Geological Association of Canada, St. John's, Newfoundland, pp. 14–247.

Kemp, B., Harrison, R., Creasy, G., 2011. Effect of mechanical leaf removal and its timing on flavan-3-ol composition and concentrations in *Vitis vinifera* L. cv. Pinot Noir wine. Aus. J. Grape Wine Res 17 (2), 270–279.

Ker, K., Pickering, G.J., 2006. Biology and control of the novel grapevine pest—the multicolored Asian lady beetle *Harmonia axyridis*. In: IN Dris, R. (Ed.), Crops: Growth, Quality and Biotechnology. IV. Control of Pests, Diseases and Disorders of Crops. WFL Publisher, Meri-Rastilan tie 3 C, Helsinki, Finland, 952-91-8601-0pp. 991–997.

Kliewer, W.M., Torres, R.E., 1972. Effect of controlled day and night temperatures on coloration of grapes. Am. J. Enol. Vitic. 23, 71–77.

Kögel, S., Botezatu, A., Hoffmann, C., Pickering, G.J., 2015. Methoxypyrazine composition of Coccinellidae-tainted Riesling and Pinot noir wine from Germany. J. Sci. Food Agric. 95, 509–514.

Lacey, M.J., Allen, M.S., Harris, R.L.N., Brown, W.V., 1991. Methoxypyrazines in Sauvignon blanc grapes and wines. Am. J. Enol. Vitic. 42 (2), 103–108.

Li, J.-C., Li, S.-Y., He, F., Yuan, Z.-Y., Liu, T., Reeves, M.J., et al., 2016. Phenolic and chromatic properties of Beibinghong red ice wine during and after vinification. Molecules 21 (4), 431.

López de Lerma, N., Moreno, J., Peinado, R.A., 2013. Determination of the optimum sun-drying time for *Vitis vinifera* L. cv. Tempranillo grapes by E-nose analysis and characterization of their volatile composition. Food Biopro. Tech. 1935–5130.

Lukić, I., Budić-Leto, I., Bubola, M., Damijanić, K., Staver, M., 2017. Pre-fermentative cold maceration, saignée, and various thermal treatments as options for modulating volatile aroma and phenol profiles of red wines. Food Chem. 224, 251–261.

Manns, D.C., Coquard Lenerz, C.T.M., Mansfield, A.K., 2013. Impact of processing parameters on the phenolic profile of wines produced from hybrid red grapes Maréchal Foch, Corot noir, and Marquette. J. Food Sci. 78 (5), C696–C702.

Mansfield, A.K., 2015. Yeast assimilable nitrogen (YAN) optimization for fermentation of cold climate cultivars. The Northern Grape Project, USA. Available from: http://northerngrapesproject.org/.

Mercurio, M.D., Dambergs, R.G., Herderich, M.J., Smith, P.A., 2007. High throughput analysis of red wine and grape phenolics—adaptation and validation of methyl cellulose precipitable tannin assay and modified Somers color assay to a rapid 96 well plate format. J. Agric. Food Chem 55, 4651–4657.

Miller, D.P., Howell, G.S., 1989. The effect of various carbonic maceration treatments on must and wine composition of Marechal Foch. Am. J. Enol. Vitic. 40, 170–174.

Nesbitt, A., Kemp, B., Steele, C., Lovett, A., Dorling, S., 2016. Impact of recent climate change and weather variability on the viability of UK viticulture—combining weather and climate records with producers' perspectives. Aus. J. Grape Wine Res. 22 (2), 324–335.

Nicolle, P., Marcotte, C., Angers, P., Pedneault, K., 2018. Co-fermentation of red grapes and white pomace: a natural and economical process to modulate hybrid wine composition. Food Chem. 242, 481–490.

Panceri, C.P., Gomes, T.M., De Gois, J.S., Borges, D.L.G., Marilde., Bordignon-Luiz., T., 2013. Effect of dehydration process on mineral content, phenolic compounds and antioxidant activity of Cabernet Sauvignon and Merlot grapes. Food Res. Int. 54 (2), 1343–1350.

Pedneault, K., Dorais, M., Angers, P., 2013. Flavour of cold hardy grapes: impact of berry maturity and environmental conditions. J. Agric. Food Chem. 61, 10418–10438.

Pedneault, P., Provost, C., 2016. Fungus resistant grape varieties as a suitable alternative for organic wine production: benefits, limits, and challenges. Sci. Hortic 208, 57–77.

Pickering, G., Inglis, D., 2013. Vintning on thin ice: the making of Canada's iconic dessert wine. In: Ripmeester, Michael, Gordon, Philip, Mackintosh, Christopher Fullerton (Eds.), The World of Niagara Wines, Chapter 13. Wilfrid Laurier University Press, Waterloo, Ontario, Canada, pp. 229–246.

Pickering, G.J. 2006. Icewine—the frozen truth. Proceedings of the Sixth International Cool Climate Symposium for Viticulture and Oenology, Christchurch, New Zealand, February 5–10, 2006, 84–99.

Pickering, G.J., Pour Nikfardjam, M.S., 2007. Influence of variety and commercial yeast preparation on red wine made from autochthonous Hungarian and Canadian grapes. Part II. Oral sensations and sensory: instrumental relationships. Eur. Food Res. Technol. 227 (3), 925–931.

Pickering, G.J., Yong, L., Reyonolds, A., Soleas, G., Riesen, R., Brindle, I., 2005. The influence of *Harmonia axyridis* on wine composition and aging. J. Food Sci. 70 (2), S128–S135.

Pickering, G.J., Lin, J., Reynolds, A., Soleas, G., Riesen, R., 2006. The evaluation of remedial treatments for wine affected by *Harmonia axyridis*. IJFST 41, 77–86.

Pickering, G.J., Karthik, A., Inglis, D., Sears, M., Ker, K., 2007a. Determination of ortho- and retronasal detection thresholds for 2-isopropyl-3-methoxypyrazine in wine. J. Food Sci. 72 (7), S468–S472.

Pickering, G.J., Ker, K., Soleas, G.J., 2007b. Determination of the critical stages of processing and tolerance limits for *Harmonia axyridis* for 'ladybug taint' in wines. Vitis. 46 (2), 85–90.

Pickering, G.J., Spink, M., Kotseridis, Y., Inglis, D., Brindle, I.D., Sears, M., et al., 2008. Yeast strain affects 3-isopropyl-2-methoxypyrazine concentration and sensory profile in Cabernet Sauvignon wine. AJGW. 14 (3), 230–237.

Pirie, A.J.G., 1977. Phenolics Accumulation in Red Wine Grapes (*Vitis vinifera* L.). PhD Thesis, University of Sydney, Australia.

Pour Nikfardjam, M.S., Pickering, G.J., 2008. Influence of variety and commercial yeast preparation on red wine made from autochthonous Hungarian and Canadian grapes. Part I: Phenolic composition. Eur. Food Res. Technol. 227 (4), 1077–1083.

Provost, C., Zerouala, L., Bastien, R., DHauteville, J. 2012. Evaluation of the agronomic and oenological characteristics of promising varieties in Quebec. Vitinord Conference 2012, Neubrandenberg, Germany, November 28–December 1, 2012.

Reynolds, A.G., Wardle, D.A., 1989. Effects of timing and severity of summer hedging on growth, yield, fruit composition, and canopy characteristics of Dechaunac. 2. Yield and fruit composition. Am. J. Enol. Vitic. 40 (4), 299–308.

Ribéreau-Gayon, P., Glories, Y., Maujean, A., Dubourdieu, D., 2006. Varietal aroma. In Handbook of Enology. John Wiley & Sons, Ltd, Chichester, UK, pp. 205–230.

Rice, S., Koziel, J.A., Dharmadhikari, M., Fennell, A., 2017. Evaluation of tannins and anthocyanins in Marquette, Frontenac and St. Croix cold-hardy grape cultivars. Fermentation. 3, 47.

Rio Segade, S., Torchio, F., Giacosa, S., Ricauda Aimonino, D., Gay, P., Lambri, M., et al., 2014. Impact of several pre-treatments on the extraction of phenolic compounds in wine grape varieties with different anthocyanin profiles and skin mechanical properties. J. Agric. Food Chem. 62, 8437–8451.

Rio Segade, S., Torchio, F., Gerbi, V., Quijada-Morin, N., Garcia-Estervez., I., Giacosa, S., et al., 2016. Impact of postharvest dehydration process of wine grapes on mechanical and acoustic properties of the seeds and their relationship with flavanol extraction during simulated maceration. Food Chem. 199, 893–901.

Roujou de Boubee, D.R., 2004. Research on the vegetal green pepper character in grapes and wines. Rev. des Oenol. 110, 6–10.

Roujou de Boubée, D., Van Leeuwen, C., Dubourdieu, D., 2000. Organoleptic impact of 2-methoxy-3-isobutylpyrazine on red bordeaux and loire wines. Effect of environmental conditions on concentrations in grapes during ripening. J. Agric. Food Chem. 48 (10), 4830–4834.

Roujou de Boubée, D., Cumsille, A., Pons, M., Dubourdieu, D., 2002. Location of 2-methoxy-3-isobutylpyrazine in Cabernet Sauvignon grape bunches and its extractability during vinification. Am. J. Enol. Vitic. 53 (1), 1–5.

Ryona, I., Pan, B.S., Intrigliolo, D.S., Lakso, A.N., Sacks, G.L., 2008. Effects of cluster light exposure on 3-isobutyl-2-methoxypyrazine accumulation and degradation patterns in red wine grapes *Vitis vinifera* L. Cv. Cabernet Franc. J. Agric. Food Chem. 56, 10838–10846.

Ryona, I., Reinhardt, J., Sacks, G.L., 2012. Treatment of grape juice or must with silicone reduces 3-alkyl-2-methoxypyrazine concentrations in resulting wines without altering fermentation volatiles. Food Res Int. 47, 70–79.

Sacchi, K.L., Bisson, L.F., Adams, D.O., 2005. A review of the effect of winemaking techniques on phenolic extraction in red wines. Am. J. Enol. Vitic. 56, 197–206.

Sala, C., Busto, O., Guasch, J., Zamora, F., 2004. Influence of vine training and sunlight exposure on the 3-alkyl-2-methoxpyrazines content in musts and wines from the *Vitis vinifera* variety Cabernet sauvignon. J. Agric. Food Chem. 52, 3492–3497.

Setford, P.C., Jeffrey, D.W., Grbin, P.R., Muhlack, R.A., 2017. Factors affecting extraction and evolution of phenolic compounds during red wine maceration and the role of process modelling. Trends Food Sci. Technol 69 (Part A), 106–117.

Shaw, A., 2005. The Niagara Peninsula viticultural area: a climatic analysis of Canada's largest wine region. J. Wine Res 16 (2), 85–103.

Sidhu, D., Lund, J., Kotseridis, Y., Saucier, C., 2015. Methoxypyrazine analysis and influence of viticultural and enological procedures on their levels in grapes, musts and wines. Crit. Rev. Food Sci. Nutr. 55 (4), 485–502.

Sivčev, B.V., Sivčev, I.L., Vasić, Z.Z.R., 2010. Plant protection products in organic grapevine growing. J. Agric. Sci. 55 (1), 103–122.

Slegers, A., Angers, P., Ouellet, É., Truchon, T., Pedneault, K., 2015. Volatile compounds from grape skin, juice and wine from five interspecific hybrid grape cultivars grown in Québec (Canada) for wine production. Molecules 20 (6), 10980–11016.

Smith, P.A., McRae, J.M., Bindon, K.A., 2015. Impact of winemaking practices on the concentration and composition of tannins in red wine. Aus. J. Grape Wine Res. 21 (S1), 601–614.

Sparrow, A.M., Smart, R.E., Dambergs, R.G., Close, D.C., 2016. Skin particle size affects the phenolic attributes of pinot noir wine: proof of concept. Am. J. Enol. Vitic. 67, 29–37.

Spayd, S.E., Tarara, J.M., Mee, D.L., Ferguson, J.C., 2002. Separation of sunlight and temperature effects on the composition of *Vitis vinifera* cv. Merlot berries. Am. J. Enol. Vitic. 53, 171–182.

Springer, L.F., Sacks, G.L., 2014. Protein-precipitable tannin in wines from *Vitis vinifera* and interspecific hybrid grapes (*Vitis* ssp.): differences in concentration, extractability, and cell wall binding. J. Agric. Food Chem. 62 (30), 7515–7523.

Springer, L.F., Chen, L.-A., Stahlecker, A.C., Cousins, P., Sacks, G.L., 2016. Relation of soluble grape-derived proteins to condensed tannin extractability during red wine fermentation. J. Agric. Food Chem. 64 (43), 8191–8199.

Stewart, A.C. 2013. Nitrogen Composition of Interspecific Hybrid and *Vitis Vinifera* Wine Grapes from the Eastern United States. PhD Thesis. Purdue University, Indiana, USA.

Sun, Q., Gates, M.J., Lavin, E.H., Acree, T.E., Sacks, G.L., 2011. Comparison of odor-active compounds in grapes and wines from *Vitis vinifera* and non-foxy American grape species. J. Agric. Food Chem. 59 (19), 10657–10664.

Tonietto, J. 1999. Les macroclimats viticoles mondiaux et l'influence du mésoclimat sur la typicité de la Syrah et du Muscat de Hambourg dans le sud de la France. Méthodologie de caractérisation. Thesis Doctorat. Agro Montpellier, France.

Van Leeuwen, C., Seguin, G., 2006. The concept of terroir in viticulture. J. Wine Res. 17 (1), 1–10.

Van Leeuwen, C., Roby, J.-P., Alonso-Villaverde, V., Gindro, K., 2012. Impact of clonal variability in *Vitis vinifera* Cabernet franc on grape composition, wine quality, leaf blade stilbene content and downy mildew resistance. J. Agric. Food Chem. 61 (1), 19–24.

Van Sluyter, S.C., Marangon, M., Stranks, S.D., Neilson, K.A., Hayasaka, Y., Haynes, P.A., et al., 2009. Two-step purification of pathogenesis-related proteins from grape juice and crystallization of thaumatin-like proteins. J. Agric. Food Chem. 57 (23), 11376–11382.

Versari, A., Patrizi, C., Parpinello, G.P., Mattioli, A.U., Pasini, L., Meglioli, M., et al., 2016. Effect of co-inoculation with yeast and bacteria on chemical and sensory characteristics of commercial Cabernet Franc red wine from Switzerland. J. Chem. Technol. Biotechnol. 91, 876–882.

Willwerth, J. 2016. Grapevine cold hardiness and innovative strategies to mitigate freeze injury in Ontario, Canada. International Cool Climate Wine Symposium, Brighton, UK, May 26–28, 2016.

Willwerth, J., Ker, K., Inglis, D., 2014. Best management practices for reducing cold injury in grapevines. Cool Climate Oenology and Viticulture Institute (CCOVI). Brock University, Ontario, Canada. Available from: http://www.brocku.ca/webfm_send/33923.

CHAPTER

24

Red Winemaking in Cold Regions With Short Maturity Periods

Ma Tengzhen[1], Kai Chen[2], Hao Yan[3], Han Shunyu[1] and Bi Yang[1]

[1]Gansu Key Laboratory of Viticulture and Enology, College of Food Science and Engineering Gansu Agricultural University, Lanzhou, P.R. China [2]College of Food Science and Nutritional Engineering, China Agricultural University, Beijing, P.R. China [3]Institute of Fruit and Floriculture Research, Gansu Academy of Agricultural Science, Lanzhou, P.R. China

24.1 INTRODUCTION

When the winter gets too cold, such as if the temperature is lower than −17°C, there should be some methods for viticulture in case of freeze damage. In this case, buried viticulture would be carried out if the extreme temperature is too low for vines during winter. It is well known that extreme low temperature will result in growth retardation of grape vines in the next year. The level of cold resistance depends on the grape variety, such as the rootstock used. *Vitis amurensis* Rupr. is a typical domestic cultivar regarded to have potential for winemaking in China. On the contrary, some cultivars of *Vitis vinifera* and local cultivars planted in northern China are comparatively sensitive against cold climate. Generally speaking, temperature lower than −17°C will cause freeze injury for *V. vinifera*. Regarding the viticulture experience and rules in northern China, a region where the average winter temperature is lower than −15°C, buried viticulture must be carried out before the winter in order to prevent freeze damage to vines.

24.1.1 Buried Viticulture

Frozen injury is the dominant natural winter disaster in Gansu Hexi Corridor, Wu wei, Jinchang, Zhangye, Jiayuguan, Jiuquan, and other wine grape planting areas also face this problem. Furthermore, the vineyard regions in the northwest of China include Gansu, Xinjiang, and Ningxia province, which account for more than 67% of China's wine grape production, should adopt buried viticulture to help vines live through the winter and avoid cold damage.

Between the stage of harvest and winter pruning, water should be poured completely, vine is pruned when the ground is not muddy anymore (Fig. 24.1). As the frost-free period of Gansu Hexi Corridor was generally around 160 days, the winter pruning cannot wait for the leaves to fully fall down after harvest, thus pruning usually occurs with leaves still present. When the vine pruning and the vineyard cleaning is completed, branches and tendrils are sprayed carefully and considerately with 3−5°. Be of lime-sulfur to eliminate the harmful germs and worm eggs parasitized on grapevine and the ground, before the vines are buried under the ground.

The buried time is from when the lowest temperature at night declines to 0°C leading to the soil freezing, which takes place from middle October to early November in Gansu Hexi Corridor, generally speaking young trees at the age of 1−2 years should be buried earlier. If the vines are buried too early, buds tend to rot due to the high temperature and humidity in the soil. However, when buried too late the branches are vulnerable to early frost, besides, the soil is difficult to pick after freezing, and soil clod is also too large to cover around. Both manual and mechanical methods are used to completely bury viticulture in the wine grape producing area in

FIGURE 24.1 Winter pruning.

Hexi Corridor. The approach of manual burying is to remove the pruned and sprayed branches from the trellis, straighten them out in the same direction and press down on the ground slowly, the branch can also be bundled by using a thin rope, and then the soil is buried and immobilized (Fig. 24.2A). In this process, excessive pressure or bending are forbidden to avoid the cleavage of the roots on main branches.

Mechanical burying is assisted by labor, when the branches are put down on the ground, the burying machines rotate and plow the interrow soil on vines to achieve the required thickness (Fig 24.2B). Workers assist the burying if the branches are not covered in soil. In mechanical burying, the original application of organic fertilizer cannot be plowed out to ensure the safety of cement columns and seedlings. Mechanical burying can greatly improve the working efficiency and is an effective way to solve the problem of labor shortage in this area. However, when being buried either by machinery or by human beings, the soil should be removed as far as possible from the roots to avoid the fertilized ditch. Generally, the soil is removed from the middle ground of the two rows, 80–100 cm away from the plant.

The thickness of buried soil generally requires exceed 20–30 cm of the upper branch. The width for young trees covered with soil should reach 80–100 cm and the elder trees should reach 100–120 cm to protect roots and branches from freezing injury. When the plants are buried, famers need to check if there are gaps that are not covered up. The lumped soil should break down and be covered with fine soil in order to prevent freezing damage or stripping caused by low wind temperature.

24.1.2 Vine Unearthed in Spring

The optimum unearthed period is when the temperature is higher than 10°C, between the vine sap moving and the sprouting period. The unearthed time for wine grapes in the Gansu Hexi Corridor generally begins from middle of the April, any earlier and it is likely to be subjected to late frost. On the other hand the unearthed time should not be too late for the growing season, so usually it is before the bud expansion period, in order to prevent the bud from germinating in the soil. However, the best time should be decided according to the local climate and the phenology of the cultivar, the unearthed time for the vine should no later than 1st of May in Gansu.

FIGURE 24.2 (A) Manual burying process. Branches are folded and buried on earth. (B) Mechanical burying of the vines by using a tractor.

Artificial or mechanical methods could be used to complete the unearthing work, this operation should be undertaken carefully to prevent damage to branches and buds. Firstly, artificial or mechanical methods are used to remove the thick soil layer around the branches, and then all soils are removed (Fig. 24.3). The branches are pulled out following the direction of burying, then the soil is shaken off and the irrigation ditch is cleaned up. For the grafted plants it should be ensured that graft is over the ground to prevent the scion of the roots from breaking off.

FIGURE 24.3 Discovering vine branches. (A) Handmade. (B) Mechanically assisted.

After the unearthing of the vine branches, the branches are tied to the trellis surface, and watered for a week to prevent the spring branch from drying (Fig 24.4).

24.1.3 Effect of Buried Viticulture on Vines

The burying system helps the vines to safely survive temperatures below −17°C in the winter (Fig. 24.5). However, the grape branches need to be removed from the trellis surface in winter; the branches and vines are

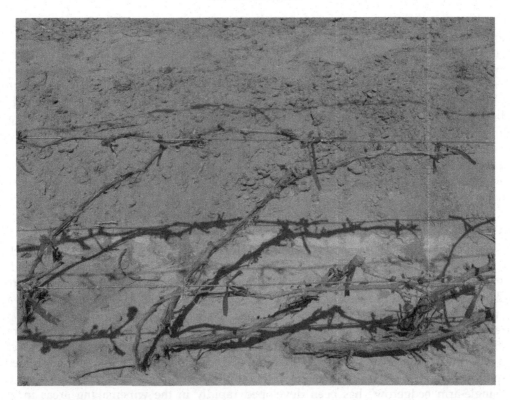

FIGURE 24.4 Unearthed branches disposed in the trellising system.

FIGURE 24.5 Vineyards covered with snow in winter at the foot of the Qilian mountains (Gansu, China).

tied again to the trellis in spring. This is one of the main reasons why the shape of vine is difficult to be fixed and thus is inconvenient to be pruned.

In addition, burying vines into soil and unearthing them is a heavy workload, and it is difficult to find a large number of laborers willing to bury into the soil in a short time, thus causing a lot of problems in practice. The application of mechanized production has effectively alleviated the shortage of the labor force in the case of buried viticulture. However branches and buds are easily damaged in this process, even if the buried soil is used to prevent cold damage, the freezing injuries may still occur in winter (Yan et al., 2014), and the branches have to be renewed after frostbite. As a result, old vines are very rare in this viticulture. Burying under soil to prevent cold damage is a time-consuming work (25% of the annual management cost of vineyard) at present. Some researchers are trying to use other materials instead of buried soil for winter, such as the preservation quilt in a greenhouse, colorful cloth or plastic film to cover vines from cold winter (Fig. 24.6), but they are still in the pilot stage and are not widely used (Li Pengcheng et al., 2011).

24.1.3.1 *The Cultivation and Management Technique of Vineyard in Cold Area*

To adapt to the cold climate conditions, it is necessary to choose grape varieties and rootstocks with strong frost resistance and late germination characteristics, Beta rootstocks and "Bei Bing Hong" are widely planted in Gansu because of their strong cold resistant and enjoyed high popularity in cool climate areas in recent years. In the middle of April, healthy and disease-free seedlings are planted, the seedlings are tilted at 45° along the row, with a spacing at 100 cm × 300 cm on average.

The main trellis system for wine grapes in Gansu at present is a system named as "single-arm hedgerow," also called as "⌐" form, which is a useful cultivation formed by combining the cultivation characteristics of the buried soil and cold prevention area (Table 24.1). In order to increase labor saving, light simplified pruning can produce high-quality grape berries. It is conducive to the application of mechanization and other characteristics. Recently, the "single-arm hedgerow" has been developed rapidly in the winemaking areas in western China, especially in the newly built wineries, most of which adopt this trellis system.

Raising the fruiting parts and balancing the distribution of branches increase the photosynthesis, since the bunch can receive light directly and thus the quality of the berry is improved. Timely weed out the redundant bud can promote branch distribution and increase its maturity in the autumn, thus also improving the resistance

FIGURE 24.6 Covered viticulture.

TABLE 24.1 The height parameter of "single-arm hedgerow" system

	The height of puncheon		The height of zinc-coated wire from ground		
Total (cm)	Underground (cm)	Overground (cm)	First wire (cm)	Second wire (cm)	Third wire (cm)
230	50	180	70	120	170

to winter freezing. The modified trellising system helps vines to grow better, making them easier to be managed, and greatly reducing the labor force in the whole growing period. Furthermore the single-arm hedgerow also increased line spacing, which is easier for mechanism cultivation.

In order to save labor and simplify the pruning process, the single-arm hedgerow shape was tested in the wine grape production area of Hexi Corridor, the fruit parts of the branches are the same. By differentiating the fruiting parts with ventilation and nutritional parts, berry coloring and sugar content become consistent, and the effect was remarkable in improving and enhancing berry quality (Yan et al., 2017; Guo et al., 2015). Rootstocks were also used to improve the stress resistance of grape vines: Beta, 5BB, 3309C, and 1103P were wildly used in the Hexi Corridor (Guo et al., 2015; Gao Xiuping et al., 1993; Huiyun and Xiuwu, 2008; Wei et al., 2008).

The phenological period for grape vine in Gansu Hexi Corridor can be divided into several parts, the unearthed period from the middle of April and ending in early May, then blooms at the beginning of June and the veraison period where grape berries begin to turn color in late July. The harvest period generally falls between early September to the middle of October, but most varieties are harvested in late September. This is followed by the winter irrigation and pruning immediately after harvest. The vines need eight times of irrigation to meet the needs of growth. The climate is rainless, with low incidence of diseases and pests, powdery mildew is one of the common diseases, and some years downy mildew can also be found. However, the spraying of 5% of sulfur mixture to sanitize before winter cutting and burying under soil and after being unearthed, or spraying fungicides three times in the growing season can prevent all disease throughout the year.

The winter pruning adopts the short-tip pruning, the annual branch leaves two to three buds, the number of remaining branches should be 20% more to avoid freezing damage. During the summer management, the bud is not usually carried out. When the length of the annual branch reaches about 30 cm, the branch is bound to the first wire to prevent it being blown around by the wind. When the branch grows to the third wire, it is necessary to pick the core of all the new branches in early August and remove some of the old leaves half a month before autumn harvest. In summer, the grapes are weeded three times in rows and ditches, which is a labor cost, and thus rotary cultivators and manual labor are combined together to complete this in the field.

Late frost freezing injury in spring is one of the main factors restricting the development of the viticulture industry in this area followed by low temperature freezing injury in winter. The most common way to prevent this is to burn compost and branches to release smoke to raise the temperature in the field. Affected by late frost, the frost-free period in the Hexi Corridor is only about 160 days. Because of the faster temperature rise in summer and the larger temperature fluctuation between day and night, the total heat can meet the needs of grape growth, and most varieties can mature to reach the required sugar content. However, because of the short growing period of the grape and the rapid accumulation of nutrients such as pigment in grape fruit, the color of red wine is not stable enough, and it is easy for it to fade in the course of storage. Furthermore, due to the speed of ripening, the aroma of some grape varieties cannot be fully displayed.

24.2 WINTERING ADAPTABILITY AND COLD RESISTANCE OF GRAPE VINE

A grape vine not only needs to be planted in a suitable environment and has to be given good cultivation, but also has to be able to withstand some adversities, such as drought, low temperature, wind, and hail damage. These adverse situations are commonly faced in winter, so they are all known as wintering adaptability, namely a resistance against the temperature lower than 0°C. Haishan (2007) has shown The wintering adaptability of grape vine is characterized by the decrease of temperature, that is, grapes move to a dormant state from a growth state as temperature falls. As a result, the cold resistance can be enhanced with an increase of low temperature adaptation. It is a result of the accumulation of plant protective compounds and a change of protoplasm. Two phases of winter adaptability of grape vine were found: the first phase is vine adaptation through 6–0°C and the second range is from −2 to −5°C.

24.3 INFLUENCE OF LOW TEMPERATURE ON DIFFERENT TISSUES OF GRAPE VINE

24.3.1 Adaptability of Shoots to Low Temperature

The water content of the cell influences the cold resistance of grape shoots. The water content of shoots is higher than 45% and there is 80%–90% water content in leaves. These shoots and leaves are sensitive to low temperature and easily result in freeze damage when the temperature is lower than 0°C. In fact, a cold resistance cultivar normally has larger xylem where the tissues contains more cells and thicker cell walls, which can effectively reduce the risk of the transformation of xylem and the breaking of membrane system as a result of freezing. On the other hand, the tissue structure and physiology of grape vines will undertake a series of changes with temperature decreasing: the activity of the vine's vascular cambium and efficiency of photosynthesis is going to decrease to some extent, leaves suffer from dehydration and naturally fall, the medulla shrink, cells of the medulla die, and the water content of shoot cells become lower and lower (Jinxing, 2013).

24.3.2 Adaptability of Buds to Low Temperature

Grape shoots develops on the bud pad, which is made up of parenchyma cells. Buds can be divided into winter buds, summer buds, and kryptoblasts. Herein, the winter buds include main and secondary buds. Mostly starch and a few chloroplasts exist in the linkage between buds and the diaphragm of the internode. During the growth of the grape berry, normally smaller shoots have the strongest cold resistance; the main shoots would be the most sensitive to cold, and then the more resistant are the secondary shoots and the third shoots (Jinxing, 2013).

On the other hand, with a temperature decrease and shortening of irritation time, the free water and starch of shoots decline, solvable sugar is accumulated, proteins and malondialdehyde increase, and hence the cold resistance of grape vines can be enhanced. Meanwhile, grapes generally have the following characteristics to be suitable for cold climate: summer shoots only have one piece of scale, while the winter shoots have two pieces of scale which have dense hair, so the main shoots and secondary shoots can be protected.

24.3.3 Adaptability of Roots to Low Temperature

Roots are sensitive to cold influence. This is because the roots have abundant tissues which contain a lot of nutritional compounds, which thus could boost the growth of vines. When winter is coming, the grape can feel the difference of the atmospheric temperature and cells will adjust so that the vines can still absorb the accumulated nutrition for consumption during winter (Jinxing, 2013).

24.4 INFLUENCE OF LOW TEMPERATURE ON GRAPE CELLS

24.4.1 Plasma Membrane Permeability

A cell membrane is a very thin membrane which consists of membrane lipid and membrane protein. The function of the cell membrane is to keep the comparative stability of the intracellular environment, to facilitate material exchange from extracellular environment, and energy and signal dissemination. In addition, the plasma membrane is the first part of cell which is easily damaged by extracellular factors. With decreasing temperature, proteins in cells are deactivated, the plasma membrane develops some cracks, and the active transport function will reduce. The change of osmotic pressure can result in the disorder of the physiological metabolism and the impediment of functions. Particularly, osmosis through the cell membrane will be enhanced in some cold resistant grape cultivars when the temperature gets lower to some extent. However, a climate that is too cold will lead to irreversible damage to plant cells. Osmosis resistance is related to grape cultivar—a cold durable cultivar has lower sensibility to environmental temperature—and some change of cell injury can be reversible.

24.4.2 Respiration in Cold Climate

Practically, if the grape could keep smooth respiration, the adaptation to atmosphere climate can be improved. A cold resistant cultivar is characterized with deep dormancy during winter. As a result, the respiration of these cultivars is weaker than normal grape varieties; it is an advantage to maintain water content in shoots and branches which are able to adapt to larger changes of temperature.

24.5 THE REASONS FOR FREEZE DAMAGE

24.5.1 Grape Variety Resistance to Freeze Damage

The resistance of a grape variety to freezing damage is variable. Under the same temperature, the level of freeze damage also has differences which relate to genetics. On the other hand, the cold resistance of shoots, stems, and roots on a plant has a difference. The least sensitive part to cold damage is an old stem, and the root is the most sensitive to cold climate. A winter bud is more sensitive to freeze damage than secondary buds. When a winter bud gets freeze damage, it is possible to use secondary buds for fruit in order to reduce the loss (Rui, 2008).

24.5.2 Humidity

Another important factor that influences vine cold resistance is humidity. Cold damage also includes a physiological drought resulting from dehydration in vines that is related to environmental low humidity. Northern China has a continental climate which is characterized with a low level of precipitation, as well as being windy and dry in winter. Without buried viticulture, some high cold resistance varieties could be sprouting (Rui, 2008).

24.5.3 Vine Physiological Limit

A practice of cold resistance means the vine experiences a series of physiological changes to adapt to low temperature and improves the ability of cold resistance to the shortening of the radiation time and the decreasing temperature. During the practice procedure, there are the following physiological changes in shoots: water content decreases, respiration weakens, the content of soluble sugar and amino acids increases to elevate the concentration of cellular fluid, condensation point decreases, and the water retaining capacity in protoplasm gets improved, so the capacity of cold resistance is improved as well.

The shortage of nutrients in shoots and stems results from different factors, such as the maturity of the stem and shoot, and can affect the cold resistance of vines in winter. Some disadvantages could lead to negative results of cold resistance, such as excessive production, late fruit maturation, early frozen climate after harvest, and no foliage trimming. These could all affect nutrient accumulation in stems and shoots. For example, although *V. amurensis* Rupr. should be a strong variety in a cold climate, the new shoot maturity can be affected if the harvest is carried out after the fruit gets too ripe.

Nevertheless, Rui (2017) found that during the dormancy season in the soil 20 cm underground the lowest temperature can be at -7 to $-8°C$, as a result, the root system of *V. vinifera* will be extremely damaged in this temperature; in the soil 30 cm underground, the lowest temperature can be -3 to $-6°C$, the root system of *V. vinifera* will be partially damaged; in the soil 40 cm underground, it is safe for the root system to overwinter. Therefore, the depth of burying should be more than 40 cm.

24.6 MATURITY ANALYSIS

The Gansu Hexi Corridor belongs to a cold climate region, thus the grape berries harvested as well as the wine brewed comprise a light body, high acidity, and low tannin content. Cold temperature during the growth period prolongs the physiological cycle of the grape berries and postpones the ripening time, which is beneficial for the aggregation of flavor substances as well as the accumulation of sugar. Thus the wine will have a fresh fruit aroma. However, if the maturity is not enough in some vintages, a green grass smell occurs, especially in some late harvested varieties such as *Cabernet Sauvignon* and *Cabernet Gernischt*.

Currently, the widely used methods for detecting grape maturity are Brix Degree and Sugar—Acid ratio, etc. However, in practice, it can be always found that although sugar—acid ratio has reached the harvesting requirement, the grape can show a poor chroma and low tannin content phenomenon. If the phenols maturity is measured on the basis of sugar content and total acid, the maturity of the grape could be evaluated more accurately.

Phenolic substances are distributed in skins, seeds, and stems of grapes and are extracted into the wine during the maceration stage—these could affect wine quality, physical—chemical composition, and physiological activity (Pardo-García et al., 2014).

Color, one of the most important organoleptic indicators for red wine, is an important determinant of wine quality and style, providing the basic characteristics such as quality, type and storage stability (Gordillo et al., 2014). The shades and hue characteristics of color are dictated by the type, state, and content of the coloring matter present in the wine. At the same time, as a key basis for most consumer purchases, color also affects the evaluation of other characteristics (Gordillo et al., 2013). Compared with high-quality red wines from European countries, some Chinese wines face problems of light color and poor stability, especially in north-western areas, where the largest-scale wine industry is located (Cai et al., 2014). These phenomena have been supposed to be one of the key issues restricting the quality of local premium wine for many years.

Anthocyanins, which are extracted from grape skins during maceration stage, could affect wine quality, composition, and bioactive functions (Pardo-García et al., 2014). Although there is a "terroir" advantage of growing high-quality wine grapes in the Hexi area of Gansu, the red wine here has the disadvantages of poor color stability and easy aging.

24.7 ANTHOCYANIN ACCUMULATION BY VITICULTURE PROCESS

The grapes are nonclimacteric fruits, and the growth of the berries shows a double S curve. The quality of grape and wine could be affected by light intensity, duration of sunlight, and spectral composition (Spayd and Tarara, 2002). When light intensity is low, anthocyanin concentration increases with the increase of light intensity. On the one hand, light affects photosynthesis, activates the light-sensitive pigment, promotes the synthesis or activation of enzymes, and affects the synthesis of metabolites such as sugar and phenylalanine (Evaghelia et al., 2010). On the other hand, light can regulate the expression of related genes (PAL, CHS, CHI, UFGT, F3H, DFR, ANS, and so on) in the anthocyanin biosynthesis pathway. Some studies on shading showed that shading treatment could significantly reduce the content of anthocyanin at fruit ripening. This is consistent with the reduced expression of structural genes and regulatory genes associated with anthocyanin synthesis (Koyama and Gotoyamamoto, 2008; Li et al., 2013; Scafidi et al., 2015). In the fruit setting stage, the anthocyanin content in the fruits of "Merlot" and "Cabernet Sauvignon" was significantly increased by leaf picking (Kotseridis et al., 2012).

In addition, the exposure rate under interfruit microclimate was changed by shaping, row orientation, leaf density, and ear position (Xiaohong et al., 2016). The study found canopy microclimate changes the maturity of the fruit quality compared with an upright canopy, with reducing sugar levels under canopy, whereas grain weight, total phenols, flavonoids, and anthocyanins content were improved. However, the analysis of anthocyanin monomers in the mature stage showed that 13 of the 21 anthocyanin monomers could be detected under the treatment of a horizontal leaf curtain, and the modification degree of anthocyanins was improved obviously. The effects of multimain vines sector shape and "Γ" shape on the photosynthetic characteristics and fruit quality of Cabernet Sauvignon grape were compared. The results showed that the "Γ" shape not only increased the photosynthetic rate of leaves but also improved the fruit quality of the grape. The soluble solids content and anthocyanin concentration were significantly higher than those in the multimain vines sector.

In addition to light, temperature is another important environmental factor that affects anthocyanin synthesis (Mori and Hashizume, 2007). In general, low temperatures favor the synthesis of anthocyanins, but lower temperatures can disrupt anthocyanin formation. Similarly, high temperatures also affect the synthesis of anthocyanins, which at higher temperatures can make grapes difficult to get color (Yanhui et al., 2012; FengDan et al., 2011; Tarara et al., 2008).

Compared with the control group, the anthocyanin content of Cabernet Sauvignon treated at 35°C decreased by 50%, which made the fruit of Cabernet Sauvignon promote anthocyanin degradation and inhibit the synthesis of anthocyanin (Jenkins et al., 1998; Mori et al., 1998). In the course of anthocyanin synthesis, low temperature during nighttime can promote anthocyanin accumulation, and the activity of PAL in fruit peel is also higher.

High temperature can reduce endogenous ABA level of grape, affecting the expression of transcription factor VvMybA1, thus inhibiting the biosynthesis of grape anthocyanin (Yamane et al., 2008). Higher night temperature inhibits the expression of VvCHS, VvF3H, VvDFR, VvANS, and VvUFGT genes in mature early fruits, decreasing the anthocyanin synthesis (Yamane et al., 2006). The decrease of color intensity may also be affected by the degree of anthocyanin degradation (Mori and Hashizume, 2007). From a macroscopic point of view, the concentration of anthocyanins is lower in the hot years, while in cold years more anthocyanins are synthesized and accumulated in the berries (Cohen et al., 2008). Spayd and Tarara (2002) reported that lower temperature can reduce the proportion of coumaroyl anthocyanin derivatives, while heating can lead to a substantial increase in their proportions. Downey et al.'s (2008) research showed that the proportion of coumaroyl acylation anthocyanins in grape fruits increased in a high temperature season, producing similar results to those of light processing (Tarara et al., 2008).

Moisture plays an important role in fruit quality formation of the grape. The water condition of the grape can affect the accumulation of flavonoids, and moderate water deficiency can induce anthocyanin synthesis (Kennedy and Jones, 2001; Kennedy et al., 2001).

Wei (2017) have reported that MeJA application could promote the increase of monoterpene compounds in grape and wine.

At the same time, foliar application of MeJA promoted the β-Glu activity of grape fruit, which favored the change and release of the bound aroma substance in grape fruit. For the "snake Dragon Ball" grape the release of grape wine aroma substances has a positive effect (Wei, 2017). D'Onofrio et al. (2018) has reported that the MeJA application led to a delay in grape technological maturity and has a significant increase in the concentration of several berry aroma classes (about twice of the total aroma: from 3 to 6 $\mu g/g$ per berry). Monoterpenes showed the most significant increase. An analysis of the expression of terpenoid biosynthesis genes confirmed that the MeJA application activated the related biosynthetic pathway (D'Onofrio et al., 2018). Shanshan et al. (2012) showed that 100–200 mg/L ABA particularly promoted some indexes of grape, and improved the quality of the grape. 400 mg/L GA3 treatments had a stimulative effect on the coloring of the grape.

In addition, anthocyanins contents in grapes could be promoted by improving the viticulture posture (Jinghan et al., 2011a,b), regulating the planting tree vigor (Cortell et al., 2008; Filippetti et al., 2013), optimizing the pruning methods of grapes (Prajitna et al., 2007), regulating the field grazing in the field of grape plants (Huzhumei et al., 2004), limiting the root zone (Wang et al., 2012, 2013), or using plant growth regulator (Wei, 2017; D'Onofrio et al., 2018).

24.8 WINEMAKING TECHNOLOGY

The influence of prefermentation cold maceration (Cai et al., 2014; Pace et al., 2014; Gordillo et al., 2014, 2013; Alcalde-Eon et al., 2014) and aging techniques affecting aging conditions (Espitia-Lopez et al., 2015; Han et al., 2015; Oberholster et al., 2015) can increase anthocyanin content in wine.

Anthocyanin content and stabilization property could be increased by changing the maceration temperature (Gomezmiguez et al., 2007). Cold maceration was a good method to improve the extraction of anthocyanin and other nonanthocyanin phenols. Studies showed that after 7–8 days of cold maceration at 10°C, the final wine was more structured and had unique aroma characteristics (Cejudo-Bastante et al., 2014). By using the electron microscope, it could be found that the structure of berry cells was broken by low temperature treatment, the cell wall was ruptured and the cytoplasm was leaked, thus, cold maceration was a good technique to extract the pigment substances from grape berries and to improve the quality of wine (Guodong et al., 2017). Fig. 24.7 shows the usual degree of phenolic ripeness in the Hexi corridor (Gansu province) and the effect of cold maceration in Cabernet Sauvignon.

At the same time, by increasing temperature to promote the thermal motion of molecules could also improve color intensity and stability of the wine—this is known as thermovinification (Tudosesanduville et al., 2012). Studies suggest that at maceration temperature, the corresponding oxidase activity in the grape fruit is passivated, which could preventing the loss of anthocyanins. And the hot-dipping process can accelerate the dissolution of anthocyanin or total phenols and increase the extraction rate of polyphenols (Aguilar et al., 2016). In particular, the flash evaporation process uses pressure differences to cause instantaneous fluid cooling and concentration can quickly vaporize some of the juice in fruit at a high temperature around 108°C, and rapidly expand the fruit particles under a negative pressure operating environment to promote peel rupture and strengthen pigment,

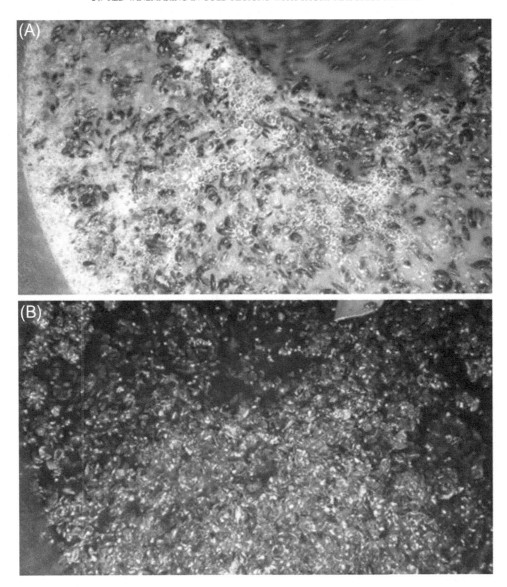

FIGURE 24.7 Effect of cold soak at 4–6°C in Cabernet Sauvignon grapes with low phenolic ripeness. (A) Initial situation with difficult skin extractability and green seeds. (B) After 4 days at low temperature together with two daily punch downs.

polyphenols, and aroma substances (Ho et al., 2008; Kekelidze et al., 2015). A relevant experiment showed that after the flash-process the juice contained abundant anthocyanins, tannins, and other substances, and the color of the wine was more stable and thus suitable for long-term aging in the later period (Hua et al., 2012).

Some grape varieties are related to a significantly higher content of some phenolic compounds, while in other varieties these are less or absent (Gordillo et al., 2012). Therefore, corresponding beneficial components can be obtained from these grape varieties that contain excessive cofactors by means of cofermentation (Gordillo et al., 2014). Lorenzo et al. (2005) studied the effect of the combined fermentation of red wine on the color of the phenols and wine by cofermentation with Monastrell and Cabernet Sauvignon and Merlot. By calculating the chromaticity and hue of the cofermented wines, the sample was found to have a significant color-increasing effect at the 530 (red band) and 620 nm (blue band) of the spectrum and remained in the aging stage, resulting in cofermentation of anthocyanins with a more stable conclusion. In the meantime, they observed that the role of co-color in fresh wines was more pronounced than that in aged wines (Lorenzo et al., 2005) by observing the color development of the sample wines. Among them, the "self-polymerization" reaction among anthocyanins in fresh dry red wine dominates, whereas the dominant color between anthocyanins and other phenolic cofactors dominates in aged wines (Aleixandre Tudó et al., 2013). Currently, cofermentation is widely used in the Côte du Rhone region of France, Chianti in Italy, and Rioja in Spain. From the beginning of alcoholic fermentation, the extraction rate of free anthocyanins rose rapidly and reached the highest level after 2–3 days of vigorous fermentation, then decreased slowly. At the same time, the concentration of polymerized anthocyanins increased gradually (García-Marino et al., 2013). Related research results show that adding to color protection substances, such as caffeic acid, quercetin 5′-sulfonate, 5′-sulfonate Moran salt, quercetin, tannic acid, chlorogenic acid, or rutin, to fermented grape juice can promote and stabilize the color of dry red wine by promoting the copigmentation of anthocyanins (Bakowska et al., 2003). In addition, research data confirm that hydroxycinnamic acid and hydroxybenzoic acid in the phenolic acids are also potent accessory pigments (Zhang et al., 2015b,c).

The use of enzymes (mainly pectinase, cellulase, and hemicellulase) in the processing of grape and wine can destroy the structure of the grape pericarp and degrade the cell wall in order to release the substance in the vacuole. This improves the degree of clarity, stability, and shortens the processing time (Fubing and Yanlin, 2013). For red wine, the effect of enhancing pigment and tannin extraction was also enhanced. Compared the wine without enzyme treatment, the color stability of red wine treated by enzyme was greatly improved during storage (He et al., 2010). After analyzing the wine treated with enzymes during alcohol fermentation, Parley et al. found that it could not only increase the chroma of red wine, but also increased the content of anthocyanin precursors in wine (Parley et al., 2008). These substances can promote the formation of proanthocyanidins in the aging stage, and play an important role in the stability of the color during the late aging of red wine. At present, these enzymes are being widely used.

The addition of acetaldehyde, color-preserving tannins, and oak products during fermentation can also improve the color of dry red wine. Sims et al. used acetaldehyde and two different commercial tannins (wine tannin and natural grape tannin) to compare muscadine wine. The results showed that the degree of browning of wine treated with acetaldehyde was lighter and the color of red was enhanced (Sims and Morris, 1986). In addition, in view of the antioxidant effect of tannins and the effect of forming stable tannin–pigment complexes with pigments, tannin was added to the fermentation to protect anthocyanins. It has also become one of the common methods in production to supplement the deficiency of condensed tannin in wine. The addition of oak products can also increase the color of wine and meet the quality requirements of aging red wine. The technology has been approved by the European Union and is considered to be a safe and reliable way to improve wine quality (Ruirui et al., 2016). The study shows that the earlier the oak product is added, the more obvious the effect is. This is mainly due to the effect of soluble tannins (tanned flower tannins and gallic acid tannins) and aldehydes in oak to promote the aggregation of wine pigments and accelerate the color stability of wine (Fredrickson, 2015).

Many of the condensation compounds in wine are thought to be formed by the direct reaction of anthocyanins and flavanols with oak phenols and aldehydes as intermediates, or anthocyanins and flavanols react directly with oak phenolic aldehyde (Bénédicte and Victor, 2010). Ellagic tannins from oak barrels can consume large amounts of oxygen and play an important role in this process. They are also involved in many chemical reactions, such as interfering with acetaldehyde-mediated anthocyanins and flavanols in wine, or directly with flavanols (Jingren et al., 2015). The effects of aging in oak barrels on wine quality are manifold, not only providing a rich aroma component to the wine but also ellagic tannins for the wine. This not only gives the wine a soft feel but also provides a stable color for the wine (Xiangyun et al., 2010). And oak barrels can provide a microoxygen environment for wine, accelerating the wine maturation. Oak structural characteristics (wood grain, porosity, and permeability) affect the physicochemical and biochemical reactions of wine during the aging of oak barrels (Zhang et al., 2015a,b,c). In addition, polyphenols extracted from oak will react with polyphenols in wine, which directly or indirectly affect the color stability, sensory or physical and chemical properties of wine. They participate in the oxidation process with anthocyanins and flavonoids during wine aging (Nevares et al., 2016). In addition, the physical and chemical characteristics of oak barrels are also important factors affecting wine aging, because they affect the interaction between oak and wine, such as oxygen diffusion, leaching of compounds, and oxidation process of wine. Although oak barrels have a lot of positive effects on wine, with the prolongation of service life, the rate of oxygen diffusion into oak barrels is affected by the clogging of wine sludge to oak pores. And with the extension of service life, there will be some harmful microbes attached to the wall of the barrel, which has a negative impact on the sensory quality of wine, so the production life of oak barrels is 3–5 years. This is a huge expense for winemaking (Yan and Guowei, 2007). So in order to simplify the aging of wines and to ensure that wines contain tannins and volatile ingredients from oak barrels, many oak barrels are substituted for the process. One of the techniques is to add oak chips to steel cans or old oak barrels to complement the lack of oak barrels (Zhou and Zhuang, 2013). However, this process cannot form a microoxygen environment, so the researchers add oak products to the steel tank supplemented with a device to control the amount of oxygen. To form a simulated microoxygenation system, the phenolic composition and color stability of wine produced by this treatment system is better than that of using different oak slices alone (Gómezplaza and Canolópez, 2011).

24.9 FINAL COMMENTS

Hexi region of Gansu Province is located in the temperate continental climate area, summer temperature is high, sunshine intensity is high, soil condition is different, soil quality is relatively poor, and the precipitation of grape fruit growth stage is limited. The dry red wine produced generally has problems such as high potassium ion content and high pH level. In addition, the unsuitable phenolic ripeness produced by adverse climatic conditions is the reason for the color defects of dry red wine in this region. This chapter reviews most of the factors affecting the viticulture of cold regions in northern China and the winemaking practices to optimize wine quality.

References

Aguilar, T., Loyola, C., Bruijn, J.D., Bustamante, L., Vergara, C., Baer, D.V., et al., 2016. Effect of thermomaceration and enzymatic maceration on phenolic compounds of grape must enriched by grape pomace, vine leaves and canes. Eur. Food Res. Technol. 242 (7), 1149–1158.

Alcalde-Eon, C., García-Estévez, I., Ferreras-Charro, R., Rivas-Gonzalo, J.C., Ferrer-Gallego, R., Escribano-Bailón, M.T., 2014. Adding oenological tannin vs. overripe grapes: effect on the phenolic composition of red wines. J. Food Compos. Anal. 34, 99–113.

Aleixandre Tudó, J.L., Álvarez, I., Lizama, V., García, M.J., Aleixandre, J.L., Toit, W.J.D., 2013. Impact of caffeic acid addition on phenolic composition of Tempranillo wines from different winemaking techniques. J. Agric. Food Chem. 61 (49), 11900–11912.

Bakowska, A., Kucharska, A.Z., Oszmianski, J., 2003. The effects of heating, UV irradiation, and storage on stability of the anthocyanin–polyphenol copigment complex. Food Chem. 81 (3), 349–355.

Bénédicte, B., Victor, D.F., 2010. A colorimetric study of oenin copigmented by procyanidins. J. Sci. Food Agric. 87 (87), 260–265.

Cai, J., Zhu, B.Q., Wang, Y.H., Lu, L., Lan, Y.B., Reeves, M.J., et al., 2014. Influence of pre-fermentation cold maceration treatment on aroma compounds of Cabernet Sauvignon wines fermented in different industrial scale fermenters. Food Chem. 154, 217–229.

Cejudo-Bastante, M.J., Gordillo, B., Hernanz, D., Escudero-Gilete, M.L., González-Miret, M.L., Heredia, F.J., 2014. Effect of the time of cold maceration on the evolution of phenolic compounds and colour of Syrah wines elaborated in warm climate. Int. J. Food Sci. Technol. 49 (8), 1886–1892.

Cohen, S.D., Tarara, J.M., Kennedy, J.A., 2008. Assessing the impact of temperature on grape phenolic metabolism. Anal. Chim. Acta 621 (1), 57–67.

Cortell, J.M., Sivertsen, H.K., Kennedy, J.A., Heymann, H., 2008. Influence of vine vigor on pinot noir fruit composition, wine chemical analysis, and wine sensory attributes. Am. J. Enol. Vitic. 59 (1), 1–10.

Downey, M.O., Harvey, J.S., Robinson, S.P., 2008. The effect of bunch shading on berry development and flavonoid accumulation in Shiraz grapes. Aust. J. Grape Wine Res. 10 (1), 55–73.

D'Onofrio, C., Matarese, F., Cuzzola, A., 2018. Effect of methyl jasmonate on the aroma of Sangiovese grapes and wines. Food Chem.

Espitia-Lopez, J., Escalona-Buendia, H.B., Luna, H., Verde-Calvo, J.R., 2015. Multivariate study of the evolution of phenolic composition and sensory profile on mouth of Mexican red Merlot wine aged in barrels vs wood chips. CyTA J. Food 13, 26–31.

Evaghelia, C., Silvia, G., Alessandra, F., Vittorino, N., 2010. Effect of different cluster sunlight exposure levels on ripening and anthocyanin accumulation in Nebbiolo grapes. Am. J. Enol. Vitic. 61 (1), 23–30.

FengDan, G., XiaoZhong, W., XueYing, L., Han, X., XingJun, W., 2011. Metabolic regulation of plants anthocyanin. Chin. Bull. Life Sci. 10, 938–944.

Filippetti, I., Allegro, G., Valentini, G., Pastore, C., Colucci, E., Intrieri, C., 2013. Influence of vigour on vine performance and berry composition of cv. Sangiovese ("*Vitis vinifera*" L.). Int. J Vine Wine Sci. 47 (1), 21–33.

Fredrickson, A., 2015. Cool Climate Winemaking: Exogenous Tannin Additions in Red Hybrid Cultivars.

Fubing, L., Yanlin, L., 2013. Application of enzymes in wine production. Food Sci. 34 (9), 392–398.

García-Marino, M., Escudero-Gilete, M.L., Heredia, F.J., Escribano-Bailón, M.T., Rivas-Gonzalo, J.C., 2013. Color-copigmentation study by tristimulus colorimetry (CIELAB) in red wines obtained from Tempranillo and Graciano varieties. Food Res. Int. 51 (1), 123–131.

Gomezmiguez, M., Gonzalezmiret, M.L., Heredia, F.J., 2007. Evolution of colour and anthocyanin composition of Syrah wines elaborated with pre-fermentative cold maceration. Int. J. Food Eng. 79 (1), 271–278.

Gómezplaza, E., Canolópez, M., 2011. A review on micro-oxygenation of red wines: claims, benefits and the underlying chemistry. Food Chem. 125 (4), 1131–1140.

Gordillo, B., González-Miret, M., Heredia, F., 2012. Importance of Anthocyanic Copigmentation on the Color Expression of Red Wine in Warm Climate.

Gordillo, B., Cejudo-Bastante, M.J., Rodriguez-Pulido, F.J., Gonzalez-Miret, M.L., Heredia, F.J., 2013. Application of the differential colorimetry and polyphenolic profile to the evaluation of the chromatic quality of Tempranillo red wines elaborated in warm climate. Influence of the presence of oak wood chips during fermentation. Food Chem. 141, 2184–2190.

Gordillo, B., Cejudo Bastante, M.J., Rodríguez Pulido, F.J., Jara Palacios, M.J., Ramírez Pérez, P., González Miret, M.L., et al., 2014. Impact of adding white pomace to red grapes on the phenolic composition and color stability of Syrah wines from a warm climate. J. Agric. Food Chem. 62 (12), 2663–2671.

Guo, Liye, C., Jun, W., Wu, C., Zhenwen, Z., 2015. Effect of training system on photosynthesis and fruit characteristics of Cabernet Sauvignon. J. Fruit Sci. 32 (2), 215–224.

Guodong, T., Yu, T., Qian, W., Jihong, Y., 2017. Effect of cryogenic treatment on the structure of grape skin cells. Food Sci. 38 (05), 191–196.

Haishan, H., 2007. Study on Cold-Resistance Characteristics of Physiology and Biochemistry of Grapevines. Gansu Agricultural University.

Han, G., Ugliano, M., Currie, B., Vidal, S., Dieval, J.-B., Waterhouse, A.L., 2015. Influence of closure, phenolic levels and microoxygenation on Cabernet Sauvignon wine composition after 5 years' bottle storage. J. Sci. Food Agric. 95, 36–43.

He, J.J., Han, F.L., Yu, Q.Q., Pan, Q.H., Duan, C.Q., Cheng, G.L., 2010. Effects of the maceration enzymes on evolution of pyranoanthocyanins and cinnamic acids during the Cabernet Gernischt (*Vitis vinifera* L. cv.) red wine making. Food Sci. Biotechnol. 19 (3), 603–610.

Ho, K.T.V., Sebastian, P., Nadeau, J.P., 2008. Multicriteria-oriented preliminary design of a flash evaporation process for cooling in the winemaking process. Int. J. Food Eng. 85 (4), 491–508.

Hua, Y., Cui, Y., Wang, L., Chai, J., 2012. Effect of maceration time on polyphenol and colour features of Cabernet Sauvignon wine produced with flash evaporation. Chin. Brew. 16 (3), 23–31.

Huiyun, L., Xiuwu, G., 2008. Influence of NaCl on activities of protective enzymes and MDA content in grape rootstock leaves. J. Fruit Sci. 25 (2), 240–243.

Jenkins, S.D., Klamer, T.W., Parteka, J.J., Condon, R.E., 1998. A comparison of prosthetic materials used to repair abdominal wall defects. Pediatr. Surg. Int. 13 (7), 487–490.

Jinghan, H., Kerui, W., Qiuhong, P., Changqing, D., Jun, W., 2011a. Influence of environmental conditions and viticulture technics to biosynthesis of grape anthocyanins(I) Impact of environmental conditions to grape anthocyanins biosynthesis. Sino Overseas Grapevine Wine 09, 71–76.

Jinghan, H., Kerui, W., Qiuhong, P., Changqing, D., Jun, W., 2011b. Influence of environmental conditions and viticulture technics to biosynthesis of grape anthocyanins(II) Impact of viticulture technics to grape anthocyanins biosynthesis. Sino-Overseas Grape vine &Wine (11), 60–65.

Jingren, H., Minjie, K., Minyu, Q., Gang, L., Shuyi, L., Nao, W., et al., 2015. Recent progress in research on pyranoanthocyanins derivatives. Food Sci. 36 (7), 228–234.

Jinxing, L., 2013. Relationship of Physiological-Biochemical Characteristics, Anatomical Structures and Cold Tolerance in Different Wine Grape. Gansu Agricultural University.

Kekelidze, I.A., Ebelashvili, N., Chankvetadze, L., Japaridze, M., 2015. Determination of monophenolics in red dessert wines produced with different maceration techniques. In: International Symposium on Pharmaceutical and Biomedical Analysis.

Kennedy, J.A., Jones, G.P., 2001. Analysis of proanthocyanidin cleavage products following acid-catalysis in the presence of excess phloroglucinol. J. Agric. Food Chem. 49 (4), 1740.

Kennedy, J.A., Hayasaka, Y., Vidal, S., Waters, E.J., Jones, G.P., 2001. Composition of grape skin proanthocyanidins at different stages of berry development. J. Agric. Food Chem. 49 (11), 5348–5355.

Kotseridis, Y., Georgiadou, A., Tikos, P., Kallithraka, S., Koundouras, S., 2012. Effects of severity of post-flowering leaf removal on berry growth and composition of three red *Vitis vinifera* L. cultivars grown under semiarid conditions. J. Agric. Food Chem. 60 (23), 6000–6010.

Koyama, K., Gotoyamamoto, N., 2008. Bunch shading during different developmental stages affects the phenolic biosynthesis in berry skins of 'Cabernet Sauvignon' grapes. J. Am. Soc. Hortic. Sci. 133 (6), 743–753.

Li, J.H., Guan, L., Hua, S., Wu, B.H., 2013. Effect of sunlight exclusion at different phenological stages on anthocyanin accumulation in red grape clusters. Am. J. Enol. Vitic. 64 (3), 349–356.

Lorenzo, C., Pardo, F., Zalacain, A., Luis Alonso, G., Rosario Salinas, M., 2005. Effect of red grapes co-winemaking in polyphenols and color of wines. J. Agric. Food Chem. 53 (19), 7609–7616.

Mori, K., Hashizume, K., 2007. Loss of anthocyanins in red-wine grape under high temperature. J. Exp. Bot. 58 (58), 1935–1945.

Mori, T., Nishimura, H., Okabe, M., Ueyama, M., Kubota, J., Kawamura, K., 1998. Cardioprotective effects of quinapril after myocardial infarction in hypertensive rats. Eur. J. Pharmacol. 348 (2-3), 229.

Nevares, I., Mayr, T., Baro, J.A., Ehgartner, J., Crespo, R., Alamo-Sanza, M.D., 2016. Ratiometric oxygen imaging to predict oxygen diffusivity in oak wood during red wine barrel aging. Food Bioprocess Technol. 9 (6), 1049–1059.

Oberholster, A., Elmendorf, B.L., Lerno, L.A., King, E.S., Heymann, H., Brenneman, C.E., et al., 2015. Barrel maturation, oak alternatives and micro-oxygenation: Influence on red wine aging and quality. Food Chem. 173, 1250–1258.

Pace, C., Giacosa, S., Torchio, F., Segade, S.R., Cagnasso, E., Rolle, L., 2014. Extraction kinetics of anthocyanins from skin to pulp during carbonic maceration of winegrape berries with different ripeness levels. Food Chem. 165 (3), 77–84.

Pardo-García, A.I., Martínez-Gil, A.M., Cadahía, E., Pardo, F., Alonso, G.L., Salinas, M.R., 2014. Oak extract application to grapevines as a plant biostimulant to increase wine polyphenols. Food Res. Int. 55 (55), 150–160.

Parley, A., Vanhanen, L., Heatherbell, D., 2008. Effects of pre-fermentation enzyme maceration on extraction and colour stability in Pinot Noir wine. Aust. J. Grape Wine Res. 7 (3), 146–152.

Pengcheng, L., Shaojie, G., Ming, L., Xuede, S., 2011. Comprehensive evaluation on cold effects of the special covers for red globe grape. Chin. Agric. Sci. Bull. 1, 206–210.

Prajitna, A., Dami, I.E., Steiner, T.E., Ferree, D.C., Scheerens, J.C., Schwartz, S.J., 2007. Influence of cluster thinning on phenolic composition, resveratrol, and antioxidant capacity in Chambourcin wine. Am. J. Enol. Vitic. 58 (3), 346–350.

Rui, G., 2008. The Study and Investigation of Freezingin Injury in the Northwest Gobi Region of China. Northwest A&F University.

Rui, S., 2017. Study of the Cold Tolerance of Wine Grape Roots in Helan Mountain Region. Ningxia University.

Ruirui, L., Xinjie, Z., Yuxia, S., 2016. Research advances on tannins in grape berries and wine. Food Ferment. Ind. 42 (4), 260–265.

Scafidi, P., Maria, G., Barbagallo, Downey, M.O., 2015. Influence of light exclusion on anthocyanin composition in 'Cabernet Sauvignon'. Front. Plant Sci. 06, 382.

Shanshan, H., Zhumei, X., Lina, M., Liying, L., 2012. Effect of plant growth regulator on the quality of Cabernet Sauvignon grape. J. Northwest A&F Univ. (Nat. Sci. Ed.) 40 (1), 183–189.

Sims, C.A., Morris, J.R., 1986. Effects of acetaldehyde and tannins on the color and chemical age of red Muscadine (*Vitis rotundifolia*) wine. Am. J. Enol. Vitic. 37 (2), 163–165.

Spayd, S.E., Tarara, J.M., 2002. Separation of sunlight and temperature effects on the composition of *Vitis vinifera* cv. Merlot berries. Am. J. Enol. Vitic. 53 (3), 171–182.

Tarara, J., Lee, J., Spayd, S., Scagel, C., 2008. Berry temperature and solar radiation alter acylation, proportion, and concentration of anthocyanin in Merlot grapes. Am. J. Enol. Vitic. 59 (5), 235–247.

Tudosesanduville, S.T., Cotea, V.V., Colibaba, C., Nechita, B., Niculaua, M., Codreanu, M., 2012. Phenolic compounds in merlot wines obtained through different technologies in Iași vineyard, Romania. Cercet. Agron. Mold. 45 (4), 89–98.

Wang, B., He, J., Duan, C., Yu, X., Zhu, L., Xie, Z., et al., 2012. Root restriction affects anthocyanin accumulation and composition in berry skin of 'Kyoho' grape (Vitis vinifera L. × Vitis labrusca L.) during ripening. Sci. Hortic. 137, 20–28.

Wang, B., He, J., Yang, B., Yu, X., Li, J., Zhang, C., et al., 2013. Root restriction affected anthocyanin composition and up-regulated the transcription of their biosynthetic genes during berry development in 'Summer Black' grape. Acta Physiol. Plant. 35 (7), 2205–2217.

Wei, L., 2017. Effects of Exogenous Methyl Jasmonate on Monoterpenes and β-Glucosidase in Cabernet Gernischt Grapes. Gansu Agricultural University.

Wei, Q., Enmao, L., Heng, Z., Xiaofang, W., Yuanpeng, D., 2008. Physiological and biochemical responses of different scion/rootstock combinations grapevine to partial root zone drought. Chin. J. Appl. Ecol. 19 (2), 306–310.

Xiangyun, C., Jianjun, H., Qiuhong, P., Changqing, D., 2010. Comparison of anthocyanin components and color of red wines aged in new and old oak barrels. Sino Overseas Grapevine Wine. 7, 18–22.

Xiaohong, L., Yongjiang, S., Hong, S., Heng, Z., 2016. Effect of canopy types on the cluster micro-environment and fruit quality of the 'moldova' grapes. Sci. Agric. Sin 49 (21), 4246–4254.

Xiuping, G., Xiuwu, G., Ke, W., Wangheng, F., 1993. The Research of resistance to cold and grown call in different grape rootstocks. Acta Hortic. Sin. 20 (4), 313–318.

Yamane, T., Jeong, S.T., Gotoyamamoto, N., Koshita, Y., Kobayashi, S., 2006. Effects of temperature on anthocyanin biosynthesis in grape berry skins. Am. J. Enol. Vitic. 57 (1), 54–59.

Yamane, T., Date, T., Tokuda, M., Aramaki, Y., Inada, K., Matsuo, S., et al., 2008. Prevalence, morphological and electrophysiological characteristics of confluent inferior pulmonary veins in patients with atrial fibrillation. Circ. J. 72 (8), 1285–1290.

Yan, H., Qilong, M., Kun, Z., 2017. Effects of different rootstocks on the growth and fruit quality of Italian Riesling in Hexi corridor. J. Fruit Sci. 10, 1286–1293.

Yan, L., Guowei, L., 2007. Oak barrel and grape wine. Liquor Making Sci. Technol. 12, 40–42.

Yan, H., Kun, Z., Qilong, M., Yaodong, B., 2014. Causes and remedial measures of Late frost damage on wine in Gansu Hexi corridor. Gansu Agric. Sci. Technol. 2, 60–61.

Yanhui, G., Chunhong, H., Yuqiu, Z., Zaikang, T., 2012. Progress on plant anthocyanin biosynthesis and regulation. Chin. Biotechnol 32 (8), 94–99.

Zhang, B., Cai, J., Duan, C.Q., Reeves, M.J., He, F., 2015a. A review of polyphenolics in oak woods. Int. J. Mol. Sci. 16 (4), 6978–7014.

Zhang, B., He, F., Zhou, P.P., Liu, Y., Duan, C.Q., 2015b. Copigmentation between malvidin-3-O-glucoside and hydroxycinnamic acids in red wine model solutions: Investigations with experimental and theoretical methods. Food Res. Int. 78, 313–320.

Zhang, B., Liu, R., He, F., Zhou, P.P., Duan, C.Q., 2015c. Copigmentation of malvidin-3-O-glucoside with five hydroxybenzoic acids in red wine model solutions: experimental and theoretical investigations. Food Chem. 170 (170), 226–233.

Further Reading

Zhumei, X., Hua, L., Yanlin, L., Yulin, F., 2004. The effect of vineyard green cover on wine quality in grape cultivar Cabernet Sauvignon. Sci. Agric. Sin. 37 (10), 1527–1531.

Author Index

A

Aagaard, O., 151
Abbott, N., 134
Abca, E.E., 170
Abguéguen, O., 245
Abrahamse, C., 100t, 102, 104, 108
Abrahamse, C.E., 102, 104, 107–108
Acuña, G., 274t
Adams, D., 262
Adams, M.R., 229–230
Agnolucci, M., 119
Agosin, E., 276–277
Aguilar, T., 367–368
Akdemir Evrendilek, G., 170
Akiyoshi, M., 264
Akkurt, M., 349
Alarcon-Mendez, A., 243
Alberts, P., 277t
Alcalde-Eon, C., 199, 201–202, 209, 367
Alderson, B., 153–155, 187–188
Aleixandre-Tudo, J.L., 1–2, 368–369
Alexandre, H., 79, 104, 107–108
Ali, K., 257
Allegro, G., 179–181
Allen, M.S., 276–277, 298t, 299
Alston, J.M., 62
Altmann, J., 244
Alvarez, M.A., 93
Amerine, M., 341–342
Amerine, M.A., 17, 21t, 25, 264
Amic, D., 197
Ancín, C., 105–106
Ancín-Azpilicueta, C., 134, 150, 319
Andaur, J.E., 12
Andorrà, I., 99, 117
Andrews, S., 229–230
Anness, B.J., 273–275
Anocibar Beloqui, A., 273–275
Antalick, G., 90–91, 99–101, 100t, 107, 300
Apetrei, I., 156
Arapitsas, A., 150, 152–153
Arapitsas, P., 212–213, 315
Araque, I., 94
Araujo, P., 211–212
Armstrong, W.G., 258
Arneberg, P., 258
Arnink, K., 79, 104, 107–108
Arnold, R.A., 1
Artés-Hernández, F., 47
Asenstorfer, R.E., 73, 201, 214–215
Asmundson, R.V., 87
Asproudi, A., 179–181
Atanasova, V., 156, 207–210
Augustin, M., 2

Auw, J.M., 265, 351t
Avbelj, M., 57t, 64
Avila, F., 12, 13f
Azen, E.A., 258
Azevedo, J., 213f
Azzolini, M., 59, 100t, 102, 104, 318

B

Back, W., 326
Bağder Elmacı, S., 318–319
Bajec, M.R., 258
Bakker, J., 73, 149, 197, 201, 207–209, 211
Bakowska, A., 368–369
Balakian, S., 245–246
Balík, J., 180t
Bamforth, C.W., 273–275
Bañuelos, M.A., 45
Baranac, J.M., 181, 198
Barata, A., 51–54, 115–116, 220, 222–223, 225, 225t, 230–231
Barba, F.J., 169
Barbanti, D., 348
Barrajón-Simancas, N., 104
Barrera-García, V.D., 187
Barron, J.J., 30
Bartowsky, E., 100t, 102, 104, 108
Bartowsky, E.J., 69, 85, 89–90, 102–104, 107–108, 117, 318
Bate-Smith, E.C., 259
Bauer, R., 87–88, 101–102
Baumes, R., 273–275, 285–289
Bautista-Ortín, A., 264
Bautista-Ortín, A.B., 58–59, 153–154, 154f, 158, 179, 201, 265
Baxter, I.R., 259
Baxter, N.F., 257–258, 260
Baxter, N.J., 257–258, 260
Bayonove, C., 276–277
Becker, N., 341
Beelman, R.B., 99–101, 103–104, 107–108
Bejaoui, H., 105
Belancic, A., 276–277
Belda, I., 64, 89, 276
Bell, S.-J., 220, 226
Belleville, M., 244
Bellincontro, A., 47, 348
Belviso, S., 3
Benavent-Gil, Y., 105
Benavides, J.A., 55
Bénédicte, B., 369
Beneduce, L., 311
Benito, A, 61–62
Benito, S., 59, 61, 73
Bennick, A., 258–259

Berbegal, C., 85, 93, 99–106, 100t
Berg, H.W., 17, 20–21, 245–246, 264
Berké, B., 207–209, 213–214
Berkelmann-Löhnertz, B., 105
Bertrand, A., 295–296
Bester, L., 117–118
Betteridge, A., 87–89, 87t
Beuchat, L., 224
Bindon, K., 262, 344
Bindon, K.A., 263
Bisotto, A., 76
Bisson, L., 220
Bisson, L.F., 56–58, 220
Blaauw, D., 157–158
Blackburn, C., 224
Blackwood, C.B., 117
Blake, A., 347
Blanchard, L., 274t, 276–277
Blanco-Vega, D., 1
Blättel, V., 119
Bleve, G., 107
Blouin, J., 247, 314–315, 316t
Blüml, Susanne, 330–332
Boase, S., 120
Boekhout, T., 219, 224
Bogicevic, M., 181, 183
Bogs, J., 181, 262
Boido, E., 75, 91, 106, 177, 179
Boidron, J.N., 132, 134
Boissier, B., 244
Boivin, S., 72
Bokulich, N.A., 51, 54–55, 69–70, 115–116, 118–119
Bonorden, W.R., 30
Bordeu, E., 99–104, 100t
Bordiga, M., 178–181
Bordons, A., 99, 117
Borgogno Mondino, E., 4
Borneman, A.R., 86
Boselli, E., 257
Bosia, P.D., 23, 314, 315t
Bosso, A., 43, 184, 264–265
Botezatu, A., 298–299, 345, 345f, 346f, 347
Bouchilloux, P., 274t, 276, 296
Boulet, J.C., 40, 263
Boulton, B., 243
Boulton, R.B., 17, 20, 21t, 25, 72, 76–77, 198, 209, 245–247
Bourzeix, M., 72, 177
Bousbouras, G.E., 17
Bradshaw, M.P., 319
Bramley, R.G.V., 4
Brandao, E., 258
Brás, N.F., 210f

Breslin, P.A.S., 257
Brezna, B., 118
Bridle, P., 185, 197, 200, 207–210
Brizuela, N.S., 99–103, 100t, 106
Brossaud, F., 1, 259
Brouillard, R., 17, 134, 197–199, 207–209, 208f, 315, 316t
Brown, S.L., 60
Buck, R.P., 20
Buelga, C.S., 211f
Bueno, M., 286t, 291, 294–295, 294t
Buffon, P., 266
Burroughs, L.F., 213–214, 314–315, 316t
Busse-Valverde, N., 39
Butler, L.G., 261
Buttery, R.G., 277t
Butzke, C.E., 220

C
Caboni, P., 290–291
Cabras, P., 89
Cabrita, M.J., 91
Cáceres-Mella, A., 260–261
Cacho, J., 285, 291
Cadahía, E., 129, 132, 134, 137–140
Cadot, Y., 2–3, 181, 244, 344
Cagnasso, E., 2
Cai, J., 366–367
Callejon, R.M., 299
Callejón, S., 105–106
Calvi, J.P., 197
Camara, J.S., 296
Cameira dos Santos, P., 244
Cameira-dos-Santos, P.J., 201
Cameron, M., 117–118
Campo, E., 295–296, 300t
Campos, F.M., 89
Camus, A., 125–126
Canal-Lalaubères, R.-M., 241
Canals, R., 43, 183–184, 263
Candiago, S., 4
Canolópez, M., 369
Cano-López, M., 149–153, 154f, 155–156, 158
Cantos-Villar, E., 85
Capece, A., 55
Capone, D.L., 278, 279t, 296, 298
Capozzi, V., 54–55, 63–64, 105–106
Cappello, M.S., 90, 107
Capucho, I., 103–104, 107–108
Caputi, L., 277, 299
Carew, A., 348
Carew, A.L., 108
Caridi, A., 72, 104, 165
Carpenter, G.H., 258
Carpinteiro, I., 290–291
Carrascon, V., 291–292, 291t
Carrau, F., 71–72, 74–75
Carrera, M., 119
Carreté, R., 88, 103, 107–108
Carrillo, E., 5, 5f
Carrillo, J.D., 290–291
Caruso, M., 105
Carvalho, E., 60
Carvalho, E.C.P., 25
Casassa, L.F., 184

Caspritz, G., 101–102
Castellari, M., 156, 317
Castino, M.R., 72
Castro-López, Ld.R., 264
Castro-Vázquez, L., 125
Castura, J.C., 283
Cavalieri, D., 55
Cavin, J.F., 91–92
Cavuldak, O., 152
Cejudo-Bastante, M.J., 152, 156–157, 367
Celotti, E., 2
Cenis, J.L., 117
Cerovic, Z.G., 5
Cerpa-Calderón, F.K., 183–184
Champagnol, F., 23
Chandra, M., 231
Charlton, A.J., 260
Charpentier, C., 60, 101–102
Chassagne, D., 72
Chassaing, S., 134–135, 187, 212
Chatonnet, D., 165
Chatonnet, P., 63, 71–72, 125–129, 132, 134, 137–140, 150, 187–188, 223, 290, 312
Chauvet, S., 314
Chemat, F., 45, 158
Chen, K., 60
Chen, L.J., 197
Chen, S., 12
Chen, Y., 293, 348
Cherry, N., 341–342
Chescheir, S., 106
Cheung, K., 158
Cheynier, V., 1, 151, 181, 185, 200, 239, 259–260
Chinnici, F., 188
Chira, K., 132, 134, 137, 139–140, 177–178, 185, 260
Ciani, M., 55, 60–62, 101–102, 104–105
Clark, A.E., 119
Claus, H., 99
Clavijo, A., 52
Clifford, M.N., 260–261
Climent, J., 125
Cocolin, L., 117–118
Coelho, J.M., 292
Cohen, S.D., 366–367
Collins, T.S., 139–140
Combina, M., 226
Comitini, F., 57t, 59, 64, 79, 104–105, 107–108
Comuzzo, P., 25, 165, 167, 169t
Conde, C., 343
Condelli, N., 258
Conn, S., 262
Connel, G.E., 258
Connel, L., 223, 226
Conterno, L., 72
Contreras, A., 57t, 62, 63t
Coquard Lenerz, C., 350
Cordero-Bueso, G., 52–54, 59
Corrales, M., 44
Correa-Gorospe, I., 246
Correia, A.C., 178f, 179–181, 182f, 184t
Corte, V., 104
Cortell, J.M., 183, 367

Cosme, F., 177–179, 259
Costa, A., 225t, 231–232, 318
Costantini, A., 89, 93
Costello, P.J., 79, 90–92, 107–108
Coton, E., 93–94
Coton, M., 93–94, 105
Coulon, J., 89
Couto, J.A., 248–249
Cozzolino, D., 74
Cravero, M.C., 1–2
Creasy, L.L., 259
Crouch, A.M., 290
Cruz, L., 201, 213f, 214–215
Cuadros-Inostroza, A., 213
Cubero, S., 8, 11
Cueva, C., 94
Cuinier, C., 72
Cullere, L., 290, 293–294, 294t, 299
Culleré, L., 276–278
Cus, F., 57t, 59
Cutzach, I., 150, 295–296

D
Da Silva, C., 74
da Silva, M.A., 210f
Dagan, L., 273–275
Dahmen, M., 327
Dalié, D.K.D., 105
Dallas, C., 185–186, 207–209
Damberg, R., 11–12
Dangles, O., 199, 207
Danilewicz, J.C., 317
Darriet, P., 75–76, 274t, 276–279, 277t, 296
Dartiguenave, C., 18, 20, 22–23
Daudt, C.E., 231–232
Davenport, R., 229
David, V., 117–119
Davies, J.M., 260, 262
Davis, C.R., 87, 99, 104
Davis, R., 341
de, Mora, S.J., 273–275
de Azevedo, L.C., 294, 316t
De Beer, D., 58–59
De Coninck, G., 188
de Cort, S., 99
de Freitas, V., 186–187, 200–201, 211–212, 211f, 239, 258, 261
De Freitas, V.A.P., 177, 179
de Klerk, J.-L., 19–21, 25
de las Rivas, B., 93–94
De las Rivas, B., 91–92
De Nobel, J., 72
De Nobel, J.G., 163
de Orduna, R.M., 291
de Revel, G., 99
de Rosissart, H., 104
de Rosso, M., 188, 212–213
de Simon, B.F., 286t
De Vero, L., 117–118
De Vos, P., 99–103
de Wijk, R.A., 258
Deak, T., 224, 229
Degand, P., 258
del Alamo-Sanza, M., 136–138
Del Alamo-Sanza, M., 149, 155, 158

del Barrio-Galán, R., 167, 169t, 170, 265–266
Del Fresno, J.M., 57t, 61
Del Llaudy, M.C., 183
Del Mónaco, S.M., 349–350
Del Prete, V., 94, 105
de-la-Fuente-Blanco, A., 285
Delaherche, A., 118
Delaporte, B., 17, 207–209
Delaquis, P., 99
Delcambre, A., 213
Delcourt, F., 69
Delcroix, A., 75–76
Delfini, C., 225t, 226, 314, 318
Delgado-Vargas, F., 196–197
Delia, L., 188
Dellaglio, F., 103–104
Demiglio, P., 258
Depinho, P.G., 295–296
Dequin, S., 103
Desmazeaud, M.J., 104
Devatine, A., 152
Devéze, M., 232
Devi, A., 107
Dharmadhikari, M., 349–350
Di Gennaro, S.F., 4, 8
Di Lecce, G., 179
Di Stefano, R., 1–2, 75–76, 198, 212–213
Diago, M.P., 4
Diako, C., 240
Dias, L., 225t
Diaz-Maroto, M., 150
Dick, K.J., 107–108
Dick, R., 105
Dicks, L., 101–102
Dicks, L.M.T., 86–88
Dietrich, H., 243
Dinnella, C., 257–258
Divol, B., 119–120
Dixon, R.A., 260, 262
Dizy, M., 56–58
Dlauchy, D., 117
Dominguez-Perles, R., 366
Domizio, P., 57t, 60–61, 166
Dong, X., 103–104
Donnelly, D., 223–224
D'Onofrio, C., 367
Doria, F., 117
Doussot, F., 126, 129, 138–140
Downey, M.O., 178–182, 366–367
Draper, D.E., 264
Drawert, F., 20–21
Drysdale, G.S., 103
Du Plessis, C.S., 273
Du Plessis, G.J., 273–275
du Plessis, H., 100t, 101, 107–108
du Toit, M., 85–86, 99, 100t, 106, 152, 248–249
Du Toit, W., 150–151, 156
Du Toit, W.J., 115–116, 119, 231, 295–296
Dubernet, M., 31–32
Dubourdieu, D., 274t, 276, 297
Ducruet, J., 264
Dueñas, M., 200, 210
Duerksen, J.D., 75–76
Dugelay, I., 91

Dumitriu, G., 153
Dunbar, J., 117
Dunn, G.M., 5, 8
Dunsford, P., 245–247
Dupin, I.V.S., 73, 163
Dupuy, P., 17
Dykes, S., 150, 158
Dziezak, J.D., 101–102

E
Eastwood, A., 125
Ebeler, S.E., 75, 293
Edinger, W.D., 318
Edwards, C., 220, 225, 230–232
Edwards, C.G., 79, 99–101, 100t, 103–104, 107–108
Eggers, N.J., 290–291
Eglinton, J., 73
Ehrhardt, C., 212–213, 349
El Hage Chahine, J.M., 315, 316t
Elias, R.J., 17, 291, 294
Endo, A., 86, 104
Engel, L., 295–296
Englezos, V., 57t, 62, 63t
Erasmus, D.J., 104
Ercolini, D., 117
Ertan Anli, R., 152
Escalona, H., 187
Esch, F., 226
Eschenbruch, R., 311
Escot, S., 73, 163, 202
Escott, C., 47, 57t, 60–61
Escribano-Bailón, M.T., 177, 198, 200, 209, 214, 259
Escribano-Bailón, T., 198, 200
Escudero, A., 275–278, 278t, 283, 290, 293–294, 294t, 296, 299
Escudier, J.L., 40
Espinosa, J.C., 117
Espitia-Lopez, J., 367
Es-Safi, N.E., 200–201, 209
Esteban-Torres, M., 90–91, 107
Esteve-Zarzoso, B., 99, 117
Etienne, A., 349–350
Evaghelia, C., 366
Ewart, A.J.W., 343–344

F
Falcao, L.D., 295t, 296
Falcão, L.D., 276–278
Falcetti, M., 341
Fang, Y., 75, 290
Farhadi, K., 180t
Fariña, L., 74–75, 77, 279, 286t, 290–291
Farris, G., 224
Fay, J.C., 55
Fazzalari, F.A., 132, 134
Fell, A., 158
Fell, J., 225
Fell, J.W., 51–52
FengDan, G., 366
Fernandes, A., 199, 202
Fernández, O., 169t, 202

Fernández de Simón, B., 125–126, 134, 137, 153–155, 172, 174, 188
Fernández-González, M., 117–118
Ferrandino, A., 5
Ferreira, A.C.S., 293–294, 295t
Ferreira, V., 74, 274t, 276–278, 285, 286t, 289–290, 292–299
Ferreira da Silva, P., 198
Ferrer, S., 101–102
Ferrer-Gallego, R., 260
Feuillat, F., 126–127, 131, 136–137, 187–188
Feuillat, M., 73
Fia, G., 60
Figueiredo-González, M., 6
Filippetti, I., 367
Fine, A.M., 177
Fischer, Sven, 330–332
Fischer, U., 39–40
Flamini, R., 212–213
Flanzy, C., 76–77
Flecknoe-Brown, A., 157
Fleet, G., 220, 222–224, 228
Fleet, G.H., 51–52, 71, 103, 115, 118
Floch, M., 274t, 276, 296, 297t
Florenzano, G., 226
Foo, L.Y., 260
Formica, J., 226
Formisyn, P., 101–102
Fornachon, J.C.M., 107–108
Fors, S., 132, 134
Fournand, D., 2–3, 263
Francesca, N., 58–59
Francia-Aricha, E.M., 201, 209–210
Francis, F.J., 197
Francis, I.L., 75, 90t, 107
Franco-Luesma, E., 283, 292–293
Frangipane, M.T., 155
Fredes, C., 12, 344
Fredrickson, A., 369
Freeman, B., 226
Friedel, M., 10–11, 343–344
Friedman, R., 258
Fröhlich, J., 115
Fubing, L., 369
Fuchs, S., 105
Fugelsang, K., 220, 225, 230–232
Fugelsang, K.C., 63, 99–101
Fulcrand, H., 73, 149, 200–201, 207–209, 211–213, 239
Fuleki, T., 179, 181, 350–353, 351t
Fumi, M.D., 86

G
Gabrielli, M., 296
Gadanho, M., 117–118
Gagne, S., 262
Gajetti, M., 4
G-Alegría, E., 88
Galgano, F., 105
Galland, D., 88
Gallander, J.F., 61
Gallardo-Chacón, J.J., 163
Gallender, J.F., 187–188
Galvan, T., 346

Gambuti, A., 6, 183–184, 187–188, 241, 261, 320
Gamero, A., 299
Gandova, V., 199
Ganeva, V., 170
Ganga, M.A., 52
Gao, L., 264
García, J.F., 170
Garcia-Estevez, I., 258, 263, 265–266
García-Marino, M., 201, 368–369
García-Moruno, E., 105
García-Ruiz, A., 89, 94, 105–106
Garcia-Viguera, C., 209
Garde-Cerdán, T., 45, 132, 134, 141, 150, 319
Gardini, F., 94
Garofalo, C., 100t, 101–102
Gastón Orrillo, A., 91
Gatti, M., 6, 8
Gautier, B., 240, 242
Gawel, R., 257–259, 261
Geffroy, O., 39–40, 290, 298–299
Genebra, T., 177, 181
Gény, L., 181–182, 262
Gerbaud, V., 240, 245–246
Gerbaux, V., 104
Gernep, 336
Ghanem, C., 73, 240
Ghosh, S., 116, 118
Giannakopoulos, P.I., 351t
Gil, M., 62, 183–184
Gil Cortiella, M., 263
Giovani, G., 60
Giudici, P., 117–118, 228
Glabasnia, A., 135, 137–138, 140–141
Gladstones, J., 341–342
Glemser, E.J., 346
Glories, Y., 2, 129, 134, 136, 139, 149, 151, 177, 188
Gobbi, M., 57t, 61–62, 63t
Goddard, M.R., 71
Godden, P., 320
Gombau, J., 136
Gómez, J., 27–28
Gomez-Alonso, S., 207–209, 211, 215
Gómez-Míguez, M., 183, 199, 367
Gómez-Plaza, E., 150–152, 154f, 155, 369
Gonçalves, F.J., 178f, 184t, 202
Goniak, O.J., 273
Gonzales-Neves, G., 240
González, M.C., 168f
Gonzalez, R., 62, 89
González-Arenzana, L., 117–118
Gonzalez-Centeno, M.R., 86
Gonzalez-Manzano, S., 207
González-Manzano, S., 183–184, 198
González-Neves, G., 184–185
Gonzalez-Ramos, D., 60
González-Royo, E., 57t, 59, 265–266
González-Saiz, J.M., 153
González-SanJosé, M.L., 187–188, 262
Gonzalo-Diago, A., 1
Gordillo, B., 201–202, 366–369
Goto, S., 273–275
Goto, T., 198
Gotoyamamoto, N., 366

Grangeteau, C., 115
Grbin, P., 87t
Grimaldi, A., 91, 99, 107
Grossmann, M., 75–76
Grube, M., 118
Guadalupe, Z., 60–61, 186, 202
Guerra-Sánchez Simán M.T., 198
Guerrero, R.F., 85
Guerrini, S., 93
Guerzoni, E., 220, 222–223, 226
Guidoni, S., 5
Guillamon, J.M., 117
Guilloux-Benatier, M., 79, 108
Günata, Z., 75, 91
Günata, Z.Y., 75
Guo, C., 363
Guo, Y.Y., 311
Guowei, L., 369
Guth, H., 69, 74, 133–134, 296
Gutiérrez, C., 119
Gutteridge, C., 225t
Guzev, L., 6
Guzzo, F., 318
Guzzo, J., 88

H
Hagerman, A.E., 260–261
Haggerty, J., 289–290
Haishan, H., 363–364
Halama, R., 28
Hall, A., 341–342
Halvorson, H., 75–76
Han, G., 367
Han, G.M., 291
Hanlin, R.L., 178–182
Harbertson, J.F., 184, 201
Harrison, R., 347–348
Hashizume, K., 346, 366
Haslam, E., 185, 240
Hayasaka, Y., 201
He, F., 35, 39, 185, 199–200, 260
He, J., 199, 201, 207–209, 211
He, J.J., 369
Heard, G., 228
Heard, G.M., 51–52, 115
Hébert, E.M., 88–89
Hemingway, R.W., 260
Henick-Kling, T., 79, 86–88, 100t, 101–104, 107–108
Henschke, P.A., 69, 71–72, 92, 220, 226
Herbst-Johnstone, M., 297–298
Herderich, M.J., 261, 277, 277t
Heredia, F.J., 196
Heredia, N., 72
Heresztyn, T., 63
Hernández, L.F., 76
Hernández-Hierro, J.M., 12, 202
Hernández-Jiménez, A., 183
Hernandez-Orte, P., 57t, 58–59, 91, 153, 156
Hernéndez, M.M., 181–182
Herrero, P., 290
Herszage, J., 293
Herve, E., 278–279, 279t
Heuft, 328
Hiemenz, P.C., 237–238

Hierro, N., 52, 117–118
Hittinger, C., 219
Hjelmeland, A.K., 75, 299
Ho, K.T.V., 367–368
Ho, P., 187–188
Hocking, A., 219, 224–225
Hoffmann, C., 346f
Hofmann, T., 135, 137–138, 140–141, 257, 260
Horneber, Annette, 329
Hoshino, T., 199
Howell, G.S., 351t
Hrazdina, G., 195–197
Hu, K., 56–59, 57t
Hua, Y., 367–368
Huang, A.-C., 277
Hufnagel, J.C., 257, 260
Huglin, P., 341–342
Huiyun, L., 363
Humbel, B.M., 166
Humbert, C., 232

I
Ibeas, J.I., 224
Ichikawa, M., 184
Iland, P.G., 12
Iñigo-Leal, B., 224
Inglis, D., 345f, 348–349
Inglis, D.L., 104, 348
Inquireo-Cañas, P.M.I., 59
Iorizzo, M., 91, 99–103
Ison, R., 225t
Israelachvili, J., 238
Izquierdo Cañas, P.M., 93, 100t, 102, 104, 106–107, 319

J
Jackowetz, J.N., 291, 315–316
Jackson, D., 341–342
Jackson, D.I., 262, 341–343
Jackson, R., 152
Jackson, R.S., 20, 349–350
James, S., 226
Jami, E., 118
Jarauta, I., 295–296
Jastrzembski, J.A., 292
Jenkins, S.D., 366–367
Jenko, M., 57t, 59
Jiao, Z., 94
Jindra, J.A., 187–188
Jinghan, H., 367
Jinxing, L., 364
Jiranek, V., 71, 87t
Johnson, H., 342
Johnson, L.F., 6
Jolly, N., 118
Jolly, N.P., 55, 115
Jones, E.T., 125
Jones, G., 341
Jones, G.P., 215, 367
Jones, G.V., 61–62, 341–342
Jordão, A.M., 134–135, 177–181, 178f, 180t, 182f, 183, 184t, 187–188
Jorge Parola, A., 210f

Joseph, L., 230
Jourdes, M., 134–135, 155
Juega, M., 86
Jun, X., 158
Jurd, L., 185, 209–210, 213–214
Jussier, D., 79, 100t, 102, 107

K

Kadim, D., 154–155
Kallithraka, S., 258
Kapsopoulou, K., 61–62, 103
Karvela, E., 153–155
Kassara, S., 263
Kassemeyer, H.-H., 105
Kawamoto, H., 260
Keding, K., 28
Keefer, R.M., 17, 20–21, 245–246
Kekelidze, I.A., 367–368
Keller, P.J., 258
Keller, R., 126–127, 136–139
Kelly, M., 136, 157
Kelly, W.J., 87
Kemp, B., 341
Kennedy, J., 151
Kennedy, J.A., 178–184, 259–263, 367
Ker, K., 346
Khan, M.K., 45
Kheir, J., 290
Kim, H.M., 285
King, A., 229
King, S.W., 104
Kitts, C.L., 118
Kliewer, W.K., 17
Kliewer, W.M., 344
Knoll, C., 100t, 102–104, 107
Knorr, D., 169t
Kocabey, N., 183–184
Koegel, S., 299
Kögel, S., 346, 346f
Kolor, M.G., 274t
Kondo, T., 198
König, H., 115
Kontoudakis, N., 1–2, 17, 24
Kopeckà, M., 166
Kosmerl, T., 152
Kotseridis, Y., 278, 278t, 285–289, 345f, 366
Kourkoutas, Y., 101–102
Kovac, V., 201–202
Koyama, K., 182, 366
Kozlovic, G., 188
Kreitman, G.Y., 292–293
Krieger, S., 104, 107–108
Krieger-Weber, S., 99–103, 100t
Krones, 335
Kuhn, F., 220
Kulkarni, P., 166–168, 168f, 169t
Kunkee, R., 220
Kunkee, R.E., 17, 27, 87, 99–101
Kunsági-Máté, S., 198
Kurtzman, C., 225–226
Kurtzman, C.P., 51–52, 116–117
Kyraleou, M., 106, 180t, 181

L

Labbe, D., 283
Lacampagne, S., 262
Lacey, M.J., 276–277, 298–299, 343–344
Lacoste, P., 125
Laffort, 104
Lafon-Lafourcade, S., 17, 86, 104
Lafontaine, M., 12
Laguna, L., 1
Lallemand, 104
Lambrechts, M.G., 74
Lamontanara, A., 86
Landaud, S., 92–93
Landete, J.M., 86, 93–94, 105
Landon, J.L., 257
Lang, J., 207–209
Lanxess, 327
Larcher, R., 290–291
Larue, F., 72
Lasanta, C., 2–3, 27–30
Lasik, M., 99–101, 104
Lasik-Kurdys, M., 107
Laszlavik, M., 187–188
Lau, M.N., 294
Lawless, H.T., 257, 265
Le Moigne, M., 2
Leaw, S.N., 117
Ledoux, V., 73
Lee, C.B., 257, 265
Lee, D.F., 73
Legrum, C., 299
Lehtonen, P., 105
Lelieved, H., 230
Lencina, A., 154f
Lencioni, L., 57t, 60
Lerena, M.C., 99–103, 100t
Lerm, E., 86, 90–91, 99–104, 100t
Lesica, M., 152
Letaief, H., 3
Levine, M., 258
Levine, M.J., 258
Li, H., 244, 317
Li, J.-C., 348–349
Li, J.H., 366
Li, M., 260
Liao, H., 207, 210
Lide, D.R., 18–19
Little, D., 149
Liu, B., 103–104
Liu, L., 167–168, 169t
Liu, M., 92–93
Liu, S., 8
Liu, S.Q., 85–86, 89–90, 99, 209
Liu, Y., 117, 183
Llaudy, M.C., 157, 261
Lleixà, J., 59
Loira, I., 163, 165, 168f, 188
Loiseau, G., 72
Lombard, P.B., 262, 341–343
Lomolino, G., 183
Longo, E., 74
Lonvaud-Funel, A., 63, 79, 89, 91–94, 99, 103, 107–108, 119–120, 222–223, 226, 230–231, 248–249
Loos, M.A., 88
López, M.C., 56–58
López, N., 45
Lopez, R., 290, 295–296, 299
López, R., 134
López de Lerma, N., 348
Lopez Leiva, M.H., 28
López-Cisternas, J., 261
López-Roca, J.M., 154f
López-Solís, R., 265–266
Lopez-Vazquez, C., 285
Lorenzo, C., 368–369
Lorrain, B., 179, 180t, 181–182
Loscos, N., 273–275, 292
Loubser, G.J., 273–275
Loureiro, V., 219–220, 222–226, 225t, 228–231, 229t
Louw, L., 285
Lu, Y., 258
Lubbers, S., 72–73
Lucio, O., 99–104, 100t
Luck, G., 258
Lukic, I., 39, 348
Luo, L., 8, 9f
Lutz, M., 180t
Lytra, G., 274t, 275, 278, 283, 285, 292

M

Ma, W., 258, 261
Maarman, B.C., 107
Maarse, H., 318
MacAvoy, M.G., 62
Maccarelli, F., 55
Maccarone, E., 197
Machado de Castilhos, M.B., 43
Macheix, J.-J., 24
Macías, M.M., 133–134
Maeda, N., 258
Maga, J.A., 150
Maggu, M., 276–277
Magnelli, P., 163
Magyar, I., 62
Maicas, S., 89, 99, 106
Mainente, F., 239
Malfeito-Ferreira, M., 219–220, 222–226, 228–231, 229t, 290
Manconi, B., 258
Mannheim, C., 154–155
Manns, D.C., 350, 351t
Manos, P.S., 125
Mansfield, A.K., 349–350
Marais, J., 264, 273–275
Marchal, R., 241
Marchetti, R., 220, 222–223, 226
Marcobal, A., 93–94
Margalef-Català, M., 88
Markakis, P., 195
Marrufo-Curtido, A., 291
Martin, B., 295–296, 295t
Martin, S.R., 5, 8
Martín-ÁLvarez, P.J., 94
Martineau, B., 69, 90
Martínez, C., 52
Martínez, J.M., 167, 169t, 170
Martini, A., 70
Martini, G., 55
Martins, G., 117–118
Martorell, N., 290–291
Martorell, P., 118
Marzano, M., 115–116, 119–120
Masino, F., 167, 169t
Masqué, M.C., 104, 106

Massera, A., 100t, 102, 104, 106–107
Masson, G., 137, 150
Mateo, E.M., 100t, 105
Mateo, J., 99
Mateo, J.J., 75–76
Mateos, J.A., 74
Mateo-Vivaracho, L., 296–298, 297t
Matese, A., 4, 8
Mateus, N., 179, 180t, 200–201, 207–212, 210f, 211f, 213f, 239, 258, 261
Mato, I., 31
Matthews, A., 90, 107
Mattivi, F., 277, 277t, 299, 315
Maturano, Y.P., 99
Maujean, A., 246–247
Maury, C., 240–241, 266
Mayer, K., 20–21
Mayr, C.M., 294, 296
Mazza, G., 38, 134, 198–199
McGovern, P.E., 55
McKinnon, A.J., 245–246
McRae, J.M., 259
Medina, K., 57t, 58–59, 61, 71–74, 77
Mehlomakulu, N.N., 57t, 64
Meillon, S., 265, 283
Mekoue Nguela, J., 239, 265
Mendes-Pinto, M.M., 278
Mendoza, L., 107–108
Mendoza, L.M., 107–108
Mercurio, M.D., 344f
Merritt, A.D., 258
Meschter, E.E., 197
Messana, I., 258
Michel, J., 134–135, 150
Mietton-Peuchot, M., 27–28
Milanovic, V., 52
Miller, B.J., 100t, 102–103
Miller, D.P., 351t
Millet, V., 119–120
Mills, D.A., 51–52, 63, 88, 90, 117–118
Minarik, E., 225
Mira de Orduña, R., 3, 61–62, 103–104, 311, 315–316
Moine-Ledoux, V., 248
Moio, L., 90
Monagas, M., 73, 177–179, 183, 260
Monge, M., 89
Montealegre, R.R., 180t
Montgomery, M.W., 199–200
Morata, A., 44–46, 61, 72–73, 164–166, 168f, 201, 265, 317
Moreau, L., 314–315
Moreira, N., 57t, 58–59
Morel-Salmi, C., 184
Moreno, P.I., 57t, 61–62
Moreno-Arribas, M.V., 85, 89, 93–94, 201
Morgan, H.H., 117
Mori, K., 366
Mori, T., 366–367
Morris, J.R., 369
Morrissey, W.F., 63
Mortimer, R., 55
Mosedale, J.R., 125–126, 137
Moss, M.D., 229–230

Mouls, L., 239
Moutonnet, M., 73
Moutounet, M., 129, 134–136, 149–150, 155, 157, 187, 240, 244, 246–248
Mozell, M.R., 103–104
Mtshali, P.S., 86
Muccilli, S., 63–64
Müller, K., 258
Müller, L., 329
Muller-Späth, H., 246–247
Muñoz, R., 85, 105
Muñoz, V., 99–101, 108
Munyon, J.R., 27, 61
Münz, T., 27
Murat, M.-L., 276
Murat, M.L., 276, 296
Murray, K.E., 277t

N
Nagel, C.W., 27, 61
Naouri, P., 85
Nascimento, R.F., 294
Naumova, E.S., 55
Naurato, N., 258–259
Navarro, M., 132, 134, 136–141, 150, 155
Nave, F., 210, 214
Nayak, A., 258
Nehme, N., 79, 104, 107–108
Neradt, F., 223–224
Nesbitt, A., 341–342
Nettenbreijer, A.H., 293–294
Neuser, F., 225
Nevares, I., 136–138, 149, 369
Newton, J.L., 90t, 107
Nicolini, G., 290–291
Nicolle, P., 350, 351t
Nida, R.H., 167, 169t
Nielsen, J.C., 90, 102–103
Niero, R., 11
Nieto-Rojo, R., 319
Nieuwoudt, H., 69
Nixon, K.C., 126
Noble, A.C., 1, 17, 257, 259, 273
Nogales-Bueno, J., 12
Noguerol-Pato, R., 3, 295–296
Nonier, M.F., 187
Nordstrom, K., 74–75
Nunes, P., 187–188
Nuñez, M., 105
Nuñez, V., 180t
Nünning, Jürgen, 332
Nurgel, C., 104
Nygaard, M., 79
Nykänen, L., 74–75, 150

O
Oberholster, A., 266, 367
Obreque-Slier, E., 178–179, 180t, 181–182, 261, 265
Ochiai, N., 299
Ohloff, G., 134
Okada, S., 104
Okuda, T., 134

Olarte Mantilla, S.M., 2
Oldfield, S., 125
Oliveira, J., 207–209, 210f, 211–212, 213f, 214–215
Oliveira, M.C., 317
Ollat, N., 2–3
Ó-Marques, J., 179–181, 183
Oms-Oliu, G., 47
Onetto, C., 99–104, 100t
Ontanon, I., 293
Oppenheim, F.G., 258
Oreglia, F., 69
Oro, L., 57t, 64
Ortega, C., 285
Ortega, L., 290
Ortega Meder, M.D., 198
Ortega-Heras, M., 155–157
Osborne, J.P., 79, 99, 103–104, 294
Ough, C.S., 20–21, 21t, 71, 231–232, 264, 318
Ozawa, T., 186
Ozmianski, J., 264

P
Pace, C., 367
Padayachee, A., 202
Padilla, B., 58, 99
Palomero, F., 57t, 60–61, 163–167, 169t, 171–172, 174, 202
Pan, W., 100t, 104, 107
Panceri, C.P., 348
Pandya, S., 120
Pantelic, M.M., 179
Pardo, I., 226
Pardo-García, A.I., 366
Pardo-Minguez, F., 154f
Parenti, A., 11
Parish, M., 150
Park, S.K., 71
Park, Y.H., 79, 100t, 103–104, 107–108
Parker, M., 276
Parley, A., 369
Parpinello, G., 152
Pascal, C., 261
Pascual, O., 136, 319
Passos, C.P., 177
Pasteur, L., 309
Patel, K.K., 12
Paterson, A., 134
Pati, S., 209
Patrignani, F., 106
Paul, R., 150, 157
Pavez, C., 297t
Payne, C., 258
Pedneault, P., 349–350, 351t
Peinado, R.A., 286t
Pellenc, R., 11
Peña-Neira, Á., 263
Peng, C.-T., 278, 278t
Peng, S., 140, 187
Peng, Z., 240
Pengcheng, L., 362
Pérez, G., 76
Pérez Prieto, L.J., 149–150
Pérez-Coello, M.S., 150

Pérez-Magariño, S., 156, 262
Pérez-Martín, F., 90–91, 105
Perez-Prieto, L.J., 187
Pérez-Través, L., 265
Peterson, R.G., 136
Petka, J., 296
Petrova, B., 318–319
Petrova, I., 199
Petruzzi, L., 99–101, 105–106
Peynaud, E., 17, 19, 76–77
Peyrot des Gachons, C., 261
Phister, T.G., 63, 120
Piano, F., 298
Piao, H., 99
Picard, M., 274t, 275–276, 279, 279t, 300, 300t
Pickering, G., 258, 348–349
Pickering, G.J., 103–104, 258, 345–350, 345f, 346f, 351t
Pigeau, G.M., 104
Piggott, J., 134
Pilatte, E., 99–103
Pilone, B.F., 246
Pilone, G.J., 209
Pineau, B., 278
Pingret, A., 158
Pino, J.A., 134
Pinto, C., 118–119
Pinto, M., 51
Pinzani, P., 86
Piotrowska, M., 105
Pirie, A.J.G., 344
Pisciotta, A., 6
Pissarra, J., 187, 200, 209–210, 211f, 214
Pissarra, J.I., 209
Pitt, J., 219, 224–225
Pittet, A., 133–134
Pizarro, C., 157, 290–291
Plata, C., 58–59
Poitou, X., 279, 344
Polgue, H., 127, 136–139
Pollnitz, A.P., 150, 290–291
Polo, M.C., 85, 93–94
Polsinelli, M., 55
Poncet, C., 238–239
Poncet-Legrand, C., 239
Pons, A., 276–279, 277t, 279t, 295–296
Popper, L., 169t
Porep, J.U., 11–12
Pour Nikfardjam, M.S., 350, 351t
Pozo-Bayón, M.A., 90–91, 99–101, 106
Prahl, C., 79, 86, 99–103
Prajitna, A., 367
Prakash, S., 239–240
Prakitchaiwattana, C.J., 117–118
Prenesti, E., 20
Pretorius, I.S., 51–52, 55, 63, 74, 85–86, 104, 115–116, 248–249
Prida, A., 132, 134, 137
Prieur, C., 177–178
Prinz, J.F., 258
Pripis-Nicolau, L., 92–93
Proffitt, A.P.B., 4
Provost, C., 349–350, 351t
Puech, J.L., 134, 137
Puertas, B., 44
Pulvirenti, A., 228

Q
Qian, M.C., 75, 290
Qiang, Z., 369
Queris, O., 134
Querol, A., 219, 224, 228
Quesada, M.P., 117
Quideau, S., 134, 187, 200
Quijada-Morín, N., 201, 258, 262
Quintela, S., 105

R
Radler, F., 101–102
Ramey, D.D., 71
Ramos, M.J., 210f
Ranjard, L., 118
Rankine, B.C., 61, 69
Raposo, R., 319
Rapp, A., 74, 295–296
Raspor, P., 52
Rauhut, D., 292–293
Raut, P., 273–275
Rayess, Y.E., 27–28
Rayne, S., 290–291
Reazin, G., 74–75
Reguant, C., 117
Regueiro, L.A., 52
Reichart, O., 229
Remy, B., 127
Remy, S., 210
Remy-Tanneau, S., 209–210
Renault, P., 59
Renouf, V., 63, 104, 117–118, 222–223, 226, 230–231, 318
Rentzsch, M., 201
Restuccia, C., 63–64
Revilla, I., 187–188
Rey-Caramés, C., 4, 6–8, 7f
Reynolds, A., 265
Reynolds, A.G., 341
Reynou, G., 243
Rhein, Stephan, 323
Ribéreau-Gayon, J., 136
Ribéreau-Gayon, P., 1, 17, 19–25, 27–28, 30–31, 99–104, 127, 131–132, 134, 136, 139, 141, 149, 151, 183, 220, 231–232, 241–242, 264, 266, 274t, 314–315, 350–353
Ricardo-da-Silva, J.M., 177, 179, 180t, 181, 261, 266
Rice, S., 179–181, 349
Richelieu, M., 90
Richter, M., 11
Rigou, P., 274t, 276
Rinaldi, A., 180t, 261
Rio Segade, S., 2, 11, 347–348
Riou, V., 239, 260
Ripperger, S., 244
Ristic, R., 12, 263
Rivas-Gonzalo, J., 211f
Rivas-Gonzalo, J.C., 207–209
Rizk, Z., 104, 107–108
Robert, V., 224
Robichaud, J.L., 1
Robnett, C.J., 116–117
Rodas, A.M., 103–104
Rodrigues, A., 165, 186, 202
Rodrigues, N., 63, 227–228
Rodriguez, H., 107
Rodríguez, M.E., 59
Rodríguez-Bencomo, J.J., 157, 163, 165, 297–298
Rodriguez-Clemente, R., 246
Rodríguez-Pulido, F.J., 12
Rodríguez-Rodríguez, P., 153–155
Rojas, V., 57t, 58–60
Roland, A., 274t, 276, 298
Rolle, L., 1–3, 5, 11, 62, 63t
Romano, A., 134, 285
Romano, P., 74, 220
Romat, H., 243
Romero, C., 149, 197
Romero-Cascales, I., 2
Romeyer, F.M., 2–3, 259
Rosi, I., 60, 73, 75–76, 79, 100t, 103–104, 108
Rosini, G., 51–52
Ross, C.F., 240
Rossouw, D., 108
Roujou de Boubée, D., 276–277, 277t, 343–344, 346
Roujou de Boubee, D.R., 346
Rous, C., 153–155, 187–188
Rouse Jr., J.W., 4
Rowan, N.J., 47
Royer Dupré, N., 275–276
Rui, G., 365
Rui, S., 365
Ruirui, L., 369
Ruiz, P., 117–118
Ruiz-Garcia, Y., 184–185
Ruiz-Hernández, M., 76–77
Russo, P., 105
Ryan, D., 299
Ryona, I., 276–277, 345, 347

S
Sabon, I., 278, 278t
Sacchi, K.L., 60, 184, 264, 347–348
Sacks, G.L., 350
Saerens, S., 99–101, 100t, 103–104
Saguir, F.M., 91
Sahu, L., 105
Saint-Cricq de Gaulejac, N., 141
Sala, C., 276–277, 298t, 299, 347
Salagoity-Auguste, M.H., 134
Salas, E., 200, 207–210, 214
Salma, M., 119–120
Salmon, J.-M., 163
Salvetti, E., 52–54
Sampaio, J.P., 117–118
San Juan, F., 286t, 290, 292t, 296, 297t, 300t
San Româno, M.V., 103–104, 107–108
San-Juan, F., 274t, 275–276, 283, 285, 286t, 293–294, 295t
Santisi, J., 103–104
Santos, A., 64
Santos, A.O., 8
Santos, H., 207–209
Santos, M.C., 89, 317
Santos-Buelga, C., 198, 209–210
Santos-Lima, M., 184–185
Sanz, M., 125, 139–140, 165, 172, 188
Sarni-Manchado, P., 197
Sarrazin, E., 274t, 296

Sasaki, K., 295–296
Sasamoto, K., 299
Sato, H., 117
Saucier, C., 73, 149, 187, 209, 213
Scafidi, P., 366
Scalbert, A., 134
Scharbert, S., 260
Schlich, P., 283
Schmarr, H.-G., 299
Schmid, F., 36, 118
Schmidtke, L.M., 103–104
Schmoll, W., 329
Schneider, R., 295–297
Schöllhammer, L., 333
Schöltz, M., 107
Schreier, P., 273–275
Schroën, C.G.P.H., 266
Schroën, K., 266
Schuller, D., 55, 228
Schumaker, M., 219
Schumaker, M.R., 63
Schwab, W., 2–3
Schwarz, B., 260–261
Schwarz, M., 72, 201, 207–209, 211
Scollary, G.R., 240, 260, 266
Sefton, M., 75
Sefton, M.A., 150
Seguin, G., 341
Ségurel, M.A., 273–275, 292
Seifert, R.M., 133–134
Serpaggi, V., 119
Sestelo, A.B.F., 107
Setati, M.E., 118
Setford, P.C., 348
Shah, M., 6
Shankar, P.S., 117–118
Shanshan, H., 367
Shaw, A., 341
Shenoy, V.R., 199–200
Sheridan, M.K., 17
Shpritsman, E.M., 28
Sidhu, D., 345–347
Siebert, T.E., 289–290, 292–293, 292t, 299
Siebrits, L., 117–118
Sierra, Lisa, 332
Sierra, Tony, 332
Sies, A., 69, 74
Silla Santos, M.H., 93
Silva, A.M.S., 213f
Silva, F.V.M., 317
Silva, M.A., 328
Silva, S., 61
Simó, G., 101–102
Simões, S., 178f, 184t
Simpson, R.F., 273–275, 298
Sims, C.A., 369
Singleton, V.L., 125, 259, 264
Sivcev, B.V., 349
Skouroumounis, G.K., 75–76
Slegers, A., 350
Smit, A., 71–72
Smit, A.Y., 93, 100t, 104, 106
Smith, P.A., 184–185, 261, 266, 348
Smyth, H.E., 74
Snow, P.G., 61

Soares, S., 241, 260
Somers, T.C., 185, 210
Soufleros, E., 93
Souquet, J.M., 177–178, 180t, 259
Sousa, A., 207–209
Sousa, C., 187, 200, 210, 214
Sowalsky, R.A., 17
Spagna, G., 76
Spano, G., 86, 107
Sparks, A.H., 314–315, 316t
Sparks, T.C., 21t
Sparrow, A.M., 347
Spayd, S.E., 344, 366–367
Spedding, D.J., 273–275
Spedding, F.H., 20
Spillman, P.J., 134, 137, 165
Splittstoesser, D.F., 225, 318
Springer, L.F., 350, 351t
Srihanam, P., 180t
Stafford, P.A., 318
Stahl, W.H., 132, 134
Stamp, C., 132
Steele, J.T., 27
Steels, H., 225
Sterckx, F.L., 134
Stewart, A.C., 350
Stratford, M., 219–220
Strauss, M.L.A., 56–58, 167
Styger, G., 274t
Su, C.T., 259
Suárez, R., 150, 226, 290
Suárez-Lepe, J.A., 44, 57t, 61, 164–166, 168f, 224
Sudraud, P., 314
Sumby, K.M., 85–86, 90–91
Sun, B.S., 177–179, 180t, 183
Sun, D., 158
Sun, Q., 278, 279t
Sun, S.Y., 99–103
Sun, Y., 117
Suriano, S., 104
Suzzi, G., 311
Swain, T., 259
Swiegers, J.H., 60, 69, 71–72, 74, 89, 90t, 91–93

T
Taillandier, P., 232
Takamuro, R., 273–275
Tanahashi, H., 246
Tanaka, T., 151
Tao, Y., 44–45, 154–155, 158
Taransaud, J., 125, 128–129
Tarara, J., 366–367
Tarara, J.M., 366–367
Tarter, M.E., 3
Tavares, M., 188
Taylor, M.W., 118
Teissedre, P.L., 132, 134, 137, 139–140
Teixeira, N., 213f
Ten Brink, B., 93
Tena, M.T., 290–291
Teresa, F.-E.M., 55
Terrier, N., 260
Tesfaye, W., 168f

Tetrapak, 328
Thach, L., 103–104
Thomas, D.S., 220, 229
Thorngate, J.H., 259
Thoukis, G., 19–21
Timberlake, C.F., 73, 185, 197, 200–201, 207–211
Tiwari, B.K., 45
Tolin, S., 241
Tominaga, T., 274t, 276, 296–297, 297t
Tonietto, J., 341–342
Torchio, F., 2–3, 11
Torrea, D., 105–106
Torrens, J., 106
Torres, R.E., 344
Toth, T., 62
Tounsi, M.S., 181
Tredoux, A., 290
Tristezza, M., 57t, 106–107
Trouillas, P., 198, 207
Troxler, R.F., 258–259
Tudosesanduville, S.T., 367–368

U
Ugliano, M., 90–91
Ulbricht, M., 266
Umeda, N., 346
Upadhyay, R., 258
Usbeck, J.C., 119
Usseglio-Tomasset, L., 17–20, 22–23, 314, 315t

V
Vaclavik, L., 213
Vaillant, H., 101–102
Val, P., 103
Valente, I.M., 290–291
Valero, A., 55
Valero, E., 52, 55, 70, 181, 222–223
Vallée, D., 247–248
Vallet, A., 92–93
Valls, J., 1
Vallverdu-Queralt, A., 213, 215
Vally, H., 89
Van der Westhuizen, L.M., 88
van Gemert, L.J., 293–294
Van Leeuwen, C., 341
van Oss, C.J., 238
van Vyk, S., 317
Varela, C., 55, 57t, 59, 62, 63t
Vasserot, Y., 72, 74, 265
Vazallo-Valleumbrocio, G., 177
Vela, E., 292
Vendrame, M., 99, 228
Verginer, M., 51
Vernhet, A., 240, 244
Vernocchi, P., 73
Versari, A., 319, 343
Versini, G., 74
Viana, F., 58–59, 99
Vichi, S., 298
Victor, D.F., 369
Vidal, S., 1, 60, 151, 209, 239, 259, 265–266
Villalba, M.L., 57t, 64
Villamor, R.R., 240

Villen, J., 285
Villettaz, J.-C., 241, 243
Vincenzini, M., 94, 106
Vinet, E., 314–315
Viriot, C., 187
Vivar-Quintana, A.M., 155
Vivas, N., 125–129, 133–134, 136–139, 141, 149, 188
Volschenk, H., 20, 25

W

Wallbridge, P.J., 317
Walling, E., 86
Wang, B., 367
Wang, H., 11–12
Ward, S.C., 9
Wardle, D.A., 341
Waterhouse, A.L., 291–292, 316, 319
Waters, E.J., 239–240
Watrelot, A.A., 263
Watson, B.T., 264–265
Wedler, H.B., 278
Wedral, D., 290
Weeks, S., 150
Wei, L., 367
Wei, Q., 363
Wejnar, R., 20, 22
Weldegergis, B.T., 290
Wen, Y., 283
Werner, M., 311
Wesche-Ebeling, P., 199–200
Whitty, M., 8
Wibowo, D., 17, 19, 25–26
Wigand, P., 239
Wilhelm, F.G., 28
Wilkinson, K.L., 134
Willwerth, J., 341, 349
Winkler, A., 341–342
Winterhalter, P., 75–76, 211
Wollan, D., 136, 157
Wongnarat, C., 180t
Wood, C., 277, 277t, 299, 300t
Wucherpfennig, K., 28, 243
Wu-Dai, Z., 262
Wüst, M., 2–3

X

Xiangyun, C., 369
Xiuping, G., 363
Xiuwu, G., 363

Y

Yamane, T., 366–367
Yan, H., 363
Yan, L., 369
Yan, Q., 258–259
Yang, N., 45, 169, 169t
Yanhui, G., 366
Yanlin, L., 369
YanLong, L., 125
Yeramian, N., 103
Yuan, L.M., 11

Z

Zakowska, Z., 105
Zamora, F., 19, 125, 128–129, 131–134, 136, 139, 141, 264
Zanchi, D., 239, 258
Zanoni, B., 1
Zapata, J., 294
Zapparoli, E., 100t, 103
Zapparoli, G., 117
Zarraonaindia, I., 54–55
Zea, L., 209
Zhang, B., 134, 368–369
Zhao, J., 259–260, 262
Zhao, P., 277t
Zhao, P.T., 290, 299
Zhao, Y., 8
Zimmerli, B., 105
Zironi, R., 25
Zoecklein, B.W., 63
Zott, K., 58–59
Zouid, I., 2
Zuñiga, A., 12
Zupan, J., 57t, 64

Subject Index

Note: Page numbers followed by "*f*" and "*t*" refer to figures and tables, respectively.

A

Abscisic acid, 182
Acacia (*Robinia pseudoacacia*), 125
Accelerated AOL, 167–171
   lees impregnation with wood-extract procedure, 170*f*
   ultrasound treated lees, 168*f*
Accelerated withering, 52–54
Accentuated Cut Edges (ACE), 347
Acetaldehyde, 73, 209, 291–292, 291*t*
Acetates, 72
Acetic acid, 11–12, 19
   bacteria, 115–116
Acetovanillone, 156
Acid, 343
   hydrolysis, 291
   increase in sugar level and decrease in acidity levels, 343*f*
Acid–base equilibrium of wine, 20, 22–23
Acidic fraction
   laboratory techniques for measuring, 30–32
   in red wines, 17
Acidic PRPs (aPRPs), 258
Acidification, 23
   activities, 61–62
   by blending with musts or wines, 24
   by electromembrane techniques, 27–29
   by supplementation with organic acids, 25
Acidity, 101–102, 349–350
   variations during winemaking, 20–21
Acutissimins, 187
Acylated anthocyanin monomers, 72
Acylated monoglucosides, 195
Additive-free winemaking approach, 24
Additives, 248
Advanced chromatographic techniques, 1
Advanced optical berry sorting systems, 12
Adventitious yeast, 219
AEM. *See* Anion exchange membranes (AEM)
AF. *See* Alcoholic fermentation (AF)
A–F pigments. *See* Anthocyanin–flavanol pigments (A–F pigments)
Aging, 79–80
   A-type vitisin-derived pigments formed in red wines during, 211–212
   anthocyanin-derived pigments formed in red wines during, 209–210
   natural evolution of proanthocyanidins during, 185
   of red wines, 185

wine
   effect of adding oak chips on wine characteristics, 153–155
   in barrels, 149–150
   combined used of MOX + chips, 156–157
   comparing effect of chips or MOX with aging wine in barrels, 155–156
   in French or American oak barrels, 165
   innovations in MOX and chips application, 157–159
   MOX technique, 150, 151*f*
   MOX technique application, 152–153
   positive factors of using MOX, 151–152
   use of oak chips, 153
Aging on lees (AOL), 163, 169*t*
   accelerated AOL, 167–171
   LC-RID chromatograms of polysaccharides, 164*f*
   lees aromatization, 171–174
   non-*Saccharomyces* yeasts, 166–167
   plasmatic membrane and cell wall structure, 164*f*
"Aglianico di Taurasi" red grape, 58–59
Alcohol, 342–343
   content in red winemaking, 71–72
Alcoholic fermentation (AF), 79, 85, 87, 99, 115, 153, 311. *See also* Malolactic fermentation (MLF)
   incidence on yeast development during, 151
Aldehydes, 294, 294*t*
Alkaline hydrolysis, 291
Alkylmethoxypyrazines, 298–299, 298*t*
Allier oak (*Quercus petraea*), 137–138
American oak (*Quercus alba*), 126, 137, 188
Amino acids, 298
Ammonium hydrogen sulfite (NH_4HSO_3), 313
Anion exchange membranes (AEM), 247–248
Anthocyanidin reductase (ANR), 262
Anthocyanin-alkyl/aryl-flavanol
   adducts, 211*f*
   pigments, 214
Anthocyanin-derived pigments
   formed in red wines during aging, 209–210
   found in red grapes and wines, 209
Anthocyanin–flavanol pigments (A–F pigments), 199
Anthocyanins, 1–3, 35, 38, 60, 149, 195, 202, 207, 208*f*, 265, 350–353, 366. *See also* Proanthocyanidins (PA)

accumulation by viticulture process, 366–367
compounds formation by yeast fermentation, 73–74, 73*f*
derivative pigments formation, 200–201
   flavanol–anthocyanin condensation products, 200
   pyranoanthocyanins, 200–201
oxidation, 199–200
stability, 195–198
   effect of bisulfite, 197
   chemical structure of anthocyanins, 195–197
   effect of oxygen, 198
   effect of pH, 197
   effect of temperature, 197
Antimicrobial activities, 63–64
   substitutes, 318–319
Antioxidant, 309
   activity, 163
   substitutes, 319–320
Antioxidase, 309
Antiseptic activities, 309
AOL. *See* Aging on lees (AOL)
Appassimento-style red wines, 348
aPRPs. *See* Acidic PRPs (aPRPs)
Aroma, 130, 150, 156, 343–344, 350–353
   aroma descriptors, wine concentrations, and odor thresholds of esters, 90*t*
   aroma-active carbonyls, 294–295
   compounds, 155, 284
      miscellaneous, 278–279
      structure, odor, concentration in wine, 279*t*
   profile, 58–60
   of putrefaction, 256
   skin and seed tannin concentrations, 344*f*
Aroma-active compounds
   alkylmethoxypyrazines, 298–299, 298*t*
   "analysis of difficult" aroma compounds in red wine, 291–300
   "analysis of easy" aroma compounds, 285–290
   analytes and analytical classification, 284–285
   compounds
      ethyl 4-methylpentanoate, 299–300
      ethyl cyclohexanoate, 299–300
      piperitone, 00020#p0545
      rotundone, 299
   highly polar compounds, 295–296, 295*t*
   hydroxyesters, 296

383

Aroma-active compounds (*Continued*)
  polyfunctional mercaptans, 296–298, 297t
  specific analysis of volatile phenols, 290–291
  strecker aldehydes and odor-active aldehydes, 293–295
  VSCs, 292–293, 292t
Aromatic compounds in red varieties
  aromatic compounds selection with distinct impact
    sulfur compounds, 273–279
Aromatic yeasts, 219
Artificial drying, 129
  comparison between natural seasoning and artificial drying, 139t
  influence, 138–139
Ascorbic acid, 319
Ash (*Fraxinus excelsior*), 125
Aspergillus carbonarius, 6
Aspergillus species, 105
Astringency, 35, 38, 257, 261, 267
  improvement of, 151
  influence of winemaking technology, 261–266
  mechanisms, 258–261
  in wines, 258–261
A-type vitisin-derived pigments formed in red wines, 211–212, 212f
Aureobasidium pullulans, 220
Automated grape cluster selection and harvest, 5, 8
Automated image analysis, 11
Automated ribosomal IGS analysis, 118
Automatic bottle burst programs, 328
Automatic sorting tables, 10
Automation standards mechanic fillers, 331

B

B-ring substitution pattern, 260–261
BA. *See* Biogenic amines (BA)
Back-pulsing, 244
Bacteria strains, 256
Bacterial identification, 119
Bag in box (BIB), 332–333
Balloon-style bottling fillers, 329
Barrel(s), 125
  aging, 143–144, 143f
    assembly and toasting of barrel, 129–130, 130f
    barrels types and barrel parts, 130–131, 131f
    drying systems, 129
    forests providing wood for cooperage, 126
    influence of botanical and geographic origin, 137–138, 137t
    influence of natural seasoning and artificial drying, 138–139
    influence of repeated use of barrels, 141–143, 142f
    obtaining staves, 127–129, 128f
    oxygen permeability of oak wood, 136
    phenolic compounds released by oak wood during, 134–136
    toasting level influence, 139–140, 140f, 141f
    tree species used in cooperage, 125
    volatile substances released by oak wood during, 132–134
    wine during, 131–132
    wines, 149–150
    wood grain concept in cooperage, 126–127
    wood grain influence, 136–137, 136f
  comparing effect of chips or MOX with aging wine in, 155–156
Basic PRPs (bPRPs), 258
BEDD index. *See* Biologically effective degree-day index (BEDD index)
Beech (*Fagus sylvatica*), 125
Bending, 129–130
Benzylmercaptan (BM), 296
Benzylthiol, 276
Berry sorting, 9
β-damascenone, 278
β-glucanase enzymes, 167
β-glycosidase activity (β-Glu), 73
  activity of grape fruit, 367
β-ionone, 278
β-irradiation, 46
β-methyl-γ-octalactones, 134, 140
BIB. *See* Bag in box (BIB)
Binding agent, 309
Biocontrol, 63–64
Biogenic amines (BA), 93–94, 101, 105–106, 220, 350
Biological deacidification, 101–102, 349–350
Biologically effective degree-day index (BEDD index), 341–342
Biopreservation, 63–64
Bipolar membranes, 28
Bisulfite
  addition to anthocyanins, 214f
  effect, 197
Blending, 79–80
BM. *See* Benzylmercaptan (BM)
Botanical origin influence, 137–138, 137t
  influence of French and American oak, 138f
Bottled wine, 222
Bottling
  hazards in bottling red wine, 325–333
  lines, 324–325
  plants, 337
  process in winery, 325f
Bousinage, 129–130
Boxing and wrapping machines for bottles, 336–337
  filling equipment of 12 ha winery, 337f
bPRPs. *See* Basic PRPs (bPRPs)
BR-forms, 293
Brettanomyces, 323
  B. bruxellensis, 91–92, 115–116, 166, 166f
  in wines, 45
  yeasts, 37, 91–92
Brownian motion, 237
Bud pad, 364
Buds adaptability to low temperature, 364
Buffer capacity in wine, 22–23
Bulk wine, 222
  monitoring, 227
Buried viticulture, 357–358
  effect on vines, 360–363
Burnt sugar family, 295

C

C_{13}-norisoprenoides, 278
Cabernet franc, 342, 348–349
Cabernet Sauvignon, 260, 348–349
  seeds, 179
Cadaverine, 256
CAgC. *See* Colloidal silver complex (CSC)
Calcium carbonate ($CaCO_3$), 101–102
Calcium sulfate ($CaSO_4$), 101–102
Calcium tartrate (CaT), 245
Canary pine (*Pinus canariensis*), 125
Candida, 51–52, 56–59
  C. guilliermondii, 58–59
  C. pyralidae, 53t, 64
  C. tropicalis, 52
  C. zemplinina, 108
Capillary electrophoresis single-strand conformation polymorphism (CE-SSCP), 118
Carbon dioxide release (CO_2 release), 240–241
Carbonate salts, 349–350
Carbonyl compounds, 89–90
Carboxypyranoanthocyanins, 211–212
Carton packaging, 332
CaT. *See* Calcium tartrate (CaT)
Catechins, 61, 177, 178f, 195, 259
Cation exchange resins, 30
CCPs. *See* Critical control points (CCPs)
CE-SSCP. *See* Capillary electrophoresis single-strand conformation polymorphism (CE-SSCP)
Cell lysis effects on anthocyanins, 72–73
Cell membrane, 364
Cell wall
  anthocyanins adsorption, 72
  polysaccharides, 166–167, 167f
Centrifugation, 241–242
Ceramic cross-flow membrane, 266
Certified management systems, 327–328
Château barrels, 131
Chemical acidification, traditional strategies for, 23–25
  by blending with musts or wines, 24
  by supplementation with organic acids, 25
Chemical composition of grapes in cool and warm climates
  acid, 343
  flavor and aroma, 343–344
  green compounds sources, 345
  "greenness" in red wines, 344
  preventing greenness in vineyard, 345–346
  remediating must and wine with elevated MP levels, 346–347
  sugar/alcohol, 342–343
Chemical contaminations, 327–328
Chemical deacidification, 101–102, 349–350
  traditional strategies for, 25–27
Chemical preservatives, 231–232
Chemical stabilization, 327
Chemical structure of anthocyanins, 195–197

SUBJECT INDEX

Cherry (*Prunus avium*), 125
Chestnut (*Castanea sativa*), 125
   aromatized biomasses, 174
Chips
   comparing effect with aging wine in barrels, 155–156
   innovations in chips application, 157–159
      continuous ultrasound system, 159f
   in treatment with chips, 158–159
Chitosan, 232
Chlorine, 333
Chromatic characteristics, 1
1,8-Cineole, 278
Cintrage, 129
CIRWG index, 5
Citric acid, 19, 25, 89–90
Clarification of red wine, 230–231, 240–245, 242f
   centrifugation and, 241–242
   filtration, 242–245
   by settling, with or without fining aids, 240–241
Classic electrodialysis, 27–28
Climate changes, 17
Clonal effects, 277
Closed bottles treatment, 334
Closures
   cork, 333–334
   screw caps, 334
   vinolok, 334
Coccinella septempunctata, 345
Coccoid species, 89
Coinoculation, 107
Cold regions with short maturity periods, red winemaking in
   anthocyanin accumulation by viticulture process, 366–367
   buried viticulture, 357–358
      effect on vines, 360–363
      freeze damage, 365
      grape vine
         influence of low temperature on different tissues, 364
         wintering adaptability and cold resistance, 363–364
      influence of low temperature on grape cells, 364–365
      maturity analysis, 365–366
      vine unearthed in spring, 358–360
      winemaking technology, 367–369
Cold soak, 38
Cold stabilization, 247
Colloidal
   aggregates, 239
   instabilities in red wines and prevention, 239–240
   interactions, 237–239
Colloidal silver complex (CSC), 319
Colloids, 237–239
Color, 195
   evolution, 199–201
      anthocyanin oxidation, 199–200
      formation of anthocyanin derivative pigments, 200–201
   stability, 60–61

Commercial yeasts, 78–79
Compound mass in cap, 334
Concentration product (CP), 245
Concrete tanks, 38
Condensed tannins, 257
Controlled fermentation, 55
Cool climates, red winemaking in
   chemical composition of grapes in cool and warm climates, 342–347
   cool climate grape varieties in Northern and Southern hemisphere, 342
   innovations in cool climate winemaking, 347–349
      appassimento-style red wines, 348
      red icewine, 348–349
   making wine from red IH varieties, 349–350
   region classification, 341–342
   tannins and anthocyanin, 350–353
   YAN, 350
Cooperage
   forests providing wood for, 126
   tree species used in, 125
   wood grain concept in, 126–127
Copigmentation, 198–199
   factors affecting, 199
   technological tools to enhancing, 201–202
Copiotrophic oxidative ascomycetes, 51–52
Copiotrophic strongly fermentative yeasts, 51–52
Cork, 333–334
Corker, 223–224
Coumarates, 72
Coumarins, 134
Covalent chromatography, 297
CP. *See* Concentration product (CP)
Critical control points (CCPs), 327–328
Cross-flow microfiltration, 243–245
Cryptococcus, 51–52
Crystallization nuclei, 27
CSC. *See* Colloidal silver complex (CSC)
Culture-dependent *vs.* culture-independent methods, 119–120
   identification and fingerprinting methods, 120f
Cyanidin, 181, 195
Cysteamine, 294

D
DAP. *See* Di-ammonium phosphate (DAP)
Deacidification, 25–26, 349–350
   activities, 61–62
   biological, 101–102, 349–350
   chemical, 101–102, 349–350
   by electromembrane techniques, 27–29
   by using processing aids, 26–27
Deacidification, physical, 101–102
Dead-end filtration, 242–243
Debaryomyces, 51–52, 56–58
Debaryomyces carsonii, 53t
Degree of instability, 246–247
Degree of polymerization, 177
Dekkera bruxelensis, 227–228
Dekkera/Brettanomyces, 51–52, 63
Dekkera/Brettanomyces bruxellensis, 226

Délestage. *See* Rack and return technique
Delphinidin, 181, 195–196
Delta Rflow, 10
Di-ammonium phosphate (DAP), 71
Diacetyl, 256
Diacetyl (2,3-butanedione), 89–90
Dichloromethane, 285–289
Digital imaging, 8
Dihydroxylated group, 260–261
Dimers, 177
Dimethyl dicarbonate (DMDC), 231–232, 318
Dimethyl sulfide (DMS), 273–276, 292
   structure, odor, concentration in wine, 274t
Dimethyl sulfoxide (DMSO), 273–275
3,5-Dimethyl-2-methoxypyrazine (MDMP), 298
Dimethyl-MP (DMMP), 345
Disinfectants, 327–328
Disk-stack centrifuges, 241–242
Dispersions, 237
4,4'-Dithiodipyridine (DTDP), 298
DMDC. *See* Dimethyl dicarbonate (DMDC)
DMMP. *See* Dimethyl-MP (DMMP)
DMS. *See* Dimethyl sulfide (DMS)
DMSO. *See* Dimethyl sulfoxide (DMSO)
DNA-based methods, 116–119
   automated ribosomal IGS analysis, 118
   CE-SSCP, 118
   gradient gel electrophoresis, 117–118
   NGS, 118–119
   PCR-RFLP, 117
   QPCR and RT-qPCR, 118
   RAPD-PCR fingerprints, 117
   T-RFLP, 117
Drying, 257
   methods, 348
   systems, 129
DTDP. *See* 4,4'-Dithiodipyridine (DTDP)

E
E-beam irradiation, 46
EC. *See* Ethyl carbamate (EC)
Eiswein in Germany and Austria, 348–349
Electrochemical micro-oxygenation (ELMOX), 157
Electrodialysis, 247–248
Electrolyzed water, 35, 47
Electromagnetic flow meters, 331
Electromembrane techniques, acidification and deacidification by, 27–29
Electronic fillers, 331
Electrophilic characteristics, 207–209
Electrostatic interactions, 238
Ellagic acid, 174
Ellagitannins, 134–135, 150, 200
ELMOX. *See* Electrochemical micro-oxygenation (ELMOX)
Emerging technologies, 35
Environmental factors, 182
Enzymatic
   activities, 56–58, 57t
   analyzers, 31
Enzymes, 369
   microbial diversity assessment through enzymes detection, 119

(−)-Epicatechin, 259
  monomers, 177, 178f
(−)-Epicatechin-3-O-gallate, 259
Esterification index (IE), 248
Esters, 75, 90–91, 278
Ethanethiol (EtSH), 292
Ethanol, 87
  content reduction, 62, 63t
  wood-extract, 171
  yield reduction, 103–104
Ethyl 4-methylpentanoate, 299–300
Ethyl and methyl thioacetates (MeSAc and EtSAc), 292
Ethyl carbamate (EC), 85, 94, 105–106, 220
Ethyl cyclohexanoate, 299–300
Ethyl esters, 163, 299–300
Ethyl phenols, 256
Ethylacetate, 227
4-Ethylcatechol, 290–291
4-Ethylguaiacol, 290
Ethylphenol, 134, 165, 227, 290
EtSH. See Ethanethiol (EtSH)
Eucalyptol. See 1,8-Cineole
Eucalyptus sp. trees, 279
Eugenol, 150
Exothermic process, 199
Extracellular enzymatic activities, 56–58

F

Fatty acids, 152–153
FERARI index, 5
Fermentation, 70, 183, 220, 263–266. *See also* Alcoholic fermentation (AF); Malolactic fermentation (MLF)
  alcoholic, 151
  biotechnologies, 44
  controlled, 55
  flavor compounds produced by yeasts, 74f
  grape must, 115
  maceration process during, 58–59
  in red wines, 35
  spontaneous must, 55
  wine, 220–221
  yeast, 73–74, 73f
Fermentative maceration of red wines, proanthocyanidins during, 183–185
FFT. *See* Furfurylthiol (FFT)
Field sampling, 253
Filling
  line processes, 324–325
  oxygen avoidance with, 328–329
  speed, 331
Film-forming species, 224
  yeast tolerance to antimicrobial agents in winemaking, 225t
Filtration, 230–231, 242–245, 323, 326
  cross-flow microfiltration, 243–245
  dead-end, 242–243
  and microbial stabilization of wines, 245
Fining
  agent, 266, 309
  clarification by settling, with or without fining aids, 240–241
  yeast populations control in wines, 230–231
"Flash détente" technique, 40
Flash pasteurization, 326
FLAV_UV index, 5
Flavan-3-ol monomers, 259
Flavanol–anthocyanin condensation products, 200
Flavanols, 1, 199, 350
Flavor, 92, 343–344. *See also* Aroma-active compounds
  synthesis of flavor compounds, 74–75, 74f
Flavylium cation, 197, 207
Flavylium ion, 195
Flexitainer system, 333
Forests
  forest for oak production, 126f
  providing wood for cooperage, 126
Freeze damage
  grape variety resistance to, 365
  humidity, 365
  vine physiological limit, 365
French oak (*Quercus petraea*), 126, 188
FRG. *See* Fungus-resistant grapes (FRG)
Frozen injury, 357
FTIR analyzers, 31–32
Functional redundancy, 258
Fungus-resistant grapes (FRG), 349
Furaneol, 295–296
Furanic aldehydes, 132
Furans, 132
Furfuryl compounds, 155
Furfurylthiol (FFT), 276, 296
Fusel alcohol acetates, 163

G

Gallic acid, 174
Galloylation, 260
Gamma rays, 334
GC–MS quadrupole analysis, 299
GDD. *See* Growing degree-days (GDD)
Geographic origin influence, 137–138, 137t
"Geranium taint", 231
Glass bottles, 323, 331–332
Glass recycling, 338
Gluconic acid, 11–12
3-O-Glucosides, 195
Glutathione (GSH), 163
Glycerol, 11–12
Glycosidic conjugates, 91
Gradient gel electrophoresis, 117–118
Grain, 126
Grape berry, yeast ecology of, 51–55
Grape cells, influence of low temperature on
  plasma membrane permeability, 364
  respiration in cold climate, 365
Grape maturity/maturation, 177–183
  evolution of different seed and skin proanthocyanidin fractions, 182f
  grape selection in winery, 9–12
  physicochemical characteristics of enological interest, 1–3
  proanthocyanidins evolution during, 179–183
  vineyard approaches to grape selection and harvest date determination, 3–8
Grape must
  fermentation, 115
  microbiota, 115–116

Grape vine
  influence of low temperature on different tissues
    buds adaptability to low temperature, 364
    roots adaptability to low temperature, 364
    shoots adaptability to low temperature, 364
    wintering adaptability and cold resistance, 363–364
Grape-originated aroma, 2–3
Grape(s), 195, 196f
  chemical composition in cool and warm climates, 342–347
  composition and content of proanthocyanidins, 177–179, 180t
  and grape juices, 220
    monitoring, 227
    organic acids in, 17
  pulp, 177
  ripening, 262–263
  seeds, 254
  skin polysaccharides, 202
  tasting, 253–254
    field sampling, 253
  variety resistance to freeze damage, 365
Green compound sources, 345
Greenness
  prevention in vineyard, 345–346
  in red wines, 344
Growing degree-days (GDD), 341–342
Growing Season Temperature index (GST index), 341–342
GSH. *See* Glutathione (GSH)

H

H^+-ATPases, 259
Hand splitting techniques, 127–129, 128f
Hanseniaspora, 56–59, 61
  H. guilliermondii, 52, 53t, 58–59
  H. osmophila, 52
  H. uvarum, 52, 53t, 56–58
  H. vineae, 53t, 56–59
Hanseniaspora/Kloeckera, 51–52
Hansenula subpelliculosa, 58–59
Harmonia axyridis, 345
Hazards in bottling red wine
  adjusting filling level, 330–331
  automation standards mechanic fillers, 331
  avoidance of oxygen with filling technology, 328–329
  BIB, 332–333
  carton packaging, 332
  chemical contaminations, 327–328
  filling speed, 331
  glass bottles, 331–332
  microbiology, 325–327
  packaging materials, 331
  PET bottles, 332
  physical contamination, 328
HDPE. *See* High density polyethylene (HDPE)
Heartwood, 126, 127f
Heat-tests, 240

SUBJECT INDEX

Henderson—Hasselbach equation, 22
Heterofermentative activity, 86
Heterofermentative conversion, 19
HHP. *See* High hydrostatic pressure (HHP)
HI. *See* Huglin Index (HI)
High density polyethylene (HDPE), 157
High hydrostatic pressure (HHP), 35, 44—45, 158, 167, 317
High performance liquid chromatography—diode array detection (HPLC—DAD), 209
High pressure homogenization (HPH), 167
High pressure processing (HPP), 317
High-resolution mass spectrometry (HRMS), 213
Homofermentative activity, 86
Homofuraneol, 295
Homogeneity, 163
Homomolecular copigmentation, 198
Hot filling, 326—327
Hot glue labeling, 336
HPH. *See* High pressure homogenization (HPH)
HPLC—DAD. *See* High performance liquid chromatography—diode array detection (HPLC—DAD)
HPP. *See* High pressure processing (HPP)
HRMS. *See* High-resolution mass spectrometry (HRMS)
HRPs, 258—259
Huglin Index (HI), 341—342
Humidity, 365
Hybrids, 342
Hydrogen peroxide (H_2O_2), 158
Hydrogen sulfide (H_2S), 292
Hydrogen sulfite adducts in wines, 316t
Hydrogen sulfite ion (HSO_3^-), 315
Hydrolysis, 187
Hydrolyzable tannins, 259
Hydrophilic repulsion, 238
"Hydrophilic/hydrophobic" polar interaction forces, 238
Hydroxybenzoic acids, 1
Hydroxycinnamic acids, 1, 91, 350
Hydroxyesters, 296
Hygiene, 230
Hygienic standard, 326
Hyperchromic effect, 198
Hyperspectral imaging, 12

I
IBMP. *See* 3-Isobutyl-2-methoxypyrazine (IBMP)
IE. *See* Esterification index (IE)
IGSs. *See* Intergenic spacers (IGSs)
IH. *See* Interspecific hybrid (IH)
Image analysis, 8, 11
Innocent yeast, 219
Innocuous yeast, 219
Interflavan bonds, 259
Intergenic spacers (IGSs), 116
Intermolecular copigmentation, 198
Internal transcribed spacer (ITS), 116
International Agency for Research on Cancer, 106
International Code of Oenological Practices, 23
International Nucleotide Sequence Database Collaboration, 116
International Oenological Codex, 27—28
International Organization of Vine and Wine (OIV), 25—30, 93
Interspecific hybrid (IH), 349
Intramolecular copigmentation, 198
Ion exchange resins, 29—30
IPMP. *See* 3-Isopropyl-2-methoxypyrazine (IPMP)
Irradiation, 35, 46
3-Isobutyl-2-methoxypyrazine (IBMP), 276—277, 298, 345
D-Isomer, 31
3-Isopropyl-2-methoxypyrazine (IPMP), 298, 345
Issatchenkia orientalis, 52
ITS. *See* Internal transcribed spacer (ITS)

K
Ketones, 134
KHT. *See* Potassium hydrogen tartrate (KHT)
Kieselguhrs, 242—243
Killer toxins, 64
Kloeckera, 56—58
Kloeckera apiculata, 53t, 58—59
Kluyveromyces, 51—52, 56—58
   K. marxianus, 58—59
   K. thermotolerans, 52
   K. wickerhamii, 64

L
LAB. *See* Lactic acid bacteria (LAB)
Labeling systems, alternative, 336
Lachancea thermotolerans, 53t, 108
Lactic acid, 19
Lactic acid bacteria (LAB), 19, 85, 99—101, 115
   coinoculation with yeast starters, 106
   implications in wine safety, 93—94
      biogenic amines, 93—94
      EC, 94
   production of off-flavors by, 91—93
      volatile sulfur compounds, 92—93
   in winemaking, 85—86
      Lactobacillus sp., 86
      Oenococcus oeni, 86
      Pediococcus sp., 86
D-Lactic acid, 108
Lactic bacteria, 256
Lactobacillus sp., 85—86
   L. mali, 103—104
   L. pantheris, 103—104
   L. paracasei, 103—104
   L. plantarum, 85—86, 91, 101—104
   L. satsumensis, 103—104
   L. vini, 103—104
Lalvin D21, 347
LAR. *See* Leucoanthocyanidin reductase (LAR)
Latitude Temperature Index (LTI), 341—342
Law of mass action, 22—23
LC. *See* Liquid chromatography (LC)
LC-RID. *See* Liquid chromatography coupled with refractive index detection (LC-RID)
Lees
   aromatization, 171—174
   polysaccharides, 163
Leucoanthocyanidin reductase (LAR), 262
Level controlled fillers, 330
Lifshitz—van der Waals forces, 238
"Light Transport" barrel, 131
Limousin, 126—127
Limousin oak (*Quercus robur*), 126, 137
Limpidity, 237
Linear alcohols, 59
Liquid chromatography (LC), 209
Liquid chromatography coupled with refractive index detection (LC-RID), 163—164, 164f
Long tube fillers, 329
Low molecular weight proanthocyanidins, 318
Low vacuum fillers, 330
LTI. *See* Latitude Temperature Index (LTI)
Lysozyme, 318

M
2M3F. *See* 2-Methyl-3-furanthiol (2M3F)
Maceration, 35, 183, 184t
   extended, 44
   and fermentation, 58—59, 263—266
   kinetics of extraction, 38—40
   mechanical processes
      extended maceration, 44
      punch downs and pump overs, 40—42
      rack and return, 42—43
      submerged cap, 43
   microwave, 348
   new extraction technologies, 44—47
      HHP, 44—45
      irradiation, 46
      ozone and electrolyzed water, 47
      PEFs, 45
      PL, 47
      USs, 45—46
   tank design for red winemaking, 35—36
   vessel materials in red winemaking, 36—38
Macromolecules, 237
Magnetic resonance imaging (MRI), 12
Maillard reaction, 133—134
Malate, 343
MALDI Biotyper system, 119
MALDI—TOF MS. *See* Matrix-assisted laser desorption/ionization—time of flight mass spectrometry (MALDI—TOF MS)
Malic acid, 18, 20—21, 254, 343
Malolactic bacteria, 104
Malolactic deacidification. *See* Malolactic fermentation (MLF)
Malolactic enzyme (MLE), 101—102
Malolactic fermentation (MLF), 69, 79, 85, 99—102, 115, 152, 343. *See also* Alcoholic fermentation (AF)
   factors impacting lab at winery, 87—88
   impact on wine organoleptic properties, 89—91

Malolactic fermentation (MLF) (Continued)
  implications in wine safety, 93–94
  LAB in winemaking, 85–86
  production of off-flavors by LAB, 91–93
  tasting during, 256
  technological strategies for managing MLF performance, 88–89
Maltol, 295
Malvidin, 195
Malvidin-3-glucoside trimer isomer, 210f
Malvidin-3-O-(6-coumaroylglucoside), 199
Mannoproteins, 60–61, 202
Manual grape selection in vineyard, 5–6
Manual sorting tables, 9, 10f
Mass action law, 315
Mass spectrometry techniques (MS techniques), 209
Matrix-assisted laser desorption/ionization–time of flight mass spectrometry (MALDI–TOF MS), 119
Matter other than grape (MOG), 9
MDGC. See Multidimensional chromatography (MDGC)
MDMP. See 3,5-Dimethyl-2-methoxypyrazine (MDMP)
Medium degree of polymerization (mDP), 260
Medium factors effects on proanthocyanidin evolution, 185–186
MeJA application, 367
Membrane filtration, 326
Mercaptanoethanol, 255
4-Mercapto-4-methylpentan-2-one (4MMP), 296
3-Mercaptohexan-1-ol (3MH), 296
3-Mercaptohexyl acetate (3MHA), 296
MeSAc and EtSAc. See Ethyl and methyl thioacetates (MeSAc and EtSAc)
MeSH. See Methanethiol (MeSH)
Metabolic products, 92–93
Metagenome sequencing approach, 52–54
Metatartaric acid, 248
Methanethiol (MeSH), 292
2-Methoxy-3-isobutylpyrazine (IBMP). See 3-Isobutyl-2-methoxypyrazine (IBMP)
Methoxypyrazines (MPs), 276–277, 343–345, 345f
  remediating must and wine with elevated MP levels, 346–347
  structure, odor, concentration in wine, 277t
3-Methyl-2,4-nonanedione, 279
2-Methyl-3-furanthiol (2M3F), 276, 296
4-Methyl-4-sulfanylpentan-2-one (4MSP), 276
Metschnikowia, 51–52, 61
Metschnikowia pulcherrima, 52, 53t, 59
3MH. See 3-Mercaptohexan-1-ol (3MH)
3MHA. See 3-Mercaptohexyl acetate (3MHA)
Micro-oxygenation technique (MOX technique), 150, 151f, 152, 231
  application, 152–153
    innovations in, 157–158
    combined used of MOX + chips, 156–157
    comparing effect of MOX with aging wine in barrels, 155–156
    positive factors, 151–152
      improvement of astringency and mouthfeel, 151

  incidence on yeast development during alcoholic fermentation, 151
  wine aroma improvement and vegetal characteristics reduction, 152
  wine chromatic characteristics and stability, 151
  system, 153
Microbial
  controlling microbial spoilage, 104–105
  diversity assessment through enzymes detection, 119
  identification, 116, 119
  stability, 79–80
"Microbial terroir", 54–55
Microbiological contamination, 325–326
Microbiological control, 227–228
  bulk wine monitoring, 227
  grape and grape juice monitoring, 227
  peculiar case of D. bruxellensis, 227–228
  wine bottling, 228
  in wineries, 228
Microbiological stabilization of red wines, 248–249
Microorganisms, 103, 115, 325–326
Microwaves (MW), 167
  maceration, 348
Mistral sorting line, 10
MLE. See Malolactic enzyme (MLE)
MLF. See Malolactic fermentation (MLF)
4MMP. See 4-Mercapto-4-methylpentan-2-one (4MMP)
Modern enology, 1
Modern volumetric fillers, 331
MOG. See Matter other than grape (MOG)
Molded corks, 333
Molecular tools to analyzing microbial populations
  classical and phenotypic methods, 116
  culture-dependent vs. culture-independent methods, 119–120
  DNA-based methods, 116–119
  MALDI–TOF MS, 119
  microbial diversity assessment through enzymes detection, 119
Monascus purpureus, 118
Monomeric flavan-3-ols, 259
Monoterpenes, 91
Most Probable Number technique (MPN technique), 228
Mousy odor or flavor, 92
Mouthfeel
  improvement of, 151
  sensations, 261
MOX technique. See Micro-oxygenation technique (MOX technique)
MPN technique. See Most Probable Number technique (MPN technique)
MPs. See Methoxypyrazines (MPs)
MRI. See Magnetic resonance imaging (MRI)
4MSP. See 4-Methyl-4-sulfanylpentan-2-one (4MSP)
MS techniques. See Mass spectrometry techniques (MS techniques)
Multi-Layer Perceptron method, 12

Multidimensional chromatography (MDGC), 296
Multiperspective simultaneous imaging approach, 11
Multipiece natural stoppers, 333
Multispectral imaging, 12
Multivariate calibration models, 11–12
Multivariate internal standards, 289–290
Must
  organic acids in, 17–19
  remediating must and wine with elevated MP levels, 346–347
    reduction in methoxypyrazine content, 346f
  techniques for acidification, 23t
  yeast diversity in, 70
MW. See Microwaves (MW)
Mycocins, 64

N

Natural evolution of proanthocyanidins during aging, 185
Natural seasoning, 129
  comparison between natural seasoning and artificial drying, 139t
  influence, 138–139
NDVI. See Normalized different vegetation index (NDVI)
Near-infrared spectrometry (NIR spectrometry), 3
Next generation sequencing (NGS), 118–119
NHTP. See Nonahydroxyterphenoyl (NHTP)
NIR spectrometry. See Near-infrared spectrometry (NIR spectrometry)
NIR-Online, 11–12
Nitrogen, 313
  competition during winemaking, 71
  droppers, 329
NMR spectroscopy. See Nuclear magnetic resonance spectroscopy (NMR spectroscopy)
Non-Saccharomyces yeasts, 44, 51, 53t, 119, 164–167, 265
  controlled fermentation, 55
  features in red wine, 56–64
    acidification and deacidification activities, 61–62
    antimicrobial activities, 63–64
    aroma profile, 58–60
    enzymatic activities, 56–58, 57t
    polysaccharides production and color stability, 60–61
    reduction of ethanol content, 62, 63t
  yeast ecology of grape berry, 51–55
Nonahydroxyterphenoyl (NHTP), 187
Nonanthocyaninc phenolic compounds, 201–202
Nontoasted oak chips, 170–171
Norisoprenoids, 75
Normalized different vegetation index (NDVI), 4
Nuclear magnetic resonance spectroscopy (NMR spectroscopy), 209
Nucleation, 245
Nuclei, 245
Nutrients, 89

O

Oak, 127
  barrels, 128
  species, 140
  structural characteristics, 369
Oak chips, 153
  effect of adding oak chips on wine characteristics, 153–155
  oak-related volatile compounds, 154f
Oak wood, 37, 150
  oxygen permeability of, 136
  phenolic compounds released by, 134–136
  volatile substances released by, 132–134
Ochratoxin, 105–106
Ochratoxin A (OTA), 6, 101, 165, 220
Odor
  Odor-active aldehydes, 293–295
  properties, 295
Oenococcus oeni, 79, 85–86, 90–91, 99, 102–103, 102f, 106
OIV. *See* International Organization of Vine and Wine (OIV)
Oligomers, 177
Oligotrophic oxidative basidiomycetous yeasts, 51–52
On-line quality assessment, 11–12
Optical sorting, 11
Oral sensation, 257
Orbitrap-HRMS, 213
Organic acids, 31, 152–153
  acidification by supplementation with, 25
  in must and wine, 17–19, 18f
    acetic acid, 19
    citric acid, 19
    lactic acid, 19
    malic acid, 18
    succinic acid, 19
    tartaric acid, 18
Organoleptic properties of red wine, 199–200
OTA. *See* Ochratoxin A (OTA)
OTR. *See* Oxygen transfer rate (OTR)
Oxidation, 187
Oxidative stress, 151
Oxidative yeasts, 115
Oxonium ion, 195
Oxygen, 151–152, 157
  avoidance with filling technology, 328–329
  effect, 198
  permeability of oak wood, 136
  and storage temperature, 231
Oxygen transfer rate (OTR), 136, 149
Ozone, 35, 47, 327

P

PA. *See* Proanthocyanidins (PA)
Packaging, 336–337
  boxing and wrapping machines for bottles, 336–337
  materials, 331
Paper labels, 335
Partial least squares (PLS), 8
Pasteurization of filled bottle, 327
PCR. *See* Polymerase chain reaction (PCR)
PCR-denaturing gradient gel electrophoresis (PCR-DGGE), 117–118
PCR-restriction fragment length polymorphism (PCR-RFLP), 116–117
PDMS. *See* Potential DMS (PDMS)
Pectolytic enzymes, 264
Pediococcus sp., 85–86
  P. damnosus, 86
  P. inopinatus, 86
  P. parvulus, 86
  P. pentosaceus, 86
Pedunculate oak. *See* Limousin oak (*Quercus robur*)
PEFs. *See* Pulsed electric fields (PEFs)
Penicillium citrinum, 94
Penicillium species, 105
2,3,4,5,6-Pentafluorobenzyl bromide (PFBBr), 297–298
Peonidin, 195
Permeability, 242
PET bottles. *See* Polyethylene terephthalate bottles (PET bottles)
Petunidin, 195
PFBBr. *See* 2,3,4,5,6-Pentafluorobenzyl bromide (PFBBr)
pH, 19–20, 87, 349–350
  control
    new technologies for, 27–30
    in red wines, 17
  effect, 197
  laboratory techniques for measuring, 30–32
Phenolic
  acids, 91–92, 317
  aldehydes, 134
  composition, 172, 173t, 195
  maturity, 2
  parameters, 188
  ripeness, 262
Phenolic compounds, 1, 35, 38, 91, 151, 171f, 177, 185, 207–209, 259–261, 344
  proanthocyanidins, 259–261
  released by oak wood during barrel aging, 134–136
    astringency and bitterness thresholds of ellagitannins, 135t
    gallic acid, ellagic acid, and ellagitannins, 135f
Phenylpropanoid, 75
Photosynthesis, 366
Pichia, 51–52, 56–59
  P. kudriavzevii, 53t
  P. membranaefaciens, 58–59
  P. membranifaciens, 53t, 56–58
Pigments, 39
Pilzwiderstandsfähige (PIWI), 349
Pinot noir winemaking, 348
Piperitone, 279, 300
PIWI. *See* Pilzwiderstandsfähige (PIWI)
PL. *See* Pulsed light (PL)
Plasma membrane permeability, 364
Plate heat exchangers, 326
PLS. *See* Partial least squares (PLS)
Polar compounds, highly, 295–296, 295t
Polyethylene terephthalate bottles (PET bottles), 332

Polyfunctional mercaptans, 296–298, 297t
Polymerase chain reaction (PCR), 116
Polymeric pigments in red wines, 209–212
  A-type vitisin-derived pigments formed in red wines, 211–212
  analysis, 212–213
  anthocyanin-derived pigments formed in red wines during aging, 209–210
  found in red grapes and wines, 209
  stability in solution and influence in red wine color, 213–215
Polymerization, 187
Polymers, 177
Polyphenoloxidase (PPO), 24, 199–200, 309
Polyphenols, 89, 172, 324
  content in red winemaking, 71–72
  polyphenolic compounds, 207
Polysaccharide–anthocyanin interaction effect, 202
  grape skin polysaccharides, 202
  yeast mannoproteins, 202
Polysaccharides, 163
  production, 60–61
  release, 73
Porosity, 242
Post-harvest grape drying/dehydration, 348
Potassium, 349–350
Potassium bicarbonate ($KHCO_3$), 101–102
Potassium bitartrate (KHT), 22–23
Potassium carbonate (K_2CO_3), 101–102
Potassium hydrogen tartrate (KHT), 245
Potassium hydroxide (KOH), 101–102
Potassium metabisulfite application, 346
Potential DMS (PDMS), 273–275
PPO. *See* Polyphenoloxidase (PPO)
Precision Viticulture approaches, 3–5, 8
Prefermentation juice runoff, 264
Pressure gradient, 242
Proanthocyanidins (PA), 2–3, 35, 177–183, 195, 259–261. *See also* Anthocyanins
  composition and content in grapes, 177–179
  evolution during grape maturation, 179–183
  during fermentative maceration of red wines, 183–185
  during red wine aging in contact with wood, 185–188
Process Analyzer X–Three industrial spectrometer, 11–12
Procyanidins, 260–261
Prodelphinidins, 260–261
Proline-rich proteins (PRPs), 258
Proteins, 186
  families, 258
  salivary, 258–259
Proteomic techniques, 119
Proximal sensing techniques, 4
PRPs. *See* Proline-rich proteins (PRPs)
Puckering, 257
Puckery, 257
Pulp's sweetness, 254
Pulsed electric fields (PEFs), 35, 45, 167
  inactivation mechanism, 169

Pulsed light (PL), 35, 47
Pump over method, 40–42
Punch down method, 40–42, 41f, 184
Putrecine, 256
Putrescine, 256
Pyrano anthocyanin–phenol pigments, 201
Pyranoanthocyanins, 200–201, 211–212
Pyrazines, 255
Pyruvate, 89–90
Pyruvic acid, 73

Q

q-PCR. *See* Quantitative PCR (q-PCR)
Q-TOF-HRMS, 213
QPCR. *See* Quantitative real-time PCR (QPCR)
Quality indicators, 3
Quantitative PCR (q-PCR), 86
Quantitative real-time PCR (QPCR), 86, 118
Quercus, 125
   Q. macrocarpa, 125–126
   Q. mongolica, 125
   Q. muehlenbergii, 126
   Q. pedunculata. *See* Limousin oak (*Quercus robur*)
   Q. petraea, 127, 138
   Q. pyrenaica, 125–126

R

Rack and return technique, 42–43, 43f, 264–265
Racking, 79–80
Randomly amplified polymorphic DNA PCR fingerprints (RAPD-PCR fingerprints), 117
Raulí (*Nothofagus alpina*), 125
Reactive oxygen species (ROS), 88
Real acidity in wine, 20
Red grapes, yeast diversity in, 70
Red icewine, 348–349
Red IH varieties, making wine from, 349–350
Red pomace (RP), 350
Red wine(s), 195, 196f, 237, 312, 323–324. *See also* Wine(s)
   acid–base equilibrium and wine buffer capacity, 22–23
   acidic fraction and pH control in, 17
   "analysis of difficult" aroma compounds in, 291–300
   aroma
     analytical problem of, 283
     compound families, 284t
     quality, 156–157
   bottling lines, 324–325
   clarification, 240–245, 242f
   closures, 333–334
   colloids and colloidal instabilities in, 237–240
   color. *See* Color
   ecology, 338
   economy, 337
   glass bottles, 323
   "greenness" in, 344
   hazards in bottling red wine, 325–333
   laboratory techniques for measuring pH and acidic fraction, 30–32
   making production scheme, 77f
   microbiological stabilization, 248–249
   modern ready to drink wines and shelf life, 324
   new technologies for pH control, 27–30
   non-*Saccharomyces* yeasts, 56–64
   organic acids in must and wine, 17–19
   packaging, 336–337
   preparing wine for market, 334–336
   proanthocyanidins, 177–183
     during fermentative maceration, 183–185
     during red wine aging in contact with wood, 185–188
   production, 51
   SO_2 replacement products for production, 318–320
     antimicrobial activity substitutes, 318–319
     antioxidant activity substitutes, 319–320
     considerations on SO_2 replacement additives, 320
   stabilization to crystallization of tartaric salts, 245–248
   tasting in production, 255–256
   total acidity and wine pH, 19–21
   traditional fashioned red wines ripening in bottle, 324
   traditional strategies
     for chemical acidification, 23–25
     for chemical deacidification, 25–27
Redox effects in red winemaking, 71
Reductive off-odors, 292
"Regional" yeasts, 219
Remote sensing techniques, 4, 8
Remotely piloted aerial systems, 6–8
Respiration in cold climate, 365
Reverse transcription quantitative real-time PCR (RT-qPCR), 118
Rhizopus oryzae, 118
Rhodotorula, 51–52, 56–58
Ripeness, 262
Ripening process, 179–181
Roots adaptability to low temperature, 364
ROS. *See* Reactive oxygen species (ROS)
Rosing, 129
Rotten grapes, 222–223, 232
Rotundone, 277, 299
Roughing, 257
RP. *See* Red pomace (RP)
16S rRNA gene, 116
RT-qPCR. *See* Reverse transcription quantitative real-time PCR (RT-qPCR)

S

S-methylmethionine (SMM), 273–275
Saccharomyces, 51–52
Saccharomyces cerevisiae, 51, 55, 58–59, 69–70, 99–101, 102f, 115, 219
   aging and microbial stability, 79–80
   hydrolysis of disaccharide glycosides, 76f
   and red wine color, 72
   and related species, 226
Saccharomyces enzymes effects on flavor, 75–76
Saccharomyces-lactic acid bacteria interactions, 79
   styles of fermentation tanks, 78f
   synthesis of flavor compounds, 74–75, 74f
   use of commercial yeasts, 78–79
Saccharomyces species, 52–54
Saccharomyces strains, 69
   alcohol and polyphenol contents in red winemaking, 71–72
   cell wall adsorption and cell lysis effects on anthocyanins, 72–73
   derived anthocyanin compounds formation by yeast fermentation, 73–74, 73f
   domination in wine ecosystem, 71
   nitrogen competition during winemaking, 71
   redox and temperature effects in red winemaking, 71
Saccharomyces yeasts, 119, 265
Saccharomycodes, 51–52, 53t, 60–61
Saccharomycodes ludwigii, 166, 166f, 226
Saccharomycopsis fibuligera, 118
Saignée process. *See* Prefermentation juice runoff
Saliva, 258
Salivary proteins, 258–259
Saturation temperature, 246
Sawing techniques, 127–129, 128f
SBMP. *See* 3-Secbutyl-2-methoxypyrazine (SBMP)
Schizosaccharomyces, 51–52, 60–61
Schizosaccharomyces pombe, 53t, 60, 89, 166, 166f, 226, 349–350
Screw caps, 334
3-Secbutyl-2-methoxypyrazine (SBMP), 298, 345
Secondary fermentation. *See* Malolactic fermentation (MLF)
Self-adhesive labels, 335
Self-association. *See* Homomolecular copigmentation
Sensory analysis, 59
   of red wines for winemaking purposes
     grape tasting, 253–254
     tasting during malolactic fermentation, 256
     tasting in production of red wine, 255–256
Sensory perception, 283
Sequential Forward Selection algorithm, 12
Sesquiterpene, 277, 278t
Settling velocity of spherical particle, 240–241
3SH. *See* 3-Sulfanylhexanol (3SH)
3SHA. *See* 3-Sulfanylhexyl acetate (3SHA)
Shoots adaptability to low temperature, 364
Short tube fillers, 329
Silica-C18, 290
Simple Perceptron method, 12
"Single-arm hedgerow" system, 362, 363t
Skin tasting, 254
Sleeves, 336

SMM. *See* S-methylmethionine (SMM)
Solid Phase Extraction (SPE), 290
Solubility product (SP), 245
Soluble "macromolecular complexes" formation, 239
Soluble solids (SS), 342–343
Sorbic acid, 231, 318
Sotolon, 295–296
SP. *See* Solubility product (SP)
Spanish *Tinaja*, 36–37, 37f
SPE. *See* Solid Phase Extraction (SPE)
Spectrophotometric analysis of phenolic compounds, 1–2
Spectrophotometric indexes, 2
Spoilage yeasts in red wines, 45
    control of yeast populations in wines, 230–232
    significance and occurrence of wine-related yeast species, 220–222
    yeast genera/species involving in wine spoilage, 224–226
    yeast monitoring, 227–230
    yeast taxonomy, 219
Spontaneous crystallization, 246
Spontaneous must fermentation, 55
Spouts, 131
Spring wood, 126
SS. *See* Soluble solids (SS)
Stabilization
    to crystallization of tartaric salts, 245–248
    mechanisms and stability assessment, 245–247
    technologies, 247–248
Stagnant water, 256
Stainless steel tanks, 38
Starmerella bacillaris, 53t
Starmerella bombicola, 53t
Staves, 127–129, 128f
Steam disinfection, 223–224
Sterilization, 230–231
Stilbenes, 1
Stokes' law, 240–241
Strecker aldehydes, 293–295
Stress-inducing factors, 88
Submerged-cap method, 43, 184
Succinic acid, 19
Sugar, 342–343
3-Sulfanylhexanol (3SH), 276
3-Sulfanylhexyl acetate (3SHA), 276
Sulfur combustion, 311
Sulfur compounds
    C_{13}-norisoprenoides, 278
    DMS, 273–276
    esters, 278
    methoxypyrazines, 276–277
    miscellaneous aroma compounds, 278–279
    sesquiterpene, 277
    thiols, 276
Sulfur dioxide (SO_2), 63–64, 88–89, 213–214, 231, 291–292, 291t, 309–311, 316f
    action mechanisms, 314–317
    antioxidant mechanisms in wine, 317f
    dosage in wine from pressurized tanks, 313f
    equilibriums in aqueous solution, 314f
    forms, 311–314
    limits for red wine in winemaking countries, 310t
    molecular SO_2 fraction, 315t
    molecular structure and resonance system, 310f
    replacement products for red wine production, 318–320
Summer wood, 126
Sweat aromas, 256
Syringaldehyde, 156, 172

T

T-RFLP. *See* Terminal restriction fragment length polymorphism (T-RFLP)
TA. *See* Titratable acidity (TA)
Tank design for red winemaking, 35–36
Tannat, 74
    grape seeds, 177
Tannin specific activity (TSA), 261
Tannins, 35, 38, 60, 257, 259, 317, 319, 350–353
    aroma, 350–353
    impact of different winemaking processes, 351t
Tannins, condensed, 257
Tartaric acid (H_2T), 18
Tartaric salts, stabilization to crystallization of, 245–248
Tartrate cream, 247
Tasting
    during malolactic fermentation, 256
    in palate, 254
    in production of red wine, 255–256
TCA. *See* 2,4,6-Trichloranisol (TCA)
TCEP. *See* [Tris(2-carboxyethyl)phosphine] (TCEP)
Technical stoppers, 333
Temperature effect, 197
    in red winemaking, 71
Temperature gradient gel electrophoresis (TGGE), 117–118
Temporal temperature gradient electrophoresis (TTGE), 117–118
Tempranillo wines, 174
Terminal restriction fragment length polymorphism (T-RFLP), 117
Tetra Pak system, 332
TGGE. *See* Temperature gradient gel electrophoresis (TGGE)
Thermovinification, 39–40, 232, 346, 348, 367–368
Thin layer chromatography (TLC), 31
Thiols, 276
Tightening, 129
Time-differential harvest, 8
Titratable acidity (TA), 31, 348–349
    in wine, 20
TLC. *See* Thin layer chromatography (TLC)
TMP. *See* Transmembrane pressure (TMP)
Toasting level influence, 139–140, 140f, 141f
Toasting of barrel, assembly and, 129–130, 130f
Tonnage, 125
Torulaspora, 58–59
Torulaspora delbrueckii, 53t, 58–59
Total acidity and Wine pH, 19–21
Traditional fashioned red wines ripening in bottle, 324
Traditional withering, 52–54
Transmembrane pressure (TMP), 243–244
Transport barrels, 131
Transport package, 336
Tree species used in cooperage, 125
Tribaie system, 11
2,4,6-Trichloranisol (TCA), 333
Trihydroxylated group, 260–261
Trimers, 177
[Tris(2-carboxyethyl)phosphine] (TCEP), 293
TSA. *See* Tannin specific activity (TSA)
TTGE. *See* Temporal temperature gradient electrophoresis (TTGE)
Turbid wine, 241–242
Tyloses, 129

U

UAVs. *See* Unmanned aerial vehicles (UAVs)
Ultrasound technique, 158
Ultrasounds (USs), 35, 45–46, 168
Ultraviolet (UV), 327
Unmanned aerial vehicles (UAVs), 4, 6–8
Urethane. *See* Ethyl carbamate (EC)
USs. *See* Ultrasounds (USs)
UV. *See* Ultraviolet (UV)
UV–visible spectroscopy, 214

V

Vanillin, 140, 155–156, 172
Vanillylthiol, 276
VBNC. *See* Viable but nonculturable (VBNC)
Vegetal aromas, 255
Vegetal characteristics reduction, 152
Vescalagin, 187
Vesicle trafficking, 262
Vessel materials in red winemaking, 36–38
Viable but nonculturable (VBNC), 117
Vigor maps, 8
Vine
    buried viticulture effect on, 360–363
    cultivation and management technique of vineyard in cold area, 362–363
    physiological limit, 365
    unearthed in spring, 358–360
Vineyard approaches, 3–8
    cultivation and management technique of vineyard in cold area, 362–363
    grape harvest and selection in vineyard, 4–8
        automated grape cluster selection and harvest, 8
        manual grape selection in vineyard by visual inspection, 5–6
        selective harvest based on vineyard area, 6–8
        selective harvest of different parts of cluster, 6
        time-differential harvest, 8
    preventing greenness in vineyard, 345–346

Vineyard approaches (*Continued*)
  spatial variability in vineyard and precision viticulture tools, 3–4
Vinolok, 334
Vintners Quality Alliance (VQA), 348–349
Vinyl phenols, 256
Vinylphenolic pyranoanthocyanin, 60
Vinylphenols, 73, 134
VIS spectrometry. *See* Visible infrared spectrometry (VIS spectrometry)
VIS/NIR spectroscopy, 11–12
Visible infrared spectrometry (VIS spectrometry), 3
Visual inspection technique, 5–6
Viticulture techniques, 349–350
  anthocyanin accumulation by, 366–367
Vitis amurensis, 349
Vitis riparia, 349
Vitis rupestris, 349
Vitis vinifera, 195, 342
Volatile
  acidity reduction, 103–104
  compounds, 172, 172t, 173t
  phenols, 134, 222
    specific analysis, 290–291
  substances
    released by oak wood during barrel aging, 132–134
    released from oak into wine, 133t
Volatile sulfur compounds (VSCs), 92–93, 292–293, 292t
Volumetric filling systems, 330
VQA. *See* Vintners Quality Alliance (VQA)
VSCs. *See* Volatile sulfur compounds (VSCs)

W

Warm climates, chemical composition of grapes in, 342–347
Weakly fermentative yeasts, 115
Whiskey lactones. *See* β-methyl-γ-octalactones
Wickerhamomyces anomalus, 53t, 56–58
Wine-related yeast species, 220–222, 221f, 221t
  factors promoting dissemination of spoilage yeasts, 222–224, 222f
  grapes and grape juices, 220
  wine fermentation, 220–221
Wine(s), 88–89, 91. *See also* Red wine(s)
  acidity control, 101–103
  aging, 165, 209
  aroma, 58. *See also* Aroma
    compounds, 284t, 286t
    improvement, 152
  during barrel aging, 131–132
  bottling, 228
  buffer capacity, 22–23
  chromatic characteristics and stability, 151
  color evolution and stability
    anthocyanin stability, 195–198
    copigmentation, 198–199
    red wine color evolution, 199–201
    winemaking practices for stabilizing red wine color, 201–202
  consumption, 283
  fermentation, 220–221
    bulk and bottled wine, 222
  interactions between wine microorganisms, 107–108
  lactobacilli, 86
  macromolecules effect, 186
  MLF impact on wine organoleptic properties, 89–91
  organic acids in, 17–19
  oxygenation, 157
  preparation for market
    alternative labeling systems, 336
    checking filling level, 334
    hot glue labeling, 336
    labeling, 335
    paper labels, 335
    self-adhesive labels, 335
    sleeves, 336
    treatment of closed bottles, 334
  quality, 1
  from red IH varieties, making, 349–350
    acidity, pH and potassium, 349–350
  SO$_2$ in, 309–311
  techniques for acidification, 23t
  toxin reduction, 105–106
  yeast populations control in, 230–232
Winemakers, 155
Winemaking, 99, 183, 213–214, 319
  influencing in wine astringency, 261–266
  LAB in, 85–86
  and legal limits, 309–311
  practices for stabilizing red wine color
    effect of polysaccharide–anthocyanin interaction, 202
    technological tools to enhance copigmentation in wines, 201–202
  techniques, 264, 367–369
Winery
  factors impacting lab at, 87–88
    ethanol, 87
    inhibitors of MLF in wines, 87t
    pH, 87
    SO$_2$, 88
    temperature, 88
  grape selection in, 9–12
    new perspectives for direct grape quality evaluation and selection, 11–12
    size, density, and image analysis sorting equipment, 10–11
    sorting tables, 9–10
Winkler Index, 341–342
Winter pruning, 357, 358f
Wintering adaptability of grape vine, 363–364
Wood
  forests providing wood for cooperage, 126
  pieces, 155
  proanthocyanidins during red wine aging in contact with
    effects of medium factors on proanthocyanidin evolution, 185–186
    influence on wine proanthocyanidin evolution, 187–188
    natural evolution of proanthocyanidins during aging, 185
  wood-aromatized yeasts, 172
Wood grain
  concept in cooperage, 126–127
  cross-section of oak trunk, 127f
  influence, 136–137, 136f
Wooded wines, 163
Wooden boxes, 336
World Health Organization, 309–310
Wrapping machines for bottles, 336–337

Y

YAN. *See* Yeast-assimilable nitrogen (YAN)
Yeast, 69, 88–89
  acceptable levels, 228–230
  autolysis, 163–164
  biomass, 164–165
  cell disruption, 170
  cell wall adsorption capacity, 165
  community, 115
  counts, 229–230
  diversity in red grapes and musts, 70, 70f
  ecology of grape berry, 51–55
  growth, 163
  involving in wine spoilage, 224–226
    Dekkera/Brettanomyces bruxellensis, 226
    film-forming species, 224
    S. cerevisiae and related species, 226
    S. ludwigii and *S. pombe*, 226
    Z. bailii and related species, 224–226
  killer toxins, 64
  mannoproteins, 202
  monitoring
    acceptable levels of yeasts, 228–230
    microbiological control, 227–228
    tools used in microbiological control in wineries, 228
  performance, 76–80
  populations control in wines, 230–232
    chemical preservatives, 231–232
    clarification, fining, and filtration, 230–231
    hygiene, 230
    oxygen and storage temperature, 231
    thermal treatments, 232
  producing fatty acids, 75
Yeast-assimilable nitrogen (YAN), 71, 349–350
Yeast-bacteria coinoculation, 44
  interactions between wine microorganisms, 107–108
  objectives
    controlling microbial spoilage, 104–105
    controlling wine acidity, 101–103
    impact on organoleptic characteristics, 106–107
    reducing ethanol yields and volatile acidity, 103–104
    reducing wine toxins, 105–106
    timing of inoculation of LAB, 100t

Z

Zygosaccharomyces, 51–52, 56–59
  Z. bailii, 53t, 115–116, 224–226
  Z. bisporus, 52
  Z. lentus, 225
  Z. rouxii, 224–225
Zygotorulaspora florentina, 53t, 60
Zymocins, 64

Printed in the United States
By Bookmasters